A
SOURCEBOOK OF
TITANIUM ALLOY
SUPERCONDUCTIVITY

A
SOURCEBOOK OF
TITANIUM ALLOY
SUPERCONDUCTIVITY

E. W. COLLINGS

Senior Research Scientist
Battelle Memorial Institute
Columbus, Ohio, U.S.A.

PLENUM PRESS · NEW YORK AND LONDON

Library of Congress Cataloging in Publication Data

Collings, E. W.
  A sourcebook of titanium alloy superconductivity.

  Bibliography: p.
  Includes indexes.
  1. Titanium alloys—Electric properties. 2. Superconductivity. I. Ti-
tle.
TN693.T5C64 1983   620.1′89322973   83-2180

  ISBN-13: 978-1-4613-3705-8   e-ISBN-13: 978-1-4613-3703-4

  DOI: 10.1007/978-1-4613-3703-4

©1983 Plenum Press, New York
Softcover reprint of the hardcover 1st edition 1983
A Division of Plenum Publishing Corporation
233 Spring Street, New York, N.Y. 10013

To **Betty Collings,** sculptor,
— *"the poetry is in the process"*
(studio notes: 1982)

first draft and tables : **Carole L. Owens**
final text typing : **Anita L. Maynard**
technical illustration and layout : **Judith S. Ward**

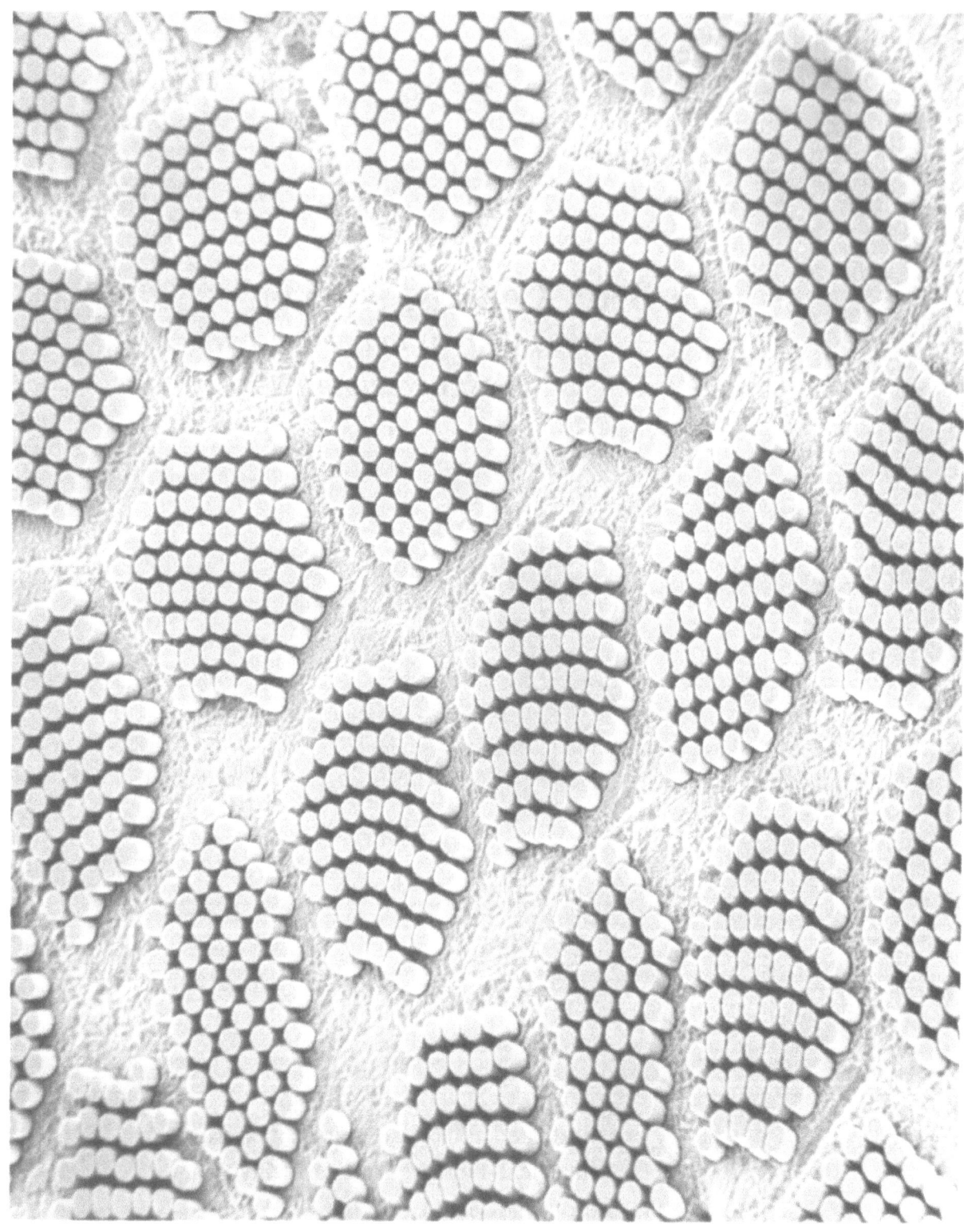

THE GOAL OF TITANIUM-ALLOY SUPERCONDUCTOR DEVELOPMENT HAS BEEN THE MANUFACTURE OF STABLE CONDUCTORS CAPABLE OF PASSING DENSE ELECTRICAL TRANSPORT CURRENTS IN THE PRESENCE OF STRONG MAGNETIC FIELDS. A portion of the cross-section of such a conductor, intended in this case for the windings of an energy-storage magnet, is shown above. The superconductive filaments are of the alloy $Ti_{61}-Zr_6-Nb_{27}-Ta_6$ (trade name: "cryozitt", Kobe Steel, Ltd., Japan), whose properties are discussed in Sect. 13.11. The specifications of the full conductor are: O.D., 3.0 x 1.5 mm^2; filament arrangement, 61 x 61; filament diameter, 18 μm; Cu/SC ratio, 3.2; twist pitch, 35 mm; 5-T critical current, 2.1 kA. The conductor was manufactured by drawing, accompanied by several intermediate heat treatments to improve its mechanical properties. *The sample depicted was courteously supplied to Battelle by T. Horiuchi, Asada Research Laboratory, Kobe Steel, Ltd. The metallographic mount was prepared by R. D. Smith, polished and deep etched (40% HNO$_3$) by K. L. Hammond, and photographed using scanning electron microscopy by A. Skidmore, all of the Battelle-Columbus Laboratories.*

FOREWORD

In less than two decades the concept of superconductivity has been transformed from a laboratory curiosity to usable large-scale applications. In the late 1960's the concept of filamentary stabilization released the usefulness of zero resistance into the marketplace, and the economic forces that drive technology soon focused on niobium-titanium alloys. They are ductile and thus fabricable into practical superconducting wires that have the critical currents and fields necessary for large-scale devices. More than 90% of all present-day applications of superconductors use titanium alloys. The drive to optimize these alloys resulted in a flood of research that has been collected, condensed, and analyzed in this volume. From the vagaries of precipitation annealing to the synergistic effects in ternary alloys, it is all here for us to use, peruse, and digest.

Dr. Collings has also created an excellent companion monograph, "The Applied Superconductivity, Metallurgy, and Physics of Titanium Alloys", that will appear in the International Cryogenic Monograph Series. The monograph provides the basic understanding in an easy to digest manner, which I find essential to sorting out the many complexities of these titanium alloy superconductors.

In every field of science there are one or two individuals whose dedication, combined with an innate understanding, permits them to be able to grasp, condense, and explain to the rest of us what that field is all about. For the field of titanium alloy superconductivity, such an individual is Ted Collings. His background as a metallurgist has perhaps given him a distinct advantage in understanding superconductivity in titanium alloys because the optimization of superconducting parameters in these alloys has been almost exclusively metallurgical. Advantages in training and innate abilities notwithstanding, it is the author's dedication that is the essential component that brought forth this pair of books that condense 20 years and thousands of publications. If we use these books, we will be led to the information and understanding needed to pursue our own interests in superconductivity. In a field fraught with alchemists and blacksmiths, this piece of articulate insight is a welcome aid.

Electromagnetic Technology Division   A.F. CLARK
National Bureau of Standards
Boulder CO, USA
September 1982

PREFACE

DESCRIPTION OF THE BOOK

PURPOSE

Although superconductors based on A15 intermetallic compounds such as Nb_3Sn, V_3Ga, and so on possess desirable high-field superconducting properties, manufacturing and handling difficulties have tended to restrict their use to fairly straightforward, usually small-scale, solenoidal-magnet applications. On the other hand conductors using the binary alloy Ti-Nb, or multicomponent alloys based on it, because of their relative ease of manufacture and excellent mechanical properties are being pressed into service in numerous large-scale devices. Such superconductors are being wound into magnets for use in energy storage, energy conversion (i.e. generators and motors), and in high-energy particle detectors and beam-handling magnets. A few representative large-scale machines which employ (or were intended to employ) stabilized Ti-Nb superconductors are listed in the following table.

The use of cold-rolled or drawn Ti-Nb-alloy wire for superconducting magnet applications was first proposed in 1961. During the ensuing ten years, while progress was being made in the development of Cu-clad filamentary-Ti-Nb-alloy conductors, Ti-Nb and other Ti-base binary transition-metal (TM) alloys were being employed as model systems in the fundamental study of type-II superconductivity. The second decade saw the development of improved current-carrying capacity in the binary Ti-Nb alloy, its incorporation into multifilamentary monoliths, cables, and braids capable of carrying tens of thousands of amperes, coupled with improvements to both the economic and technical attractiveness of the binary alloy through the incorporation into it of ternary and quaternary transition-metal additions.

But Ti-Nb, as a binary transition-metal alloy is not, conceptually, unique. In phase equilibrium it is a member of the so-called β-isomorphous class of Ti-TM alloys which includes Ti-V, Ti-Ta, and Ti-Mo; when quenched its metallurgical and physical properties are comparable to those of other quenched Ti-TM alloys. These properties can, moreover, be systematized in terms of average group number or electron/atom ratio, e/a; for example, it has been shown that in alloys quenched to room temperature the hcp-structured phase gives way to what is essentially a bcc phase at e/a = 4.15 ± 0.03. Thus to fully understand the properties of Ti-Nb alloy superconductors it is necessary to discuss them within the framework of Ti-TM alloys in general. Indeed during the early years of the practical development of type-II superconductivity, all such alloys were investigated on an equal footing. One of the goals of this work has been to provide a full description of the genesis of titanium alloy superconductivity. Following historical precedence, the review commences with a complete survey

Magnet System Name	Application	Operating Specifications		Magnet Configuration	Stored Energy, MJ	Magnet Weight, Tonnes
		Current kA	Field T			
Fermilab Energy Double/Saver[†]	High-energy physics	4.5	4.5	774 dipoles	387	---
Isabelle Ring Magnets[†]	High-energy physics	4	5	732 dipoles 352 quadrupoles	622	110
Mirror Fusion Test Facility (MFTF)[††]	"Magnetic fusion"	6	7.7	yin-yang pair	409	300
Torus II Supra[†††]	Tokamak fusion	1.4	8	24 circular coils	440	144
Large Coil Program (LCP)[†††*]	Tokamak fusion	10	8	single D-coil	1200	---
Coal-Fired Flow Facility (CFFF)[†††**]	MHD	4	6	dipole	168	157
Component Development and Integration Facility (CDIF)[†††**]	MHD	6	6	dipole	184	164

[†] R.M. Scanlan, "Superconducting Materials", Ann. Rev. Mater. Sci., 10 113 (1980); with regard to the status of Isabelle, cf. Science, 216 158 (1982).

[††] C.D. Henning, "Superconductivity for Mirror Fusion", IEEE Trans. Magn., MAG-15 525 (1979).

[†††] R. Aymer et al, "Conceptual Design of a Superconducting Tokamak: Torus II Supra", IEEE Trans. Magn. MAG-15 542 (1979).

[†††*] J.P. Heinrich et al, "Conceptual Studies of Toroidal Field Magnets for an Experimental Power Reactor", Seventh Symposium on the Engineering Problems of Fusion Research, 1977, p.931.

[†††**] P.G. Marston et al, "Superconducting MHD Magnet Engineering Program", Adv. Cryo. Eng., 25 1 (1980).

of superconductivity in all of the binary alloys of Ti with the other transition elements. After dealing extensively with Ti-Nb itself (fully up to the mid-1960's and critically thereafter) the multicomponent alloys of it are considered. Presentation of the material in such a format accomplishes a second goal, namely to guide the prospective researcher away from unnecessarily repeating experiments already adequately performed some 15 or so years ago and to help him assess which experiments, in the light of new techniques or evaluation criteria, may be worthy of repetition. The primary goal, embodied within Chapters 7 through 13, is of course to provide up-to-date information on the superconducting properties of practical Ti-Nb-base alloys, to present quantitatively the

effects of thermomechanical processing on them, and to enable intercomparisons to be made between the critical current densities of Ti-Nb and its ternary and quaternary derivatives.

SCOPE

When a group-IV transition metal (electron/atom ratio, e/a, = 4.0) such as Ti or Zr is alloyed with a transition metal of higher group number, a maximum in the superconducting transition temperature occurs somewhere between e/a = 4 and e/a = 5. In the case of alloys of Ti with Nb, this and other considerations have led to a choice of composition in which equal parts by weight of the constituent elements are present. The common ingredient of many superconductors of practical interest, whether they be intermetallic compounds (A15 or B1) or bcc-phase alloys, is Nb or occasionally V -- both group-V TM's. Thus it could perhaps be argued that Nb plays a central role in type-II superconductivity and that conventional practical superconducting compounds and alloys are "modified Nb". In the development of applied superconductivity, the alloy Zr-Nb, although it possessed several attractive superconducting properties, soon gave way to Ti-Nb with its 20-kOe higher maximum 4.2-K upper critical field and better workability. It is probably safe to say that most of the large superconducting magnet systems constructed up till now have used composite conductors incorporating Ti-Nb as the superconducting component. From time to time, but particularly recently, it has been recognized that improvements to the basic binary alloy could be achieved by substituting Zr for some of the Ti, and/or Ta for some of the Nb, yielding the ternary and quaternary alloys: Ti-Zr-Nb, Ti-Nb-Ta, and Ti-Zr-Nb-Ta. Although as indicated above, from one point of view Nb could be regarded as the dominant component of Ti-Nb, from a metallurgical standpoint the practical superconducting alloys of Ti and Nb are clearly members of the Ti-TM class of binary alloys. The equilibrium microstructures of Ti-Nb alloys are similar to those of the other so-called "β-isomorphous" binary Ti-TM alloys, e.g. Ti-V and Ti-Mo, and quenched Ti-Nb alloys possess microstructures and microstructural responses to aging comparable to those exhibited by quenched Ti-TM alloys in general. For these reasons, and also since the superconducting properties of all the Ti-TM alloys were originally explored as part of the experimental development of type-II superconductivity, it was decided to treat the subject of Ti-Nb- and Ti-Nb-base-alloy superconductivity within the general framework of Ti-TM alloys and to refer to the materials so reviewed as TITANIUM ALLOY SUPERCONDUC-

TORS. This book deals only with their electrical and magnetic properties. It does not treat topics such as "stress effects", "radiation effects", "stability", and so on, or is it intended to be a design handbook. These aspects are, however, fully addressed in a companion volume entitled *Applied Superconductivity, Metallurgy and Physics of Titanium Alloys* (E.W.C., Plenum Press, in preparation) which is referred to throughout this book by way of appropriate variants of the general citation [Mon00.00] -- see REFERENCES. This volume is referred to as a SOURCEBOOK since its purpose is to describe the properties of Ti-Nb-base-alloy superconductors within a historically based materials context. The steady progress of superconductive-alloy research from basic scientific enquiry to a developmental activity in preparation for the design of large magnet systems is reflected in the type of publication in which the results have been presented: Physical Review and similar publications at first, then tending towards applied journals and the proceedings of applied-physics and materials-and-engineering conferences in recent years. All of the Ti-alloy literature, except that dealing with binary Ti-Nb, is covered. The review treats Ti-Nb fully only up to the mid-1960's during which period the research was basic and the results not too voluminous, but restricts itself to critically selected literature from that time up to the present day. The subject is covered in fourteen chapters. Chapter 1, which can be regarded as an expanded table of contents, summarizes all the literature to be reviewed in the remainder of the book. Chapter 2 deals with unalloyed Ti; Chapter 14 treats amorphous alloy superconductors based on Ti-Nb. The crystalline binary, ternary, and quaternary alloys are all discussed in the intervening eleven chapters. In them, for the reasons given above, Ti is regarded as the solvent element, and, as will be explained in the introduction to Chapter 1, the order of presentation follows a logic based on e/a ratio and the periodic table.

ACKNOWLEDGMENTS

ADMINISTRATIVE CREDITS

The idea of preparing a review of the literature of titanium alloy superconductors arose in November 1976 during conversations with Professor U. ZWICKER, Universität Erlangen-Nürnberg, who had earlier prepared a comprehensive treatise on their metallurgical properties. In the following year a search for financial support took place, as a result of which joint funding was obtained from Battelle Memorial Institute's Corporate Technical Development office and from the then Energy Research and Development Administration's (ERDA) Division of Magnetic Fusion Energy. The combined backing of F.G. DAWSON (CTD, Battelle) and E. ZIURYS (ERDA) gave the project its initial impetus. As the work proceeded, ERDA became the Department of Energy (DOE) and the host division became the Office of Fusion Energy; officers within it who continued to lend their support to the project were G.M. HAAS, E.N.C. DALDER, and subsequently D.S. BEARD. In Battelle, continued work on the review was strongly advocated by L.A. RANCITELLI during his association with CTD, while completion of this book and its companion volume have been made possible by F.J. MILFORD, Associate Director for Research, and by D.E. NIESZ, Manager of the Materials Department, who has given generous and whole-hearted practical support to the project at all phases of it, but particularly during the final writing and production stage.

Although a preliminary literature collection had been on hand for several years prior to 1977, the updating and expansion of it began in earnest at the beginning of 1978. A comprehensive bibliography was prepared in January 1978 by the Cryogenic Data Center of the Institute for Basic Standards, National Bureau of Standards, Boulder, CO, under the direction of N.A. OLIEN. The result of this computer survey conducted by the Center, now unfortunately defunct, was of inestimable value and provided another essential link in the chain of operations which needed to be performed before any serious writing could commence. Most of the literature identified by the search was already available in Battelle's Main and Foreign Science Libraries, respectively. In the latter, the Soviet literature was sought and copied by K. COTTRILL. In the Main Library, many hundreds of articles were acquired and copied under the direction of L.S. VAHEY to whom another debt of gratitude is owed. Literature to be collected consisted of: (a) regular journal articles, (b) review articles, (c) symposia proceedings, (d) government reports, (e) Ph.D. and other theses, (f) books, (g) patent documents. All the interlibrary arrangements were conducted again by L.S. VAHEY. Reports were obtained through our Report Library by D.L.E. WELCH and L.A. MASTERS; patents were acquired from or through our Patent Library by S.L. WAKEFIELD.

At the very beginning, a scheme of chapters, sections, and subsections was intuitively developed forming a complex set of categories into which the contents of the literature was to be sorted. The task of collecting, scanning, sorting, and indexing was completed by the end of January 1979, at which point the writing commenced. One of the chapters was to deal with "materials", all the rest with "properties" (i.e. superconductive, metallurgical, and physical). Writing commenced on the "materials" aspects; this manuscript was completed in first-draft form by the end of August 1979, and was set to one side. Work then commenced on the "properties" manuscripts. It had already become apparent that the first manuscript was far too large to be absorbed as a chapter in a single book. As work on the second manuscript proceeded it was decided, after consultation with A.F. CLARK of the National Bureau of Standards, Boulder, CO, that a pair of conjoint books should be prepared. The "materials" manuscript was to become the present book, while the "properties" manuscript was to be developed into a companion volume entitled *Applied Superconductivity, Metallurgy and Physics of Titanium Alloys* to join Plenum Press' Cryogenic Monograph Series (Consulting Editor, A.F. CLARK, NBS). During this period an agreement was reached with Plenum Publishing Corporation (L.S. MARCHAND, Senior Editor) to publish both books as a related pair.

Work on this volume recommenced in October 1980; the original manuscript was rewritten and enlarged and all relevant journal and conference literature that had appeared in the interim (including preprints of articles scheduled to appear in 1981) were incorporated into it. Final typing was started on a series of large tables which summarize the literature of the field, hence the book, and which constitute Chapter 1 of it. The finished manuscript, minus Chapter 1, was then sent out for critical review. Sincere thanks are due to J.D. SCUDIERE, K. HEMACHALAM, and R.E. SCHWALL of the Intermagnetics General Corporation who each carefully read the entire text. At about the same time, another copy of Chapters 2 through 14 was being reviewed by A.F. CLARK, NBS, Boulder, to whom I am also deeply indebted. Chapter 1 was painstakingly proofread at Battelle by V.E. WOOD whose valuable help in this regard, as well as for general and technical advice and encouragement, is hereby acknowledged. Incorporating the comments of all these reviewers, and armed with additional information provided by others (to be mentioned below), the manuscript was again fully revised, updated, and parts of it were again rewritten. The volume as it stands may be regarded as current as of May 1982.

Throughout the period of this writing, which has voraciously encroached not only into evenings and weekends but also into vacation time, my wife BETTY

COLLINGS has provided constant moral support and encouragement. As a practicing sculptor she has been able to appreciate the sacrifices demanded by intense creative labor of which book-writing is just one example.

TECHNICAL ACKNOWLEDGMENTS

During the writing of this review and its companion volume, numerous scientists world-wide have submitted additional information, helped solve difficulties, or contributed samples or photographs. Some of them are able to be recognized only in the preface to the companion volume; others, whose contributions were in one way or another also applicable to this book, are acknowledged (in alphabetical order) below: P.B. ALLEN, SUNY-Stony Brook (USA), for critical comment on the subject of transition temperature in strong-coupled superconductors; M.R. BEASLEY, Stanford University (USA), for helpful discussions of recent insights into paramagnetic limitation and, in that context, of spin-orbit-coupling effects; C.W. CURTIS, W.K. McDONALD, and M. SHIELDS, Teledyne Wah Chang Albany(USA), for general discussions on superconductor fabrication, and micrographs of cold-drawn and recrystallized Ti-Nb alloy (Fig.7-25); H. HILLMANN, Vacuumschmelze GmbH (W. Germany), for discussions on flux pinning and anisotropy effects, reprints, unpublished reports, and several micrographs (Figs.7-31 through 7-33); T. HORIUCHI, Kobe Steel Ltd (Japan), for a wealth of valuable information on the properties of Ti-Zr-Nb-Ta alloys, for conductor specifications (Table 13-6), information on J_c-measurement criteria, some photographs, and some samples of conductor to be photographed (Fig.13-7); Y. ISHIGAMI, Hitachi Cable Ltd (Japan), for discussions of the X-type and Z-type Ti-Zr-Nb alloy superconductors developed by Hitachi Ltd and information on their present status with regard to applications and production levels, also for supplying a copy of JAERI-M 8785 [Tad80]; C.C. KOCH, Oak Ridge National Laboratory (USA), for unpublished research results and data relating to precipitation in Ti-Nb alloys; D. LARBALESTIER, University of Wisconsin (USA), for discussions and copies of recent research results dealing with critical fields, critical current density, and flux pinning in Ti-Nb-base alloys; M.A.R. LeBLANC, University of Ottawa (Canada), for helpful discussions and information on "longitudinal" (i.e. "force-free" or "nearly force-free") current flow in superconducting wire; T. MASUMOTO, Tôhoku University (Japan), for introducing me to the subject of amorphous superconductors; H. NAKAMOTO, JAERI (Japan), for permission to reproduce material from JAERI Report M 8785 for use in Tables 7-22 and 11-9, and Figs.7-54 and 11-14; D.F. NEAL, IMI (England), for micrographs of transverse and longitudinal sections of Ti-58Nb wire (Figs.7-34 and 7-35); D.H. POLONIS, University of Washington (USA), for numerous reprints on metallurgical subjects and for helpful conversations and correspondence particularly in regard to the ω-phase reversion effect; R.W. ROLLINS, Ohio University (USA), for helpful information on the critical state in superconductors; H.R. SEGAL, MCA (USA) and, more recently Holec Draad (Netherlands), for information on critical-current criteria as applied to the measurement of certain Ti-Nb and Ti-Nb-Ta alloys, thereby enabling valid comparisons between their current-carrying capacities to be deduced (cf. Fig.12-9); S. SHIMAMOTO, JAERI (Japan), for supplying glossy prints (Figs.7-53) from JAERI-M 8785 of cross-sections of some of the conductors whose properties are reported in Figs.7-54; E. TADA, JAERI (Japan), for information concerning the H_r-measurement criterion adopted by the JAERI group (Table 11-9); B.I. VERKIN, Institute for Low-Temperature Physics and Engineering (USSR), for reviewing my list of Soviet alloy compositions and approving the composition ranges; H.WADA, National Research Institute for Metals (Japan), for information on the properties of various ternary alloy systems and on the criteria used in determining their H_r's and J_c's, also for supplying the previously unpublished H_r-data for the Ti-Hf-Nb alloys given in Fig.12-2; A.W. WEST, University of Wisconsin (USA), for helpful discussions on the range of α-phase precipitation in Ti-Nb alloys and related assistance; J.J. WHITE III, Battelle-Columbus (USA), for help in the interpretation of the results of low-temperature calorimetric measurements of mechanically deformed alloys; J. WILLBRAND, Krupp Forschungsinstitut (W. Germany),

for details of the thermomechanical process depicted in Fig.7-49(a) and for micrographs of Ti-50Nb (Figs.7-29 and-30), Ti-50Nb-Ge and Ti-50Nb-Cu (Fig.9-10); U. ZWICKER, Universität Erlangen-Nürnberg (W. Germany) for copies of doctoral theses unavailable in this country and micrographs of cold-swaged and swaged-and-heat-treated Ti-Nb alloys (Figs.7-26 through 7-28).

PRODUCTION CREDITS

Finally in this preface I am pleased to acknowledge the people who have collaborated in preparing this manuscript in camera-ready form. The thirty-nine tables which comprise Chapter 1, many of the other tables, and the Reference List, were typed in draft and final forms by C.L. OWENS who was also responsible for the first-drafting of Chapters 2 through 14. The completion of other tables and ancillary components, and the final typing and correcting of the entire main text, were performed conscientiously and ably by A.L. MAYNARD. I am also particularly grateful to J.S. WARD, technical illustrator, who provided invaluable technical advice during the production of this book and who inked and lettered the 180-odd line drawings; an essential and time-consuming task expertly and tastefully accomplished. J.S. WARD was also responsible for the layout of the book. All the lettering and numbering for the figures was composed by C.L. CONRAD who, with S. EHASZ, also composed the figure captions. The section-heads and running-heads were type-set by B.S. DOERMAN. Numerous other members of Battelle's Report and Photographic units also contributed their skills to the production of this book in its pre-publication form.

Materials Department   E.W. COLLINGS
Battelle Memorial Institute
Columbus OH, USA
May 1982

EDITORIAL PREFACE

GENERAL NOTE ON UNITS AND CONVENTIONS

UNITS

The cgs-practical system of units is generally, but not exclusively, used. Particular units are, however, specified locally enabling substitutions in any of the equations to be conveniently made. In magnetization experiments a c.g.s. applied field (oersteds) gives rise to a magnetization of, or induction within, the sample (gauss). If critical current is being measured it varies in response to that same applied field which, for the sake of consistency, should also be expressed in Oe (c.g.s) or A m^{-1} in SI units. The latter is hardly ever employed; the tesla, the SI unit of induction, being used instead since a convenient and numerically recognizable conversion to and from kG is then possible. In this book the unit kOe is preferred for applied magnetic field, H_a, although the tesla, $\mu_o H_a$, (equal to 10 kG) is occasionally used, especially when quoting from some of the more recent literature.

REFERENCES

References to the literature are listed at the end of the book. The format used is derived from the first three letters of the first author's name and the year of publication -- for example SMITH (1975) would be referred to as [Smi75]. As mentioned above, reference is frequently made to supportive information contained in a companion volume entitled *Applied Superconductivity, Metallurgy and Physics of Titanium Alloys* (E.W.C., Plenum Press, *in preparation*) by way of appropriate variants of the general citation [Mon00.00] -- see that entry under REFERENCES.

EQUATIONS

Equations are numbered thus: (P-Q), where P is the section number, and Q is the serial number. Occasionally it is convenient to repeat an equation; if so, its previously assigned number is also given, but in square brackets.

COMPOSITIONS

In designating alloy compositions, Ti is taken as the solvent element, and solutes are generally listed primarily in order of increasing group number (e.g. Ti-V-Cr, Ti-Nb-Fe) and secondarily in order of increasing atomic number (e.g. Ti-Zr-Nb, Ti-Nb-Ta) in such a way that the elements of the same periodic group are grouped together; an exception to this rule obviously occurs when Ti-Nb is treated as the solvent for the "interstitial elements" B, C, N, and O, or the "simple metals". Then depending on the point of view taken in the original publication, alloy compositions are expressed either in weight percent in formats such as Ti-X(n wt.%) or, for brevity, Ti-nX, or in atomic percent in formats such as Ti-X(n at.%), Ti-X$_n$, or Ti$_\ell$-X$_m$-Y$_n$ (where $\ell+m+n$ = 100). With regard to the binary Ti-Nb alloys, the research and technical literature combine to present compositions in terms of either: *(a)* at.% Nb, *(b)* wt.% Nb, or *(c)* wt.% Ti (e.g. Nb-46.5Ti). Accordingly, Table 7-1 has been constructed to enable the reader, whatever his preference with regard to compositional format, to conveniently convert between representations *(a)*, *(b)*, and *(c)*. The short-hand notation [M] is generally used used to denote "concentration of M".

SYMBOLS AND ABBREVIATIONS

LOCATION OF GREEK SYMBOLS

English	Greek	English	Greek
A	α	L	λ
B	β	M	μ
C	γ	N	ξ
D	Δ,δ	R	ρ
E	ϵ	S	σ
F	ζ	T	τ
G	η	U	Φ,ϕ
H	θ	X	χ
K	κ	Z	Ω,ω

ALPHABETICAL LIST OF SYMBOLS AND ABBREVIATIONS

A

AC   alternating current, alternating.

am   amorphous.

atm   atmosphere (pressure).

at.%   atomic percent.
"-X(n at.%)" indicates n at.% of element X, "-X$_n$" also indicates n at.% of element X.

α

α angle between the current direction and the rolling direction in rolled strip.

α exponent of M in the isotope effect.

α hexagonal-close-packed (hcp) crystal structure.

α pinning-force parameter, especially in tube-magnetization experiments.

(α—continued)

α the MAKI-WHH paramagnetic limitation parameter $\alpha = \sqrt{2}\, H_{c20}^{*}/H_{p0}$.

α_s the above calculated from superconductive quantities.

α_n the above calculated from normal-state quantities.

α' hexagonal martensite.

α'' orthorhombic martensite.

α^m generalized martensite.

α^H a magnetoresistance, $\alpha^H \equiv \rho^{-1}(\partial\rho/\partial H)_T$.

α^T a temperature coefficient of resistance, $\alpha^T \equiv \rho^{-1}(\partial\rho/\partial T)_{H=0}$.

B

B   magnetic induction (may be subscripted).

\<B\>   mean value of above within sample (wall).

B_0 magnetic induction just below sample surface.

B' magnetic induction at the interior wall of a tube.

BCS   BARDEEN, COOPER, and SCHRIEFFER [Bar57] (theory).

b_0 a constant in tube-magnetization theory.

bcc   body-centered cubic.

bct   body-centered tetragonal.

β

β angle between the applied-field direction and the rolling plane in rolled strip.

β bcc crystal structure.

β the MAKI paramagnetic-limitation mixed parameter defined by: $\beta^2 \equiv \alpha^2/1.78\lambda_{so}$.

β' a β-phase leaner in solute than the average.

β'' a β-phase richer in solute than the average.

C

C   Curie constant in magnetism.

C   total low-temperature specific heat.

C_{es} superconductive electronic specific heat.

c   concentration, as in dT_c/dc, etc. The "concentration of element A" is also occasionally represented by [A].

c   velocity of light.

cw   cold work.

γ

γ electronic specific heat coefficient.

γ_{bs} "bare" or band-structure-predicted electronic specific heat coefficient.

D

DC   direct current, steady.

DTA   differential thermal analysis.

d   subband diameter.

Δ, δ

$\Delta\sigma_{f\ell}$ extra conductivity due to superconducting fluctuations.

$\Delta\rho_{f\ell}$ $\Delta\rho_{f\ell} = -\Delta\sigma_{f\ell}$.

Δ_{00} BCS energy half-gap at zero K.

E

EDAX   energy dispersive analysis of x-rays.

e   charge on the electron.

e/a   electron/atom ratio -- referring to the "valence" or "s+d" electrons.

ε

ε — percent deformation (strain).

F

F_p — bulk or specific pinning force.

f — fraction (atomic-, mass-, mole-, volume-) of one of the components of a mixture, usually with subscripts.

f_p — elementary pinning force.

f_ϕ — individual-fluxoid driving force (includes the elementary Lorentz force, f_L).

ζ

ζ — valence.

G

G — gauss.

GLAG — GINZBURG, LANDAU, ABRIKOSOV, and GOR'KOV.

g — free energy per unit volume.

η

η — atomic-parameter in the HOPFIELD *et al* coupling-constant formula.

η — flux-flow viscosity.

H

H_a — the sample-surface value of an externally applied magnetic field.

H_m — amplitude of an applied AC field.

H_s — self-field due to a transport current.

H' — field in the bore of a tube in a tube-magnetization experiment.

H^* — value of a slowly increasing external field applied to a wall when the flux in the wall has just penetrated to the far side.

H^{**} — value of a dynamic applied field under the above conditions.

H_c — thermodynamic critical field.

H_{c1} — lower critical field.

H_{c1}^* — field at the "first" maximum of the $4\pi M$ versus H_a loop.

H_p — the CLOGSTON paramagnetically imposed upper critical field limit.

H_r — the resistively determined upper critical field.

H_u — the magnetically or calorimetrically determined upper critical field.

H_{c2} — upper critical field -- either as an abbreviation or to signify the paramagnetically limited value.

H_{c2}^* — upper critical field in the absence of para-magnetic limitation.

H_{c3} — the "third", or surface, critical field.

$H_{()0}$ — all of the above critical fields at zero K.

HW — HELFAND and WERTHAMER [Heℓ64, Heℓ66] (theory).

h^* — a reduced upper critical field defined by $h^* \equiv H_{c2}/(-dH_u/dt)_{t=1}$.

h — reduced magnetic field; $h = H_a/H_{c2}$.

h — hour(s).

hcp — hexagonal close packed.

h — Planck's constant.

$\hbar$ — $h/2\pi$

θ

θ_D — the Debye temperature.

θ_W — the Weiss temperature in magnetism.

I

I.D. — inside diameter.

(I—continued)

Intl   "interstitial element" such as B, C, N, and O.

in   inch(es).

J

J   current density.

J_c critical current density.

K

k_B Boltzmann's constant.

ksi 10^3 pounds per square inch (pressure).

κ

κ_{GL} the Ginzburg-Landau parameter (valid near t = 1). Also used with subscript-1 to emphasize its t = 1 value.

κ_{GL}^c the intrinsic or "clean" component of κ_{GL}.

κ_{GL}^d the electron-scattering or "dirty" component of κ_{GL}.

κ_1 a MAKI "Ginzburg-Landau" parameter; used also with the subscripts '2' and '3'.

L

L_p twist pitch.

ℓ as a superscript, the length of an object.

ℓ electronic mean free path length; also specialized to s-wave-scattering mfp when a distinction is needed, as in EILENBERGER theory.

ℓ_{tr} in EILENBERGER theory, the average transport mean free path.

λ

λ electron-phonon coupling constant.

λ s-electron scattering frequency parameter of HW theory.

λ_{tr} s- and p-electron-lattice (anisotropic) scattering frequency parameter of EILENBERGER theory.

λ_{so} the spin-orbit-scattering frequency parameter.

M

M   atomic mass.

M   magnetization.

MHD   magnetohydrodynamic (electrical generation).

m   mass of the electron.

mfp   mean free path.

min   minute(s).

μ

μ^* electron-electron Coulomb pseudopotential.

μ_B Bohr magneton.

μ_o permeability of space to permit the conversion of magnetic field units from oersteds to tesla.

μ_{eq} differential permeability of an unpinned (equilibrium or reversible) superconductor.

N

N_{eff} an effective atomic-size-effect-corrected electron/atom ratio.

$n(E_F)$ Fermi density of states.

n   volume density of conduction electrons.

n_p number density of precipitated particles.

n subscript signifying normal-state property

ξ

ξ superconducting coherence length.

0

O.D.   outside diameter.

Q

$\dot{Q}$ power dissipated per unit volume of a composite conductor.

$\dot{Q}_h$ hysteretic loss of the superconducting component of a composite.

$\dot{Q}_e$ eddy-current loss in the matrix of a composite.

R

RF   radio frequency.

ρ

ρ electrical resistivity in general.

ρ_f flux-flow resistivity.

ρ_n residual (normal-state) electrical resistivity.

$\rho_\perp$ transverse resistivity of a composite conductor.

S

S   entropy.

S   area of the Fermi surface.

S_F area of the free-electron Fermi sphere.

SM   "simple metal" such as Al, Ga, In, Sn, Pb, etc. or a metalloid.

SAD   selected area diffraction; a special technique of transmission electron microscopy.

SEM   scanning electron microscopy.

STEM   scanning transmission electron microscopy.

s   second(s).

s   spin, usually of a localized magnetic moment.

s subscript signifying superconducting-state property.

s/n superconducting-to-normal state

σ

σ_B stress at point of fracture; i.e. ultimate tensile strength.

σ_n normal-state residual electrical conductivity.

T

T_c superconducting transition temperature.

T_{c0} above in zero magnetic field when a distinction is useful.

$T_{c,()}$ superconducting transition temperature of the () phase; () may represent: α, β, or ω.

T_1 nuclear spin relaxation time.

TM   transition metal.

TEM   transmission electron microscopy.

T-T-T   "time, temperature, and transformation"; a type of diagram representing the kinetics of phase transformation.

t   reduced superconducting transition temperature; $t \equiv T/T_c$.

t   as a superscript, the thickness of a strip, ribbon, or layer.

τ

τ total electron-scattering relaxation time.

(τ—continued)

τ_{so} spin-orbit-scattering relaxation time.

τ_{tr} normal electron-scattering relaxation time.

U

Φ, φ

ϕ_0 the flux quantum.

ϕ as a superscript, the diameter of a precipitate or wire.

ϕ diameter of a small cylinder or wire.

V

V   BCS electron-phonon pairing potential.

V_g gap-voltage in a tunneling experiment.

v   unit-cell volume.

v_F Fermi velocity.

v_f flux-flow velocity.

W

WHH   WERTHAMER, HELFAND, and HOHENBURG [Wer66] (theory).

w   thickness of a superconducting slab or cylinder wall.

wcch   water-cooled copper hearth (used for arc melting).

wt.%   weight percent.
"-X(n wt.%)" indicates n wt.% of element X,
"-nX" also indicates n wt.% of element X.

X, Z

Z   atomic number.

χ, Ω, ω

χ magnetic susceptibility.

Ω atomic volume.

ω phonon frequency (expressed as a temperature).

ω ω-phase.

CONTENTS

2. UNALLOYED TITANIUM   123

6. TERNARY ALLOYS OF TITANIUM WITH SIMPLE METALS AND TRANSITION METALS (EXCEPT NIOBIUM)

7. TITANIUM-NIOBIUM BINARY ALLOYS

8. TITANIUM-NIOBIUM AND TITANIUM-NIOBIUM-BASE ALLOYS CONTAINING SMALL ADDITIONS OF BORON, CARBON, NITROGEN, OR OXYGEN   299

1

TITANIUM ALLOY SUPERCONDUCTORS—
A TABULATED REVIEW

Chapter 1 is a summary in tabular form of all of the literature to be considered in Chapters 2 through 14. Accordingly this introduction to it serves as an introduction to the entire volume. Although quite valid arguments have been advanced for regarding Nb as the solvent in superconducting alloys of Nb and Ti and related multicomponent systems [Lar81], for other valid and fundamental reasons this review treats Ti as the host element. The most important system to be considered is Ti-Nb, an alloy upon which most of the large-scale superconducting magnet systems constructed to date have been based. Thus the largest number of table and text pages have been allocated to it. But *equilibrium* Ti-Nb is a member of the class of so-called β-isomorphous Ti-TM alloys which include Ti-V, Ti-Ta, and Ti-Mo; the *quenched* Ti-Nb alloys support the same microstructures that are to be found in quenched Ti-TM alloys in general, while during the aging of them common precipitational effects are encountered. Thus to fully understand the properties of Ti-Nb super-conductors it is necessary to have them presented within the context of Ti-TM alloys in general. This is one of the functions of the present work. Only electronic and magnetic properties are treated, together with sufficient metallurgical information to assist in their interpretation. All other properties, such as stress effects and so on, are considered in a companion volume to which some 250 references are made by calling on appropriate variants of the general citation [Mon00.00] (see REFERENCES). This chapter

and the text to follow are systematically laid out on an e/a-ratio basis and according to the logic of the periodic table. Thus the first alloys to be considered are those of Ti with each of the 3d-TM elements in succession, viz: V, Cr, Mn, etc. The next binary solutes are taken in pairs from the second (4d) and third (5d) long periods. This leads to the sequence: Ti-Zr,Hf; Ti-(Nb),Ta; Ti-Mo,W; and so on. The systematically treated ternary alloys of Table 1-26 and Chapter 6 do not contain Nb. Because of its importance, Ti-Nb is accorded special treatment in Tables 1-27 through 1-29 and Chapter 7, and multicomponent alloys based on it are considered after the binary alloy has been reviewed. When considering the Ti-Nb-Intl alloys (Table 1-30, Chapter 8) and the Ti-Nb-SM alloys (Table 1-31, Chapter 9) the Interstitial elements and the Simple Metals are treated in order of increasing atomic number. In the all-TM multi-component Ti-Nb-base alloys, the elements Zr and Hf are regarded as substituting for Ti with which they are isoelectronic; likewise V and Ta are regarded as substituting for Nb. This philosophy leads to alloy designations such as Ti-Zr-Nb, Ti-Nb-V, Ti-Zr-Nb-Ta, with total disregard for the relative amounts of each constituent. In both this chapter and the main text, opportunities are frequently taken to subdivide the properties into the *primary, secondary, and tertiary* superconductive categories: Transition Temperature, Critical Field (or Mixed State), and Critical Current Density.

TABLE 1-1 UNALLOYED TITANIUM — ALPHA-PHASE (hcp) TITANIUM

Sample Details	Refrigeration	Measurement Technique	Remarks, Discussion	Transition Temperature, K	Literature
Single crystal of purity 99.75%. Principal impurities: Zr (0.2%); Pb (0.03%) and possibly some Fe.	Pumped liquid He	Electrical resistance	This sample, and those referred to in in the following three citations, were prepared by the "iodide process" in the Philips Laboratory at Eindhoven. The sample was not completely superconducting at 1.13 K.	<1.13	W. Meissner (1930) [Mei30]
Unstated purity	--------------	Electrical resistance	Broad transition region.	1.72	W.J. de Haas and P.M. van Alphen (1931) [Deh31]
Two new samples from the same source as that referred to in the first entry.	Pumped liquid He	Electrical resistance	Several temperatures noted for various samples, the highest being 1.77 K.	<1.77	W. Meissner *et al.* (1932) [Mei32]
Thick wire, of purity 99.9%. Principal impurity: Ca.	Pumped liquid He	Magnetic permeability	Down to 1.0 K no complete superconductive transition was noted. A small diamagnetic anomaly was observed at 1.5 K.	<1.0	D. Shoenberg (1940) [Sho40]
Impurities: Mg < 0.25%; Si, 0.02%; Fe, 0.03%; H < 0.01%	Pumped liquid He	Electrical resistivity	Sample prepared by the Kroll process (i.e. the reduction of titanium chloride by Mg at about 1000°C in the presence of pure argon). Resistance dropped by 50% between 3.2 and 1.5 K and began to level off at 1.1 K, the lower limit of the apparatus. The term "incipient superconductivity" was applied to the results of these observations.	No complete transition noted above 1.1 K.	R.T. Webber and J.M. Reynolds (1948) [Web48]
Purity: 99.95%	Temperatures below 1 K were obtained by adiabatic demagnetization from a field of 8 kOe.	Magnetic susceptibility of salt pill with imbedded sample using ballistic mutual inductance.	$(dH_c/dT)_{T=T_C} = 470$ Oe K^{-1}	0.527 ± 0.006 (Curie scale)	J.G. Daunt and C.V. Heer (1949) [Dau49]
Same sample as above but heat treated at 800°C for 2-1/2 hours.	Same as above	Same as above	Results practically unchanged by heat treatment. (Iodide Ti is unstrained in the as-prepared condition.) Rockwell hardness: as-received, B32; annealed, B28. $(dH_c/dT)_{t=1} = 450$ Oe K^{-1}.	0.558	T.S. Smith and J.G. Daunt (1952) [Smi52]

TABLE 1-1 UNALLOYED TITANIUM — ALPHA-PHASE (hcp) TITANIUM -*continued*

Sample Details	Refrigeration	Measurement Technique	Remarks, Discussion	Transition Temperature, K	Literature
Purity: 99.99% vacuum annealed at 670°C for 3 hours	Same as above	Same as above	Iodide process Ti obtained from the Foote Mineral Co. Vickers hardness: as-received, 76 kg mm^{-2}; annealed, 67 kg mm^{-2}. $(dH_c/dT)_{T=T_C}$ = 89.5 Oe K^{-1}; H_{c0} ∼ 20 Oe	0.387	T.S. Smith *et al.* (1953) [Smi53]
Ti crystal bar, purity: 99.99% Impurities: O ∼ 0.005%; Si ∼ 0.0001%	Temperatures below 1 K were obtained by adiabatic demagnetization from field of a Bitter solenoid.	Same as above	Iodide process Ti (commonly referred to as "crystal bar") origin unknown. Vickers hardness: ∼70 kg mm^{-2}. $(dH_c/dT)_{T=T_C}$ = 400 Oe K^{-1}	0.49 ± 0.01	M.C. Steele and R.A. Hein (1953) [Ste53]
Unannealed wire from Ti crystal bar, purity: 99.98% Impurities: O ∼ 0.01%; Cr ∼ 0.01%.			Wire prepared by cold swaging crystal bar Ti. Vickers hardness: ∼70 kg mm^{-2}. $(dH_c/dT)_{T=T_C}$ = 465 Oe K^{-1}	0.37 ± 0.01	
Review paper	-------------	-------------	Results of a critical review of the literature through 1953. H_{c0} = 20-130 Oe	0.387-0.56	J. Eisenstein (1954) [Eis54]
Ti^{46} Purity: 99.6%	Initial temperature reduction to 1.15 K by pumping followed by adiabatic demagnetization of a separate but thermally coupled salt from a maximum field of 11 kOe.	Mutual inductance at 96 Hz, with the sample acting as the core of the transformer.	Abundance Ti^{46}, 86.4 at.%; mean isotopic mass, 46.3	0.36 ± 0.02	R. Netzel (1960) [Net60]
Ti^{48} Purity: 92.5%			Abundance Ti^{48}, 99.19 at.%; mean isotopic mass, 48.0	0.22 ± 0.01	
Ti^{50} Purity: 99.6%			Abundance Ti^{50}, 84.1 at.%; mean isotopic mass, 49.7. Samples Ti^{46} and Ti^{50} were obtained from the Oak Ridge National Laboratory (ORNL). Spectroscopic and isotopic analyses were made at ORNL. The measurements were carried out in order to investigate the "isotope effect".	1.11 ± 0.01	
325-mesh powder Quoted purity: 99.5%	Same as above	Same as above	Obtained from A. D. Mackay Co. X-ray analysis revealed approximately equal amounts of Ti and TiC.	Essentially no transition observed down to 0.16 K.	Same as above

TABLE 1-1 UNALLOYED TITANIUM — ALPHA-PHASE (hcp) TITANIUM—*continued*

Sample Details	Refrigeration	Measurement Technique	Remarks, Discussion	Transition Temperature, K	Literature
100-mesh powder Quoted purity: 99.5%	Same as above	Same as above	Obtained from above supplier. Yielded an x-ray powder pattern for pure Ti.	Same as above	R. Netzel (1960) [Net60]
USBM powder size between 8- and 34-mesh Brinell hardness: 78 Purity: 99.88%	Same as above	Same as above	Electrolytically refined Ti obtained from U.S. Bureau of Mines (USBM), Boulder City, Nevada. Large crystallites deposited from fused salts at 850°C and cooled slowly to room temperature. Crystallites distorted and highly strained.	Same as above	Same as above
USBM powder size between 34- and 48-mesh Brinell hardness: 81 Purity: 99.82%	Same as above	Same as above	Same preparation procedure as above.	Same as above	Same as above
Ti crystal bar; purity at least 99.92%.	Same as above	Same as above	Obtained from the Foote Mineral Co; turned down to a cylinder 1.84 in. long and 0.12 in. in diameter.	0.27 ± 0.01	Same as above
Same as above	Same as above	Same as above	Turnings from the above machining operation consisting of 0.0015 in. cuts taken at slow speed.	No evidence of a transition down to 0.25 K.	Same as above
Electron beam melted specimen of selected high-purity Ti Impurities: Mn < 5 ppm; Fe < 2 ppm.	Adiabatic demagnetization	Ballistic mutual inductance	Sample yielded a reversible magnetization curve -- hence no trapped flux. Influence of trace Mn and Fe impurities studied. $H_c = 56[1-(T/0.42)^2]$	0.42	R.L. Falge (1963) [Fa63]
Series of argon-arc-melted α-Ti-Ir alloys. Starting material: iodide Ti from Foote Mineral Co.	Above 1.2 K, He[4] Below 1.2 K, He[3]	Above 1.2 K, ballistic mutual inductance during the collapse of a 10 Oe field. Below 1.2 K the Schawlow-Devlin resonance frequency shift technique was used.	Transition temperature of α-Ti was determined by extrapolating the data for α-Ti-Ir solid solutions to zero solute concentration.	0.4	C.J. Raub and G.W. Hull (1964) [Rau64c]

TABLE 1-1   UNALLOYED TITANIUM — ALPHA-PHASE (hcp) TITANIUM *-continued*

Sample Details	Refrigeration	Measurement Technique	Remarks, Discussion	Transition Temperature, K	Literature
Arc melted sample Impurities: Fe, 9.2 ppm; Mn, 4.0 ppm.	Adiabatic demagnetization	Specific heat in temperature range 0.2 to 1.2 K.	Reference point with which data on Ti-Fe alloys could be compared. The ±0.03 K uncertainty in T_c refers to the half-width of the transition.	0.33 ± 0.03	R.H. Batt (1964) [Bat64]
Iodide Ti Purity: 99.99%	Temperatures in the range 0.06-0.6 K were obtained by adiabatic demagnetization.	Mutual inductance method	T_c was measured at pressures of about 9, 16, 19, 23, 24 and 25 kbar. Hydrostatic compression was accompanied by increase in both $(\partial H_c/\partial T)_{T_c}$ and T_c. $(\partial T_c/\partial P) = 6\times10^{-3}$ K kbar^{-1}	Before pressurization: $T_c = 0.23$ K After pressurization to 25 kbar: $T_c = 0.32$ K	N.B. Brandt and N.I. Ginzburg (1965) [Bra65]
Iodide Ti Purity: 99.99%	Same as above	H_c determined magnetically as function of temperature. T_c determined by extrapolating $H_c(T)$ to zero field.	Various samples measured in what was essentially an experiment to determine the pressure dependence of T_c for pressures of up to about 25 kbar. $(\partial T_c/\partial P) = 7\times10^{-3}$ K kbar^{-1}	0.23 0.26 0.32-0.34	N.B. Brandt and N.I. Ginzburg (1966) [Bra66]
Electron-beam plus levitation melted iodide Ti Impurities: C and 0 < 0.01, Al < 0.001 at.%; S, N, Cu, Fe, V, Mn, <300 ppm (at.); others, <10 ppm (at.). each	Adiabatic demagnetization	Ballistic method	Experiment aimed at comparing the T_c's of the α and ω modifications of Ti -- the latter formed under hydrostatic pressure (120kbar).	0.45 ± 0.03	V.F. Degtyareva *et al.* (1974) [Deg74]
Estimated critical value	---------	-----------------	---------------------------	0.387	T.S. Smith *et al.* (1953) [Smi53]
				0.37±0.01	M.C. Steele and R.A. Hein (1953) [Ste53]
				0.33±0.03(half-width)	R.H. Batt (1964) [Bat64]
			Mean	0.36±0.03(std.dev.)	

TABLE 1-2 UNALLOYED TITANIUM — BETA PHASE (bcc) TITANIUM

Sample Details	Refrigeration	Measurement Technique	Remarks, Discussion	Transition Temperature, K	Literature
Ti-Rh alloys were prepared from Ti sponge (purity: 99.5%) and Rh powder (purity: 99.98%).	Liquid He in the temperature range 1.8-4.2 K	Electrical resistance	By extrapolating the T_c of Ti-Rh alloys to zero Rh concentration a value of about 4 K was obtained. It was suggested that this was a reduced value due to inclusions of α-phase. As a consequence a T_c value for β-Ti as high as 6 K was assumed to be possible. Similar results were obtained for both as-cast and annealed (60h/700°C plus 30 h step-cooled to room temperature) samples. The obtaining of a "good" value for T_c (β-Ti) was fortuitous.	4-6	W. Buckel *et al.* (1962) [Buc62]
Ti-Rh alloys were prepared from iodide Ti obtained from Foote Mineral Co. and Rh (99.8%, Johnson and Matthey).	Liquid He	Change of inductance of coil surrounding sample	The T_c *vs* composition curve was one of the earliest true representations of the influence of quenched microstructure (α^m-martensite, ω-phase, β-phase) on T_c. But the "β-Ti value" obtained was the result of an extrapolation from the ω-rich-segment of the T_c curve, and consequently is more nearly representative of T_c (ω-phase).	$< T_c(\alpha\text{-Ti})$	C.J. Raub and C.A. Anderson (1963) [Rau63]
Ti-Mo alloys were prepared from Ti sponge (Titanium Metals Corp., purity: 99.88%) and Mo bar (Climax Molybdenum Co., purity: 99.98%).	Liquid He	Low temperature specific heat, 1.8 ∿ 6 K	A $T_{c,\beta}$ was obtained by extrapolating the single-phase β segment of the curve of T_c *vs* Mo concentration.	6.4	E.W. Collings *et al.* (1972) [Col72]

TABLE 1-3 UNALLOYED TITANIUM — OMEGA PHASE, THIN FILMS, and AMORPHOUS

Sample Details	Refrigeration	Measurement Technique	Remarks, Discussion	Transition Temperature, K	Literature
Electron-beam plus levitation melted iodide Ti Impurities (at.%): C and O < 1.10^{-2}; Al < 1.10^{-3}; S, N, Cu, Fe, V, Mn < 3.10^{-4}; others <1.10^{-5} each	Adiabatic demagnetization	Ballistic method	The ω-phase structure was retained at 1 atm after quasihydrostatic compression to 120 kbar for 30 min only in the sample containing least oxygen (≈0.01 at.%). ω-Ti remained normal down to 0.06 K.	<0.06	V.F. Degtyareva *et al.* (1974) [Deg74]
Film deposited by Xe ion beam sputtering Gas purity: >5N; Target purity: >4N	-------------	-------------	Deposition rate 50-300Å/min over an area of 1 sq.in. Typical sample thickness 1-5 μm. Increase in lattice parameters above those of the bulk material: Δa, 0.8%; Δc, 1.0%. In general T$_c$ was a strong function of atomic radius of bombarding ion.	2.52	P.H. Schmidt (1973) [Sch73]
Amorphous transition metals including Ti	-------------	Theoretical	T$_c$ calculated using McMillan equations upon the assumption that the atomic volume change ΔΩ between crystalline and amorphous phases was similar to that encountered upon melting. The effect of ΔΩ upon the McMillan electron-phonon parameters is central to the calculation.	0.52	G. Kerker and K.H. Bennemann (1973) [Ker73]

TABLE 1-4 TITANIUM-VANADIUM ALLOYS — THE SUPERCONDUCTING TRANSITION

V Concentration Range	Starting Materials, Sample Preparation	Properties and Procedures	Remarks, Discussion	Other Systems Studied	Literature
From published curve: 50, 90, 97, 98.5, 100 at.%	Argon-arc melted on wcch; annealed near the mp for several days under electron bombardment at about 10^{-7} Torr.	T_c studied magnetically as function of average electron/atom ratio (e/a) for bcc Ti-V and other alloys.	In one of the first papers to display T_c vs e/a, T_c whose maximal value was 6.7 K for e/a = 4.5, was plotted for e/a = 4.5, 4.9, 4.97, 4.985 and 5.0.	Ti-Mo V-Mo	E.Bucher et al. (1959) [Buc59]
From published curve: 1, 2, 3, 4, 5, 6, 7, 8, 10, 15 at.%	Argon-arc melted and quenched. The alloys with 10 and 15 at.% V were also annealed for several days at 680°C.	T_c measured magnetically vs composition for dilute amounts of "nonmagnetic" solutes in response to a previously claimed enhancement of T_c [Mat59] by "magnetic" solutes.	It was determined that V in Ti-V raised T_c just about as strongly as Cr, Mn, Fe and Co, thus invalidating the postulated "magnetic interaction" mechanism for superconductivity.	Ti-Nb	E.Bucher and J.Muller (1961) [Buc61]
From published curve: 10, 14, 29, 39, 49, 59, 69, 79, 90 at.%	Iodide Ti (Fe<100ppm,etc.). Iodide V (Fe<280ppm,etc.). From Battelle Memorial Institute. Arc melted on wcch under Zr-gettered argon.	T_c measured magnetically (using the Hardy-Hulm method) on a comprehensive series of alloys.	It was shown that T_c passed through two maxima as compositions move along a transition metal row in the periodic table. The availability of data for Ti-V-Cr from the work of Cheng, Wei and Beck [Che60] enabled the first, albeit qualitative, comparison to be made between experimental T_c and that calculated from BCS-Goodman.	Comprehensive matrix of binary transition metal alloys within and between the three long periods, including Ti-Zr, Nb.	J.K.Hulm and R.D.Blaugher (1961) [Hu61]
20, 30, 50, 75, 85 at.%	Iodide Ti and 99.8% V; argon-arc melted on wcch. Annealed 3 days/1200°C and oil quenched.	Low temperature specific heat measured in temperature range ~1.6 K to ~8 K enabling a comparison to be made with the predictions of the Goodman-modified BCS expression for T_c, viz. $T_c/\theta_D = 0.855 \exp[-1/n(E_F)V]$	Metallographically single phase, but presence of ω-phase suspected in Ti-V (20 and 30 at.%). Averaged over all alloys (≥30 at.% V) measured, the BCS parameter V had the value $0.128_8 \pm 0.002$ eV-atom.	------	C.H.Cheng et al. (1962) [Che62]
Review including Ti-V	Homogenized by careful vacuum annealing 50-100h/1500-2500°C and either furnace cooled or quenched -- depending on composition range.	Review of the superconductive literature and discussion with reference to metallurgical microstructure.	The collected results for binary Ti-TM alloys are presented in diagrams of T_c vs at.% solute, in which the boundary of the quenched α^m-martensitic field (at which T_c makes a pronounced dip) is indicated.	Alloys of Ti with V and ten other transition elements and numerous compounds.	U.Zwicker (1963) [Zwi63] see also [Zwi74, p.81]

V Concentration Range	Starting Materials, Sample Preparation	Properties and Procedures	Remarks, Discussion	Other Systems Studied	Literature
4 at.%	Argon-arc melted repeatedly; quenched from slightly below solidus temperature in jet of cold argon.	The superconducting transition in hcp Ti-V,-Nb, -Fe alloys was studied in a further investigation of the validity of Matthias' "magnetic interaction" model. Low temperature specific heat fitted to a function which assumes the two-fluid model for superconductivity for $T_c/2 < T < T_c$ (better than BCS), and a normal distribution of the volume fraction, x, undergoing the superconducting transition. Ballistic magnetic measurements (1 Oe field) also carried out.	Plots of $\ell n(T_c/\theta_D)$ vs $1/\gamma$ were in essential agreement with those for the isoelectronic Ti-Zr, thus substantiating a BCS-type electron-phonon picture for superconductivity. A comparison was made between magnetically detected and calorimetrically detected transitions. It was noted, for example, that in "many cases complete flux exclusion may be observed as soon as 10% of the volume is superconducting". An inexplicable, reduced, calorimetric anomaly was noted in Ti-Fe.	Ti-Fe Ti-Nb	J.Heiniger and J.Müller (1964) [Hei64]
See above	See above	Same as above, but with the addition of some extra alloys.	Same as above	Ti-Cr,-Fe,-Nb,-Mo Ti-V-Cr Ti-Zr-TM (9 alloys)	E.Bucher *et al.* (1965) [Buc65]
2.5 at.%	Iodide Ti (Foote Mineral Co.) and V (99.8%; Vickers hardness -- 60 kg mm^{-2}). Argon-arc melted three times. Cold rolled 40% and annealed 2h/650°C in Ti cylinder in evacuated silica tube.	T_c's of Ti-V,-Nb and -Ta, were measured in an investigation of superconductivity in *equilibrium* α-phase Ti-base alloys.	Homogeneous α-phase attained $T_c = 1.4$ K. This was the first study of *equilibrium - hcp* Ti-V,-Nb and -Ta alloys. It was claimed that the previous work on quenched, supposedly single-phase alloys, may have been perturbed by the presence of threads of β-phase. Such an objection can scarcely be leveled in good faith against the results of calorimetric measurements. Nevertheless, from a comparison of this work with the results of the previous measurements on martensitic-α' (hcp) alloys it appears that the latter do in fact exhibit an enhanced T_c.	Ti-Nb Ti-Ta	C.J.Raub and U.Zwicker (1965) [Rau65]
Review	------	The significance of a solute-size-corrected "effective" electron/atom ratio, N_{eff}, was studied in an attempt to universally parameterize the positions of T_c and H_{c2} maxima.	The systematic investigation of T_c in binary alloys from Groups IV, V, VI and VII show that maxima occur at $N_{eff} \sim 4.4$ and 6.6. The positions of the peaks were claimed to be the same in all the (3d, 4d, 5d) periods. N_{eff} was claimed to be a better descriptor than simple average e/a, and is of course, proportional to the "electron density" investigated later by Matthias and Jensen. H_{c2} also peaked at $N_{eff} \sim 4.4$, in verification of the relationship $H_{c20} = 3.1 \times 10^4 \rho_n \gamma T_c$ (Oe).	Ti-Zr Ti-Nb,-Mo	W.DeSorbo (1965) [Des65]

TABLE 1-4 TITANIUM-VANADIUM ALLOYS — THE SUPERCONDUCTING TRANSITION *–continued*

V Concentration Range	Starting Materials, Sample Preparation	Properties and Procedures	Remarks, Discussion	Other Systems Studied	Literature
60, 70 at.%	Sample ground to powder, 40-100 $\mu m^{\diamond}$ for Ti-V$_{0.70}$, otherwise 1-10 $\mu m^{\diamond}$.	T_1, the nuclear spin-lattice relaxation times of V and Nb in Ti-V and Zr-Nb were measured. The spin-echo experiment was performed in the temperature range 1.4-4.2 K in a field of about 10 kOe.	The first measurement of T_1 in a *hard* superconductor. The linearity of log T_1 *vs* T_c/T verified that $T_1 \propto \exp(\Delta/k_B T)$ with $2\Delta = 3.5\ k_B T_c$, consistent with any gap model for superconductivity. Agreement with BCS theory was implied by drawing attention to the fact that the BCS gap was also $3.5\ k_B T_c$.	Zr-Nb	K.Asayama and Y.Masuda (1965) [Asa65]
~10-100 at.%	Iodide Ti (99.9%) and "carbothermal" V (99.766%) arc melted repeatedly on wcch under 0.7 atm of Ti-gettered He. Cast samples processed by rolling, drawing, annealing.	T_c was studied as a function of alloying; the effect of annealing time and temperature on T_c and J_c in cold-worked Ti-V was also investigated.	T_c is the same for as-cast and cold-worked alloys. The T_c-composition curve has a maximum at about 70 at.% V. High temperature annealing may cause T_c to rise or fall depending on the compositions of the decomposition products.	-----	Yu.V.Efimov *et al.* (1967/1970) [Efi70]
30, 40, 50, 60, 70, 80, 90, 100 at.%	Ti, 99.9%; major impurity: 0.01-0.02% Zr. V, 99.9%; major impurity: 0.03% Fe plus O, C, N. Arc melted twice, homogenized 24h/1600°C in 10^{-6} Torr, furnace cooled. Cold rolled 15:1 to sheet, hence strip.	Transition temperatures were measured using a self-inductance technique in an exploration of the ternary Ti-Nb-V system.	Some inconclusive references were made to Fermi density-of-states, n(E$_F$), variation with composition. The highest transition temperatures in the Ti-Nb-V system were to be found within Ti-Nb itself (10.0 K, centered at about 60 at.% Nb).	Ti-Nb Ti-Nb-V	P.H.Bellin *et al.* (1969) [Bel69]
24.4 at.%	Iodide Ti and 99.9% V were levitation melted together and cast. Solution heat treatment was followed by isothermal aging and also up-quenching.	T_c was measured by monitoring, as a function of temperature, the impedance of a coil surrounding the specimen (essentially responsive to AC susceptibility at 1 kHz). A comparison was made of the T_c's of Ti-V (and Ti-Cr) in the aged (ω+β structure) and up-quenched (β+β' structure) conditions. The effects of aging on the *microstructural* and *superconductive* properties were investigated.	Ti-V (24.4 at.%) T_c: (a) quenched (β) 4.382 K, (b) 35h/300°C (ω+β) 3.875 K, (c) above, plus 3min/450°C (β+β') 5.089 K. The higher T_c of the up-quenched alloy is a result of the V-enriched β phase. Since the Ti-rich phase is stable it may be possible to deduce $T_{c,\beta}/T_{c,\omega}$ for one concentration of V. T_c of Ti-V (24.4 at.%) decreased rapidly with aging time at 300°C as ω-phase precipitation developed. T_c dropped from 4.43 K at 10 min to 3.97 K after about 10-1/2 h then to 3.88 K after 35 h.	Ti-Cr,-Fe Ti-Nb	T.S.Luhman (1970) [Luh70] see also [Luh71]

V Concentration Range	Starting Materials, Sample Preparation	Properties and Procedures	Remarks, Discussion	Other Systems Studied	Literature
3, 9, 12, 15, 17.5, 19, 20, 25, 30, 40, 50, 60, 80 at.%	Electrorefined Ti sponge (>99.8%) from Titanium Metals Corp. "vp" grade V (99.5%) from Materials Research Corp. Argon-arc melted repeatedly. Annealed 1h/1000°C and quenched. Some alloys additionally annealed 8h/1350°C and quenched.	The low temperature specific heat parameters γ, θ_D and T_c of a series of Ti-V alloys in the microstructural ranges α', $\omega+\beta$ and β were measured.	Plots of C/T vs T^2 for each alloy; also plots of T_c, γ, θ_D against composition. The presence of ω-precipitation depresses T_c; however, the transition appears to be complete (or "bulk") indicating that a superconductive proximity effect must be operative.	------	P.E.Upton (1972) [Upt72]
1, 2 at.%	Ti from Titanium Metal Corp. (purity: >99.8%), V (99.93%) from Bureau of Mines, Boulder City. Arc melted under elaborately cleaned conditions in Ti-gettered argon; frequently melted. Samples for measurement annealed and ice-brine quenched.	The low temperature specific heats, in the temperature range of about 1.3-4.7 K, and the electrical resistivities of dilute Ti-TM (3d) α^m-phase alloys were measured in a partially successful attempt to clarify the discrepancies among earlier resistive, magnetic and calorimetric studies.	Neither of the Ti-V alloys were super-conducting above 1.6 K, consistent with interpolations between the Heiniger and Müller [Hei64], Raub and Zwicker [Rau65] and pure Ti (Table 1-1) data. The results for Ti-Cr,-Fe and -Co were anomalous and inexplicable at the time.	Ti-Sc,-Cr Ti-Mn,-Fe,-Co Ti-Ni Ti-Hf	K.L.Agarwal (1974) [Aga74]
3-80 at.% (See above)	See above	The low temperature specific heats, in the temperature range of about 1.8-9.8 K, of a series of Ti-V alloys were measured after quenching, and in the case of Ti-V (15 and 19 at.%) "during" aging at 300°C for times up to 2000 h.	Growth of ω-phase, with its lower density of states caused the average (over $\omega+\beta$) γ and χ to decrease with aging. T_c increased as a result of solute enrichment of the β combined with a growing incompleteness (with respect to ω) of the superconducting transition.	------	E.W.Collings et al. (1975) [Col75^b] see also [Upt72]
38.5, 58.5, 79.0, 89.4 at.%	Johnson Matthey Co. "specpure" starting materials. Arc melted twice in Ti-gettered argon.	Thermal conductivity within temperature range of about 2-16 K was measured in an investigation of an anticipated (based on Ti-Nb results) anomalously predominant phonon thermal conductivity.	Although calculated and experimental values of C (which characterizes "lattice" as distinct from "electron" scattering) were in satisfactory agreement in Ti-V (58.5 and 79.0 at.%) alloys a large discrepancy existed for Ti-V (38.5 at.%). This of course (unbeknown to these authors) is near the V composition below which ω-phase-related fluctuation or precipitation sets in as discussed in detail by Collings [Col74] in connection with electrical resistivity studies.	------	N.Morton et al. (1977) [Mor77]

TABLE 1-4 TITANIUM-VANADIUM ALLOYS — THE SUPERCONDUCTING TRANSITION *–continued*

V Concentration Range	Starting Materials, Sample Preparation	Properties and Procedures	Remarks, Discussion	Other Systems Studied	Literature
19 at.%	----------	Nonlinear least-squares fitting procedure applied to earlier [Whi78] calorimetric data.	The T_c of bcc Ti-V (19 at.%) obtained by rapid quenching is suppressed by some 2 K by athermal ω-phase precipitation through the operation of an almost complete proximity effect. Subsequent aging at 300°C progressively broadens the transition, decreases its completeness, and increases T_c. These results contrast with equivalent experiments on Ti-Mo and Ti-Fe alloys where the transition remains sharp and complete while T_c decreases with aging time. The data analysis involves fitting with a BCS model broadened by an asymmetrical Gaussian distribution of T_c's.	------	J.J.White and E.W.Collings (1978) [Whi78]

TABLE 1-5 TITANIUM-VANADIUM ALLOYS — THE MIXED STATE

V Concentration Range	Starting Materials, Sample Preparation	Properties and Procedures	Remarks, Discussion	Other Systems Studied	Literature
15, 20, 25, 28.7, 40.5, 50.0, 60.0, 70.0, 80.0, 84.0, 90.0, 92.0, 96.0 at.%	Starting material purity >99.9%. Argon-arc melted on each at least six times. Reduced to sample size with diamond wheel, grinding and polishing, and by cold rolling if sufficiently ductile. Measured as-machined or as-rolled.	The upper critical field, H_r, was measured resistometrically (at low current densities, ~10 A cm^{-2}) in pulsed fields of up to about 160 kOe, as a function of sample e/a, in an experimental evaluation of the GLAG theory as modified by paramagnetic limitation (upper critical field, H_p).	H_r (~10 A cm^{-2}) was found to be nearly independent of cold working, and of the relative orientations of magnetic field, current and defect texture. The results were used to underscore a discussion of current stabilization according to "Gorter-Anderson model" and the Mendelssohn filamentary-mesh model. Attention was drawn to a very early observation of the "peak effect".	Ti-Nb,-Mo,-Ta Zr-Nb Hf-Nb,-Ta U-Nb,-Mo	T.G. Berlincourt and R.R. Hake (1963) [Ber63a] see also [Ber63] [Ber62]
60, 67, 75, 85, 90, 93 at.%	Iodide Ti and 99.95% V. Arc melted in an inert atmosphere. Ingots cold swaged to 0.065 in.² rods: rolled to strip 0.008 in.† without intermediate or final anneal.	A comparison of the measured critical field with predictions from the Gor'kov-Shapoval expression; and of the measured critical current with predictions of a lamellar model for the mixed state. Critical current was measured resistometrically in fields of up to 50 kOe. Critical field was determined using current densities down to 10 A cm^{-2} followed by extrapolation to zero current. Normal-state conductivity measured.	In this very early study of current transport in superconductors, the detailed effects of cold work were not considered. In fact J_c was regarded as being calculable using a thermodynamic expression involving only H_c and H_{c2}. A laminar model (in particular, Goodman [Goo62]) was used to describe the current-carrying mixed state. With regard to critical field good agreement with Gor'kov-Shapoval was claimed on the basis of measurements made at 4.2 K. Berlincourt, however, stated that a valid conclusion could be drawn only on the basis of measurements made as function of temperature, as he had done. Calculations of upper critical field were carried out according to: *Gor'kov-Goodman:* $H_{c2}/H_c = (1.77-0.43\,t^2 + 0.07\,t^4)\,\kappa_{GL}(t=1)$ *Abrikosov-Ginzburg:* $H_{c2}/H_c = 2\sqrt{2} \cdot \kappa_{GL}(t=1)(1+t^2)^{-1}$ with $\kappa_{GL} = 2.43\,\gamma^{1/2} T_c (1-t^2)$, and $t = T/T_c$ $H_c(BCS) = 1.84\times10^4\,T_c\,(1-t^2)$, and that A comparison was also made with Clogston's paramagnetically limited upper critical field, $H_p = 1.84\times10^4\,T_c$, and that due to Maki and Tsuneto [Mak64b].	V Zr-Nb	A. El Bindari and M.M. Litvak (1964) [Elb64] see also [Elb63a]

TABLE 1-5 TITANIUM-VANADIUM ALLOYS — THE MIXED STATE *-continued*

V Concentration Range	Starting Materials, Sample Preparation	Properties and Procedures	Remarks, Discussion	Other Systems Studied	Literature
Calculations based on the low-temperature specific heat data of Cheng *et al.* [Che62] -- hence Ti-V (20-85 at.%).	---------	For a series of Ti-V alloys at 1.2 K the upper critical fields, computed according to the theories of Gor'kov-Goodman, Abrikosov-Ginzburg and Clogston were intercompared.	------------	------	Y.Shibuya and T.Aomine (1965) [Shi65]
20-95 at.% (9 alloys)	----------	The resistance ratio, R/R_n, for well-annealed samples was measured as function of H (0-40kOe) at temperatures between 1.5 and 4.2 K in a study of flux-flow resistivity, hence to obtain an experimental value for the hypothetical upper critical field, H^*_{C20} of GLAG theory by means of $$\rho_f/\rho_n = H/H^*_{C20}. {}^\dagger$$	The specific goals of the work were to study the relationship of flux-flow dissipation to the upper critical field. The work also contains some reference to H_{C3}. The following fields were intercompared: H^*_{C20}, from flux-flow-resistivity measurements, H^*_{C20}, calculated from Maki's expression in the dirty limit, H_{p0}, calculated from the Clogston formula, $$H_{p0} = 1.84 \times 10^4\, T_c,$$ $H_{c,0}$, obtained by smooth extrapolation of resistively measured upper critical field to T = 0, H_{C20}, calculated from Maki's (non-"spin-orbit") theory [Mak64b] of paramagnetic limitation, $$H_{C20} = H^*_{C20}/\sqrt{1 + \alpha^2}, \alpha = \sqrt{2} \cdot H^*_{C20}/H_{p0}.$$	Ti-Nb Zr-Nb Nb-Ta	Y.B.Kim *et al.* (1965) [Kim65]
38.5, 48.5, 58.5 at.%	40x3x3 mm³ rod prepared by alternate cold rolling and vacuum annealing.	In a study of the temperature dependence of the lower critical field, magnetization measurements between 1.5 and 4.2 K were made in fields of up to 11 kOe by moving a pair of opposing coils about a fixed sample. Resistivity measured using a four-probe method.	Strongly irreversible magnetization was noted. The measured H^*_{C1} at (t=0.564) was plotted against V concentration and compared with H_c, a bulk thermodynamic critical field calculated from the work of Cheng *et al.* [Che62]. The curves intersect at about 49 at.% V. Also calculated for comparison was the Abrikosov $$H_{C1} = (H_c/\sqrt{2}\,\kappa_{GL})(\ell n\,\kappa_{GL} + 0.08)$$ with κ_{GL} given by the Gor'kov-Goodman relationship $$\kappa_{GL} = \kappa^c_{GL} + 7.49 \times 10^3\,\gamma^{1/2}\,\rho_n.$$	Ti-Mo	R.D.Blaugher (1965) [Bla65]

† The "*" anticipates the realization that the GLAG-calculated upper critical field is not paramagnetically limited.

TABLE 1-5 TITANIUM-VANADIUM ALLOYS — THE MIXED STATE *-continued*

V Concentration Range	Starting Materials, Sample Preparation	Properties and Procedures	Remarks, Discussion	Other Systems Studied	Literature
35, 65, 75, 85 at.%	See Reference [Kim65]	The upper critical field, H_r, was measured resistometrically for comparison with theoretical values based on the Helfand and Werthamer [He264] and Maki [Mak64] extensions to GLAG theory.	Paramagnetic lowering of upper critical field is taken into consideration.	Nb Nb-Ta	Y.B.Kim and A.R.Strnad (1966) [Kim66]
58 at.%	Vacuum-annealed	The upper critical field was measured resistometrically in the temperature range, $1.3 \text{ K} \gtrsim T \gtrsim T_c$, in steady fields of up to 150 kOe using current densities in the range 0.1-50 A cm^{-2}. The goal was to investigate experimentally the influence of spin-orbit scattering on a Pauli-paramagnetically reduced upper critical field as function of Z, the mean atomic number, which was varied by alloying.	Due to Pauli spin paramagnetism $H_{c_2}^*$ (Maki) < $H_{c_2}^*$ (GLAG). But past measurements showed that $H_{c_2}^* < H_r < H_{c_2}^*$; the elevation of H_r was supposedly due to spin-orbit scattering. The latter also influences the temperature dependence of H_{c_2}. An interesting conclusion was the postulate that the depressive influence of Pauli paramagnetism on H_{c_2} may be able to be counteracted by addition of high-Z solute elements.	Ti-Nb,-Ta	L.J. Neuringer and Y. Shapira (1966) [Neu66]
Experimental reference data from Berlincourt and Hake (1963) [Ber63^a]	Theoretical	Experimental and theoretical values of paramagnetically limited upper critical field were intercompared.	It is shown that H_u should be some function of $H_{c_2}^*$, the GLAG upper critical field, and H_p the Clogston limiting field, of the form $$H_u = H_p \left(\sqrt{4 H_p^2 + 4 H_{c_2}^{*2}} - H_p \right)/2 \, H_{c_2}^* \, .$$ Good agreement between this and the experimental H_u or H_r is achieved if (as should be the case) H_p is increased by some 30-40% above the simple Clogston value. This work pre-dates that of WHH and Maki which provides a justification in terms of spin-orbit effects, for that enhancement.	V$_3$Ga (brief note)	R.Hancox (1966) [Han66]

TABLE 1-5 TITANIUM-VANADIUM ALLOYS — THE MIXED STATE –continued

V Concentration Range	Starting Materials, Sample Preparation	Properties and Procedures	Remarks, Discussion	Other Systems Studied	Literature
88, 91, 94, 97, 100 at.%	Argon-arc melted, cut to 15x5x0.2 mm³, homogenized 6h/300°C. Surfaces electrolytically polished. Purity: 0 and H < 1%	The surface critical field H_{C3}, the ratio H_{C3}/H_{C2} and the critical surface current were studied for comparison with Abrikosov's theory. H_{C2} was determined magnetically. H_{C3} was obtained from the results of $J_c(H)$ measurement in the following way: with surface and transport currents parallel to the magnetic field, J_c was measured as function of applied field. H_{C3} is the field for which J_c becomes independent of H -- i.e. the field at the foot of the curve.	J_s, the critical surface current, so obtained was related to H_{C3} through Abrikosov's relationship $$J_s = (5H_{C2}/3\sqrt{\sigma\pi}\cdot\kappa_{GL})(1-H/H_{C3})^{3/2} \text{ A cm}^{-2} ,$$ with κ_{GL} given by the usual Gor'kov formula $7.49\times10^3 \gamma^{1/2} \rho_n$ (dirty limit) . H_{C3} was found to be independent of the smoothness of the surface and of copper plating.	------	K.Kwasnitza and G.Rupp (1966) [Kwa66]
40 at.%	Numerical analysis	The limiting (as distinct from achievable) values of upper critical field for several materials were estimated from fundamental physical measurables.	Comprehensive tabulations of physical property data were presented. The zero temperature upper critical field limits, $H_{C20,max}$, were calculated taking into account Maki paramagnetic limitation, and the Werthamer et al. [Wer66] (WHH) spin-orbit-coupling/spin-flip-scattering effect.	B-W compounds B1 compounds Ti-Nb Zr-Nb	R.R.Hake (1967) [Hak67a]
22.5, 25 at.%	Arc melted and cast. Studied as-cast (ac); ac plus cold rolled 2:1 (cr); ac plus cr plus solution heat treated 1h/0.8 mp and cooled 7°C/min; 70°C/min, 30°C/min, 15°C/min to 800°C, 500°C, 350°C, 20°C, respectively.	The magnetizations of high-κ type-II superconductors were measured by moving specimen between a pair of oppositely-wound 18,000-turn coils connected in series with a fluxmeter in a quantitative evaluation of the Pauli paramagnetic spin-orbit-effect theories of Maki and WHH. Electrical resistivity was also measured.	Reversible paramagnetic superconductivity was noted in association with a second-order transition to the high-field normal state. Good agreement between experimental reduced upper critical field temperature dependence, $h^*(t)$, and reduced temperature, t, is obtained with Maki-WHH theories for $\lambda_{so} \sim 0.5$. Fine theoretical details were not substantiated. Reference [Hak65] was predominantly a Ti-Mo paper. With increasing field the mixed state magnetization decreased monotonically and reversibly, thus smoothly entering in turn: (a) a paramagnetic mixed state, (b) the paramagnetic normal state. The experimental results supported the existence of paramagnetic free energy terms, but the lack of a full set of physical property data hindered quantitative analyses.	Ti-V-Cr Ti-Mo Hf-Nb	R.R.Hake (1967) [Hak67b] see also [Hak67] [Hak65]

TABLE 1-5 TITANIUM-VANADIUM ALLOYS — THE MIXED STATE *—continued*

V Concentration Range	Starting Materials, Sample Preparation	Properties and Procedures	Remarks, Discussion	Other Systems Studied	Literature
50 at.%	Annealed	In order to study the M(h) curve at 4.2 K, to confirm Hake's results and to compare H_u (magnetic upper critical field) with H_r (resistive upper critical field), magnetization measurements by the moving sample technique (using a pair of 1,000-turn coils) and resistometric measurements were carried out.	No trace of first order transition noted, in agreement with Hake [Hak65]: $H_r \approx 1.3 \, H_u$.	------	Y. Shibuya and T. Aomine (1967) [Shi67]
30, 40, 50 at.%	The sample, a thin flat plate, formed the end wall of an undercoupled rectangular cavity.	Microwave surface impedance at 14.4 GHz was measured in a study of: (a) flux-flow resistivity and its implications in terms of mixed state vortex structure and motion; (b) the microscopic details of mixed state paramagnetism and (c) the thermodynamic order of the s/n transition.	The advantages of microwave, over DC flux-flow measurements were mentioned. Mixed-state paramagnetism, in terms of the "hard vortex core" model was discussed. No unequivocal evidence for a first-order s/n transition was obtained.	------	W. H. Hackett *et al.* (1967) [Hac67]
25 at.%	Non-experimental	The essential characteristics of thermodynamic relationships in types-I and II superconductors were reviewed. Fluctuation superconductivity was considered.	Extremely useful review of magnetic properties, pressure effects and the status of fluctuation superconductivity.	Ti-Mo, -Ru V-Ta	R. R. Hake (1969) [Hak69]
28 at.%	Ti-V heat treatments: solution treated (st) at 1000°C and; step-quenched to 540°C and held for 1 h, 10 h and 24 h; st at 1000°C/water quenched and (a) aged 10, 24h/350°C, (b) aged 24h/350°C plus 3min/550°C/quenched.	The utility of magnetization in the superconductive state as a microstructural diagnostic tool was explored. The experimental technique consisted of motorized transfer of the specimen between oppositely-wound 10,000-turn coils with electronic integration of the output. The field range employed was 0 ∿ 40 kOe.	See also Reference [Pol71] (Ti-V,-Nb and Zr-Nb). Magnetization data interpreted in terms of anticipated microstructural changes.	------	T. S. Luhman *et al.* (1972) [Luh72] see also [Pol71]
36, 70, 82 at.%	Iodide Ti (deliberately not zone refined) and high-purity V. Argon-arc melted, homogenized and cold-rolled.	Resistometric studies of critical field (at 10 mA; ≈ 10 A cm^{-2}) and critical current density were undertaken in investigations of the effects of (normal-state) paramagnetism on the mixed state in high-field superconductors.	A simple, phenomenological, model of the paramagnetic mixed state was established. A detailed investigation of flux pinning in the paramagnetic mixed state showed that both a local interaction, f_p, and a macroscopic (Lorentz) force, F_p, acting over the flux lattice needed to be taken into consideration. A breakdown of scaling laws for $\kappa_{GL} \gtrsim 30$ (as for *all* the Ti-V alloys), due to the effect of paramagnetic limitation, was noted.	Ti-Nb	R. A. Brand (1972) [Bra72] see also [Bra75] for summary

TABLE 1-5 TITANIUM-VANADIUM ALLOYS — THE MIXED STATE *–continued*

V Concentration Range	Starting Materials, Sample Preparation	Properties and Procedures	Remarks, Discussion	Other Systems Studied	Literature
72.5 at.% and other un-specified compositions	A specimen 1 cm³ and 0.03 cm thick was an end wall of a cylindrical microwave (33GHz) cavity.	A study of the flux-flow resistance minimum effect initially discovered by Kim *et al.* [Kim65].	A minimum in $\rho_f(t)$ was noted whose position (i.e. t), but not depth, could be reasonably well accounted for by existing theory.	------	T.Akachi *et al.* (1976) [Aka75]
23.75, 70 at.%	Sample glued into a recess in the end plate of a cylindrical TE.011 cavity resonating at 35.5 GHz.	An investigation of: (a) the thermodynamic order of the *s/n* transition in paramagne-tically limited superconductors and (b) Maki's dynamic microscopic theory of flux flow [Mak68]. Reflected power from a microwave cavity containing the sample was measured by a crystal detector, yielding information re-lating to ρ_f, the flux-flow resistivity.	The search for evidence of a first-order *s/n* transition parallels earlier work [Hac67]. Again, the results and their relationship to theory seemed inconclusive.	------	K.S.Kim and Y.B.Kim (1976) [Kim75]

TABLE 1-6 TITANIUM-VANADIUM ALLOYS — CURRENT TRANSPORT EFFECTS

V Concentration Range	Starting Materials, Sample Preparation	Properties and Procedures	Remarks, Discussion	Other Systems Studied	Literature
29, 50, 70 at.%	Arc-melted repeatedly on wcch. Samples cold rolled from the as-cast condition.	The short sample J_c was measured at 4.2 K in fields up to 30 kOe in a study of $J_c(\parallel)$ and $J_c(\perp)$ as function of thickness reduction by cold rolling.	Cold working to reductions of 67 to 92% increased J_c by factors of 1.9 to 88 over those for the as-cast alloys. Thus the influence of *cold work* on J_c was recognized apparently for the first time; and it was predicted that larger J_c values could be obtained by further cold working and/or heat treatment. The concept of flux pinning by dislocations had not been considered at this stage, and an explanation for the effect of cold work was sought with recourse to the "filamentary model". Highest current value was $J_c(\parallel) = 1.5 \times 10^4$ A cm^{-2} at 30 kOe at 4.2 K for Ti-V$_{70}$ rolled 93%. Attention was drawn to a positive dJ_c/dH in certain alloys in one of the earliest discussions of what later became known as the "peak effect".	Ti-Mo,-Ta Zr-Nb,-Mo Nb-Sc,-Hf	R.R.Hake (1962) [Hak62^a]
45 at.%	Sputter-deposited films. Target purity, initially 99.8%. Films prepared in low-pressure, hot-cathode discharge tubes. Deposition rates of 40-400Å/min onto heated (300-450°C) glass substrates.	A study of critical current *vs* field for sputter-deposited films of hard superconductors. Four-terminal measurements were made at 4.2 K in pulsed magnetic fields of up to 85 kOe (transverse fields) and 136 kOe (longitudinal fields).	J_c(30kOe) of sputter-deposited 1000 Å Ti-V: $J_c(\perp)$, 2.0×10^4 A cm^{-2}; $J_c(\parallel)$, 5.4×10^5 A cm^{-2}.	Nb Ti-Nb Nb-Zr V-Si Nb-Sn	J.Edgecumbe *et al.* (1964) [Edg64]
26, 32, 52, 56, 86, 95.3 wt.%	Iodide Ti (99.9 wt.%) and "carbothermal" V (99.766 wt.%, Vickers hardness, 110 kg mm^{-2}). Arc melted repeatedly on wcch in 0.7 atm of gettered He. Three sets of samples prepared corresponding to three thermomechanical processing sequences, Ti-V alloys only.	Mechanical properties, microstructure and critical current density in Ti-V alloys were studied. Short-sample critical current density was measured at 4.2 K in fields of up to 22.2 kOe.	First published by Nauka (1965) as part of the proceedings of the first (Soviet) conference on the Metal Science and Metal Physics of Superconductors. The effects of deformation and heat treatment on the hardness and critical current density of Ti-V was considered in detail on pp. 48-54 of the translation. It was recognized that an increase in J_c with annealing was associated with microstructural effects such as phase separation and "re-distribution of impurities".	-----	Yu.V.Efimov *et al.* (1965) [Efi65]

TABLE 1-6 TITANIUM-VANADIUM ALLOYS — CURRENT TRANSPORT EFFECTS –continued

V Concentration Range	Starting Materials, Sample Preparation	Properties and Procedures	Remarks, Discussion	Other Systems Studied	Literature
The same as above	Additional thermomechanical processing sequences were studied.	Short-sample critical current density was measured at 4.2 K in fields of up to 26.6 kOe.	Part of translation of second and third (Soviet) conferences on the Metal Science and Metal Physics of Superconductors held in 1965 and 1966, respectively. Essentially a summary of the earlier work, it was concluded that the maximum J_c at 26.6 kOe (viz. 6.1×10^4 A cm^{-2}) was achieved in Ti-26V after annealing 5h/400°C. The Nb-33Zr of the day was capable of about 2×10^5 A cm^{-2} in that field [Hak62^a].	-----	Yu.V.Efimov *et al.* (1967/1970) [Efi70] see also E.M.Savitskii *et al.* [Sav73, p.244]
20 wt.%	Triple arc-melted buttons sectioned to 0.1 in. thickness. Homogenized 6h/1000°C, quenched; cold-rolled cut and shaped; recrystallized 2h/800°C, quenched. Samples aged in salt bath for 1/4, 1, 4 and 20 h, at temperatures of 200, 300, 400, 500 and 600°C.	Short-sample J_c measured at 4.2 K (>400°C aging) in fields of up to about 45 kOe in a study of the effect of ω-phase and α-phase precipitation on critical current density.	α-phase is an effective flux pinner, but less so than ω-phase. Current densities (e.g. about 10^4 A cm^{-2} at 25 kOe after 4-1/4h/400°C) much less than those for optimized Zr-Nb and Ti-Nb. The pioneer work on the effect of *heat treatment* on J_c was carried out by Vetrano and Boom prior to 1965 using Ti-Nb (20.7 at.%).	-----	J.B.Vetrano *et al.* (1968) [Vet68]
22.5 at.%	---------	In a study of the occurrence of fluctuation superconductivity at temperatures above T_c, electrical resistivity was measured as function of temperature. Zero-field results were plotted in the format $\sigma(t)/\sigma(0)$ *vs* t.	The appearance of high-temperature resistivity tails was suggestive of weak superconductivity up to temperatures of at least 2 T_c. The effect was not strongly dependent upon mechanical surface polish, surface-volume ratio, annealing, quenching, or cold work. Further discussion of this effect appears in the context of Ti-Fe alloys.	Ti-Fe Ti-Mo,-Ru Ti-Os	R.R.Hake (1969) [Hak69^a]
25 at.%	---------	Electrical resistivity measurements were made at temperatures between 10 and 20 K in fields of up to 40 kOe and the results presented in the format $[\rho(H)-\rho(0)]/\rho(0)$ *vs* H in a study of fluctuation superconductivity in high magnetic fields.	Kohler's-law behaviour was excluded on temperature-dependence grounds -- the increase in $\Delta\rho(H)$ with H being attributed to "magnetoparaconductance", i.e. the H-induced quenching of superconductive fluctuations. Fluctuation superconductivity up to 3 T_c was claimed. Anomalous scattering mechanisms were not considered.	-----	R.R.Hake (1970) [Hak70]

TABLE 1-6 TITANIUM-VANADIUM ALLOYS — CURRENT TRANSPORT EFFECTS –continued

V Concentration Range	Starting Materials, Sample Preparation	Properties and Procedures	Remarks, Discussion	Other Systems Studied	Literature
17-100 at.% (more than 60 alloys)	DC triode co-sputtering from two separate targets of high purity material (5N, excluding gases) onto fused quartz substrates, under argon, at a rate of about 100Å/min to a thickness of about 3000 Å. Samples annealed several hours at 700°C, rapidly cooled.	The T_c and critical current densities of sputtered films were studied. J_c vs H characteristics were measured using a pulsed-field, pulsed current technique at 4.2 K in liquid He with the field parallel to the surface of the film.	T_c vs V composition rises monotonically with increasing V content to a maximum of 12.8 K at 99.99 at.% V, dropping to 5.3 K for pure V (about 0.2 K above the bulk value). Very high J_c's were reported, for example, $\sim 6.5 \times 10^5$ A cm^{-2} in Ti-V (99.8, 99.96 at.%) and $\sim 3.5 \times 10^5$ A cm^{-2} in Ti-V (88.4, 92.7 and 99.4 at.%) at 60 kOe. The enhanced T_c and J_c performances were suspected to relate to the presence of impurities or precipitates in an otherwise single-phase-β Ti-V matrix.	Ti-Nb,-Ta Nb-Zr,-V,-Ta and the compounds NbN and NbC	H.J.Spitzer (1974) [Spi74^a] Summarized in [Spi74^b]
3-80 at.% (14 compositions)	Multiple arc melted on wcch using Ti sponge (>99.8%) from Titanium Metals Corp. and "vp" grade V (99.95%) from Materials Research Corp.	Four-terminal measurements of electrical resistivity vs composition at 77, 200, 298 K in a study of the relationship between ω-phase precipitation and the electrical resistance anomaly.	The results were discussed within the context of transition-metal binary-alloy phase stability, expressed in terms of soft phonons which become more and more localized as the solute concentration increases in the average e/a-ratio range 4.1 to 5.0. A connection was made with superconductivity in alloys formed between groups IV and V transition metals. It was pointed out that the phonons of instability, which are responsible for the anomalous resistivity, also favor superconductive coupling and consequently a relatively "high" T_c; but that at the same time, this effect must compete with a proximity-effect-induced lowering of the average T_c by the product of the lattice instability (viz. ω phase), which itself has a very low T_c.	-----	E.W.Collings (1974) [Col74] see also [Col78]
20, 25, 30, 40, 50, 60, 70, 80, 90, 97, 100 at.%	Water quenched from 1100°C.	Zero field resistometric measurements between He^4 temperatures and the ice point were undertaken in studies of the electrical resistivity temperature dependences of the Ti-(group V) alloys Ti-V,-Nb,-Ta. Of particular interest in the case of Ti-V was a negative dρ/dT and an anomalously broad superconducting transition.	An additional spin-flip scattering mechanism was assumed. As a pair-breaking mechanism this broadens the superconducting transition. Ti-V's negative dρ/dT was supposed to be the low-temperature part of a localized-spin-fluctuation/Friedel-Anderson resistivity minimum. These effects were supposed to be operative in varying degrees, in both Ti-Nb and Ti-Ta. The relationship between the observations of Prekul and of Hake are discussed in connection with Ti-Fe alloys.	Ti-Nb Ti-Ta	A.F.Prekul et al. (1974) [Pre74] see also [Pre73]

TABLE 1-7 TITANIUM-CHROMIUM ALLOYS — THE SUPERCONDUCTING TRANSITION

Cr Concentration Range	Starting Materials, Sample Preparation	Properties and Procedures	Remarks, Discussion	Other Systems Studied	Literature
0-30 at.% (13 alloys)	Argon arc melted repeatedly. Measured in as-cast condition.	T_c measured by a mutual inductive technique as function of composition. The properties of Ti-TM (3d) and Ti-TM (4d, 5d) alloys were intercompared.	Cold working changed T_c in the vicinity of the α/β boundary. The width of the superconducting transition was 0.1-0.25 K.	Ti-Mn,-Fe,-Co Ti-----Ru,-Rh Ti-Re	B.Matthias et al. (1959) [Mat59]
Review including Ti-Cr	---------	-------------	See Table 1-4	-----	U.Zwicker (1963) [Zwi63]
2.5 at.%	---------	-----------	See Table 1-4	Ti-V,-Fe,-Nb,-Mo Ti-V-Cr Ti-Zr-TM, nine TM-elements one at a time	E.Bucher et al. (1965) [Buc65]
0.005-75 at.% (~24 alloys)	Iodide or sponge Ti and vacuum out-gassed electrolytic Cr. Pure argon arc melted on wcch. Cut or swaged samples (depending on composition) were homogenized.	T_c measured by an inductive technique as function of composition and aging with due regard to optical microstructure.	Samples were studied in the conditions: (a) as-cast, (b) 1/2h/1000°C/water quenched -- if [Cr] > 25 at.%, an additional 1h/ (1250-1325°C)/oil quenched, (c) aged at 450°C up to 10,000 min, (d) deformed by swaging up to 95%.	-----	C.J.Raub et al. (1966) [Rau66]
10.3, 13.2, 18.5, 24.4 at.%	Iodide Ti and 99.99% Cr. Levitation melted and cast in atmosphere of purified He. Homogenized 72h/ 1040°C in 10⁻⁶ Torr. Solution treated 1-1/2h/1040°C (24.4%) and 1-1/2h/980°C (others) and quenched.	T_c was studied inductively as function of composition and heat treatment (hence microstructure).	Ti-Cr (10.3 at.%), solution treated (ST) 1.5h/980°C/quenched and ST plus up to 28min/196°C. Ti-Cr (24.4 at.%), ST 1.5h/1040°C/quenched and ST plus up to 28min/300°C.	Ti-V,-Fe Ti-Nb	T.S.Luhman et al. (1969) [Luh69]
7.9, 9.5, 10.3, 11, 13.2, 15, 18.5, 24.4 at.%	Same as above	Same as above	An extensive study of the effects of aging at ~300°C and aging plus upquenching.	Same as above	T.S.Luhman (1970) [Luh70]
0.85, 2.09 at.%	---------	The low temperature specific heat was measured in temperature range 1.4-4.7 K. Electrical resistance measurements were also conducted over the range of the expected superconducting transition.	The calorimetric superconducting transition in Ti-Cr was sufficiently broad as to occupy the entire measuring temperature range and extremely incomplete. A broad resistive transition was noted with Ti-Cr (0.85 at.%); a somewhat sharper one with Ti-Cr(2.09 at.%). For further details refer to Table 1-4.	Ti-Sc,-V,-Mn Ti-Fe,-Co,-Ni Ti-Hf	K.L.Agarwal (1974) [Aga74]

TABLE 1-8 TITANIUM-CHROMIUM ALLOYS — CURRENT TRANSPORT and MAGNETIC EFFECTS

Cr Concentration Range	Starting Materials, Sample Preparation	Properties and Procedures	Remarks, Discussion	Other Systems Studied	Literature
1.15 at.%	Ti crystal bar and Cr (>99.4 wt.%) argon arc melted repeatedly on wcch. Sectioned for measurement in the as-cast condition.	A resistometric study of low temperature electrical transport properties.	Resistivity, magnetoresistivity and Hall coefficient were measured. "Resistivity minima" were found near 18 K in Ti-Cr,-Mn, -Fe but not in Ti-Co,-Ni. Only the dilute Ti-Mn alloys exhibited negative magneto-resistance.	Ti Ti-Mn,-Fe Ti-Co,-Ni	R.R.Hake *et al.* (1962) [Hak62[b]]
10, 13, 15, 20 at.%	Iodide Ti (99.97%) and Cr (99.9%) levitation melted in purified, dry positive-pressure He atmosphere. Alloys homogenized 36h/1050°C in 5×10^{-6} Torr and swaged to rods 0.1 in.$^{\phi}$. Measurement conditions for Ti-Cr$_{20}$: as-quenched (β) aged 1140min/300°C $(\omega+\beta)$ aged 4min/435°C $(\beta'+\beta)$ quenched + deformed 1% (β) aged 6min/300°C (β) above + deformed 1% (β) aged 320min/300°C $(\omega+\beta)$ above + deformed $(\omega+\beta)$.	Electrical resistance measured using standard potentiometric techniques in the temperature range 0 to -196°C in a study of negative $d\rho/dT$ in quenched Ti-TM alloys. The effects of heat-treatment and deformation on it were also investigated.	Selected-area electron diffraction was conducted at temperatures from 25 to -180°C. The results of this study, taken together with the (reviewed) results of earlier workers established that the negative $d\rho/dT$ in Ti-TM alloys is associated with the metastability of the β phase.	Reviewed were: Ti-V,-Mn Ti-Nb,-Mo Ti-V,-Al Ti-V-Cr-Al Ti-O,-N Zr-Nb,-O	V.Chandrasekaran *et al.* (1974) [Cha74]
1.15 at.%	Ti crystal bar and Cr (>99.4 wt.%) argon arc melted repeatedly on wcch. Measured in the as-cast condition.	An investigation of the existence of localized magnetic states.	The principal thrust of this paper was a study of the "local-moment" magnetic properties of Ti-Mn alloys. Ti-Cr, Ti-Fe and Ti-Co were not Curie-Weiss paramagnetic -- see Table 1-10.	Ti-Mn,-Fe,-Co	J.A.Cape (1963) [Cap63]

TABLE 1-9 TITANIUM-MANGANESE ALLOYS — THE SUPERCONDUCTING TRANSITION

Mn Concentration Range	Starting Materials, Sample Preparation	Properties and Procedures	Remarks, Discussion	Other Systems Studied	Literature
0-25 at.% (about 13 alloys)	---------	T_c was studied as a function of composition.	See Table 1-7	Ti-Cr,------Fe,-Co Ti--------Ru,-Rh Ti-----Re	B.Matthias *et al.* (1959) [Mat59]
Review including Ti-Mn	---------	-------------	See Table 1-4	-----	U.Zwicker (1963) [Zwi63]
Mn < 5 ppm; Fe < 2 ppm Mn, 30 ppm; Fe, 20 ppm Mn, 30 ppm; Fe, 0.15 at.% Mn, 100 ppm; Fe, 20 ppm	Argon arc melted on wcch.	A study of superconductivity in pure Ti and in Ti with known levels of impurity, especially Mn. Refrigeration was accomplished by magnetic cooling down to 0.06 K; superconductivity was monitored by means of ballistic mutual inductance.	The presence of Fe increased T_c at the rate ~0.1 K per 200 ppm; Mn reduced it at the rate ~0.1 K per 10 ppm (at these levels no distinction was drawn between wt.% and at.%).	-----	R.L.Falge (1963) [Fa£63]
0.39, 1.71 at.%	---------	Low temperature specific heat and resistometric study of superconductivity in dilute Ti-TM alloys.	See Table 1-4. Results, in the format C/T vs T^2, were quite similar to those reported by Hake and Cape for Ti-Mn (0.36 and 1.7 at.%, and others).	Ti-Sc,-V,-Cr Ti-Fe,-Co,-Ni Ti-Hf	K.L.Agarwal (1974) [Aga74]

TABLE 1-10 TITANIUM-MANGANESE ALLOYS — CURRENT TRANSPORT and MAGNETIC EFFECTS

Mn Concentration Range	Starting Materials, Sample Preparation	Properties and Procedures	Remarks, Discussion	Other Systems Studied	Literature
0.02, 0.11, 0.21, 0.40, 1.0, 1.1, 2.0 at.%	--------	A study of low temperature electrical transport properties.	See Table 1-7	Ti-Cr,-Fe Ti-Co,-Ni	R.R.Hake *et al.* (1962) [Hak62b]
0.28, 1, 2, 4, 14 at.%	Ti crystal bar (T_c=0.14K) and Mn >99.4 wt.%, argon arc melted on wcch. Annealed; (a) 1000°C, and (b) 8h/690°C, respectively, and quenched to 77 K.	The superconducting transition was monitored resistometrically and the magnetic susceptibility was measured using the Gouy technique (temperature range, 4.2~50K) in a study of the relationship between the occurrences of localized magnetic moments and superconductivity in Ti-TM (3d) alloys.	Quenched^m α alloys and annealed equilibrium (α+β)-phase alloys were investigated. Alloys of up to 4 at.% Mn had "localized" (i.e. Curie-Weiss) moments; the β-phase alloys exhibited no Curie-Weiss behavior.	Ti-Cr,-Fe,-Co briefly referred to	J.A.Cape (1963) [Cap63]
0.17, 0.36, 0.85, 1.7, 14 at.%	Ti crystal bar and Mn >99.4 wt.%. Argon arc melted repeatedly on wcch. Measured in as-cast condition.	Low temperature specific heat measured in the temperature range 1.2-4.5 K in a study of the influence of a localized magnetic moment on the superconducting transition.	Magnetic susceptibility studies, 4-300 K, on on Ti-Mn (≈0.2-0.4 at.%) by Cape [Cap63] (see below) indicated that the dilute Ti-Mn (≳1.7 at.%) alloys are local moment systems and nonsuperconducting. Only β-Ti-Mn (14 at.%) was superconductive.	Ti-Co Ti-Zr-Mn	R.R.Hake and J.A.Cape (1964) [Hak64]
16 at.%	--------	Resistometric studies in the temperature range $1.2 \lesssim T \lesssim 4.2$ K were made in magnetic fields $H \lesssim 140$ kOe in studies of negative normal-state resistivity temperature dependence parameterized by $\alpha^T \equiv (1/\rho)(\partial\rho/\partial T)_H$ and magnetoresistance $\alpha^H \equiv (1/\rho)(\partial\rho/\partial H)_T$.	Several possible origins of the negative α^T and α^H were suggested: (a) magnetic spin-disorder scattering, (b) influence of sharp structure in density of states plus Zeeman splitting of the state density, (c) electron scattering from soft-phonon modes plus H-induced mode stiffening.	Ti------Fe Ti-Mo,-----Ru Ti-------Os Ti-V-Cr	R.R.Hake *et al.* (1975) [Hak75]
0.6, 1.1, 1.9, 3.0, 3.9, 5.0, 5.8, 7.0, 7.8, 11.8, 14.8, 24.9 at.%	Electrorefined Ti sponge grade ELXX from Titanium Metals Corp. and high-purity electrolytic Mn. Argon arc melted repeatedly on wcch; annealed 5h/1000°C/brine quenched.	Magnetic susceptibility measured using the Curie technique in the temperature range 77 ~ 450 K in a study of the relationship between the occurrence of Curie-Weiss susceptibility and composition-dependent crystal structure.	Ti-Mn (5 and 7 at.%) were selected for aging studies -- up to about 1000h/300°C. The (α+β)-phase and β-phase alloys Ti-Mn (≳5 at.%) exhibited no Curie-Weiss behavior.	-----	E.W.Collings [Col83]

TABLE 1-11   TITANIUM-IRON ALLOYS — THE SUPERCONDUCTING TRANSITION

Fe Concentration Range	Starting Materials, Sample Preparation	Properties and Procedures	Remarks, Discussion	Other Systems Studied	Literature
0-20 at.% (12 alloys)	See Table 1-7	A study of T_c as function of composition in order to examine "magnetic solute" effects in Ti-TM alloys.	Results for Ti-Fe and Ti-Ru are juxtaposed. The work was also discussed in [Bar61] (1961) and [Mat63] (1963), in which a magnetic electron-electron coupling mechanism was proposed. No reason was advanced as to why this should lead to superconductivity -- see Table 1-7.	Ti-Cr,-Mn,-----Co Ti-------Ru,-Rh Ti-----Re	B.Matthias et al. (1959) [Mat59]
Review including Ti-Fe	----------	------------	See Table 1-4	------	U.Zwicker (1963) [Zwi63]
0.15 at.%	Impure Ti	Magnetic study of the effect of various impurities on the T_c of Ti -- see Table 1-9.	The addition of fractional percentages of Fe was found to increase T_c at the rate 5.5 K per at.% Fe.	------	R.L.Falge (1963) [Fal63]
0.05, 0.25, 1.0 at.%	Ti (9.2 ppm Fe, 4.0 ppm Mn) arc melted with Fe. Conditions: 0.05%: cold worked and annealed 14days/700°C; 0.25%: quenched; 1.0%: annealed 6days/700°C *without* cold work.	Low temperature specific heat investigation within the temperature ranges 0.25-4.2 K (0.05, 0.25 at.% Fe) and 0.25-20 K (1.0 at.% Fe) of the apparently anomalous superconducting transition in Ti-Fe. The superconducting transition was also examined magnetically.	The heights of the specific heat jumps at T_c compared to the BCS prediction of $\Delta C/T = 1.43 \gamma$ are: 0.05 at.% Fe, 0%; 0.25 at.% Fe, 4.1% and 1 at.% Fe, 12.5%. Temperatures of specific heat anomalies -- 0.05% ∼ 3 K (possibly an impurity effect); 0.25% ≲ 0.25 K; 1.0% ∼ 3 K. The calorimetric and magnetic results were in conflict and inconclusive as to the roles played by solute and microstructure in influencing T_c in dilute Ti-Fe alloys.	Unalloyed Ti	R.H.Batt (1964) [Bat64]
0.13, 0.4, 0.9 at.%	----------	Superconducting transition studied using the technique of AC magnetic susceptibility.	It was noted that the addition of small amounts of Fe raised T_c more than did equivalent amounts of its electronic isomer Ru. But from the nature of the zero-field a.c. transition it was thought that the observed superconductivity originated in a small fraction (≲3%) of some phase other than hcp Ti-Fe -- possibly β phase.	Several inter-metallic compounds	M.Strongin et al. (1964) [Str64]

Fe Concentration Range	Starting Materials, Sample Preparation	Properties and Procedures	Remarks, Discussion	Other Systems Studied	Literature
1.0, 1.5 at.%	-----------	-------------	See Table 1-4. By fitting a distributed specific heat function to the data the volume fraction, x, undergoing transition was determined. Whereas Ti-V,-Nb yielded $0.95 < x < 1.0$, Ti-Fe (1.5 at.%) yielded $0.3 < x < 0.35$, i.e. a transition only about 30% complete -- inexplicable in terms of microstructural analysis.	Ti-V,-Nb	F.Heiniger and J.Müller (1964) [Hei64] see also E.Bucher et al. (1965) [Buc65]
--------	Theoretical analysis		The development of a theoretical model involving a new electron-electron interaction to explain the increase in T_c with composition in Ti-Fe alloys.	-----	B.N.Ganguly et al. (1966) [Gan66]
0.01, 0.025, 0.035, 0.05 at.%	Iodide Ti and electrolytic Fe, arc melted in purified argon. Cold rolled 60%, homogenized 2weeks/700°C and 3weeks/600°C under gettered conditions, water quenched.	Resonant-frequency shift method of Schawlow and Devlin (at 16 kHz) used to detect the superconducting transition. Also determined were the equilibrium α-phase solubility limits.	Samples were examined metallographically after annealing at various temperatures such as 850°C, 700°C, 600°C, 500°C. Transitions beginning at 0.44 K were detected in the homogeneous α-solid solutions Ti-Fe (0.035 and 0.05 at.%). Heterogeneous samples of Ti-Fe (0.025, 0.035 and 0.05 at.%) yielded very broad transitions beginning at 3.2 K.	-----	E.Raub et al. (1967) [Rau67]
4, 9.5 at.%	Ti and purified Fe melted in rod-shaped crucibles. Ingot hot rolled, turned down to thin rods, vacuum annealed 1/2h/900°C/water quenched.	A study of T_c, electrical resistivity, and hardness as functions of aging time at 400 and 500°C.	Electrical resistivity temperature dependence was measured, and a detailed metallographical study was undertaken.	Ti-Nb Ti-Nb-Fe	T.Nishimura and U.Zwicker (1968) [Nis68]
0.53, 1.47 at.%	-------------	Calorimetric study of superconductivity in dilute Ti-TM alloys in an attempt to clarify discrepancies among earlier results on Ti-Fe and other Ti-TM (magnetic) alloys.	See Table 1-4. Extremely broad and incomplete calorimetric transitions were noted (as was the case with Ti-Cr and -Co). Both Ti-Fe alloys had the same superconductive onset temperature, T_c = 3.6 K.	Ti-Sc,-V,-Cr Ti-Mn,-,-Co,-Ni Ti-Hf	K.L.Agarwal (1974) [Aga74]
2.5, 5, 7.5, 10 at.%	Ti (Titanium Metals Corp. sponge, 99.9%) and Fe (Materials Research Corp., 99.99%) argon arc melted repeatedly on wcch. Annealed 1h/1000°C in Ti-gettered environment, ice-brine quenched.	Low-temperature specific heat measurements in the temperature range $1.8 \sim 6$ K were undertaken in a study of the influence of aging on the T_c of Ti-Fe (7.5 at.%). Room temperature magnetic susceptibility was also measured.	The results on quenched alloys were compared with magnetic susceptibility data. Calorimetric and magnetic studies of the aging of Ti-Fe (7.5 at.%) for various times up to about 3000 h at temperatures of 175 and 300°C were undertaken.	-----	J.C.Ho and E.W.Collings (1974) [Ho73a]

TABLE 1-12 TITANIUM-IRON ALLOYS — CURRENT TRANSPORT and MAGNETIC EFFECTS

Fe Concentration Range	Starting Materials, Sample Preparation	Properties and Procedures	Remarks, Discussion	Other Systems Studied	Literature
0.96 at.%	See Table 1-8	Low temperature electrical transport properties including electrical resistivity, magnetoresistivity and Hall coefficient.	See Table 1-8	Ti Ti-Cr,-Mn Ti-Co,-Ni	R.R.Hake (1962) [Hak62b]
0.96 at.%	See Table 1-8	A study of local-moment properties in dilute Ti-TM alloys.	Although yielding a resistivity minimum [Hak62b], Fe does not possess a localized magnetic moment in Ti.	Ti-Cr,-Mn,-Co	J.A.Cape (1963) [Cap63]
1.46, 2.15, 3.27, 4.0, 6.5, 8.0, 10, 12, 15, 20 at.%	Helium arc melted four times from iodide Ti and carbonyl Fe. Homogenized 5h/1200°C in 10^{-6} Torr. Water quenched from 1200°C after cutting.	A study of electrical resistivity temperature dependence from 300 K into the liquid He range, with particular attention to the transition to the superconducting state.	Magnetic susceptibility(χ) composition dependence also measured. It is claimed that since χ and T_c scale for [Fe] > 0.05 at.% the superconductivity is a normal density-of-states effect and not unusually enhanced by phonon softening. Secondly (quoting Raub et al. (see above)), the T_c of Ti-Fe (<0.05 at.%) is less than that of pure Ti.	-----	A.F.Prekul et al. (1976) [Pre76]
8 at.%	---------	See Table 1-10	------------	Ti-------Mn Ti------Mo,----Ru Ti---------Os	R.R.Hake et al. (1975) [Hak75]

TABLE 1-13 TITANIUM-COBALT ALLOYS

Co Concentration Range	Starting Materials, Sample Preparation	Properties and Procedures	Remarks, Discussion	Other Systems Studied	Literature
0-20 at.% (10 alloys)	---------	T_c determined as function of composition in an examination of the influence of "magnetic" solutes on the T_c of Ti-TM alloys.	Results for Ti-Co and Ti-Rh juxtaposed, see Table 1-7.	Ti-Cr,-Mn,-Fe Ti------Ru,-Rh Ti----Re	B.T.Matthias et al. (1959) [Mat59]
1.3 at.%	Argon arc melted and cast.	Low temperature electrical transport properties such as resistivity magneto-resistivity and Hall effect.	See Table 1-8	Ti-Cr,-Mn Ti-Fe,-Ni	R.R.Hake et al. (1962) [Hak62b]
1.3 at.%	---------	An investigation of the possible existence of localized magnetic states.	See Table 1-8	Ti-Cr,-Mn,-Fe	J.A.Cape (1963) [Cap63]
Review including Ti-Co	---------	------------	See Table 1-4	-----	U.Zwicker (1963) [Zwi63]
1.0 at.%	---------	An investigation of localized magnetic moments and superconductivity.	See Table 1-10	Ti-Mn Ti-Zr-Mn	R.R.Hake and J.A.Cape (1964) [Hak64]
0.5, 2.0, 13.5 at.%	Iodide Ti (99.9% from Foote Mineral Co.) and Co ≳ 99.8%. Argon arc melted on wcch.	An intercomparison of the T_c composition dependences of alloys of Ti with the electronic isomers Co, Rh and Ir. Above 1.2 K, ballistic mutual inductance was measured during the collapse of a 10 Oe field; below 1.2 K, the Schawlow-Devlin resonance-frequency-shift technique was used.	T_c data, together with some from Reference [Mat59], were plotted vs at.% solute, and e/a. An extrapolation of data for α-phase solid solutions yielded $T_{c,\alpha-Ti}$ ~0.4 K.	Ti-Rh,-Ir	C.J.Raub and G.W.Hull (1964) [Rau64c]
0.71, 2.2 at.%	---------	A calorimetric (and resistometric) study of superconductivity in dilute α^{m} Ti-TM (3d) alloys.	See Table 1-4	Ti-Sc,-V,-Cr Ti-Mn,-Fe,-Ni Ti-Hf	K.L.Agarwal (1974) [Aga74]

TABLE 1-14 TITANIUM-NICKEL ALLOYS

Ni Concentration Range	Starting Materials, Sample Preparation	Properties and Procedures	Remarks, Discussion	Other Systems Studied	Literature
Through 9 at.%	Measured in as-cast condition.	------------	See Table 1-4. Results limited to the observation that at least 9 at.% Ni was needed to raise T_c above 1 K in the hexagonal phase.	Ti-Cr,-Mn,-Fe,-Co Ti-------Ru,-Rh Ti-----Re	B.T.Matthias *et al.* (1959) [Mat59]
1.0 at.%	Measured in as-cast condition.	------------	See Table 1-8	Ti-Cr,-Mn,-Fe	R.R.Hake *et al.* (1962) [Hak62[b]]
0.43, 1.04 at.%	40h/1000°C/quenched.	------------	See Table 1-4	Ti-Sc,-V,-Cr Ti-Mn,-Fe,-Co Ti-Hf	K.L.Agarwal (1974) [Aga74]

TABLE 1-15 TITANIUM-ZIRCONIUM and TITANIUM-HAFNIUM ALLOYS

Zr Concentration Range	Starting Materials, Sample Preparation	Properties and Procedures	Remarks, Discussion	Other Systems Studied	Literature
35, 50, 70, 83, 100 at.%	Annealed and quenched	T_c determined magnetically.	See Table 1-4 for preparation, etc. details. T_c yields a well-defined maximum at about 50 at.%.	Numerous binary-TM alloy systems.	J.K.Hulm and R.D.Blaugher (1961) [Hul61]
25, 50, 75 at.%	----------	Low temperature specific heat studies of the variation of T_c with $n(E_F)$ at constant e/a ratio.	First calorimetric study of isoelectronic binary-TM alloys. Data for Ti and Zr end-points from Smith et al. [Smi53] and [Smi52] (Ti) and Hulm and Blaugher [Hul61] (Zr). Plots of T_c, γ and θ_D were presented. The strong parallelism between T_c and γ was accepted as support for the BCS model which calls for a dependence of T_c on $n(E_F)$γ.	Nb-Re,-Ru	E.Bucher et al. (1964) [Buc64] see also [Buc63]
50 at.%	As-cast	Low temperature specific heat studies of equiatomic Ti-Zr for use as a "Ti-like" hcp base alloy with $T_c > 1$ K for studying the influence of Mn additions on superconductivity.	See Table 1-9 for preparation, etc. details. The paper described a magnetic and calorimetric study of Mn in: (a) Ti, and (b) Ti-Zr (50 at.%). Mn, which has four times the moment in Ti as in Ti-Zr (50 at.%) lowers the T_c of Ti at a much faster rate.	Ti-Mn,-Co Ti-Zr-Mn	R.R.Hake and J.A.Cape (1964) [Hak64]
Review including Ti-Zr	---------	An investigation of the effectiveness of a solute-size-corrected "effective" e/a ratio.	See Table 1-4 for more details.	Ti-V,-Nb,-Mo and numerous other binary-TM alloy systems.	W.DeSorbo (1965) [Des65]
9.6, 30.6, 39.5, 49.4, 50.2, 59.5 at.%	Iodide Ti and Zr, argon arc melted repeatedly on wcch. Either annealed ~1200°C/quenched, or annealed 500-750°C/quenched, depending on alloy.	Low temperature specific heat study within temperature range 1.1-4.5 K aimed at elucidating factors which influence fundamental superconductive interactions. The transition temperatures of α^m and α structures were intercompared.	Alloys quenched from a high-temperature β-anneal were martensitic (α^m), whereas moderate-temperature annealing yielded the equilibrium α phase. In general $T_{c,\alpha^m} > T_{c,\alpha}$, although the difference was less than the transition width itself, typically ± 0.1 K. The results were discussed within the framework of BCS theory, and analyzed with reference to electron-electron and electron-phonon interactions and the results of band-structure calculations.	Zr-Hf Zr-Sc	J.O.Betterton and J.O.Scarbrough (1968) [Bet68]

TABLE 1-15   TITANIUM-ZIRCONIUM and TITANIUM-HAFNIUM ALLOYS –continued

Zr Concentration Range	Starting Materials, Sample Preparation	Properties and Procedures	Remarks, Discussion	Other Systems Studied	Literature
Theoretical study	---------	A calculation of λ, the electron-phonon coupling , using CPA methods and a examination of the results by comparing experimental T_c's with those calculated using the simple weak-coupling expression $T_c = 1.13\, \theta_D \exp(-1/\lambda)$.	------------	Ta-V Ta-Nb Nb-V	H.Lustfeld (1975) [Lus75]

Hf Concentration Range	Starting Materials, Sample Preparation	Properties and Procedures	Remarks, Discussion	Other Systems Studied	Literature
35.4, 64.6 at.%	---------	A low temperature specific heat study of Ti-Hf as an isoelectronic system for comparison with the results of a similar study on Ti-Zr.	γ passed through a maximum and θ_D through a minimum near 40 at.% Hf paralleling the behavior of Ti-Zr. Unfortunately, the lower temperature limit of the measurement (1.7K) was too high to enable the variation of T_c to be followed.	-----	K.L.Agarwal (1974) [Aga74]

TABLE 1-16 TITANIUM-TANTALUM ALLOYS — THE SUPERCONDUCTING TRANSITION

Ta Concentration Range	Starting Materials, Sample Preparation	Properties and Procedures	Remarks, Discussion	Other Systems Studied	Literature
Review including Ti-Ta	----------	----------	See Table 1-4. In this paper the hitherto unpublished data of Blaugher and Joiner [Bla63] for Ti-Ta (26-60 at.%) were presented for the first time.	Alloys of Ti with ten other TM elements and numerous compounds.	U.Zwicker (1963) [Zwi63]
2.5, 5 at.%	Iodide Ti (Foote Mineral Co.) electron-beam melted and de-gassed, Ta (99.8%, DPH 45 kg mm^{-2}). Argon arc melted three times; cold rolled 40%; annealed 2h/650°C in Ti cylinder in evacuated silica tube.	A study of T_c as function of composition in equilibrium α-phase Ti-TM (group V) alloys.	The microstructure was homogeneous α phase. For intercomparison, the T_c's of Ti-V,-Nb,-Ta (2.5 at.%) were 1.4, 1.5 and 1.3 K, respectively.	Ti-V,-Nb	C.J.Raub and U.Zwicker (1965) [Rau65]
30.1, 34.2, 42.3, 46.2, 47.3, 50.8, 53.5, 56.5, 63.6, 68.5, 77.7, 86.3, 97.7 at.% (analyzed)	Iodide Ti (Foote Mineral Co.), 99.92% (75-125 wt. ppm O) and Ta (Fansteel Metallurgical Corp.), 99.96%. Arc melted 14 times in gettered argon on wcch. Annealed 13h/1000°C; machined, stainless steel clad, swaged to 0.140 in.⬦, drawn to 0.010 in.⬦.	T_c was measured resistometrically at a current of 0.5 mA ($\sim$1 A cm^{-2}). T_c was taken as the temperature of the 1/2-residual resistance point.	One alloy, Ti-Ta (50.9 at.%) made with Ti sponge containing 249 wt. ppm O. T_c vs composition was claimed to be similar to that reported by Blaugher and Joiner [Bla63] for β-Ti-Ta alloys. Annealing 1h/1000°C reduced T_c in Ti-Ta (31.5-53.8 at.%) alloys. The effects on T_c of annealing, aging, phase separation and oxygen content were considered.	------	D.A.Colling et al. (1966) [Col66a]
7, 16, 32, 52, 70, 90 at.% and others	DC triode sputtering in argon at 100Å/min onto fused quartz substrate at 700°C.	T_c measured resistometrically in a study of T_c-composition dependence in sputtered films.	See Tables 1-4 and 1-6.	Ti-V,-Nb	H.J.Spitzer (1974) [Spi74], see also [Spi74a]
18, 20, 28, 40, 50, 71, 80, 91 at.% (from plotted data)	----------	----------	See Table 1-4	Ti-V,-Nb	A.F.Prekul et al. (1974) [Pre74]

TABLE 1-17 TITANIUM-TANTALUM ALLOYS — THE MIXED STATE

Ta Concentration Range	Starting Materials, Sample Preparation	Properties and Procedures	Remarks, Discussion	Other Systems Studied	Literature
10.0, 19.8, 25.0, 29.5, 37.5, 45.0, 50.0, 60.0, 65.0, 70.0, 75.0, 80.0, 90.0 at.% (nominal)	---------	H_r measured resistometrically in pulsed fields in order to establish a data base for an experimental critique of the GLAG theory as modified by paramagnetic limitation.	See Table 1-5	Ti-V,-Nb,-Mo Zr-Nb Hf-Nb,-Ta U-Nb,-Mo	T.G.Berlincourt and R.R.Hake (1963) [Ber63^a], see also [Ber63]
30.1-97.7 at.% (analyzed)	---------	H_r measured resistometrically for comparison with the Clogston H_p and an H_{c2} computed with the aid of an equation due to Gor'kov. J_c was also studied.	See Table 1-16 for an experimental note. Also studied was the influence of final-size heat treatment on the H_r in Ti-Ta (50.8 at.%).	-----	D.A.Colling et al. (1966) [Col66]
52 at.%	---------	H_r measured resistometrically in steady fields in an investigation of the predicted spin-orbit-scattering induced alleviation of paramagnetic limitation of H_{c2}.	See Table 1-5	Ti-V,-Nb	L.J.Neuringer and Y.Shapira (1966) [Neu66]
53 at.%	Arc melted repeatedly on wch. Swaged and drawn to 0.010 in. φ.	H_u measured by observing the rate of change of sample magnetization in a pulsed field. The purpose of the study was to improve upon the simple electron/atom ratio systematics previously used for displaying T_c and H_{c2} data.	The technique used was related to that of Suenaga and Ralls [Sue69].	25 other alloys, mostly ternary and quaternary such as Ti-Nb-TM and Ti-Nb-TM$_1$-TM$_2$.	W.DeSorbo et al. (1967) [Des67]

TABLE 1-18   TITANIUM-TANTALUM ALLOYS — THE CRITICAL CURRENT DENSITY

Ta Concentration Range	Starting Materials, Sample Preparation	Properties and Procedures	Remarks, Discussion	Other Systems Studied	Literature
20, 25, 29, 29, 60, 70, 75 at.%	Reduced from as-cast slabs about 75% by: (a) grinding, and (b) rolling.	A study of $J_c(\parallel)$ and $J_c(\perp)$ on cold-rolled alloys.	See Table 1-6 for experimental details. It was stated that the alloys were predominantly β, except Ti-Ta₂-Nb (20 at.%) which contained some hcp martensite. However, on the basis of later work [Ber63[a]] it is to be concluded that the 25 and 29 at.% alloys also possessed traces of hcp.	Ti-V₂-Mo Zr-Nb₂-Mo Nb-Hf₂-Sc	R.R.Hake *et al.* (1962) [Hak62[a]]
90 at.%	Reduced in thickness 140:1 by cold rolling.	A study of cold-rolling induced critical current anisotropy in fields of up to 30 kOe.	J_c *vs* H was measured for three values (9°, 90°, 170°) of β, the angle between H and the rolling plane. Noted were: (a) a peak effect, and (b) evidence for variation of effectiveness of "pinning centers" with angle.	Ti-Nb	R.R.Hake *et al.* (1963) [Hak63[a]]
75 at.%	Cold drawn wires of 0.010 in. and 0.005 in.ᵠ.	A study of longitudinal critical current density.	Unlike the usual transverse J_c *vs* H characteristics, longitudinal current goes through a pronounced maximum at intermediate values of the field. These maxima, at about 22 kOe correspond to 3.8×10^5 A cm⁻² (0.010 in.ᵠ) and 4.3×10^5 A cm⁻² (0.005 in.ᵠ); 4.2 K.	Nb-Zr Mo-Re	S.T.Sekula *et al.* (1963) [Sek63]
70, 75, 80 wt.%	Wire samples	Measurements at 4.2 K in fields of up to about 90 kOe were made in a comparison of J_c *vs* H for several TM_1-TM_2 alloys and "powder-core" Nb₃Sn wires.	Coil tests were carried out upon Zr-Nb and Ti-Nb wires.	Ti-Nb Zr-Nb	R.C.Wolgast *et al.* (1963) [Wol63]
30.1-97.7 at.% (analyzed, 13 alloys)	--------	Measurements were made at 4.2 K transverse to the axis of 2 in. air-core Bitter solenoid in a study of J_c *vs* composition in cold-worked alloys. Also investigated was the influence of heat treatment on $J_c(H)$.	See Table 1-16 for sample preparation. J_c was investigated as function of heat treatment; a peak effect noted; the highest J_c at H > 50 kOe was obtained in response to a 1h/400°C heat treatment. The influence of 0 on the J_c of Ti-Ta (~50 at.%) was considered.	------	D.A.Colling *et al.* (1966) [Col66]
60 at.%	Samples were heat treated to "optimize" J_c.	Part of a study of the influence on J_c of ternary and quaternary additions to basic Ti-Nb (and Ti-Ta) alloys.	Precipitation heat treatment significantly increased J_c in the Ti-Ta alloy, but additions of 0.08 to 0.27 at.% 0 had no significant effect. This alloy had been homogenized at 1250 to 1500°C.	Ti-Nb Ti-Nb-0 Ti-Nb-Fe Ti-Nb-0-Fe Ti-Ta-0	J.O.Betterton *et al.* (1966) [Bet66]

TABLE 1-18 TITANIUM-TANTALUM ALLOYS — THE CRITICAL CURRENT DENSITY –*continued*

Ta Concentration Range	Starting Materials, Sample Preparation	Properties and Procedures	Remarks, Discussion	Other Systems Studied	Literature
Ti-Ta	Heat treated to optimize J_c.	A study of the influence on J_c of interstitial element additions and precipitation heat treatment.	Due regard was paid to the difficulty of drawing alloys with excessive interstitial content.	Ti-V Ti-Nb Ti-Nb-Ta Ti-Nb-Hf the above with interstials (O, N, C)	French Patent No. 1,517,216 March 15, 1968 [Ass68[a]]
7, 16, 32, 52, 70, 90 at.% and some 34 other compositions	DC triode sputtering in argon at 100Å/min onto fused quartz substrate at 700°C.	J_c measured using a pulsed-current, pulsed-field technique was studied as a function of composition in sputtered Ti-Ta films.	See Tables 1-4 and 1-6	Ti-V,-Nb	H.J.Spitzer (1974) [Spi74[a]], see also [Spi74[a]]

TABLE 1-19 TITANIUM-MOLYBDENUM ALLOYS — THE SUPERCONDUCTING TRANSITION

Mo Concentration Range	Starting Materials, Sample Preparation	Properties and Procedures	Remarks, Discussion	Other Systems Studied	Literature
30, 35, 40, 45, 48.5 at.%	See Table 1-4	T_c studied magnetically as a function of average electron/atom ratio.	See Table 1-4	Ti-V, V-Mo	E.Bucher et al. (1959) [Buc59]
6.25, 6.50, 7.54, 8.60 at.%	Iodide Ti (99.92%) (Foote Mineral Co.), Mo (99.9%). Arc melted six times in gettered argon on wcch. Measured as-cast.	Low temperature specific heats (1.1-4.3K) of a number of Ti-Mo alloys were measured in a fundamental study of a representative series of well-categorized Ti-TM alloys.	One of the first detailed experimental studies of the superconducting transition in a short mfp alloy superconductor. The presence of ω phase was denied.	----	R.R.Hake (1961) [Hak61], see also E.Bucher et al. (1964) [Buc64]
1, 1.5, 2, 2.5, 3, 4, 5.5, 6, 8.5, 9.5, 10, 10.5, 21, 29, 33, 40, 50, 55, 60 at.%	Levitation melted and cast into rods, annealed by electron bombardment 30h/1100°C/water quenched.	T_c was measured over a wide electron/atom ratio range using the ballistic method of Hardy and Hulm.	The results showed a maximum in T_c at an electron/atom ratio of 4.32.	----	R.D.Blaugher et al. (1961) [Bla61] see also U.Zwicker (1963) [Zwi63]
16 at.%	Annealed 1h/900°C/water quenched.	A study of the influence of phase transformation on the T_c and hardness.	The alloy, Ti-Mo (8.7 at.%) was aged up to 1000min/400°C.	Ti-Nb, Ti-V-Al	C.J.Raub and U.Zwicker (1964) [Rau64^a]
2.5, 4 at.%	See Table 1-4	T_c was measured using calorimetric and magnetic techniques in a study of the superconductivity of dilute Ti-TM alloys.	See Table 1-4	Ti-V,-Cr,-Fe Ti-Nb	E.Bucher et al. (1965) [Buc65]
13, 15, 20, 30, 40, 75, 85 at.%	Iodide Ti (99.95%) and Mo (99.9%) argon arc melted on wcch. Homogenized at 1400°C in <3x10^{-6} Torr for 50 h (Ti rich) to 250 h (Mo rich), oil quenched.	The low temperature specific heats (1.4-4.2K) of a wide range of bcc alloys were measured.	The maximum in T_c occurred within the electron/atom ratio range 4.3 ∿ 4.4. γ vs composition was compared with those for Ti-TM (3d) and Ti-TM (4d) series of alloys.	----	A.K.Sinha (1968) [Sin68]
1, 2, 3, 3.5, 4, 4.5 at.% (nominal)	Ti sponge (99.9%) from Titanium Metals Corp. with Mo (99.9%) from Climax Molybdenum Co., argon arc melted repeatedly on wcch. Annealed in gettered argon 8h/1300°C/ice-brine quenched.	Low temperature specific heats (1.5∿6K) were measured in a study of the superconducting transition of low-concentration martensitic (α^m) Ti-Mo alloys.	As a result of a comparison of the rates of increase of γ and T_c in the α^m alloys with the dependence of T_c on γ in the bcc alloys, it was claimed that T_c was enhanced about 2.5 times by some property of the martensitic structure.	----	E.W.Collings and J.C.Ho (1969) [Co69]

TABLE 1-19 TITANIUM-MOLYBDENUM ALLOYS — THE SUPERCONDUCTING TRANSITION –continued

Mo Concentration Range	Starting Materials, Sample Preparation	Properties and Procedures	Remarks, Discussion	Other Systems Studied	Literature
5, 7, 8-1/2, 10, 15, 20, 25 at.% (nominal)	See above	The low temperature specific heats (1.5-6K) of a series of quenched ($\omega+\beta$)-phase and β-phase alloys were measured in a study of the influence of ω phase as a suppressor of T_c on the Ti-rich side of $T_{c,max}$.	A large volume fraction of ω-phase was noted in Ti-Mo (4-1/2, 5 and 10 at.%).	-----	J.C.Ho and E.W.Collings (1969) [Ho69]
4-1/2 at.%	As in [Coℓ69] above, except that samples were measured in the: (a) as-cast, (b) quenched, and (c) quenched plus deformed, conditions.	Low temperature specific heat study of the response of the superconducting transition in a quenched Ti-Mo alloy to mechanical deformation.	T_c increased following a transformation of the structure from $\omega+\beta$ to α_m at constant composition.	-----	E.W.Collings and J.C.Ho (1970) [Coℓ70]
5, 7 at.%	Prepared as in [Coℓ69] above and: (a) quenched, and (b) deformed.	In an extension of the above study the low temperature specific heats of quenched and quenched + deformed Ti-Mo (5, 7 at.%) alloys were measured and intercompared.	Samples were deformed by compression such that $\Delta\ell/\ell = 23\%$ (Ti-Mo, 5 at.%) and 47% (Ti-Mo, 7 at.%).	-----	J.C.Ho and E.W.Collings (1971) [Ho71], see also J.C.Ho and E.W.Collings (1973) [Ho73] and E.W.Collings et al. (1971) [Coℓ71]
10 at.%	As in [Coℓ69] above	Low temperature specific heat (1.5-5K) measurements used in a study of bulk superconductivity in a two-phase (in particular, $\omega+\beta$-phase) Ti-TM alloy.	Annealed 2h/1200°C/quenched (Q) then: (i) 1h/300°C/Q; (ii) above plus 8h/350°C/Q; (iii) above plus 150h/350°C/Q; (iv) above plus 720h/350°C/Q.	-----	J.C.Ho et al. (1970/1971) [Ho70]
-----	Nonexperimental	A procedure for the empirical determination of the electronic specific heat enhancement factor in a series of TM superconductors was developed.		-----	E.W.Collings and J.C.Ho (1971) [Coℓ71a]
1, 2, 3, 5, 7, 8-1/2, 10, 15, 20, 25, 40, 70 at.% (nominal)	As in [Coℓ69] above	Superconductivity and structural instability in "absolutely unstable" β-Ti-Mo (<4-1/2 at.%) and "virtual" β-Ti-Mo (5<at.% Mo<15) alloys were investigated.	The analysis and the discussion were assisted by availability of magnetic susceptibility and elastic constant data.	-----	E.W.Collings et al. (1972) [Coℓ72]

TABLE 1-19 TITANIUM-MOLYBDENUM ALLOYS — THE SUPERCONDUCTING TRANSITION *-continued*

Mo Concentration Range	Starting Materials, Sample Preparation	Properties and Procedures	Remarks, Discussion	Other Systems Studied	Literature
As in [Coℓ72] above	As in [Coℓ69] above	-------------	Review of magnetic and calorimetric studies in Ti-Mo alloys treated under the heads: (a) quenched real alloys, (b) postulated T_c's for dilute β-Ti-Mo alloys, (c) aging of ω+β alloys and equilibration in the α+β field.	-----	E.W.Collings (1972) [Coℓ72a]
As in [Coℓ72] above	As in [Coℓ69] above	-------------	A review of the properties of quenched, aged and deformed Ti-Mo alloys including a study of the aging for 880h/350°C followed by equilibration at 550°C of a quenched Ti-Mo (10 at.%) alloy.	-----	J.C.Ho and E.W.Collings (1973) [Ho73]
1, 1.5, 2, 2.5, 3, 3.5, 4, 4.5 at.%	In some cases following [Coℓ69] above	T_c in quenched α^m-Ti-TM alloys was reviewed.	It was agreed, with particular reference to the Ti-TM system, that superconductivity in the quenched α^m-alloys was a property of the martensitic structure itself rather than that of any included residual β-phase.	Ti-V,-Cr,-Fe,-Co Ti-Nb,-Mo	E.W.Collings and J.C.Ho (1975) [Coℓ75a]
As in [Coℓ72] above	As in [Coℓ69] above	T_c of Ti-TM alloys was discussed within the framework of lattice stability.	It was concluded: (a) that the stabilization of the β-Ti-Mo lattice through alloying with a solution-strengthening solute such as Al results in a depression of T_c, the extent of which depends on the degree to which the binary lattice has already been stabilized by ω-phase precipitation, and (b) the extreme structural disorder inherent in a martensitic structure leads to an enhanced value of T_c.	Ti-Mo$_5$-Al$_{1,3}$ Ti-Mo$_{10}$-Al$_{1,3}$ Ti-Mo$_{25}$-Al$_{1-6}$	E.W.Collings and J.C.Ho (1976) [Coℓ76]
5 at.%	See above	A two-part seven-parameter nonlinear least-squares fitting program was developed and applied to the low-temperature specific heat data for the deformed Ti-Mo (5 at.%) alloy referred to in [Ho71] above. A Gaussian-rounded exponential superconductive specific heat function was employed.	It was concluded: (a) that the bulk of the material is responsible for the observed broad transition, (b) that an enhancement in T_c is brought about by the presence of lattice defects (cf. [Coℓ69] above), and (c) that T_c broadening is the result of a wide variation of defect environments.	Ti-Mo$_5$-Al$_{1,3}$	J.J.White and E.W.Collings (1976) [Whi76]
As in [Whi76] above	As in [Whi76] above	An analysis of the calorimetric data for quenched and quenched-plus-deformed Ti-Mo alloys.	In an extension of the above work the data were analyzed by applying a Gaussian-rounded BCS-specific heat function to the sub-T_c specific heat.	-----	E.W.Collings and J.J.White (1978) [Coℓ78a]

TABLE 1-20   TITANIUM-MOLYBDENUM ALLOYS — THE MIXED STATE

Mo Concentration Range	Starting Materials, Sample Preparation	Properties and Procedures	Remarks, Discussion	Other Systems Studied	Literature
6.25, 9.07, 11.1, 16.3, 25.0, 37.5, 50.0 at.%	Sample prepared from arc-melted ingot by machining or cold rolling.	Resistometric measurements of the upper critical field, H_*, in pulsed magnetic fields $0 \lesssim H \lesssim 160$ kOe, in a comparison of the H_{c2}^* of GLAG theory with the Clogston paramagnetically limited H_p.	See Table 1-5 for further details. The calorimetric data of Hake [Hak61] were used to calculate H_{c2}^* (T=1.2K) and the T_c data of Blaugher et al. [BZa61] were used to deduce H_p (T=1.2K). H_r and H_p were in good accord within the range $4.1 < e/a < 4.6$.	Ti-V,-Nb,-Ta; Zr-Nb; Hf-Nb,-Ta; U-Nb,-Mo	T.G.Berlincourt and R.R.Hake (1963) [Ber63a], see also T.G.Berlincourt (1963) [Ber63]
16 at.%	Annealed 1h/0.8 mp (~1500°C).	Ballistic magnetization measurements were undertaken by moving the sample between a pair of 18,000-turn search coils in fields $0 < H \lesssim 27$ kOe in a study of the transition between the diamagnetic and paramagnetic mixed states.	See Table 1-5 for further details. The paramagnetic mixed state $M_s(H)$ line smoothly contacts at H_u that for the normal state $M_n(H)$, indicating a second, or higher, order s/n transition.	Ti-V; Ti-V-Cr; Hf-Nb	R.R.Hake (1965) [Hak65]
16 at.%	Annealed 1h/1440°C, cooled 6°C/min.	Vibrating sample magnetometer measurements in fields $0 < H \lesssim 50$ kOe in a study of the transition between the paramagnetic-superconducting and normal states at H_u in order to compare H_u and $H_t(T)$ with the predictions of paramagnetic-limited and spin-orbit-inclusive theory.	H_u is less than H_{c2}^* but larger than expected if spin paramagnetism is allowed but spin-orbit scattering is not. The Pauli spin parameter, α, is calculated from atomic data, while τ_{so} is an adjustable parameter enabling the fitting of $H_u(T)$ and $H_{c2}(T)$. The quantity $\kappa_2(T) \propto 1/\sqrt{d(M_s-M_n)/dH}$ at H_u *decreases* with decreasing T.	-----	J.A.Cape (1966) [Cap66]
16 at.%	Annealed 1h/0.8 mp (~1500°C) and cooled 7°C/min, 70°C/min, 30°C/min, 15°C/min, to 800°C, 500°C, 350°C, 20°C, respectively.	Ballistic magnetization measurements as in [Hak65] in fields of $0 < H < 55$ kOe. A more detailed repetition of the above experiment using wider field and temperature ranges and an extensive set of alloys.	Detailed comparisons were made between the experimentally derived values of H_u. $\kappa_1 = H_u/\sqrt{2} H_c$ and $\kappa_2 \propto 1/\sqrt{d(M_s-M_n)/dH}$ (at H=H_u), and the theories of Maki (e.g. [Mak66]) and Werthamer et al. [Wer66].	Ti-V; Ti-V-Cr	R.R.Hake (1967) [Hak67b], see also R.R.Hake (1967) [Hak67] and L.J.Barnes and R.R.Hake (1966) [Bar66]

TABLE 1-20 TITANIUM-MOLYBDENUM ALLOYS — THE MIXED STATE *–continued*

Mo Concentration Range	Starting Materials, Sample Preparation	Properties and Procedures	Remarks, Discussion	Other Systems Studied	Literature
16 at.%	Annealed as in [Hak65] above.	Low temperature specific heat study ($0.33 \lesssim t \lesssim 1$) ($T_c = 3.2$-$4.2$K) in fields $0 \lesssim H \lesssim 29$ kOe of the second-order transition between the paramagnetic mixed and normal states. Calorimetric measurements were also able to provide data for the calculation of $H_{c0} = 2.42 \gamma^{1/2} T_c$, and other parameters.	The second-order nature of the transition implies that the Pauli susceptibilities (dM/dH) of the paramagnetic mixed and normal states must be comparable, regardless of other possible contributions to the superconducting and normal state magnetizations.	-----	L.J.Barnes and R.R.Hake (1967) [Bar67], see also L.J.Barnes and R.R.Hake (1966) [Bar66]
16 at.%	Annealed 1h/0.8 mp ($\sim$1500°C).	Longitudinal and transverse voltage measurements were made in fields of up to 39 kOe at 1.2 K in a study of the mixed-state Hall effect.	------------	-----	R.R.Hake (1968) [Hak68]
Review	--------	An examination of the core model for mixed state Hall effect and resistivity.	Model calculations were made in the case of TM alloys.	Nb, V In-Pb Nb-Ta	C.H.Weijsenfeld (1968) [Wei68]
16 at.%	--------	A review of the thermodynamics of types-I and II superconductivity.	Thermodynamic relationships, upper critical field limits and fluctuation superconductivity were discussed.	Ti-V V-Ta	R.R.Hake (1969) [Hak69]
16 at.%	--------	A re-analysis of the data of Barnes and Hake [Bar67] above within the context of a search for evidence of one-dimensional behavior in the specific heat transition.	------------	-----	R.F.Hassing *et al.* (1973) [Has73]

TABLE 1-21 TITANIUM-MOLYBDENUM ALLOYS — CURRENT TRANSPORT EFFECTS

Mo Concentration Range	Starting Materials, Sample Preparation	Properties and Procedures	Remarks, Discussion	Other Systems Studied	Literature
16 at.%	Samples either sawed and ground to shape or cold rolled.	Four-terminal J_c measurements at 1.2 and 2.4 K as function of the orientation (∥ or ⊥ to surface of strip) at a transverse magnetic field 0 < H < 30 kOe.	It was noted, for example, that cold rolling increased J_c (30kOe, H∥) by a factor of 12 at 1.2 K, and 24 at 2.4 K, but did not appreciably alter the J_c (30kOe, H⊥) values. The high current-carrying capacity of hard super-conductors, and especially its enhancement by cold work was interpreted in terms of the "sponge" or "filamentary" model.	Nb-Zr	R.R.Hake et al. (1962) [Hak62]
9, 16 at.%	---------	Four-terminal J_c measurements at 1.2 K in fields of up to 30 kOe.	See Table 1-6 for details and discussion.	Ti-V,-Ta Zr-Nb Nb-Sc,-Hf	R.R.Hake et al. (1962) [Hak62a]
6, 9, 16 at.%	---------	Four-terminal J_c measurements at 1.2 K in fields 0 < H < 30 kOe of cold rolled Ti-TM alloys, with emphasis on the critical current anisotropy. In the companion paper [Ber63] the field range is extended (by pulsing) to about 130 kOe.	See Table 1-20. High current-carrying capacity was interpreted in terms of a filamentary model, i.e. a model based on interconnected metallic defects such as "dislocations and/or their strain effects". As a consequence of the extended field range, a high-field peak was revealed in J_c(H⊥ RP) -- an interesting example of "peak effect".	-----	R.R.Hake and D.H.Leslie (1963) [Hak63], see also T.G.Berlincourt (1963) [Ber63]
16 at.%	---------	Resistivity temperature-dependence measurements in the temperature range $1 < t ≳ 3.8$ in a study of fluctuation superconductivity above T_c.	See Table 1-6	Ti-V,-Ru Ti-V-Cr	R.R.Hake (1969) [Hak69a]
16 at.%	---------	Resistivity temperature-dependence and field-dependence measurements undertaken in a study of fluctuation superconductivity and its quenching by magnetic fields.	------------	-----	R.R.Hake (1971) [Hak71]
5.09, 6.86, 10.30, 14.92, 16.59, 19.38 at.%	Ti sponge (99.9%)(Titanium Metals Corp.) and Mo (99.9%)(Climax Molybdenum Co.) arc melted frequently on wcch. Heat treated 8h/1000°C (5 and 7 at.%) 2h/1200°C (10-20 at.%).	Resistivity temperature-dependence in the temperature range 4.2-620 K in a study of the negative $d\rho/dT$ effect.	A negative $d\rho/dT$ (4-300K) was noted in the alloys Ti-Mo (7, 10 and 15 at.%). That of Ti-Mo (10 at.%) could be accounted for in terms of a negative $d\rho_{def}/dT$ attributed to the reversible precipitation of a small volume fraction of athermal ω-phase.	-----	J.C.Ho and E.W.Collings (1972) [Ho72]

TABLE 1-21 TITANIUM-MOLYBDENUM ALLOYS — CURRENT TRANSPORT EFFECTS –continued

Mo Concentration Range	Starting Materials, Sample Preparation	Properties and Procedures	Remarks, Discussion	Other Systems Studied	Literature
16 at.%	--------	A study of negative normal-state resistivity temperature dependence parameterized by $\alpha^T \equiv 1/\rho(\partial\rho/\partial T)_H$ and magnetoresistance parameterized by $\alpha^H \equiv 1/\rho(\partial\rho/\partial H)_T$.	See Table 1-10	Ti-----Mn,-Fe Ti--------Ru Ti--------Os Ti-V-Cr	R.R.Hake et al. (1975) [Hak75]

TABLE 1-22 TITANIUM-TUNGSTEN ALLOYS

W Concentration Range	Starting Materials, Sample Preparation	Properties and Procedures	Remarks, Discussion	Other Systems Studied	Literature

No entries

TABLE 1-23 TITANIUM-TECHNETIUM and TITANIUM-RHENIUM ALLOYS

Tc Concentration Range	Starting Materials, Sample Preparation	Properties and Procedures	Remarks, Discussion	Other Systems Studied	Literature
15, 25, 35, 50, 67, 75, 85, 87.5, 95, 97 at.%	Iodide Ti (Zr < 700; Mg < 100; Fe, 200; O, 100, C, 190 ppm, etc.); Tc (Al, 330; Cu, 50 ppm, etc.). Arc melted repeatedly and cast into copper mold. Measured as-cast and also after annealing at 700-1500°C depending on composition.	T_c measured inductively on as-cast cylindrical shapes over the temperature range >1.70 K.	Of particular interest was the composition dependence of T_c. A local T_c maximum was noted in the bcc alloys about $e/a \sim 4.7$; however the largest T_c's occurred in the hcp T_c-rich alloys.	-----	C.C.Koch (1976) [Koc76a]

Re Concentration Range	Starting Materials, Sample Preparation	Properties and Procedures	Remarks, Discussion	Other Systems Studied	Literature
2, 4, 5.5, 10 at.%	--------	T_c studied as function of solute concentration.	See Table 1-7. Owing to high melting point of Re (3180°C) it is likely that, due to Ti loss, the Re levels are higher than nominal. Structures are probably $\omega+\beta$ for all alloys.	Ti-Cr,-Mn,-Fe,-Co Ti-------Ru,-Rh	B.T.Matthias et al. (1959) [Mat59]
Review including Ti-Re	--------	------------	See Table 1-4. Large $\partial T_c/\partial c$ in Ti-rich α^m region assumed by analogy with data for Ti-Rh.	----------	U.Zwicker (1963) [Zwi63]

TABLE 1-24 TITANIUM-RUTHENIUM and TITANIUM-OSMIUM ALLOYS

Ru Concentration Range	Starting Materials, Sample Preparation	Properties and Procedures	Remarks, Discussion	Other Systems Studied	Literature
5, 10, 15 at.%	---------	A study of T_c vs e/a.	The pioneering study on the relationship between T_c and valence electron/atom ratio -- statement of Matthias' rule.	Ti-Rh	B.T.Matthias (1957) [Mat57]
Same as above	--------	A study of T_c vs e/a and a comparison of the effects of "ferromagnetic" (i.e. late 3d) solutes with those of 4d and 5d solutes.	See Table 1-7. Same data as above but compared with that for Ti-Fe.	Ti-Cr, -Mn, -Fe, -Co; Ti------Rh; Ti------Re	B.T.Matthias et al. (1959) [Mat59]
Same as above	--------	A review of superconductivity in transition elements.	Although data for the α^m region, plotted for Ti-Fe, were missing for Ti-Ru, it was assumed that there was a fundamental difference between $\partial T_c/\partial c$ for Ti-Fe and that for Ti-Ru, and that fundamentally different superconductive pairing interactions were responsible. A model was suggested.	Ti-Fe	B.T.Matthias (1963) [Mat63]
Review	--------	------------	See Table 1-4. Large $\partial T_c/\partial c$ in Ti-rich α^m region was assumed by analogy with data for Ti-Rh.	------	U.Zwicker (1963) [Zwi63]
8 at.%	--------	Studies of fluctuation superconductivity in the vicinity of T_c, negative magnetoresistance in high fields and anomalous normal-state resistivity temperature dependence.	[Hak69] is a review of type-I and II superconductivity which goes on to deal with fluctuation superconductivity and magnetoresistance. [Hak69a] deals with fluctuation superconductivity and its quenching by high fields and large transport current densities. [Lue75] is an extension of [Hak69a] to higher fields and the discovery of a weakly negative high-field magnetoresistance. [Hak75] associates this negative magnetoresistance $\alpha^H \equiv \rho^{-1}(\partial\rho/\partial H)_T$ with a spin-fluctuation model for the negative $\alpha^T \equiv \rho^{-1}(\partial\rho/\partial T)_H$ noted in several Ti-TM alloys.	Ti-V,-Mo; Ti-V-Cr; Ti-V,-Mo; V-Ta; Ti------Mn,-Fe; Ti-Mo; Ti------Os; Ti-V-Cr	R.R.Hake (1969) [Hak69a], see also; R.R.Hake (1969) [Hak69] and; J.W.Lue et al. (1975) [Lue75] and; R.R.Hake et al. (1976) (1975) [Hak75]

Os Concentration Range	Starting Materials, Sample Preparation	Properties and Procedures	Remarks, Discussion	Other Systems Studied	Literature
8 at.%	--------	------------	Member of a large group of superconductive Ti-TM alloys investigated in a study of low-temperature anomalies in the normal-state resistivity.	See [Hak75] above	R.R.Hake et al. (1975) [Hak75]

TABLE 1-25   TITANIUM-RHODIUM, -IRIDIUM, -PALLADIUM, and -PLATINUM ALLOYS

Rh Concentration Range	Starting Materials, Sample Preparation	Properties and Procedures	Remarks, Discussion	Other Systems Studied	Literature
5, 10, 15 at.%	----------	------------	See Table 1-7	Ti-Cr,-Mn,-Fe,-Co; Ti-------Ru; Ti-----Re	B.T.Matthias (1957) [Mat57], see also B.T.Matthias *et al.* (1959) [Mat59]
0.063, 1.3, 5, 10, 12.5, 15 at.%	Ti sponge (>99.5%) and Rh powder (99.98%). Helium arc melted on wch. Strips 10x2x0.5 mm³ were prepared for resistance and x-ray study. Some turnings and drillings were also taken.	Resistometric study of T_c as function of composition.	Heat treatment conditions; (a) as-cast, and (b) 60h/700°C in vacuum and step-cooled (30h) to room temperature. The resistometrically measured T_c was uniformly high (~4K; cf. that of Ti-Rh$_{15}$ in above study) as a consequence of the presence of included β-phase.	-----	W.Buckel *et al.* (1962) [Buc62]
0.5, 1, 2, 3, 4, 12 at.%	Iodide Ti (Foote Mineral Co.) and Rh (99.8%). Arc melted in Zr-gettered argon. Ingots cut up and remelted several times and measured in as-cast and annealed conditions.	T_c measured by the Hardy-Hulm inductive method, and also resistometrically as a function of composition in order to shed further light on the results of the preceding pair of papers.	The quench-rate during hearth cooling, which naturally decreases with increasing sample size, was judged to be ~10³ °C s⁻¹. The heat treatment, when administered, was 100h/700°C/quench. A series of measurements with Zr as solvent was part of this study.	Zr-Rh	C.J.Raub and C.A.Anderson (1963) [Rau63]
A review which in the case of Ti-Rh referred to the above study.	---------	------------	See Table 1-4. A review in which T_c *vs* composition for Ti-TM alloys were assembled for the first time and discussed with reference to metallurgical microstructure.	Ti-V,-Cr,-Mn,-Fe,-Co; Ti-Nb,-Mo----Ru; Ti-Ta------Re	U.Zwicker (1963) [Zwi63]
0.5, 1, 2, 3, 4, 5.5, 7, 8, 9, 10.5, 12, 13 at.%	Same as in [Rau63]. Alloys measured as-cast.	The Hardy-Hulm inductive method above 1.2 K and the Schawlow-Devlin frequency-shift method (in a He³ bath) below that temperature were employed to measure T_c as a function of composition in order to shed further light on Matthias' postulated "magnetic-interaction" model for superconductivity in the electronic isomer Ti-Fe [Mat59].	The transition widths were 0.1-0.2 K. The work enabled a comparison to be made between the superconducting properties of the related (solute) systems Ti-Co,-Rh,-Ir.	Ti-Co,-Ir	C.J.Raub and G.W.Hull (1964) [Rau64c]
Review, with reference to [Rau64c] and [Rau63] in case of Ti-Rh.	---------	------------	A review of superconductivity in alloys and compounds involving noble and platinum-group metals.	Ti-Ir; Ti-Rh-M with M = Ru,( ),Pd,Ag; Os, Ir,Pt,Au	C.Raub (1964) [Rau64]

TABLE 1-25 TITANIUM-RHODIUM, -IRIDIUM, -PALLADIUM, and -PLATINUM ALLOYS -continued

Rh Concentration Range	Starting Materials, Sample Preparation	Properties and Procedures	Remarks, Discussion	Other Systems Studied	Literature
1, 2, 3, 6, 8, 10 at.%	Electron-beam melted Ti sponge and Rh powder, or Ti (99.5%) and Rh (99.89%) wire. Arc melted together repeatedly and hearth quenched (20 seconds).	Low temperature specific heat measurements were carried out down to 0.9 K, electrical resistance and magnetic inductance down to 1.2 K, in studies of the composition- and structure-dependence of T_c in Ti-Rh alloys spanning the hcp and bcc structural ranges.	As-cast alloys were measured calorimetrically and resistometrically. Heat-treated alloys were measured inductively and resistometrically. The results were discussed in terms of measuring technique and quench-rate-controlled microstructural states.	-----	G. Dummer and E. Oftedal (1968) [Dum68]
0.5, 1, 1.5, 2, 2.5, 3 at.%	Same as above. One sample, Ti-Rh (1 at.%) was cooled from the mp to room temperature in about 1 min.	Specific heat measurements were performed between 0.4 and 4 K in a simple He^3 cryostat. Comparisons were made between calorimetrically and resistometrically determined superconducting transitions in a detailed study of T_c in quenched martensitic Ti-Rh alloys.	The calorimetric experiment was described. A slowly-cooled sample of Ti-Rh (1 at.%) exhibited a double transition.	-----	S. Danner and G. Dummer (1969) [Dan69]

Ir Concentration Range	Starting Materials, Sample Preparation	Properties and Procedures	Remarks, Discussion	Other Systems Studied	Literature
0.4, 0.5, 0.8, 1, 1.5, 1.9, 3, 4, 5, 6, 7, 8.5, 10, 13.5 at.%	---------	---------	See above entry for Ti-Rh. Although dT_c/dc are similar for α^m-Ti-Co, -Rh, that for α^m-Ti-Ir is, inexplicably, much less.	-----	C. J. Raub and G. W. Hull (1964) [Rau64^C], see also C. Raub (1964) [Rau64]
5, 10, 12.5, 15, 17.5 at.% (solid solutions)	Ti (99.9%) and Ir powder (99.999%). Arc melted; samples measured (a) as-cast, and (b) after homogenization at 1400°C.	Primarily a low temperature specific heat study of superconductivity in A15 Ti-Ir-Pt compounds.	Ti-Ir composition range investigated (including compounds): up to 32 at.%. The T_c measurements indicated that the solid solution range extends up to about 18 at.% Ir.	Ti-Pt Ti-Ir-Pt A15 compounds	A. Junod *et al.* (1976) [Jun76]

Pd Concentration Range	Starting Materials, Sample Preparation	Properties and Procedures	Remarks, Discussion	Other Systems Studied	Literature
--------------	---------	---------	No information available.	-----	--------

TABLE 1-25 TITANIUM-RHODIUM, -IRIDIUM, -PALLADIUM, and -PLATINUM ALLOYS –continued

Pt Concentration Range	Starting Materials, Sample Preparation	Properties and Procedures	Remarks, Discussion	Other Systems Studied	Literature
5, 10 at.% (solid solutions)	----------	------------	See above entry for Ti-Ir. Ti-Pt composition range (including compounds): up to 30 at.%. The terminus of the solid solution range in the as-cast alloys is, according to the T_c measurements, between 10 and 15 at.% Pt.	Ti-Ir Ti-Ir-Pt A15 compounds	A.Junod *et al.* (1976) [Jun76]

TABLE 1-26   TITANIUM-BASE TERNARY ALLOYS (excluding alloys with niobium)

Alloy Classification	Compositions	Properties and Procedures	Remarks, Discussion	Other Systems Studied	Literature
(a) Ti-TM-SM	Ti-16V-3Al	Influence of aging-induced phase transformations on T_c and hardness of the initially 1h/900°C/quenched alloy.	Transition temperature plotted vs aging time to 1000 min at 250, 350, 400, 450, 500, 550, 650°C.	Ti-Nb Ti-Mo	C.J.Raub and U.Zwicker (1964) [Rau64[a]]
	Ti-V$_{50}$-Al$_{10}$ Ti-V$_{50}$-Sn$_{10}$	Hardness, ductility, strength, T_c and J_c of cold-worked wire.	Additions of Al and Sn brought about *rapid reductions* in T_c and also J_c of the cold-worked wire.	Ti-V-TM	Yu.V.Efimov *et al.* (1967/1970) [Efi70[a]]
	Ti-4V-6Al	AC measurements of complex susceptibility and T_c as function of aging-induced microstructure.	AC susceptibility regarded as useful diagnostic tool.	-----	E.G.Wolff *et al.* (1973) [Wol73], see also R.Lepper *et al.* (1972) [Lep72]
	(Ti-Mo$_5$)-Al$_{1,3}$ (Ti-Mo$_{10}$)-Al$_{1,3}$	Calorimetric studies of the superconducting transition.	The quenched (Ti-Mo$_5$)-Al$_{1,3}$ alloys are martensitic. Their T_c's extrapolate back to that of deformed-to-α^m Ti-Mo$_5$. The quenched (Ti-Mo$_{10}$)-Al$_{1,3}$ alloys are wt%. The T_c of quenched Ti-Mo$_{10}$ is only weakly depressed by the addition of Al.	Ti-Mo	E.W.Collings and J.C.Ho (1976) [Col76]
	(Ti-Mo$_{25}$)-Al$_{1,2,4,6}$	Resistometric studies of T_c.	The addition of Al to Ti-Mo$_{25}$ reduces T_c at the rate 0.38K/at.%Al.	-----	E.W.Collings and H.L.Gegel (1975) [Col75], see also E.W.Collings and J.C.Ho (1976) [Col76]
(b) Ti-V-TM	Ti-V$_{50}$-TM$_{10}$ TM = Zr, Hf, Nb, Ta, Mo, W, Re	A study of microstructures and phase stability, hardness and tensile strength, critical temperature and critical current.	All the transition elements cited reduced T_c by amounts which increased with the group number of the alloying element; thus the suppression rate increased in the sequence Nb, Zr, Ta, Mo, Hf, W and Re. Critical current density of cold-worked wires was also adversely affected by addition of these elements.	-----	Yu.V.Efimov *et al.* (1967/1970) [Efi70[a]]

TABLE 1-26   TITANIUM-BASE TERNARY ALLOYS (excluding alloys with niobium) — *continued*

Alloy Classification	Compositions	Properties and Procedures	Remarks, Discussion	Other Systems Studied	Literature
	Ti-V_{50}-$Zr_{0,1,5,10}$	A study of transition temperatures and alloy phases.	In the series Ti-V_{50}-Zr_{0-10}, the transition temperature *decreased monotonically* from 7.8 to 5.5 K.	Review	E.M.Savitskii *et al.* (1973) [Sav73, pp.337-8]
	Ti-V_{50}-$Ta_{0,0.5,1,5,10}$	A study of transition temperatures and alloy phases.	In the series Ti-V_{50}-Ta_{0-10}, the transition temperature *decreased monotonically* from 7.8 to 6.7 K.	Review	E.M.Savitskii *et al.* (1973) [Sav73, pp.339-40]
	Ti-V_{50}-$Mo_{0,1,5,10}$	A study of transition temperatures and alloy phases.	In the series Ti-V_{50}-Mo_{0-10}, the transition temperature *decreased monotonically* from 7.8 to 6.2 K.	Review	E.M.Savitskii *et al.* (1973) [Sav73, pp.340-1]
	Ti-V-Ta system	A detailed study of structures and properties of the ternary system.	Quantities plotted on ternary diagrams were: microhardness, 12 and 77 K resistivity and T_c. The highest T_c's were found along the binary axes Ti-V and Ti-Ta. The equiatomic minimum in V-Ta generates a trough connecting that alloy to the Ti apex.	-----	E.M.Savitskii *et al.* (1971) [Sav71][a]
	$(60.5-40.5)Ti-(0.01-0.95)W-(bal.)V$	Deals with transition temperature, critical current and tensile strength.	Transition temperature claimed 7-8 K; critical current density claimed $5 \times 10^3 - 1 \times 10^4$ A cm^{-2}; tensile strength 130-150 kg mm^{-2}.	-----	E.M.Savitskii *et al.* (1968) [Sav68]
(b₂) Ti-V-Cr	Ti-$V_{2.2}$-$Cr_{1.1}$	Investigations of magnetic and calorimetric superconductive transitions.	Very broad superconductive transitions were noted, indicative of the complicated metallurgical structures of these essentially martensitic alloys.	Ti-V,-Cr,-Fe Ti-Nb,-Mo, etc.	E.Bucher *et al.* (1965) [Buc65]
	Ti-V_{60}-Cr_{10}	An investigation of the field and temperature dependences of the mixed state magnetization.	The magnetically determined upper critical fields of these bcc alloys were compared with the predictions of Maki and WHH theories: cf. Table 1-20.	Ti-V Ti-Mo	R.R.Hake (1965) [Hak65], see also R.R.Hake (1967) [Hak67][b] and R.R.Hake (1967) [Hak67]
	Ti-V_{60}-Cr_{10}	Electrical resistivity temperature dependence measurements were undertaken, particular attention being paid to the temperature range just above T_c.	Evidence was cited for the existence of superconductive fluctuations.	Ti-V Ti-Mo Ti-Ru	R.R.Hake (1969) [Hak69][a]

TABLE 1-26

TABLE 1-26 TITANIUM-BASE TERNARY ALLOYS (excluding alloys with niobium) –continued

Alloy Classification	Compositions	Properties and Procedures	Remarks, Discussion	Other Systems Studied	Literature
	Ti-V$_{60}$-Cr$_{10}$	Studies of the temperature dependence and field dependence of normal-state resistivity.	The possible origins of an observed negative $\alpha^m \equiv 1/\rho(\partial\rho/\partial T)_H$ and negative $\alpha^H \equiv 1/\rho(\partial\rho/\partial H)_T$ were discussed.	Ti-,-Mn,-Fe Ti-Mo,-,-Ru Ti-,-,-Os	R.R.Hake et al. (1975) [Hak75]
(c$_1$) Ti-Zr-TM	Ti-Zr$_{49.9}$-Mn$_{0.2}$	Magnetic susceptibility temperature dependence and calorimetric studies of superconducting transitions in as-cast alloys.	The absence of a calorimetric transition was attributed to the existence of a localized magnetic moment on the Mn as evidenced by the Curie-Weiss-like susceptibility temperature dependence.	Ti-Mn,-Co Ti-Zr	R.R.Hake and J.A.Cape (1964) [Hak64]
	Ti-Zr$_{50}$-TM$_{\sim0.2}$ with TM = V , Cr, Mn, Fe Nb, Mo, --, Ru --, W , Re, --	Calorimetric and magnetic studies of superconducting transition in: (a) quenched, and (b) annealed 72h/500°C, alloys.	*All solutes* except Cr and Mn *slightly increased* T_c. Cr slightly reduced T_c and Mn produced a drastic reduction.	Ti-V ,-Cr,-Fe ---,-Nb,-Mo,---	E.Bucher et al. (1965) [Buc65]
(c$_2$) Ti-Zr-Ta	Ti-Zr$_7$-Ta$_{63}$	Resistometric studies of upper critical fields using pulse-field technique.	H_r data plotted *vs* an atomic-volume-corrected electron/atom ratio, N_{eff}.	Numerous Ti-base binary, ternary and quaternary alloys.	W.DeSorbo et al. (1967) [Des67]
(d) Ti-Rh-TM	Ti-Rh$_4$-TM$_5$ with TM = Sc, Ti, V , Cr, Mn, Fe, Co, Ni, Cu Y , Zr, Nb, Mo, --, Ru, Rh, Pd, Ag --, Hf, Ta, W , Re, Os, Ir, Pt, Au	Transition temperature measured inductively.	1-1.5 g samples were produced by argon arc melting and pouring into a water-cooled copper mold. The small amounts of martensite, observed only in the Hf, V and Ta alloys were removed by annealing 2h/950°C and quenching. T_c plotted *vs* electron/atom ratio was *maximum for* TM ≡ Rh (binary alloy) at 4.35 K, dropping on one side: to about 3.5 K for TM ≡ Ru, Os (e/a = 4.4) and on the other to 2.0, 3.0 K, respectively, for "TM" ≡ Ag, Au (e/a = 4.55). In the secondary reference some noble-metal compounds were considered.	-----	C.J.Raub (1964) [Rau64b] see also [Rau64]

Alloy Classification	Compositions	Properties and Procedures	Remarks, Discussion	Other Systems Studied	Literature
(e) Ti–$\left(\frac{V}{Ta}\right)$–Int'l	Ti–Ta$_{50.7}$–O$_{0.2}$	An investigation of normal-state resistivity, H_r, and J_c (4.2K) *vs* H.	The oxygen concentration in Ta–Ti (49.1 at.%) alloy from Ti sponge was claimed to be 249 ppm. This apparently low level of O amounts to 0.17 at.% when the relative molar weights of Ti–Ta (~114) and O (=16) are considered. A similar alloy based on iodide Ti served as "oxygen-free" control.	-----	D.A.Colling *et al.* (1966) [Col66]
	Ti–Ta$_{60}$–O$_{0.08,0.27}$	An investigation of J_c at 4.2 K, 30 kOe with and without heat treatment.	No significant effect of O on J_c was noted in these studies.	Ti–Ta Ti–Nb and Ti–Nb plus O and/or Fe	J.O.Betterton *et al.* (1966) [Bet66]
	Ti–$\left(\frac{V}{Ta}\right)$–$\left(\frac{C}{N}\atop O\right)$	Patent claims relating to critical current density.	It was claimed that the critical current densities of ductile alloys of Ti such as Ti–V,–Nb,–Ta can be raised by the addition of the elements C, N and/or O, in interstitial solid solution.	Ti–Nb–Int'l	French Patent No. 1,517,216 [Ass68^a]

TABLE 1-27 TITANIUM-NIOBIUM ALLOYS — THE SUPERCONDUCTING TRANSITION

Nb Concentration Range	Procedures	Other Properties Studied	Remarks, Discussion	Other Systems Studied	Literature
2.5,9,11,18,27,35,44, 50,60,68,78,83,95 and 97 at.%	T_c was studied as function of composition on arc melted, annealed and quenched alloys. The Hardy and Hulm magnetic technique was used.	-------------	See Table 1-4	A comprehensive matrix of binary transition metal alloys including Ti-Zr,-V.	J.K.Hulm and R.D.Blaugher (1961) [Hul61]
2.5,5,7,11 and 15 at.%	Alloys were arc melted repeatedly and quenched; two alloys were annealed for several days at 680°C. T_c was studied magnetically as function of composition.	-------------	See Table 1-4	Ti-V	E.Bucher and J.Muller (1961) [Buc61]
1 at.%	T_c of arc melted buttons was studied resistively.	Numerous electrical transport property measurements.	A very broad resistive transition was encountered. A current density of 100 A cm^{-2} was used.	Ti-Cr,-Mn,-Fe,-Co,-Ni	R.R.Hake et al. (1962) [Hak62b]
2-97 at.% (see Hulm and Blaugher above)	A review of published data.	Some J_c results were reviewed.	This was a review of superconductivity in alloys and compounds based on Ti, in which the parallelisms, in T_c vs composition and α-β microstructure, among various Ti-TM alloys were emphasized.	Alloys of Ti with other transition metals and numerous intermetallic compounds.	U.Zwicker (1963) [Zwi63]
33 at.%	T_c was studied as function of aging time ($<10^4$ min) at 400°C for 1h/900°C quenched alloy.	Hardness as a function of aging time at 400°C.	The influence of precipitation and solute diffusion on T_c was discussed. The emphasis was on Ti-V-Al, but the binaries Ti-Nb,-Mo were included for comparison. The influence of ω-phase precipitation was considered for the first time.	Ti-Mo, Ti-V-Al	C.J.Raub et al. (1964) [Rau64a]
4 at.%	T_c was investigated both magnetically and calorimetrically.	-------------	Heat capacity was used to investigate the completeness of the transition, in a study of "dilute transition element effects" on the T_c of Ti. The experiments were carried out in response to Matthias' suggestion of a superconductive mechanism different from electron-phonon. The results were not inconsistent, as far as they went, with BCS.	Ti-V,-Cr, V-Cr,-Fe, Ti-Mo, Ti-Zr-TM with TM = V,-Cr,-Mn,Fe, Nb,Mo,--,Ru, --,W,Re,--	E.Bucher et al. (1965) [Buc65], see also F.Heiniger and J.Müller (1964) [Hei64]
Compositions corresponding to e/a ≈ 4.3-5.0	T_c was correlated with a size-effect-corrected electron/atom ratio.	Resistive critical field.	The purpose of the work was to investigate the systematical variation of T_c and H_{c2} with a corrected electron/atom ratio.	Ti-V,-Zr,-Mo, Zr-Nb, Hf-Ta, Mo-Re and some others	W.DeSorbo (1965) [Des65]

TABLE 1-27 TITANIUM-NIOBIUM ALLOYS — THE SUPERCONDUCTING TRANSITION –*continued*

Nb Concentration Range	Procedures	Other Properties Studied	Remarks, Discussion	Other Systems Studied	Literature
2.5 at.%	A measurement of the transition temperature of a single-phase α-Ti-Nb alloy.	-------------	The triple-arc-melted as-cast samples showed a Widmannstätten structure. They were therefore cold rolled >40%, annealed 2h/650°C, yielding a homogeneous α-structure. This was claimed to be the first study of Ti-TM alloys whose compositions did not exceed the α-stability limit.	Ti-V, Ti-Ta	C.J.Raub and U.Zwicker (1965) [Rau65]
20,25,30,35,40 and 50 at.%	The T_c's of 0.25 mm$^\phi$ wire were measured magnetically after 1h-aging at temperatures between 400 and 700°C.	Tensile strengths, electrical resistivity and critical current density.	In some cases T_c increased with aging time as a result of the appearance of the Nb-rich β-phase during phase decomposition.	-------------	M.I.Bychkova *et al.* (1967/1970) [Byc70]
4,9,20 and 25 at.%	The T_c's of 1/2h/900°C/quenched alloys were measured resistively as function of aging time at 400 and 500°C.	Hardness, optical microstructures, dρ/dT, as functions of aging time at 400 and 500°C.	A study of the influence of phase transformation (micro-structures) on T_c. Measurements were made on 1.75 mm$^\phi$ wires.	Ti-Fe Ti-Nb-Fe	T.Nishimura and U.Zwicker (1968) [Nis68]
25 at.%	The low temperature specific heat of an unannealed sample was measured between 4.2 and 17 K. The sample had not been subjected to a homogenizing heat treatment; such an alloy in the as-cast condition would tend to be rich with precipitates and microscopic compositional inhomogeneities.	-------------	As a consequence of the lack of homogeneity, the transition was rather broad, affording the authors an opportunity to develop and apply a model claimed to be generally suitable for dealing with second-order phase transitions in inhomogeneous samples.	-------------	B.Ya.Sukharevskii *et al.* (1968) [Suk68]
30,40,50,60,70,80 and 90 at.%	Bcc samples were prepared by arc melting; annealed 24h/1600°C, furnace cooled 1 h to room temperature, then cold rolled 15:1. Transition temperatures were measured inductively.	-------------	This was one part of a study of superconductivity in the ternary system Ti-Nb-V.	Ti-V Nb-V Ti-Nb-V	P.H.Bellin *et al.* (1969) [Bel69]
10,25,50,75 and 80 at.%	T_c was studied as function of composition using low-temperature specific heat in the temperature range 2.5-20 K.	-------------	The results were interpreted from the standpoint of a partly-overlapping two-band (s- and d-band) BCS model.	-------------	B.Ya.Sukharevskii *et al.* (1971) [Suk71]

Ti-Nb ALLOYS — Transition Temperature

TABLE 1-27 TITANIUM-NIOBIUM ALLOYS — THE SUPERCONDUCTING TRANSITION *-continued*

Nb Concentration Range	Procedures	Other Properties Studied	Remarks, Discussion	Other Systems Studied	Literature
92.8 at.%	The low temperature specific heat of an annealed sample was measured between 2.5 and 18 K (about 200 data points).	----------	A deviation-function plot placed this alloy between Hg and Pb -- i.e. in the strong-coupling regime. A plot of $\ln(C_{es}/\gamma T_c)$ vs T_c/T could be separated into two straight lines. As in the work of Sukharevskii $et\ al.$ (above) this result was interpretable in terms of a two-band (hence two-gap) modified BCS model.	----------	I.S.Shchetkin and T.N.Kharchenko (1973) [Shc73]
9.5-86 at.%	The transition temperatures of sputtered alloy films were measured resistometrically; results were plotted as function of composition.	J_c was measured as a function of applied field up to 100 kOe.	When plotted vs composition the transition temperatures of the films appeared to be consistently higher by some 0.6 ∿ 3.5 K across the composition range, than those of bulk Ti-Nb alloys.	Ti-V,-Ta Nb-Zr,-V,-Ta NbN, NbC	H.J.Spitzer (1974) [Spi74^a], see also [Spi71]
25,30,40,50,60,70,80,90, 95 and 97 at.%	A study of electrical resistivity temperature dependence.	----------	The $d\rho/dT$ results were interpreted in terms of fluctuation effects associated with the superconducting transition.	Ti-V,-Ta	A.F.Prekul $et\ al.$ (1974) [Pre74]
50.1,52.1 and 61 wt.%	T_c values determined according to the following criteria: (1) $R = R_n/2$, (2) $H_{c2} \to 0$, (3) $J_c \to 0$, were intercompared.	H_{c2} vs temperature and composition; J_c vs temperature, cold deformation; pinning force vs composition, field; and numerous other properties.	The subject article is a comprehensive review of commercial composite superconductor development and manufacture, with particular reference to high-field conductors carrying currents greater than 3000 A for application in magnet systems supplied with non-steady currents.	----------	H.Hillmann $et\ al.$ (1979) [Hil79]
20,25,50,75,90 and 92.8 at.%	Low temperature specific heat in the temperature range 2.5 ∿ 18 K.	----------	With Ti-Nb$_{25}$ [Suk68] a pronounced smearing of the transition was noted as a result of inhomogeneities. The results for Ti-Nb$_{92.8}$ [Shc73] were interpreted in terms of a two-band model.	----------	B.I.Verkin (1980) [Ver80], see also [Shc73] and [Suk68] above
35.8 and 37 at.%	T_c was taken from the mid-point of the resistive s/n transition at a current density of 5 A cm^{-2} (cf. H_r studies by the same group -- next Table).	Resistive upper critical field at $J = 5$ A cm^{-2}.	The primary goal of the study was to investigate the influence of Hf and Ta additions on the H_{c2} of technically important Ti-Nb alloys.	Ti-Hf-Nb Ti-Nb-Ta Ti-Hf-Nb-Ta	D.G.Hawksworth and D.C.Larbalestier (1980) [Haw80]
Review	Review	Upper critical field vs composition, etc.; see Tables 1-28 and 1-29.	The subject paper is a review of the present status of Ti-Nb and Ti-Nb-base alloy superconductors.	Ti-TM TM$_1$-TM$_2$ Ti-Nb-Ta	D.C.Larbalestier (1980) [Lar80], see also [Lar81]

TABLE 1-28 TITANIUM-NIOBIUM ALLOYS — CRITICAL FIELDS and the MIXED STATE

Nb Concentration Range	Procedures	Other Properties Studied	Remarks, Discussion	Other Systems Studied	Literature
10.0,20.0,25.0,30.0,34.1, 38.7,42.5,43.7,52.0,60.0, 70.0,85.0,90.0 and 95.0 at.%	H_r was studied as function of composition at 1.2K. Resistometric measurements were made in pulsed fields of up to 160 kOe using a measuring current density of 10 A cm^{-2}.	Normal state resistivity at 1.2 K vs composition.	The results of resistive critical field, H_p, studies were compared with the Clogston paramagnetically limited field, $H_p = 18.4\, T_c(1-t^2)$ kOe and the prediction of GLAG theory.	Ti-V,-Ta,-Mo Nb-Zr,-Hf Hf-Ta U-Nb,-Mo	T.G.Berlincourt and R.R.Hake (1963) [Ber63[a]]
68,83 and 96.3 at.%	Critical fields were studied as function of temperature 0.1<t<1.0. Resistometric measurements were made on rolled strips using a current density of 10 A cm^{-2}.	-----------	The purpose of the experiment was to study the temperature dependence of H_{c2} expressed in terms of a normalized temperature function of the form: $W(t) = H_{c2}/K_0 H_{c0}$, where $K_0 = \kappa_{GL}(t=0)$.	Nb-Ta	B.S.Chandrasekhar $et\ al.$ (1963) [Cha63]
90 at.%	Critical fields of the (a) cold worked and (b) heat treated (1h/1100°C) alloy were measured at 4.2 K. $H_r(H)$ was determined by extrapolating $J_c(H)$ to a current density of 10 A cm^{-2}.	Normal state electrical resistivity as function of temperature.	Shapoval, calculating H_{c20} in the dirty limit concluded that: $H_{c2}/H_{c20} = (1-2t\,\ell n2)\ (t<1)$ $= (1-t)8/\pi^2\ (t\cong1)$ with $H_{c20} = (3/2\pi)\,ec\gamma T_c/k_B\sigma$ (esu) $= 5.53 \times 10^4\ \gamma\rho\,T_c$ (Oe). These it was claimed adequately described H_{c2} and its temperature variation, including the effects of cold work, with the aid of accompanying resistivity measurements.	Zr-Nb	A.El Bindari and M.M.Litvak (1963) [ElB63]
15,30,35,40,50,65,74.5, 80,85,90 and 95 wt.%	Critical fields were measured resistometrically at 4.2 K using current densities of 10^3, 10 and 1 A cm^{-2}.	Critical current density vs field at 4.2 K; also room temperature resistivity.	H_r was plotted vs Nb concentration for heavily cold worked wire (0.4 → 0.01 in.ϕ, 99.94%). H_r data are also available for Ti-65Nb final heat treated 1h/400,600,800,1000°C.	Ti-Zr-Nb	K.M.Ralls (1964) [Ra£64], see also R.L.Ricketts (1969) [Ric69]
20 at.%	Pulsed-field magnetization measurements of cold worked and annealed wire.	Critical current.	Trapped current (flux) reduces the critical current; thus I_c is related to the magnetization. This may be regarded as an early study of stability and the effects of rapidly changing fields.	-----------	I.Dietrich and R.Wey] (1964) [Die64]
44,68,83 and 96 at.%	Critical fields were measured resistometrically, at a current density of 10 A cm^{-2}, as function of temperature 0.1<t<1.0.	-----------	An extension of [Cha63]; the purpose was to experimentally evaluate H_{c20} and $H_{c2}(t)$ for comparison with the predictions of the Shapoval, Gor'kov and Abrikosov-Ginzburg theories. As κ_{GL} varied, agreement in $H_c(t)$ shifted from A-G to G, the cross-over occurring for most alloys at $\rho_n \cong 30\ \mu\Omega$ cm, i.e. $\ell \cong \xi_0$.	Zr-Nb	C.K.Jones $et\ al.$ (1964) [Jon64]

TABLE 1-28 TITANIUM-NIOBIUM ALLOYS — CRITICAL FIELDS and the MIXED STATE -continued

Nb Concentration Range	Procedures	Other Properties Studied	Remarks, Discussion	Other Systems Studied	Literature
Westinghouse alloy HI 120 (33 at.% Nb) 75% cold worked	Tube magnetization studies as function of temperature. The bore field B' was measured as function of applied field, H_a (<27kOe).	----------	The tube magnetization results were interpreted in terms of the models of Bean [Bea62] and Kim et al. [Kim63]. Flux jumping was investigated.	Zr-Nb Nb$_3$Sn	R.Hecht and C.M.Harper (1965) [Hec65]
90 at.%	The relative resistance R/R$_n$ was measured as a function of field (H$_a$<40kOe) and temperature (1.5<T<4.2K) in a study of flux-flow resistivity.	----------	It was argued that, for the field range within which the flux-flow resistivity, ρ_f, increased linearly with H (viz. $H/H_{c2} \gtrsim 0.7$), $\rho_f/\rho_n = H/H^*_{c20}$ where H^*_{c20} is the neo-GLAG nonparamagnetically limited upper critical field. Thus by extrapolation of the lower field flux-flow data a hitherto unobtainable experimental estimate of H^*_{c20} could be extracted. Also reported was a new type of resistance-minimum effect; it was noted that R/R$_n$ vs T (<T$_c$) passed through a minimum prior to increasing towards 1.0. This behavior was interpreted in terms of spacial nonuniformity associated with the dissipative mechanism -- i.e. dissipation occurring outside the vortex core.	----------	Y.B.Kim et al. (1965) [Kim65]
44 and 63 at.%	The upper critical field was measured resistively at temperatures down to 1.3 K, and compared with the the neo-GLAG upper critical field H^*_{c20} as determined using the flux-flow resistivity technique.	Ultrasonic attenuation at 10 MHz; superconducting transition temperature.	This was an important experimental study dealing with the role of Pauli spin paramagnetism in influencing H_{c2}. The effect was found to be smaller than that predicted by Maki. Possible counter influence of spin-orbit interaction was considered.	----------	Y.Shapira and L.J.Neuringer (1965) [Sha65]
44 at.%	The upper critical field was measured resistively in transverse fields (<150kOe) within the temperature range 1.3K<T<T$_c$ using current densities within the range 0.1-30 A cm^{-2}.	Normal state resistivity at 4.2 K (H>H$_{c2}$).	An experiment intended to examine the influence of spin-orbit scattering on H$_{c2}$; hence the choice of three isoelectronic solutes of increasing atomic number (Z), viz: V(23), Nb(41) and Ta(73). Comparisons of H$_{c2}$(T) with Maki spin-orbit-neglecting theory and the full WHH theory [Wer66] were made.	Ti-V$_x$-Ta	L.J.Neuringer and Y.Shapira (1966) [Neu66], see also N.R.Werthamer . . . [Wer66]
Westinghouse alloy HI 120 (33 at.% Nb) 75% cold worked	J$_c$ was determined from the results of a Kim-type tube magnetization experiment -- the magnetization of a hollow superconducting cylinder -- in fields of up to 30 kOe at temperatures within the range 4.2<T<T$_c$.	----------	In a continuation of [Hec65], J$_c$ was derived from tube magnetization measurement of an alloy in the conditions: (a) as-received (cold worked); (b) annealed 2.5h/400°C/10^{-6} Torr; (c) copper plated internally and externally to a thickness of 2x10^{-3} cm.	Nb Nb$_3$Sn	D.A.Gandolfo and C.M.Harper (1966) [Gan66a]

TABLE 1-28 TITANIUM-NIOBIUM ALLOYS — CRITICAL FIELDS and the MIXED STATE *—continued*

Nb Concentration Range	Procedures	Other Properties Studied	Remarks, Discussion	Other Systems Studied	Literature
75 at.%	A study of flux-flow velocity using a pulsed-applied-field tube-magnetization method.	Normal state resistivity in zero field at $T > T_c$, H_{c1}(4.2K).	A study of flux motion. Ti-Nb$_{75}$ was annealed 2h/1200°C plus 30min/800°C in 10^{-7} Torr, radiation cooled. The flux pinning was thus sufficiently reduced as to eliminate flux jumping, hence the conditions of Wipf and Lubell [Wip65] were not applicable. A simple viscous flux-flow analysis, starting with an elementary force-balance equation of the form: $$\vec{f}_L + \vec{f}_p + \vec{f}_v = 0 \text{ (Lorentz, pinning and viscous forces,}$$ respectively) and leading to: $$\eta v_f = \vec{f}_p + \vec{f}_L, \text{ where } v_f \text{ is the fluxoid velocity,}$$ was employed.	Zr-Nb	D.M.Kroeger (1966) [Kro66]
46 and 76 at.%	Upper critical fields were determined magnetically (hence H_u), using the pulse-field technique of Hart and Lawrence, on swaged and drawn alloy wires 0.25 mm in diameter.	--------------	The goal of the research was to improve the systematics used by previous authors to display H_{c2} (and elsewhere, I_c). In place of a simple valence-electron-to-atom ratio, the authors recommended an "effective" electron/atom ratio which took atomic volume into account. Such an electron density approach for T_c had previously been discussed by DeSorbo [Des63] and Jensen *et al.* [Jen65].	Numerous binary ternary and quaternary TM alloys such as: Ti-Zr-Nb (10 alloys); Ti-Nb-TM$_1$-TM$_2$ (5 alloys), etc.	W.DeSorbo *et al.* (1967) [Des67]
87.5,91,95.5,98.5 and 99.5 at.%	The temperature- and mean-free-path dependences of the magnetic properties near H_{c2} were investigated. Magnetization of a long cylindrical sample was measured in fields $H \gtrsim 40$ kOe and within the temperature range $1.2 < T < 15$K	Superconducting transition temperature; also normal state resistivity near T_c (with or without the aid of quenching field).	This was intended as a study of $H_{c2}(T)$ and $(dM/dH)_{H_{c2}}$ related, respectively, to the generalized Ginzburg-Landau parameters κ_1 and κ_2, on alloys of nearly constant T_c and $H_c(T)$ but of widely varying mean-free-path, ℓ, $(0.1\xi_0 < \ell < 15\xi_0)$ where ξ_0 is the pure-Nb coherence length. Also considered were flux pinning, magnetization-derived J_c, flux motion, and instability.	Nb-Ta	W.A.Fietz (1967) [Fie67], see also W.A.Fietz and W.W.Webb (1967) [Fie67^a] and (1969) [Fie69]
22 at.%	Resistive critical fields (H_r) measured in pulsed fields at current densities as low as 10 A cm^{-2}. H_r was studied as function of aging temperature (200–650°C) and aging time (typically 10 min to several tens of h).	Microstructures, normal-state resistivity at 4.2 and 300 K, hardness. T_c was studied as function of aging; J_c *vs* H was measured for various aging conditions.	H_r *vs* aging temperature passed through the usual broad maximum situated between 400 and 500°C (cf. Ricketts [Ric69] below).	--------------	L.C.Salter (1966) [Sa£66] see also [Sa£65]

TABLE 1-28 TITANIUM-NIOBIUM ALLOYS — CRITICAL FIELDS and the MIXED STATE -continued

Nb Concentration Range	Procedures	Other Properties Studied	Remarks, Discussion	Other Systems Studied	Literature
30 at.%	A numerical study (calculation) of the upper limit to H_{C2U}.	-------------	A comprehensive tabulation of physical property data was presented, together with calculations of upper critical field limits, $H_{C2U,max}$ taking into account Maki paramagnetic limitation [Mak66] and the Werthamer et al. [Wer66] spin-orbit-coupling spin-flip-scattering effect.	Ti-V, Zr-Nb, NbCN, NbN, Nb_3Sn, V_3Si, V_3Ga	R.R.Hake (1967) [Hak67[a]]
52 %	A theoretical analysis of stability and an investigation of the possible role played by solute segregation in flux jump stability.	X-ray and microprobe study of banding.	It was concluded that flux jumps can arise as a result of the presence of long-range variations in κ_{GL} due to compositional variation.	------------	J.R.Moon (1968) [Moo68]
40 wt.% (25.6 at.%)	H_r was studied resistometrically at 4.2 K as a function of 1-h aging at about 250, 300, 350, 400, 450, 500, 550 and 600°C.	Electrical resistivity, J_c vs H, x-ray and metallographic studies.	Ti-40Nb was the base for a series of quaternary alloys with Er + 0 or Sc + 0.	Ti-Nb-$\begin{pmatrix}Er\\Sc\end{pmatrix}$-0	R.L.Ricketts (1969) [Ric69]
30,40,50,60,70,80 and 90 at.%	Resistive critical field, H_r, was measured at 4.2 K on cold-rolled (15:1) strips 0.4x0.4x2.5 mm^3 in a 150 kG magnet at the National Magnet Laboratory.	T_c vs composition in a companion paper [Bel69].	Sample material was prepared by arc melting; homogenized 24h/1600°C/10^{-6} Torr and furnace cooled prior to rolling. Part of a comprehensive study of upper critical field in the Ti-Nb-V system. H_r data for the binary Ti-Nb and Ti-V alloys and several families of ternary alloys were plotted vs conventional electron/atom ratio.	Ti-Nb-V	P.H.Bellin et al. (1970) [Bel70]
75,90 and 95 at.%	An important study of experimental and calculated values of the lower critical field. Experimentally, H_{C1} was taken to be the highest field for which the diamagnetic magnetization was reversible; i.e. the highest Meissner field in this case. H_{C10} was obtained by extrapolating H_{C1} to zero temperature.	Normal state resistivity was measured at 4.2 K in $H>H_{C2}$; T_c was determined magnetically; H_{C20}^{*} was obtained from flux-flow-resistivity measurements (cf. Kim et al. [Kim65] above).	A test of the range of applicability of a relationship: $H_{C10} = (74.6\ T_c/\rho_n)\ \ell n\ (4.26\ H_{C20}^{*}\ \rho_n/T_c)$, based on Ginzburg-Landau theory and a Maki [Mak64[a]] expression for H_{C10}. The results suggested that Maki theory was not applicable for $\kappa_{GL} < 26$. Other theories of the lower critical field were discussed.	------------	M.S.Lubell and R.H.Kernohen (1971) [Lub71]
33,39 and 44 at.%	Resistive critical fields, H_r, as gauged by the appearance of 1 μV across the 1 mmℓ x18 mmℓ sample during the passage of a 10 mA current, were measured at 4.2 K in a Bitter solenoid ($H_{a,max}$=150kOe) at the Naval Research Laboratory.	Hardness, electrical resistivity, T_c and J_c.	H_r was measured as a function of aging time at 400 and 700°C.	Ti-Nb-Cu	R.Lohberg (1971) [Loh71]

TABLE 1-28 TITANIUM-NIOBIUM ALLOYS — CRITICAL FIELDS and the MIXED STATE — *continued*

Nb Concentration Range	Procedures	Other Properties Studied	Remarks, Discussion	Other Systems Studied	Literature
60 at.%	Magnetic hysteresis was investigated as a function: (a) of 1-h aging at 300, 500, 600, 700 and 1100°C of 90% cold-worked alloy; (b) of aging time (1/2–5h) at 500°C. The magnetization was measured at 4.2 K in fields ≳30 kOe on alloy strips of dimensions 10x2x0.25 mm.³	Transmission electron microscopy.	A study of flux pinning mechanisms.	------------	M.J.Witcomb and A.V.Narlikar (1972) [Wit72]
32,55,62 and 83 at.%	Flux pinning ($F_p = J_c H/10$) was systematically studied in fields sufficiently strong that paramagnetic effects had to be considered. H_{c2} was measured for use as a field normalization, and pinning force, parameter. H_{c2} was measured resistometrically as function of temperature, 2≲T≲10 K on strips cold rolled about 2000:1 (99.95%).	J_c was measured as function of field and temperature. Also determined were transition temperature and normal-state electrical resistivity.	Critical fields in the paramagnetic mixed state were discussed at length in terms of the Maki [Mak66] and WHH [Wer66] extensions to GLAG theory. Detailed discussion of flux pinning theory was presented. It was concluded that pinning in a wide range of cold-worked Ti-Nb (and Ti-V) alloys was due to the cellular dislocation structure, and consequent modulations of normal-state-resistance (the "mfp effect") and thus of κ_{GL}.	Ti-V	R.A.Brand (1972) [Bra72] see also [Bra75]
50.1,52.1 and 61 wt.%	Upper critical field for the alloys listed was measured *vs* temperature within the range 1.8–8.8 K.	T_c for the alloys listed, J_c *vs* temperature, cold deformation, etc. (see Tables 1-27,1-29).	H_{c2} for the alloys listed decreased linearly with T^2.	------------	H.Hillmann *et al.* (1979) [Hil79]
31.6,33.7,34.3,35.9 and 37.2 at.%	Resistive critical field, H_r, was measured at 4.2 K and below in a Bitter solenoid at the National Magnet Laboratory at $J = 5$ A cm⁻². Samples were laboratory-made Cu-clad monofilaments whose cores (~0.5mm$^\phi$) had experienced >97% cold reduction by swaging and drawing.	------------	H_r in these studies was defined in terms of the onset of the normal state rather than the first appearance of resistance with increasing H_a. The influence of such criteria, and the measuring current density, on the apparent value of H_r was discussed.	Ti-Hf-Nb Ti-Nb-Ta Ti-Hf-Nb-Ta	D.G.Hawksworth and D.C.Larbalestier (1980) [Haw80]
Review	Review	T_c *vs* composition; subband geometry in Ti-Nb *vs* cold reduction; critical current density and pinning forces as function of area reduction, subband diameter, alloy composition, temperature and applied field where applicable.	A review of the contemporary status of Ti-Nb and Ti-Nb-base alloy superconductors.	Ti-TM TM₁-TM₂ Ti-Nb-Ta	D.C.Larbalestier (1980) [Lar80] see also D.C.Larbalestier (1981) [Lar81]

TABLE 1-29 TITANIUM-NIOBIUM ALLOYS — CRITICAL CURRENT DENSITY, FLUX PINNING

Nb Concentration Range	Procedures	Remarks, Discussion	Other Properties Studied and Systems Investigated	Literature
25,50 and 75 at.%	J_c *vs* composition was measured at 5 K, without and with the presence of a 5 kOe magnetic field.	Very early study by Zwicker *et al.* of the formation by cold work (>99.5% with and without intermediate heat treatment) of a lamellar structure favorable to high critical current density. By application of heat treatment (insufficient to initiate recrystallization) a "high critical current density was achieved".	Cold worked microstructure; transition temperatures of Ti-Nb (15,20,25,50,67,75 at.%). Pb Pb-Bi Nb-Zr	I.Dietrich *et al.* (1962) [Die62]
20,35,50 and 72 at.%	A study of the influence of cold rolling on J_c. Measurements were made at 4.2 K, 30 kOe.	Ti-Nb alloys were cold reduced 89-92%. The concept of flux pinning by microstructure had not been considered; and an explanation for the effect of cold rolling in terms of the "filamentary model" was offered.	Ti-V,-Ta Zr-Sc,-Nb,-Mo Nb-Sc,-Hf Ta-Hf	R.R.Hake *et al.* (1962) [Hak62a]
30,35,39 and 50 at.%	The influence of cold work, of various kinds and extents, on J_c was measured at 1.2 and 4.2 K as a function of H_a ($\lesssim$140kOe).	In this paper it was suggested that the current-carrying state should be considered in terms of the Abrikosov vortex structure, rather than a current-filamentary structure.	Resistive critical field, H_r. Ti-V,-Ta Zr-Nb	T.G.Berlincourt (1963) [Ber63]
60 and 90 at.%	J_c at 4.2 K, 30 kOe *vs* applied field orientation and J_c (Ti-Nb$_{90}$, 4.2K) *vs* H_a were measured in a study of critical current anisotropy in cold-rolled strip.	A "giant anisotropy" was observed and discussed in terms of microstructural anisotropy. The appearance of a "peak-effect" was noted in J_c *vs* H_a.	Rolled microstructure. Ti-Ta	R.R.Hake *et al.* (1963) [Hak63a]
43 at.%	Short-sample J_c (4.2K) *vs* H (<90kOe) was measured in a comparison of short-sample and coil behavior.	A non-optimized Ti-Nb alloy sample was used; it was recognized as a "readily fabricable ductile alloy". The work represents an early comparison of short-sample and coil performances.	Ti-Ta Nb-Zr Nb$_3$Sn (Nb sheath)	R.C.Wolgast *et al.* (1963) [Wo63]
26 and 33 wt.%	A review of metallurgical and superconductive properties.	The original contribution included studies of the microstructures of deformed and deformed-plus-heat treated Ti-Nb alloys. It was suggested that critical current of useful density was favored by a lamellar structure with an optimal width of about 0.1 μm.	Transition temperature. Electrical resistivity. Vickers hardness, microstructures. Numerous alloys and compounds referred to.	U.Zwicker (1964) [Zwi64]
34 at.%	An examination of the cold rolled microstructure and a correlation of it with the observed increases in J_c.	It was claimed that high J_c is due to a well-developed system of thin layers and filaments. The cold deformation of metastable alloys was stated to result in fine dispersions of precipitates.	Electron microscopy. Zr-Nb	B.G.Lazarev *et al.* (1964) [Laz64]

Nb Concentration Range	Procedures	Remarks, Discussion	Other Properties Studied and Systems Investigated	Literature
Not specified	A study of the influence of AC frequency (10 ~ 300 Hz) and wire diameter on J_c.	In short-wire tests it was discovered that the critical AC current density decreased with $\log_{10}$ (wire diameter). Several coils were also tested for AC critical current density.	Zr-Nb Nb$_3$Sn	F.J.Young and H.L.Schenk (1964) [You64]
33 %	An investigation of the heating effects produced by current pulses in order to investigate the validity of a current impulse method for measuring J_c when large currents are to be used and their heating effects avoided.	The impulse method enabled intrinsic sample heating to be distinguished from heating via the current lead-ins. The influence of the rise-time of the current pulse was small.	-----------	E.W.Berkl and R.Weyl (1964) [Ber64]
15,30,35,40,50,65,74.5, 80,85,90 and 95 wt.%	J_c (4.2K) vs H to current densities ≤ 3 A cm^{-2} was measured in a study of the influence of 1-h aging time on J_c(H).	An investigation of superconductivity in cold-worked-plus-aged alloys. No final cold work was administered. No metallography or TEM was carried out.	Normal-state resistivity at 4.2 K and RT; critical fields. Zr-Nb Ti-Nb-Zr	K.M.Ralls (1964) [Ra64]
40 at.%	In a study of critical current density in sputtered films J_c (4.2K) was measured vs H (<100kOe) with the applied field directed both parallel and transverse to the current. A sensing current density of 10^3 A cm^{-2} was used. Errors in thickness determination tended to predict thicker films than were actually deposited.	As subsequent studies of critical-current anisotropy have shown to be generally true J_c($J \parallel H$) was considerably greater than J_c($J \perp H$). For example, in the former case, J_c(40kOe, $\parallel$) = 5×10^5 A cm^{-2}; while J_c (40kOe, $\perp$), the usual orientation; = 6.3×10^4 A cm^{-2} comparable to the values subsequently expected for the unoptimized bulk alloy. It was suggested that hard thin-film superconductors may possess promise in superconducting device applications.	Nb Ti-V Nb-Zr V-Si Nb-Sn	J.Edgecumbe et al. (1964) [Edg64]
20.7 at.%	Recrystallized and cold worked samples were aged in the two-phase $\alpha+\beta$-region in order to: (a) induce flux pinning precipitation; (b) to enrich the matrix with Nb, thus raising H_{c2}. The influence of post-work final heat treatments of 3h/(up to 460°C) was investigated.	The deformation plus final aging of Ti-Nb$_{20.7}$ for (3/4-6)h/400°C (no final cold work) yielded: J_c (4.2K, 30kOe) > 10^5 A cm^{-2}. As function of aging temperature the hardness peaked at 350°C (attributed to ω-phase) while J_c peaked at 410°C (attributed by x-ray diffraction to α-phase of particle diameter 2000 Å).	H_{c2}, Vickers hardness, relative room-temperature resistivity.	J.B.Vetrano and R.W.Boom (1965) [Vet65]
20 at.%	Rectangular current pulses (5μs,rise time/length,400μs) were applied to short samples at 27 kOe in a study of the influence of field- and current-application procedures on the measured J_c. Short-sample and coil results were intercompared.	The samples were annealed after cold work in order to achieve compositional segregation. The difference in I_c between DC and pulse current was attributed to the existence of circulating currents trapped in multiply connected regions of a superconducting "sponge".	Magnetization field dependence.	R.Weyl and I.Dietrich (1965) [Wey65], see also I.Dietrich and R.Weyl (1964) [Di64]

TABLE 1-29 TITANIUM-NIOBIUM ALLOYS — CRITICAL CURRENT DENSITY, FLUX PINNING — *continued*

Nb Concentration Range	Procedures	Remarks, Discussion	Other Properties Studied and Systems Investigated	Literature
22 at.%	Short-sample tests at 4.2 K were carried out on wires which had experienced a matrix of aging times/temperatures in a study of the response of steady-state $J_c(H)$ to sample heat treatment. Instability in response to sweeping of the applied field according to: $dH_a/dt(\lesssim 2kOe\ s^{-1},\ \lesssim 35kOe)$ was also investigated.	This was an extension and amplification of the Vetrano and Boom study ([Vet65], above). In the sweep tests no current instabilities were noted in any Ti-Nb sample up to the maximum ramp rate of 2 kOe s^{-1}. The high J_c coupled with the excellent stability noted were attributed to a high number density of weak pinning sites.	Metallurgical studies, normal state resistivity at 4.2 and 300 K, T_c, H_r. Zr-Nb	L.C.Salter (1966) [Sal66], see also [Sal65a] [Sal65] [Sal64]
54 at.%	The short-sample J_c (4.2K, 16kOe) was measured as a function of starting ingot annealing time ($\lesssim 5h$) at 1500°C in a study of the influence of prior microstructure (as-cast or homogenized) on the critical current density of cold deformed 99.96% wire.	J_c was found to decrease with increasing degrees of ingot homogeneity.	Nb-Zr 65 BT	V.Ya. Pakhomov et al. (1966) [Pak66]
20 at.%	Short-sample J_c (4.2K) vs H (<40kOe) tests were made on strips spark cut from quenched/heat-treated sheet in a study of the effects of quenched microstructure and subsequently heat-treated microstructure on the J_c of un-worked alloy.	The aging at 334 ± 2°C of the as-quenched structure (α''-martensite with fine, $\lesssim 1000$ Å, internal structure) yielded precipitation of plates $\gtrsim 100$ Å diameter and $\gtrsim 25$ Å thick; these we now know to be ω-phase precipitates. A pronounced "peak effect" was observed, the origin of which was thought to be fluxon/precipitate "matching".		J.Sutton and C.Baker (1966) [Sut66]
75 at.%	Short-sample *longitudinal* (H_a parallel to the current direction) J_c (4.2K) vs H (<100kOe) measurements were undertaken in a study of deformation-induced microstructures and their influence on superconductive properties.	The development and flux-pinning properties of a dislocation cell structure was emphasized, and the concept of "κ_{GL}-pinning" was discussed.	TEM, magnetization. Nb Zr-Nb	A.V.Narlikar and D.Dew-Hughes (1966) [Nar66]
30 and 35 at.%	Short-sample J_c (4.2K) vs H (<30kOe) was measured in a study of the effects of heat treatment (and interstitial impurities) on J_c.	In preliminary studies, J_c was measured after cold work 99.85%, and cw plus 1h/400°C.	Ti-Nb-O, O plus Fe Ti-Ta Ti-Ta-O Nb-Zr	J.O.Betterton et al. (1966) [Bet66]
22 at.%	Short-sample J_c (4.2K) vs H (<40kOe) measurements were made and the results reconciled with the observed microstructures produced by aging 1h/350,400,450°C, 48h/400°C, and 24h/450°C.	Ti-Nb$_{22}$, air cooled in an evacuated capsule from 3h/800°C exhibited ω-phase precipitation ($\sim 10^{17}$ cm^{-3}) with size and interparticle spacing ~ 50 Å. It was concluded that flux pinning by ω-phase was responsible for high critical currents in Ti-Nb$_{22}$ and that the highest J_c corresponded to a precipitate number density of 10^{17} cm^{-3}.	Transmission electron microscopy.	D.Kramer and C.G.Rhodes (1967) [Kra67], see also [Ato67], see also [Ato66], see also N.G.Brammer and C.G.Rhodes (1967) [Bra67]

Nb Concentration Range	Procedures	Remarks, Discussion	Other Properties Studied and Systems Investigated	Literature
20,25,30,35,40 and 50 at.%	J_c (26kOe) was studied *vs* 1 h aging temperature for all compositions. The effects of intermediate- and final-heat treatment were investigated.	An important paper dealing with thermomechanical processing of Ti-Nb. The effects of varying intermediate and final heat treatment, heat treatment temperature (300-800°C) and alloy composition, were compared. Under the conditions of the study the final heat treatment option was more successful. A maximal J_c (prob. 26kOe) of 1.02×10^5 A cm^{-2} was achieved with Ti-Nb25 (1h/450°C). Flux pinning in terms of ω- or α-phases was discussed.	Tensile strength Electrical resistivity Transition temperature	M.I.Bychkova *et al.* (1967/70) [Byc70]
22 at.% Nb	Short-sample J_c (4.2K) *vs* H (≲90kOe) was measured in an investigation of the influence of wire diameter (0.0024-0.05in.) on J_c and in order to derive an empirical relationship between J_c and H_a.	J_c was not dependent on wire diameter in the range $0.0024 \lesssim \phi \lesssim 0.03$in. For one example the relationship $J_c = (A_0/b_0)\exp(-B/b_0)$ with $(A_0/b_0) = 1.1 \times 10^6$A cm^{-2} and $b_0 = 25$kOe was established. Note that for small B/b_0, the above reduces to $A_0/(b_0+B)$ -- an equation of Kim *et al.* [Kim62].	-----------	A.D.McInturff *et al.* (1967) [Mci67]
47 wt.%	J_c (4.2K) on 0.1 mm† specimens *vs* H (<23kOe) was measured in a study of the influence on J_c of: (a) quenched-and-aged, and (b) cold-worked-and-aged microstructures.	The *quenched alloys* aged 10h/400, 2-4h/500 and ≳1h/600°C, yielded similar formations of ω-phase (viz. 200-250 Å diameter, 1000-2000 Å separation). The *deformed alloys* aged 2h/400 and 40h/500°C, respectively exhibited cellular structures, with dislocations and free of dislocations, respectively, but no precipitation. α-phase occurred in both quenched *and* deformed alloys after 150h/500°C.	TEM studies of deformation and precipitation structures.	V.A.Vozilkin *et al.* (1968) [Voz68]
35 and 50 wt.%	J_c *vs* H (≲80kOe) for numerous mechanical and thermal processing conditions was measured in a study of the influence of heavy cold work followed by heat treatment and final cold work on the microstructure and J_c of drawn wire and rolled ribbon.	Considered in detail were: subband structure, conditions for ω- and α-phase precipitation, martensite formation, substructure recovery and the effects of these on J_c. Ti-35Nb: α-phase precipitation as a result of heat treatment was possible either before or after deformation. In the strongly deformed alloy ω-phase, appearing after 10min/380°C, gave way to α-phase after a 2h/380°C anneal. Ti-50Nb: α-phase precipitation was only possible after cold deformation and longer heat treatment times were required than for the Ti-rich alloy..	TEM studies of deformation and precipitation structures.	I.Pfeiffer and H.Hillmann (1968) [Pfe68]

TABLE 1-29   TITANIUM-NIOBIUM ALLOYS — CRITICAL CURRENT DENSITY, FLUX PINNING *—continued*

Nb Concentration Range	Procedures	Remarks, Discussion	Other Properties Studied and Systems Investigated	Literature
50,60 and 80 wt.%	J_c (4.2K) vs H ($\gtrsim$110kOe) was measured in a study of the influence of heat-treatment-induced structure on J_c and other properties.	Low-oxygen Ti-50Nb, Ti-60Nb and Ti-80Nb alloys, 99.9% cold worked and aged 1h to 600, 700 and 750°C, respectively, were investigated. The influence of heat treatment on J_c was discussed in terms of *deduced* precipitation. Powder samples prepared for x-ray analysis were nonrepresentative, being heavily contaminated with O (~5000ppm).	Knoop hardness; relative resistivity vs aging; upper critical field, H_r. (X-ray investigation was unsuccessful.) Ti-Nb-O	G.C.Rauch (1968) [Rau68] see also [Rau68a]
40 wt.%	J_c (4.2K, 50kOe) vs 1 h aging temperature was measured in a study of the influence of heat-treatment-induced structure on J_c and other properties.	As a result of x-ray diffractometry the temperature regimes favorable for ω-phase precipitation (400°C) and α-phase precipitation (500°C) were able to be distinguished. As part of a study of the effect of O, etc., the subject alloy contained a typical residual O level of 440 ppm. J_c(H) data were also presented in the format J_c.H (pinning strength) vs H.	Electrical resistivity at 77 and 300 K; resistive critical field, H_r; x-ray and metallographic studies. Ti-Nb-O Ti-Nb-$\left(\begin{smallmatrix}Er\\Sc\end{smallmatrix}\right)$-O	R.L.Ricketts (1969) [Ric69] see also [Ric70]
Review	A discussion of ω-phase occurrence and growth during heat treatment with particular reference to its effectiveness as a flux pinner.	A useful model for athermal and isothermal nucleation and growth, respectively, was presented. Reference was made to the critical current results of Rauch ([Rau68], above).	-----------------	T.H.Courtney and J.Wulff (1969) [Cou69]
22 at.%	Short-sample J_c (4.2K) vs H (<88kOe) was measured in a study of the influence on J_c(H) of the following processing conditions, and their consequent microstructures: (a) cold work (cw); (b) cw plus recrystallization (1h/800°C) and aging (390 and 425°C); (c) cw plus aging.	A parabolic functional relationship between J_c and H was developed.	-----------------	Y.F.Bychkov *et al.* (1969) [Byc69]
20 at.%	J_c (4.2K) vs H (<45kOe) was measured in an important study directed towards interrelating some superconducting and microstructural properties.	The following conditions were investigated: (a) quenched from 800°C (α") plus 78% cold worked (α"), also cw plus aged at 330°C (to α+β); (b) recrystallized 10h/1300°C, annealed 30min/900°C and quenched (α"); (c) above plus aged 330°C (α"+β+ω+β). The emphasis of the paper was placed on conditions (b) and (c); in fact no critical current data were presented for the cold-worked-plus-aged (α-precipitated) condition.	T_c, magnetization (4.2K), normal state resistivity at 4.2 and ~300 K; hardness, optical metallography and TEM.	C.Baker and J.Sutton (1969) [Bak69]

TABLE 1-29   TITANIUM-NIOBIUM ALLOYS — CRITICAL CURRENT DENSITY, FLUX PINNING –*continued*

Nb Concentration Range	Procedures	Remarks, Discussion	Other Properties Studied and Systems Investigated	Literature
45 at.%	Short-sample J_c (4.2K) vs H (<45kOe) was measured, the results being frequently presented in the format ΔJ_c vs H — where ΔJ_c is the increase in J_c above that of the cold-worked alloy — in a study of the effect of heat-treatment-induced microstructure on J_c.	Test wire 0.24 mm$^\phi$, cold reduced from 6.7 mm$^\phi$ (99.87%), was aged (20min-200h)/(195-450°C). X-ray investigations after heat treatment revealed no evidence of second phases. Consequently the emphasis was placed on the development of dislocation cells, and the migration of interstitial impurities to the cell walls. An activation-energy model was developed.	Microhardness, metallography, x-ray diffractometry, normal-state resistivity at 20 K.	J.P.Charlesworth and P.E.Madsen (1970) [Cha70]
33,39 and 44 at.%	Short-sample J_c (4.2K) vs H (<100kOe) measurements were undertaken in an investigation whose primary goal was a study of the influence of Cu additions on the superconducting properties of Ti-Nb alloys. The binary alloys were measured for reference.	Alloys were measured in the 97.2% cold-deformed-plus-aged condition, and in the homogenized-quenched-and-aged (at 400°C and 700°C) conditions. Numerous studies of the effects of prolonged aging were undertaken.	Hardness, electrical resistivity at LN and RT; x-ray structure determinations; T_c and H_{c2} measurements. Ti-Nb-Cu	R.Löhberg (1971) [Loh71]
13,20,25,31,33,35,39 and 44 at.%	Short-sample J_c (4.2K) vs H (<100kOe) measurements were undertaken in an investigation whose primary goal was a study of the influence of Ge additions on the superconducting properties of Ti-Nb alloys. The binary alloys were measured for reference. Investigated were the effects of cold deformation plus aging, Nb content, aging time and temperature on the superconducting properties of selected alloys.	Alloys listed, cold deformed 97.2% and aged 1h/400°C. Ti-49Nb itself, and the ternary Ti-Nb-Ge alloys, cold deformed 97.2% and aged: (a) 1/4 to 500h/400°C; (b) 300, 400, 500, 600°C for times up to ≳100h; (c) aged various times at 700°C; (d) aged 10h/400°C after 93, 95, 97 and 99% cold work; (e) homogenized, quenched and aged.	Hardness, elastic modulus, electrical resistivity, microfluorescence analysis, x-ray structure and TEM. Ti-Nb-Ge	W.Heller (1971) [He271]
58 wt.% (nominally Ti-56Nb)	The results of short-sample J_c (4.2K) vs H (<60kOe) measurements were plotted in the format J_c.H vs H in a systematic study of the effect on flux pinning of: (a) cold reduction of 33-99.999% followed by 1h/385°C; (b) 99.998% cold work followed by 1h/300-500°C, etc.	No precipitation was observed of a size or distribution that would be suitable for flux pinning (Pfeiffer and Hillmann, it will be recalled, had noted the presence of α-phase in both Ti-35Nb and Ti-50Nb). Hardness measurements too, provided no evidence for dispersed precipitation. The optimal J_c was related entirely to an optimization of the elongated cell geometry.	TEM, electron diffraction, hardness, electrical resistivity at room temperature and in liquid H.	D.F.Neal *et al.* (1971) [Nea71]
58 wt.% (same alloy as above)	The results of short-sample J_c (4.2K) vs H (<100kOe) measurements, undertaken on material that had experienced 99.998% (4x10^4:1) cold work plus 1h/(250-600°C), were reported. The purpose of the work was to develop a flux-pinning model appropriate to these, and the above, results.	The applicability of contemporary flux pinning theory to the experimental results was discussed. For the subject material a modulated-κ_{GL} wall-pinning theory was proposed.	Measurement of dislocation cell size.	R.G.Hampshire and M.T.Taylor (1972) [Ham72]
20,30,40,50,60 and 70 at.%	J_c (45kOe) measured as function of composition for various processing conditions, also vs aging time and temperature for selected alloys.	One purpose of the study was the establishment of base data against which the effects of third-element additions could be compared. Iodide- and Kroll-process Ti as starting materials were intercompared.	Hardness	G.Rassmann and L.Illgen (1972) [Ras72]

TABLE 1-29 TITANIUM-NIOBIUM ALLOYS — CRITICAL CURRENT DENSITY, FLUX PINNING —continued

Nb Concentration Range	Procedures	Remarks, Discussion	Other Properties Studied and Systems Investigated	Literature
50 wt.%	Short-sample J_c (4.2K) vs H (<80kOe) measurements provided data in an investigation of the influence of mixed cold work and heat treatment on the flux pinning structure, hence J_c. Useful data were acquired relating J_c to solute concentration for various levels of final cold work. The optimization of J_c during final cold deformation of a previously heat-treated 61-filament composite was demonstrated.	In contrast to the conclusions of Neal *et al.* (above) with regard to Ti-58Nb, favorable flux pinning in Ti-50Nb with an attendant high J_c ($2-3 \times 10^5$ A cm^{-2}, 50 kOe), was attributed to a high number per unit area ($\sim 10^{10}$ cm^{-2}) of α-phase precipitates. Alternate cold work and heat treatment was claimed to optimize the precipitate distribution; on the other hand, excessive precipitation, which reduced the volume fraction of superconducting matrix, reduced J_c. A flux-lattice/precipitate-spacing matching mechanism was considered.	TEM	H.Hillmann and D.Hauck (1972) [Hil72] see also H.Hillmann (1974) [Hil74]
32,55,62 and 83 at.%	Short-sample J_c was measured as a function of temperature (t = 0.2 ~ 0.8 depending on sample) and field (H < 100 kOe) on 0.15 mm^t cold rolled (2000:1, 99.95%) strips. The purpose of the research was to extend contemporary pinning theories (which were discussed) to include the effects of Pauli paramagnetic limitation.	It was concluded that pinning in a wide range of cold worked Ti-Nb (and Ti-V) alloys -- with certain exceptions -- was due to a cellular dislocation structure; hence mfp-, and consequently κ_{GL}-, modulations. This may be regarded as a logical extension of the work of Neal *et al.* and Hampshire and Taylor (above). The cellular model which is traceable to the work of Narlikar and Dew-Hughes (above) on precipitate-free alloys leads to a pinning force function of the form: $$F_p \sim h^{3/2}(1-h)H_{c2}^{5/2}/\kappa_2^3$$	H_{c2} Ti-V	R.A.Brand (1972) [Bra72], see also [Bra75]
40 at.% (Soviet alloy T 60)	Short-sample J_c measurements in fields up to 60 kOe were under taken in a study of the influence on J_c of the thickness of a Cu cladding.	An investigation of Soviet Alloy 60 T, initially Cu plated to a thickness of 40 μm. The plating was: (a) reduced to 4 μm; (b) built up to 120 μm.	Soviet Alloy SS-2 (viz. Ti$_{25}$-Zr$_{25}$-Nb$_{50}$)	B.G.Lazarev and S.I.Goridov (1973) [Laz73]
22,25,27.6,30 and 37 at.%	Short-sample measurements of J_c (4.2K, 30kOe) vs precipitation heat treatment, and other aspects of thermomechanical processing (TMP) were undertaken. The investigation focussed attention on the influence of final heat treatment parameters on the J_c of previously cold-worked Ti-Nb alloys; also on details of the TMP sequence: cold work/ heat treatment/final cold work.	Flux pinning was discussed with reference to the literature. Some principles of commercial conductor design were outlined.	Resistivity at 300 K Hardness Workability T_c vs heat treatment	A.D.McInturff and G.G.Chase (1973) [Mci73]
50 and 55 wt.%	Short-sample J_c (4.2K) vs H (<80kOe) measurements were made during the course of a program intended to develop a superconductor for use in pulsed magnets, requiring a rise-time of about 1 s.	The development work concerned multifilamentary Ti-Nb-base composite conductors, with pure Cu, Cu-Ni mixed substrates, filament numbers up to 1200, filament diameters down to 10 μm. Conductor optimization consisted of cold work with numerous intermediate heat treatments presumably at 380°C.	Thermal expansion, mechanical properties at room temperature and 4.2 K.	H.Hillmann (1973) [Hil73a]

Nb Concentration Range	Procedures	Remarks, Discussion	Other Properties Studied and Systems Investigated	Literature
9.5 and 86 at.% (>20 alloys)	In a study of critical current density in sputtered films J_c (4.2K) was measured *vs* H (<100kOe) with the applied field transverse to the current direction.	It was concluded that the J_c(H) characteristics were generally 1-1/2 orders of magnitude higher than those of the bulk. As an example, J_c (4.2K, 40kOe) in Ti-Nb$_{43}$ = 3.7x10^5 A cm^{-2}, a remarkably high value and, incidentally, considerably higher than a value reported earlier for sputtered Ti-Nb$_{40}$ (Edgecumbe [Edg64], above).	Transition temperature *vs* composition. Ti-V,-Ta Nb-Zr,-V,-Ta and the compounds NbN and NbC.	H.J.Spitzer (1974) [Spi74^a] see also [Spi71]
50 wt.%	Short sample measurements of J_c (4.2K, 50kOe, 60kOe) *vs* aging (e.g. 3h/(350-500°C) and J_c (4.2K) *vs* H (<70kOe) for wires heat treated according to various prescriptions. Heat treatments took place for various times at 200, 300, 390 (optimal), 450 and 500°C. Reported were the results of a comprehensive study of the influence of microstructures induced by deformation and heat treatment and the particular properties of precipitates on the critical current density.	Cu-clad single and multifilamentary conductors were considered. With regard to : (a) Ti-50Nb, *excessively heat treated* either with respect to time or temperature, and in which the subbands and precipitates were large; (b) Ti-Nb-Ge *for all heat treatments*, and which exhibited small subbands without noticeable precipitation, a fit to the general pinning function h^m (1-h)^2 could not be obtained. But for *optimized* Ti-50Nb (and Ti-Nb-Cu) a universal fit could be achieved with a reduced pinning function of the form h$^{1/2}$ (1-h). Some useful results of a study of the influence of heat treatment on subband diameter were included.	TEM, electron diffraction. Ti-Nb-Cu Ti-Nb-Ge	J.Willbrand *et al.* (1975) [Wil75^a] see also R.Arndt and R.Ebeling (1974) [Arn74]

50 wt.%	Studies of 61-filament Cu/Ti-Nb (1.3:1) composites. Processing had consisted of several heat treatments at 390°C during the wire drawing followed by 80 to 95% final cold deformation to a fixed final diameter (hence fixed overall reduction for all samples) of 0.25 mm. By varying the process parameters two series of samples were produced:
	
	Series A: subband φ 250 Å;
	          precipitate φ 35 Å;
	
	Series B: subband φ 350-400 Å;
	          precipitate φ 50-130 Å.
	
	J_c (4.2K) *vs* H was measured. For both series J_c (4.2K, 5T) for example lay between 2.3 and 3.6x10^5 A cm^{-2}.
	
	J_c (4.2K, 5T) was studied as function of: (a) final cold deformation; (b) number density of precipitates, n$_p$; (c) precipitate diameter, φ; (d) the product n$_p$φ^2.
		One of the conclusions to be drawn from the study was that subband diameters of between 250-400 Å were a necessary but not sufficient condition for high current density -- precipitate density and particle diameter are extremely important. It was shown that J_c increased with the product n$_p$φ^2.	TEM studies of deformation structure.	J.Willbrand and W.Schlump (1975) [Wil75]

Nb Concentration Range	Procedures	Remarks, Discussion	Other Properties Studied and Systems Investigated	Literature
53.5 and 55 wt.%	Studies included the commercial-scale fabrication of multifilamentary composites, Cu/SC ratios 1.25:1 to 5.5:1, up to 2046 filaments, from billets up to 10 in. in diameter. Billets were extruded to 2.5 in.ϕ and drawn to wire size with two or at most three heat treatments including the major precipitation heat treatments (PHT) and a final heat treatment to anneal the Cu. J_C was measured (and the 4T, 5T and 6T results plotted) as function of total reduction in area from pre-extrusion diameter. Total area reductions were: $5 \times 10^3 \sim 10^6$. Final reduction following PHT: 90-97%. Current densities of: 2.4×10^5 A cm^{-2} (4T), 2.0×10^5 A cm^{-2} (5T) and 1.6×10^5 A cm^{-2} (6T) were attained.	In contrast to the results of Hillmann and Hauck (above) and Willbrand and Schlump (above), J_C did not maximize with cold reduction but tended to a saturation value. Satisfactorily high J_C values were attained only after some minimum level of total cold reduction ($\sim 5 \times 10^4$:1) had been exceeded. The transition from "low" to "high" J_C with area reduction was more pronounced at 4T than at 6T. Highest J_C's are reached only for reductions in excess of 10^5:1 (i.e. 99.999%). According to West and Larbalestier (see above) Ti-53.5Nb derives about one-third of its J_C from α-precipitation with appropriate PHT. Alloys as Nb-rich as Ti-55Nb have not been examined in this regard.	Room-temperature mechanical properties with tabulations of yield strength, ultimate strength, % elongation and % area reduction at fracture.	D.A.Colling *et al.* (1977) [Col77]
53.5 wt.%	J_C (4.2K) *vs* H (<75kOe) measured in order to establish a reference against which the properties of Ti-Nb-Ta could be compared.	The basic alloy is the commercial (U.S.) "Nb-46.5Ti". Samples were Cu-clad monofilaments, the processing of which are to be considered in Table 1-36.	Hardness, room-temperature tensile properties. Ti-Nb-Ta	C.W.Curtis and W.K.McDonald (1979) [Cur79]
Same as above	The Cu-clad monofilaments prepared by Curtis and McDonald (above) were tested at 4.2 K and 1.9 K in fields of 9 and 10T continuing on to 11 through about 13T at 1.9 K.	Again the binary Ti-Nb was used as reference against which the properties of small additions of Ta could be estimated.	Ti-Nb-Ta	Z.J.J. Stekly *et al.* (1978) [Ste78] see also [Seg80]
50 wt.%	Studies included the simulation of commercial-scale fabrication of a multifilamentary composite, Cu/SC ratio 2.3:1, 294 filaments, from a 10-cm diameter billet. Processing variables were: heat treatments and cold reductions between them. Two to five heat treatments of 8 to 75 h each (probably at 375°C) were administered.	The goal of the study was the development of a conductor for use in 12T tokamak toroidal field coils. Ti-50Nb was the binary alloy selected for development for two reasons: (a) it possesses superior performance at low fields; (b) its higher Ti content than another "standard" alloy Ti-53.5Nb offered the possibility that heat treatment would result in a high density of flux-pinning precipitates, whose presence would enhance J_C, and whose Ti-rich content would release additional Nb to the β-phase matrix and increase its critical field. It turned out that it was not possible to measure useful current densities in Ti-50Nb above 13T at 2 K, in contrast to Ti-53.5Nb which has performed satisfactorily in a coil at 14T.	The conceptual design parameters of a reactor-compatible toroidal field coil using a Ti-Nb-base conductor were outlined. Suitable compound conductors were discussed. Ti-Nb-Ta	H.R.Segal *et al.* (1980) [Seg80]

Nb Concentration Range	Procedures	Remarks, Discussion	Other Properties Studied and Systems Investigated	Literature
45 and 53.5 wt.%	J_c (4.2K) vs H (≳10T) data were presented for both alloys. J_c was studied as a function of total cold reduction in area from pre-extrusion to final size, and also with regard to area reduction following the "major" heat treatment (75h/375°C was selected as an optimal heat treatment). For Ti-53.5Nb, J_c (4.2K, 10T) and J_c (1.9K, 13T) were plotted vs total area reduction to 10^6:1; for Ti-45Nb, J_c (1.9K, 12T) was plotted vs total area reduction to 5×10^5:1. The temperature dependences of J_c (Ti-53.5Nb) at 10-14T were also investigated. J_c did not maximize with cold reduction (cf. Collings et al., above).	Studies included the commercial-scale fabrication of multi-filamentary composites (up to 2046 filaments) from billets up to 10 in. in diameter. Ti-Nb, optimally processed for high-field performance, was shown to be of practical use in fields of up to 13.5T (at temperatures < 4.2K). It was claimed that the reduced-temperature current densities of Ti-Nb are comparable to those of Nb_3Sn at 4.2 K. A small coil insert was able to be operated at 95% of the quench current in fields as high as 14T thus verifying the feasibility of using Ti-Nb as a high-field magnet insert.	Some small-coil testing was undertaken.	H.R.Segal et al. (1979) [Seg79] see also [Ste78]
50 wt.%	A monofilamentary Cu-Ti-Nb composite conductor was subjected to from 0 to 4 intermediate heat treatments of 5h/380°C each prior to a final cold reduction of 84%. The optimal conductor was flattened by rolling to various aspect ratios a/b ≳ 13:1; J_c vs H (3-8T) was measured with the applied field directed either parallel or perpendicular to the flat face in a study of anisotropy.	The anisotropy was discussed in terms of a redistribution of precipitates; the anisotropy factor f = $J_c(\parallel)/J_c(\perp)$ was studied as function of: the number of intermediate heat treatments, the applied field, and the aspect ratio.	————————	K.J.Best et al. (1979) [Bes79]
50 wt.%	In what was essentially a development of the above study the influence of flattening on underoptimized and overoptimized 60-filament Cu/Ti-50.9Nb (1.37:1) composite conductors was studied.	The influence of flattening on $J_c(\parallel)$ and $J_c(\perp)$ was interpreted by superposing the new data over a final-cold-work-optimization curve for round conductor.	The influence of twist was investigated.	K.J.Best et al. (1979) [Bes79a]
50 ~ 60 wt.%	J_c (4.2K) data for optimized 1- and 60-filament Cu/Ti-Nb composites were presented in the format $p_{,max}$ vs Ti content. Attention was focussed on Ti-50Nb which was subjected to intense study of practical interest. Useful results presented included: J_c vs final conductor diameter; J_c vs field for various mixed-matrix conductors.	The article referred to is a valuable report describing in great detail the development of superconductor materials and processes directed towards the fabrication of monolithic and cabled conductors with current carrying capacities of more than 3000 A for use in magnet systems for fields with alternating components. Conductor design and measurement, and production techniques, were dealt with. The conductor-flattening studies referred to above are included.	Mechanical properties, T_c vs composition, H_{c2} vs composition, temperature.	H.Hillmann et al. (1979) [Hil79]

Nb Concentration Range	Procedures	Remarks, Discussion	Other Properties Studied and Systems Investigated	Literature
47.3,49.6,50.3 and 53.5 wt.%	J_c vs H measurements were made on 20 cm$^\ell$ hairpin samples at Madison (<7T) and at the National Magnet Laboratory (>7T) on as-received commercial composite conductors, and the same conductors after additional heat treatments at temperatures between 300 and 500°C. The results were presented in both J_c vs H and $F_p/F_{p,max}$ vs h = H/H$_{c2}$ formats.	The manufacturers were: Ti-47.3Nb -- Airco; Ti-49.6,50.3Nb -- Vacuumschmelze; Ti-53.5Nb -- MCA. Compositions of the nominally Ti-47Nb and Ti-50Nb alloys were determined by electron probe; the tolerance on the Ti-53.5Nb alloy was stated to be ±1.5 wt.%. In the range 300-500°C the effect of heat treatment was always deleterious indicating that the condition of the as-received material was certainly not underoptimized. Surprisingly, of the two Ti-50Nb alloys that with the smaller filament diameter (larger cold work) had the lower J_c.	H_r vs T^2 (a linear decrease).	K.F.Hwang and D.C.Larbalestier (1979) [Hwa79]
47.3,49.6,50.3 and 53.5 wt.%	TEM and selected-area electron diffraction studies were carried out on specimens taken from commercial multifilamentary Cu/Ti-Nb composites. Ti-47.3Nb was also studied after additional heat treatments of (1 and 100)h/350°C, 1h/425°C and 1h/500°C. The metallographic results were discussed with reference to the J_c results reported earlier by Hwang and Larbalestier (above).	An important result was the discovery of α-phase precipitates in all samples whatever their heat treatment. It had not been generally recognized that precipitation would occur in an alloy as Nb-rich as Ti-53.5Nb. It will, however be recalled that Neal *et al.* (above) had noted precipitates in their Ti-58Nb, although not of a size or distribution that would be suitable for flux pinning. Of course before categorical statements can be made with regard to composition ranges favorable for precipitation the complete absence of coring must be ascertained.	----------------	A.W.West and D.C.Larbalestier (1980) [Wes80]
47,49.6 and 53.5 wt.%	J_c (4.2K and ~2K) vs H (<14T) data were reported. The results were also presented in the pinning-force vs reduced-field, and reduced-pinning-force vs reduced-field formats. Sample materials were commercial multifilamentary composites measured as-received except for Ti-49.6Nb which was given an additional 3h/350°C.	A useful discussion of critical current measuring criteria was presented. An intercomparison of results obtained according to the 0.3 μV cm^{-1}, 10^{-13} Ω m, 5x10^{-14} Ω m, and 10^{-14} Ω m criteria, respectively, was presented. In this paper the latter was used. Some typical results: Ti-47Nb : J_c (4.2K, 5T) = 1.48x10^5 A cm^{-2}; Ti-49.6Nb: J_c (4.2K, 5T) = 2.51x10^5 A cm^{-2}; Ti-53.5Nb: J_c (4.2K, 5T) = 1.77x10^5 A cm^{-2}.	Resistive upper critical field, H_r, (J = 5 A cm^{-2}) results were reported. Deformation-subband sizes were measured and reported. Ti-Nb-Ta Ti-Zr-Nb-Ta	D.G.Hawksworth and D.C.Larbalestier (1980) [Haw80a]
45.0,47.3,49.6,50.3 and 53.5 wt.%	J_c (4.2K) data for the five alloys referred to, in the form of optimized Cu/Ti-Nb multifilamentary conductors, were collected vs applied field up to about 10.5T. The same results were also recast in the form of a plot of pinning force ($\sim J_c$,H) vs applied field. The temperature dependences of the maximum pinning force for the same alloys (except Ti-45Nb) were presented.	A review of the contemporary status of Ti-Nb and Ti-Nb-base alloy superconductors. Numerous critical-current- or pinning-force-related data from the literature were reviewed.	T_c vs composition; subband geometry in Ti-Nb vs cold reduction; electrical resistivity vs composition for numerous Ti-TM and TM$_1$-TM$_2$ alloys. Ti-TM TM$_1$-TM$_2$ Ti-Nb-Ta	D.C.Larbalestier (1980) [Lar80] see also [Hwa79]

TABLE 1-29 TITANIUM-NIOBIUM ALLOYS — CRITICAL CURRENT DENSITY, FLUX PINNING –*continued*

Nb Concentration Range	Procedures	Remarks, Discussion	Other Properties Studied and Systems Investigated	Literature
40 at.%	*Sample preparation:* ingots cold worked to 3 mm$^\phi$; Cu sheathed and drawn (99%) to 0.5 mm$^\phi$; 7-filament composites were prepared by rebundling at 1.6 mm^2 and drawing to 0.5 mm$^\phi$ (reduction, 99.94%). J_c (4.2K, 3-9T) measured as function of 50h-aging temperature, 350-450°C. J_c (4.2K, 3-8T) measured as function of 400°C-aging time to 200 h. J_c (4.2K) *vs* H for 1-filament (99% reduction) and 7-filament (99.94% reduction) composites aged 50h/375°C.	Ti-Nb$_{40}$ was prepared and measured for use as a reference alloy in a study of the properties of the Ti-Hf-Nb system. The results focussed attention, of course, on the influence of Hf additions on the superconducting properties -- see Table 1-34. With regard to Ti-Nb$_{40}$ itself, J_c (4.2K, 5T) for 7 filament wire aged 50h/375°C was only 3.3×10^4 A cm^{-2}, an order of magnitude lower than contemporaneously measured experimental values.	Hardness *vs* aging temperature T_c *vs* (24h) aging temperature H_r (4.2K) *vs* composition. Ti-Hf-Nb	H.Wada *et al.* (1981) [Wad81]
30 at.%	I_c (4.2K, 4T) *vs* aging time to 10^4 min at 350°C.	The paper was primarily reporting on the superconducting properties of quaternary Ti-Zr-Nb-Ta alloys conforming to the general specification: $[(Ti-Nb_\ell)-Zr_m]-Ta_n$. Ti-Nb$_{30}$ was included as a reference system.	Vickers hardness as a function of cold work to 10^6:1 area reduction ratio. Guinier radius (a measure of precipitate particle size) and relative interparticle distance in wire specimens of Ti-Nb$_{36}$ (cold worked to 40 μm$^\phi$ (99.994%)) as function of aging time at 380°C. Ti-Zr-Nb-Ta	T.Horiuchi *et al.* (1980) [Hor80]
30 through 50 at.% -- in the form of seventeen different Cu/Ti-Nb composite conductors	Short-sample tests of spool-wound conductor (separation of voltage taps, 95 ± 48 mm) in transverse fields of up to more than 10T, in an intercomparison of the properties of commercial multifilamentary composite conductors from five Japanese, one British and two American manufacturers. Results for all conductors were presented as J_c (4.2K) *vs* H plots (criteria, $10^{-10}\ \Omega$ cm and $10^{-11}\ \Omega$ cm).	Plots of J_c (8,9,10T) and H_r *vs* Ti concentration were presented. Equipment for performing tests on the "large-coil" conductor was described and some results for two conductors were presented.	H_r data given for each conductor.	E.Tada *et al.* (1980) [Tad80]
45, 50, and 53.5 wt.%	Optimization studies were performed on three Ti-Nb alloys and the five Ti-Nb-Ta alloys discussed in Table 1-36. All the samples studied were produced from 10 cm to 25 cm diam. billets and standard wire-production procedures were followed. J_c *versus* H$_a$ measurements were undertaken at temperatures down to about 2 K, and data were presented which enabled the performances of the three classes of Ti-Nb-Ta alloy to be intercompared with those of Ti-Nb-Ta considered elsewhere. Testing of a small coil wound from Cu/Ti-53.5Nb was also undertaken.	(a) At fields of 5 T and below, the J_c of Ti-45Nb was within 13% of that of Ti-53.5Nb. However, at 10 T, on account of the higher critical field, the J_c of Ti-53.5Nb was almost five times greater than that of Ti-45Nb. At 12 T and 2 K, Ti-53.5Nb carried more than twice the current of Ti-45Nb. (b) The performance of optimized Ti-50Nb was also compared with that of Ti-53.5Nb which turned out, at high fields, to be the slightly better alloy. For example, J_c(2K, 12T, Ti-50Nb) = 4×10^4 A cm^{-2}, while J_c(2K, 12T, Ti-53.5Nb) = 4.5×10^4 to 5.5×10^4 A cm^{-2}. (c) The paper under discussion is a summary of [Ste78].	(Ti-53.5Nb)-0.5Ta (")-1.0Ta. (")-1.5Ta. Ti-48Nb- 8Ta Ti-32Nb-25Ta An intercomparison between the properties of the binary and ternary alloys is given in Table 1-36.	H.R.Segal *et al.* (1981) [Seg81][a] cf.[Ste78].

TABLE 1-29 TITANIUM-NIOBIUM ALLOYS — CRITICAL CURRENT DENSITY, FLUX PINNING –continued

Nb Concentration Range	Procedures	Remarks, Discussion	Other properties Studied and Systems Investigated	Literature
53.5 wt.%	TEM at 120 kV on samples removed from a multifilamentary Cu/Ti-Nb composite, in particular the Fermilab Ti-53.5Nb conductor taken from an intermediate stage in the manufacturing process. Strand diam., 3.66 mm; filament number, 2000. The influence of further cold drawing and heat treatment on subband size and precipitation was examined. J_c measurements were made on long samples at the final strand diameter of 0.66 mm; *criterion:* 10^{-12} Ω cm.	The influences of heat treatment for times of up to 200 h at 300, 375, and 400°C on the subband size and 5-T J_c were examined and the results presented in tabular form. It was concluded that a high J_c is favored by a small subband size, not as a consequence of the subband size itself but because the fine structure is conducive to a high density of α-phase precipitates. Thus the highest J_c was found in samples not necessarily with the smallest subband diameter but with the highest density of precipitation.	------------------	A.W. West and D.C. Larbalestier (1981) [Wes81]
45 and 53.5 wt.%	(a) TEM at 120 kV on samples of Ti-45Nb rod and rolled strip after heat treatment for 100h/450°C. (b) STEM and EDAX investigations of a sample of Fermilab conductor (cf. [Wes81]) measured at a diameter of 0.67 mm after undergoing the following thermomechanical treatment: 3.66 mm$^\phi$(80h/375°C) -----> 1.5 mm$^\phi$(40h/375°C) ----> 0.67 mm$^\phi$	(a) The TEM studied demonstrated that although in the rod sample the α-phase precipitates which form during the heat treatment bore a definite orientational relationship to the bcc matrix, the same was not true of the worked material. (b) The STEM investigation shed additional light on the morphology of the α-phase precipitate and enabled it to be concluded that: (i) the α-phase which nucleates in heavily cold worked alloys forms as equiaxed particles with no definite orientation relationship to the matrix; (ii) during subsequent wire drawing the precipitates deform with the matrix and elongate in the drawing direction.	------------------	A.W. West and D.C. Larbalestier (1982) [Wes82]

TABLE 1-30 TITANIUM-NIOBIUM-BORON, -CARBON, -NITROGEN, and -OXYGEN ALLOYS

Binary Host Alloy	Solute Concentrations	Sample Preparation, Properties Investigated	Remarks, Discussion	Literature
(a) BORON Additions				
Ti-56Nb	0.05 wt.% B	B was introduced in form of TiB_2. J_c (4.2K,50kOe) vs heat treatment temperature after 99.9% cold work.	Micrograph showed needle-like precipitates, ~30 μm long; presumed to be the preferentially etched ends of dendritic branches. Only small B levels are technologically permissible on account of possible work hardening.	G. Rassmann and L. Illgen (1972) [Ras72b]
(b) CARBON Additions				
Ti-Nb$_{50}$	0.1, 0.15, 0.4 wt.% C	C was introduced in form of NbC. J_c (25kOe) vs C content in cold-worked wire, and J_c vs annealing temperature with 0.15 wt.% C.	J_c of cold-worked alloy increases with C content; and with 0.15 wt.% C is maximized for anneals between 350 and 450°C. That of Ti-Nb$_{35}$ was maximized within 350-400°C.	Ya.N. Kunakov et $al.$ (1967/1970) [Kun70]
Ti-20Nb Ti-40Nb Ti-60Nb Ti-80Nb	0.013, 0.026, 0.031, 0.078 0.013, 0.025, 0.044, 0.085 } wt.% C 0.007, 0.020, 0.043, 0.098 0.006, 0.016, 0.046, 0.083	Samples prepared by arc melting using consumable electrode containing C in the form of cloth. Vacuum annealed 16h/1550±20°C in 10^{-4} Torr. J_c (4.2K,20kOe) vs 1 h heat treatment temperature after ~99% (area reduction) cold work.	Sample preparation, metal working, J_c (as function of heat treatment temperature) was described in considerable detail.	British Patent No. 1,089,786 (1967) [Wes67]
(c) NITROGEN Additions				
Ti-56Nb	0.11 wt.% N	N was introduced in form of TiN to augment N already present in starting alloy which contained 405 ppm C, 50 ppm N, 645 ppm O. Comparisons were made between the J_c's of: (a) forged alloys, and (b) N-doped and quenched alloys. Detailed microstructural studies were undertaken. J_c (4.2K,30-90kOe) vs 1 h heat treatment temperature was measured on: (a) forged plus 99.93% deformed wire. (b) N-doped, quenched plus 99.93% deformed wire.	During 1h/1100°C/quench only about half of the N remained in solution. For the wire-drawn alloys in the optimal aged condition the forged material yielded the higher J_c above 75 kOe, while the N-doped alloy was superior below 75 kOe. A well-defined cell structure with carbide precipitation and N in solution was recommended.	C. Baker (1970) [Bak70]

TABLE 1-30 TITANIUM-NIOBIUM-BORON, -CARBON, -NITROGEN, and -OXYGEN ALLOYS –continued

Binary Host Alloy	Solute Concentrations	Sample Preparation, Properties Investigated	Remarks, Discussion	Literature
		(d) OXYGEN Additions		
Ti-60Nb	0.239 wt.% 0	0 was introduced as Nb_2O_5. Arc melted, machined to 0.415 in. φ, jacketed in stainless steel, cold swaged to 0.117 in.², unclad, swaged to 0.051 in.² and drawn to 0.010 in.² (hence 99.9% cold worked). J_c (4.2 K) versus H_a (0–110kOe), metallography.	The following conditions were compared: (a) as-worked, (b) 1h/290, 370, 500, 600, 700, 800, 900, 1000°C, (c) 1h/500°C at 0.113, 0.078, 0.030, 0.010 in. φ with intermediate heat treatment, (d) 500°C for 1/2, 1, 2, 8, 32 h and 1h/1000°C plus 1h/500°C. J_c is influenced by the heat-treatment-induced fine-scale precipitates, in the formation of which the presence of 0 plays a role.	F.W. Reuter et al. (1966) [Reu66]
Ti-Nb₃₀ Ti-Nb₅₅	0.08 ⎫ at.% 0 0.2 ⎭	(a) As-cast plus swaged to 0.220 in. φ, homogenized 1h/1250°C, cold drawn (99.9%) to 0.0086 in.². (b) Above (99.8% at 0.009 in.²) plus 1h/400°C.	0 increased the J_c of cold-drawn and precipitation-heat-treated Ti-Nb, possibly, it was thought, as a result of local 0-induced lattice expansion and resulting lattice strain.	J.O. Betterton (1966) [Bet66]
Ti-39Nb	"low oxygen" and "high oxygen"	Two levels of 0, "low" and "high", introduced as Nb_2O_5 powder. Samples were prepared by arc melting, clad in 304 stainless, swaged. Unclad, wrapped in Nb foil, 8h/1450°C/5x10⁻⁶Torr/slow cooled. Rewrapped, 4h/1000°C/ quenched. Clad in stainless steel, swaged to 0.100 in., unclad, swaged to various diameters. Recrystallized 2h/1000°C, swaged or drawn to 0.010 in final diameter. Cold work ranged from 0 to 99.9%. Optical microstructure, resistive critical field H_r (4.2K), and J_c (4.2K,H) studied. Field capability, 150 kOe.	After cold work, final aging heat treatments were administered consisting of 1h/425°C (low oxygen) or 1h/525°C (high oxygen). Both were thought to represent overaging. Under conditions of 99.9% cold worked and 99.9% cold worked plus aging, the low-oxygen alloy exhibited the higher J_c. It was recognized that optimal aging conditions should be sought.	K.R. Comey (1967) [Com67]
Ti-40Nb Ti-50Nb Ti-60Nb Ti-80Nb	0.14 , 0.16 ⎫ 0.051 ⎪ wt.% 0 0.052, 0.11, 0.27 ⎪ 0.034 ⎭	0 was introduced as Nb_2O_5. Samples were prepared by multiple arc melting, machined and jacketed in stainless steel, swaged >50° RA. Unclad, wrapped in Nb, homogenized to remove coring 7–12h/1500°C. Cold swaged to 0.042 in. and cold drawn to 0.010 in. -- no heat treatment. H_r (4.2K) and J_c (4.2K,H) measured in same equipment as above.	An extension of the above studies. J_c(H) studied as function of 1 h aging temperature (250–many steps-1000°C). In an appendix the effect of varying aging time was explored in an optimization study of four Ti-Nb-0 alloys. Five variables governing J_c in such systems are: Nb content, 0 content, cold work prior to aging, aging temperature and aging time. Although it was stated that in Ti-60Nb, aged 1h/450°C, J_c decreased with increasing oxygen content, the conclusions were generally discursive, and tended to deal with the effect of 0 and process variables on x and ω precipitation, rather than intrinsic properties of dissolved or precipitated 0.	G.C. Rauch (1968) [Rau68ᵃ] see also [Rau68]

TABLE 1-30 TITANIUM-NIOBIUM-BORON, -CARBON, -NITROGEN, and -OXYGEN ALLOYS *–continued*

Binary Host Alloy	Solute Concentrations	Sample Preparation, Properties Investigated	Remarks, Discussion	Literature
Ti-50Nb Ti-60Nb	0.051 ⎱ wt.% O 0.239 ⎰	-------------	Review of superconductivity and metallurgy in Ti-Nb alloys. The addition of O can accelerate transformation kinetics, thus compensating for the retardation encountered at higher Nb levels. The transformation temperature is also reduced.	U. Zwicker *et al.* (1970) [Zwi70]
Ti-Nb$_{60}$	0.17, 0.68, 1.17, 1.66, 2.17, 2.60, 3.13 at.% O	O was introduced as Nb$_2$O$_5$ prior to arc melting. *Transition temperature* examined as function of O level in Nb and Ti-Nb$_{60}$. The O levels were determined using fast-neutron activation analysis; T$_c$ was measured using the Schawlow-Devlin technique.	T$_c$ decreases at the linear rate, over the range studied, of -0.56K/at.%O.	F.A.Rodriguez-Gonzalez (1970) [Rod70]
Ti-40Nb	0.044, 0.24 wt.% O	-------------	Pinning-force parameter, J$_c$H, plotted *vs* H ($\gtrsim$100kOe) for 1 h anneals at 250-1000°C (8 steps) (0.24wt.%O) and 400, 500, 1000°C (0.04wt.%O).	R.L. Ricketts *et al.* (1970) [Ric70]
Ti-Nb$_{30}$ Ti-Nb$_{55}$	(a) 176 ppm O (b) 352 ppm O plus 537 ppm Fe (c) 44 ppm O	Samples were prepared by arc melting from Ti (RRR,42) and Nb (RRR,80). O was introduced as TiO$_2$. Alloys were cold swaged from 0.220 in. to final diameters of 0.007 $\sim$ 0.010 in. and measured in the conditions: (a) 99.9% cold work (cw) 99.8% cw + 3h/375°C, (b) 99.8% cw 99.8% cw + 3h/375°C, (c) 99.6% cw 99.6% cw + 1h/300°C. J$_c$ (4.2 K, 0 < H $\lesssim$ 32 kOe)	As expected the effect of the "precipitation heat treatment" is greatest at the lower Nb content. In each case the critical current density was enhanced by the addition of 44 to 285 ppm (wt.) O.	D.S. Easton *et al.* (1971) [Eas71]
Ti-Nb$_{35}$	0.30 wt.% O	Short sample J$_c$ (4.2 K, 0 < H $\lesssim$ 95 kOe), mixed state magnetization (4.2K), and normal-state resistivity.	The main emphasis of the paper was on the α-phase/TiO equilibrium -- its establishment and possible effects on flux pinning. It was concluded that the degree of deformation necessary to produce filaments could not be achieved when interstitials were present in sufficient quantity to produce significant precipitation. On the other hand the very fine dislocation cell size resulting from high reductions gave flux pinning and critical current densities comparable to those due to precipitation. A possible corollary might be the suggestion that precipitation could be an *alternative* to cold work under some circumstances.	M.J. Witcomb and D. Dew-Hughes (1973) [Wit73]

TABLE 1-30 TITANIUM-NIOBIUM-BORON, -CARBON, -NITROGEN, and -OXYGEN ALLOYS –continued

Binary Host Alloy	Solute Concentrations	Sample Preparation, Properties Investigated	Remarks, Discussion	Literature
	(e) NITROGEN and/or OXYGEN Additions			
Ti-Nb	N and O	------------------	The total amount of N and O claimed in the invention does not exceed the residual quantity of up to about 1500 ppm. The contribution of this to J_c was regarded as minor.	British Patent No. 1,089,786 (1967) [Wes67]
Ti-33Nb	0.20, 0.41 at.% N 0.32 at.% O	Samples were prepared by arc melting and cast into rods 6 mm$^\phi$, cold swaged about 92% to 1.75 mm$^\phi$, quenched. Drawn to diameters of 1.0 to 0.5h/950°C, annealed 0.20 mm without annealing.	The influence of interstitial level, time and temperature of heat treatment was studied. Under the conditions of this particular investigation N and O generally had a deleterious effect on J_c.	D. Bachmann et al. (1968) [Bac68]
Ti-40Nb	0.21 at.% N 0.17, 0.32 at.% O 0.21 at.% N plus 0.17 at.% O	The J_c (4.2 K, 0 < H < 100 kOe)'s of N- and O-containing alloys were compared with that of the "pure" binary. The transition temperatures of N-, O-, and N+O-containing Ti-33,40Nb alloys were studied as function of interstitial content and heat treatment.		
Ti-60Nb	0.049, 0.118, 0.16, 0.24 wt.% O 0.097 wt.% N plus 0.187 wt.% O	Samples arc melted, hot forged, rolled and cold drawn. The density and arrangement of dislocations were studied in the electron microscope. J_c was investigated as function of oxygen content and heat treatment.	Under the conditions of this study, J_c increased with interstitial content corresponding to an increasing degree of refinement of the internal structure which takes the form of subdivided filamentary or lenticular bands. For example, at 50 kOe (5T) the following critical current densities were recorded: 490 ppm O, 1x10^5 A cm^{-2}; 2400 ppm O, 1.5x10^5 A cm^{-2}; 1870 ppm O + 996 ppm N, 1.7x10^5 A cm^{-2}.	M. Bidault and J. Dosdat (1970) [Bid70], see also E. Adam et al. (1970) [Ada70]
	(f) CARBON, NITROGEN and/or OXYGEN Additions			
Ti-Nb$_{50}$	C, N, O and H	Interstitials were introduced in the form of TiC, NbN, TiO$_2$ and NbH. Transition temperature was measured as function of interstitial content.	T_c decreased continuously with N and O concentration in the range 0-0.44 at.%, decreased slightly upon the addition of H, and increased following the addition of C, up to ~0.3 wt.%.	I.A. Baranov et al. (1967) [Bar67a]
Ti-Nb$_{40}$	Initial C, N, O (viz. 204 ppm C, 56 ppm N, 3169 ppm O) plus 1000 ppm N. Initial C, N, O plus 1000 ppm C.	Interstitials were introduced in the form of TiC or Nb$_2$C, TiN or Nb$_2$N, Nb$_2$O$_5$ or TiO$_2$ to augment the already high interstitial content of the commercially pure starting material, viz: Ti: 80 ppm C, 50 ppm N, 800 ppm O; Nb: 300 ppm C, 60 ppm N, 5000 ppm O. J_c (4.2K) measured in fields of up to 70 kOe.	Critical currents were reported for various interstitial contents and thermomechanical processing sequences involving various combinations of hot (650°C) and cold forging, swaging, wire drawing and 1h/400°C heat treatments. The invention also claimed, separately or in combination, additions in the following ranges: C, 500-1500 ppm; N, 500-2000 ppm; O, 500-4000 ppm to Ti-V,-Nb,-Ta,-Nb-Hf.	French Patent No. 1,517,216 (1968) [Ass68a], see also French Patent No. 1,512,971 (1968) [Ass68]

TABLE 1-30 TITANIUM-NIOBIUM-BORON, -CARBON, -NITROGEN, and -OXYGEN ALLOYS *-continued*

Binary Host Alloy	Solute Concentrations	Sample Preparation, Properties Investigated	Remarks, Discussion	Literature
Ti-Nb$_{20}$	300 ppm, 1500 ppm O, 2000 ppm O plus 1500 ppm O, as Y$_2$O$_3$	It was pointed out that O may be introduced into the final wire from the annealing atmosphere, in the form of Nb$_2$O$_5$ powder or by pre-oxidizing iodide-process Ti. N was added by pre-nitriding iodide Ti, and C was added to the Ti-Nb alloy in the form of acetylene soot. During processing the wire was clad in Fe, which acted as an oxidation protection for the core material during heat treatment. Its advantage over Cu as a protective coating was discussed.	Also studied was J$_c$ vs annealing temperature (1h at up to 500°C) following 99.9% cold work for Ti-Nb$_{40}$ (Kroll Ti) and Ti-Nb$_{40}$ (iodide Ti) plus 500 ppm N, 500 ppm O, and 500 ppm N plus 500 ppm O and Ti-52Nb plus 1500 ppm C. All curves peaked at 400°C except that for Ti-Nb-C (350°C).	G. Rassmann and L. Illgen (1972) [Ras72^a]
Ti-Nb$_{40}$	225 ppm, 500 ppm O, 1500 ppm O, 1500 ppm O plus 1500 ppm O, as Y$_2$O$_3$			
Ti-Nb$_{40}$	500 ppm N, 500 ppm N plus 500 ppm O			
Ti-52Nb	1500 ppm C	J$_c$ vs H (<60kOe) for cold worked (99.9%) and cw plus 500°C heat treated alloys with various interstitial contents.		

Ternary Host Alloy	Solute Concentrations	Sample Preparation, Properties Investigated	Remarks, Discussion	Literature
(g) OXYGEN or NITROGEN Additions				
Ti-Nb$_{30}$-Fe$_{0.1}$	0.16 at.% O	J$_c$ (4.2K,30kOe) compared as-drawn and after optimal heat treatment.	Under the conditions of this study additional O plus Fe did not improve the J$_c$ of Ti-Nb$_{30}$-O$_{0.08}$, and the addition of 0.13 at.% O made a slight improvement to that of Ti-Nb$_{55}$-Fe$_{0.52}$.	J.O. Betterton $et\ al.$ (1966) [Bet66]
Ti-Nb$_{55}$-Fe$_{0.52}$	0.13 at.% O and without added O			
Ti-Nb$_{30}$-Ta$_5$	500 ppm N	N was added as TiN. Samples were prepared by cold swaging to 1.8 mm^b, drawing to 0.25 mm^b; final anneal 1h/400°C.	The addition of N improved J$_c$ in both cases. The invention also claimed a range of interstitial combinations and levels as mentioned above in connection with additions to binary Ti-Nb.	French Patent No. 1,517,216 (1968) [Ass68^a]
Ti-Nb$_{30}$-Hf$_5$	500 ppm N	J$_c$ (4.2K) measured in fields of up to 70 kOe.		
Ti-Nb$_{50}$-Zr$_{40}$	0.1, 0.3, 0.95 wt.% O	Optical and electron microscopy accompanied measurements of J$_c$ vs H (0 < H $\lesssim$ 70 kOe) in studies of solution treated and aged (4h,100h/700°C) oxygen-doped Zr-Nb-rich alloys.	Improvements noted in J$_c$ were attributed to both direct and indirect influences of O, viz: (a) flux pinning by dissolved O, and (b) influence of O on the kinetics of the $\beta \rightarrow \beta' + \beta''$ reaction and refinement of the resulting $\beta' + \beta''$ lamellar structure.	M. Kitada and T. Doi (1970) [Kit70^b]

TABLE 1-30 TITANIUM-NIOBIUM-BORON, -CARBON, -NITROGEN, and -OXYGEN ALLOYS –*continued*

Ternary Host Alloy	Solute Concentrations		Sample Preparation, Properties Investigated	Remarks, Discussion	Literature
	(h) OXYGEN Additions				
	Total (wt.%)	Excess Oxygen			
Ti-39.4Nb-1.0Y	0.45	0.18	Y, Gd, Th were added in the form of metal chips or powder; O as Nb_2O_5 powder. Samples arc melted, homogenized 6h/1450°C.	"Excess" O was computed by assuming complete oxidation of minor additions to Y_2O_3, Gd_2O_3, ThO_2, respectively. The J_c's were claimed to be superior to those of all other undoped Ti-40Nb alloys previously studied.	T.H. Courtney and J. Wulff (1967) [Cou67]
-1.0Gd	0.45	0.298			
-1.0Th	0.45	0.312	An examination, assisted by optical metallography, of the influence of interstitial and minor element additions on the J_c(H) of cold-worked and cw + aged 1h/425°C alloys.		
Ti-40Nb-Er$_{0.72}$,0.66	0.3, 0.29 wt.% O, respectively		Er, Sc added as chips, O as Nb_2O_5 powder; samples were prepared by multiple arc melting. Some of the Er, Sc was in solution as was some of the O. Some M_2O_3 oxide was also present.	Results for the "reference" alloys Ti-40Nb-0.044, 0.24 O were presented. J_c (50kOe) was plotted for all alloys as function of 1 h annealing temperature. A pinning-force-parameter, J_cH, was plotted vs H for all alloys corresponding to the conditions: as-drawn, 1h/400,500,1000°C. It is concluded that ω-phase is the most effective flux pinner in these alloys, that O and RE additions diffuse to the α-phase at elevated temperatures. The optimal 1 h aging temperatures for α- and ω-phase are 500 and 400°C, respectively.	R.L. Ricketts *et al.* (1970) [Ric70]
Ti-40Nb-Sc$_{0.67}$,0.82	0.38, 0.45 wt.% O, respectively		J_c(H) was measured vs heat treatment.		
Ti-Nb$_{20}$-Y	0.35 wt.% O		O in the form of Nb_2O_5 was added to Ti-Nb alloys containing Y which is itself insoluble in both Ti and Nb, but reacts with released O to form Y_2O_3 precipitates.	A finely divided Y_2O_3 precipitate was formed, whose volume fraction was unaffected by heat treatment.	G. Rassmann and L. Illgen (1972) [Ras72^a]
Ti-Nb$_{40}$-Y	0.30 wt.% O		J_c(H) was investigated in cold worked and heat treated alloys.		

TABLE 1-31   TITANIUM-NIOBIUM-SIMPLE-METAL ALLOYS

Host Alloy	Solute Concentrations	Properties Investigated	Remarks, Discussion	Other Systems Studied	Literature
	(a) ALUMINUM Additions				
Ti-36.4Nb (22.0 at.%)	1.2 wt.% (2.0 at.%) Al	Electrical resistivity and resistance-ratio vs aging, up to 10,000 min/350°C. T_c of cold deformed wire.	An overview of some properties of numerous Ti-Nb and Ti-Nb-base ternary alloys. Samples for measurement had been arc cast, cold swaged ∼92% to 1.75 mm², vacuum annealed 30min/950°C; and drawn (with/without annealing) to wire 0.35 to 0.20 mm².	Ti-Nb (∼8 compositions) with O, N and eleven simple metal substitutional elements	B. Bachmann et $al.$ (1968) [Bac68]
Ti-Nb$_{33}$	1, 1.5, 2 and 2.5 at.% Al	T_c, H_r and J_c (100kOe) vs Al concentrations.	A review paper dealing with the interrelationship between metallurgical and superconductive properties.	Ti-Nb-Cu,-Ge,-Au Ti-Nb-Zr Ti-Nb-O	U. Zwicker et $al.$ (1970) [Zwi70]
Ti-49Nb (33 at.%)	0.9 wt.% (2 at.%) Al	H_r measured resistometrically on 97% cold deformed plus 10h/400°C wire.	An overview of the properties of numerous Ti-Nb-base ternary alloys; detailed results were presented only for Ti-Nb-Cu and -Ge. Additions of 2 at.% Al, Ga, Be, Ag, In, Sn, Cr, Mn, Fe and Zr were found to decrease H_r; while 2 at.% Cu did not change the H_r (=114kOe) of Ti-50Nb.	Ti-Nb-Cu,-Ga,-Ge,-Ag,-In,-Sn Ti-Nb-Cr,-Mn,-Fe,-Zr	W. Heller et $al.$ (1971) [He71a]
Ti-50.6Nb	2.8 wt.% Al	J_c vs (0 < H ∼ 48 kOe) for various heat treatments.	Results for ∼99.9% cold worked wire were compared with the J_c of an un-heat-treated Ti-39Nb sample.	Ti-Nb-C,-N,-O	G. Rassmann and L. Illgen (1972) [Ras72a]
Ti-Nb (5.00-46.11 at.%) (18 compositions)	4.98-25.03 at.% Al (18 compositions)	T_c vs composition and heat treatment.	Results were displayed as function of e/a ratio.	-----	W.L. Cotton et $al.$ (1974) [Cot74]
Ti-Nb$_{30}$ Ti-Nb$_{50}$ Ti-Nb$_{75}$	2, 6, 10 at.% Al	Resistive measurements of T_c and upper critical field (the latter at 4.2K using 10 and 17.5 T superconducting magnets) on samples which had received the following thermomechanical processing: 5h/1000°C/cold rolled 50%/aged 24h at temperatures between 300 and 800°C.	Al is a strong α-stabilizer; consequently the superconducting properties respond markedly to heat treatment. Small amounts of Al (and Mo) result in only small reductions of the T_c and H_r of Ti-Nb$_{30}$ and increase its J_c when aged.	Ti-Nb-Cu Ti-Nb-Cr Ti-Nb-Mo	H. Wada et $al.$ (1980) [Wad80]

TABLE 1-31   TITANIUM-NIOBIUM-SIMPLE-METAL ALLOYS *–continued*

Host Alloy	Solute Concentrations	Properties Investigated	Remarks, Discussion	Other Systems Studied	Literature
	(b) SILICON Additions				
Ti-Nb$_{40}$	0.5 ∼ 2 at.%	The effect of Si additions on the J$_c$ of Ti-Nb$_{40}$ was examined. J$_c$ (40k0e) was measured as function of aging time at 500°C; J$_c$ of the 3h/500°C aged alloy measured in fields up to about 90 kOe. The influences of Si additions on grain size and tensile strength were also investigated.	The J$_c$ of Ti-Nb$_{40}$ aged 350-550°C increased markedly with addition of 0.5 ∼ 2 at.% Si as a consequence, it was believed, of more uniformly distributed and finer α-phase precipitation. After 98.4% cold drawing and 3h/500°C aging the alloy with 2 at.% Si attained a maximal J$_c$ of 6.0x10^4 A cm^{-2} at 70 kOe.	Ti-Nb$_{40}$ control	F. Ishida *et al.* (1970) [Ish70]
Ti-Nb$_{19.0}$ Ti-Nb$_{20.5}$ Ti-Nb$_{33.6}$ Ti-Nb$_{35.2}$	2.1 0.3 } at.% Si 6.2 2.4	J$_c$ briefly referred to.	cf. [Ras72, Ras72^a, Ras72^b, Ras73]	Ti-Nb-V,-Ta,-Cu	G. Rassmann and L. Illgen (1973) [Ras73]
Ti-Nb$_{40}$	2 at.% Si	J$_c$ at 70 kOe *vs* aging time up to about 1.8x10^3 h at 350, 400 and 500°C.	Comparisons were made with a Ti-Nb$_{32}$ control alloy. Alloys and numerous compounds reviewed.	Nb-Zr, Nb-Ti, Nb-Ti-Ta,-Zr Numerous compounds (β-W, NaCl and Laves types).	M. Watanabe (1974) [Wat74]
	(c) COPPER Additions				
Ti-36.6Nb (23.0 at.%)	1.1 wt.% (1.0 at.%) Cu	T$_c$ of a (deformed) wire was measured.	An overview paper -- cf. [Bac68] above.	See Al entries	D. Bachmann *et al.* (1968) [Bac68]
Ti-Nb$_{33}$	0.5, 1, 3 at.% Cu	T$_c$, H$_r$ and J$_c$ (100kOe) *vs* Cu concentration.	Cf. [Zwi70] above. The J$_c$ (100kOe) of Ti-Nb$_{33}$-Cu, was measured as function of 400°C aging time up to about 90 h.	See Al entries	U. Zwicker *et al.* (1970) [Zwi70]

TABLE 1-31 TITANIUM-NIOBIUM-SIMPLE-METAL ALLOYS −continued

Host Alloy	Solute Concentrations	Properties Investigated	Remarks, Discussion	Other Systems Studied	Literature
Ti-Nb$_{33}$ (48.5,48.3 wt.%) Ti-Nb$_{39}$ (55.0,54.7 wt.%) Ti-Nb$_{44}$ (60.0,59.7 wt.%) } 3, 5 at.% Cu		H_r measured resistometrically vs aging time.	H_r of homogenized (3h/1000°C) and quenched alloys were measured vs aging time at 400°C (Ti-Nb$_{33}$-Cu$_{3,5}$) and 700°C (Ti-Nb$_{33}$-Cu$_3$); and of deformed (~97%) Ti-Nb$_{33,39,44}$-Cu$_{3,5}$ vs aging time at 400°C.	See Al entries	W. Heller et $al.$ (1971) [Hel71d], see also
		Study of H_r as mentioned above plus an investigation of microstructures and J_c vs H (0<H<100kOe) as function of thermomechanical processing. Also given is T_c vs aging for: (a) homogenized, and (b) deformed samples.	This investigation provided material for several publications. Wire samples were prepared by 97.2% deformation of cast rod.	Ti-Nb controls	R. Löhberg (1971) [Loh71] and
		J_c (H<100kOe) and microstructures were studied as function of thermomechanical processing.	Data from R. Löhberg (1971) [Loh71] and W. Heller (1971) [Hel71], see below.	Ti-Nb-Ge	W. Heller et $al.$ (1972) [Hel72] and R. Löhberg et $al.$ (1973) [Loh73]
Ti-Nb$_{18.5}$ Ti-Nb$_{21.3}$ Ti-Nb$_{32.1}$ Ti-Nb$_{35.2}$ Ti-Nb$_{36.1}$	1.8 0.2 11.0 } at.% Cu 1.8 1.8	J_c (45kOe) vs 1 h aging temperature (≥500°C) for the alloys Ti-Nb$_{21.3}$-Cu$_{0.2}$ and Ti-Nb$_{35.2}$-Cu$_{1.8}$.	Hardness data were presented for all alloys; metallurgical conditions: (a) as-cast, and (b) 24h/500°C/ water quenched.	See [Ras73] above	G. Rassmann and L. Illgen (1973) [Ras73]
Ti-Nb$_{43}$ Ti-Nb$_{10}$	5 } at.% Cu 40	J_c vs H (≥79kOe) for various structures and growth conditions; also T_c vs growth conditions.	The paper deals with in $situ$ composite superconductors and largely with solidification problems associated with their growth.	-----	A. Olivei (1974) [Oli74]
Ti-46.1Nb Ti-47.1Nb Ti-46.8Nb Ti-45.7Nb Ti-46.3Nb	0.48 0.90 1.85 } wt.% Cu 2.90 4.05	Detailed studies of microstructures and critical current densities (4.2 K, 0 < H < 70 kOe).	Closely related to the work of the Erlangen group (Zwicker, Löhberg and Heller, see above). Flux pinning mechanisms were considered in detail with reference to comparative plots of (J_cH)$_{norm}$ vs H$_{norm}$ for Ti-50Nb, Ti-46Nb-Cu and Ti-46Nb-Ge. The report contains considerable technical detail.	Ti-Nb control Ti-Nb-Ge	R. Arndt and R. Ebeling (1974) [Arn74], see also J. Willbrand et $al.$ (1975) [Wil75a]
Ti-Nb$_{40}$ Ti-Nb$_{38}$ Ti-Nb$_{38}$ Ti-Nb$_{34}$	0.15 2.6 } at.% Cu 5.4 8.5	T_c and J_c studied as function of composition (and composition/heat-treatment-controlled microstructure).	In response to the addition of Cu, the rate of linear decrease of T_c is 0.14K/at.%, and that of H_{c2} is 4kOe/at.%. Ti$_2$Cu particles, the precipitation of which is relatively rapid at 400°C, provide flux pinning.	Review of superconducting materials	E.M. Savitskii et $al.$ (1976) [Sav76, p.175]

TABLE 1-31 TITANIUM-NIOBIUM-SIMPLE-METAL ALLOYS –continued

Host Alloy	Solute Concentrations	Properties Investigated	Remarks, Discussion	Other Systems Studied	Literature
Ti-Nb$_5$ Ti-Nb$_{10}$ Ti-Nb$_{15}$ Ti-Nb$_{20}$	5.2 5.0 4.7 4.4 } at.% Cu	Resistive measurements of T_c and upper critical field. J_c was measured as function of applied field up to about 90 kOe (9T) for various thermomechanical treatments -- hot working at 500 or 600°C followed by aging at 400, 500 or 600°C.	The optimal 48 h aging temperature for maximizing T_c was 500°C. J_c responded strongly to aging following hot working; cold-work optimization was not part of the study. Precipitation of Ti_2Cu and α-phase, which occurs most rapidly at ~500°C are responsible for flux pinning. It was claimed that Ti-Nb-Cu alloys with up to 10 at.% Cu can be hot worked readily at 600°C to achieve superconducting properties close to those of commercial Ti-Nb alloys.	Ti-Nb-Al Ti-Nb-Cr Ti-Nb-Mo	H. Wada et al. (1980) [Wad80]

__(d) GALLIUM Additions__

Host Alloy	Solute Concentrations	Properties Investigated	Remarks, Discussion	Other Systems Studied	Literature
Ti-35.1Nb (22 at.%)	2.5 wt.% (2.0 at.%) Ga	T_c of deformed wire.	An overview paper -- cf. [Bac68] listings above.	See Al entries	D. Bachmann et al. (1968) [Bac68]
Ti-49Nb (33 at.%)	2.2 wt.% (2.0 at.%) Ga	H_r measured resistometrically on 97% cold deformed plus 10h/400°C wire.	See [Hel71a] above. H_r = 103.5 kOe compared to 114.0 kOe for the binary Ti-50Nb control alloy.	See Al entries	W. Heller et al. (1971) [Hel71a]

__(e) GERMANIUM Additions__

Host Alloy	Solute Concentrations	Properties Investigated	Remarks, Discussion	Other Systems Studied	Literature
Ti-36.4Nb (20 at.%)	2.1 wt.% (2.0 at.%) Ge	T_c of deformed wire.	--------------------------	See Al entries	D. Bachmann et al. (1968) [Bac68]
Ti-Nb$_{33.3}$ Ti-Nb$_{39}$ Ti-Nb$_{44}$	0.5,1,1.5,2,3,4,5 2, 4 2, 4 } at.% Ge	T_c, H_c and J_c (100kOe) studied *vs* Ge concentration and annealing time.	Only brief mention of the Ti-Nb$_{33}$-Ge series was made in this review paper.	See Al entries	U. Zwicker et al. (1970) [Zw170]
		H_r *vs* aging time.	These alloys were prepared on the basis of constant Nb content.	See Al entries	W. Heller et al. (1971) [Hel71a]
		Complete study of H_r, J_c ($H \gtrsim 100$kOe), metallurgical details.	Additional ternary alloys, based on constant Ti content (66.7 at.%) were: Ti-Nb$_{32.8}$-Ge$_{0.5}$, Ti-Nb$_{32.3}$-Ge$_1$, Ti-Nb$_{31.8}$-Ge$_{1.5}$, Ti-Nb$_{31.3}$-Ge$_2$, Ti-Nb$_{30.3}$-Ge$_3$, Ti-Nb$_{29.3}$-Ge$_4$, Ti-Nb$_{28.3}$-Ge$_5$.	Ti-Nb controls	W. Heller (1971) [Hel71]

TABLE 1-31   TITANIUM-NIOBIUM-SIMPLE-METAL ALLOYS –*continued*

Host Alloy	Solute Concentrations	Properties Investigated	Remarks, Discussion	Other Systems Studied	Literature
Ti-Nb$_{33.3}$ Ti-Nb$_{39}$ Ti-Nb$_{44}$	0.5,1, 1.5, 2,3,4,5 2, 4 } at.% Ge 2, 4	See above	A summary of results from the above was made together with a comparison between Ti-Nb-Ge and Ti-Nb-Cu.	Ti-Nb Ti-Nb-Cu	W. Heller *et al.* (1972) [Hel72]
		See above	For English language version, see above	Ti-Nb Ti-Nb-Cu	R. Löhberg *et al.* (1973) [Loh73]
Ti-31.4Nb Ti-32.8Nb Ti-51.3Nb	0.63 0.45 } wt.% Ge 1.49	J_c (50kOe) of ~99.9% cold-deformed wire was measured as function of 1 h aging temperature below 500°C and compared with the corresponding binary alloy.	-----------------------	Ti-Nb-Sn Ti-Nb-B Ti-Nb-Zr	G. Rassmann and L. Illgen (1972) [Ras72b]
Ti-Nb$_{34,39,44}$	1 at.% Ge	J_c (4.2K) *vs* applied field (≥90kOe) for rolled and aged (at 500°C) alloy sheets.	The J_c *vs* aging characteristics of Ti-Nb-Zr and Ti-Nb-Ge were intercompared.	Ti-Nb-Zr	M. Kitada and T. Doi (1972) [Kit72]
Ti$_{75}$-Nb Ti$_{70}$-Nb Ti$_{65}$-Nb Ti$_{60}$-Nb Ti$_{55}$-Nb	0.5, 1, 2 at.% Ge	J_c (4.2K) *vs* applied field (≥90kOe) for numerous heat treatment conditions.	Oxygen content was varied within the range 427–882 ppm.	-----	M. Kitada (1973) [Kit73]
Ti-47.8Nb Ti-46.2Nb Ti-46.9Nb Ti-46.0Nb Ti-45.7Nb	0.50 0.98 2.10 } wt.% Ge 2.90 3.90	See above	See above	Ti-Nb control Ti-Nb-Cu	J. Willbrand *et al.* (1975) [Wil75a], see also R. Arndt and R. Ebeling (1974) [Arn74]

__(f) YTTRIUM, SILVER, INDIUM, TIN and Other Additions__

Host Alloy	Solute Concentrations	Properties Investigated	Remarks, Discussion	Other Systems Studied	Literature
Ti-Nb$_{25,38.2}$	2 at.% Y	J_c *vs* applied field (≤66kOe).	This work is related to the Ti-Nb-Y-O studies of Rassmann and Illgen [Ras72a]; to the Ti-Nb-Y-O, Ti-Nb-Gd-O, Ti-Nb-Th-O studies of Courtney and Wulff [Cou67] and to the Ti-Nb-Er-O, Ti-Nb-Sc-O studies by Ricketts *et al.* [Ric70] -- see Table 1-30.	Ti-Nb control	C.C. Koch and J.O. Scarbrough (1976) [Koc76]

TABLE 1-31 TITANIUM-NIOBIUM-SIMPLE-METAL ALLOYS –continued

Host Alloy	Solute Concentrations	Properties Investigated	Remarks, Discussion	Other Systems Studied	Literature
Ti-36.3Nb (23.0 at.%)	1.8 wt.% (1.0 at.%) Ag	T_c for 1.75 mm^2 wire deformed from as-cast condition.	-------------	See Al entries	D. Bachmann et al. (1968) [Bac68]
Ti-Nb$_{33}$	1, 2 at.% Ag	H_r and J_c (100kOe) in 97% cold deformed wire aged 10h/400°C.	-------------	See Al entries	U. Zwicker et al. (1970) [Zwi70], see also W. Heller et al.
Ti-34.6Nb (22.0 at.%)	3.9 wt.% (2.0 at.%) In	T_c for 1.75 mm^2 wire deformed from as-cast condition.	-------------	See Al entries	D. Bachmann et al. (1968) [Bac68]
Ti-48Nb (33 at.%)	3.6 wt.% (2 at.%) In	H_r in 97% cold deformed wire aged 10h/400°C.	-------------	See Al entries	W. Heller et al. (1971) [Hel71][a]
Ti-31.9Nb (20 at.%)	4.1 wt.% (2.0 at.%) Sn	T_c of deformed wire.	-------------	See Al entries	D. Bachmann et al. (1968) [Bac68]
Ti-48Nb (33 at.%)	3.7 wt.% (2.0 at.%) Sn	H_r of 97% cold worked and aged 10h/400°C wire.	-------------	See Al entries	W. Heller et al. (1971) [Hel71][a]
Ti-31.5Nb Ti-51.8Nb	0.78 0.79 } wt.% Sn	J_c vs 1 h aging temperature (≤500°C) for 99.9% cold-worked wire.	-------------	Ti-Nb control Ti-Nb-Zr Ti-Nb-Ge,-B	G. Rassmann and L. Illgen (1972) [Ras72][b]
Ti-15.5Nb (10.0 at.%) Ti-25.6Nb (16.0 at.%) Ti-29.1Nb (18.0 at.%)	10.0 4.0 2.0 } at.% Sb				
Ti-35.8Nb (23.0 at.%) Ti-30.9Nb (20.0 at.%) Ti-15.0Nb (10.0 at.%)	1.0 at.% Au 2.0 at.% Pb 5.0 at.% U	T_c of deformed wire.	-------------	See Al entries	D. Bachmann et al. (1968) [Bac68]
Ti-32.4Nb (20.1 at.%) Ti-32.1Nb (20.0 at.%) Ti-31.9Nb (20.0 at.%) Ti-14.6Nb (10.0 at.%)	Al(1.0) + Sb(1.0 at.%) Ga(1.0) + Sb(1.0 at.%) In(1.0) + Sb(1.0 at.%) Sn(2.5) + U (5.0 at.%)				

TABLE 1-32 THE SOVIET ALLOYS

Soviet Alloy Type	Compositions and Descriptive Data	Preparation and Properties Investigated	Remarks, Discussion	Other Systems Studied	Literature
SS 2	-------------	Arc melted, homogenized (0-5h)/1500°C, cold forged to bars 11x11 mm^2 and drawn 99-96% to 0.25 mm^2. J$_c$ was measured.	The effect of final annealing on the short-sample J$_c$ (16kOe) of unhomogenized and homogenized alloys was studied. J$_c$ was highest in the initially-unhomogenized (i.e. cored) alloy.	Ti-Nb$_{35,54}$ Zr-Nb$_{25}$ Ti$_{39}$-Zr$_8$-Nb$_{53}$ Ti-Nb-C	Ya.N. Kunakov *et al.* (1967/1970) [Kun70]
65 BT	Ti-65Nb plus a "small quantity" of Zr.	J$_c$ was measured.	Short-sample and solenoid properties of 65 BT were studied. This type of wire was developed and manufactured by the Institute of Precision Alloys of the I.P.Bardin Central Scientific Research Institute of Ferrous Metallurgy. Stabilization of solenoids using the "transformer method" was investigated, leading to a design for a copper-stabilized monofilament.	-----	N.E.Alekseevskii *et al.* (1967) [Ale67]
65 BT	-------------	Magnetizations of single-layer coils were studied.	Samples of 65 BT were subjected to differing, but unspecified, amounts of deformation and heat treatment in order to alter the pinning micro-structure. Magnetization hysteresis was able to be correlated with J$_c$.	-----	A.F.Prekul and N.V.Volkensteyn (1967) [Pre67]
SS 2 T 60	Ti$_{25}$-Zr$_{25}$-Nb$_{50}$ Ti-Nb$_{40}$	J$_c$ was measured.	Alloys and wires were prepared by the Girednet Experimental Works and measured as short samples and sample coils. Some wires were copper plated, others enameled.	-----	B.G.Lazarev *et al.* (1968) [Laz68]
65 BT	-------------	Cold drawn to 0.25 mm$^\phi$, and heat treated 3h/1000°C. J$_c$ was measured.	Curves were presented of sample mV *vs* sample current (A) in fields of 59.3-82.6 kOe, 4.2 K (cold drawn) and 35.4-67.7 kOe, 4.2 K (3h/1000°C).	Pb-In Zr-4%Nb	N.E.Alekseevskii *et al.* (1968) [Ale68]
KSMI-6	Six copper-plated strands of 65 BT and one copper strand, twisted, and impregnated with indium.		A solenoid wound with KSMI-6 was used in the short-sample testing of Ti-Nb$_{22}$.	Ti-Nb$_{22}$	Yu.F.Bychkov *et al.* (1969) [Byc69]

TABLE 1-32 THE SOVIET ALLOYS –continued

Soviet Alloy Type	Compositions and Descriptive Data	Preparation and Properties Investigated	Remarks, Discussion	Other Systems Studied	Literature
65 BT	Ti-65Nb plus "other components".	Drawn from 8 mm$^\phi$ to 0.25 mm$^\phi$ (99.9%) with intermediate low-temperature anneals at 3, 1.5 and 0.8 mm. Aged (0.006-10)h/ (100-600°C), (0-24)h/(250 and 400°C). J$_c$ was measured.	Properties investigated were: thickness and micro-structure of the oxide layer and microhardness after heat treatment (5-120)min/(500-1200°C) in air. J$_c$ and relative-J$_c$ were measured as a func-tion of final heat treatment. Stability to moderate temperature aging (such as in enameling) was investigated. A summary was presented of the mechanical and electrical properties of 65 BT.	-----	N.B.Gorina et al. (1969) [Gor69]
65 BT	Ti-Zr-Nb	The effective DC resistance in the presence of an alternating magnetic field was measured.	A bifilar coil of 65 BT wound on an insulating form was positioned along the axis of a super-conducting solenoid.	-----	V.V.Sychev et al. (1969) [Syc69]
65 BT	Ti-Zr-Nb	Deformed in grooved rolls, homogenized 4h/1550°C, quenched from 1250°C and aged: 1/2h/300-880°C 45h/600°C; 2h/700°C 6h/800°C; 20h/880°C. A detailed metallographic investigation was carried out.	Samples were studied by electron microscopy and Laue x-ray diffractometry. It was discovered that: (a) Several hours at 550-600°C yielded small precipitates tens of Å in size; (b) 45h/600°C or 2h/700°C yielded coherent precipitate particles; (c) 1/2h/880°C yielded precipitates surrounded by dislocations.	-----	N.N.Buynov et al. (1970) [Buy70]
65 BT T 60 NT 50	Ti-Zr-Nb Ti-Nb	J$_c$ was measured.	An intercomparison was made of the performances of welded joints formed in the alloys 65 BT, T 60 and NT 50.	-----	I.S.Krainskii and I.F.Shchegolev (1971) [Kra71]
35 BT	Ti-7.5Zr-30Nb Ti-7.5Zr-35Nb	Arc melted; hot worked to bars 80 mm$^\phi$, or sheet 3.5 mm^t; then cold worked ∿99.999% to wire 0.25-0.3 mm$^\phi$ with intermediate anneals at 3, 1.5 and 0.8 mm (cf. [Gor69] above), or 97% to strip 0.1 mm^t, respectively.	The BT's are an alloy type containing Nb whose approximate concentration (wt.%) is given by the numerical prefix. They also contain "small amounts" (meaning ∿7 wt.%) of Zr. For example, according to this reference, 35 BT may represent Ti-7.5Zr-(30-35)Nb.	-----	G.N.Kadykova and L.N.Fedotov (1972) [Kad72]
50 BT 65 BT	Ti-7.5Zr-50Nb Ti-7.5Zr-65Nb	Transmission electron microscopy and electron and x-ray diffractometry were undertaken.	The results of a detailed study of the effects of cold work and precipitation heat treatment in the three alloy classes were presented.		

TABLE 1-32 THE SOVIET ALLOYS –continued

Soviet Alloy Type	Compositions and Descriptive Data	Preparation and Properties Investigated	Remarks, Discussion	Other Systems Studied	Literature
T 60 / SS 2	Ti-Nb$_{\sim40}$ / Ti$_{\sim25}$-Zr$_{\sim25}$-Nb$_{\sim50}$	Measurements conducted on wire 0.23 mm$^\phi$ coated with Cu (30-45 μm^t) and enamel ($\sim$15 μm^t). J$_c$ was measured.	Critical currents were measured on (a) short samples (20-30 cm$^\ell$), (b) loosely wound coils, (c) one-layer tightly wound coils, and (d) multi-layer coils tightly wound with either SS-2 or 60 T.	-----	B.G.Lazarev et al. (1972) [Laz72]
T 60	Ti-Nb$_{40}$	Heavily deformed wire 0.25 mm$^\phi$, heat treated 2h/350,400,450,500,550°C. Carbon replica electron microscopy was performed; J$_c$ measurements were made.	Critical current (25, 40, 60 kOe) displayed vs 2-h annealing temperature exhibited the usual pronounced maximum at 400°C. Carbon replica micrographs were presented for the conditions: (a) cold deformed, (b) 2h/400°C, and (c) 1h/550,700°C.	-----	B.G.Lazarev et al. (1974) [Laz74]
T 60 / SS 2	Ti-Nb$_{40}$ / Ti$_{25}$-Zr$_{25}$-Nb$_{50}$	Starting material: optimally heat treated 60 T with 40 μm^t copper coating. J$_c$ measurements were made.	The original 40 μm coating was either reduced by chemical or electrochemical etching or built up by electroplating. In this way the thickness of the copper layer was adjusted from 0 to 120 μm.	-----	B.G.Lazarev and S.I.Goridov (1973) [Laz73]
T 60 / 35 BT / 50 BT / 65 BT	Ti-54Nb / Ti-(several %)Zr-30Nb, Ti-(small amount)Zr-35Nb / Ti-(small amount)Zr-50Nb / Ti-(small amount, e.g.10%)Zr-65Nb	Review article. Critical temperature, field and current measurements were made. Electron microscopy of cold-deformed and 4h/400°C aged 50 BT was carried out; α-phase precipitates were revealed.	A review was presented of flux-pinning microstructures (dislocations, α^m, ω- and α-phase precipitates, and G.P. zones, etc.) in Ti-Nb-base alloys.	Ti-Nb, Ti-Nb-O, Nb-Ta, Nb-Zr	G.N.Kadykova et al. (1973) [Kad73]
35 BT (with variable Zr and addition of Fe)	(see composition table below)	No superconductivity data were given. X-ray diffractometry was carried out; electrical resistivity, dilatometric and lattice parameter measurements were undertaken as function of heat treatment.	Physical properties were measured as diagnostic of heat treatment induced precipitation.	-----	G.N.Kadykova et al. (1974) [Kad74]
T 60 / NT 50	Ti-Nb$_{40}$ / Ti-Nb$_{50}$	No superconductivity data were given. Low temperature mechanical properties were measured.	The mechanical properties of superconducting cable of 6 and 24 twisted strands of copper-clad T 60 and NT 50, respectively, were compared with those of the individual wires.	-----	B.I.Verkin et al. (1976) [Ver76]

Composition table for 35 BT (with variable Zr and addition of Fe):

Ti	Zr	Nb	Fe	O (wt.%)
bal.	3.1	34.1	0.10	0.017
bal.	3.2	33.8	0.71	--
bal.	3.1	33.9	--	--
bal.	7.7	34.4	0.02	--
bal.	7.5	34.0	0.66	0.013

TABLE 1-33 TITANIUM-ZIRCONIUM-NIOBIUM ALLOYS — (a) RESEARCH ALLOYS

Compositions	Preparation, Procedures, Properties	Remarks, Discussion	Other Systems Studied	Literature
(Ti$_{50}$-Zr$_{50}$)$_{100-x}$-Nb$_x$ with x = 11,20,33,50,56,70,88	Arc melted, induction homogenized, cold reduced from ~0.36 in.⌀ to ~0.051 in.⌀ (99.92%). Resistometric measurements (at 1, 10, 10³ A cm⁻²) of H$_r$; J$_c$ (4.2K) vs H.	As a result of this work a comparison was made possible between the composition dependences of H$_r$ for the systems Ti-Nb, Nb-Zr and Ti-Nb-Zr whose maximal H$_r$'s (99.92% cw) were in the sequence 113, 95, 90 kOe at 4.2 K. For x = 70, J$_c$ (4.2K) at 30 kOe and 50 kOe (peak) were 1.7 and 2.0x10⁴ A cm⁻², respectively.	Ti-Nb Zr-Nb	K.M.Ralls (1964) [Ral64]

Ti	Zr	Nb (at.%)
62.0	5.0	33.0
52.4	5.9	41.7
75.0	6.9	18.1
10.0	10.0	80.0
20.0	10.0	70.0
38.6	11.1	50.3
62.0	11.2	26.8
48.0	13.0	39.0
28.0	16.0	56.0
46.9	17.3	35.8
10.0	20.0	70.0
20.0	20.0	60.0
30.0	20.0	50.0
23.8	23.9	52.3
28.0	24.0	48.0
10.0	27.0	63.0
27.4	32.5	40.1
10.1	40.0	49.9

Compositions	Preparation, Procedures, Properties	Remarks, Discussion	Other Systems Studied	Literature
(above composition block)	The ternary alloys were prepared by levitation melting. The critical temperature was measured inductively. J$_c$ was measured at 4.2 K as function of applied field up to 100 kOe. Electrical resistivities were measured at 77 and 293 K and the results plotted on a composition triangle. Optical microstructure was studied.	Optimal critical current performance was displayed by Ti-Zr$_{40}$-Nb$_{50}$. Some results were: J$_c$ = 4.0, 2.7, 1.5 x 10⁴ A cm⁻² at 60, 70 and 80 kOe, respectively. Further heat treatment was expected to result in improvements.	---------	T.Doi et al. (1966) [Doi66]
Ti-Zr$_{70}$-Nb$_{20}$ Ti-Zr$_{50}$-Nb$_{40}$ Ti-Zr$_{40}$-Nb$_{50}$	Normal-metal/superconductor intermetallic junctions were formed under pressure in excess of the normal-metal yield strength. I-V characteristics were determined using standard AC and DC techniques in studies of the gap energy.	Current-voltage characteristics were measured as a function of temperature and magnetic field, and interpreted in terms of the energy gap. The voltage, V$_g$, at which the curvature d²V/dl² is maximum, when extrapolated to zero K, is close to half the BCS energy gap.	Ta, Nb Zr-Nb	D.B.Sullivan and C.E.Roos (1967) [Sul67]
Ti-Zr$_{50}$-Nb$_{35}$ to Ti-Zr$_5$-Nb$_{47}$ Ti-Zr$_{22}$-Nb$_{48}$ and in particular	H$_{c1}$ was measured magnetically at 4.2 K in fields up to 40 kOe. H$_{c2}$ was measured in pulsed fields. Data were presented for cast, cold-worked and annealed Ti-Zr$_{22}$-Nb$_{48}$.	For alloys between the compositional limits indicated H$_{c1}$ depended only slightly on composition and lay between 500 and 600 Oe. H$_{c2}$ for Ti$_{30}$-Zr$_{22}$-Nb$_{48}$ was 78 kOe for all three conditions.	---------	B.G.Lazarev et al. (1967/1970) [Laz70a], see also E.M.Savitskii et al. (1973) [Sav73, p.311]

Compositions			Preparation, Procedures, Properties	Remarks, Discussion	Other Systems Studied	Literature
Ti	Zr	Nb (at.%)				
15	10	75	Samples were measured in the conditions: (a) as-cast, (b) deformed and annealed (usually) 24h/520°C and 120h/560°C. T_c and H_{c2} were determined resisto-metrically. Also measured was $(dH_{c2}/dT)_{T_c}$.	The H_{c2} surface was found to be concave with a fairly wide plateau in the center of the concentration triangle where H_{c2} = 70-80 kOe. H_{c2} changed little with annealing.	----------	B.G.Lazarev et al. (1968) [Laz68a], see also [Laz70]
15	20	65				
14	24	62				
16	32	52				
48	5	47				
30	22	48				
27	30	43				
15	44	41				
15	50	35				
Ti	Zr	Nb (at.%)				
46	--	54	Alloys multiply arc melted; homogenized 3-5h/1500°C; forged; cold rolled and wire drawn; annealed 1-3h/550°C. Critical fields were measured resistometrically at 300 A cm^{-2}. X-ray structural analysis was performed.	Critical fields were generally higher than those obtained by Lazarev, above. Particularly anomalous, within the context of Lazarev's work, was the value of 135.8 kOe for the alloy Ti-Zr$_{21}$-Nb$_{40}$.	----------	N.E.Alekseevskii et al. (1968) [Al68a]
36	6	58				
25	14	61				
13	21	66				
--	30	70				
53	10	37				
40	21	39				
21	34	45				
10	43	47				
64	16	20				
45	32	23				
28	45	27				
9	62	29				
Ti	Zr	Nb (at.%)				
30	35	35	Upper critical field, H_u, was measured magnetically -- see Table 1-17.	H_u was plotted vs a size-effect-corrected electron density, N_{eff}, as well as the conventional e/a. The latter plot reveals out-of-line H_u values for alloys (a) and (b).	Numerous alloys	W.DeSorbo et al. (1967) [DeS67]
45	20	35				
45	5	50				
30	12	58				
30	10	60				
15	19	66				
15	16	69				
10	13	77 (a)				
21	1	78				
10	10	80 (b)				

Compositions	Preparation, Procedures, Properties	Remarks, Discussion	Other Systems Studied	Literature
Ti-11%Zr-58%Nb	Magnetization at 4.2 K was measured as function of field ≈20 kOe for this single alloy subjected to eleven thermomechanical processing conditions.	The purpose of the investigation was to interrelate the M(H) behavior with current-carrying capacity. Whether the composition referred to is at.% or not was not specified.	--------	A.F.Prekul (1967) [Pre67a]
Ti$_{20}$-Zr$_y$-Nb$_x$ (10 ternary alloys) Ti$_x$-Zr$_{20}$-Nb$_y$ (10 ternary alloys) Ti$_x$-Zr$_y$-Nb$_{50}$ (6 ternary alloys)	Blanks, 6x6 mm^2, cut from arc melted ingot rolled to 0.15 mm. t. J$_c$ was measured at 17 kOe with the applied field directed perpendicular to the rolling plane.	This was the report of an early Soviet study dealing with alloy phases, hardness, and J$_c$, as function of composition in Ti-Zr-Nb. No heat treatment was administered and the field direction selected is well known to be the least favorable from the standpoint of critical current.	--------	A.M.Grigor'yev et al. (1968) [Gri68]
Ti Zr Nb (at.%) 74 7 19 61 18 21 51 30 19 25 53 22 56 8 36	Alloys arc melted and cast into a finger-mold 6 mm$^\phi$; swaged about 91.5% to 1.75 mm$^\phi$. T$_c$ was measured using an inductive technique -- *sample conditions:* (a) cold worked (cw); (b) cw plus 1/2h/900°C plus final deformation; (c) cw plus 160h/500°C; (d) cw plus 1/2h/900°C plus deformation plus 160h/500°C. Some critical current density measurements were made.	Optical metallography was carried out, and electrical re- sistivity measurements made. Such J$_c$ measurements that were undertaken upon cold-deformed material did not reveal prac- tically useful current-carrying capacities. It was noted that the "addition of Ti to Nb-Zr improved its deformability".	--------	U.Zwicker et al. (1968) [Zwi68]
Ti-Zr$_4$-Nb$_{31,33,35}$ Ti-Zr$_{2,4,6}$-Nb$_{33}$	Material cold worked plus: (a) 17h/400°C; (b) 34h/400°C; (c) 51h/400°C. Some upper critical field and critical current density measurements were made.	This was part of a study of the influence of small per- centages of third-element additions to Ti-Nb$_{\sim 33}$.	Ti-Nb Ti-Nb-Ge Ti-Nb-Al Ti-Nb-Ag Ti-Nb-Cu	U.Zwicker et al. (1970) [Zwi70]
Ti-Zr$_2$-Nb$_{33}$	Alloys were electron-beam melted and cast to 6 mm$^\phi$ rods, swaged 97.2% to 1 mm$^\phi$, annealed 10h/400°C and water quenched.	The purpose of the investigation was to study the influence of third-element additions on the critical field of Ti-Nb. Zr was only one of numerous solutes investigated.	Ti-Nb$_{33}$ plus numerous $\sim$2% additions	W.Heller et al. (1971) [Hel71a]

Compositions			Preparation, Procedures, Properties	Remarks, Discussion	Other Systems Studied	Literature
Ti	Zr	Nb (at.%)				
76	5	19	The alloys were cold deformed ~99.9% to wire of ~0.25 mm$^\phi$. J_c was measured at 45 kOe. Sample conditions: (a) cold worked; (b) 1 h at temperatures up to 500°C. The influence of oxygen, administered in the form of Nb_2O_5, was also considered.	The alloy, which when in the optimally heat-treated condition had the highest J_c ($>10^5$ A cm^{-2} at 62 kOe) was Ti-Zr$_{18}$-Nb$_{36}$. It was noted that although the substitution of Zr for Ti to the extent of about 15 at.% in alloys with Nb contents above 35 at.% resulted in superconducting properties better than those of (iodide) Ti-Nb, Ti-Zr-Nb should not automatically be considered as a substitute for binary Ti-Nb with its superior drawing characteristics (workability).	Ti-Nb Ti-Nb-Ge Ti-Nb-Sn	G.Rassmann and L.Illgen (1972) [Ras72[b]]
69	5	26				
60.5	5	34.5				
55	5	40				
47	5	48				
76.5	10	13.5				
72	10	18				
64.5	10	25.5				
57.5	10	32.5				
52.0	10	38				
67	16	17				
55	15	30				
45	18	37				
Ti	Zr	Nb (at.%)				
--	75	25	Transition temperature data were presented.	The transition temperatures of this set of compositions do not appear to have been published previously in the journal literature.	Comprehensive review book	E.M.Savitskii *et al.* (1973) [Sav73, p.311]
10	65	25				
20	55	25				
30	45	25				
40	35	25				
50	25	25				
65	10	25				
70	5	25				
74	1	25				
75	--	25				
0	50	50				
5	45	50				
10	40	50				
25	25	50				
(Ti-Nb$_{50}$)$_{100-x}$-Zr$_x$			Specimens for measurement were prepared in the form of wire 0.25 mm$^\phi$ using 99.8% cold reduction in area. Critical field measurements were made resistometrically at 4.2 K using a pulsed magnetic field upon both as-drawn and 380°C-aged wires. Critical current densities were measured.	The measurements referred to were simply a minor part of a study whose primary goal was the investigation of quaternary Ti-Zr-Nb-Ta alloys (see Table 1-38). The presence of Zr was seen to play an important role in the establishment of flux-pinning precipitation during aging.	Ti-Nb Numerous Ti-Zr-Nb-Ta alloys	T.Horiuchi *et al.* (1980) [Hor80]
(Ti-Nb$_{40}$)$_{100-x}$-Zr$_x$						
(Ti-Nb$_{30}$)$_{100-x}$-Zr$_x$						
with x = 5,10,15						

Compositions (at.%)			Processing Parameters	Properties Investigated	Remarks, Discussion	Literature
Ti	Zr	Nb [Nb]$_{incr}$ →				
62.0	5.0	33.0	Alloy ingots 4.5 mm$^\phi$ (5g) prepared by levitation melting; cold rolled 95% to 1.0 mm$^\phi$, annealed 1h/600°C, and cold drawn 93.75% to 0.25 mm$^\phi$.	An intercomparison of the J$_c$'s of the Ti-Zr-Nb alloys, and a comparison of them with those of Ti-Nb(33-61 at.%) and Zr-Nb(43-96 at.%).	Critical temperatures were measured magnetically on 4.5 mm$^\phi$x20 mm ingots. Electrical resistivities at 77 and 293 K of 0.25 mm$^\phi$ wires were measured; optical micro-structures were examined. Alloy compositions covered the entire range from X-type (Zr-Nb-rich) to Z-type (Ti-Nb-rich). Uniform processing was administered to all alloys, no alloy being selected for individual optimization. Of the binary and ternary alloys the best *relative* per-formance was achieved with Ti$_{10}$-Zr$_{40}$-Nb$_{50}$ for which: J$_c$ (4.2K) = 4.0, 2.7, 1.5x10^4 A cm^{-2} at 60, 70 and 80 kOe, respectively. Subsequently, as a result of process opti-mization better performances were achieved -- see T. Doi *et al.* (1968) [Doi68^a].	T. Doi *et al.* (1966) [Doi66]
48.0	13.0	39.0				
52.4	5.9	41.7				
28.0	24.0	48.0				
10.1	40.0	49.9				
38.6	11.1	50.3				
10.0	27.0	63.0				
10.0	20.0	70.0				
Ti$_{10}$-Zr$_{40}$-Nb$_{50}$ (X-type)			Material cold rolled 4.5 mm$^\phi$ to 2.0 mm$^\phi$, anneal-ed 5h/1100°C and: (a) cold drawn 98.44% to 0.25 mm$^\phi$ and annealed 1h/500,570,700,800,900,1100°C; (b) cold drawn 75% to 1.0 mm$^\phi$ and annealed 1h/570,700,800,900,1100°C then cold drawn 93.75% to 0.25 mm$^\phi$; (c) cold drawn 75% to 1.0 mm$^\phi$ and annealed 0.5,1,3h/700°C then cold drawn 93.75% to 0.25 mm$^\phi$; (d) cold drawn 23%, 75%, 93.75% and 98.44% to 1.75, 1.0, 0.5, 0.25 mm$^\phi$, annealed 1h/700°C, then cold drawn 97.96, 93.75 and 75% respec-tively to 0.25 mm$^\phi$.	A detailed comparison was undertaken of the effects of cold work before and after intermediate heat treatment on the J$_c$'s of Zr-Nb$_{75}$ and Ti$_{10}$-Zr$_{40}$-Nb$_{50}$. Procedures varied were: (a) final heat treatment temperature; (b) inter-mediate heat treatment temperature; (c) intermediate heat treatment time; and (d) degree of cold work before and after heat treatment.	The J$_c$ of the Zr-Nb responded to cold work rather than heat treatment. The converse was true for the Ti-Zr-Nb alloy, J$_c$ increasing by about 100 A cm^{-2} as a result of heat treatment at 700°C which resulted in decomposition of the β structure to β'+β". The best Ti$_{10}$-Zr$_{40}$-Nb$_{50}$ alloy performances (viz. 10^5 A cm^{-2}, 4.2K at fields of 58, 55 and 50kOe), were obtained under the following respective conditions: (1) 2.0mm$^\phi$(5h/1100°C) $\frac{CW}{75\%}$→1.0mm$^\phi$(3h/700°C) $\frac{CW}{93.75\%}$→0.25mm$^\phi$; (2) 2.0mm$^\phi$(5h/1100°C) $\frac{CW}{98.44\%}$→0.25mm$^\phi$(1h/570°C); (3) 2.0mm$^\phi$(5h/1100°C) $\frac{CW}{93.75\%}$→0.5mm$^\phi$(1h/700°C) $\frac{CW}{75\%}$→0.25mm$^\phi$;	T. Doi *et al.* (1966) [Doi66^b]
Ti$_{10}$-Zr$_{40}$-Nb$_{50}$ (X-type)			Sample conditions prior to measurement were: (a) no cold work (cw); (b) 93.75% cw (1.0 → 0.25 mm$^\phi$); (c) 98.44% cw (2.0 → 0.25 mm$^\phi$); (d) 98.44% cw plus 1h/570°C; (e) 75% cw (1h/700°C) plus 93.75% cw.	Process optimization was carried out on this alloy whose properties were then compared with those of Zr-Nb$_{75}$	In what was essentially a continuation of the previous paper [Doi66^b] the authors compared the effects of thermo-mechanical processing on the flux-pinning microstructures in Zr-Nb$_{75}$ and Ti$_{10}$-Zr$_{40}$-Nb$_{50}$. In the former, although precipitation was found to make an important contribution to flux pinning at low fields, dis-location structures introduced by cold work were the prin-cipal high-field flux pinners. On the other hand the authors attributed flux pinning in Ti$_{10}$-Zr$_{40}$-Nb$_{50}$ princi-pally to the effects of localized internal stress on heat-treatment-induced β-phase decomposition (in particular to β'+β").	T. Doi *et al.* (1966) [Doi66^c]

TABLE 1-33   TITANIUM-ZIRCONIUM-NIOBIUM ALLOYS — (b) A COMMERCIAL WIRE DEVELOPMENT PROGRAM –continued

Compositions	Processing Parameters	Properties Investigated	Remarks, Discussion	Literature
Ti_{45}-Zr_{15}-Nb_{40} (Z-type)	Material cold rolled 4 mmϕ to 2 mmϕ, annealed 5h/1100°C, and: (a) cold drawn 98.44% 2 mmϕ to 0.25 mmϕ and annealed 1h/500°C; (b) above reduction plus 99h/500°C; (c) above reduction plus 1h/400°C; (d) cold drawn 75% to 1 mmϕ and annealed 1h/500°C, drawn 93.75% to 0.25 mmϕ; (e) cold drawn 98.44%, no anneal.	A study of thermomechanical-processing-induced flux pinning mechanisms and, in particular, the peak effect.	After 1h/500°C α-phase precipitates were seen to form along the fibers of the cold-worked structure. At the longer aging time (99h) the structure tended to become globular. Also, an initially coherent precipitate (ω-phase?) grew and became incoherent. The (high-field) peak effect was supposed to be a feature of the pinning of a dense, and consequently rigid and interacting, flux lattice by high randomly spaced barriers.	T. Doi *et al.* (1967) [Doi67]
Numerous alloys	-------------	-----------	A conference paper (in English) summarizing references [Doi66, Doi66b, and Doi66c].	S. Maeda *et al.* (1967) [Mae67]

	Ti	Zr	Nb	(at.%)
(1)	70	5	25	
(2)	65	5	30	
(3)	60	5	35	
(4)	50	5	45	
(5)	45	5	50	

All Z-type alloys

Processing Parameters	Properties Investigated	Remarks, Discussion	Literature
Cold rolled 4 mmϕ to 2 mmϕ and annealed at 1100°C followed by: (a) cold reduction 98.44% to 0.25 mmϕ and aged: 10h/400°C [(1) and (2)], 100h/400°C [(3)], 12h and 20h/500°C [(4) and (5) respectively]; (b) alloy (3) aged: 2h/1100°C, 1h/400,500,600°C, 20h/400°C, also 1,3,6,99h/500°C.	A study of J_c in Ti-Zr-Nb alloys close to Ti, and the Ti-Nb edge of the composition triangle. The goal was the discovery of the best composition and heat treatment in the temperature range 400-600°C for obtaining a suitable α-phase precipitation structure.	Preliminary experiments showed Ti_{60}-Zr_5-Nb_{35} to be a favorable composition. After 98.44% cw plus 100h/400°C some typical J_c values were: 1.7×10^5, 1.2×10^5, 6.4×10^4, 4.4×10^4 A cm^{-2} at 40, 50, 70 and 80 kOe respectively. Aging 1h/500°C was more beneficial than 1h/400°C, but J_c decreased with heat treatment time at 500°C. Flux pinning by α-precipitates was most effective in the early stages of precipitation. High levels of prior cold work were not investigated.	F. Ishida *et al.* (1968) [Ish68]
(1) Ti_{10}-Zr_{40}-Nb_{50} (X-type) (2) Ti_{60}-Zr_5-Nb_{35} (Z-type) Common starting condition: cold rolled in grooved mill 4 to 2 mmϕ (75%), homogenized 5h/1100°C 10^{-6} torr, water quenched. Alloy (1)-(a) cold drawn 98.44% to 0.25 mmϕ and solution treated 5h/1100°C then aged 0,10,50,91,503h/700°C; (b) cold drawn to 0.25 mmϕ and aged 3,10,50,100h/700°C; also 5h/1100°C (c) cold drawn 75% to 1 mmϕ; aged 3h/700°C or 5h/1100°C; cold drawn 93.75% to 0.25 mmϕ. Alloy (2)-(a) cold drawn to 0.25 mmϕ and aged 0,1,6,99h/500°C.	A study of the optimized properties of: (1) X-type Ti-Zr-Nb, (2) Z-type Ti-Zr-Nb, flux pinning in terms of the nature, size and distribution of the precipitates, and the microstructure of the matrix.	Microstructures were investigated using optical metallography and replica electron microscopy. Alloy(1) was: (a) solution treated (1100°C) and aged, (b) cold worked and aged with and without intermediate heat treatment. Alloy(2) was: (a) cold worked and aged. *Optimal processing conditions were found to be:* Alloy (1) - 4mm $\xrightarrow{CW}$ 2mm (5h/1100°C) $\xrightarrow{CW}$ 1mm (3h/700°C) $\xrightarrow{CW}$ 0.25mmϕ, Alloy (2) - 4mm $\xrightarrow{CW}$ 2mm (5h/1100°C) $\xrightarrow{CW}$ 0.25mmϕ (1h/500°C), yielding the following 40, 80 and 90-kOe J_c's: Alloy (1) - 3.6×10^5, 5.0×10^4 and 1.0×10^4 A cm^{-2}, Alloy (2) - 1.0×10^5, 4.2×10^4 and 3.0×10^4 A cm^{-2}.	T. Doi *et al.* (1968) [Doi68a]

TABLE 1-33 TITANIUM-ZIRCONIUM-NIOBIUM ALLOYS — (b) A COMMERCIAL WIRE DEVELOPMENT PROGRAM – continued

Compositions	Processing Parameters	Properties Investigated	Remarks, Discussion	Literature
Ti-Zr-Nb (57 alloys including six Ti-Nb and six Zr-Nb binary alloys) With special attention to: Ti_{10}-Zr_{40}-Nb_{50} (X-type) Ti_{25}-Zr_{30}-Nb_{45} Ti_{60}-Zr_5-Nb_{35} (Z-type)	Common starting condition: cold rolled 4 to 2 mm$^\phi$, homogenized 5h/1100°C/1x10^{-6}Torr/water quenched: (a) $0 < [Ti] \lesssim 20$ at.%: $\mathrm{CW} \xrightarrow{75\%} 1mm^\phi (3\sim5h/700°C) \xrightarrow{93.75\%} 0.25mm^\phi$; (b) $20 \lesssim [Ti] \lesssim 35$ at.%: $\mathrm{CW} \xrightarrow{75\%} 1mm^\phi (1\sim10h/600°C) \xrightarrow{93.75\%} 0.25mm^\phi$; (c) $35 \lesssim [Ti]$: $\mathrm{CW} \xrightarrow{98.44\%} 0.25mm^\phi (1\sim10h/500°C)$(except $[Nb] \lesssim 30$ at.%, $1\sim10h/400°C$); and for the special alloys: Ti(10at.%): $\mathrm{CW} \xrightarrow{75\%} 1mm^\phi (3h/700°C) \xrightarrow{\mathrm{CW}}_{93.75\%} 0.25mm^\phi$; Ti(25at.%): $\mathrm{CW} \xrightarrow{75\%} 1mm^\phi (10h/600°C) \xrightarrow{\mathrm{CW}}_{93.75\%} 0.25mm^\phi$; Ti(60at.%): $\mathrm{CW} \xrightarrow{98.44\%} 0.25mm^\phi (1h/500°C)$.	A study of the joint effects of composition and processing on critical current density in Ti-Zr-Nb.	The detailed study of Ti_{10}-Zr_{40}-Nb_{50} duplicated that of Reference [Doi68[a]] above in which a complete discussion of precipitation effects in that alloy, and the Z-type alloy Ti_{60}-Zr_5-Nb_{35} is given. It was concluded that high J_c's are to be found in alloys joining Ti-Nb_{33} and Zr-Nb_{53} in the composition triangle. Alloys (1) and (2) above lay on that line which is close to that derived by Rassmann and Illgen [Ras72[b]] as a result of their 1972 investigation. Closer examination of the results suggests the existence of two zones of high J_c, one in the $\beta'+\beta''$ field and the other paralleling the Ti-Nb axis, at about 5 at.% Zr, and centered on $[Nb] \cong 38$ at.%.	T. Doi et al. (1968) [Doi68]
Ti_{10}-Zr_{40}-Nb_{50} (X-type) plus 0.1, 0.3, 0.95 wt.% O	Cold rolled to 2 mm$^\phi$ and heat treated 5h/1100°C; water quenched, in the usual way. Reduced either to 1 mm or 0.25 mm$^\phi$ whereupon O was introduced by heating to 500°C. The O content was analyzed by vacuum fusion (1mm$^\phi$) or gravimetrically (0.25mm$^\phi$). J_c measurements were conducted on material in the conditions: (a) solution treated at 1100°C; (b) solution treated and aged 4 h and 100h/700°C.	The effect of O on precipitation and consequently on critical current density in Ti_{10}-Zr_{40}-Nb_{50} was investigated.	Due to the acceleration of the $\beta \to \beta'+\beta''$ decomposition reaction at 700°C, and consequent refinement of its characteristic lamellar structure, as a result of the addition of O, the flux pinning, hence J_c, was improved. Unfortunately the brittleness of the specimen also increases with addition of O.	M.Kitada and T.Doi (1970) [Kit70[b]]
Ti-$Zr_{2.5}$-$Nb_{34.8}$ (a) Ti-$Zr_{2.5}$-$Nb_{35.0}$ (b) Ti-$Zr_{2.5}$-$Nb_{39.9}$ (c)	From alloys (a) and (b) the following commercial conductors were prepared: <table><tr><td>Alloy</td><td>Wire$^\phi$ mm</td><td>Fil.$^\phi$ μm</td><td>Fil. No.</td><td>Cu/SC Ratio</td></tr><tr><td>(a)</td><td>0.778</td><td>45.1</td><td>127</td><td></td></tr><tr><td>(b)</td><td>2.3</td><td>58.6</td><td>271</td><td>4.6</td></tr><tr><td>(c)</td><td>0.778</td><td>46.8</td><td>127</td><td></td></tr></table>	The critical current densities of three Hitachi conductors (a), (b) and (c), were measured at fields between 8 and about 11 T in an intercomparison with the properties of seventeen different Cu/Ti-Nb commercial composites.	J_c vs H plots were presented corresponding to transition criteria of 10^{-10} and 10^{-11} Ω cm.	E. Tada et al. (1980) [Tad80]

TABLE 1-33 TITANIUM-ZIRCONIUM-NIOBIUM ALLOYS — (c) PROPERTIES OF ROLLED STRIP

Compositions	Thermomechanical Processing ("t" Refers to Thickness)	Current (α) and Field (β) Orientations, Degrees	Remarks, Discussion	Literature
Ti_{10}-Zr_{40}-Nb_{50} (X-type)	$1mm^t(1h/950°C) \longrightarrow 0.2mm^t(1h/1100°C) \longrightarrow 0.1mm^t(HT1)$ HT1 = as-rolled, 1h/570,700,800,1100°C, 3h/700°C $1mm^t(1h/950°C) \longrightarrow 0.2mm^t(HT2) \longrightarrow 0.1mm^t$ HT2 = 1, 3.75, 5, 10h/700°C $\longrightarrow 0.2mm^t(5h/1100°C) \longrightarrow 0.1mm^t(HT3)$ HT3 = 1h/1100°C, also 1h/1100°C + 1h/700°C $\longrightarrow 0.2mm^t(1h/1100°C) \longrightarrow 0.1mm^t(HT4)$ HT4 = as-rolled, also 1h/700°C $1mm^t(1h/950°C) \longrightarrow 0.2mm^t(HT5) \longrightarrow 0.1mm^t$ HT5 = 1, 3, 3.75h/700°C $1mm^t(1h/950°C) \longrightarrow 0.2mm^t(3.75h/700°C) \longrightarrow 0.1mm^t$	$\alpha = 0, \quad \beta = 0$ $\alpha = 0, \quad \beta = 0$ $\alpha = 0, \quad \beta = 0,90$ $\alpha = 0, \quad \beta = 0,90$ $\alpha = 0, \quad \beta = 0,30,45,60,90$ $\alpha = 0, \left\{ \begin{array}{l} 0,15,30,60,90, \quad \beta = 0 \\ 0,15,30,60,90, \quad \beta = 90 \end{array} \right.$	Critical current densities were measured 'n applied fields of up to 80 kOe. With $\alpha = 0°$ and β varying from 0 to 90° the critical-current anisotropy was about 6:1 to 3:1 depending on field, J_c being of course greatest with H_a parallel to the rolling plane ($\beta=0°$). With $\beta = 0°$, J_c was independent of α at all fields, with $\beta = 90°$ (the unfavorable orientation) and α varying from 0 to 90°, J_c decreased by from 40 to 70% depending on the field. Optimal processing included a (1-5h)/700°C intermediate heat treatment. For *strip* processed according to: $1mm^t(1h/950°C) \longrightarrow 0.2mm^t(1h/700°C) \longrightarrow 0.1mm^t$, J_c (4.2K, 60kOe) = $\underline{1.6\times10^5}$ A cm^{-2}. For *wire* processed according to: $2mm^\phi(5h/1100°C) \longrightarrow 1mm^\phi(3h/700°C) \longrightarrow 0.25mm^\phi$, J_c (4.2K, 60kOe) = $\underline{1.9\times10^5}$ A cm^{-2}. Cold deformations in the above rolling experiment were of course: 80% |HT| 50%, while those in the wire drawing were: 75% |HT| 93.75%.	M. Kitada *et al.* (1970) [Kit70]
Ti_{10}-Zr_{40}-Nb_{50} (X-type)	---------------------------	---------------------------	Metallurgical studies of rolled and heat treated microstructures and texture as they relate to critical current flow in an applied magnetic field.	M. Kitada and T. Doi (1970) [Kit70^a]
Ti_{60}-Zr_5-Nb_{35} (Z-type)	$5mm^t(3h/1000°C) \longrightarrow 0.05mm^t(HT)$ $HT = \left\{ \begin{array}{l} 250,360,880h/300°C \\ 5,20,50,75,125,250,400h/350°C \\ 5,25,75,150h/400°C \\ 5,50,115,325h/450°C \\ 3,50,100h/500°C \end{array} \right.$	All Cases $\alpha = 0, \quad \beta = 0,90$	The level of cold work administered prior to final heat treatment is 99% -- considerably higher than that for the X-type β-alloys. Optimal final heat treatments were considered to be 300 or 350°C for at least about 500 or 100 h, respectively. For comparison with the above results: J_c (4.2K, 60kOe, $\beta=0$)(125h/350°C) = $\underline{1.4\times10^5}$ A cm^{-2}.	M. Kitada and T. Doi (1970) [Kit70^f]

Ti-Zr-Nb ALLOYS

TABLE 1-33 TITANIUM-ZIRCONIUM-NIOBIUM ALLOYS — [C] PROPERTIES OF ROLLED STRIP *–continued*

Compositions	Thermomechanical Processing ("t" Refers to Thickness)	Current (α) and Field (β) Orientations, Degrees	Remarks, Discussion	Literature
Ti_{60}-Zr_5-Nb_{35} (Z-type)	Cold rolled and aged, HT = (0-400)h/350°C, (0-250)h/400°C, (0-320)h/450°C	------------------	A study of critical current anisotropy ratio, $J_c(\parallel)/J_c(\perp)$, and a quantity $(a+d)/a$ representing rolling-induced deformation banding, *vs* aging time/temperature.	M. Kitada (1972) [Kit72a]
Ti_{75}-Zr_5-Nb_{20} (Z-type)	5mm^t(3h/1000°C)⟶0.05mm^t(HT), HT = 50,100h/350°C, 1/2,2,5,50h/400°C, 2h/450°C, 1h/500°C		99% cold work was administered prior to final heat treatment. Of the alloys Ti-Zr$_5$-Nb20-40 that with the highest 60 and 80 kOe J$_c$'s was Ti-Zr$_5$-Nb$_{30}$. For that alloy in the form of strip, aged 50h/350°C: J_c (4.2K, 60kOe, β=0) = 1.9×10^5 A cm^{-2}, J_c (4.2K, 80kOe, β=0) = 6.3×10^4 A cm^{-2}, J_c (4.2K, 80kOe, β=90°) = 3.4×10^4 A cm^{-2}.	M. Kitada and T. Doi (1972) [Kit72]
Ti_{70}-Zr_5-Nb_{25} (Z-type)	5mm^t(3h/1000°C)⟶0.05mm^t(HT), HT = 150h/350°C, 25h/400°C, 10h/450°C, 3h/500°C	All Cases $\alpha = 0$, $\beta = 0.90$	The maximal 60 kOe and 80 kOe J$_c$ was plotted *vs* Ti concentration. The curve had a broad maximum covering 60 and 65 at.% Ti.	
Ti_{65}-Zr_5-Nb_{30} (Z-type)	5mm^t(3h/1000°C)⟶0.05mm^t(HT), HT = 25,50h/300°C, 25,50h/350°C, (0-150)h/400°C		Also plotted was $J_c(\parallel)/J_c(\perp)_{min}$ *vs* Ti concentration. Values less than 1.5 occurred for Ti concentrations (at.%Zr=5) greater than about 60 at.%.	
Ti_{55}-Zr_5-Nb_{40} (Z-type)	5mm^t(3h/1000°C)⟶0.05mm^t(HT), HT = 250,400,600,1000h/350°C, 25,400h/400°C, 325h/450°C, 330h/500°C			

Compositions		Processing, Conductor Details	Properties and Procedures	Results, Discussion	Literature
Ti$_{0-15}$-Zr$_{25-50}$-Nb	(X-type)	18mm$^\phi$(annealed)$\overset{cw}{\overline{99.69\%}}$1mm (3h/700°C)$\overset{cw}{\overline{93.75\%}}$0.25mm$^\phi$	Magnetic hysteresis (AC) loss, its relationship to J$_c$ and its dependence on alloy composition, and AC frequency were investigated using bare wires and some multifilamentary composites.	Within the composition ranges considered the AC losses, of both X-type (Zr-Nb-rich) and Z-type (Ti-Nb-rich) alloys, passed through minima:	M. Kudo et al. (1972) [Kud72]
Ti$_{50-75}$-Zr$_{0-15}$-Nb	(Z-type)	18mm$^\phi$(annealed)$\overset{cw}{\overline{99.98\%}}$0.25mm$^\phi$(100h/350°C)		(1) of the X-type alloys Ti$_8$-Zr$_{40}$-Nb$_{52}$ had the lowest loss;	
Ti$_{62.5}$-Zr$_{2.5}$-Nb$_{35}$	(Z-type)	Monofilament Composite: 0.25mm$^\phi$ with 25-125μm^t of Cu	The AC loss was measured using a He boil-off calorimeter. Samples were noninductively wound and exposed to an alternating magnetic field (peak surface, H$_m$ = 1kOe) of frequency 20-500 Hz.	(2) of the Z-type alloys Ti$_{70}$-Zr$_{2.5}$-Nb$_{27.5}$ had the lowest loss;	
		Multifilament Composite: 16-19 filaments, 40-100μm$^\phi$ Cu-SC area ratio: 1.5-3.0:1 Twist pitch: 2-20mm and ∞		(3) the losses incurred were about 30% lower than those of Zr-Nb$_{75}$ and Ti-Nb$_{36}$; (4) The AC loss of the ternary alloys decreased as the (composition-dependent) J$_c$ (zero field) increased; (5) bare superconductor loss per cycle could be expressed as $\hat{Q}_h/f \propto H_m^{3.0-3.5}/J_c(0)^{1.6}$ where H$_m$ is the peak surface field; (6) the total loss of single- and multi-core composites equaled the sum of the linear (SC core) and quadratic (matrix) frequency components.	
Ti$_{60.6}$-Zr$_{2.4}$-Nb$_{37}$	(Z-type)	Multifilament Composite: 37 close-packed (1+6+12+18) filaments Cu/SC area ratio: 2.17:1 Twist pitch: 2,5,20,50mm and ∞	Magnetic hysteresis (AC) loss was measured as function of the electromagnetic coupling between the filaments of a twisted multifilamentary composite conductor. The twist pitch, which was given the values L$_p$ = ∞, 50, 20, 5 and 2 mm, was a measure of the coupling. Loss was measured as function of: (a) H$_m$; (b) twist pitch; and (c) frequency. The AC loss was measured using a He boil-off calorimeter. Samples were noninductively wound and exposed to a transverse sinusoidal magnetic field (H$_m$≲4kOe) of frequency 35-250 Hz.	(1) With a twist pitch L$_p$ = 2 mm the filaments were uncoupled; (2) in the uncoupled case, the AC loss was the sum of the hysteretic losses of the filaments and the eddy current losses of the Cu matrix; (3) when the coupling was weak (L$_p$≲5mm) loss increased with the coupling -- conversely when coupling was strong (L$_p$≳20mm). AC loss thus passed through a maximum.	K. Shiiki et al. (1974) [Shi74]
Ti$_{60.2}$-Zr$_{2.8}$-Nb$_{37}$	(Z-type)	Monofilament of 252μm$^\phi$, bare and with 78μm^t of Cu			
Ti$_{62.5}$-Zr$_{2.5}$-Nb$_{35}$	(Z-type)	Monofilament of about 250μm$^\phi$, bare and plated with 34, 51, 78, 126μm^t of Cu	Alternating transport current (AC) loss in bare and Cu-plated monofilaments was studied in the frequency range 20-500 Hz. The AC loss associated with a sinusoidal 100 A rms transport current was measured using a He boil-off calorimeter.	A relationship of the form $\hat{Q}=\hat{Q}_h+\hat{Q}_e$, was developed, where $\hat{Q}/f$ is the total loss per cycle, af refers to the hysteretic loss in the superconductor, and bf^2 represents loss associated with the Cu cladding. Both a and b were functions of the magnitude of the AC current. As before, $a\equiv\hat{Q}_h/f$ (for the bare wire)$\propto H_m^{3.5}$, $b\equiv\hat{Q}_e/f^2\propto H_m^2 t^2/\rho$, where t and ρ are the thickness and resistivity, respectively, of the copper.	K. Shiiki et al. (1974) [Shi74^c]

Compositions	Processing, Conductor Details	Properties and Procedures	Results, Discussion	Literature
$Ti_{4.7}-Zr_{43.1}-Nb_{52.2}$ (X-type)	$10mm^\phi$(annealed)$\frac{CW}{99.93\%}$261μm^ϕ(2.5h/290°C) $10mm^\phi$(annealed)$\frac{CW}{99.74\%}$508μm^ϕ(100h/350°C) $10mm^\phi$(annealed)$\frac{CW}{99.94\%}$250μm^ϕ(100h/350°C)	*Magnetic hysteresis* (AC) loss was studied with a view to examining an anomalous increase in loss, believed to be due to flux jumping, which occurs at certain values of H_m.	In the plot of $\dot{Q}_h/f$ *vs* field, at particular values of field the loss exhibited a catastrophic increase (anomalous loss) upon the occurrence of a flux jump.	K. Shiiki and M. Kudo (1974) [Shi74[d]]
$Ti_{60.2}-Zr_{2.8}-Nb_{37.0}$ (Z-type)	$10mm^\phi$(annealed)$\frac{CW}{99.99\%}$45μm^ϕ(100h/350°C) $10mm^\phi$(annealed)$\frac{CW}{99.99\%}$101μm^ϕ(100h/350°C) $10mm^\phi$(annealed)$\frac{CW}{99.94\%}$250μm^ϕ(100h/350°C) $10mm^\phi$(annealed)$\frac{CW}{99.74\%}$508μm^ϕ(100h/350°C) $10mm^\phi$(annealed)$\frac{CW}{99.9\%}$319μm^ϕ+34μm^tCu(100h/350°C)	AC loss was measured calorimetrically in alternating transverse fields (H_m<4kOe) of 35, 50 and 100 Hz. Loss per cycle, $\dot{Q}_h/f$, was plotted as function of (a) peak applied field, H_m, and (b) wire diameter.	By reducing the wire diameter, the catastrophic loss increase was able to be deferred to higher and higher fields until complete stability was achieved. In $Ti_{60.2}-Zr_{2.8}-Nb_{37.0}$ the critical diameter necessary for adiabatic stability was about 27 μm. Swept field (118 Oe s^{-1}) and AC field (35Hz) results were compared.	
$Ti_{61.8}-Zr_{2.2}-Nb_{36.0}$ (Z-type)	$10mm^\phi$(annealed)$\frac{CW}{99.94\%}$242μm^t (no HT)			
$Ti_{62.5}-Zr_{2.5}-Nb_{35.0}$ (Z-type)	Four series of multifilamentary composite conductors each containing 37 close-packed filaments about 24 μm^ϕ were investigated. They are describable as follows: (1) SC in copper; Cu/SC area ratio, 2.17:1; (2) SC sheathed in 1-2 μm^t Cu, again in Cu-Ni, and the whole in Cu. Cu/Cu-Ni/SC area ratio, 1.6:0.34:1; (3) the same but with a ratio, 0.51:1.82:1; (4) SC in Cu-Ni; Cu-Ni/SC area ratio, 1.87:1. The configurations of the filaments in the matrices were the same for each composite. Overall wire diameter was 0.25 mm; twist pitches were: 2, 3, 5, 20, 50, ∞ mm.	*Magnetic hysteresis* (AC) loss and *coupling between filaments* was studied as a function of *matrix resistivity* (ρ) and *twist pitch* (L_p). AC loss was measured calorimetrically in alternating transverse field (H_m<4kOe) of 35, 50, 100 and 250 Hz.	(a) Results for Series-(1) conductors were discussed in [Shi74] above. (b) Eddy current losses were given by: $$\dot{Q}_e \propto f^2t^4H_m^2\rho^{-1} \text{ -- high resistivity limit,}$$ $$\dot{Q}_e \propto f^{1/2}tH_m^2\rho^{1/2} \text{ -- low resistivity limit,}$$ where: f = frequency; t = thickness of the matrix of resistivity ρ; H_m = peak value of the magnetic field. (c) With no coupling, AC loss is given by $\dot{Q}$=a+bf so that $\dot{Q}/f$ *vs* f is linear with the slope b=$\dot{Q}_e/f^2$, representing the eddy current loss. In a family of curves for various values of L_p, departure from linearity signified the onset of coupling. (d) When the *matrix resistivity was low* (ρ ≈ 5x10^{-8} Ω cm) the AC loss of the untwisted composite was lower than that of the twisted composite. With a *high resistivity matrix*, the composites with perfect twist had the lowest loss. The critical twist pitch is approximately proportional to ν$^{1/2}$.	K. Shiiki (1974) [Shi74[a]]

TABLE 1-33   TITANIUM-ZIRCONIUM-NIOBIUM ALLOYS — (d) AC EFFECTS IN X-TYPE and Z-TYPE ALLOYS –continued

Compositions			Processing, Conductor Details	Properties and Procedures	Results, Discussion	Literature
Ti	Zr	Nb (at.%)				
9.1	39.4	51.5 (All X-type)		*Magnetic hysteresis* (AC loss) anisotropy was investigated.	The relationship between the critical current density, which depends on pinning force, and hysteresis loss, was considered.	K. Shiiki and K. Aihara (1974) [Shi74b]
7.0	39.1	53.9	$10\text{mm}^{\phi}(2\text{h}/1000°\text{C})\xrightarrow[\sim 99.9\%]{\text{CW}}\begin{cases}\rightarrow 270\mu\text{m}^{\phi}(100\text{h}/350°\text{C})\\\rightarrow 280\mu\text{m}^{\phi}(100\text{h}/350°\text{C})\\\rightarrow 263\mu\text{m}^{\phi}(1\text{h}/500°\text{C})\\\rightarrow 261\mu\text{m}^{\phi}(2.5\text{h}/290°\text{C})\end{cases}$	AC loss was measured calorimetrically in external AC fields <4 kOe at frequencies of 35, 50 and 100 Hz: (1) directed parallel to the wire; (2) directed transverse to the wire; (3) with AC currents ~100 A of frequencies 20, 35, 50, 100, 250 and 500 Hz passing through the wire.	That same microstructural anisotropy which is responsible for J_c (long.) > J_c (trans.) also led to hysteresis anisotropy as the applied field direction was redirected from perpendicular to the wire to parallel to it.	
5.0	44.9	50.1				
4.7	43.1	52.2				
60.2	2.8	37.0 (Z-type)	$10\text{mm}^{\phi}\xrightarrow[99.7\%]{\text{CW}} 508\mu\text{m}^{\phi}(100\text{h}/350°\text{C})$ $10\text{mm}^{\phi}\xrightarrow[99.94\%]{\text{CW}} 250\mu\text{m}^{\phi}(100\text{h}/350°\text{C})$ $10\text{mm}^{\phi}\xrightarrow[99.99\%]{\text{CW}} 101\mu\text{m}^{\phi}(100\text{h}/350°\text{C})$ $10\text{mm}^{\phi}\xrightarrow[99.998\%]{\text{CW}} 44.5\mu\text{m}^{\phi}(100\text{h}/350°\text{C})$			
54.7	2.2	43.1 (Z-type)	$10\text{mm}^{\phi}(2\text{h}/1000°\text{C})\xrightarrow[\sim 99.9\%]{\text{CW}}\begin{cases}\rightarrow 253\mu\text{m}^{\phi}(100\text{h}/350°\text{C})\\\rightarrow 240\mu\text{m}^{\phi}(100\text{h}/350°\text{C})\end{cases}$			
67.6	2.1	30.3				
Ti$_{4.7}$-Zr$_{43.1}$-Nb$_{52.2}$ (X-type)			$10\text{mm}^{\phi}\xrightarrow[99.93\%]{\text{CW}} 261\mu\text{m}^{\phi}(2.5\text{h}/290°\text{C})$	*Critical AC currents*, I_{mc} and I_q were measured in *zero applied field*, where I_{mc} is the peak value of the AC current just prior to a $s \rightarrow n$ transition, and I_q is the instantaneous current at the transition threshold.	This and the subsequent paper in this group [Shi77] *did not deal with AC loss measurement.* The equipment used in both papers was a "wave-memory"-type current-and-voltage recording system.	M. Kudo and K. Shiiki (1976) [Kud76]
Ti$_{60.2}$-Zr$_{2.8}$-Nb$_{37.0}$ (Z-type)			$10\text{mm}^{\phi}\xrightarrow[99.99\%]{\text{CW}} 101\mu\text{m}^{\phi}(100\text{h}/350°\text{C})$ $10\text{mm}^{\phi}\xrightarrow[99.94\%]{\text{CW}} 250\mu\text{m}^{\phi}(100\text{h}/350°\text{C})$	AC currents of frequency 20-500 Hz, swept at rates 0.1-10 A s^{-1} up to more than 700 A, were applied to 3 cm long samples of superconductor. Sample current and voltage were monitored, stored and recorded.	If I_c is the DC critical current, stabilized superconductors exhibit "normal" transitions characterized by $I_c=I_{mc}=I_q$. Two classes of anomalous transitions, distinguished by: (a) $I_c > I_{mc} > I_q$; (b) $I_c >> I_{mc} \sim I_q$; occur as a result of flux jumps.	
Ti$_{60.6}$-Zr$_{2.4}$-Nb$_{37.0}$ (Z-type)			$10\text{mm}^{\phi}\xrightarrow[99.999\%]{\text{CW}} 23\mu\text{m}^{\phi}(100\text{h}/350°\text{C})$			
Ti$_{61.8}$-Zr$_{2.2}$-Nb$_{36.0}$ (Z-type)			$10\text{mm}^{\phi}\xrightarrow[99.94\%]{\text{CW}} 242\mu\text{m}^{\phi}(\text{no heat treat})$ $10\text{mm}^{\phi}\xrightarrow[99.94\%]{\text{CW}} 250\mu\text{m}^{\phi}(100\text{h}/350°\text{C})/\text{A,B}$ A = Cu-clad 34μm^t to 318μm^{ϕ} B = Cu-clad 78μm^t to 406μm^{ϕ}		The following conductors exhibited normal transitions: (a) the bare 23μm^{ϕ} wire for which $I_c=I_{mc}=I_q$; (b) The Cu-clad (78μm^t) monofilament for which $I_c=I_{mc}=I_q$; (c) The Cu-clad (34μm^t) monofilament for which $I_c > I_{mc} \sim I_q$.	
Ti$_{60.2}$-Zr$_{2.8}$-Nb$_{37.0}$ (Z-type)			$18\text{mm}^{\phi}\xrightarrow[99.997\%]{\text{CW}} 101\mu\text{m}^{\phi}(100\text{h}/350°\text{C})$	*Critical direct current* was measured in *longitudinal AC* magnetic fields. DC current (≲200A) swept at rates 10-100 A min^{-1} was applied to a 3 cm long wire exposed to longitudinal magnetic AC fields (H_m<4kOe) of frequencies 35, 50 and 100 Hz.	In DC experiments longitudinal critical currents are much higher than transverse giving rise to the concept of the "force-free configuration" of magnetic flux lines. In longitudinal AC fields the critical currents, which are smaller than those in DC fields, decrease monotonically as the field is raised. The longitudinal AC condition tends to be unstable.	K. Shiiki and M. Kudo (1977) [Shi77]

TABLE 1-33 TITANIUM-ZIRCONIUM-NIOBIUM ALLOYS — (e) THE PATENT LITERATURE

Compositions	Preparation, Procedures, Properties	Remarks, Discussion	Other Systems Studied	(Inventor) Assignee Priority or Filing-Issuance Dates	Listing
26 alloys within the range: $Ti_{0-70}-Zr_{0-50}-Nb_{25-95}$	Cold drawn to wire with intermediate heat treatment. $I_c(4.2K)$ vs $H \gtrsim 100$ kOe for six alloys, also $I_c(4.2K)$ at 50, 70 and 80 kOe, were given for wires 0.25 mm$^\phi$.	The suitability of such wires for small coil fabrication was discussed. $I_c(4.2K)$ vs H was given for 0.25 mm$^\phi$ wires of compositions: $Ti_{40}-Zr_{10}-Nb_{50}$. $Ti_{30}-Nb_{20}-Nb_{50}$. $Ti_{20}-Zr_{30}-Nb_{50}$. $Ti_{10}-Zr_{40}-Nb_{50}$.	Ti-Nb Zr-Nb	Hitachi Ltd., Japan 1963-1965	French Patent No. 1,410,055 September, 1965 [Hit65]
Alloys within the range: $Ti_{1-79}-Zr_{1-79}-Nb_{20-63}$	Alloys were electron beam melted and cast. Ingots were reduced to 2.0 mm$^\phi$ and homogenized 5h/1100°C; cold drawn to 1.0 mm$^\phi$ (75%), aged 1h/600°C. Critical temperature data available. I_c (4.2K, 0.25mm$^\phi$) data were presented for fields of 50, 70, 80 kOe.	This and the above patent possess common 1963 origin. Some refinement of the composition range was specified. Also considered were: (a) workability, ductile-brittle transition; (b) influence of tramp elements; (c) insulation and coatings; (d) small coils and effects such as degradation and training. Suggested applications were superconducting magnets for electron microscopes or physical research; saddle magnets in MHD, magnets for particle accelerators, and motors for submarine propulsion.	Ti-Nb Zr-Nb	(T. Doi *et al.*) Hitachi Ltd., Japan 1963-1968	U.S. Patent No. 3,408,604 October, 1968 [Doi68[b]]
(5-80)Ti-(5-80)Zr-(≲25)Nb (5-80)Ti-(5-80)Zr-(≳20)Nb (5-80)Ti-(5-80)Zr-(≲15)Nb (5-50)Ti-(20-60)Zr-(≲25)Nb (5-50)Ti-(20-60)Zr-(≳20)Nb (5-50)Ti-(20-60)Zr-(≲15)Nb	Ingots were forged at 1000°C and drawn to filaments 0.25 mm$^\phi$ without intermediate heat treatment. No numerical data were made available.	A preferred composition was 15Ti-≳25Zr-25Nb. It was recommended that at least 5 wt.% Ti, and preferably 15 wt.%, should be present in Ti-Zr-Nb in the interests of cold workability.	Ti-Zr-Nb-Ta Ti-Zr-Nb-Hf	Imperial Metal Industries Ltd., Great Britain 1964-1965	Belgian Patent No. 659,033 July, 1965 [Imp65]
Ti-Zr-Nb		Some very general statements were made with regard to materials, wire and magnets. No data were offered.	Ti-Nb Zr-Nb Quaternary alloys	(P.F. Chester) Central Electricity Generating Board, Great Britain 1964-1966	French Patent No. 1,452,977 September, 1966 [Che65]

TABLE 1-33 TITANIUM-ZIRCONIUM-NIOBIUM ALLOYS — (e) THE PATENT LITERATURE *-continued*

Compositions		Preparation, Procedures, Properties	Remarks, Discussion	Other Systems Studied	(Inventor) Assignee Priority or Filing-Issuance Dates	Listing
Ti-7Zr-63Nb			Cladding in stainless steel was mentioned.	Ti-Nb Ti-Nb-Ta	Compagnie Francaise Thomson-Houston, Paris, France 1965-1967	Netherlands Patent No. 6,614,471 April, 1967 [Com67a]
Ti$_{10}$-Zr$_{40}$-Nb$_{50}$	(X-type)	Superconducting wire was coated with stabilizer by passage through a bath of molten (a) Sn, (b) Al, or (c) Cd. Critical current densities in fields of up to about 90 kOe were intercompared.	Results were presented for wires that had been aged and coated according to the following prescriptions: (a) cold drawn; (b) 1h/500°C plus coated with Al at 700°C; (c) 1h/550°C plus coated with Cd at 350°C; (d) 1h/500°C plus coated with Sn at 300°C; (e) 1h/500°C or 550°C plus imbedded in Cu by rolling plus final aging 10h/500°C.	Ti-Nb Zr-Nb Nb$_3$Sn, Nb$_3$Al	(T.Doi and M.Kudo) Hitachi Ltd., Japan 1967-1973	U.S. Patent No. 3,710,844 January, 1973 [Doi73]
Ti$_{60}$-Zr$_5$-Nb$_{35}$	(Z-type)					

TABLE 1-34 TITANIUM-HAFNIUM-NIOBIUM ALLOYS

Compositions	Properties Investigated	Remarks, Discussion	Other Systems Studied	Literature
$Ti_{64}-Hf_4-Nb_{32}$	Swaged and drawn from cast ingot.	H_u determined magnetically in a study of H_{c2} vs effective electron concentration.	Numerous ternary and quaternary Ti-Nb-base alloys	W. DeSorbo et al. (1967) [Des67]
$Ti_{65}-Hf_{10}-Nb_{25}$ $Ti_{50}-Hf_{25}-Nb_{25}$	Homogenized at 1500°C in vacuum furnace; annealed 200h/1100°C/water quenched; 99.9% cold deformed. T_c determined (presumably magnetically) on bundles (>3 mm†) of 7 mm long wires. J_c measured at 4.2 K vs H ≈ 26 kOe.	Normal state resistivities, tensile strengths and lattice parameters were also investigated.	Ti-Nb Ti-Nb-Mo Ti-Nb-Re	M.I. Bychkova et al. (1967/1970) [Byc70], see also E.M. Savitskii et al. (1973) [Sav73, p.325]

Ti	Hf	Nb	(at.%)
bal.	6,18,24	70	
bal.	20,40	50	
bal.	14,56	30	

T_c data provided.

Comprehensive review of superconductive materials.

E.M. Savitskii et al. (1976) [Sav76, p.174]

Ti	Hf	Nb	(at.%)
60	2.5	37.5	
62.5	2.5	35	
65	2.5	32.5	
67.5	2.5	30	
70	2.5	27.5	
52.5	5	42.5	
57.5	5	37.5	
60	5	35	
62.5	5	32.5	
65	5	30	
67.5	5	27.5	
55	10	35	
57.5	10	32.5	
60	10	30	
65	10	25	

Arc melted repeatedly in gettered argon, homogenized 8h/1350°C; cold rolled >70% and recrystallized 1h/875°C. Swaged to 3 mm$^\downarrow$, sheathed in Cu and drawn to 1 mm‡ (core ~0.5 mm$^\downarrow$). T_c measured resistometrically; H_r measured at 1.7, 2, 3, 4.2 K and other temperatures with current of 10 mA ($J ≡ 5$ A cm^{-2}).

The work was partly motivated by the need to move H_r closer to H_{c2}^*, the GLAG limit by reducing the paramagnetic susceptibility difference between the s and n states. Thus the purpose of the heavy-element addition to Ti-Nb was to spin-decouple the Cooper pairs (thus raising the mixed-state susceptibility and bringing it closer to the normal-state value) by means of high-Z electron scattering, following the speculation of Neuringer, Shapira [Neu66] and others.

Ti-Nb Ti-Nb-Ta Ti-Hf-Nb-Ta

D.G. Hawksworth and D.C. Larbalestier (1980) [Haw80]

see also

D.G. Hawksworth and D.C. Larbalestier (1981) [Haw81]

TABLE 1-34 TITANIUM-HAFNIUM-NIOBIUM ALLOYS –continued

Compositions	Properties Investigated	Remarks, Discussion	Other Systems Studied	Literature
$Ti-Hf_{0.15}-Nb_{25}$ $Ti-Hf_{0.15}-Nb_{30}$ $Ti-Hf_{0.4}-Nb_{35}$ $Ti-Hf_{0.3}-Nb_{37.5}$ $Ti-Hf_{0.15}-Nb_{40}$ $Ti-Hf_{0.4}-Nb_{45}$ $Ti-Hf_{0.15}-Nb_{50}$ $Ti-Hf_{0.15}-Nb_{60}$	Argon-arc melted, homogenized at 1200°C to remove coring segregation. Materials for *preliminary studies* were cold rolled 80% to 0.2 mm thickness and aged in argon 24h/(300-800°C). Several high-Hf alloys received 2-stage processing: 24h/800°C + 80% cold work + 24h/400°C. T_c was measured resistively; H_r (4.2K) was measured at a current density of 1 A cm^{-2} in 13 and 17 T (170kOe) superconducting magnets; J_c (4.2K) was measured mainly on the high-H_r alloys in the 17 T magnet.	Based on the preliminary results the alloys $Ti-Nb_{40}$ (control) and $Ti-Nb_{40}-Hf_3$ were selected for fabrication into single-core Cu/SC composites (total reduction, 99%) and seven-core Cu/SC composites (total reduction, 99.94%). J_c was measured as function of thermomechanical processing and the results discussed. It was concluded that T_c, H_r and J_c for the ternary alloy were slightly higher than those of the binary control; and that the results were sufficiently promising to warrant further study and attempts at process optimization.	-----	H. Wada *et al.* (1980) [Wad81]

TABLE 1-35 TITANIUM-NIOBIUM-VANADIUM ALLOYS

Compositions (at.%)			Properties Investigated	Remarks, Discussion	Other Systems Studied	Literature
Ti	**Nb**	**V**				
20	50	30	Arc melted samples were homogenized 24h/1000°C, and reduced 15:1 (93%) by cold rolling. Samples were exposed to a flux of 3.7×10^{19} neutrons cm^{-2}. T_c, H_{c2} and J_c measurements were made before and after irradiation.	A degradation occurred in all samples as a result of irradiation: H_{c2} was reduced 8-15%; T_c and J_c decreased.	-----	J.T.A. Pollock et al. (1969) [Po£69]
30	50	20				
40	40	20				
40	10	50				
Ti-Nb-V (some 35 ternary alloys)			Arc melted samples (cored) were homogenized 24h/1600°C at 10^{-6} Torr and furnace cooled to room temperature (1h) and cold rolled 15:1 to sheet. T_c was measured magnetically; H_r measurements were associated with the J_c investigation.	T_c results expressed in formats: T_c (V =10,20,30 and 40 at.%) vs e/a. T_c (Nb=10,20,30 and 40 at.%) vs e/a. H_r results expressed in a similar format. Some interesting systematics are noticeable.	-----	P.H. Bellin et al. (1969) [Be£69] see also [Be£70]
Ti	**Nb**	**V**				
50	--	50	Arc melted, cold worked 50%, homogenized 3h/1400°C at 10^{-5} to 10^{-6} Torr, and quenched. All structures bcc.	Ti-Nb-V$_x$-Ta were used as subjects for a T_c vs composition multiple regression analysis.	Ti-Nb-Ta [Sav71]	E.M. Savitskii et al. (1971) [Sav71] see also [Sav73, p.326]
49	1	50				
45	5	50				
40	10	50				
75	25	--				
70	25	5				
65	25	10				
50	25	25				
Ti	**Nb**	**V**				
75	20	5	J_c measurements were made at 50 kOe on cast plus cold-deformed (ca 99.9%) 0.25 mm^2 wire. An intercomparison in graphical form was presented between J_c vs Nb content (20, 30, 40 at.%) of the binary Ti-Nb alloys and the ternary alloys in the cold-worked and in the cw plus 1h/500°C heat-treated conditions.	-------------------------	Ti-Nb-Ta,-Mo; Ti-Nb-Cu,-Si; Ti-Nb; Ti-Nb-all elements (summary)	G. Rassmann and L. Illgen (1973) [Ras73]
65	30	5				
55	40	5				
75	15	10				
61	23	16				
72	11	17				

TABLE 1-36 TITANIUM-NIOBIUM-TANTALUM ALLOYS

Compositions Ti	Nb	Ta (at.%)[†]	Properties and Procedures	Remarks, Discussion	Literature
75	25	0	T_c was measured magnetically on alloys in the as-cast state, and after homogenization 4h/1500°C.	In constant-(50 wt.%) Nb alloys, T_c passed through a maximum with increasing Ta content. From the standpoint of substituting Ta for Nb, with which it is isoelectronic, T_c always decreased -- see [Sue68] and [Sue69] below. In the most recent paper [Sav71] multiple linear regression analysis was applied to define a relationship between T_c and composition (see Ti-Nb-V, above).	I.I. Baranov et al. (1967/1970) [Bar70], see also E.M. Savitskii et al. (1973) [Sav73, p.329] and E.M. Savitskii et al. (1971) [Sav71]
74	25	1			
70	25	5			
65	25	10			
50	25	25			
75	24	1			
75	20	5			
75	15	10			
75	0	25			

[†]Nominal -- see literature for analyzed compositions

Compositions	Properties and Procedures	Remarks, Discussion	Literature
Ti$_{60}$-Nb$_{36}$-Ta$_4$	See Ti-Hf-Nb, Table 1-34.		W. DeSorbo et al. (1967) [Des67]

Compositions Ti	Nb	Ta (at.%)	Properties and Procedures	Remarks, Discussion	Literature
70	bal.	0 A series	Arc melted samples were swaged and remelted; swaged or drawn to 9.00, 3.96, or 1.95 mm²; fully recrystallized at 900°C; cold drawn 98.35%, 99.60% or 99.93%, respectively, to 0.25 mm². Properties measured included Vickers hardness, T_c and J_c (99.93% cw) vs: (a) composition for 1h/400°C-aged alloys, (b) 1h-annealing temperature up to 600°C, (c) annealing time to 80 or 40 h at 400 or 500°C, respectively. In addition, J_c vs H (≲55kOe) for (i) fully annealed, (ii) 98.35% cw, (iii) 99.60% cw, and (iv) 99.93% cw alloys were intercompared.	The J_c (4.2K,40kOe) of Ti-Nb$_{25}$-Ta$_5$ vs 1h-aging temperature maximized at the usual 400°C. With the A-series alloys J_c peaked at 5 ∿ 10 at.% Ta. With the B-series J_c increased steeply and monotonically with increasing Ti content, becoming largest for Ti-Nb$_{20}$-Ta$_5$ (the last alloy of the sequence) at higher Ti levels than which the alloys were too brittle to be cold drawn (cf. the usual problems with ω-phase).	Y. Hashimoto et al. (1968) [Has68]
70	bal.	2.5			
70	bal.	5			
70	bal.	10			
70	bal.	15			
70	bal.	25			
70	bal.	30			
50	45	5 B series			
60	35	5			
65	30	5			
75	20	5			

Compositions	Properties and Procedures	Remarks, Discussion	Literature
Ti-Nb-Ta (27 ternary alloys)	Arc melted alloys, homogenized 2h/1200°C <10⁻⁵ torr, cold rolled typically 90% to 0.030 in. T_c was measured magnetically; H_u and J_c were measured in pulsed field (10ms to 130kOe) at 4.2 K. Prior to J_c measurement some heat treatment, e.g. 1,4,10,40h/500°C was administered.	It was anticipated that the addition of Ta, isoelectronic with Nb but with a higher electronic number, would increase H_{c2} as a result of spin-orbit-scattering-induced increase of superconducting state paramagnetism (thus offsetting Pauli limitation).	M. Suenaga and K.M. Ralls (1968) [Sue68], see also M. Suenaga and K.M. Ralls (1969) [Sue69]

TABLE 1-36 TITANIUM-NIOBIUM-TANTALUM ALLOYS *–continued*

Compositions			Properties and Procedures	Remarks, Discussion	Literature
Ti	Nb	Ta (at.%)			
bal.	20	0	See Ti-Nb-V, Table 1-35	Other systems investigated were: Ti-Nb	G. Rassmann and
bal.	30	0		Ti-Nb-V,-Mo	L. Illgen
bal.	40	0		Ti-Nb-Cu,-Si	(1973)
				Ti-Nb-all elements	[Ras73]
bal.	19	5		(summary).	
bal.	25	10			
bal.	33	10			
bal.	40	10			
Ti	Nb	Ta (wt.%)			
40	bal.	0	Repeatedly arc melted, homogenized 3h/1500°C. All cooled alloys bcc. Wire (0.25mm^2) and ribbon (0.1x1 mm^2) prepared by 99.9% cold drawing or rolling with no intermediate anneal. Vacuum annealed 3h/300,350,375,400,450,500,600°C and quenched.	Suenaga and Ralls (see above) had noted the high critical fields obtainable through the addition of Ta to Ti-Nb alloys (in fact it was claimed that they have the highest H_{c2} of all ductile super- conductors) and had measured J_c in unoptimized material. In what is, therefore, a sequel to that work Bychkov *et al.* have measured the J's of partially optimized alloys. J_c (4.2K,50kOe) ≈ 2x10^4 A cm^{-2}.	Yu.F. Bychkov *et al.* (1974) [Byc74]
40	bal.	10			
40	bal.	20			
40	bal.	40			
46.5Ti-Nb-(0,0.5,1,2)Ta			During fabrication the alloy rods were annealed at 0.56 in.‡ and cold worked to 0.117 in.‡, Cu-clad and further cold re- duced 99.7% to 0.045 in.‡ or 99.9% to 0.018 in.‡; all alloys were then heat treated 24h/380°C, cold reduced 95% and final heat treated 2h/300°C. Final core diameters were 0.010 in. (0.25mm) and 0.004 in. (0.10mm), respectively. One of the alloys (2 wt.% Ta) did not receive the anneal at 0.56 in.‡, instead it was "extra cold worked" from the recryst- allization + quench accorded all alloys at 1.5 in.‡. J_c (4.2K) was measured at fields of up to 75 kOe (7.5T) by three independent laboratories.	The purpose of the study was to examine the effect on the electrical and mechanical properties of the wire of the presence of low levels of Ta in the starting Nb. It was concluded: (a) that for equi- valent cold work the Ta-bearing alloys were stronger than the Ti-Nb reference, while retaining excellent ductility, (b) that levels of up to 2 wt.% Ta in 46.5Ti-Nb had no adverse influence on J_c (30 < H < 75 kOe).	C.W. Curtis and W.K. McDonald (1979) [Cur79]
46.5Ti-Nb-(0,0.5,1,2)Ta			Cu-clad wire samples of 0.01 cm‡ (0.004 in.) and 0.025 cm‡ (0.010 in.) prepared by Curtis and McDonald (as described in [Cur79] above) were tested at 4.2 K and 1.9 K in fields of 9 and 10 T, continuing on to 11 through 13.5 T at 1.9 K.	The ternary alloys frequently performed better than the Ti-Nb control alloy. As is generally noted for Ti-Nb-type alloys J_c improved with the extent of the cold work. With regard to the Ta (2 wt.%) alloy, since J_c (13T) increased in response to "extra cold work" it was clear that the process was not yet optimized.	Z.J.J. Stekly *et al.* (1978) [Ste78], see also H.R. Segal *et al.* (1980) [Seg80]

TABLE 1-36 TITANIUM-NIOBIUM-TANTALUM ALLOYS –continued

Compositions (Ti / Nb / Ta, at.%)	Properties and Procedures	Remarks, Discussion	Literature
57.5 bal. 5 62.5 bal. 5 65 bal. 5 67.5 bal. 5 70 bal. 5 57.5 bal. 15 60 bal. 15 62.5 bal. 15 65 bal. 15 67.5 bal. 15 57.5 bal. 10 60 bal. 10 62.5 bal. 10 65 bal. 10 67.5 bal. 10 70 bal. 10	See Ti-Hf-Nb, Table 1-34.	In this re-examination of the results of Suenaga and Ralls (see above) the rather high values of H_r reported by those authors were not confirmed. Other systems investigated were: Ti-Nb, Ti-Hf-Nb, Ti-Hf-Nb-Ta.	D.G. Hawksworth and D.C. Larbalestier (1980) [Haw80], see also D.C. Larbalestier (1980) [Lar80]
Ti-48Nb-8Ta (i.e. Ti-Nb$_{35}$-Ta$_3$) Ti-32Nb-25Ta (i.e. Ti-Nb$_{25}$-Ta$_{10}$)	In the first series of tests 10-cm diameter billets consisting of 300 Ti-Nb-Ta elements, with a Cu/SC-ratio of 2.3:1, were reduced to wire of 0.25 mm². J_c (1 μV cm⁻¹ criterion) at 4.2 and 1.9 K was measured in fields of up to 13 T. In a second series of tests the influence of cold reduction was investigated by reducing billets of both alloys to wires of diameter 0.76 to 0.25 mm. For both alloys J_c was measured vs T (2-4K) at fields of 5, 7, 10, 11, 12 and 13 T.	Studies of 0.25 mm² wires showed: (a) that at 10 T both alloys had the same J_c; (b) that at 12 T and 2 K the 25 wt.% Ta alloy carried 10% more current than the 8 wt.% alloy; and (c) that under those same conditions the J_c's of the Ti-Nb-Ta alloys are about "75% higher than that of Ti-Nb". In [Seg80] it was noted that if, for example, 5×10^4 A cm⁻² in a peak field of 12 T is needed, then the maximum operating temperature is 1.8 K if Ti-50Nb is used and 2.5 K with Ti-Nb-25Ta. Thus the higher fixed-temperature operating field of the Ta-bearing alloy translates into a higher operating temperature at a given design field. It was concluded that all the alloys tested (Ti-Nb, Ti-Nb-low Ta and Ti-Nb-25Ta) could be used in a 12 T, 2 K conductor, and that the high-Ta alloy had the best performance.	H.R. Segal et al. (1981) [Seg81], see also H.R. Segal et al. (1980) [Seg80]
(Ti-53.5Nb)-0.5Ta (")-1.0Ta (")-1.5Ta Ti-48Nb-8Ta Ti-32Nb-25Ta	The high-field J_c's of commercially produced multifilamentary composites based on Ti-Nb-Ta were measured, as were composites based on Ti-45Nb and Ti-53.5Nb, and their relative performances were discussed. Data were presented enabling a direct comparison between the properties of Ti-53.5Nb (actually with 0.5 wt.% Ta, which had a null effect on J_c), Ti-48Nb-8Ta, and Ti-32Nb-25Ta to be made.	The alloys with large amounts of Ta had significantly higher current densities than did the binary alloys at the high fields. For example, the J_c of Ti-32Nb-25Ta is up to 75% greater than that of the binary alloys at 12 T and 2 K. Moreover, the useful range of the current densities can be extended to at least 13 T for some of the ternary alloys.	H.R. Segal et al. (1981) [Seg81a] Cf. preceding literature.

TABLE 1-36 TITANIUM-NIOBIUM-TANTALUM ALLOYS –*continued*

Compositions			Properties and Procedures	Remarks, Discussion	Priority or Filing and Issuance Dates	(Inventor) Assignee	Listing
Ti	Nb	Ta	(wt.%)				
20	25	55	Electron beam melted, cold deformed 99% and annealed at 500 and 900°C. T_c and J_c were measured, the latter at 4.2 K in a field of 50 kOe.	In the interests of workability it was found advantageous to replace the Zr in Ti-Nb-Zr alloys with Ta. Such a replacement was found to result in improved superconducting properties.	1964-1967	(R.R. Reinbach) Vacuumschmelze GmbH, Hanau, BRD	German Patent No. 1,237,786 (March, 1967) [Re167]
25	45	30					
30	60	10					
Ti-11Nb-25Ta			No properties presented.	A process for producing magnet wire was claimed in which various alloys were swaged, clad in annealed Cu, or other metals specified, reduced to wire with intermediate annealing and coated with an organic insulator. Other alloys mentioned were: Ti-64Nb Ti-Zr-Nb.	1965-1967	Compagnie Francaise Thomson-Houston, Paris, France	Netherlands Patent No. 6,614,471 (April, 1967) [Com67ª]
Ti$_{20-80}$-Nb$_{2-80}$-Ta$_{1-80}$			Arc melted, forged and cold drawn 99% to a filament 0.25 mm². heat treated 1h/400°C. J_c (4.2K) measured *vs* H ≈ 58 kOe.	The ternary alloys Ti-Nb-Ta were claimed to have several advantages over the then available Ti-Nb alloys. Performance was compared with those of Ti-Nb and Zr-Nb.	1966-1968 (Fr) 1969-1972 (US)	Mitsubishi Electric Corporation, Japan	French Patent No. 1,512,769 (February, 1968) also issued as U.S. Patent No. 3,671,226 (June, 1972) [Mit68]

TABLE 1-37 TITANIUM-NIOBIUM-(GROUPS VI, VII, and VIII) TRANSITION-METAL-TERNARY ALLOYS

Compositions	Properties and Procedures	Remarks, Discussion	Other Systems Studied	Literature
CLASS: Ti-Nb – Group VI				
$Ti_{65}-Nb_{33}-Cr_2$	H_r measured resistively on samples deformed 97% to wire (1mm[ϕ]) and annealed 10h/400°C.	The paper deals principally with upper critical fields in Ti-Nb-Cu,-Ge ternary alloys.	Ti-Nb Ternary alloys formed with 2 at.% of: Sn, Cu, Ge, Ga, In, Ag, Al, also Mn, Fe and Zr in Ti-Nb_{33}.	W. Heller *et al.* (1971) [Hel71[a]]
$(Ti-Nb_{75})-Cr_{2,6,10}$ $(Ti-Nb_{50})-Cr_{2,6,10}$ $(Ti-Nb_{30})-Cr_{2,6,10}$	Arc melted samples were homogenized 15min/1200°C, cold rolled to 0.5 mm thickness with intermediate 15min/1100°C anneal. Final anneal 5h/1000°C, cold rolled 50%, aged 24h/up to 300°C. T_c was measured by the four-probe method and followed as function of 24 h aging temperature (0-800°C). H_r was determined resistometrically at 4.2 K on as-rolled and optimally aged alloys and plotted *vs* solute concentration.	In the plot of J_c (4.2K,8T) *vs* 24 h aging temperature (0-800°C) the usual 400°C J_c-peak was noted. The addition of Cr reduced T_c and H_r; but when added to Ti-Nb_{30}, improved the J_c. Process optimization was not carried out.	Ti-Nb-Al Ti-Nb-Cu Ti-Nb-Mo	H. Wada *et al.* (1980) [Wad80]
$Ti_{65}-Nb_{34}-Mo_1$	The H_u's of cold worked wire samples were measured using a pulse-field magnetization technique.	Part of a systematic study of H_{c2} *vs* electron concentration.	Numerous Ti-Nb-base ternary and quaternary alloys	W. DeSorbo *et al.* (1967) [Des67]
Ti Nb Mo (at.%)[†] 74.3 25.4 0.44 69.6 25.4 5.0 63.6 27.77 8.8 46.3 27.7 26.1 † By chemical analysis	Homogenized at 1500°C in vacuum furnace; annealed 200h/1100°C/water quenched; 99.9% cold deformed. T_c determined (presumably magnetically) on bundles (<3 mm[ϕ]) of 7 mm long wires, J_c measured at 4.2 K *vs* H ≃ 26 kOe.	Normal state resistivities, tensile strengths and lattice parameters were also investigated. All the Ti-Nb-Mo alloys were single phase in the cold-worked (99.9%) state.	Ti-Nb Ti-Nb-Hf Ti-Nb-Re	M.I. Bychkova *et al.* (1967/1970) [Byc70[a]], see also E.M. Savitskii *et al.* (1973) [Sav73, p.331]

TABLE 1-37 TITANIUM-NIOBIUM-(GROUPS VI, VII, and VIII) TRANSITION-METAL-TERNARY ALLOYS –continued

Compositions (at.%) Ti	Nb	Mo	Properties and Procedures	Remarks, Discussion	Other Systems Studied	Literature
47.5	47.5	5	T_c was determined using the self-inductive technique and H_r measured by the standard four-probe method at 4.2 K on ribbon samples cold rolled from annealed (24h/1300°C) arc-melted buttons.	J_c measurements were also undertaken although the results were not reported. The (conventional) electron/atom ratio range of the alloys studied was $4.40 < e/a < 4.80$.	-----	V. Sadagopan et al. (1970) [Sad70]
55.0	40.0	5				
64.7	30.3	5				
45.0	45.0	10				
60.0	30.0	10				
42.5	42.5	15				
65.0	20.0	15				
40.0	40.0	20				
74	21	5	J_c measurements were made at 37.5 kOe on cast plus cold-deformed (ca 99.9%) 0.25 mm² wire. A comparison was made between the as-deformed and 1h/500°C annealed wire, and also between the ternary alloys and Ti-Nb (20,30,40 at.%) samples prepared under the same conditions.	Hardness measurements and metallography were carried out; the cast structures were homogeneous β-phase excepting for the three most-Ti-rich alloys which exhibited some deformation martensite.	Ti-Nb-V,-Ta Ti-Nb-Cu,-Si Ti-Nb-all elements (summary)	G. Rassmann and L. Illgen (1973) [Ras73]
66	29	5				
54	41	5				
80	10	10				
74	16	10				
65	25	10				
93	4	3	T_c was measured magnetically on samples prepared by levitation melting. Metallurgical conditions included: (a) quenched, (b) quenched plus aged at 350°C, (c) quenched plus ω-reverted at 500°C; (d) slow-cooled, (e) slow-cooled plus aged at 750, 650 or 550°C.	T_c was investigated as a function of conventional electron/atom ratio.[+] The addition of Mo to Ti-Nb depressed T_c well below that of the equivalent-e/a binary Ti-Nb alloy. [+] $e/a = 4 + 1 \cdot [\mathrm{Nb}] + 2 \cdot [\mathrm{Mo}]$, where $[N]$ represents the atom fraction of element N.	Ti-Nb-Al Ti-Nb-Mo-Al	W.L. Cotton et al. (1974) [Cot74] see also [Cot73] and V. Chandrasekaran (1973) [Cha73]
86	8	6				
82.5	13	4.5				
66	19	15				
51	28	21				
37	36	27				
bal.	2-62	9 (7 alloys)				
bal.	12	14-39 (6 alloys)				
(Ti-Nb75)-Mo5,10			See Ti-Nb-Cr above	Both T_c and H_r were reduced by the addition of Mo. The 15 at.% Mo alloys missing from the list were not able to be cold rolled suggesting the presence of brittle phases. $J_c(H)$ measurements, which were confined to the alloys with favorable H_r's, were not carried out on these alloys.	Ti-Nb-Al Ti-Nb-Cu Ti-Nb-Cr	H. Wada et al. (1980) [Wad80]
(Ti-Nb50)-Mo5,10						
(Ti-Nb30)-Mo5,10,15						

TABLE 1-37 TITANIUM-NIOBIUM-(GROUPS VI, VII, and VIII) TRANSITION-METAL-TERNARY ALLOYS -continued

Compositions	Properties and Procedures	Remarks, Discussion	Other Systems Studied	Literature
$Ti_{64}-Nb_{35}-W_1$	See Ti-Nb-Mo$_1$ above	See Ti-Nb-Mo$_1$ above	See Ti-Nb-Mo$_1$ above	W. DeSorbo et al. (1967) [Des67]
$Ti-(42,74,85,88)Nb-10W$ $Ti-(24,47,60,78)Nb-20W$ $Ti-(40,41,62)Nb-30W$	Superconducting transition temperature was measured on alloys prepared by arc melting, annealed at 1000°C, and, presumably quenched.	The transition temperature results were presented in the format: T_c vs wt.% Ti at constant [W] = 10, 20 and 30 wt.%, respectively. In the alloys referred to here: $T_{c,max}$ = 7.5 K for Ti-(42-85)Nb-10W $T_{c,min}$ = 4.5 K for Ti-24Nb-20W.	Comprehensive survey	E.M. Savitskii et al. (1973) [Sav73, p.333]
CLASS: Ti-Nb - Group VII				
$Ti_{65}-Nb_{33}-Mn_2$	See Ti-Nb-Cr$_2$ above	See Ti-Nb-Cr$_2$ above	See Ti-Nb-Cr$_2$ above	W. Heller et al. (1971) [Hel71a]

Ti	Nb	Re	(at.%)				
74	25	1		A study of the T_c and J_c of homogenized, quenched and 99.9% cold deformed wires. Electrical, mechanical and microstructural properties were also investigated.	Although the Ti-Nb-Re alloys seemed to be single phase when homogenized and quenched a second hexagonal phase appeared during cold deformation.	Ti-Nb Ti-Nb-Hf Ti-Nb-Mo	M.I. Bychkova et al. (1967/1970) [Byc70a]
70	25	5					
65	25	10					
50	25	25					

CLASS: Ti-Nb - Group VIII

Ti	Nb	Fe	(at.%)				
72	25	3		The transition temperatures of hot-rolled, annealed 1h/900°C and water quenched 1.75 mm$^\phi$ wires were measured as function of aging time >10^4 min at 400 and 500°C.		The electrical and metallurgical properties were studied.	Ti-Nb Ti-Fe	T. Nishimura and U. Zwicker (1968) [Nis68]
86	10	4					
86.9	4.9	8.2					
33.3	33.3	33.3					

$Ti_{19}-Nb_{80}-Fe_1$	See Ti-Nb-Mo$_1$ above	See Ti-Nb-Mo$_1$ above	See Ti-Nb-Mo$_1$ above	W. DeSorbo et al. (1967) [Des67]

Compositions	Properties and Procedures	Remarks, Discussion	Other Systems Studied	Literature
Ti_{57}-Nb_{38}-Fe_5	The critical current density was measured on 97.7% cold rolled strip, in the as-rolled condition and after being annealed 10h/350-800°C.	J_c was measured perpendicular to the rolling plane -- the least favorable orientation. The influence of thermo-mechanical-processing-induced structural change on J_c and the Mössbauer spectra were discussed.	-----	A.K. Prokoshkin and I.M. Puzei (1970) [Pro70]
Ti_{65}-Nb_{33}-Fe_2	See Ti-Nb-Cr_2 above	See Ti-Nb-Cr_2 above	See Ti-Nb-Cr_2 above	W. Heller *et al.* (1971) [Hel71a]
$Ti_{87.5}$-Nb_{10}-$TM_{2.5}$ $TM = \begin{cases} Ru, Rh, Pd \\ Os, Ir, Pt \end{cases}$	Superconducting transition temperature was measured both inductively and resistively on water-quenched samples.	Electrical and metallurgical properties were studied.	Ti-Nb	U. Zwicker (1965) [Zwi65]
Ti_{65}-Nb_{33}-Ni_2	See Ti-Nb-Cr_2 above	See Ti-Nb-Cr_2 above	See Ti-Nb-Cr_2 above	W. Heller *et al.* (1971) [Hel71a]

TABLE 1-38 TITANIUM-NIOBIUM-BASE QUATERNARY ALLOYS

Compositions	Preparation, Procedures, Properties	Remarks, Discussion	Other Systems Studied	Literature
ALLOY CLASS: Ti-Nb-SM$_1$-SM$_2$				
Ti-Nb$_{20}$-Al$_1$-Sb$_1$ Ti-Nb$_{20}$-Ga$_1$-Sb$_1$ Ti-Nb$_{20}$-In$_1$-Sb$_1$ Ti-Nb$_{10}$-Sn$_{2.5}$-U$_5$	Cold deformed from 6 mm$^\phi$ cast cylinders to 1.75 mm or 1 mm$^\phi$ (91.5 or 97.2%, respectively) wires. T_c measured magnetically on cast-plus-deformed 1.75 mm$^\phi$ wire using a pulse technique. J_c measured on cast-plus-deformed 1 mm$^\phi$ wire. Measurement conditions: 4.2 K, H < 100 kOe.	Simultaneous additions of Al and Sb led to a T_c intermediate between those of the individual additions. (Ga+Sb) and (Sn+U) led to a decrease in T_c. The (Sb+In) addition raised T_c relative to those of the individual ternary alloys. It was stated that the "critical fields and currents of Ti-Nb alloys can be improved through additions of metals or semimetals".	Ti-Nb-O-N Ti-Nb-simple metal ternary alloys	D. Bachmann *et al.* (1968) [Bac68]
ALLOY CLASS: Ti-Nb-TM-SM				
Ti-Nb$_{12}$-Mo$_4$-Al$_{10}$ Ti-Nb$_{12}$-Mo$_4$-Al$_{20}$	(a) Quenched, (b) quenched and aged at 350°C, (c) above plus "reverted" at 500°C, (d) isothermally transformed ($\alpha+\beta$) and furnace cooled, (e) above plus aged 40 h at 750, 650 or 550°C. Transition temperature measured magnetically.	Variation in T_c with heat treatment could be justified in terms of anticipated solute redistribution.	Ti-Nb-Mo Ti-Nb-Al	W.L. Cotton (1973) [Cot73]
ALLOY CLASS: Ti-Nb-TM$_1$-TM$_2$				
Ti$_{11}$-Nb$_{67}$-Zr$_{11}$-Hf$_{11}$ Ti$_{20}$-Nb$_{40}$-Zr$_{20}$-Hf$_{20}$ Ti$_{50}$-Nb$_{42}$-Ta$_4$-Hf$_4$ Ti$_{52}$-Nb$_{40}$-V$_4$-Hf$_4$ Ti$_{53}$-Nb$_{39}$-Zr$_4$-Ta$_4$	Arc melted, cast, cold swaged and drawn to wire, 0.010 in.$^\phi$. Critical field measured by pulse-field technique.	Part of a study of the dependence of H_{c2} on the solute-size-corrected "effective electron/atom ratio".	Numerous binary and ternary alloys	W. DeSorbo *et al.* (1967) [Des67]

Ti-Zr-Hf-Nb$_{70}$ (6 quaternaries)
Ti-Zr-Hf-Nb$_{50}$ (6 quaternaries)
Ti-Zr-Hf-Nb$_{30}$ (6 quaternaries)
	Arc melted, cast, homogenized 3h/1400°C, hot forged in stainless steel), sectioned and cold rolled to 1.2x1.2 mm^2; cold drawn to 0.3 mm$^\phi$ through a series of dies (cw: 99.95%). Heat treated:
	   3h/550°C (Nb = 70 at.%)
	   2h/500°C, 3h/380°C (Nb = 50 at.%)
	   1h/500°C (Nb = 30 at.%).
	
	T_c was measured (probably resistometrically) on wire samples.
	
	H_r was determined from $J_c(H_r) = 10^2$ A cm^{-2}.
	
	J_c (4.2K) was measured (ten specimens at a time) in the 1 mm gap of the permendur-concentrated field (<82kOe) of a superconducting solenoid.
		A detailed systematic study of the influence of quaternary composition on alloy phases, and superconducting properties. A complete discussion is complicated. If a single conclusion is to be drawn, it is that the addition of Hf to Ti-Zr-Nb usually has a deleterious influence on the superconducting properties.	-----	I.I. Rayevskii *et al.* (1969)
				[Ray69]
				(Nb = 50, 70 at.%)
				
				I.I. Rayevskii *et al.* (1971)
				[Ray71]
				(Nb = 30 at.%)

TABLE 1-38 TITANIUM-NIOBIUM-BASE QUATERNARY ALLOYS –continued

Compositions Ti	Hf	Nb	Ta (at.%)	Preparation, Procedures, Properties	Remarks, Discussion	Other Systems Studied	Literature
55	2.5	32.5	10	Samples were multiply arc melted, cold rolled recrystallized 1h/875°C, swaged to 3 mm$^\phi$, sheathed in Cu and drawn to 1 mm‡ (core, ~0.5 mm$^\phi$).	The alloy group studied included what might be regarded as Ti-Zr$_5$-Nb$_{25}$-Ta$_{10}$ in which the Zr had been replaced by an equal amount of the other group IV element, Hf.	-----	D.G. Hawksworth and D.C. Larbalestier (1980) [Haw80]
57.5	2.5	30	10				
60	2.5	27.5	10	H_r was measured at 1.7, 2, 3 and 4.2 K at the Francis Bitter National Magnet Laboratory. The measuring current density was 5 A cm^{-2}.	The highest values of H_r (4.2K) were 109-113 kOe compared to 113-115 for some Ti-Nb alloys. At 2 K the quaternary alloys showed some H_r (2K) enhancement (e.g. 153 kOe compared to 143 kOe for Ti-Nb) but not as much as for example Ti$_{65}$-Nb$_{25}$-Ta$_{10}$ which, according to the same authors, yielded H_r (2K) = 154 kOe.		
62.5	2.5	25	10				
70	2.5	17.5	10				
52.5	5	32.5	10		Considering this result in the light of Rayevskii et al. [Ray69, Ray71], and comparing it with those of Horiuchi et al. to follow, we do not recommend Hf as an addition to either Ti-Nb-Ta or Ti-Nb-Zr.		
55	5	30	10				
60	5	25	10				
47.5	10	32.5	10				
50	10	30	10				
55	10	25	10				
60	10	20	10				
62.5	10	17.5	10				

Compositions Ti	Zr	Nb	TM	(at.%)	Preparation, Procedures, Properties	Remarks, Discussion	Other Systems Studied	Literature
10	40	40-50	(0,2,5,10)	V	(a) 0, 2, 5, 10 at.% V 4 mm$^\phi$ → 2 mm$^\phi$ (5h/1100°C) → 1 mm$^\phi$ (3h/700°C) → 0.25 mm$^\phi$ (b) 5 at.% V 4 mm$^\phi$ → 2 mm$^\phi$ (5h/1100°C) → 1 mm$^\phi$ (A) → 0.25 mm$^\phi$ A = 1h/600°C, (1,3,10h)/700°C (c) 2 at.% Mo Same as above (d) 1, 5 at.% Ta Same as above (without,10-h heat treatment) T_c plotted vs TM concentration for all alloys solution treated 5h/1100°C. J_c (4.2K) vs H ≤ 80 kOe determined for the cold worked and intermediate heat treated 0.25 mm$^\phi$ wires.	Optical metallography reported for quenched (from 1100°C) and aged 300h/700°C V, Mo and Ta alloys. The solubilities of V, Ta and Mo in Ti$_{10}$-Zr$_{40}$-Nb are about 5, 5 and 2 at.%, respectively. All additions depress T_c. The severity of the *depression* increases in the order Ta < V < Mo. The J_c (H = 40, 50 kOe) of the 5 at.%-Ta alloy, after 3h/700°C, is equal to or a little less than that of the Ti$_{10}$-Zr$_{40}$-Nb base. The addition of V or Mo always depresses the J_c of the base under consideration.	-----	M. Kitada and T. Doi (1970) [Kit70^9]
10	40	45-50	(0,2,5)	Mo				
10	40	45-50	(0,1,5)	Ta				

TABLE 1-38 TITANIUM-NIOBIUM-BASE QUATERNARY ALLOYS *–continued*

Compositions	Preparation, Procedures, Properties	Remarks, Discussion	Other Systems Studied	Literature
Ti-Nb-V-Ta (matrix of alloys)	Transition temperatures were plotted on a quaternary tetrahedron.	Alloys with highest T_c's lay close to the Ti-V and Ti-Ta edges of the tetrahedron. Alloying with a third or fourth component led to a reduction in T_c.	Review monograph	E.M. Savitskii *et al.* (1973) [Sav73, p.351]
Ti-Zr$_{30}$-Nb$_{30}$-Ta$_7$	Alloys were solution treated 1h/900°C, cold worked 0, 50, 90, 99% and aged at temperatures from 400 to 700°C. A metallurgical study of the effect of cold work and heat treatment on phase transformations, using: (a) electrical resistivity, (b) microhardness, (c) X-ray (powder) diffraction.	A discussion of phase equilibria and kinetics of transformation. T-T-T diagrams for 90 and 99% cold worked and aged alloys were constructed.	-----	T. Horiuchi *et al.* (1973) [Hor73^a]
Ti$_{20-63}$-Zr$_{5-35}$-Nb$_{24-60}$-Ta$_5$ (18 alloys) Ti$_{19-31}$-Zr$_{5-30}$-Nb$_{24-60}$-Ta$_{10}$ (13 alloys)	Alloys were arc melted, hot forged, cold swaged, solution heat treated and cold drawn to 0.25 mm$^\phi$. T_c, H_r, and microhardness were measured and the results plotted on pairs of Ti-Zr-Nb Gibbs triangles (one representing [Ta] = constant = 5 at.%; and the other, [Ta] = 10 at.%). Residual resistivities, ρ_n were measured and critical fields calculated using the Clogston and WHH-Maki-modified GLAG relationship between H_{c2} and the product $\rho_n \gamma T_c$.	T_c was greatest along the Ti-Nb-Ta (5, 10 at.%) edges of the Gibbs triangles; maximal values of 9.8 K (Ta: 5 at.%; Ti: 30, 40 at.%) and 9.6 K (Ta: 10 at.%; Ti: 30, 40 at.%) were attained. Resistive critical field: according to Suenaga *et al.* [Sue69] Ti-Nb-Ta exhibits highest H_r's for [Ta] = 5 ∿ 10 at.%, i.e. along the Ti-Nb edges of the Ti-Nb-Zr-Ta (= 5, 10 at.%) composition triangles. This paper shows that even higher H_r's exist in the interiors of these triangles accompanying the addition of some 5 at.% Zr. The alloy with the highest known critical field is Ti-Zr$_5$-Nb$_{25}$-Ta$_{10}$ for which H_r (4.2K) was reported to be 131 kOe.	-----	T. Horiuchi *et al.* (1973) [Hor73]
Ti-Zr$_6$-Nb$_{27}$-Ta$_6$ (Cryozitt)	Properties of multifilamentary composites were presented. Specifications of the conductors covered the ranges: (a) conductor φ (mm) : 0.35 to (1.6x3.8) (b) number of filaments: 1 to 3721 (c) filament φ (μm) : 18 to 250 (d) Cu/SC ratio : 1:1 to 4:1 (e) twist pitch (mm) : 10 to 50 Critical current *vs* field (up to 7T) was given for (a) as heat treated, and (b) heat treated and final cold worked wire. Mechanical properties given were (a) 4.2 K tensile strength *vs* Cu/SC ratio, (b) static stress degradation, and (c) dynamic stress degradation.	The J_c (4.2 K, 5 T) of the as heat treated material was given as 1.85x10^5 A cm^{-2}; with final cold work this could be raised to 2.4x10^5 A cm^{-2}. Under static stress the critical current degradation of a 61-filament, 2:1 Cu/SC-ratio conductor at 30%, 50% and 70% of σ_B at 4.2 K were respectively 0, 2% and 5%; at 64% σ_B (4.2K) no further degradation was noted during repeated application (up to 100 times) of the stress. In dynamic tests at 5 T of a 61-filament, 0.9:1 Cu/SC-ratio conductor with I/I$_c$ ∿< 0.95 no stress effect was noted at stress levels up to 70% σ_B (4.2K).	-----	Kobe Steel Technical Data Sheet October, 1978 [Kob78] see also [Kob80]

TABLE 1-38 TITANIUM-NIOBIUM-BASE QUATERNARY ALLOYS –continued

Compositions	Preparation, Procedures, Properties	Remarks, Discussion	Other Systems Studied	Literature
Ti-Zr$_6$-Nb$_{27}$-Ta$_6$ (Cryozitt)	Properties documented were: (a) J$_c$ (4.2K, 5T, 6T, 7T) vs 370°C aging time up to 50 h for wire in the conditions: (i) 99.998% total cw (ii) cw plus 12 and 38% cw after aging, (ii) 99.99998% total cw (ii) cw plus 16 and 40% cw after aging, (b) J$_c$ (4.2K, 5T, 6T, 7T) vs cold reduction up to 47% after aging 50h/370°C. Total cold reductions: 99.998% and 99.99998%, (c) J$_c$ (4.2K, 7T) vs total cold reduction in area of up to 10^6:1, (d) J$_c$ (4.2K) vs H (≤7T) for wire in the conditions: 99.998% cw plus (i) 50 h/340°C and (ii) 50h/380°C and these plus 40% final cold work, (e) several J$_c$(H) intercomparisons among various wires.	The results indicate that: (a) final cold reduction (~40%) after final heat treatment is beneficial to J$_c$; (b) a total cold reduction of 99.998% is preferable to 99.99998% in the presence of 50h/370°C final heat treatment and final cold reductions of up to 47%; (c) the J$_c$ of the quaternary alloy is to first approximation relatively insensitive to total area reduction (in contrast to Ti-Nb$_{37}$); (d) the best quaternary alloy performance was achieved with 99.998% total cold reduction in association with 50h/370°C plus 40% final cold reduction; (e) Although the 4.2K J$_c$ of the quaternary alloy is inferior to those of several Ti-Nb alloys, at about 2 K in fields of less than about 8 T the quaternary alloy is superior to both Nb-46.5Ti (Ti-Nb$_{37}$) and Nb-50.5Ti (Ti-Nb$_{34}$).	Ti-Nb	Kobe Steel Technical Data Sheet May, 1980 [Kob80^a] (An extended version of the above)
[(Ti-Nb$_{50}$)$_{95}$-Zr$_5$]-Ta$_{0,5,7}$ (10 alloys) [(Ti-Nb$_{40}$)$_{95}$-Zr$_{5,10,15}$]-Ta$_{4-10}$ (11 alloys) [(Ti-Nb$_{30}$)$_{95}$-Zr$_{5,10,15}$]-Ta$_{3-10}$	Critical fields were measured at 4.2 K, resistively on wire samples in a pulsed magnetic field. Samples were measured after 99.8% cold reduction by drawing, and again after a final heat treatment.	The highest 4.2 K critical fields were found in the alloys [(Ti-Nb$_{30}$)$_{95}$-Zr$_5$]-Ta$_{5,7,10}$ and of these the alloy with 10 at.% Ta gave the highest field of 13.1 T.	Ti-Nb Ti-Zr-Nb	T. Horiuchi et $al.$ (1980) [Hor80]
[(Ti-Nb$_{40}$)$_{95}$-Zr$_5$]-Ta$_6$ [(Ti-Nb$_{30}$)$_{95}$-Zr$_5$]-Ta$_{5,7,10}$ [(Ti-Nb$_{30}$)$_{90}$-Zr$_{10}$]-Ta$_6$	I$_c$ (4.2k, 4T) measured as function of 350°C aging time up to 10^4 min.	The [(Ti-Nb$_{30}$)$_{90}$-Zr$_{10}$]-Ta$_6$ alloy responded strongly to aging especially for times greater than about 24 h.		
Ti-Zr$_6$-Nb$_{27}$-Ta$_6$	J$_c$ (4.2K, 3 to 7T) measured as function of 370°C aging time up to 10^4 min. J$_c$ (4.2K, 5, 6, 7T) measured as function of cold area reductions by drawing of up to 80 ~ 90% after final heat treatment. In studies of precipitation the Guinier radius (a measure of particle diameter) and "relative interparticle distance" were measured as function of aging time up to about 560 h at 380°C. Comparisons were made between the quaternary alloy (99.994% cw, fil.;, 40 μm) and Ti-Nb$_{36}$ (99.96%, 99.986%, 99.994% cw, fil.'s, 110, 60, 40 um).	J$_c$ (4.2K) generally increased or decreased with aging time depending on whether H was less than or greater than 5 T. The 7T J$_c$ passed through a broad maximum in the vicinity of 16 h. J$_c$ (4.2K) always passed through a broad maximum centered upon 55 to 65% cold area reduction. As anticipated the more heavily cold worked alloys possessed the initially-smaller precipitate particles. During aging the differences diminished and Guinier radii tended towards 65 ~ 80 Å.		

TABLE 1-38 TITANIUM-NIOBIUM-BASE QUATERNARY ALLOYS –*continued*

Compositions	Preparation, Procedures, Properties	Remarks, Discussion	Other Systems Studied	Literature
$Ti-Zr_6-Nb_{27}-Ta_6$	----------------------	A complete description of alloy preparation and ingot examination, and the process metallurgy of composite superconductor with particular reference to Kobe Steel's quaternary alloy "Cryozitt".	-----	T. Horiuchi *et al.* (1980) [Hor80a]
$Ti-Zr_6-Nb_{27}-Ta_6$ (i.e. Ti-8Zr-36Nb-15Ta)	The quaternary alloy was a sample of "Cryozitt" from Kobe Steel Ltd., Japan. J_c (~2K, 4.2K) for Ti-Zr-Nb-Ta and several Ti-Nb alloys were intercompared (cf. Reference [Kob80a] - Item (e) - above). Bulk pinning forces, $F_p = J_c H$ were plotted *vs* applied field and also reduced field, $h(\equiv H/H_r)$. Plots of reduced pinning force, $F_p/F_{p,max}$ *vs* h for various alloys were given; also a plot was constructed of $F_p/F_{p,max}$ *vs* h for the quaternary alloy at temperatures T = 2.05 K through 4.20 K.	Four commercial composite superconductors were intercompared. $F_p/F_{p,max}$ *vs* h for $Ti-Zr_6-Nb_{27}-Ta_6$ at temperatures of 2.05, 2.55, 2.78, 3.29, 3.50, 3.83 and 4.20 K adhered closely to the usual scaling law.	Ti-Nb Ti-Nb-Ta	D.G. Hawksworth and D.C. Larbalestier (1980) [Haw80a]

TABLE 1-38 TITANIUM-NIOBIUM-BASE QUATERNARY ALLOYS –continued

Compositions	Preparation, Procedures, Properties	Remarks, Discussion	Other Systems Studied	(Inventor) Assignee Priority or Filing and Issuance Dates	Listing
$(5-80)Ti-(5-80)Zr-(\leq25)Nb$ " " $(\geq20)Nb$ " " $(\geq15)Nb$ $(5-50)Ti-(20-60)Zr-(\leq25)Nb$ " " $(\geq20)Nb$ " " $(\geq15)Nb$ Above plus either ≤10 wt.% Ta or ≤ 5 wt.% Hf	Ingot forged at 1000°C and drawn to filament of 0.25 mm² without intermediate heat treatment.	No striking advantages were claimed for additions of either Ta or Hf. It was found to be possible to add up to 10 wt.% Ta without significantly changing the properties of Ti-Zr-Nb -- in fact some advantage may accrue under certain conditions. In addition it was noted that Hf, which is occassionally associated with Zr to the extent of 2 to 3% could be present in Ti-Zr-Nb in amounts of up to 5% without significantly affecting the superconducting properties. Covered by the claim were the superconducting alloys as well as conductor made from them.	Ti-Nb-Zr	Imperial Metal Industries, Ltd., Great Britain 1964-1965	Belgian Patent No. 659,033 July, 1965 [Imp65]
Quaternary alloys containing Nb and Ti (or Nb and Zr)		General statements were made with regard to alloy types; no compositions or data were specified.	Ti-Nb, ternary alloys with Ti and Nb or Zr and Nb	(P.F.Chester) Central Electricity Generating Board, Great Britain 1964-1966	French Patent No. 1,452,977 September, 1966 [Che65]
$Ti_{33}-Zr_{30}-Nb_{30}-Ta_7$	See below	See below	$Ti_{10}-Zr_{40}-Nb_{50}$	(T.Horiuchi *et al.*) Kobe Steel, Ltd., Kobe, Japan 1973	Japanese Patent No. 85,405 November, 1973 [Hor73b]
  Ti Zr Nb Ta (at.%) (40-75) (1-10) (20-40) (1-20) 50 20 25 5 55 15 25 5 60 5 25 10 $[(Ti_{70}-Nb_{30})_{95}-Zr_5]_{100-y}-Ta_y$	H_r was measured using 0.25 mm² wire annealed 5h/1000°C. J_c was measured using 10h/400°C final heat treated wire.	The H_r's of sixteen 5 at.% Ta alloys and sixteen 10 at.% Ta alloys were plotted on composition triangles. The H_r of the 5 at.% Ta series was also plotted as a function of Ta level (0-25 at.%). J_c (4.2K) *vs* H (<100kOe) was plotted for each of the three alloys specified and compared with those of $Zr-Nb_{75}$ and a Ti-Nb alloy.	Ti-Nb $Zr-Nb_{75}$ Ti-Nb-Ta	(T.Horiuchi *et al.*) Kobe Steel, Ltd., Kobe, Japan 1973-1974	German Patent No. 2,350,199 July, 1974 [Hor74]

TABLE 1-38 TITANIUM-NIOBIUM-BASE QUATERNARY ALLOYS –continued

Compositions				Preparation, Procedures, Properties	Remarks, Discussion	Other Systems Studied	(Inventor) Assignee Priority or Filing and Issuance Dates Listing
Ti	Zr	Nb	Ta (at.%)				
(10-50)	(10-40)	(20-50)	(5-12)	Ti$_{33}$-Zr$_{30}$-Nb$_{30}$-Ta$_7$ heat treated according to:	The H$_r$'s of thirteen 5 at.% Ta alloys and eleven 10 at.% Ta alloys were plotted on composition triangles. The H$_r$ of Ti-Zr$_{30}$-Nb$_{30}$-Ta$_{0-14}$ was plotted vs Ta composition. Critical field claims for the quaternary system were modest, viz:	Ti$_{10}$-Zr$_{40}$-Nb$_{50}$	(T.Hor'uchi $et\ al.$) German Patent No. 2,347,400
33	30	30	7	$2h/850°C \xrightarrow[87\%]{CW} HT(A) \xrightarrow[90\%]{CW} (0\text{-}1000min)/550°C \xrightarrow[90\%]{CW}$			Kobe Steel, Ltd., April, 1975
40-26	30	30	0-14	with HT(A) = 2,50h/850°C;3,50h/650°C;3h/550°C.	H$_r$ ≥ 96 kOe.		Kobe, Japan [Hor75]
					For the preferred composition specified: J$_c$ (4.2K) vs H (<80 kOe) was given for the five intermediate heat treatments specified and 100h/500°C final heat treatment. J$_c$ (4.2K, 60kOe) was plotted vs 0-1000min/550°C final heat treatment.		1973-1975

TABLE 1-39 AMORPHOUS TITANIUM-BASE ALLOYS

Compositions, Preparation[+] (at.%)	Superconducting Properties Investigated	Other Properties Investigated	Results, Discussion	Literature
Ti Nb Si 70 15 15 60 25 15 55 30 15 50 35 15 45 40 15 56 30 14	Resistively measured T_c was studied: (a) as function of Nb content, (b) as function of Si content, (c) in Ti$_{70}$-Nb$_{15}$-Si$_{15}$ and Ti$_{55}$-Nb$_{30}$-Si$_{15}$ as function of reduction in thickness by cold rolling, and as function of subsequent 1 h annealing temperature to 300°C, (d) in Ti$_{55}$-Nb$_{30}$-Si$_{15}$ as function of 1 h annealing to 500°C. Resistive upper critical fields were measured, as were dH$_r$/dT. 4.2 K J$_c$ was measured in fields up to typically 34 kOe.	Structures were studied using x-ray diffractometry and TEM/SAD especially during aging. Crystallization effects were followed using DTA. Vickers microhardness and tensile strengths were measured.	T_c increased with increasing Nb content and decreased with increasing Si content. The highest T_c (for Ti$_{45}$-Nb$_{40}$-Si$_{15}$) was 5.1 K, considerably less than the 9 K of Ti-Nb$_{40}$. T_c decreased with cold rolling but recovered after annealing at 1h/300°C. T_c decreased with annealing in the glass phase at ~400°C but increased during crystallization at temperatures near 500°C and above.	A. Inoue *et al.* (1980) [Ino80]
Ti$_{85-x}$-Nb$_x$-Si$_{15}$ (x = 10, 15, 30, 40) Ti$_{85-x}$-V$_x$-Si$_{15}$ (x = 5, 10, 20, 30) Ti$_{55}$-Nb$_{30}$-Si$_4$-B$_{11}$ Ti$_{57}$-Nb$_{30}$-Si$_{10}$-B$_3$	The T_c's of Ti$_{55}$-Nb$_{30}$-Si$_{15}$ and Ti$_{70}$-Nb$_{15}$-Si$_{15}$ were reported as function of (a) cold rolling, and (b) annealing temperature to 300°C after cold rolling. The T_c of Ti$_{55}$-Nb$_{30}$-Si$_{15}$ was reported as function of 1 h annealing to 500°C. The T_c's of Ti$_{85-x}$-V$_x$-Si$_{15}$ (x = 5, 10, 20, 30) were reported as functions of 1 h annealing to about 750°C. The T_c's of Ti$_{85-x}$-Nb$_x$-Si$_{15}$ (x = 10, 15, 30, 40) were reported as functions of 1 h annealing to about 900°C. J$_c$ vs H (<90kOe) was reported for 1 h-annealed (650, 700, 750°C) Ti$_{50}$-Nb$_{35}$-Si$_{15}$ and in partially recrystallized (bcc+Am) Ti$_{55}$-Nb$_{30}$-Si$_4$-B$_{11}$ and Ti$_{57}$-Nb$_{30}$-Si$_{10}$-B$_3$.	-------------------------	The superconducting properties T_c, H$_c$ and J$_c$ improved remarkably after recrystallization; in particular, high values of these quantities were achieved when fine crystalline particles were present in the amorphous matrix. In the ternary alloys, the first stage of crystallization resulted in each case in two types of crystalline precipitate: (a) an intermetallic compound, and (b) a supersaturated bcc solid solution. Some brief reference to the possibility of A-15 intermetallic compound precipitation under certain conditions was made.	T. Masumoto *et al.* (1980) [Mas80]

Ti$_{55-x}$-Nb$_{30}$-Si$_{15}$-Mo$_x$
 (x = 0, 3, 5)
 	
 Ti$_{55}$-Nb$_{30}$-Si$_{15-x}$-M$_x$
 M = B,   x = 0, 3, 5
 M = C,   x = 0, 3, 5
 M = Ge,   x = 1.5, 3
 	
 (Ti-Nb)$_{85}$-Si$_{12}$-B$_3$
	Transition temperatures were measured resistively at a specimen current of 1 mA.
	
	 Critical fields were measured resistively.
	
	 Critical current densities were measured in fields of up to about 75 kOe.
		The composition range of the amorphous single phase within the ternary Ti-Nb-Si triangle was mapped.
		
		 The composition ranges of the amorphous single phases of the quaternary systems:
		 Ti$_{55-x}$-Nb$_{30}$-Si$_{15}$-M$_x$
		     M = Mo, Ru, Rh, Pd and Ir
		 Ti$_{55}$-Nb$_{30}$-Si$_{15-x}$-M$_x$
		     M = B, C and Ge
		 were indicated on a diagram.
			Increases of T_c took place in Ti-Nb-Si when the Nb content was increased, or the Si content decreased; and to a slight extent in the quaternary, as B replaced Si.
			
			 Decreases of T_c took place in the quaternary as Mo replaced Ti, and as C or Ge replaced Si; the T_c's were undetectably low (<3.5K) when Ti was replaced by Ru, Rh, Pd or Ir.
			
			 The T_c's of amorphous Ti-Nb-Si and Ti-Nb-Si-B lay very close to, but a little above, the Collver-Hammond curve [Col73].
			
			 In (Ti-Nb)$_{85}$-Si$_{12}$-B$_3$, T_c maximized at 5.4 K for [Nb] = 40 at.%.
				A. Inoue *et al.* (1980) [Ino80]

TABLE 1-39   AMORPHOUS TITANIUM-BASE ALLOYS *–continued*

Compositions, Preparation[†]			Superconducting Properties Investigated	Other Properties Investigated	Results, Discussion	Literature
Ti	Nb	Si (at.%)				
70	15	15	Transition temperature was measured resistometrically at specimen current of 1 mA.	Twenty-two binary and ternary alloys were prepared in an investigation of the range of amorphous compositions and their crystal-	As a logical extension of research described in the preceding three papers, the primary purpose of this work was to study the influence of aging and eventual crystallization on the superconducting properties.	A. Inoue *et al.* (1980) [Ino80[b]]
55	30	15	Upper critical field was measured resistively.	lization temperatures.	During annealing of the amorphous alloys at temper-	
45	40	15	Critical current density was measured using a 1μV/25mm criterion in both as-quenched (amorphous) and crystal-	The various stages of the crystal-	atures up to 400°C T_c decreased, but increased again as the annealing temperature was raised and crystal-	
			lized alloys.	lization process were investigated using DTA and TEM.	lization set in.	
					Upon recrystallization the T_c's of the Ti-Nb-Si alloys were *larger* than those of the respective binaries with the same Nb concentrations especially for the Ti-Nb$_{10}$-Si$_{15}$ and Ti-Nb$_{15}$-Si$_{15}$ alloys.	
					The difference was attributed to the stabilization in these materials of a low-Nb high-T_c metastable bcc phase.	

[†] Materials were pre-alloyed by arc melting on wcch before being transferred to the melt-quench apparatus.

5 g samples were levitation melted under argon and fed to a 20 cm diameter Cu roller spinning at 4000 rpm to yield continuous "melt-spun" ribbon 1∿2 mm wide and 0.03∿0.04 mm thick.

2

UNALLOYED TITANIUM

A study of the history of superconducting transition temperature measurement in pure Ti reveals the several difficulties which the experimentalist faces when attempting to make accurate determinations of low transition temperatures in pure substances. The principal difficulties encountered had to do in one way or another with sample purity and spurious effects associated with refrigeration and sample temperature determination. Such a study of course also emphasizes the fact that measurements which may be performed with extreme accuracy quite routinely today, were difficult and time-consuming three decades ago. The work up to 1953 has been summarized and discussed by EISENSTEIN [Eis54] who concluded that the most likely T_c for pure Ti was 0.387 K [Smi53] and that the wide discrepancies among the previously published results were attributable partly to differences in sample purity or condition and partly to lack of thermal equilibria between sample and paramagnetic salt capsule in those cases in which magnetic cooling was employed.

2.1 SAMPLE PURITY AND MEASURING TECHNIQUE

2.1.1 Influence of Trace Impurities

Sample purity and measuring technique must be considered jointly since one has a strong influence over the choice of the other. The three earliest studies which all used the disappearance of electrical resistance as the indicator of the onset of

superconductivity, yielded anomalously high values for the superconducting transition temperature (1.13 - 1.77 K). The magnetic measurement due to SHOENBERG [Sho40], who intended to take advantage of the MEISSNER effect, suggested that T_c was below 1.0 K, the lowest temperature attainable in that particular experiment. In the light of the earlier results, the fact that a small diamagnetic anomaly was observed at 1.5 K was most significant. As a consequence of the high current-carrying capacity of superconducting alloys, the existence of any high-T_c percolation path along the sample, formed from suitable second-phase precipitates, is sufficient to yield an anomalously high T_c if four-terminal resistivity is the measuring technique. Such a resistive T_c value would not be confirmed if a bulk measurement, either magnetic or calorimetric, were then performed on the same sample. This conclusion is graphically illustrated in a pair of studies carried out in 1962 and 1963 on Ti-Rh alloys. BUCKEL, DUMMER, and GEY [Buc62] measured the superconducting transition temperatures of a series of as-cast and also annealed Ti-Rh alloys using the electrical resistance technique, while RAUB and ANDERSON [Rau63] used magnetic inductance. The results of these studies are compared in Fig.2-1 which suggests immediately that filamentary inclusions of ß-Ti-Rh act as electrical short circuits around the lower-T_c α^m-martensite which forms the bulk of the alloy. The equilibrium α (hcp) and ß (bcc) phases of binary Ti-(transition metal), i.e. Ti-TM, alloys, and their non-equilibrium quenched structures (which in the

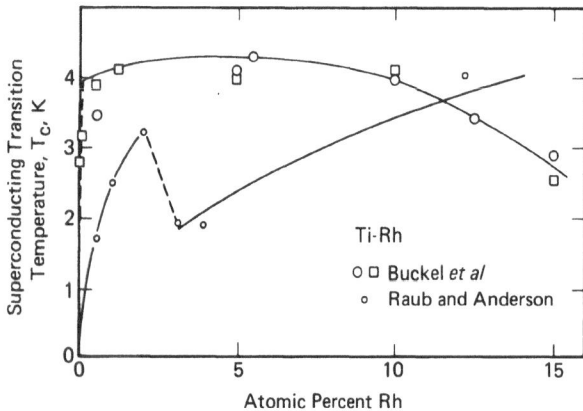

FIGURE 2-1. Superconducting transition temperatures of Ti-Rh alloys as determined resistometrically on massive unannealed samples (○) and unannealed turnings (□) by BUCKEL *et al* [Buc62]; and magnetically on as-cast samples (○) by RAUB and ANDERSON [Rau63].

low-concentration alloys are α'- or α"-martensitic -- collectively α^m, herein) are discussed in [Mon1.00].

As indicated in the subsequent chapters of this book, the superconducting transition temperatures of alloys of Ti with most other transition elements are usually higher than that of pure Ti. A notable exception is α-Ti-Mn in which the localized magnetic moment of Mn suppresses the temperature of the superconducting transition. After discussing the effects of traces of Mn and Fe on the transition temperature of Ti, FALGE [Fal63] has suggested that the presence of small amounts of dissolved transition elements may have been responsible for the variation among the values of T_c which had been reported for unalloyed Ti. As well as measuring T_c in a sample of high-purity Ti (Mn < 5 ppm, Fe < 2 ppm) FALGE also investigated the influence of small additional amounts of Mn and Fe; the results are summarized in Table 2-1.

Comparing rows 2 and 3 of the table we see that 0.15% Fe raises T_c by 0.81 K; thus a few tens of ppm of Fe would have a negligible effect on the transition temperature. Comparing rows 1 and 2 and ignoring the effect of the change in Fe content it may be concluded that ~25 ppm of Mn depresses T_c by 0.25 K; a rate of about 0.1 K per 10 ppm Mn, which of course explains why superconductivity was not detected in Ti-Mn(100 ppm).

The Mn content of unalloyed Ti according to the typical analysis supplied by Materials Research Corporation is 5 wt. ppm (high-purity grade) or 1.2 wt. ppm (zone-refined grade), that of electrolytic Ti sponge as supplied by Titanium Metals Corporation is

stated to be typically <10 wt. ppm, while that of Ti from the Foote Mineral Company has been specified as <50 wt. ppm. In the absence of further chemical analysis the latter would be suitable only as an alloying base.

The pronounced difference between the results of electrical resistance measurements on one hand, and bulk measurements on the other (Fig.2-1) is strong evidence in favor of the filamentary precipitate picture. Bulk measurements of superconducting transition temperature may be made either magnetically or calorimetrically. The latter (i.e. low temperature specific heat), although more difficult to perform, yields not only the position and width of the superconducting transition but also, in terms of the height of the specific heat jump at T_c, a measure of the fraction of the specimen participating in the transition. Of the magnetic methods, the SCHAWLOW-DEVLIN resonance frequency shift technique [Sch59] has been frequently employed. A slight disadvantage of it is the possibility that partial internal shielding by higher T_c components of an inhomogeneous sample may occur. Frequently employed in some of the earlier work was a DC or ballistic mutual inductance method involving the collapse of a magnetic field. Provided the field is kept small (≲10 Oe) errors caused by the possible lowering of the measured T_c by trapped flux would not be significant. Much more serious would be the influence of large stray magnetic fields when refrigeration is by adiabatic demagnetization.

TABLE 2-1 SUPERCONDUCTIVITY IN IMPURE TITANIUM -- after FALGE [Fal63]

Row Number	Impurity Content ppm[+]		Transition Temperature K
	Mn	Fe	
1	<5	<2	0.42
2	30	20	0.17
3	30	1500	0.98
4	100	20	<0.06

[+] Although not specified in the original article it is assumed that parts per million *by weight* were being referred to.

2.1.2 Refrigeration and Sample Temperature

Occasionally liquid He (He^4 above 1.2 K and He^3 below that temperature) has been the refrigerant, but in most cases adiabatic demagnetization was employed. The numerous difficulties which have been encountered with the use of that technique include those associated with (a) the preparation of the salt pill, (b) the establishment of thermal equilibrium between salt and sample, and (c) the possible deleterious effect of the stray field of the magnetization/demagnetization magnet.

In the experiments of DAUNT and HEER [Dau49] good thermal contact between sample and salt was achieved by randomly mixing small pieces (2-3 mm) of iodide Ti (99.95%, supplied by Battelle Memorial Institute) with the powdered chromium potassium alum, and compressing the mixture (200-300 atm) into an ellipsoidal pill of axial lengths 2.7 mm and 13 mm. As a possible result of the strain introduced into the Ti specimen during cutting, the H_c versus T curve was not reversible[+], nevertheless the trapped flux which, presumably, was responsible for the hysteresis did not appear to depress $T_c(H=0)$ to any significant extent. T_{c0} was in fact slightly higher than currently accepted values. The same experimental arrangement was again used by SMITH and DAUNT [Smi52] this time in a study of the influence of annealing on the transition temperature. It turned out that the results obtained on Ti samples measured in the as-received condition and after vacuum annealing at 800°C for 2.5 h differed insignificantly. In spite of the heat treatment the superconducting transition remained highly irreversible, an effect which in this case was apparently not due to strain.

In a final experiment in this series SMITH, GAGER and DAUNT [Smi53], using the same experimental procedures as those referred to above, again measured the properties of a sample of iodide Ti. This material, supplied by the Foote Mineral Company and stated to be 99.99% pure, yielded a reversible approximately parabolic $H_c(T)$ curve, and a superconducting transition temperature of 0.398 K, lower than any previously reported value. Since the same techniques were used in all three of the experiments by DAUNT and his co-workers, these results suggest that in the earlier work traces of impurity were raising T_c and introducing irreversibility into the $H_c(T)$ characteristic.

+ Flux pinning by defects is discussed in [Mon16.1] et seq.

STEELE and HEIN [Ste53] repeated the DAUNT experiment using small pieces of wire (3-5 mm in length) obtained by cold swaging a rather thin crystal bar (1.6 mm$^\phi$) of 99.98% Ti to a diameter of 0.6 mm. The transition temperature so obtained (0.37 ± 0.01 K) agreed excellently with the value of 0.387 K acquired in the previous work. However, when a more massive sample (13 mm$^\phi$ x 13 mm$^\ell$) mounted adjacent to the salt pill was measured a higher value of T_c (viz. 0.49 ± 0.01 K) was obtained, contrary to expectation, if poor thermal contact had previously been suspected as being the culprit. In STEELE and HEIN's measurement of Ti crystal bar, the paramagnetic salt (chrome potassium alum) was pressed around a Cu fin, silver soldered to a brass base to which the Ti sample was thermally and mechanically coupled by means of a long (35 mm) Cu screw. Since the salt pill was the local refrigerant, poor thermal contact between it and the metallic parts of the sample assembly and thermal resistance along the path to the sample, in association with eddy current heating of the sample during demagnetization, would lower the apparent transition temperature. By attaching a carbon thermometer to the Cu screw between salt pill and sample it was possible to monitor the quality of the thermal contact between the salt and the metal. As the thermal contact could always be shown to be adequate, and departures from salt-sample equilibrium led to a lowering of the measured transition temperature, the reason for the observed higher values of T_c for the massive as compared to the finely divided sample was never resolved.

In the light of the earlier comments on alloying and precipitate effects greater credence should be placed on the lower of the two transition temperature values cited above. This conclusion is reinforced by the results of the calorimetric work of BATT [Bat64] who obtained a T_c of 0.33 K in association with a transition width of 0.06 K.

2.1.3 Spurious Mechanical Effects

Deformation will not affect the superconducting transition temperature of pure Ti. On the other hand, as indicated above in connection with the preparation of a salt-base composite suitable for magnetically cooling an imbedded specimen through the superconducting transition, deformation of the sample either during cutting or during compaction of the composite into the usual ellipsoidal shape, may produce lattice defects capable of trapping flux and consequently lowering the apparent zero-field T_c. With $(dH_c/dT)_{T=T_c} \cong 400$ Oe K^{-1}, residual fields of less than 100 Oe would be adequate to lower T_c by ~ 0.2K. This will be particularly important when the samples are in the form of fine-mesh powders or lathe turnings as in the work of NETZEL [Net60], or during high-pressure studies of T_c when magnetic refrigeration is used, and plastic deformation of the sample due to departures from ideal hydrostatic conditions is encountered, as in the experiments of BRANDT and GINZBURG [Bra65, Bra66].

In the second of these papers, in which the influence of pressure cycling upon the transition temperature was considered, it was noted that T_c increased linearly with the initial application of pressure up to some 25 kbar, decreased reversibly upon reduction of pressure down to 15 kbar and then remained almost constant as the pressure dropped to zero. By comparison with the results of others, the initial 1-atm transition temperate encountered ($T_c = 0.23$ K) was too low, but the final value ($T_c = 0.32$ K) was, within reasonable limits, equal to the presently accepted critical value. It is tempting to suggest that this was a result of improved thermal contact. The hysteresis noted in the pressure-cycling studies was attributed to plastic deformation brought about as a result of departures from ideal hydrostatic conditions. This possibility was examined in detail in [Bra66] in which the effects of both pressure and deliberately administered plastic deformation upon $H_c(T)$ and T_c, were explored. Although it seems clear that an intrinsic increase of T_c accompanies the application of hydrostatic pressure, the true effect could conceivably be masked by apparent changes in the T_c of pure Ti brought about by plastic deformation. Measurements were made at practically constant temperature as the sample warmed up slowly from 0.08 K to 0.6 K in some 8 to 10 h. At each selected temperature superconductivity was destroyed using the magnetic field of a pair of Helmholtz coils straddling the sample space. Even supposing an enhanced upper critical field begins to develop under heavy deformation it is difficult to comprehend the observed spread in extrapolated zero-field critical temperatures (viz. 0.23 to 0.53 K [Bra66]).

2.2 TRANSITION TEMPERATURE—THE INFLUENCES OF PRESSURE AND ALLOTROPIC TRANSFORMATION

2.2.1 The Influence of Pressure

According to BRANDT and GINZBURG [Bra65, Bra66], T_c increases with pressure, P, at the rate of 6×10^{-3} K kbar^{-1}. Their attempts to quantitatively analyze the effect in terms of fundamental electronic parameters and the existing microscopic theory of BARDEEN, COOPER and SCHRIEFFER (BCS) [Bar57] were unable to be completed. It was, however, observed that the signs and relative magnitudes of $\partial T_c/\partial P$ for various transition elements (e.g. Ti, Ta, Mo) agreed with those of $\partial T_c/\partial(e/a)$, where e/a is the electron/atom ratio. Curves of γ (the electronic specific heat coefficient, which is proportional to the Fermi density of states) and T_c *versus* electron/atom ratio, based on data for some transition elements and their alloys, are given in Fig. 2-2.[†] Returning to the pressure dependence, it was noted that $\partial T_c/\partial P$ was positive for Ti, negative for Ta and very small for Mo. The McMILLAN formula itself predicts a decrease of T_c with decrease in atomic volume, after the responses of the phonon spectrum, and the d-electron matrix elements have been taken into account [Mcm68].

The influence of pressure on the superconducting transition temperature has recently been the subject of a comprehensive review article by GARLAND and BENNEMANN [Gar72]. Their conclusions with regard to Ti are summarized in Table 2-2 which includes a positive value for $\partial T_c/\partial P$ in accordance with the results of BRANDT and GINZBURG [Bra65, Bra66].

It is reasonable for the associated $\partial \ln T_c/\partial \ln V$ to be negative, but if so the results are in conflict with the predictions of KERKER and BENNEMANN [Ker73], considered below, that T_c should increase by 30% when the crystal lattice expands during disordering.

[†] Further discussions of electronic specific heat coefficient and its relationship to the superconducting transition are given in [Mon10.1] *et seq.*

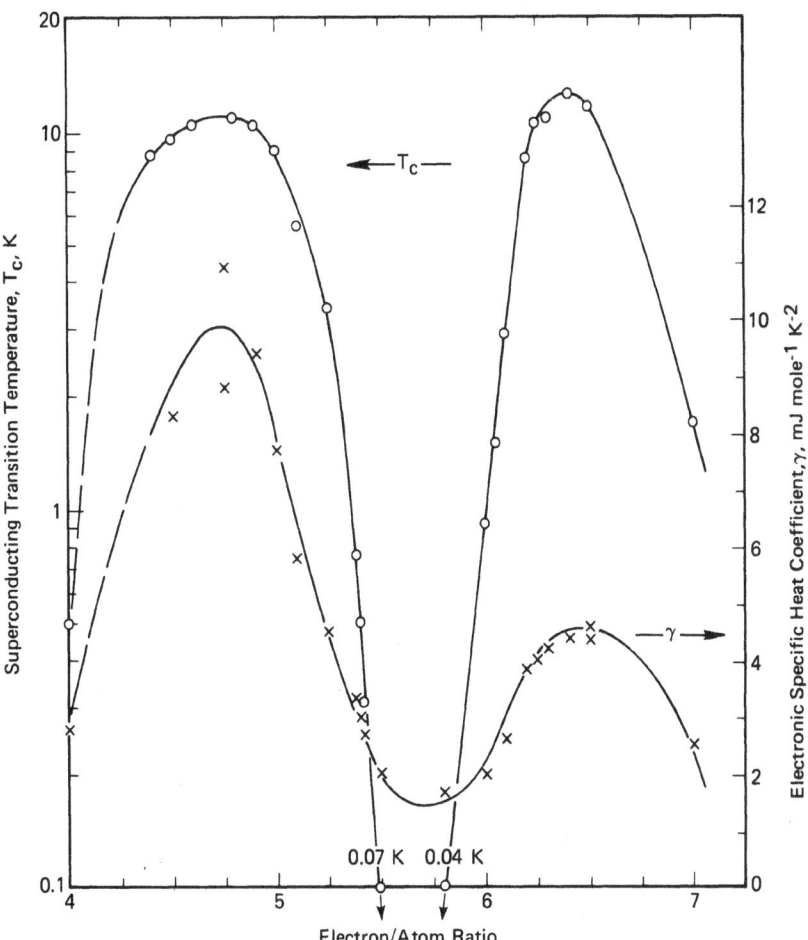

FIGURE 2-2. Superconducting transition temperature (T_c) and electronic specific heat coefficient (γ) as function of electron/atom ratio for transition elements and their binary alloys. 4d elements and alloys are represented within $4 < e/a < 6$, with Mo-Re taking over to complete the series up to e/a = 7 on account of the lack of Mo-Tc data (except for γ, Mo-Tc (50 at. %) = 4.6 mJ mole^{-1} K^{-2}; cf. γ, Mo-Re (50 at. %) = 4.4 mJ mole^{-1} K^{-2}). T_c data from GLADSTONE et al [Par69, p. 665] augmented by RALLS [Ral64]; γ-data from MORIN and MAITA [Mor63], see also HEINIGER et al [Hei66].

TABLE 2-2 VOLUME AND PRESSURE DEPENDENCES OF T_c FOR PURE TITANIUM -- after GARLAND and BENNEMANN [Gar72]

$\partial \ell n T_c / \partial \ell n V$		$\partial T_c / \partial P$	
Calc.	Expt.	Calc.	Expt.
-1.7	-1.65	0.06	0.06

2.2.2 The Influence of Structure -- Amorphous Titanium

KERKER and BENNEMANN [Ker73], who predicted the superconducting transition temperatures of fourteen amorphous transition metals including Ti, drew upon the semi-empirical theory of McMILLAN [Mcm68] and concluded that either increases or decreases in T_c were possible depending on the outcome of a competition between changes in the phonon spectrum, changes in the d-electron matrix elements, and a smearing of the electronic density of states at the Fermi level, all under the influence of lattice disorder.

The starting point of the analysis was the assumption that the increase in atomic volume $\Delta\Omega$ which accompanies complete disordering of the crystalline state is equivalent to that which occurs during melting. The influence of this volume dilatation upon $\Delta\ell n\langle\omega^2\rangle$, the relative change of the squared phonon frequencies averaged over the phonon spectrum, $\Delta\ell n\lambda$, the relative change in the electron-phonon coupling constant, and $\Delta\mu^*$, the change in the Coulomb pseudo-potential, when applied to the McMILLAN [Mcm68] equation, in the form:

$$T_c = \frac{\langle\omega^2\rangle^{1/2}}{1.2} \ \exp\ [-1.04(1+\lambda)/(\lambda-\mu^*(1+0.62\lambda))], \quad (2-1)$$

where ω, the phonon frequency or energy, is expressed in the form of a temperature, predicted the change in transition temperature. For Ti, a 30% increase in T_c was predicted as a direct result of the volume increase during disordering. The smearing-out of the electronic density of states resulting from lattice disorder was also supposed to play a role, increasing T_c when $n(E_F) < 1$ state eV^{-1} atom^{-1}, and decreasing it when $n(E_F) > 1$ state eV^{-1} atom^{-1}. For α-Ti, COLLINGS and HO [Coℓ73] have shown that $n(E_F) \cong 0.60$ states eV^{-1} atom^{-1} (for one direction of spin). Eqn.(2-1) is reintroduced in Chapter 14 in a full discussion of superconductivity in amorphous transition-metal alloys wherein attention is focused on the central role played by density-of-states-induced changes of the electron-phonon coupling constant.

2.2.3 The Influence of Structure -- Omega Phase

The detailed applications of McMILLAN theory to the study of transition temperature in transition metals, their alloys, and intermetallic compounds, undertaken by BENNEMANN and GARLAND [Ben72] and extended by them to deal with the effects of pressure [Gar72] have been referred to above. Similar methods could no doubt be brought to bear on the task of predicting the influence of allotropic transformation on T_c, and in particular the changes in T_c accompanying the transformation of α-Ti to its ω-phase and β-phase modifications.

Available experimental evidence suggests that for unalloyed Ti, $T_{c,\omega} < T_{c,\alpha}$. In studies of Ti-Mo, COLLINGS [Coℓ72^a] has demonstrated that for ω-Ti-Mo$_{4.3}$

$T_c = 0.20$ K compared with the estimated 0.36 K for α-Ti. But since T_c in quenched martensitic (α^m) Ti-Mo decreases with decrease in solute concentration we would expect the T_c of ω-Ti, which is also hexagonal in structure, to be lower than 0.2 K.

A direct measurement of T_c in ω-Ti has been made by DEGTYAREVA and co-workers [Deg74]. By exposing a sample of high-purity iodide Ti to a quasi-hydrostatic pressure of up to 120 kbar at room temperature for 30 min a transformation to the ω modification took place. As confirmed by x-ray diffraction this structure was retained after unloading to atmospheric pressure. Using a ballistic method for detecting the s/n transition, and magnetic refrigeration for controlling the temperature, it was determined that ω-Ti, if superconductive, would have a transition temperature lower than 0.06 K.

2.2.4 The Influence of Structure -- The BCC Phase[†]

Since the unperturbed bcc high-temperature allotrope of Ti cannot be retained, even in a metastable state, at temperatures below 883°C, it is possible only to estimate its transition temperature by extrapolating to zero solute concentration the transition temperatures of bcc Ti-base transition metal alloys. The results of two studies of the Ti-Rh system, those by BUCKEL, DUMMER and GEY [Buc62] and RAUB and ANDERSON [Rau63] have been referred to above and summarized in Fig.2-1. The latter authors erroneously believed that $T_{c,\beta}$ would be obtained by back-extrapolating the data for concentrations greater than 4 at.% Rh, when in fact the "extrapolated structure" is actually ω-phase. This, as mentioned above, has a transition temperature lying lower than that of the hexagonal α-phase -- a result attributed by RAUB and ANDERSON [Rau63] to β-Ti. The earlier of the two studies in fact yielded the relevant result. By measuring electrical resistivity, BUCKEL et al [Buc62] were able to detect the superconducting transitions of threads of β-Ti-Rh which short-circuited the lower T_c α^m-matrix. The extrapolated zero-Rh-concentration

---

[†] The equilibrium states of Ti-TM alloys are considered in [Mon1.1] et seq and the nonequilibrium phases, α^m-martensite and ω-phase, in [Mon1.6] et seq. See also Sect.7.19.

transition temperature was ~4 K; but under the assumption that the adjacent hexagonal phase would tend to have a depressive influence on the T_c of the β-phase, it was assumed that the "pure cubic phase" would have a transition temperature of between 4 and 6 K. This is in accord with the more recent value of 6.4 K obtained by COLLINGS et al [Coℓ72[a], Coℓ73[a]][Ho73] who used an elaborate extrapolation argument based on the results of a series of conjoint magnetic susceptibility and low temperature specific heat studies of a number of quenched Ti-Mo alloys, cf. [Mon10.4].

2.2.5 The Influence of Structure -- Thin Films

Thin films of several transition elements, including Ti, have been prepared by SCHMIDT [Sch73] using the technique of ion-beam bombardment. This method differs from other sputtering procedures in that it is carried out in vacuum. Any of the rare gases may be employed and the charge-state of the ejected atoms may be selected.

In SCHMIDT's experiments, Xe (purity > 99.999%) bombarded a Ti target (purity > 99.99%) to produce films of thickness 1-5 μm at a deposition rate of 50-300 Å min^{-1} (per square inch). The Ti as well as the Zr, Mo and W films investigated were all finely polycrystalline and possessed the crystal structures of the targets. An unexpected feature of the results, however, was a sizeable lattice dilatation (frequently several percent), and a spectacular enhancement of the transition temperature. Based on the reported evidence only noble gas ions were present as impurities (to the extent of ~0.1-1.0%) in the Ti (and other) films. The transition temperature of the Ti film was 2.52 K. The enhancement mechanism was not identified, although reference was made to the numerous previously cited phenomenological sources of T_c enhancement such as order-disorder effects, particle size effects, and so on.

2.3 THE ISOTOPE EFFECT

One of the fundamental concepts of superconductivity, traceable from the work of FRÖHLICH [Fro50] through that of BARDEEN [Bar50] to the modern theories of BCS [Bar57] and McMILLAN [Mcm68] is the existence of an electron-phonon interaction. It followed that T_c was expected to vary with atomic mass according to:

$$T_c \propto M^{-\alpha} \quad , \quad \text{with } \alpha = 0.5 \qquad (2-2)$$

a relationship that between 1950 and 1958 was verified for the elements Hg, Sn, Tl and Pb [Net60].

According to BCS theory in the form:

$$T_c = 1.14 <\omega> \exp[-1/n(E_F)V] \qquad (2-3)$$

where V represents a net electron-phonon pairing potential, cf. [Mon12.1], and the McMILLAN result as stated above, Eqn.(2-1), it follows that:

$$T_c \propto \omega \propto M^{-1/2} \qquad (2-4)$$

provided that the terms in the exponents remain invariant. PINES [Pin58] for example has argued that the $n(E_F)V$ in the BCS approximation should be independent of atomic mass even for transition elements. So strong was the belief in the $T_c M^{1/2}$ = constant law that departures from it in the cases of Ru and Os in 1961 and 1962 were taken as evidence for the existence of an alternative electron-coupling interaction, following which further experimental evidence was eagerly sought that would bolster such a train of thought. However, already in 1959 SWIHART [Swi59] had been able to show that α could depart considerably from 1/2 even within the context of a standard electron-phonon model. In particular he predicted an isotope effect for Ti varying as $M^{-0.35}$.

Subsequently, instead of attempting to compute the magnitude of an isotope effect for a particular element or compound, it became more usual to proceed in the opposite direction, and to use the isotope effect to provide information about the detailed nature of the electron-phonon interaction or the Coulomb pseudopotential μ*.

In the first study of isotope effect in a transition element, NETZEL [Net60] set out to measure the transition temperatures of Ti samples enriched to at least 85% in the isotopes Ti^{46}, Ti^{48} and Ti^{50}, respectively. Assuming $T_{c,Ti}$ = 0.36, the isotope effect $T_c M^{\alpha}$ = constant should, according to Swihart, yield a difference in T_c between adjacent isotopes of 0.008 K if α = 0.50, and 0.005 K if α = 0.35. Since in past studies the transition temperature of Ti had been measured with an uncertainty of typically ± 0.01 K for a single specimen; and the width of the transition in pure material can be as large as 0.06 K (BATT [Bat64]) and in view of the fact that the standard deviation among the "best" values from three independent experiments is ± 0.03 K (cf. Table 2-3), the undertaking was doomed to failure at the outset.

2.4 SUPERCONDUCTING TRANSITION TEMPERATURE OF UNALLOYED Ti

Table 2-3 lists chronologically the measured superconducting transition temperatures of unalloyed Ti and offers a critical value based on the results of three selected investigations. Table 1-1 (and of course the above discussion) should be consulted for further details.

2.5 THERMODYNAMIC CRITICAL FIELD OF UNALLOYED Ti

Along with some twenty-five other metallic elements, pure Ti is a type-I superconductor.[†] FALGE [Faℓ63] has measured the reversible magnetization loops of a sample of very high purity electron-beam melted Ti containing less than 5 ppm Mn and less than 2 ppm Fe. The thermodynamic critical fields, H_c, determined from these loops satisfied the relationship

$$H_c = 56[1 - (T/0.42)^2] \qquad (Oe) \qquad (2-5)$$

indicative of a zero-K value of 56 Oe in association, of course, with a T_c of 0.42 K.

TABLE 2-3 SUPERCONDUCTING TRANSITION TEMPERATURE (T_c) OF UNALLOYED TITANIUM

T_c, K	Literature		
<1.13	W.Meissner	[Mei30]	
1.72	W.J. de Haas and P.M. van Alphen	[Deh31]	
<1.77	W.Meissner *et al.*	[Mei32]	
<1.0	D.Shoenberg	[Sho40]	
<1.1	R.T.Webber and J.M.Reynolds	[Web48]	
0.527±0.006	J.G.Daunt and C.V.Heer	[Dau49]	
0.558	T.S.Smith and J.G.Daunt	[Smi52]	
0.387	T.S.Smith *et al.*	[Smi53]	
0.49±0.01 0.37±0.01	M.C.Steele and R.A.Hein	[Ste53]	
0.36±0.02 0.22±0.01 1.11±0.01 0.27±0.01	R.Netzel	[Net60]	
0.42	R.L.Falge	[Faℓ63]	
0.4	C.J.Raub and G.W.Hull	[Rau64[c]]	
0.33±0.03	R.H.Batt	[Bat64]	
0.23	N.B.Brandt and N.I.Ginzburg	[Bra65]	
0.23 0.26 0.32-0.34	N.B.Brandt and N.I.Ginzburg	[Bra66]	
0.45±0.03	V.F.Degtyareva *et al.*	[Deg74]	
Critical Values			
	T.S.Smith *et al.*	[Smi53]	0.387
	M.C.Steele and R.A.Hein	[Ste53]	0.37±0.01
	R.H.Batt	[Bat64]	0.33±0.03
Mean	(value ± std.dev.)		0.36±0.03

[†] Elemental type-II superconductors are Nb and Tc

3

TITANIUM-VANADIUM BINARY ALLOYS

Ti-V has played a dominant role in the early development of type-II superconductivity. Since the late 1950's when the results of transition temperature measurements were first reported, Ti-V has been subjected to both critical field and critical current measurement. The alloy has been investigated in both bulk and thin-film form, in the non-equilibrium rapidly-quenched state, and in equilibrium and aged conditions. Measurements of superconducting transition temperature have been made within the context of an argument over the possible existence of a magnetic, non-phonon-moderated mechanism for electron-electron coupling. The results of measurements of lower and upper critical fields, and flux-flow resistivity, have aided in the development of extensions of the GINZBURG-LANDAU-ABRIKOSOV-GOR'KOV (GLAG) theory into the realm of the paramagnetic mixed state, and have helped in defining the order of the s/n transition in extreme type-II superconductors. Critical current density investigations in Ti-V and related alloys, under way since the early 1960's have helped to define the phenomenological roles played by uniaxial cold work (wire drawing, rolling) and precipitation heat treatment in optimizing the current carrying capacities of superconducting alloys, and have aided in the development of the microscopic concepts of flux pinning. These investigations are mentioned in Tables 1-4, 1-5, and 1-6.

PART 1: THE SUPERCONDUCTING TRANSITION IN TITANIUM-VANADIUM ALLOYS

3.1 SYSTEMATICS OF THE TRANSITION TEMPERATURE

In one of the earliest studies of Ti-V alloys, BUCHER, BUSCH AND MÜLLER [Buc59] used a magnetic technique to measure the variation of T_c with composition within the concentration range 50 through 98.5 at.% V, expressing the results as a function of electron/atom ratio, e/a. T_c was shown to decrease monotonically from 6.7 K as e/a increased from its initial value of 4.50. Subsequent work [Huℓ61][Coℓ75[b]] would recommend a higher value than this for Ti-V$_{50}$ and would go on to show that a broad maximum ($T_c \sim 7.3$ K) is centered near Ti-V(60 at.%) (i.e. Ti-V$_{60}$). Electron/atom ratio (or "average group number") had been used as early as 1957 by MATTHIAS [Mat57] as a display variable against which to plot the superconducting transition temperatures of the elements. Subsequently an "electron density" was claimed to be a closely related, but more physically meaningful parameter [Jen65]. Within the context of Ti-V (and other Ti-TM) alloys DeSORBO [Des65] claimed that an "effective electron/atom ratio", N_{eff}, which took into account relative atomic-size-effect corrections, could even more successfully parameterize, for transition-metal alloys, the positions of the peaks not only in T_c (at $N_{eff} \sim 4.4$ and 6.6) but also in the related quantity H_{c2} (at $N_{eff} \sim 4.4$ only). In a landmark paper HULM and BLAUGHER [Huℓ61] explored the superconducting transition temperatures of a comprehensive matrix of well-homogenized binary transition-metal (TM) alloys selected from four columns and the three long rows (3d, 4d and 5d) of the periodic table. For Ti-V, a rapid increase in T_c with solute concentration in the low-solute-content quenched martensitic (α^{m}) regime was noted; this was followed by a minimum and a second rapid increase leading up to a maximum as the solute concentration was still further increased in the bcc-based regime. This confirmed the earlier observations of MATTHIAS and co-workers [Mat59] who had studied seven other Ti-TM alloy systems. In the vicinity of the maximum, transition temperatures of 7.5, 7.6 and 7.5 K were encountered in the bcc-phase alloys at solute concentrations of 60, 70 and 80 at.% V, respectively. These may be compared with the values of about 8.0, 8.3 and 8.2 K[+] obtained by EFIMOV et al

---

† Interpolated from a published curve.

[Efi70] for Ti-V (60, 70 and 80 at.%) and the 8.0, 8.0 and 7.8 K obtained for the same set of solute concentrations, by BELLIN et al [Beℓ69] as part of a study of the ternary Ti-Nb-V system. HULM and BLAUGHER [Huℓ61] compared their experimental transition temperatures with a set of values calculated using the BCS-MOREL [Bar57][Mor59] expression:

$$T_c = 0.855 \, \theta_D \exp[-1/n(E_F)V], \qquad (3-1)$$

(cf. [Mon12.1]) with the aid of numerous assumptions inspired by (rather than based on) the specific heat work of CHENG, WEI, and BECK [Che60] and some data derived by linear interpolation between α-Ti and V. It was encouraging at the time that reasonable qualitative agreement between theory and experiment was obtained. Subsequently, as a result of the work of CHENG, GUPTA, van REUTH and BECK [Che62] and more recently of UPTON [Upt72] and COLLINGS et al [Coℓ75b], accurate low temperature specific heat data are available for the quenched Ti-V system. CHENG et al [Che62] noted that excellent agreement between experimental and BCS-MOREL (or GOODMAN [Che62])-calculated T_c composition dependence could be obtained by setting V = 0.1288 eV atom. The calorimetrically derived superconducting transition temperatures of quenched Ti-V, and two other Ti-TM alloy systems to be considered in subsequent chapters, are displayed as functions of conventional e/a in Fig.3-1.

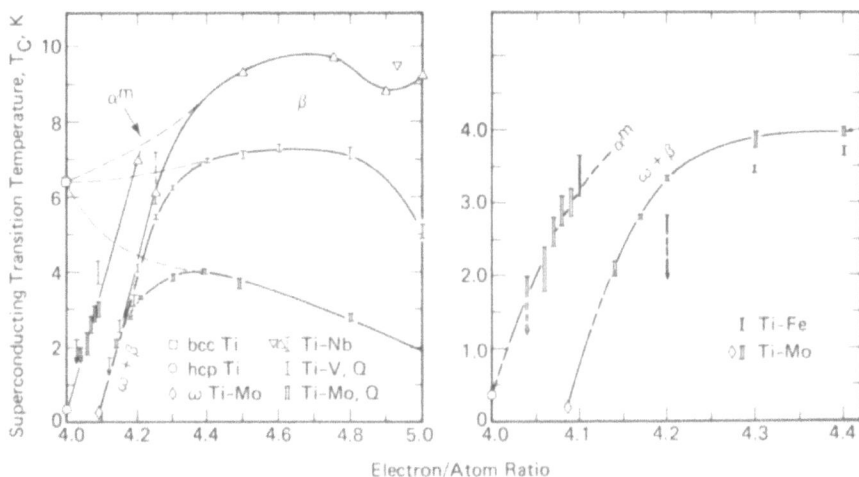

FIGURE 3-1. Calorimetrically determined superconducting transition temperatures in Ti-V, -Nb, -Mo and -Fe alloys in various microstructural states. Data sources: Ti-Mo, COLLINGS et al [Coℓ72, Ho73]; Ti-V, COLLINGS et al [Coℓ75b]; Ti-Fe, HO et al [Ho73a]; Ti-Nb, VERKIN [Ver80] (△), SHCHETKIN and KHARCHENKO [Shc73] (▽) and SUKHAREVSKII et al [Suk68] (Ī).

3.2 MICROSCOPIC MECHANISMS OF SUPERCONDUCTIVITY

3.2.1 The Electron-Phonon Interaction

Evidence for BCS-type[†] superconductivity has been found in the results of several studies on Ti-V alloys. Attention has already been drawn for example to the work of HULM and BLAUGHER [Huℓ61] who qualitatively verified the BCS-MOREL relationship with the aid of some preliminary data of CHENG, WEI, and BECK [Che60] and to the quantitative agreement subsequently achieved by CHENG, GUPTA, van REUTH and BECK [Che62].

As a result of nuclear spin-lattice relaxation time (T_1) measurements, ASAYAMA and MASUDA [Asa65] showed that:

$$T_1 \propto \exp\left(\delta/k_B T_c\right) \qquad (3-2)$$

with $2\delta = 3.5\, k_B T_c$. $\qquad (3-3)$

Agreement with BCS theory was implied by drawing attention to the fact the BCS energy gap at absolute zero, $2\Delta_{00}$ is also $3.5\, k_B T_c$ (see for example [Mon15.1] for further discussion of the BCS energy gap).

3.2.2 The Magnetic Interaction

Ti-V (along with other Ti-TM alloys) has played an important role during the development of arguments for and against the "magnetic interaction" theory of superconductivity. According to an early suggestion by MATTHIAS and co-workers [Mat59, Mat63, Mat63[a]], superconducting electrons in certain alloy systems -- notably alloys of Ti with "magnetic" (i.e. late 3d) transition metals -- instead of being coupled *via* virtual phonons as in conventional theory, were thought to have been coupled by a mechanism which in other contexts is known as "indirect exchange" or the "s-d interaction". The magnetic-interaction postulate arose as a result of a series of experiments by

MATTHIAS, COMPTON, SUHL and CORENZWIT [Mat59] on the composition dependence of superconductivity in some Ti-(3d)TM and Ti-(4d)TM alloys. It appeared, for example, that dilute additions of Mn, Fe, and Co to Ti were able to raise T_c more rapidly than were the corresponding 4d-elements Re, Ru and Rh.

This suggestion triggered a series of measurements on the rate of increase of T_c with composition in low-concentration non-magnetic Ti-TM(3d) alloys, notably Ti-V and Ti-Nb, for comparison with its rate of increase in those previously studied. The first paper to specifically address the problem was that of BUCHER and MÜLLER [Buc61] who studied a closely spaced series of alloys in the concentration range 1 through 15 at.% V, only to find that V brought about essentially the same rate of increase of T_c with concentration as did Cr, Mn, Fe and Co. Subsequently, in a continuation of that work, HEINIGER and MÜLLER [Hei64] studied the superconducting transition in Ti-V, and two other systems (Ti-Fe,-Nb) using low temperature specific heat, whereupon it was found that plots of $\ell n(T_c/\theta_D)$ *versus* $1/\gamma$ agreed reasonably well with those for Ti-Zr and with the expectations of the BCS theory. Examples of such "BCS-MOREL-MORIN-MAITA plots" [Mor63] are to be found in [Mon8.7][Mon10.3][Mon11.9]. It could therefore be concluded that since dT_c/dc in Ti-V, itself a BCS superconductor, was comparable to that in Ti-Fe, a magnetic interaction mechanism was no longer required to explain its superconductive behavior. In a subsequent paper, BUCHER, HEINIGER and MÜLLER [Buc65] reconfirmed their opinion that a special interaction mechanism was not required to explain the behavior of dilute alloys of Ti with Mn, Fe, and Co, but that *all* such dilute alloys did in fact possess anomalously large rates of increase of T_c with composition -- possibly associated in some way with details of the quenched microstructure. Further evidence in favor of this conclusion can be found in a comparison of the T_c composition dependences of quenched α'-martensitic and equilibrium-α-phase Ti-V alloys, Fig.3-2. RAUB and ZWICKER [Rau65], the authors of the latter data, believed that the higher T_c's exhibited by the quenched alloys may have been induced by the presence of extraneous phases -- particularly threads of β-phase. Indeed the T_c of α'-Ti-V appears to lie somewhere between that of α-Ti-V and what would now be expected of its β-phase allotrope (cf. Fig.3-1).

---

† Short for virtual-phonon-moderated electron-electron coupling modelled by the BCS theory [Bar57] in the weak-coupling limit.

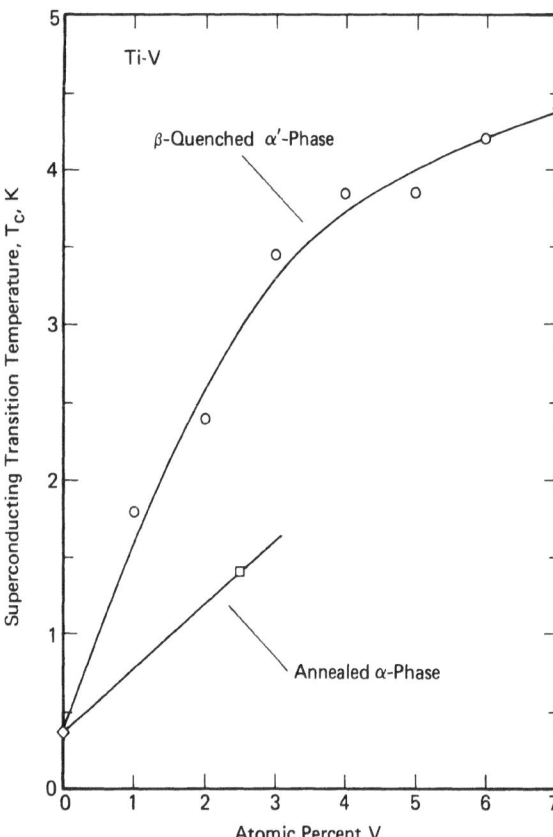

FIGURE 3-2. Intercomparison of the T_c-composition dependences of β-quenched (to α'-martensite) and α-equilibrated Ti-V alloys. Data sources: unalloyed Ti ($\Diamond$), Table 2-3; α'-Ti-V ($\circ$) BUCHER and MÜLLER [Buc61] (see also, COLLINGS and HO[Coℓ75^a]); α-Ti-V ($\square$), RAUB and ZWICKER [Rau65].

3.3 TRANSITION TEMPERATURE AND MICROSTRUCTURE

3.3.1 Properties of Annealed and Quenched Microstructures

Fig.3-1 with its independent branches corresponding to the α^m-martensitic and the bcc-base structures confirms the early observations of HULM and BLAUGHER [Huℓ61]. The measurements on quenched dilute α'-Ti-V alloys initiated by those authors were extended by MÜLLER and co-workers [Hei64][Buc65]; and the results have been summarized and compared with those for several other α^m-Ti-TM alloys by COLLINGS and HO [Coℓ75^a]. The occurrence of the hexagonal martensite, α', in quenched Ti-TM alloys within certain composition ranges, and its orthorhombic variant, α'', in others are dealt with in [Mon1.6, Mon1.7] wherein the decision was made to refer to them collectively as α^m.

It is instructive to compare, as in Fig.3-2 the results of RAUB and ZWICKER's [Rau65] measurements of an annealed single-phase-α Ti-V alloy with those of MÜLLER and co-workers [Hei64][Buc65] who studied a series of quenched samples, with the postulated magnetic mechanism in mind. In the quenched alloys, T_c did in fact appear to be enhanced, but this enhancement it was claimed appeared to have more to do with the complicated metallurgical structure than any fundamental interaction mechanism. COLLINGS and HO [Coℓ69] had suggested that the enhanced T_c was a property of the disordered nature of the martensitic structure, and postulated a localized soft-phonon mechanism within the framework of the McMILLAN [Mcm68] model. But more recently, COLLINGS and WHITE [Coℓ78^a], as a result of a detailed statistical curve-fitting of the calorimetric superconducting transitions, were led to the conclusion that the intuition of MÜLLER *et al* was correct and that the upper-temperature limb of the broad superconducting transition exhibited by quenched alloys owed its origin to the influence of small amounts of higher T_c included-β material, a conclusion which as mentioned above had been anticipated by RAUB and ZWICKER [Rau65]. The applicabilities of curve-fitting procedures to the analysis and interpretation of rounded specific heat transitions in situations such as this are discussed in [Mon10.3, Mon10.8].

As indicated in Fig.3-1, T_c *versus* composition for quenched β-phase Ti-V alloys passes through a broad maximum defined by transition temperatures of 7.1 ± 0.1, 7.3 ± 0.1, and 7.15 ± 0.15 K occurring at 50, 60 and 80 at.% V.

With reference to the metallurgical literature COLLINGS *et al* [Coℓ75^b] have claimed that the rapid decrease in superconducting transition temperature that occurred when the V concentration dropped below about 30 at.% had to do with the depressive influence on the average T_c of ω-phase precipitation of both the isothermal (already present in the quenched samples at room temperature) and athermal varieties (the latter occurring reversibly upon further cooling of the sample into the liquid-N and liquid-He temperature ranges. Particularly interesting is the fact that, in spite of the two-phase nature of the alloys, the calorimetric superconducting transitions were just as sharp, and the "jumps" just as high, as they were in the single-phase-β alloys, thus indicating the existence of a complete proximity effect in the as-quenched alloys. A discussion of proximity effect as it relates to superconductivity in two-phase ω+β-Ti-TM alloys is given in [Mon11.9]. Also noticeable in the results of the calorimetric measurements of COLLINGS et al [Col75^b] on the Ti-V(3-90 at.%) system is the parallelism that exists between the composition dependences of T_c and γ (the electronic specific heat coefficient), thus demonstrating qualitatively the expected BCS-based "proportionality" between T_c and the Fermi density-of-states, $n(E_F)$; cf. Eqn.(3-1), and further discussions of this topic in Sect.14.3 and [Mon12.00].

3.3.2 Influences of Aging and Other Heat Treatments

The effects on the superconducting transition temperature of ω+β-phase Ti-V alloys, of (a) aging and (b) aging plus "up-quenching"[+] into the β-phase field, have also been examined. COLLINGS et al [Col75^b] have taken two quenched alloys, Ti-V$_{15}$ and Ti-V$_{19}$, the choice of composition having been guided by the ready availability of metallurgical information through the work of HICKMAN [Hic68, Hic69], and aged them at 300°C for times of up to 2000 h. During aging the β-phase became enriched with V, at the expense of the ω-precipitates, which however grew in volume. In the competition between the increasing T_c of the β-matrix, and the depressive influence of the ω-precipitates, only the outer layers of which participated in the superconductivity, the bulk T_c in fact increased with aging time.

The influence of aging on the superconducting transition temperature of cold-worked Ti-V alloys has been briefly considered by EFIMOV et al [Efi70]. During the phase-separation which accompanied the aging (in their case they referred to a decomposition into two bcc phases, β' and β -- cf. [Mon2.3] for a discussion of this metallurgical topic) T_c rose or fell depending on the geometry of the precipitates, their volume fractions, and the relative values of the individual composition-dependent T_c's.

+ Up-quenching: For example, the rapid raising of the temperature of a two-phase alloy into a higher-temperature single-phase field (followed of course by rapid chilling in order to preserve the product so obtained for observation and measurement) thereby generating in that field a well-defined compositional modulation.

LUHMAN and co-workers [Luh70, Luh71] have measured, using an inductive technique, the superconducting transition temperature of a heat-treated Ti-V$_{24.4}$ alloy. Measurements were made first of all with the alloy in the as-quenched condition (T_c = 4.38 K). In that state, the sample was supposed to have been single-phase bcc; however, since alloys in that concentration range have been shown several times to possess negative resistivity temperature dependences, it is possible that athermal ω-phase was present at the temperature of measurement. Anomalous resistivity temperature dependences of Ti-V and similar alloys have been considered by COLLINGS [Col74], CHANDRASEKARAN et al [Cha74] and discussed in detail in [Mon5.5] through [Mon5.7]. The Ti-V$_{24.4}$ sample was next aged 35 h/300°C and remeasured. The ensuing decrease in T_c to 3.88 K is interpretable as the result of the overwhelming depressive influence of ω-phase precipitation. Similar behavior has been encountered in Ti-Mo alloys, the subject of a subsequent review. Up-quenching to 450°C (3 min) caused a "reversion" of the ω-phase, with no time for significant solute redistribution. In such reversions, the ω is converted back to β, leaner in solute than the parent phase (hence the appellation β') which persists in a strain-stabilized state upon the alloy's return to room temperature and below. The Ti-rich zones of the reverted ω-phase expressed the inherent tendency toward clustering of the Ti-V system. Naturally, the up-quenched all-β structure had a higher transition temperature (in particular 5.09 K) than the initial, aged, ω+β-phase alloy.

One's ability to interpret the effects on T_c of aging and up-quenching illustrates the utility of superconductive measurements as metallurgical (i.e. microstructural) diagnostic tools.

3.4 SPUTTERED FILMS

Superconducting films are generally of interest from a *processing* rather than a *fundamental* standpoint. The goal of sputtering is frequently not to produce a material with new superconducting properties (there are of course some obvious exceptions to this statement) but rather to prepare in thin-film form, with particular applications in mind, a conventional or new alloy having superconducting properties appropriate to bulk material. Film deposition techniques have been applied to the production of superconductors in ribbon form and to the coating of RF cavities.

SPITZER [Spi74[a]] has measured the superconducting transition temperatures of a wide range (17-100 at.%) of Ti-V alloys which he had prepared by DC triode sputtering onto fused quartz substrates. T_c was noted to rise monotonically with V content to a maximum value of 12.8 K at 99.99 at.% V, before dropping to a terminal value of 5.3 K, an acceptable figure for bulk unalloyed V (cf. [Rad66]). The enhanced T_c encountered among the alloys was attributed to the possible presence of impurities or precipitates in an otherwise single-phase-β Ti-V matrix. Enhanced critical current densities were also noted.

PART 2: THE MIXED STATE IN TITANIUM-VANADIUM ALLOYS

The literature of magnetic effects in Ti-V alloys, more than that of any other single system, traces an experimental accompaniment to the historical development of theories of the mixed state. Transitions into and out of that state have been detected both magnetically and resistometrically. Experimental values of the lower critical fields have been compared with predictions of ABRIKOSOV theory [Abr57]. Early upper critical field data have been compared with estimates based on GOR'KOV-SHAPOVAL theory according to which, at zero K [Elb64]:

$$H_{c20}^{GS} = (3/2\pi)\ ec\gamma T_c/k_B\ \sigma_n \qquad (esu,\ Oe) \qquad (3-4)^{\dagger}$$

where σ_n is the residual normal-state electrical conductivity. The temperature dependences of H_{c2}^{*} were compared against the GOR'KOV and ABRIKOSOV-GINZBURG expressions which are, respectively:[++]

$$H_{c2}^{*}(t)/H_c = (1.77 - 0.43t^2 + 0.07t^4)\kappa_{GL1} \qquad (3-5)$$

$$H_{c2}^{*}(t)/H_c = 2\sqrt{2}\ (1 + t^2)^{-1}\kappa_{GL1} \qquad (3-6)$$

[with $t \equiv T/T_c$

$$H_c = 2.43\gamma^{1/2}T_c\ (1-t^2) \qquad (3-7)$$

and where κ_{GL1} is the value of the Ginzburg-Landau parameter at $T = T_c$].

These equations are collectively referred to as the results of "GLAG theory". The temperature dependences

of H_{c2}^{*} were also compared with further developments of that theory by CLOGSTON [Clo62], CHANDRASEKHAR [Cha62] and MAKI [Mak64[a]], who considered normal-state Pauli paramagnetism which tended to lower the upper critical field below the H_{c2}^{*} of GLAG theory, and with more fully developed expressions for the temperature dependence of H_{c2} again due to MAKI [Mak64[b], Mak66] and WERTHAMER, HELFAND, and HOHENBERG (WHH) [Wer66] who took the paramagnetic susceptibilities of both the normal and superconducting states into account. The magnetic properties of the mixed state are treated generally in [Mon13.00][Mon14.00][Mon15.00] and, as they apply to Ti-Nb alloys, in Chapter 7 of this book. Also considered, using Ti-V as a working substance, were ABRIKOSOV's predictions with regard to the surface critical field H_{c3}.

In what follows we consider the acquisition of experimental critical field data, and go on to trace that coupling of experiment to theory which led to improvements to the initial GLAG model of the mixed state and to what might be termed "post-GLAG" theory and has in fact been referred to as "neo-GLAG" theory [Hak67[b]].

3.5 THE LOWER CRITICAL FIELD, H_{c1}

A theoretical lower critical field may be calculated using the relationship:

$$H_{c1} = (H_c/\sqrt{2}\ \kappa_{GL})(\ell n\ \kappa_{GL} + 0.08) \qquad (3-8a)$$

$$\doteq (H_c/\sqrt{2}\ \kappa_{GL})\ \ell n\ \kappa_{GL} \qquad (3-8b)$$

usually attributed to ABRIKOSOV [Abr57] -- but see [Mon13.7] for further discussion -- where H_c, the bulk thermodynamic critical field, is easily calculable from low-temperature specific heat data, cf. Eqn.(3-7), such as that referred to above [Che62][Col75[b]]. κ_{GL} is the Ginzburg-Landau parameter which, according to

---

+ In cgs-practical units Eqn.(3-4) becomes $H_{c20}^{GS} = 5.53\times10^4\ \gamma\rho_n T_c$, in error by almost a factor of 2 as compared to the more recent value, $H_{c20}^{*} = 3.06\times10^4\ \gamma\rho_n\ T_c$.

++ The use of the (*) anticipates the problem that H_{c2} is seriously overestimated due to the neglect of paramagnetic effects.

the work of GOR'KOV [Gor60] followed by GOODMAN [Goo62] is calculable in terms of atomic constants and physical measurables. Thus, as discussed in [Mon13.5], [Mon14.4], κ_{GL} may be expressed as the sum of intrinsic (or "clean") and extrinsic (or "dirty") components according to:

$$\kappa_{GL} = \kappa_{GL}^{c} + \kappa_{GL}^{d} \quad , \qquad (3-9)$$

(defined at t = 1)

where $\kappa_{GL}^{c} = 1.60 \times 10^{24} \, T_c \, \gamma^{3/2} (n^{2/3} S/S_F)^2 \quad (3-10)$

and $\kappa_{GL}^{d} = 7.49 \times 10^{3} \, \gamma^{1/2} \, \rho_n \qquad (3-11)$

where S/S_F is the ratio of the Fermi surface area to that of a free-electron gas of density, n; T_c is in K; γ, the electronic specific heat coefficient, is in erg cm^{-3} K^{-2}; and ρ_n, the residual resistivity, is in Ω cm.

In "ideal" reversible type-II superconductors, as the applied field passes across the threshold, H_{c1}, of the mixed state the (diamagnetic) magnetization, $-4\pi M$, previously increasing with increasing field, passes through a cusp before eventually dropping to zero at H_{c2}, cf. [Mon16.1]. In "non-ideal" irreversible type-II superconductors, on the other hand, in which flux pinning impedes both the ingress and egress of flux, the peak in the magnetization curve becomes rounded off and moved to higher values of $-4\pi M$ and H_a. BLAUGHER [Bla65] investigated the field, H_{c1}^{*}, at the magnetization maximum in cold-rolled-and-vacuum-annealed strips of Ti-V(38.5, 48.4, and 58.5 at.%) at temperatures of 2.1 and 4.2 K, comparing the results, as in Table 3-1, with values of H_{c1} calculated using Eqns.(3-8) through (3-11). The H_{c1}^{*}'s appeared to follow $1-t^2$ temperature dependences. Although they were much larger than the corresponding H_{c1}'s, the differences diminished with decreasing V concentration indicating that the flux-pinning centers responsible for the irreversibility were impurities that had been introduced *via* the V component.

3.6 THE UPPER CRITICAL FIELD, H_{c2}

In what follows the usual symbol-convention for the upper critical field will be adopted in which H_{c2} denotes the fully paramagnetically limited quantity; H_{c2}^{*} the same theoretical quantity in the limit of no

TABLE 3-1. CRITICAL FIELDS AT 4.2 K AND OTHER PARAMETERS FOR SOME Ti-V ALLOYS -- after Blaugher [Bla65].

At.% V	T_c K	ρ_n $\mu\Omega$ cm	$\kappa_{GL}^{\dagger}$	H_{c1} Oe	H_{c1}^{*} Oe	$H_{c}^{\dagger\dagger}$ Oe
38.5	7.07	91.5	74.1	50	450	1,190
48.4	7.20	72.3	60.0	62	1,200	1,270
58.5	7.49	58.1	49.3	78	2,000	1,370

+ Calculated from Eqn.(3-9).

†† Bulk thermodynamic critical field calculated from the results of measurements of Cheng *et al.*[Che62]

paramagnetic limitation (i.e. such that the superconducting and normal states have the same magnetic susceptibilities); H_r represents the resistometrically determined field; H_u represents the magnetic or calorimetric value, and in all cases the further subscript '0' designates a zero-K value [Ber62][Hak65].

3.6.1 Temperature Dependences -- Early Studies

The first experimental data on the upper critical field of Ti-V alloys were acquired by BERLINCOURT and HAKE [Ber62, Ber63, Ber63a] who used a resistive method at 1.2 K. The normal state electrical resistivity, ρ_n, was also measured enabling a GLAG upper critical field to be calculated from the expression:

$$H_{c2}^{*} = 2.58 \times 10^4 \, \rho_n \, \gamma \, T_c \, (1-t^2) \qquad (Oe) \qquad (3-12)$$

with ρ_n in Ω cm, and γ in erg cm^{-3} K^{-2}. As explained in [Mon14.6] the nonparamagnetic limit of MAKI theory would today yield the more familiar relationship:

$$H_{c2}^{*} \, (dirty) = 3.06 \times 10^4 \, \rho_n \, \gamma \, T_c \, (1-t^2)(Oe) \qquad (3-13)$$

This field was subsequently referred to by HAKE [Hak67b] as the "neo-GLAG" upper critical field; the (*) should strictly be confined to *it* rather than shared with the GLAG upper critical field, although it is *also* not paramagnetically limited. BERLINCOURT and HAKE [Ber63a] compared the measured H_r with the H_{c2}^{*} given by Eqn.(3-12) and with H_p, the CLOGSTON paramagnetically imposed upper critical field limit given by:

$$H_p = 18.4\ T_c\ (1-t^2) \quad . \qquad (kOe) \qquad (3-14)$$

Since the occurrence of a superconducting transition in an applied magnetic field, H_a, is controlled by the free energy balance between the normal and superconducting states, and $g_s(H_a)$ is an increasing function of H_a, whilst $g_n(H_a)$ (if, as is usually the case, the Ti-TM alloy is Pauli paramagnetic in the normal state) decreases with H_a, it follows, with reference to Fig.3-3 that the experimentally determined upper critical field lies below the smaller of H_{c2}^* and H_p. This is illustrated beautifully in Fig.3-4, in which it is seen that H_r coincides with H_p when this is

lower, and with H_{c2}^* at high V concentrations when $H_{c2}^* < H_p$. Subsequent studies of Ti-V gave experimental support to discussions of the various early formulations for the temperature dependence of H_{c2}. So-called "GLAG-theory" expressions developed during this period were: that due to GOR'KOV [Gor60], Eqn.(3-5), the ABRIKOSOV-GINZBURG [Gin56] expression, Eqn.(3-6), or the prediction by SHAPOVAL [Sha62] of a nearly linear temperature dependence between the end-points:

$$H_{c2}^*(t) = \begin{cases} H_{c20}^*(1-2t\ell n2) & t \ll 1 \qquad (3\text{-}15a) \\ H_{c20}^*(1-t)8/\pi^2 & t \cong 1 \qquad (3\text{-}15b) \end{cases}$$

with $H_{c20}^* = 3.03\ ^\kappa{GL1}\ H_{c0}$ (cf. Eqn.(3-4)). (3-15c)

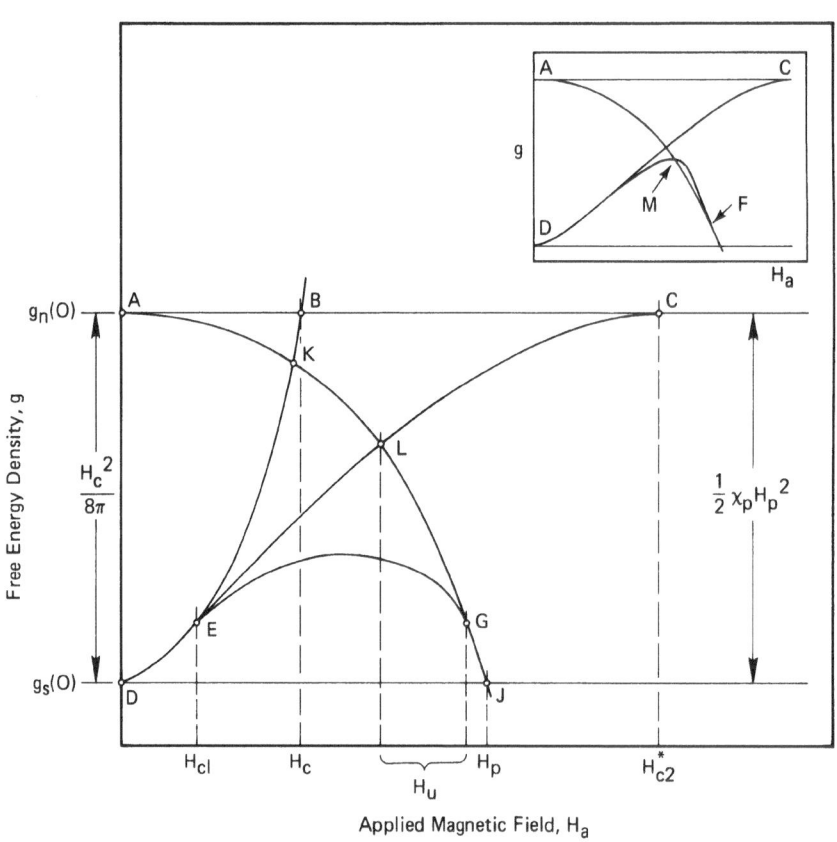

FIGURE 3-3. Schematic diagram of normal-state and superconducting-state free energies *versus* applied magnetic field, H_a. An arbitrary temperature $0 < T < T_c$ is assumed. The normal state in the absence of Pauli spin paramagnetism is represented by AC and its continuation; that in the presence of spin paramagnetism, of susceptibility χ_p, by the parabola AJ. The field-ignoring superconducting state is represented by DJ and its continuation; the point J defines the Clogston-Chandrasekhar paramagnetically limited first-order critical field, H_p. The parabola DB represents the response of a type-I superconductor to H_a, with B itself occurring at the thermodynamic critical field H_c. The magnetization of a GLAG (non-paramagnetic) type-II superconductor is represented by DEC with lower critical field, H_{c1}, and upper critical field, H_{c2}^*, at E and C, respectively. The inclusion of spin in the normal state leads to a first-order transition at an upper critical field corresponding to L. The further inclusion of spin in the superconducting state can lead to a second-order transition at an upper critical field corresponding to G. Both these upper critical fields, designated H_u in the figure, are lower than H_{c2}^*. The inset suggests that within the context of paramagnetic theory it is also possible to have a first-order *s/n* transition (at M) followed by a second-order transition at F with a metastable state lying between the two — after WERTHAMER *et al* [Wer66].

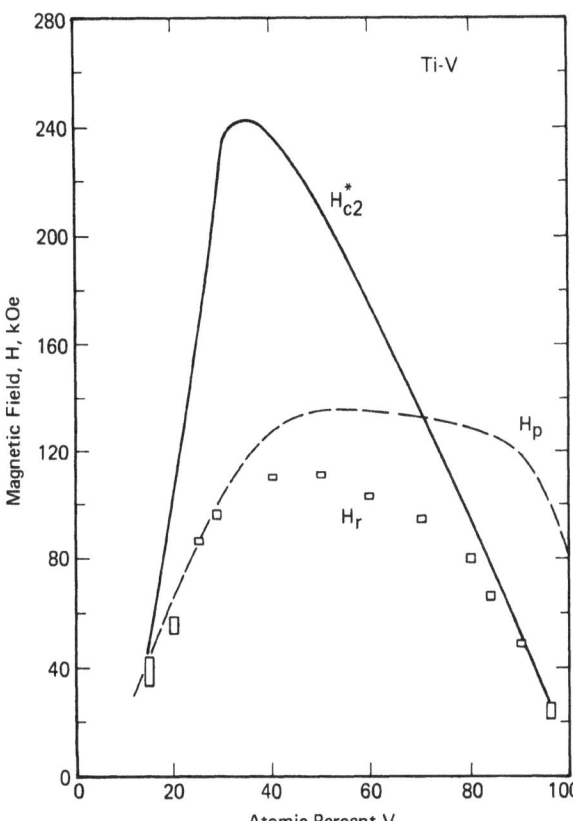

FIGURE 3-4. Upper critical fields at 1.2 K for Ti-V alloys. A comparison of the resistively measured values, H_r (from *threshold* to *full restoration* of resistivity at 10 A cm^{-2}), with the Clogston paramagnetic limit, H_p, and the GLAG-calculated upper critical field, H_{c2}^* — after BERLINCOURT and HAKE [Ber63[a]].

Tests of these expressions of course required critical fields to be measured as functions of temperature.

EL BINDARI and LITVAK [Eℓb63[a], Eℓb64] focused attention on the SHAPOVAL formulation although the measurements of six experimental Ti-V alloys (60-93 at.% V) were carried out at only a single temperature, 4.2 K. SHIBUYA and AOMINE [Shi65] intercompared the critical field theories of GOR'KOV, ABRIKOSOV-GINZBURG and SHAPOVAL at 1.2 K, using as input data the results of the experiments of BERLINCOURT and HAKE [Ber63[a]] (reported above) and CHENG et al [Che62].

3.6.2 Composition Dependences -- Early Studies

Again, using Ti-V as a model system, HANCOX [Han66] attempted to fit the BERLINCOURT and HAKE data, within the framework of the GLAG-CLOGSTON approach, to a function of H_{c2}^* and H_p. The first equation

$$H_{c2} = H_{c2}^* H_p / (H_{c2}^* + H_p) \qquad (3\text{-}16a)$$

underestimated H_r throughout the entire concentration range, and by about 30% at the peak ($\sim$ 40 at.% V). Further improvements were made, leading to

$$H_{c2} = H_p(\sqrt{H_p^2 + 4H_{c2}^{*2}} - H_p)/2H_{c2}^* , \qquad (3\text{-}16b)$$

a relationship which had already been derived by MAKI and TSUNETO [Mak64[b]] and others (cf. [Mon17.2]) from the generalized GINZBURG-LANDAU equation, and which underestimated (i.e. lay below) H_r by only 16% at the peak, Fig.3-5. Having obtained a satisfactory functional form for H_{c2} it was found that excellent agreement with the experimental composition dependence of H_r could be obtained simply by increasing H_p by some 30-40%.

Although no explanation was offered, it was realized that the application of full CLOGSTON Pauli paramagnetic limitation led to excessively severe reductions of the upper critical field. Further work by MAKI [Mak66] and WERTHAMER, HELFAND, and HOHENBERG

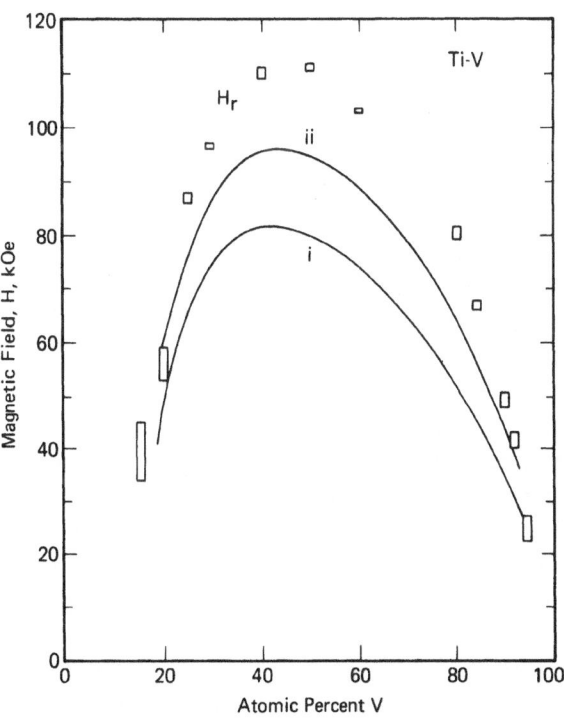

FIGURE 3-5. Upper critical fields of Ti-V alloys. A comparison of the 1.2-K experimental data of Berlincourt and Hake, H_r, with a pair of suggested (and temperature ignoring) calculated results: curve (i) represents Eqn (3-16a), and curve (ii) represents Eqn (3-16b) — after HANCOX [Han66].

(WHH) [Wer66] was to clarify this picture and to provide a basis for modern theories of the paramagnetic mixed state and the associated upper critical field. The situation is considered in greater detail in Sect.7.10 within the context of Ti-Nb alloys.

3.6.3 Temperature Dependences -- Paramagnetic Limitation

Possible modifications to the Pauli paramagnetic limitation effect can be expressed in terms of a parameter α attributed to MAKI and defined by:

$$\alpha \equiv \sqrt{2} \; H_{c20}^{*}/H_{p0} \quad , \tag{3-17}$$

in terms of which the CLOGSTON field becomes:

$$H_{p0} = \frac{\sqrt{2}}{\alpha} \cdot H_{c20}^{*} \quad . \tag{3-18}$$

In the work of KIM and STRNAD [Kim66] on a series of Ti-V alloys, α underwent a variation of from 0.5 to 2.7. Subsequent upper critical field investigations of Ti-V usually dealt with experimentally based discussions of the validities of the theory of MAKI as propounded in his later papers (e.g. [Mak66]) and that of WHH [Wer66]. Papers considering these further developments of GLAG theory with reference to data for Ti-V alloys are those by NEURINGER and SHAPIRA [Neu66] and HAKE [Hak65, Hak67, Hak67[a], Hak67[b]] and excellent reviews of the subject have been given by BRANDT [Bra72], CODY [Bre73, p.159], and BEASLEY [Bea82].

Pauli spin paramagnetism has the effect of reducing the upper critical field below the GLAG value $H_{c2}^{*}(T)$. But the theories of CLOGSTON [Cℓo62] and CHANDRASEKHAR [Cha62] and one of the 1964 papers of MAKI [Mak64[a]] overestimated this reduction. It was found necessary to offset the effect of normal-state Pauli paramagnetism by the introduction of paramagnetism into the superconducting component of the high-field mixed state. Spin-orbit scattering, again according to MAKI [Mak66], and to WHH [Wer66], did this by partially decoupling the aligned spins of the Cooper pairs thereby permitting them to respond paramagnetically to the applied field. As discussed fully in [Mon15.5], the MAKI paramagnetic limitation parameter defined in Eqn.(3-17) can be expressed in terms of atomic constants:

$$\alpha = 3\mu_{B}c/ev_{F}^{2}\tau_{tr} \tag{3-19}$$

$$= 3\hbar/2mv_{F}^{2}\tau_{tr} \quad , \tag{3-20}$$

where the symbols have their usual meanings in transport theory, or in terms of experimental superconducting (s) or normal-state (n) quantities:

$$\alpha_{s} = 5.23\times10^{-5}(-dH_{c2}/dT)_{T_c} \tag{3-21}$$

$$\alpha_{n} = 2.37 \; \rho_{n}\gamma \quad . \tag{3-22}$$

In WHH theory, according to [Mon15.6], the reduced upper critical field temperature dependence

$$h^{*}(t) \equiv \frac{H_{c2}}{(-dH_{c2}/dt)_{t=1}} \tag{3-23}$$

is a unique function of α, t, and λ_{so} the spin-orbit-scattering frequency parameter defined by

$$\lambda_{so} = 2\hbar/3\pi k_{B}T_{c}\tau_{so} \tag{3-24}$$

where τ_{so} is the spin-orbit-scattering relaxation time.

The absence of paramagnetic limitation can come about in either of two ways: (a) the normal state is not paramagnetic, or (b) the superconductive state, hence the mixed state, has the same paramagnetic susceptibility as the normal state into which it will transform at H_{c2}. In terms of the MAKI-WHH theory the first condition is satisfied by $\alpha = 0$, and the second by $\lambda_{so} = \infty$.

3.6.4 Experimental Evaluation of the MAKI-WHH Theory

As in the studies just considered, λ_{so} is generally regarded as being an adjustable parameter there being no reliable way of deducing its value from atomic parameters, especially for alloys and inter-metallic compounds. The nearest approach to an *ab initio* assignment of a value to λ_{so} was taken by HAKE [Hak67[a]] in a series of calculations of the upper critical field limits of Ti-V and several other alloys and representative compounds. In doing so, HAKE assumed the spin-orbit-scattering mean-free-path (mfp) to be equal to twice the transport-electronic mfp which in turn was taken to be comparable to the inter-atomic spacing.

The situation in 1965 and the difficulties of making an adequate experimental assessment of the status of contemporary mixed-state theory were discussed by HAKE [Hak65] with reference to Ti-V(22.5 and 25 at.%) and several other alloys. NEURINGER and SHAPIRA [Neu66] in 1966, armed with data from their experiments on Ti-V$_{58}$ (as well as Ti-Nb$_{44}$ and Ti-Ta$_{52}$), were among the first to discuss the effects of these new improvements to the theory. At the same time WHH themselves were comparing the predictions of their theory with the upper critical field temperature dependences of Ti-V$_{65}$ (and Ti-Nb$_{44}$) as measured by KIM and STRNAD [Kim66] and SHAPIRA and NEURINGER [Sha65], respectively, and plotted in the format $h^*(t)$ *versus* t, with the $h^*(t)$ defined as in Eqn.(3-23) but with H_{c2} replaced by the experimental quantity, H_u. The normal-state value of the paramagnetic limitation parameter (Eqn.(3-22)) was accepted, while λ_{so}, regarded as an adjustable parameter, was chosen to yield what seemed to be a best fit to the data. By experiment: for Ti-V$_{58}$ [Neu66] α_s = 1.56, λ_{so} = 0.7; for Ti-V$_{65}$ [Wer66], α_n = 1.37, λ_{so} = 0.75.

3.6.5 Properties of the Paramagnetic Mixed State

NEURINGER and SHAPIRA [Neu66] drew attention to the interesting result embodied in the λ_{so} = ∞ condition referred to above, that the stronger the spin-orbit-scattering the less the alloy was subject to the influence of normal-state Pauli paramagnetism, and the closer the upper critical field was able to approach its theoretical limit. Next, since

$$\tau/\tau_{so} \sim (Ze^2/\hbar c)^4 \qquad (3\text{-}25)$$

(where τ is the total scattering relaxation time and Z is the atomic number) the effect of spin-orbit-scattering should increase rapidly with increasing Z -- thus their work suggested the possibility that Pauli paramagnetic limitation could be more-or-less overcome by the use of suitable alloying elements with high atomic numbers, a philosophy which was to guide the direction of numerous superconductive-alloy development programs over the subsequent fifteen years. Unfortunately the work of NEURINGER and SHAPIRA [Neu66] contained a serious flaw, the consequences of which were overlooked by all at the time, having to do with the observation that the experimental upper critical field of Ti-Ta$_{52}$ exceeded theoretical expectations even with λ_{so} = ∞. This result implied that factors other than (or in addition to) spin-orbit-scattering were at work to elevate H_{c2}. The possi-

bility of an electron-phonon-renormalized density-of-state reduction of the Pauli susceptibility had already been considered by CLOGSTON [Clo62] and forgotten until very recently. The roles played by this and other factors in reducing the need for unrealistically rapid spin-orbit-scattering rates have been discussed in detail by ORLANDO *et al* [Orl79] and BEASLEY [Bea82], and the entire situation is reviewed in [Mon15.8].

A magnetization study of Ti-V$_{22.5}$ in the as-cast state, Ti-V$_{25}$ in the as-cast and two other heat-treatment conditions, as well as several other alloys, by HAKE [Hak67^b] as summarized in [Hak67], provided the forum for a detailed discussion of the joint effects of Pauli paramagnetic limitation and spin-orbit scattering as embodied in the MAKI-WHH approach. The paper also discussed the qualitative significance of normal-state Pauli paramagnetism in association, on one hand, with fully spin-decoupled superconducting electrons, and on the other, with completely spin-coupled Cooper pairs. In the first case, and with reference to Fig.3-6, a uniform background Pauli paramagnetism, $M_n(H_a) = M_s(H_a)$, gives rise to a negatively sloped basis for the usual superconductive $M_s(H_a)$ curves and in so doing does not influence H_{c20}^* which is simply the "GLAG-MAKI" value in the dirty limit. In the second extreme case, the diamagnetism of the Cooper pairs forces the paramagnetic excursion of $M_s(H_a)$ to fall short of its previous value, Fig.3-6(b); but since the condensation energy remains unchanged, the constant-area rule requires $H_{u0}(PP)$ to now be less than H_{c20}^*. Partial decoupling of the Cooper spins by spin-orbit-coupling-induced spin-flip scattering leads to an intermediate situation in which $H_{u0}(PP) < H_{u0}(PP + SO) < H_{c20}^*$ as illustrated in the figure. The other important matter to be dealt with in this landmark paper had to do with the order of the superconducting transition. The simple CLOGSTON theory called for a first-order transition to the normal state, whilst MAKI-WHH predicted a second-order transition, as observed. In conclusion it is noted that Ti-V$_{25}$ is not the best alloy upon which to study the improvements to CLOGSTON predictions which result from the application of MAKI-WHH theory since, as indicated already in Fig.3-4, for that alloy H_p (= 96.4 kOe) and H_r ($\sim$90 kOe) are already in good agreement.

A discussion along similar lines was also offered by BRAND [Bra72] who studied Ti-V (36, 70, and 82 at.%) and some Ti-Nb alloys in order to extend the

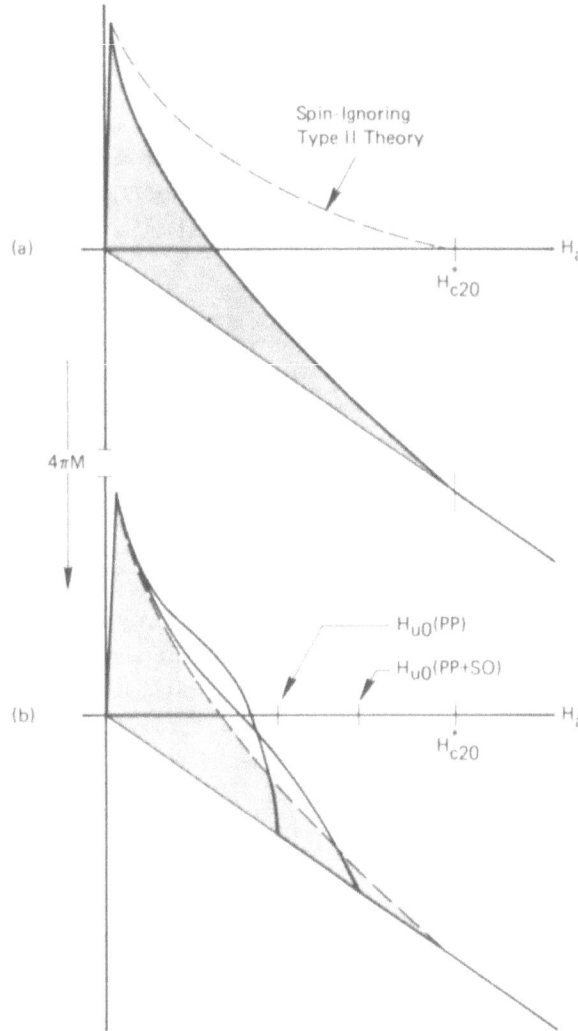

FIGURE 3-6. Schematic magnetization ($4\pi M$) *versus* applied field (H_a) curves for a dirty type-II superconductor at zero K. In (a) the susceptibilities of the normal and superconducting states are supposed equal — either zero or Pauli paramagnetic (shaded diagram). In (b) two levels of Pauli paramagnetism of the superconducting component of the mixed state are represented — one curve terminating in the upper critical field $H_{u0}(PP)$ represents complete superconductive-state Cooper pairing and full Clogston paramagnetic limitation; the other, terminating at $H_{u0}(PP+SO)$ showing the effect of some spin-depairing consequent upon the introduction of spin-orbit-induced spin-flip scattering — after HAKE [Hak67[b]].

conventional elastic-vortex-lattice theory of κ_{GL}-modulation flux pinning into the paramagnetic mixed state. The approach taken is fully discussed in [Mon21.8].

Both HAKE and BRAND have commented on the fact, referred to in Sect.3.6.3 that although two different physical situations are represented, the limits $\alpha \longrightarrow 0$

and $\lambda_{so} \longrightarrow \infty$ yield identical relationships between reduced critical field and reduced temperature. In addressing the interesting question as to whether spin-orbit scattering, which decouples spins, also depairs the superconducting electrons, BRAND [Bra72] pointed out that "although λ_{so} is a pair spin-decoupling parameter, it is not a depairing parameter", and HAKE [Hak67[b]] emphasized that "such scattering effectively acts to decouple spin pairing but not time-reversal (i.e. momentum) pairing of one-electron wave functions appropriate to the strong nonmagnetic scattering case". In the light of this separability of spin and momentum coupling it is interesting to recall the "magnetic-interaction" idea of electron pairing, considered in Sect.3.2.2, which relied entirely on *spin* interactions.

3.7 THE SURFACE SHEATH CRITICAL FIELD, H_{c3}

The properties of the surface-sheath (or "third") critical field have been reviewed in [Mon13.9]. According to predictions by SAINT-JAMES and de GENNES [Sai63] if the applied field is orientated parallel to the surface of an ideal sample, superconductivity should persist up to (or, conversely, nucleate at) a field, H_{c3}, higher than the usual upper critical field and given by

$$H_{c3} = 2.38 \, \kappa_{GL} \, H_c \quad . \qquad (3-26)$$

Then for type-II superconductors, for which according to the usual theories $H_{c2} = \sqrt{2} \, \kappa_{GL} \, H_c$ (cf. Eqn. (3-15b)), it follows that:

$$H_{c3} = 1.69 \, H_{c2} \quad . \qquad (3-27)$$

Superconductivity in fields between H_c (type-I) or H_{c2} (type-II) and H_{c3} is attributed to the existence of a surface film about one coherence length in thickness.

KWASNITZA and RUPP [Kwa66] have measured the third critical fields of a series of homogenized (6h/1300°C) alloys which, in contrast to the subjects of earlier such investigations, possessed large values (up to 25) of the Ginzburg-Landau parameter, κ_{GL}. The upper critical field was measured magnetically, while the potentiometrically determined H_{c3} was taken to be the field for which J_c became independent of H_a; i.e. the field at the foot of the $J_c(H_a)$ curve. In confirmation that the effect being measured was intrinsic, rather than artifactual, was the observation that the influence of the sample's surface condition on the measured quantities was weak and in particular that H_{c3} itself was independent of the smoothness of the surface, and of the presence of Cu plating. The measured critical fields are listed in Table 3-2; it is interesting to note that their ratios (average value = 1.53 ± 0.27) are in excellent agreement with the predictions of Eqn.(3-27).

TABLE 3-2   MEASURED CRITICAL FIELDS AT 4.2 K OF SOME Ti-V ALLOYS [Kwa66]

At.% V	$H_u^\dagger$ kOe	$H_{c3}^{\dagger\dagger}$ kOe	$H_{c3}|H_u$
88	17.3±0.2	28.1±0.5	1.64±0.03
91	14.3±0.3	16.4±0.8	1.15±0.03
94	8.2±0.1	12.7±0.3	1.54±0.03
97	3.8±0.2	6.8±0.3	1.78±0.07

† From magnetization measurements.

†† H_{c3} is the field at which J_c (criterion: 2×10^{-8} V across sample) becomes independent of H_a.

3.8 FLUX-FLOW RESISTIVITY

Although associated with the presence of transport current, flux-flow resistivity is better treated as a mixed-state effect than as a transport property. Ti-V alloys have been the subjects of flux-flow resistivity studies using either microwave loss or resistive measurements. Using the latter technique, KIM, HEMPSTEAD, and STRNAD [Kim65] in a classical paper have reported on the results of a general study of flux-flow resistivity in type-II superconductors with particular reference to the Ti-V(20-95 at.%) alloy system. Although flux-flow resistivity is an expression of energy dissipation resulting from the motion of the ABRIKOSOV flux under the influence of the Lorentz force produced as a result of its interaction with a transport current, it is interesting to note that a useful expression relating the flux-flow and normal-state resistivities, ρ_f and ρ_n respectively, can be derived if it is simply assumed that mixed-state resistivity is a result of current sharing between the superconductive and normal-core components

of the ABRIKOSOV state. Such a model leads to the expression

$$\rho_f/\rho_n = H_a/H_{c20} \quad , \qquad (3\text{-}28a)$$

where H_{c20} is some upper critical field at zero K. The derivation of this expression, which is considered in greater detail in [Mon23.8] and of course the original article, is based on the principles that: (a) the vortex core is equivalent to a cylinder of normal metal; (b) that the volume fraction occupied by such cores is given by H_a/H_{c2}. KIM et al [Kim65] then went on to show experimentally that the upper critical field to be used in their expression was not the paramagnetically limited value, but the bare H_{c20}^* of GLAG theory as given by Eqn.(3-13). Thus Eqn.(3-28a) became:

$$\rho_f/\rho_n = H_a/H_{c20}^* \quad . \qquad (3\text{-}28b)$$

With reference to Ti-V as a test alloy, Fig.3-7 intercompares an experimentally measured (and extrapolated to zero-K) upper critical field, H_{r0}, with an Eqn.(3-13)-calculated value of H_{c20}^* and a flux-flow-resistively measured (also extrapolated to zero-K) value of it. The utility of the flux-flow method in obtaining values for the non-paramagnetically limited

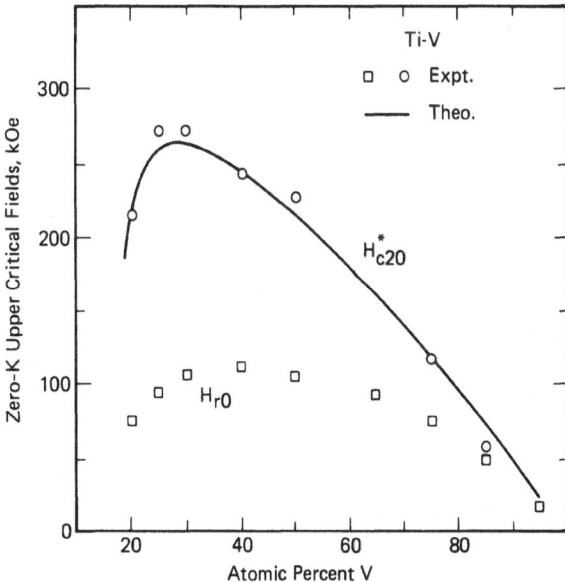

FIGURE 3-7. Zero-K upper critical fields of Ti-V alloys. Intercompared are: (a) the extrapolated (t→0) resistive upper critical field, H_{r0} (□); (b) an extrapolated flux-flow-measured critical field, H_{c20}^* (expt), obtained via Eqn (3-28) (○); (c) a theoretical GLAG upper critical field, H_{c20}^* (theo) derived from Eqn (3-13) — after KIM et al [Kim65].

fields is clearly evident. Other features of the flux-flow phenomenon are considered in Sect.7.16 within the context of a discussion of the mixed-state properties of Ti-Nb alloys.

In measurements of the temperature dependence of flux-flow resistivity it was noted that at low fields, ρ_f initially decreased as the temperature increased from near zero. This combined with the eventual increase in ρ_f with temperature led to a minimum in $\rho_f(T)$ -- the so-called flux-flow-resistivity minimum.

Subsequently HACKETT *et al* [Hac67] studied flux-flow resistivity, using a microwave surface-impedance technique at 14.4 GHz, on Ti-V (30, 40 and 50 at.%). The results in the form ρ_f/ρ_n *versus* H_a were in remarkable qualitative agreement with the earlier DC resistivity studies. Available for discussion were DC results for Ti-V$_{50}$ taken at 2.0, 3.5, and 4.2 K, and microwave results for Ti-V$_{40}$ taken at 1.18, 2.18, and 4.2 K, enabling a comparison to be made of the respective ρ_f/ρ_n ratios at, for example, 4.2 K, 60 kOe. In the former case $[\rho_f/\rho_n]_{DC} \cong 0.32$, while in the latter it was found that $[\rho_f/\rho_n]_{14.4\ GHz} \cong 0.28$. The DC technique, because of the high current densities needed to de-pin the flux lines, was expected to lead to erroneously high ($\sim$20%) values of ρ_f, for which reason the microwave method was regarded as being the more satisfactory. It was noted that the upper-field transition indicated by ρ_f/ρ_n *versus* H_a increased in sharpness as T decreased, thereby leading to questions, couched within the framework of the CLOGSTON approach and the assumption of an s/n volume ratio equal to H_a/H_{c20}^*, about the possibility of a first-order paramagnetically-limited transition to the normal state.

After an interval of ten years investigations were re-opened, again using microwave techniques, into the two important questions in flux-flow resistivity that had earlier been asked, viz: *(a)* the origin of the flux-flow-resistance minimum; *(b)* the thermodynamic order of the flux-flow-resistively-determined s/n transition at H_{c2}. In agreement with the earlier

conclusions of HACKETT, MAXWELL, and KIM [Hac67], KIM and KIM [Kim75] claimed evidence for the possibility of a first-order transition at temperatures lower than those actually attained in their experiments. AKACHI, KIM, and KIM [Aka75] as a result of experiments on a series of Ti-V alloys including Ti-V$_{72.5}$, developed empirical expressions for the position and depth of the flux-flow-resistance minimum, the origin of which it was claimed had to do with dissipative mechanisms other than simple electromagnetic Lorentz-force-instigated loss [Cℓe68].

3.9 MAGNETIZATION MEASUREMENT AS A METALLURGICAL DIAGNOSTIC TECHNIQUE

LUHMAN, TAGGART, and POLONIS [Luh70, Luh70^a], [Poℓ71] have conducted magnetization measurements on a Ti-V$_{28}$ alloy after subjecting it to various heat treatments. One such treatment consisted of an anneal at 1000°C to form the equilibrium-β structure followed by a step-quenching to 540°C and an aging at that temperature for times of 1, 10, and 24 h. An increase in the magnetic hysteresis was noted in response to the development of compositional fluctuations of the type discussed by KOUL and BREEDIS [Kou70] (precursors to the establishment of α+β-phase equilibrium which is unable to be fully attained at that temperature). A second type of aging heat treatment, (10 and 24 h)/350°C, resulted in the development of various amounts of ω-phase precipitation. The response of the magnetization field dependence to this was noted, as was that of reversion to a double-β structure after 24h/350°C followed by up-quenching to 550°C, cf. Sect.3.3.2. It may be concluded that magnetization hysteresis studies are useful diagnostic aids, after suitable calibration by means of optical and electron microscopy, to the following of changes which occur in the bulk microstructures of superconducting alloys in response to thermal treatments of various kinds.

PART 3: CURRENT TRANSPORT EFFECTS IN TITANIUM-VANADIUM ALLOYS

Current transport effects are considered under the categories: *(a)* fluctuation superconductivity -- which could just as well have been included in either of the previous major subdivisions; *(b)* critical current density; *(c)* normal-state transport properties

related to superconductivity. Electrical transport experiments on Ti-V alloys, frequently carried out since 1962 when HAKE, BERLINCOURT, and LESLIE [Hak62^a] performed their pioneering studies of critical current density in cold-rolled Ti-TM alloys, have contributed to our understanding of the pinning and de-pinning mechanisms which control critical current and flux flow in superconductors, and of the superconductive critical-fluctuation effects which manifest themselves above T_c.

3.10 FLUCTUATION SUPERCONDUCTIVITY

HAKE [Hak69] has pointed out that the upper critical field transition of type-II superconductors is second order at all points on the transition surface. Accordingly one would expect critical fluctuations into the superconductive state to extend for some distance outside that surface. With the aid of Ti-V (22.5 and 25 at.%) and some other Ti-TM alloys, HAKE [Hak69[a]] has explored these fluctuations in H_a-T space both along the temperature axis, and parallel to the H-axis at temperature below T_{c0}; and in a second paper [Hak70] has reported on the results of low current density (30 A cm^{-2}) resistive measurements carried out on Ti-V_{25} as functions of temperature ($10 \lesssim T \lesssim 20$ K) and magnetic field ($H_a \lesssim 40$ kOe). According to theory, the relative fluctuation-conductivity temperature dependences are:

$$\Delta\sigma_{f\ell}/\sigma_n \propto (\ell n\ t)^{-1} \text{ for a 2-D film, and} \quad (3\text{-}29a)$$

$$\Delta\sigma_{f\ell}/\sigma_n \propto (\ell n\ t)^{-1/2} \text{ in three dimensions.} (3\text{-}29b)$$

Measurements of the zero-field resistivity temperature dependences yielded good qualitative agreement with Eqn.(3-29b). It was noted that the *zero-field* fluctuations (expressed as reductions in the average bulk normal-state resistivity) were present to temperatures of at least 2 T_c. The possibility of this being a static rounding effect due to compositional fluctuations was considered, but rejected on the grounds that the transition temperature of any reasonable solute-rich component would be too low to explain the effect. High-magnetic-field tails were also noted in the range $H_{c2} \lesssim H_a < H_{c3}$ [Hak69[a]] and in *high fields* fluctuation superconductivity was claimed for the temperature range $T_c < T < 2T_c \sim 3T_c$ [Hak70].

3.11 CRITICAL CURRENT DENSITY

3.11.1 Bulk Alloys

Although Ti-V alloys were found to be technically unimportant from the standpoint of current-carrying superconductivity, they served as useful model systems for the study of flux pinning by microstructural features introduced through cold deformation [Hak62[a]] or cold deformation followed by heat treatment [Vet68]. The results of HAKE *et al* [Hak62[a]], although interpreted at the time in terms of a MENDELSSOHN-GOODMAN (filamentary) rather than ABRIKOSOV-ANDERSON (flux pinning) model, illustrated that a relationship existed between cold-rolling-induced microstructure and critical current density that was sensitive to the orientation of the applied transverse magnetic field with respect to the rolling plane. VETRANO *et al* [Vet68] took the next step and showed how considerable enhancements of the critical current density of their Ti-V(20 wt.%) (i.e. Ti-20V) sample could be achieved through the development by heat treatment of a suitable *precipitate* structure. The results of the transport-property investigation were discussed, with reference to those of a detailed microstructural study, in terms of the pinning of ANDERSON flux bundles. The highest critical current densities were obtained by aging at 400°C suggesting at the time that ω-phase, on account of its better size match, was a more effective flux pinner than the α-phase precipitation which was claimed to take place at temperatures above 500°C. On the other hand the 300°C aging, although it certainly produced ω-phase, left the Ti-20V alloy with a very low transition temperature (<4.2 K). Although the microstructural results themselves were instructive, the associated critical current densities were less than those obtainable with suitably processed Zr-Nb or Ti-Nb.

Comparable work had previously been carried out by the Soviet group headed by SAVITSKII who studied the alloy series Ti-V(23, 32, 52, 56, 86 and 95.3 wt.%) after subjection to various sequences of cold work and heat treatment such as, for example, cold rolling and drawing 99% plus vacuum annealing (1-25h)/(200-900°C), also 1h/1100°C [Efi70]. The effects of this thermomechanical processing on microstructure and critical current density were discussed. It was noted, for example, that during the aging of Ti-26V at 400°C, J_c (26.6 kOe) passed through a maximum of 6.1×10^4 A cm^{-2} after 5 h. More recently, BRAND [Bra72, Bra75] has investigated homogenized-plus-cold-rolled Ti-V(36, 70 and 82 at.%) alloys. Resistive measurements of $J_c(H)$ were carried out, the results being presented in terms of reduced critical pinning force density *versus* reduced magnetic field in a fundamental study of flux-pinning models appropriate to the paramagnetic mixed state. Further details of cell-wall pinning in the paramagnetic mixed state, with particular reference to this work of BRAND, are given in [Mon21.8].

3.11.2 Sputtered Films

As indicated in Sect.3.4, sputtering techniques of various kinds have been applied to produce superconductive coatings, with bulk properties, of a wide range of alloys and compounds. Within the context of Ti-V alloys EDGECUMBE et al [Edg64] were responsible for an early investigation of the current-carrying capacities of sputter-deposited Ti-V_{45} and other alloy and compound films. The results of J_c versus H_a measurements were reported in which both transverse and longitudinal orientations of the applied field (with respect to the current direction) had been arranged. Some typical data, for two values of the applied field, are given in Table 3-3. Subsequently,

SPITZER [Spi71, Spi74[a], Spi74[b]] prepared a wide range of Ti-V alloys (17-100 at.% V) by DC triode sputtering from separate targets onto fused quartz substrates. The critical current densities at 4.2 K of samples about 3000 Å thick, after annealing for several h at 700°C and rapidly cooled, were measured using a pulsed field, pulsed current technique. Unexpectedly high values of J_c were obtained, e.g. $\sim 5 \times 10^5$ A cm^{-2} at 60 kOe (cf. Table 3-3) leading to the suggestion that the films may have been contaminated by precipitation, the existence and/or nature of which was unable to be confirmed or specified at the time.

3.12 NORMAL-STATE PROPERTIES RELATED TO SUPERCONDUCTIVITY

Several recent normal-state property studies, which have a direct bearing on superconductivity in Ti-V alloys include the electrical resistivity measurements of COLLINGS [Col74, Col78] and PREKUL et al

TABLE 3-3   INTERCOMPARISON OF THE CRITICAL CURRENT DENSITIES (J_c) OF SPUTTER-PROCESSED AND CONVENTIONALLY-PROCESSED TITANIUM-VANADIUM ALLOYS

At.% V	Applied Field kOe	Condition	J_c, A cm^{-2}		Reference
			$J \perp H$	$J \parallel H$	
45	30	sputtered film	2.0×10^4	5.2×10^5	[Edg64]
45	40	" "	1.0×10^4	2.0×10^5	"
88.4	60	sputtered film	3.7×10^5		[Spi74[a]]
92.7	"	" "	"		"
99.4	"	" "	"		"
99.8	60	sputtered film	6.3×10^5		"
99.96	"	" "	6.7×10^5		"
20	30	recrystallized	6.0×10^3		[Vet68][+]
20	40	plus 20h/400°C	3.0×10^3		"
29	30	cold rolled 78%	0.032×10^3	0.08×10^3	[Hak62[a]][++]
50	"	" " 92%	0.35×10^3	4.1×10^3	"
70	"	" " 93%	0.93×10^3	15×10^3	"

[+] J_c criterion: current at quench.

[++] J_c criterion: 0.25 µV across 7 mm.

[Pre73, Pre74] and the thermal conductivity investigation of MORTON *et al* [Mor77]. The anomalous resistivity composition dependences of Ti-TM alloys which, as is well known, depart from a simple parabolic law are discussed, together with their anomalous temperature dependences in [Mon5.5].

The results of the measurements by PREKUL *et al* [Pre74] on Ti-V alloys with compositions greater than 20 at.%, at temperatures from the ice point to below T_c showed: *(a)* a marked deviation from MATTHIESSEN's Rule (cf. [Mea65]) leading to a negative $d\rho/dt$ in alloys of compositions less than 40 at.% V; *(b)* a cusp in the residual resistivity at the point (20 at.%V, 170 $\mu\Omega$ cm). Had the results been extended to lower V concentrations it would have been noted that for compositions less than 20 at.% V, $d\rho/dt$ would have gradually approached "normalcy" in magnitude and sign. In justifying the anomalous resistivity, PREKUL invoked a spin-flip-scattering mechanism. The negative $d\rho/dt$ and the anomalously broad transition to the superconducting state was explained by PREKUL *et al* [Pre74] in terms of the postulated spin-flip-scattering operating in addition to normal scattering. Such a pair-breaking effect, it was claimed, would lower the critical temperature by $T_c^{bs} - T_c^{sf} = \Delta T_c^{sf}$ (where bs and sf refer to "band structure" and "spin-flip", respectively); i.e. by an amount equal to the width of the superconducting transition. It is interesting to note than spin-orbit-induced spin-flip scattering had already appeared as an ingredient of the theory of mixed-state paramagnetism, its presence being necessary to offset the full effect of CLOGSTON-CHANDRASEKHAR paramagnetic limitation. In addition, the negative temperature dependence was claimed by PREKUL *et al* to be the low-temperature part of a resistivity minimum. The influence of ω-phase and lattice instability was alluded to, but only with reference to the sharp dip in T_c which separates the "hcp" from "bcc" regions of the T_c-composition plot. Very accurate resistivity measurements by COLLINGS

[Col74, Col78] at temperatures of 273.2, 77.3 K, and an intermediate temperature, have confirmed the existences of: *(a)* negative $d\rho/dt$ in the composition range 20 to $\sim$30 at.% V; *(b)* a cusp in the isothermal (200.0 K, and 77.3 K) resistivity-composition dependences at 20 at.% V. The results were correlated with the occurrence of ω-phase effects, and in conjunction with the results of magnetic-susceptibility composition dependence studies [Col78] were explicable in terms of the occurrence of "isothermal" and "athermal" ω-phases over certain composition and temperature ranges in these alloys.

It is well recognized by now that athermal ω-phase precipitation in metastable (quenched) β-Ti-TM alloys, especially when diffusion is *unequivocally absent* (i.e. in the cryogenic temperature range), is a result of soft-phonon lattice instability. Accordingly one might expect correspondingly anomalous behavior in the low-temperature thermal conductivity. The alloy Ti-Nb$_{35}$ had already exhibited anomalous phonon conductivity at low temperatures. In order to investigate this effect in a related system MORTON *et al* [Mor77] measured the thermal conductivities of Ti-V(38.5, 58.5, 79.0 and 89.4 at.%) over a temperature range of about 2-16 K. They found that although the calculated and experimental values of C (a quantity which characterized "lattice" as distinct from "electronic" scattering) were in satisfactory agreement in the more concentrated alloys, discrepancies became larger as the Ti concentrations increased. An anomalously high phonon component of lattice thermal resistivity (of which C was a measure), particularly noticeable in Ti-V$_{38.5}$, appeared to mirror that of the electrical resistivity which we have attributed to ω-phase effects.[+]

_____

+ By "ω-phase effects" we include the possibility of both static and dynamic ω-phase precipitation and the phonons of instability.

PART 4: TABULATED DATA

TABLE 3-4   TRANSITION TEMPERATURES (T_c) OF TITANIUM-VANADIUM ALLOYS

At.% V	T_c K	Data Sources[+]		At.% V	T_c K	Data Sources[+]	
1	1.8	[Buc61]	g	49	7.4	[Huℓ61]	g
2	2.4	"	g	50	7.3	[Che62]	n
3	3.4_5	"	g	50	6.7	[Buc59]	g
	<2.2	[Coℓ75[b]]	n	50	7.5	[Beℓ69]	n
4	3.8_5	[Buc61]	g	50	7.1	[Coℓ75[b]]	n
	2.6	[Buc65]	g	59	7.5	[Huℓ61]	g
5	3.8_5	"	g	60	8.0	[Beℓ69]	n
6	4.2	"	g	60	7.3	[Coℓ75[b]]	n
7	4.4	"	g	69	7.6	[Huℓ61]	g
8	4.3	"	g	70	8.0	[Beℓ69]	n
9	4.0	[Coℓ75[b]]	n	75	7.1_6	[Che62]	n
10	4.2_5	[Buc61]	g	79	7.5	[Huℓ61]	g
	6.3	[Huℓ61]	g	79.5	7.3_6	[Des63]	g[++]
12	<1.7	[Coℓ75[b]]	n	80	7.1_5	[Coℓ75[b]]	n
14	2.3	[Huℓ61]	g	80	7.8	[Beℓ69]	n
15	2.1	[Coℓ75[b]]	n	85	7.0_2	[Che62]	n
17.5	2.5_5	"	n	88.4	6.8_6	[Des63]	g[++]
19	3.2_5	"	n	89.8	6.7_1	"	g[++]
20	3.5	[Che62]	n	90	6.0	[Buc59]	g
20	4.1	[Coℓ75[b]]	n	90	6.8	[Huℓ61]	g
25	5.4_5	"	n	90	6.9	[Beℓ69]	n
29	6.4	[Huℓ61]	g	90	5.7_5	[Coℓ75[b]]	n
30	6.6	[Beℓ69]	n	93.3	6.3_7	[Des63]	g[++]
30	6.2_5	[Coℓ75[b]]	n	96.8	5.8_9	"	g[++]
30	6.1_4	[Che62]	n	97	5.6	[Buc59]	g
39	7.2	[Huℓ61]	g	98.5	5.4	"	g
40	7.4	[Beℓ69]	n	98.8	5.4_0	[Des63]	g[++]
40	6.9_5	[Coℓ75[b]]	n				

+ n = numerical data
   g = graphical data

+ n = numerical data
   g = graphical data

++ Normalized to T_c, vanadium = 5.1 K

TABLE 3-5 TRANSITION TEMPERATURES OF TITANIUM-VANADIUM ALLOYS -- DATA SOURCES

V Concentration Range	Condition	Structures	Procedure	Literature
50 - 98.5 at.% (4 alloys)	Homogenized near the melting point.	β	Magnetic	E. Bucher et al. (1959) [Buc59]
1 - 15 at.% (10 alloys)	As-cast (10 and 15 at.% V also annealed several days at 680°C).	1-10 at.% V, (as-cast), α' 15 at.% V, (as-cast), ω+β	Magnetic	E. Bucher and J. Muller (1961) [Buc61]
10 - 90 at.% (9 alloys)	Homogenized 1500-2500°C and either furnace cooled or quenched -- depending on composition range.	10 at.% V, α' 14-90 at.% V, ω+β tending to β at high concentrations	Magnetic	J.K. Hulm and R.D. Blaugher (1961) [Huℓ61]
20 - 85 at.% (5 alloys)	Homogenized and quenched from 1200°C.	20, 30 at.% V, possibly ω+β 50-85 at.% V, β	Calorimetric	C.H. Cheng et al. (1962) [Che62]
79.5 - 98.8 at.% (6 alloys)	-----	β	---	W. DeSorbo (1963) [Des63]
4 at.%	Quenched from just below solidus in chilled Ar.	α'	Calorimetric and magnetic	E. Bucher et al. (1964) [Buc65][Hei64]
8 - 95 at.% (10 alloys)	As-cast.	8, 12 at.% V, α' 25-95 at.% V, ω+β tending to β at high concentrations	---	Y.V. Efimov et al. (1967/1970) [Efi70]
30 - 90 at.% (7 alloys)	Homogenized 1600°C and furnace cooled and cold rolled.	ω+β tending to β at high concentrations	Magnetic	P.H. Bellin et al. (1969) [Beℓ69]
17 - 100 at.% (more than 60 alloys)	Sputtered to thickness of ∿3000 Å. Annealed several hours 700°C, rapidly cooled.	β (sputtered films)	Pulsed resistometric	H.J. Spitzer (1974) [Spi74a]
3 - 90 at.% (14 alloys)	Quenched from 1000°C or 1350°C.	3 and 9 at.% V, α' 12 at.% V, α'+ω+β 15-90 at.% V, ω+β tending to β at high concentrations	Calorimetric	E.W. Collings et al. (1975) [Coℓ75b]
20 - 97 at.% (10 alloys)	Quenched from 1100°C.	ω+β and β depending on composition	Resistometric	A.F. Prekul et al. (1974) [Pre74]

TABLE 3-6 CRITICAL FIELDS OF TITANIUM-VANADIUM ALLOYS

(a) Lower Critical Fields[†] (H_{c1}^*) at 4.2 K [Bℓa65]

At.% V	H_{c1}^* kOe
38.5	0.45
48.4	1.20
58.5	2.00

† Taken from peak of the magnetization curve.

(b) Resistometrically Determined Upper Critical Fields (H_r) at 1.2 K [Ber63[a]]

At.% V	H_r[†] kOe	
15.0	33.9	44.5
20.0	53.0	58.7
25.0	86.2	87.8
28.7	96.3	97.3
40.5	109.6	111.4
50.0	110.6	112.0
60.0	103.0	103.8
70.0	94.3	96.2
80.0	79.6	81.9
84.0	66.0	67.3
90.0	48.0	50.8
92.0	40.0	43.1
96.0	22.3	27.0

(c) Upper Critical Fields (H_r or H_u) at 4.2 K

At.% V	H_r or H_u kOe	Data Sources[†]	
22.5 (ac)	24	[Hak67[b]]	n
25 (cr)	∿52	"	n
25 (an)	>53	"	n
36	81	[Bra72]	g
58	78	[Neu66]	g
60	>50	[Eℓb64]	g
67	>50	"	g
70	69	[Bra72]	g
75	>50	[Eℓb64]	g
82	51	[Bra72]	g
85	48.0	[Eℓb64]	g
88	36.0	"	g
88	17.3±0.2	[Kwa66]	n
90	28.0	[Eℓb64]	g
91	14.3±0.3	[Kwa66]	n
93	17.5	[Eℓb64]	g
94	8.2±0.1	[Kwa66]	n
97	3.8±0.2	"	n

† Supplied as numerical data -- refers to onset of detectable resistance and full restoration of resistance, respectively; measuring current density, 10 A cm^{-2}.

† n = numerical data
  g = graphical data

ac = as-cast
cr = cold-rolled
an = annealed

TABLE 3-7 CRITICAL FIELDS OF TITANIUM-VANADIUM ALLOYS -- DATA SOURCES

V Concentration Range	Condition	Procedure	Literature
38.5, 48.4 and 58.5 at.%	Sample rod prepared by alternate cold rolling and vacuum annealing.	Magnetic (4.2 K)	R.D. Blaugher (1965) [Bℓa65]
15-96 at.% (13 alloys)	Arc melted and cold rolled.	Resistive (1.2 K, 10 A cm^{-2})	T.G. Berlincourt and R.R. Hake (1963) [Ber63[a]]
60-93 at.% (6 alloys)	Cold rolled.	Resistive (4.2 K)	A. El Bindari and M.M. Litvak (1964) [Eℓb64]
22.5 and 25 at.%	As-cast (ac), ac plus cold-rolled (2:1) (cr), ac plus cr plus solution heat treated and step cooled (an).	Magnetic (4.2 K)	R.R. Hake (1967) [Hak67[b]]
58 at.%	Vacuum annealed.	Resistive (1.3 K - T_c)	L.J. Neuringer and Y. Shapira (1966) [Neu66]
36, 70 and 82 at.%	Homogenized and cold rolled.	Resistive ($\sim$2 K - T_c)	R.A. Brand (1972) [Bra72]
88-97 at.% (4 alloys)	Arc melted, homogenized and polished.	Magnetic (4.2 K) Resistive (4.2 K)	K. Kwasnitza and G. Rupp (1966) [Kwa66]

TABLE 3-8 CRITICAL CURRENT DENSITIES (J_c) AT 4.2 K OF TITANIUM-VANADIUM ALLOYS

(a) Cold-rolled Strip [Hak62[a]]

At.% V	Thickness Reduction %	J_c(30kOe), 10^3 A cm^{-2}	
		J_c ($\perp$)[+]	J_c ($\parallel$)[+]
29	78	0.032	0.08
50	92	0.35	4.1
70	93	0.93	15

(b) Thermomechanically Processed Wire

Wt.% V	Aging Time h	J_c, A cm^{-2} Aging Temperature, °C 400	500	Applied Field kOe	Reference
20[+]	1		1.6x10^3	25	[Vet68][+++]
"	20	9x10^3	4.3x10^3	"	"
26[++]	1	5.7x10^4		26.6	[Efi70]
"	5	6.1x10^4		"	"
"	25	4.2x10^4		"	"

+ J_c criterion: 0.25 µV across 7 mm.

+ Recrystallized 2h/800°C/quenched prior to aging.

++ Cold worked 99% prior to aging.

+++ J_c criterion: current at quench.

4

BINARY ALLOYS OF TITANIUM WITH CHROMIUM, MANGANESE, IRON, COBALT, OR NICKEL

This chapter deals with superconductivity in alloys of Ti with all the other first-long-period (3d) transition elements with the exception of V, which was the subject of the preceding chapter. The solutes are considered in order of increasing group number (with reference to the periodic table) or e/a ratio.

ALLOY GROUP 1: TITANIUM-CHROMIUM BINARY ALLOYS

4.1 TRANSITION TEMPERATURE AS A FUNCTION OF COMPOSITION IN DILUTE Ti-Cr ALLOYS

In an early study by MATTHIAS *et al* [Mat59], Cr was a member of the series of 3d elements which, when added in small amounts to Ti, brought about unexpectedly rapid increases in T_c, and initiated the concept of magnetic, rather than BCS-type, superconductive-electron pairing. Subsequently, BUCHER *et al* [Buc65] used both magnetic and calorimetric techniques in investigating the superconducting transition in Ti-Cr and several other binary Ti-TM and ternary Ti-Zr-TM alloys. It was claimed as a result of this work that the observed rapid increases in T_c with composition were characteristic of all α^m-Ti-TM alloys and not confined to the Ti-("magnetic 3d-element") binary systems and consequently that, for the latter, no special interactive mechanism was required. Ti-Cr$_{2.5}$ was shown to exhibit a very broad calorimetric superconductive transition; this, and direct metallurgical evidence, suggested that the sample may have contained several phases. Stimulated by the results of these early investigations of dilute Ti-TM alloys, Ti-Cr was

again measured both resistometrically and calorimetrically, along with six other Ti-TM systems, by AGARWAL [Aga74] in a further study of dilute martensitic Ti-TM alloys. As was the case with Ti-Fe and Ti-Co alloys, Ti-Cr(0.85 and 2.09 at.%) exhibited both extremely broad and very incomplete superconductive specific-heat anomalies. For example, as T tended to zero, C/T *versus* T^2 extrapolated to >4 mJ mole^{-1}K^{-2} rather than to zero. The resistive transition exhibited by the lower concentration specific heat specimen, centered about 2.6 K, was also extremely broad; that of Ti-Cr$_{2.09}$ centered about 3.9_5 K, was reasonably sharp. The superconductive low temperature specific heats of dilute α^m-Ti-TM alloys are fully discussed in [Mon10.3].

Electrical resistivity studies by HAKE *et al* [Hak62b] had shown Ti-Cr$_{1.2}$ to possess a weak low temperature resistance minimum, Fig.4-1, generally known by now to be characteristic of the presence of a localized magnetic moment. The work of CAPE [Cap63] went on to demonstrate that Ti-Cr$_{1.15}$ (also Ti-Fe$_{0.96}$ and Ti-Co$_{1.3}$) was not a Curie-Weiss paramagnet. These observations, taken together, justify the

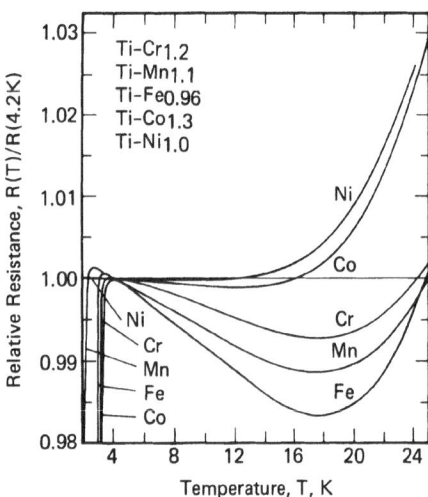

FIGURE 4-1. Relative resistivity temperature dependences, $\rho(T)/\rho(4.2K)$, of dilute (i.e., ~1 at. %) Ti-TM alloys. Resistance minima are exhibited only by Ti-Cr, -Mn and -Fe — after HAKE *et al* [Hak62[b]].

classification of dilute α^m-Ti-Cr as a spin-fluctuation system and provide a clue to the reason for the breadth and incompleteness of the superconducting calorimetric anomaly.

According to AGARWAL [Aga74], rapidly quenched foils of the Ti-Cr which had been heat treated and quenched along with the bulk specimens showed no resistive transitions whatever down to 1.1 K. The conflicting properties of the bulk and foil samples were presumed to be a result of differences in microstructure incurred in response to the differing quench rates. The possible presence of traces of ω- or β-phase precipitation in the more massive, and consequently more slowly cooled, samples was considered, but the discrepancy was not able to be properly resolved.

4.2 TRANSITION TEMPERATURE AND MICRO-STRUCTURE IN QUENCHED AND HEAT-TREATED Ti-Cr ALLOYS

4.2.1 Transition Temperatures of Quenched Alloys

Following a review by ZWICKER [Zwi63] of the transition temperature composition dependences of eleven Ti-TM alloys, in which the importance of metallurgical microstructure as a parameter was emphasized, RAUB, RÖSCHEL and ZWICKER [Rau66], in one of the first detailed studies of the influence of composition and

composition-controlled microstructure on the superconductivity of quenched-and-aged Ti-base alloys, measured the superconducting transition temperatures of an extensive series (0-75 at.%) of Ti-Cr alloys. Metallographical mapping, with the assistance of a phase diagram, enabled the influences of various as-quenched phases on T_c to be deduced. T_c *versus* composition for the quenched alloys exhibited two negative-going cusps: the first at ~8 at.% Cr (corresponding to the boundary of the α^m and $\omega+\beta$ phases) and the second at ~36 at.% Cr. The microstructural cause of the second cusp was not apparent in the optical micrographs (at 200X) and was claimed, therefore, to be the result of some submicroscopic effect. Several of the alloys were subjected to aging for times of up to 10,000 min at 450°C the influence of which on the T_c, hardness, and lattice constant was noted.

4.2.2 Influence of Aging and Other Heat Treatments on the Transition Temperature

In the work of RAUB *et al* [Rau66] just referred to, the influence of aging on the transition temperature of Ti-9Cr was interpreted, with the aid of optical metallography and x-ray diffractometry, in terms of the precipitation of phases designated $\omega+\beta$, $\omega+\alpha+\beta+\gamma'$ and $\alpha+\gamma'+TiCr_2$. The results of further detailed studies of the effects of aging and up-quenching on the microstructures and associated superconducting transitions in Ti-Cr alloys were outlined in a series of papers and reports by LUHMAN, POLONIS, TAGGART and others [Luh69, Luh70[a], Luh71][Pol69[a], Pol70] and were fully discussed by LUHMAN [Luh70]. The alloys under investigation had analyzed compositions of 7.04, 7.89, 9.04, 9.48, 10.32, 11.01, 11.99, 13.20, 14.23, 15.04, 16.98, 18.45, 19.00, 20.08, 22.03 and 24.41 at.% Cr. Magnetically determined transition temperatures were reported for eight representative samples in the as-quenched condition, while, as indicated in Table 4-1, five compositions, viz 7.9, 9.5, 10.3, 15.0 and 18.5 at.% Cr, were selected for the aging studies and one of these, 15.0 at.% Cr, for the up-quenching.

The results of superconducting transition temperature measurements were compared with the heat-treatment-induced changes in microstructure as determined by x-ray and electron-diffraction analysis. It is useful to note here the following observations regarding these microstructures: The solute-lean Ti-Cr

TABLE 4-1   LIST OF HEAT TREATMENTS ADMINISTERED TO
Ti-Cr ALLOYS by Luhman *et al.*,
e.g.[Luh70], DURING STUDIES OF
SUPERCONDUCTIVITY AND MICROSTRUCTURE.

At.% Cr		Heat Treatment[†] (time/temperature)
7.9	ST	1h/1000°C/Q plus aged 5h/350°C.
9.5	ST	1000°C plus step quenched to 540°C for 3, 6, and 30 min.
10.3	ST	1h/1000°C/Q plus aged up to 28 min./196°C.
15.0	ST	1h/850°C/Q plus:
		(a) aged up to 51 min./300°C,
		(b) aged up to 283 min./250°C,
		(c) step quenched to 450°C for 3 min., Q.
	ST	1h/1000°C/Q plus:
		(a) aged 36h/300°C,
		(b) aged 36h/300°C plus up-quenched to 450°C for 3 min., Q.
18.5	ST	1040°C/Q plus aged up to 44h/300°C.

† ST = solution treated (i.e., annealed)
  Q = water quenched

alloys (e.g. Ti-Cr$_{9.5}$) contained ω and β phases in the
as-quenched condition, whilst in more concentrated
quenched alloys (e.g. Ti-Cr$_{15.0}$) only the β phase was
detected. During moderate-temperature aging the
structure may become $\alpha+\beta$ as in Ti-Cr$_{7.9}$ after
5h/350°C, or $\omega+\beta$ as in Ti-Cr$_{18.5}$ after $\overset{\sim}{<}$ 44h/300°C.
The ω phase is not stable at 450°C, consequently
either step cooling to that temperature prior to
quenching, or up-quenching to 450°C prior to quenching
back to room temperature, left Ti-Cr with a modulated
β structure. So-called "beta-phase separation" and
the up-quenching-induced ω-reversion reactions are
described in [Mon2.3] and [Mon2.5] and also in
Sect.3.3.2 above in connection with the influence of
heat treatment on the properties of Ti-V alloys. In
the experiments with Ti-Cr alloys as reported by
LUHMAN *et al* (e.g. [Luh70]), many instances of double

superconducting transitions were noted, and it was
concluded that T_c measurements were generally useful
as diagnostic tools with which to study precipitation
effects in Ti-TM alloys.

4.3 SUPERCONDUCTIVITY IN Ti-Cr ALLOYS— TABULATED DATA

TABLE 4-2   TRANSITION TEMPERATURES (T_c) OF
TITANIUM-CHROMIUM ALLOYS

At.% Cr	T_c K	Data Sources[†]	
0.85	2.6$_4$	[Aga74]	n
1.0	2.1	[Mat59]	g
1.15	2.7$_5$	[Hak62[b]]	n
1.6	2.9	[Mat59]	g
2.09	3.9$_5$	[Aga74]	n
2.5	3.5	[Mat59]	g
2.5	3.5	[Buc65]	n
4	3.4$_5$	[Mat59]	g
6	2.3	"	g
7.9	2.2$_5$	[Luh70]	n
9.5	3.8$_7$	"	n
10	1.5	[Mat59]	g
10.3	4.0$_2$	[Luh70]	n
11.0	4.1$_7$	"	n
13.2	4.4$_2$	"	n
15	4.2	[Mat59]	g
15.0	4.5$_0$	[Luh70]	n
18.5	4.4$_1$	"	n
20	4.2	[Mat59]	g
24.4	3.8$_5$	[Luh70]	n
25	3.7$_5$	[Mat59]	g
30	3.4$_5$	"	g

† n = data from individual resistive or magnetic
       transformation curves
  g = graphical data

Cr Concentration Range	Condition	Structures	Procedures	Literature
1-30 at.% (13 alloys)	As-cast.	1 - 2.5 at.% Cr, α^m 4 and 6 at.% Cr, mixed α^m,ω,β $\geq$10 at.% Cr, $\omega+\beta$ tending to β	Mutual induction	B. Matthias *et al.* (1959) [Mat59]
1.15 at.%	As-cast.	α^m	Resistometric	R.R. Hake *et al.* (1962) [Hak62[b]]
2.5 at.%	As-cast.	α^m	Magnetic and calorimetric	E. Bucher *et al.* (1965) [Buc65]
0.005-75 at.% ($\sim$24 alloys)	Annealed and quenched from 1000°C or 1325°C also aged at 450°C.	Entire range of solid-solution phases and compounds	Magnetic	C.J. Raub *et al.* (1966) [Rau66]
7.9-24 at.% (8 alloys)	Homogenized, annealed and quenched from 1000°C; also aged.	$\omega+\beta$ and β	Magnetic	T.S. Luhman (1970) [Luh70]
0.85, 2.09 at.%	Annealed and quenched from 950°C.	α^m	Resistometric and calorimetric	K.L. Agarwal (1974) [Aga74]

ALLOY GROUP 2: TITANIUM-MANGANESE BINARY ALLOYS

4.4   TRANSITION TEMPERATURE AS A FUNCTION OF COMPOSITION IN Ti-Mn ALLOYS

Initial studies of the composition dependence of T_c in as-cast Ti-Mn(0-25 at.%) alloys, the results of which were juxtaposed against those for the "corresponding" Ti-Re system and compared with the transition temperature results for several other Ti-TM systems, seemed to indicate a rapid increase in T_c with composition in the quenched-martensitic regime. Since, as indicated elsewhere, these results formed the basis of a newly postulated principle for superconductive interaction, several workers proceeded to examine more closely the structures and superconducting properties of Ti-Mn especially in the very dilute concentration range.   By 1963, as a result of the work of HAKE *et al* [Hak62[b]], FALGE [Faℓ63], and CAPE [Cap63], it was clear: *(a)* that in the hcp-Ti-Mn alloys (either equilibrium-α or quenched-α^m), Mn carried a localized magnetic moment which, as has generally been found to be the case, suppressed the superconductivity; *(b)* that superconductivity was to be found only in the β-phase alloys, which were not magnetic, and whose T_c-composition dependence was comparable to those generally observed in β-Ti-TM alloys.

Using a magnetic technique, FALGE [Faℓ63] investigated the superconductivity of Ti both with and without the inclusion of small amounts (some 10's of ppm) of Fe and Mn.   By examining FALGE's results as summarized in Table 4-4 it can be deduced that while small amounts of Fe do indeed raise T_c quite rapidly -- at a threshold rate of some 0.1 K per 200 ppm -- the addition of Mn has a severely depressive effect on T_c -- some -0.1 K per 30 ppm.   This accounts for the fact that in the presence of Mn no superconductivity was noted above the minimum available temperature of 0.06 K.

4.5 CALORIMETRIC STUDIES OF SUPERCONDUCTIVITY IN Ti-Mn ALLOYS

Calorimetric studies of Ti-Mn(0.17 through 14 at.%) were carried out by HAKE and CAPE [Hak64] and ten years later by AGARWAL [Aga74] who measured Ti-Mn$_{0.39}$ and Ti-Mn$_{1.71}$.   The graphical analysis of

TABLE 4-4 INFLUENCE OF TRACES OF Mn AND Fe ON THE
SUPERCONDUCTING TRANSITION TEMPERATURE (T_c)
OF TITANIUM -- after Falge [Fal63]

	Concentration ppm[+]		T_c K
	Mn	Fe	
(a)	< 5	< 2	0.42
(b)	30	20	0.17
(c)	30	1500	0.98
(d)	100	20	< 0.06

Mn-effect between
  (a) and (b): -0.1 K per 10 ppm
  (b) and (d): -0.1 K per 41 ppm

Fe-effect between
  (b) and (c): +0.1 K per 183 ppm

+ It is assumed that parts per million by weight
   is being referred to.

low temperature specific heat data in the form of C/T
versus T^2 plots, which for normal metals are usually
linear with positive slopes, has been considered in
[Mon8.1] and the standard literature of the subject,
e.g. [Gop66]. In contrast to the usually expected
result, C/T for the low-concentration Ti-Mn($\gtrsim$1.7 at.%)
alloys, according to both of the above authors,
increased rapidly as the temperature *decreased*.

Using data for pure Ti as a basis, an excess
specific heat, ΔC, was extracted and plotted in the
format log (ΔC/T) *versus* log T in order to determine
the exponent, x, in the relationship $\Delta C/T \propto T^{-x}$.
According to HAKE and CAPE [Hak64], x = 1.56 (T < T_b)
and 1.35 (T > T_b), where T_b = 2.7 K (for 0.17 at.% Mn)
or 3.3 K (for 0.36 at.% Mn). AGARWAL's data, plotted
in the above format, also exhibited a change of slope
near 3 K, below which both Ti-Mn$_{0.39}$ and Ti-Mn$_{1.71}$
yielded a common slope exponent, x = 1.6. Thus for
Ti-Mn, ΔC/T obeyed neither the T^{-3} law characteristic
of nuclear hyperfine interaction, or a T^{-1} relationship (ΔC = constant) ascribable to superparamagnetic
clusters. On the other hand, the areas under the
excess specific heat curves yielded spin-entropies
S_{expt} which, when compared to S = c R ℓn (2s + 1),
where c is the concentration and s (the spin) = 3/2
(in accord with CAPE's magnetic results) gave the
relative entropies, S_{expt}/S, listed in Table 4-5.

These numerical results, which are in excellent
agreement, suggest that the specific heat anomaly
could be the result of some kind of ordering of
localized Mn spins.

TABLE 4-5 ENTROPY ASSOCIATED WITH LOW TEMPERATURE
SPECIFIC HEAT ANOMALIES IN
LOW-CONCENTRATION Ti-Mn ALLOYS

At.% Mn	Temperature Range of Integration K	Relative Entropy, S_{expt}/S %	Reference
0.17	4.5-1.2	27	[Hak64]
0.36	"	"	"
0.39	4.3-1.3	25	[Aga74]
1.71	"	12	"

4.6 TRANSPORT PROPERTY AND MAGNETIC STUDIES OF Ti-Mn ALLOYS

Particularly interesting as a result of the
transport-property measurements of HAKE *et al*
[Hak62[b]], Fig.4-1, was the discovery of low-temperature resistivity minima not only in Ti-Mn -- which is
to be expected in view of the local-moment character
of Mn -- but also in Ti-Cr and Ti-Fe which are not
Curie-Weiss paramagnetic, Fig.4-2. It would be convenient to be able to designate dilute Ti-Cr and Ti-Fe
as localized-spin-fluctuation systems analogous to
Al-Mn excepting that the latter, although it carries
a proven fluctuating moment, does not exhibit a low-
temperature resistance minimum [Hed64]. The origin
of the minima in Ti-Cr and Ti-Fe is inexplicable.
These alloys, although not paramagnetic and which
therefore gave rise to what might be termed "normally
rapid" dT_c/dc's in the α^m-regime, do appear to yield
rather broad, reduced, superconductive specific heat
anomalies, cf. HEINIGER and MÜLLER [Hei64] and
AGARWAL [Aga74].

CAPE [Cap63], who carried out susceptibility
measurements of magnetic moment and resistive measurements of the superconducting transition temperature
on both α^m (1000°C, quenched) and equilibrium α+β

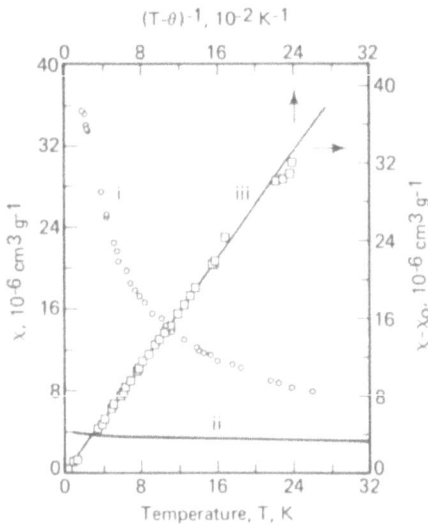

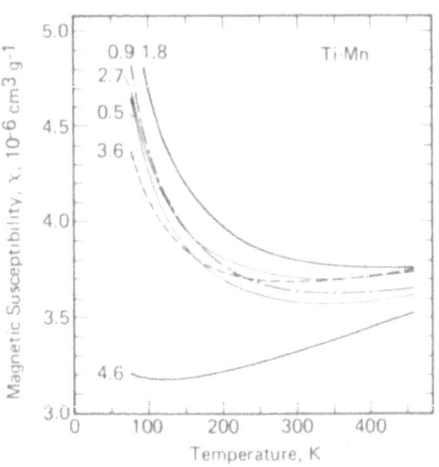

FIGURE 4-2. Magnetic susceptibility temperature dependences of some dilute Ti-TM alloys plotted in the format χ versus T in the case of (i) Ti-Mn (1.0 at.%) and (ii) unalloyed Ti, Ti-Cr (1.15 at.%), Ti-Fe (0.96 at.%), Ti-Co (1.30 at.%) and in the format χ-χ_0 versus $(T-\theta)^{-1}$ in the case of Ti-Mn (1.0 at.%) (iii) with χ_0 = 3.4×10^{-6} cm^3 g^{-1} and θ = -2.2 K. Evidently of the solutes studied only Mn carries a local moment — after CAPE [Cap63].

FIGURE 4-3. Magnetic susceptibilities of Ti-Mn (1 ~ 5 at.%) alloys over the temperature range 77 ~ 450 K. Only the hcp alloys show "Curie-Weiss" temperature dependences. The numbers refer to the analyzed compositions in at.% — after COLLINGS [Coℓ83].

(annealed 690°C, quenched) dilute Ti-Mn alloys, was able to associate the existence of an hcp structure with the appearance of a localized moment on the Mn and a corresponding depression of the transition temperature. COLLINGS [Coℓ83], in a magnetic susceptibility study of a series of Ti-Mn (0.5 through 25 at.%) alloys, Fig.4-3, confirmed that the local-moment behavior is indeed confined to the hcp structure and that not even the ω-phase, although hexagonal, is magnetic in the Curie-Weiss sense.

4.7 SUPERCONDUCTIVITY AND MICROSTRUCTURE IN Ti-Mn ALLOYS

In an optical investigation of the microstructures of Ti-Mn (1 and 2 at.%) CAPE [Cap63] was able to show that during the annealing of the quenched martensite the equilibrium β-phase began to form at the boundaries of the α-platelets. In three dimensions the β-phase can be envisioned as a cellular structure which if $T_{c,\beta} > T_{c,\alpha}$ is capable of yielding premature resistive transitions (with decreasing T) due to the establishment of short-circuit paths, and premature magnetic transitions, as a result of a screening of the interiors of the cells by the β-phase

cell-wall or crust. Incidentally, the magnetic equilibration study, with the aid of a tie-line technique, enabled a point on the α/(α+β) phase boundary (viz 690°C, 0.54 at.% Mn) to be determined.

4.8 SUPERCONDUCTIVITY IN Ti-Mn ALLOYS— TABULATED DATA

TABLE 4-6   TRANSITION TEMPERATURES (T_c) OF TITANIUM-MANGANESE ALLOYS

At.% Mn	T_c K	Data Sources[†]	
(a) Quenched Alloys			
0.25	n.d. (<1.1)	[Cap63]	r
0.39	n.d. (<1.3)	[Aga74]	r
1	n.d. (<1.1)	[Cap63]	r
1.71	n.d. (<1.3)	[Aga74]	r
2	n.d. (<1.1)	[Cap63]	r
4	n.d. (<1.1)	"	r
14	3.0	"	r
(b) As-Cast Alloys			
0.02	n.d. (<1.1)	[Hak62[b]]	r
0.11	n.d. (<1.1)	"	r
0.17	mag.[††]	[Hak64]	c
0.21	n.d. (<1.1)	[Hak62[b]]	r

At.% Mn	T_c K	Data Sources[†]		At.% Mn	T_c K	Data Sources[†]	
(b) As-Cast Alloys -- *continued*				**(c) Annealed (690°C) Alloys**			
0.36	mag.	[Hak64]	c	0.28	n.d. (<1.1)	[Cap63]	r
0.40	n.d. (<1.1)	[Hak62[b]]	c	1	2.9_8	"	r
0.85	mag.	[Hak64]	c	2	3.1_2	"	r
1.0	1.7	[Hak62[b]]	r	4	2.9_8	"	r
1.1	1.7	"	r				
1.5	1.75	[Mat59]	g				
1.7	mag.	[Hak64]	c				
2.0	2.4	[Hak62[b]]	r				
2.5	2.3_5	[Mat59]	g				
3.0	2.3	"	g				
5.0	2.1_5	"	g				
7.0	2.0	"	g				
8.5	2.7	"	g				
10.0	3.0_2	"	g				
12.0	2.9_8	"	g				
14	2.5±0.15	[Hak64]	c				
15	2.70	[Mat59]	g				
18	2.0_7	"	g				
20	1.8_5	"	g				
25	1.0_8	"	g				

† r = data from mid-point of resistive transition

  c = calorimetric result

  g = from plotted T_c *vs* composition data

†† Magnetic anomaly

TABLE 4-7 TRANSITION TEMPERATURES OF TITANIUM-MANGANESE ALLOYS -- DATA SOURCES

Mn Concentration Range	Condition	Structures	Procedures	Literature
0-25 at.% (about 13 alloys)	As-cast.	Mostly α''' through β	Magnetic	B.T. Matthias *et al.* (1959) [Mat59]
0.02-2.0 at.%	As-cast.	Mostly α^m	Resistometric	R.R. Hake *et al.* (1962) [Hak62[b]]
5, 30, 100 ppm	As-cast.	α^m	Magnetic	R.L. Falge (1963) [Faℓ63]
0.28, 1, 2, 4 and 14 at.%	Quenched from (a) 1000°C, (b) 690°C.	(a) ≲4 at.% Mn, α^m; 14 at.% Mn, $\omega+\beta$, (b) $\alpha+\beta$	Resistometric	J.A. Cape (1963) [Cap63]
0.17, 0.36, 0.85, 1.7 and 14 at.%	As-cast.	≲1.7 at.% Mn, α^m, 14 at.% Mn, $\omega+\beta$	Calorimetric	R.R. Hake and J.A. Cape (1964) [Hak64]
0.39 and 1.71 at.%	Quenched from 950°C.	α^m	Calorimetric and resistometric	K.L. Agarwal (1974) [Aga74]

ALLOY GROUP 3: TITANIUM-IRON BINARY ALLOYS

4.9 TRANSITION TEMPERATURE AS A FUNCTION OF COMPOSITION IN Ti-Fe ALLOYS—ALTERNATIVE MODELS FOR SUPERCONDUCTIVITY

4.9.1 The Magnetic Interaction Model

Whereas small amounts of the magnetic rare-earth element Gd were found to depress the superconducting transition temperature of La [Mat58] it was subsequently discovered that the transition temperature of Ti could be increased by up to nine times by the addition of up to 2 at.% Fe in solid solution. Similar results were obtained at the time with the other late-3d (or "magnetic") transition-element solutes Cr, Mn and Co, but not (erroneously as it turned out) with 4d transition elements. Since magnetic properties are frequently associated with Mn, Fe, Co, and Ni, workers at the time leapt prematurely to the conclusion that a magnetic interaction between pairs of conduction electrons, mediated by a direct d-d interaction between adjacent solute-atom spins might supplement or even replace the virtual-phonon interaction.

Stemming chiefly from a comparison of the T_c versus composition curves for Ti-Fe and Ti-Ru (Ru belonging to the same column in the periodic table as Fe) was the observation by MATTHIAS in 1959 [Mat59] of an apparently anomalous rapid increase in superconducting transition temperature accompanying the addition of small amount of Fe to Ti, Fig.4-4. A glance at the figure suffices to show, however, that had one or two more data points been acquired for dilute Ti-Ru, the need for ascribing special properties to the Ti-("magnetic element") series may have disappeared. Accepting the experimental results at face value, BARDEEN and SCHRIEFFER [Bar61] agreed that the ostensibly enhanced T_c exhibited by Ti-Fe and similar alloys could be associated with a magnetic-interaction model of the type referred to in the introductory paragraph.

In MATTHIAS' model, magnetic polarization substitutes for the deformation potential of the electron-phonon approach. Spin pairing is thereby achieved, but the essential mechanism of superconductivity, viz momentum conservation, is not a feature of the model. An equally serious problem was the fact that Curie-Weiss paramagnetism was unable to be detected in

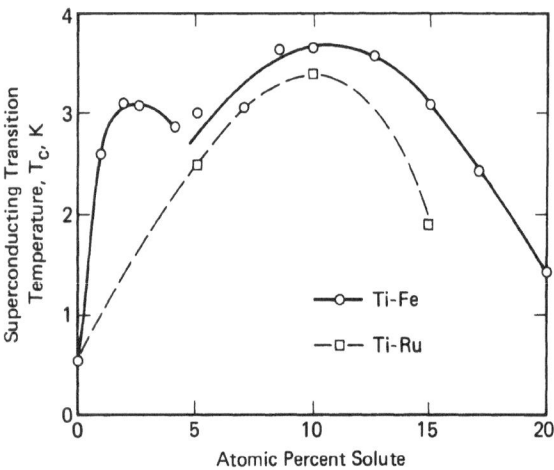

FIGURE 4-4. Magnetically determined superconducting transition temperature *versus* composition in as-cast Ti-Fe and Ti-Ru alloys — after MATTHIAS *et al* [Mat57, Mat59].

Ti-Fe($\sim$1 at.%) [Mat62]. Nevertheless, in spite of these discrepancies, a picture of Ti-Fe as an alloy with unusual superconductive and magnetic properties was beginning to emerge especially as a result of some very careful work by HAKE [Hak62[b]] to be considered below.

4.9.2 Other Localized-State Interactions

As pointed out by BATT [Bat64], MATTHIAS' interpretation of the transition temperature composition dependences of Ti-(3d)TM alloys, if able to be substantiated, would have been of considerable importance in the development of superconductivity theory. In any event the results certainly stimulated considerable further experimentation and some theoretical conjecturization. By 1966 it was realized that the alloys Ti-Cr, -Mn, -Fe, -Co, -Ru etc., and Zr-Co, -Rh, -Ir etc., were all characterized by anomalously rapid increases of T_c with solute addition, increases which in some cases clearly exceeded the accompanying rate of increase of γ, the electronic specific heat coefficient, and that only Mn in Ti carried a localized moment. With reference to Ti-Fe, GANGULY [Gan66] attempted to explain the rapid increase of T_c with solute concentration in terms of an additional attractive interaction between BCS pairs (over and above the usual phonon-induced pair-interaction) brought about "through the Coulomb and exchange type interactions between the conduction electrons and impurity electrons *via* unoccupied localized impurities states". The calculated T_c *versus* concentration curve took the form of a double inverted parabola which closely matched the experimental results for Ti-Fe. However, the validity of the model is suspect if for no other reason than that the positions of the experimentally observed humps correlate closely with the regimes of occurrence of the successive α^m, $\omega+\beta$, and β structures.

4.10 TRANSITION TEMPERATURE AND MICRO-STRUCTURE IN QUENCHED AND HEAT-TREATED Ti-Fe ALLOYS

4.10.1 Transition Temperatures of Quenched and Equilibrated Alloys

Soon after the pioneering work by MATTHIAS *et al* [Mat59], ZWICKER [Zwi63] carried out a comprehensive review of superconductivity in Ti-TM alloys and emphasized that because of the sensitivity of superconducting transition temperature to structural state, and owing to that fact that the transformation processes were not only complicated but quite responsive to the presence of impurities, conflicting results could be obtained for alloys of the same nominal composition. In studying superconductivity in hcp Ti-TM alloys ZWICKER and his colleagues, then at the Forschungs-institut für Edelmetalle und Metallchemie, Schwäbisch-Gmünd, were concerned that the results on quenched α^m-martensitic alloys could have been perturbed by the presence of untransformed β-phase. Consequently they initiated some measurements on annealed equilibrium-α-phase alloys. The paper jointly authored by RAUB, RAUB, and RÖSCHEL of the above-mentioned institute and COMPTON, GEBALLE, and MATTHIAS of the Bell Telephone Laboratories [Rau67] dealt with a determination of the boundary of the α-phase in Ti-Fe, and a study of the superconducting transition temperature appropriate to that phase. It was found that a maximum solubility limit of 0.05 at.% occurred at temperatures between about 500 and 600°C, and that the superconducting transitions for Ti-Fe(0.035 and 0.05 at.%) both began at 0.44 K. In spite of comments to the contrary, this value was in accord with the 3.10 K reported earlier by MATTHIAS [Mat59] for as-cast Ti-Fe$_{2.0}$ (Fig.4-4)

interpreted as a martensitic alloy, and the results of studies of other quenched α^m-Ti-TM alloys.

4.10.2 Influence of Aging on the Transition Temperature

NISHIMURA and ZWICKER [Nis68] studied the effects of aging on the T_c, resistivity, and hardness of Ti-Fe(4, 9.5 at.%) and other alloys. After quenching from $\frac{1}{2}$h/900°C, ω-phase precipitates were found to be present in both of the bcc alloys. The T_c of the as-quenched Ti-Fe$_4$ was immeasurably low (<2 K); that of the Ti-Fe$_{9.5}$ which was 3.7 K, as-quenched, continued to decrease with aging time at 400 and 500°C as the ω-phase precipitates became depleted in Fe through diffusion. HO and COLLINGS [Ho73^a] carried out a detailed investigation of the aging of initially annealed (1h/1000°C) and quenched Ti-Fe$_{7.5}$ (starting T_c, $3.4_5 \pm 0.07$ K). The course of the aging at 175 and 300°C (i.e. the approach to "saturation") was monitored using room temperature magnetic susceptibility. The superconducting transition, whose temperature steadily decreased as the aging proceeded, was studied using low-temperature calorimetry. The results were interpreted, with the aid of some published microstructural observations by WILLIAMS and BLACKBURN [Wil69], in terms of a proximity effect between the bcc matrix and the developing ω-phase precipitates.

4.11 CALORIMETRIC STUDIES OF SUPERCONDUCTIVITY IN Ti-Fe ALLOYS

The pioneering measurements on dilute Ti-(3d)TM alloys had used a magnetic method to determine the superconducting transition temperatures. Since the conclusion, which had to do with the possibility of magnetic superconductive interaction mechanisms, could have had an impact on the theoretical development of the subject, it was important to determine if the high transition temperatures noted by MATTHIAS *et al* [Mat59] were bulk properties of the sample. To be sure, complete diamagnetic transitions had been noted, however, as recognized by CAPE [Cap63], BATT [Bat64] and MÜLLER and co-workers [Hei64][Buc65], the possibility of enhanced transition temperatures due to the presence of high-T_c residual β-Ti-Fe, distributed in such a way that it shielded the major α^m-phase fraction of the alloy, could not be ignored. Accordingly,

BATT [Bat64] and MÜLLER et al [Hei64][Buc65] set out to study superconductivity in dilute Ti-Fe alloys using calorimetric as well as magnetic methods. Ten years later AGARWAL [Aga74] duplicated the calorimetric work and carried out companion electrical resistivity measurements, in an unsuccessful attempt to clarify some of the uncertainties which remained following the completion of the earlier work, while HO and COLLINGS [Ho73^a] studied a series of quenched Ti-Fe(2.5, 5, 7.5 and 10 at.%) alloys, the first of which was martensitic, prior to investigating some aging properties of the $\omega+\beta$ structure.

AGARWAL [Aga74] measured two alloys, Ti-Fe$_{0.53}$ and Ti-Fe$_{1.47}$, quenched from 950°C. Although no metallographic work was carried out, the alloys were believed, on the basis of literature results, to be in the α^m condition. For Ti-Fe$_{1.47}$ the mid-point of the resistive transition (2.3 K) could be said to be in agreement with that of the calorimetric transition (2.6 K), although the specific heat jump was broad (1.5-3.6 K from peak to threshold) and unusually low. No attempt was made to fit a distribution function to the transition in order to quantify, in the manner of MÜLLER et al [Hei64][Buc65], the volume-fraction of superconducting component present (cf. [Mon10.3, Mon10.8] for discussions of suitable curve-fitting procedures). The superconducting fraction may in fact have been small in the alloys measured by AGARWAL, since at 0.6 mJ mole^{-1}K^{-2} the relative height of the specific-heat jump was small compared to the BCS-predicted value[†] of 5.4 mJ mole^{-1}K^{-2}, i.e. a ratio of 11%. For Ti-Fe$_{0.53}$ the mid-point of the resistive transition was 2.1 K. The calorimetric threshold temperature was 3.6 K, as for the higher concentration alloy, but the transition was too poorly resolved to enable any further information to be extracted from it.

The best work with higher concentration α^m alloys is still that by MÜLLER et al [Hei64][Buc65] who carried out magnetic and calorimetric investigations, and analyzed the rounded specific heat jumps by combining,

in the manner detailed in [Mon10.3], a transition-temperature distribution function, $f(T_c)$, to be determined by the analysis, with a temperature variation of C_{es} (the electronic specific heat) provided by the classical GORTER-CASIMIR "two-fluid" relationship:

$$C_{es} = 3\gamma T_c (T/T_c)^3 \quad . \qquad (4\text{-}1)$$

As a result of the analysis, HEINIGER and MÜLLER [Hei64] were able to show that although Ti-V and Ti-Nb exhibited sharp transitions for which the superconducting fraction, f, lay between 0.95 and 1.0, the alloy Ti-Fe$_{1.5}$ yielded 0.3<f<0.35, indicating that only about 30% of that specimen was participating in the superconducting transition within the temperature range of the analysis. Since, however $[\Delta C_{es}/\gamma T_c]$ (GORTER-CASIMIR) = 2, while $[\Delta C_{es}/\gamma T_c]$(BCS) = 1.43, from a BCS standpoint it may have been possible to have claimed for the volume fraction a value somewhat greater than 0.4.

As a result of the work under discussion, excellent agreement was obtained between the mid-points of the magnetic and calorimetric transitions for Ti-Fe$_{1.0}$ (viz 2.3 and 2.6 K, resp.) and Ti-Fe$_{1.5}$ (viz 2.8 and 2.8 K, resp.). These values should be compared with the 2.7 K determined resistometrically upon Ti-Fe$_{0.96}$. The calorimetric transitions were broad ($\Delta T_c \sim 1$ K in each case). These results are consistent with those of HO and COLLINGS [Ho73^a] who obtained for quenched α^m-Ti-Fe$_{2.5}$ a transition of width 0.6 K centered on 3.5 K. The relative height of this somewhat rounded transition (viz $\Delta C_{es}/T_c$ = 1.15 mJ mole^{-1}K^{-2}) was 21% of that expected for a sharp BCS transition (viz $\Delta C_{es}/T_c$ = 1.43 × 3.90 = 5.6 mJ mole^{-1}K^{-2}).

The work of MÜLLER et al was accompanied by x-ray examination which failed to reveal evidence for the presence of any structure other than hexagonal. By invoking comparison with the results of the study of Ti-V, dealt with in Chapter 3 (and the results of the magnetic investigations of Ti-Fe by CAPE [Cap63] referred to in Sect.4.1, Sect.4.2, and again below), MÜLLER et al derived two conclusions from their work: (a) the apparent enhancement of T$_c$ in α^m-Ti-Fe alloys was commensurate with that observed in the "non-magnetic" alloy α^m-Ti-V, thus obviating the need for a magnetic superconductive interaction mechanism; (b) an undetermined mechanism was operative in Ti-Fe, as distinct from alloys such as Ti-V (and, as we shall see, Ti-Mo), in such a way as to smear out and reduce the height of the specific heat jump at T$_c$.

---

† According to [Mon8.4] and standard sources, the relative height of the BCS specific heat jump, ΔC_{es}, at T$_c$ is $\Delta C_{es}/\gamma T_c$ = 1.43; thus with γ = 3.8 mJ mole^{-1}K^{-2}, $\Delta C_{es}/T_c$(Ti-Fe$_{1.47}$) = 1.43γ = 5.4 mJ mole^{-1}K^{-2}.

The dilute hcp alloys, Ti-Fe(0.05, 0.25 and 1.0 at.%) have been investigated calorimetrically by BATT [Bat64] whose results are discussed here in terms of his own metallographic analyses in conjunction with those of RAUB et al [Rau67] mentioned above. Ti-Fe(0.05 at.%) was measured after cold work plus a 14 day/700°C heat treatment had been administered in an attempt to establish the equilibrium α structure. No calorimetric transition was, however, detectable although one would have been expected within the temperature range of the measurement (0.25-4.2 K). A very small fraction (<0.1%) of an Fe-rich (>15% Fe) precipitate (1-2 μm^{ϕ}) was detected within the sample; its identity is, however, unknown even in retrospect. Neither was an interpretable transition found in Ti-Fe$_{0.25}$ which had been measured in the as-cast, α^m, condition, although a magnetic transition within the temperature range 2.0-2.6 K had been claimed for this alloy (unpublished results of T.H. GEBALLE reported by BATT [Bat64]). It seems that the preparation of a reliably single-phase low-concentration Ti-Fe sample ($\lesssim$0.05 at.% Fe) is unusually difficult. With reference to the phase equilibrium studies of RAUB et al [Rau67] the greatest difficulties would have been expected with Ti-Fe$_{1.0}$ annealed 6 days/700°C, i.e. in the two-phase ($\alpha+\beta$) field. BATT concluded on the basis of metallographic examination that annealing had had no effect on the structure of this previously as-cast alloy which seemed to have remained more than 99.9% α^m. Contrary to this conclusion, and with reference to the work of CAPE [Cap63] on Ti-Mn it can now be safely concluded that the annealing of such an as-cast structure 6 days/700°C is sufficient to achieve thermodynamic equilibrium in the $\alpha+\beta$ field, and that the reported structure was probably prior α^m platelets decorated with β precipitates. CAPE had determined that 8h/690°C was adequate to achieve ($\alpha+\beta$) equilibrium in Ti-Mn$_{1.0}$ yielding α-Ti-Mn$_{0.5}$ plus β-Ti-Mn$_{11}$ with a resistive T_c, common to all of the annealed Ti-Mn(1-4 at.%) alloys, of about 3 K.

4.12 TRANSPORT PROPERTY AND MAGNETIC STUDIES OF Ti-Fe ALLOYS

As a result of resistive measurements, the mid-point superconducting transition temperature of Ti-Fe$_{0.96}$ was found to be 2.7 K; then by extending the resistivity measurements to higher temperatures (in particular, up to 24 K) HAKE [Hak62^b] also discovered that Ti-Fe (along with Ti-Mn and Ti-Cr) exhibited a low-temperature resistivity minimum, Fig.4-1, an effect usually associated with localized magnetic moments, and later to become known as the "KONDO effect". Low-temperature resistivity minima were also noted by PREKUL et al [Pre76] in the homogenized-and-quenched α^m-phase alloys Ti-Fe(1.46, 2.15 and 3.27 at.%). Of the alloys Ti-Cr, -Mn, -Fe, -Co and -Ni studied by HAKE [Hak62^b] only Ti-Mn showed negative magnetoresistance; and as a result of the studies by CAPE [Cap63] only Ti-Mn supported a Curie-Weiss paramagnetic moment on the solute atoms. Work by subsequent authors, considered above, showed that although some questions about the superconductivity of Ti-Fe remained unanswered, its behavior was not sufficiently anomalous as to require its being classified separately from the other Ti-TM alloys. The only spectacular exception was Ti-Mn, whose hcp phase is Curie-Weiss paramagnetic with an as-yet undetected superconducting transition temperature.

Resistometric studies of the transition from temperatures of 3 or 4 T_c into the superconducting state have been reported by HAKE [Hak69^a]. In that work, in which Ti-Fe$_8$ and representatives of several of the other Ti-TM alloy systems were investigated, it was found that the normal state residual resistivity began to decrease as the temperature was lowered below 2~3 T_c, dropping to zero of course at T_c itself. Arguments against this being a surface effect were advanced, after which it was able to be concluded that the drop in resistivity which became more and more pronounced as T_c was approached from above, was a manifestation of fluctuation superconductivity.

A more complete set of data was presented by PREKUL et al [Pre76] who studied the resistivity of the ten Ti-Fe(1.46-20.0 at.%) alloys in the vicinity of their T_c's. When nested together it became immediately obvious that the curves fell into two classes: (a) those with sharp transitions, viz the α^m-alloys Ti-Fe(1.46, 2.15, 3.27 at.%); (b) those with rounded transitions, viz the β-alloys Ti-Fe(6.5 through 20 at.%) with Ti-Fe$_4$ as an intermediate case. The rounded transition is clearly a property of the β-structure, and possibly related to dynamical ω-phase fluctuations which are expected to exist in quenched β-alloys, cf. [Mon1.8].

With regard to the normal-state resistivity of Ti-Fe (and related Ti-TM alloys) several phenomena must be accounted for:

(a) a resistivity minimum, as reported in
α^m-Ti-Fe$_{0.96}$ by HAKE et al [Hak62^b] and in
α^m-Ti-Fe(1.46, 2.15 and 3.27 at.%) by PREKUL
et al [Pre76];

(b) a positive magnetoresistance,
$\alpha^H \equiv \rho^{-1}(\partial\rho/\partial H)_T$, as in Ti-Fe$_{0.96}$;[+]

(c) a negative dρ/dT over certain temperature
ranges as in β-Ti-Fe(4, 6.5, 8, 10 and
12 at.%) [Pre76];

(d) a negative α^H as noted in β-Ti-Fe$_8$ and five
other β-Ti-TM alloys by HAKE et al [Hak75].

We will consider first of all the β-Ti-TM alloys.
PREKUL et al [Pre76] suggested that in these systems
the observed negative dρ/dT was the low-temperature
part of a localized-spin-fluctuation-induced resistivity minimum. HAKE et al [Hak75] in considering
negative $\alpha^T \equiv \rho^{-1}d\rho$/dT and negative α^H in Ti-Fe$_8$ and
other alloys also considered localized spin fluctua-
tions, but offered as alternative possibilities: (i)
sharp structure in the Fermi density-of-states to gen-
erate negative α^T (in the manner described in [Mon5.7])
coupled with a Zeeman splitting of such states to
confer negative α^H; (ii) electron scattering from
soft-phonon modes increasing in strength as T decreases (α^T) in association with a compensatory field-
induced mode stiffening which would result in negative
α^H. Since there is ample evidence for soft-phonon
modes in Ti-TM alloys the above explanation for α^T is
quite reasonable, and had in fact been advanced
earlier by COLLINGS [Col74] in connection with anomalous dρ/dT in Ti-V alloys. The negative α^H remains an
unexplained phenomenon.

The regime of validity of any localized-spin-
fluctuation model for Ti-TM alloys must lie among the
dilute α^m-alloys. Indeed there is ample evidence in
terms of the recent work by PREKUL et al [Pre76] and
the earlier results of HAKE et al [Hak62^b] that

[+] In a 1962 paper dealing with dilute α^m-Ti-TM alloys,
HAKE noted that of the alloys Ti-Mn$_{1.0}$, Ti-Fe$_{0.96}$,
Ti-Co$_{1.3}$, and Ti-Ni$_{1.0}$, only Ti-Mn, a local-moment
system, yielded negative α^H, while Ti-Cr, -Mn and
-Fe exhibited pronounced low-temperature resistivity
minima. In their 1976 paper, HAKE and co-workers
[Hak75] showed that the six β-phase alloys Ti-Mn$_{16}$,
Ti-Fe$_8$, Ti-Mo$_{16}$, Ti-V$_{60}$-Cr$_{10}$, Ti-Ru$_8$, and Ti-Os$_8$
possessed negative α^H's.

α^m-Ti-Fe, with its low-temperature resistance minimum,
but absence of Curie-Weiss moment, is a spin-fluctuation alloy. The same is possibly true for Ti-Cr.
This important result could conceivably be responsible for the suppression of the height of the superconductive anomaly noted by numerous authors in Ti-Fe and
by AGARWAL [Aga74] in Ti-Cr, Ti-Co, and Ti-Fe.

4.13 SUPERCONDUCTIVITY IN Ti-Fe ALLOYS— TABULATED DATA

TABLE 4-8   TRANSITION TEMPERATURES (T$_c$) OF
TITANIUM-IRON ALLOYS

At.% Fe	T$_c$ K	Data Sources[+]	
0.035	0.44	[Rau67]	qt
0.05	0.44	"	"
0.05	n.d. (<0.24)	[Bat64]	h
0.25	2.0-2.6	"	mt
"	$\gtrsim$0.25	"	h
0.53	2.1	[Aga74]	r
"	3.6	"	ht
0.96	2.7	[Hak62^b]	r
1.0	∿2.6	[Bat64]	mt
"	3.1±0.2	"	h
1	2.6	[Mat59]	g
1.0	2.3	[Buc65]	mm
"	2.6	"	h
1.46	2.6	[Pre76]	rt
1.47	2.3	[Aga74]	r
"	1.5-3.6	"	h
1.5	2.8	[Buc65]	mm
"	2.8	"	h
2	3.1	[Mat59]	g
2.15	3.0	[Pre76]	rt
2.5	3.1$_5$-3.6$_5$	[Ho73^a]	h
2.5	3.0$_9$	[Mat59]	g
3.27	2.6	[Pre76]	rt
4.0	1.7	"	"
4	2.8$_8$	[Mat59]	g
4	<2	[Nis68]	r
5	<1.5-2.7	[Ho73^a]	h
5	3.0	[Mat59]	g
6.5	2.9	[Pre76]	rt
7	3.0$_7$	[Mat59]	g

7.5	3.3_7-3.5_2	[Ho73[a]]	h	12.5	3.5_8	[Mat59]	g
8.0	3.2	[Pre76]	rt	12.5	3.3_2-3.4_2	[Ho73[a]]	h
8.5	3.6_5	[Mat59]	g	15	3.0_9	[Mat59]	g
9.5	3.7	[Nis68]	r	15.0	2.5	[Pre76]	rt
10	3.6_3-3.7_6	[Ho73[a]]	h	17	2.4_3	[Mat59]	g
10	3.6_7	[Mat59]	g	20.0	2.0	[Pre76]	rt
10.0	3.1	[Pre76]	rt	20	1.4_4	[Mat59]	g
12.0	3.0	"	"	20	n.d. (<1.5)	[Ho73[a]]	h

+ mt = tabulated magnetic result   mm = mid-point of magnetic transition

    qt = quoted magnetic threshold   h = calorimetric transition

     r = mid-point of resistive transition   ht = threshold of calorimetric transition

    rt = tabulated resistive data   g = from plotted T_c *vs* composition data

TABLE 4-9 TRANSITION TEMPERATURES OF TITANIUM-IRON ALLOYS -- DATA SOURCES

Fe Concentration Range	Condition	Structures	Procedures	Literature
0-20 at.% (12 alloys)	As-cast.	α^m through β	Magnetic	B.T. Matthias *et al.* (1959) [Mat59]
0.96 at.%	As-cast.	α^m	Resistometric	R.R. Hake *et al.* (1962) [Hak62[b]]
0.15 at.%	Not stated.	α	Magnetic	R.L. Falge (1963) [Fal63]
0.05-1.0 at.% (3 alloys)	As-cast or annealed 700°C.	0.05 at.% Fe, 99% α-phase, 0.25 at.% Fe, α^m, 1.0 at.% Fe, 99.9% α^m(reported)	Magnetic and calorimetric	R.H. Batt (1964) [Bat64]
1.0 and 1.5 at.%	As-cast.	α^m	Magnetic and calorimetric	E. Bucher *et al.* (1965) [Buc65]
0.01-0.05 at.% (4 alloys)	Cold rolled and homogenized 600-700°C.	α	Magnetic	E. Raub *et al.* (1967) [Rau67]
4 and 9.5 at.%	Quenched from 900°C.	ω+β	------	T. Nishimura and U. Zwicker (1968) [Nis68]
0.53 and 1.45 at.%	Quenched from 950°C.	α^m	Calorimetric and resistometric	K.L. Agarwal (1974) [Aga74]
2.5-20 at.% (6 alloys)	Quenched from 1000°C.	2.5 at.% Fe, α^m, 5-7.5 at.% Fe, ω+β	Calorimetric	J.C. Ho and E.W. Collings (1974) [Ho73[a]]
1.46-20 at.% (10 alloys)	Homogenized and quenched from 1000°C.	α^m through β	Resistometric	A.F. Prekul *et al.* (1976) [Pre76]

ALLOY GROUP 4: TITANIUM-COBALT AND TITANIUM-NICKEL BINARY ALLOYS

4.14 MAGNETIC AND CALORIMETRIC STUDIES OF THE SUPERCONDUCTING TRANSITION IN Ti-Co ALLOYS

Combining an explanation of the apparently universal dependence of T_c upon electron/atom ratio with an interest in examining the effects of "ferromagnetic" solute elements on the superconducting transition temperature, MATTHIAS et al [Mat59] made direct comparisons of the superconducting transition temperatures of Ti-(3d)TM alloys with those in which the solute was selected from the second (4d) and third (5d) long periods. A relatively rapid increase of T_c in the α^m-regime was noted. RAUB and HULL [Rau65] confirmed the Ti-Co results by filling in a few additional data points for the alloys Ti-Co(0.5, 2.0 and 13.5 at.%), but went on to carry out a detailed investigation of the T_c composition dependences of the "corresponding" Ti-(4d)TM and Ti-(5d)TM sequences, i.e. Ti-Rh and Ti-Ir, respectively. It is interesting to note at this point that when the data for α^m-Ti-Rh that were missing from MATTHIAS' initial paper were filled in, an almost exact parallelism between the T_c composition dependences of Ti-Co and To-Rh was obtained, thus eliminating any need to assign special properties to the former system.[†] MATTHIAS' work was quickly followed by an intense study of the superconductive and magnetic properties of Ti-Co. HAKE et al [Hak62[b]] showed that as-cast Ti-Co$_{1.3}$ unlike Ti-Cr, -Mn and -Fe did not exhibit a low-temperature resistivity minimum,[††] clear evidence for the non-existence of a localized magnetic moment. This conclusion was substantiated by the results of the magnetic susceptibility study of CAPE [Cap63] who showed that as-cast Ti-Co$_{1.3}$, as with Ti-Cr and Ti-Fe, was not a Curie-Weiss paramagnet. Thus if enhanced superconductivity did occur in the α^m-regime it must have been as a

normal alloying effect, and not as the result of magnetic-impurity-moderated interactions. Having so disposed of the magnetic issue, it then remained to establish whether the rapidly increasing transition temperatures noted by MATTHIAS et al [Mat59] were properties of bulk as-cast α^m, or whether they were the result, for example, of a network of superconducting bcc inclusions, as was shown to be the case for Ti-Mn. In the absence of metallurgical evidence to the contrary this is always a possibility since, as HEINIGER and MÜLLER [Hei64] pointed out in connection with the interpretation of magnetic results, complete flux exclusion can occur as soon as 10% of the volume is superconducting.

To settle the question of bulkness of a superconducting transition it is necessary, in order to avoid the possibility of magnetic shielding by grain-boundary precipitation, or electrical short-circuiting by that or other forms of interconnected precipitation, to use low-temperature specific heat. If the super-conducting specific heat jump is *sharp* it is possible to evaluate immediately, in a quantitative way, the relative volume of the superconducting material. In the case of a *rounded transition*, a fairly accurate estimate of the volume fraction of the superconducting component can be obtained by fitting an appropriate distribution function in the manner described in [Mon10.3] and [Mon10.8]. Calorimetric measurements of Ti-Co$_{1.0}$ have been made by HAKE and CAPE [Hak64] and of Ti-Co(0.71 and 2.2 at.%) by AGARWAL [Aga74]. In the earlier work, although an anomalous rise or "bump" appeared in the C/T *versus* T^2 plot within $2 < T^2 < 12$ K^2, it was too insignificant to qualify as the indication of a bulk superconducting transition. In fact HAKE and CAPE suggested that if it did represent a transition at all, it might have been one that was associated with some filamentary component of the alloy structure. The experiments by AGARWAL [Aga74] confirmed the results of the earlier specific heat measurement. It is almost certain that superconducting transitions of some kind were involved and if so, for both alloys, they were very broad (peak-to-peak threshold widths of ~ 1 K) and low. The heights (threshold-to-peak) of the C_{es}/T_c jumps were 0.5 mJ mole^{-1}K^{-2} for Ti-Co$_{0.71}$ and 1.1 mJ mole^{-1}K^{-2} for Ti-Co$_{2.2}$; these were 6 and 12%, respectively, of

---

† As a matter of fact the T_c in α^m-Ti-Rh rose even more rapidly than that in α^m-Ti-Co.

†† In fact a barely detectable minimum was noted, but so shallow as to be attributable to traces of Cr, Mn or Fe as impurities.

the BCS jump height. The next step in the analysis would require proper curve fitting, but in its absence it is clear that the calorimetric experiments were not detecting bulk superconductivity. In a recent contribution to this subject COLLINGS [Coℓ80] has pointed out that, since the diffusion coefficients of Fe, Co, and Ni in β-Ti at, say 1000°C, are almost two orders of magnitude higher than those of the earlier transition elements, Fig.4-5, an opportunity exists for significant levels of solute redistribution, and possibly Widmanstätten growth, to take place during the quenching of the moderately massive samples needed for the conventional low-temperature specific heat measurement.

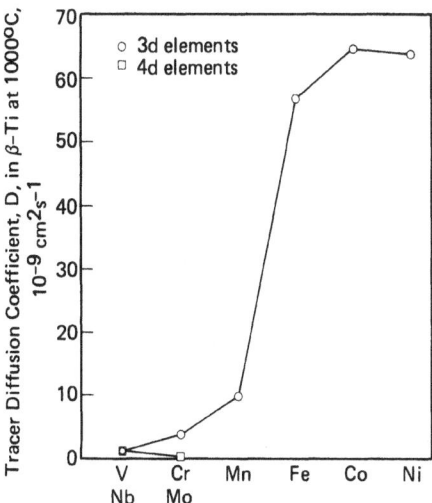

FIGURE 4-5. Tracer diffusion coefficients for the 3d solutes V, Cr, Mn, Fe, Co and Ni, and the 4d solutes Nb and Mo in β-Ti at 1000°C — computed from the frequency-factor and activation-energy data of ZWICKER [Zwi74, p. 108].

4.15 SUPERCONDUCTIVITY IN Ti-Co ALLOYS— TABULATED DATA

TABLE 4-10   TRANSITION TEMPERATURES OF
TITANIUM-COBALT ALLOYS

At.% Co	T_c K	Data[†] Sources	
0.5	1.4	[Rau64[C]]	g
0.71	2.8-3.7	[Aga74]	h
"	3.2$_8$	"	r
1.0	3.5	[Hak64]	ht
1	2.2$_9$	[Mat59]	g
1.3	2.7	[Hak62[b]]	r
2.0	2.5	[Rau64[C]]	g
2	2.5$_7$	[Mat59]	g
2.2	3.5$_2$	[Aga74]	r
"	3.0-4.0	"	h
3	2.8$_5$	[Mat59]	g
5	2.6$_4$	"	g
6	2.9$_4$	"	g
7	3.0$_7$	"	g
10	3.8$_0$	"	g
12	3.7$_5$	"	g
13.5	3.9	[Rau64[C]]	g
15	3.8$_0$	[Mat59]	g
20	3.7$_9$	"	g

† g = from plotted T_c *vs* composition data
  h = calorimetric transition
  ht = threshold of calorimetric transition
  r = mid-point of resistive transition

Co Concentration Range	Condition	Structures	Procedures	Literature
0-20 at.% (10 alloys)	As-cast.	1, 2, 3 at.% Co, α^m, 5 at.% Co, mixed, 6-20 at.% Co, $\omega+\beta$ through β	Magnetic	B.T. Matthias *et al.* (1959) [Mat59]
1.3 at.%	As-cast.	α^m	Resistometric	R.R. Hake *et al.* (1962) [Hak62[b]]
1.0 at.%	As-cast.	α^m	Calorimetric	R.R. Hake and J.A. Cape (1964) [Hak64]
0.5, 2.0 and 13.5 at.%	As-cast.	0.5, 2.0 at.% Co, α^m, 13.5 at.% Co, β	Magnetic	C.J. Raub and G.W. Hull (1964) [Rau64[c]]
0.71 and 2.2 at.%	Quenched from 950°C.	α^m	Calorimetric and resistometric	K.L. Agarwal (1974) [Aga74]

4.16 TRANSPORT PROPERTY AND CALORIMETRIC STUDIES OF Ti-Ni ALLOYS

Very little information is available on the influence of dissolved Ni on the superconductivity of Ti. Certainly, Ti-Ni is not a member of that class of Ti-TM alloys whose T_c rises rapidly with concentration in the quenched-α^m regime. MATTHIAS claimed that quenched alloys of concentrations up to 9 at.% were hcp[†] and that at least that amount of Ni was required to raise T_c above 1 K. HAKE [Hak62[b]] in studies of the electrical transport properties of several Ti-(3d)TM alloys showed that as-cast Ti-Ni$_{1.0}$, in common with dilute Ti-Co but unlike Ti-Cr, -Mn, and -Fe, did not exhibit either a low-temperature resistance minimum or a negative magnetoresistance. Ni can thus be regarded as a "nonmagnetic" addition to Ti (in the local-moment sense). All searches for superconductivity in α-phase Ti-Ni yielded null results.

Resistive studies by HAKE *et al* [Hak62[b]] revealed no decrease in the residual resistivity of as-cast Ti-Ni$_{1.0}$ down to 1.1 K, the lowest temperature available. Calorimetric measurements by AGARWAL [Aga74] on Ti-Ni(0.43 and 1.04 at.%) alloys, quenched from 1000°C, failed to reveal any traces of superconducting specific heat anomalies down to 1.6 K, the lowest temperature attainable. On the other hand, electrical resistivity measurements to 1.1 K showed that although Ti-Ni$_{0.43}$ retained 95% of its 4.2-K (i.e. residual) resistance, that of Ti-Ni$_{1.04}$ was down 80% from the residual value indicative of superconductive partial short-circuiting of the sample. The final decrease in the resistivity of this sample commenced at 2.9 K.

In summary then it appears that although Ni is a nonmagnetic addition to Ti any influence that additions of up to 1 at.% might have on the T_c of the α-phase alloys is only to be seen at temperatures below 1 K, and in this regard Ni behaves like a noble metal rather than a transition element. Such incipient superconductivity as has been detected by the work of AGARWAL [Aga74] on 40h/1000°C-annealed Ti-Ni(0.43 and 1.04 at.%) must be interpreted as being due to second-phase precipitation. We are not in a position to judge the nature of this precipitate -- β-phase or the compound Ti$_2$Ni are both likely candidates.

---

† According to MOLCHANOVA [Mol65, p.94], Ti-Ni alloys of up to 7 at.% undergo martensitic transformation; the β-phase is retainable at higher concentrations.

5

BINARY ALLOYS OF TITANIUM WITH THE SECOND-LONG-PERIOD (4d) AND THIRD-LONG-PERIOD (5d) TRANSITION ELEMENTS

This chapter deals with superconductivity in alloys of Ti with the 4d-transition elements Zr through Pd and the 5d-transition elements Hf through Pt. Whenever possible, the solutes are treated in common-group-number pairs (e.g. Zr is coupled with Hf, Tc with Re, and so on) in view of the close relationship that exists between 4d and 5d elements from the same column of the periodic table. As in the previous chapter, the treatment proceeds in order of increasing group number. There are, however, some exceptions to this system: Ti-Nb is omitted, being the subject of a separate chapter, and due to a lack of data, Ti-W and Ti-Pd are also excluded from the following discussion.

ALLOY GROUP 1: TITANIUM-ZIRCONIUM AND TITANIUM-HAFNIUM BINARY ALLOYS

5.1 SUPERCONDUCTIVITY IN Ti-Zr ALLOYS

5.1.1 Composition Dependence of the Transition Temperature

The first studies of superconductivity in Ti-Zr alloys were conducted by HULM and BLAUGHER [Huℓ61], who in an investigation of the sequence Ti-Zr(35, 50, 70 and 83 at.%) using a magnetic technique, discovered a pronounced maximum ($\sim$1.6 K) in the superconducting transition temperature near 50 at.% Zr. The alloy had been annealed at elevated temperatures in the β regime and may have been furnace cooled rather than quenched. The structure was probably single-phase α^m, although alternative possibilities were α^m with retained β which may occur in the 20-80 at.% composition range under very rapid quenching conditions (such as 8000°C

sec^{-1}) [Moℓ65, p.3] or $\alpha^m + \omega$ under slow-cooling conditions such as 200-300°C sec^{-1} [Moℓ65, p.4]. At normal quench rates such as 1000-1500°C sec^{-1}, x-ray analysis has shown that complete transformation to α^m-martensite with no retained β-phase is to be expected over the entire Ti-Zr composition range [Moℓ65, p.4]. Annealing at temperatures below the equilibrium $\alpha/(\alpha+\beta)$ transus (which sags to a minimum at 545°C, 65.6 wt.% Zr) yields the α-phase which in this system is essentially α^m with the defects annealed out.

The transition-temperature data of HULM and BLAUGHER [Huℓ61] was displayed as a function of solute concentration. Other workers, notably MATTHIAS and colleagues [Jen65][Mat70] have introduced the average electron/atom ratio (essentially an average group number) as a display parameter while DeSORBO [Des65], in a paper dealing with numerous TM-alloy systems, including Ti-V, -Nb, -Zr, and -Mo, has argued in favor of a solute-size-corrected number, N_{eff}, which, as an expression of the volume density of conduction electrons, is possibly a more physically meaningful parameter. In a like vein, the use of valence-electron density as an electronic-property display parameter has also been advocated by JENSEN *et al* [Jen65].

5.1.2 Calorimetric Studies of Superconductivity

The importance of Ti-Zr as a system for study rests in its being an isoelectronic alloy of fixed composition-independent structure. It was therefore regarded as a system upon which the influence on T_c of electronic and related atomic properties could be studied in the absence of extraneous disturbances, and in particular one in which the interdependences of T_c, $n(E_F)$, and θ_D, as implied by equations such as Eqn.(3-1), could be examined. With this in mind, BUCHER *et al* [Buc64] carried out low temperature specific heat measurements on the three alloys Ti-Zr(25, 50 and 75 at.%) as a result of which they were able to draw attention to the existence of a close parallelism between the composition dependences of T_c and γ (which in turn is proportional to $n(E_F)$).

It was also noted that, although the normal-state specific heat components were unaffected, the maximum T_c was strongly influenced by the heat treatment to which the alloy had been subjected. For example, while the superconducting transition temperature of well-annealed samples of equiatomic Ti-Zr occurred at 1.23 K, it was able to be raised to 2.0 K by very rapid quenching. Such an increase must have been due either to special properties of the martensitic-hcp structure, as opposed to the annealed-hcp, or to the inclusion of retained β-phase. The structures likely to be encountered after quenching at the rate of 1000-1500°C sec^{-1}, or cooling at the rate of 200-300°C sec^{-1}, have just been described.

Subsequent work by BETTERTON and SCARBROUGH [Bet68], as a result of which $T_{c,max}$ (at 1.3 K) was found to be significantly lower than those noted in the earlier studies (1.5-1.6 K), demonstrated that starting-material purity and heat treatment had an influence on T_c but not on γ and θ_D.

In another calorimetric study of this system, HAKE and CAPE [Hak64] selected one alloy, Ti-Zr$_{50}$, to be used as a "high-transition-temperature" pseudo-Ti

host into which 0.2 at.% Mn was to be dissolved in a study of the influence of localized magnetic moments on T_c. In this experiment the transition temperature of the binary host was 1.60±0.15 K; upon the addition of the 0.2 at.% Mn it dropped below the level of detectability (i.e. below 1.24 K).

5.1.3 Concluding Discussion

With the aid of Ti-Zr the influence of the martensitic structure on T_c is able to be isolated. As in the work of BETTERTON and SCARBROUGH [Bet68] it is possible to study, over a wide composition range, the relative influences, at fixed composition, of the α^m and α structures on the superconducting transition temperature. A typical example is presented in Table 5-1 from which it can be seen that the martensitic structure is characterized by broader superconducting transitions with slightly higher T_c's than those of the equilibrium-hcp controls.

Although the parallelism between γ and T_c noted initially by BUCHER *et al* [Buc64] is strongly suggestive of a direct influence of $n(E_F)$ on T_c, effects due to variation of the electron-phonon or Coulomb interaction at constant $n(E_F)$ cannot be ruled out. As a result of an empirical discussion of these possibilities, BETTERTON and SCARBROUGH [Bet68] concluded that

TABLE 5-1 INFLUENCE OF MARTENSITIC STRUCTURE
ON THE TRANSITION TEMPERATURES OF
Ti-Zr ALLOYS [Bet68].

At.% Zr	Structure†	Transition Temperature K	Enhancement $100\ \Delta T_c/T_{c,\alpha}$ %
30.6	α	1.15±0.02	
	α^m	1.22±0.1	6
39.5	α	1.15±0.1	
	α^m	1.15±0.1	0
49.4	α	1.08±0.15	
	α^m	1.22±0.05	13
59.5	α	1.08±0.1	
	α^m	1.11±0.1	3

$\dagger$ α signifies equilibrium hcp,
α^m signifies the martensitic structure

the variation of T_c could in fact best be described in terms of variation of $n(E_F)$ at constant electron-phonon-interaction strength. The occurrence of a maximum in $n(E_F)$ for an isoelectronic alloy (which might have been expected to have yielded a monotonic change of $n(E_F)$ with composition) is intriguing. BETTERTON and SCARBROUGH examined this effect, drawing upon the results of energy-band calculations of Ti and Zr by MATTHEISS and LOUCKS, respectively [Bet68]. Recognizing that the relative energies of the third and fourth bands near Γ-point of the Fermi surface were reversed for those two elements, BETTERTON and SCAR-BROUGH suggested that it may be possible, when Zr is alloyed with Ti, for $n(E_F)$ to peak near the composition at which these bands cross each other at the Fermi surface.

5.2 SUPERCONDUCTIVITY IN Ti-Hf ALLOYS

The equilibrium phase diagram for Ti-Hf, which is quite similar to that of Ti-Zr, has a minimum in the $\beta/(\alpha+\beta)$ transus at the point (770°C, 47 wt.% Hf). As with Ti-Zr the interest in Ti-Hf as a superconducting system lies in its being an isoelectronic alloy in which the response of T_c to atomic properties can be studied free of any possible difficulties associated with precipitation and structural transformation. The low temperature specific heat parameters, γ and θ_D, of Ti-Hf$_{50}$ have been measured by HEINIGER and reported by HEINIGER, BUCHER, and MÜLLER [Hei66]. The low temperature specific heats of Ti-Hf(35.4 and 64.6 at.%) were subsequently measured by AGARWAL [Aga74], who was able to show that the composition dependence of γ passed through a maximum, paralleling the behavior of Ti-Zr whose maximum was, however, a little higher. The Debye temperatures of the Ti-Hf alloys, which increased monotonically between Hf and Ti, were always lower than those of Ti-Zr (in keeping with those of the unalloyed end-points). Since $T_{c,max}$(Ti-Zr) = 1.3~1.6 K, it is to be expected that the maximum T_c (if such exists) in the Ti-Hf system would be less than about 1.3 K. Unfortunately, since in AGARWAL's experiment on Ti-Hf the lower temperature limit was 1.7 K, the dependence of T_c upon alloying in that system was unable to be followed.

5.3 SUPERCONDUCTIVITY IN Ti-Zr AND Ti-Hf ALLOYS—TABULATED DATA

TABLE 5-2(a) TRANSITION TEMPERATURES (T_c) OF TITANIUM-ZIRCONIUM ALLOYS

At.% Zr	T_c K	Data[†] Sources	
9.62	<1.12	[Bet68]	n
25	1.2	[Buc63]	g
30.6	1.15±0.02	[Bet68]	n
35	1.37	[Huℓ61]	g
39.5	1.15±0.1	[Bet68]	n
49.4	1.08±0.15	"	"
50	1.57	[Huℓ61]	g
50	1.5	[Buc63]	g
50	1.60±0.15	[Hak64]	n
50.2	1.30±0.02	[Bet68]	n
59.5	1.08±0.1	"	"
69.1	1.03±0.15	"	"
70	1.36	[Huℓ61]	g
75	1.2	[Buc63]	g
83	1.01	[Huℓ61]	g
89.6	<1.27	[Bet68]	n
100	0.54	[Huℓ61]	g

† n = numerically listed or quoted calorimetric result (on α-phase material in the case of [Bet68], cf. Table 5-1).

  g = from plotted T_c vs composition data

TABLE 5-2(b) TRANSITION TEMPERATURES (T_c) OF TITANIUM-HAFNIUM ALLOYS

At.% Hf	$T_c^†$ K	Data Source
35.4	n.d. (<1.7 K)	[Aga74]
64.6	n.d. (<1.7 K)	[Aga74]

† Calorimetric measurement.

TABLE 5-3 TRANSITION TEMPERATURES OF TITANIUM-ZIRCONIUM ALLOYS -- DATA SOURCES

Zr Concentration Range	Condition	Structures	Procedures	Literature
0-100 at.% (Ti, Zr, + 4 alloys)	Annealed in β, probably furnace cooled.	α^m	Magnetic	J.K. Hulm and R.D. Blaugher (1961) [Huℓ61]
25, 50 and 75 at.%	----------	α	Calorimetric	E. Bucher *et al.* (1963) [Buc63]
50 at.%	As-cast.	α^m	Calorimetric	R.R. Hake and J.A. Cape (1964) [Hak64]
9.6-89.6 at.% (8 compositions)	Quenched from ∿1200°C. Quenched from 500-750°C.	α^m α	Calorimetric	J.O. Betterton and J.O. Scarbrough (1968) [Bet68]

ALLOY GROUP 2: TITANIUM-TANTALUM BINARY ALLOYS

5.4 SUPERCONDUCTIVITY IN Ti-Ta ALLOYS

In contrast to the Ti-(3d)TM alloy series: Ti-(Cr through Ni), whose properties are primarily of scientific interest from the standpoints of metallurgy and physics, the family of Ti-(group V)TM alloys Ti-V, -Nb, and -Ta, received early attention because of their potential technical importance as superconductors with high T_c's, high upper critical fields, as well as the possibility of their possessing high critical current densities in strong magnetic fields. The transition temperatures of both α- and β-phase Ti-Ta alloys have been studied, and their resistive upper critical fields have been measured and discussed in terms of GLAG theory and extensions of it. Several useful studies of critical current density, J_c, in Ti-Ta have been carried out, while the effects on J_c of important variables such as: *(a)* solute concentration in cold-worked alloys, *(b)* precipitation-heat-treatment following cold work, *(c)* precipitation resulting from the heat treatment of alloys containing previously dissolved interstitial elements, have been investigated. Ti-Ta has been prepared and studied in the form of sputtered film. That work is considered separately, not because films of the thicknesses considered (∿3000Å) are not conceptually bulk material, but because: *(a)* coating is a potentially important preparation process for superconductors; *(b)* the

properties of films, as a result presumably of preparationally induced artifacts, are usually different from those of the bulk.

5.5 TRANSITION TEMPERATURES OF Ti-Ta ALLOYS

5.5.1 Composition Dependence of the Transition Temperature

Although the experimental results had been first reported by BLAUGHER and JOINER [Bℓa63], a plot of T_c *versus* composition for a limited range of Ti-Ta alloys (up to 60 at.% Ta) was first published by ZWICKER [Zwi63] in his review of superconductivity in Ti-base alloys. With a plateau of height $T_c \sim 8.6$ K, commencing at about 50 at.% Ta, the results were to be in reasonable accord with those subsequently obtained by COLLING *et al* [Coℓ66[a]] who investigated the entire concentration range. However, discrepancies did appear to exist at higher Ta levels due possibly to undetected losses of Ti during preparation of the earlier alloys. The data-set was completed by RAUB and ZWICKER [Rau65] who carried out the only study on equilibrium-α-phase (as distinct from quenched-α^m) low-concentration Ti-Ta alloys. By intercomparing the results for α-Ti-V, α-Ti-Nb, and α-Ti-Ta they showed (on the basis of 2.5 at.%-solute-concentration data) that dT_c/dc was 0.40, 0.44, and 0.36 K per at.% respectively, and claimed that in the light of the then-current results on the electronic specific heat coefficient, γ, the increases in T_c observed in the α-phase alloys were in conformity with increases in $n(E_F)$, deduced on the assumption of proportionality between it and γ. In their detailed investigation of β-Ti-Ta, COLLING et al [Col66[a]] measured the superconducting transition temperature of a complete series of Ti-Ta(30.1-97.7 at.%) cold-worked alloys, and went on to investigate the effect on them of heat treating for 1h/(300-1200°C).

In subsequent studies of the Ti-Ta system, PREKUL et al [Pre74] showed that at 50 at.% Ta, T_c passed through a maximum which, at 7.4 K, was some 1.3 K lower than the values previously obtained. The principal burden of PREKUL's work, however, was a description of the normal-state resistivity temperature dependence in terms of a spin-fluctuation model.

5.5.2 Influence of Aging on the Transition Temperature

Heat treatment of Ti-TM alloys in the low- to mid-temperature ranges results chiefly in precipitation, while annealing at the highest temperatures produces recrystallization, and consequently an erasing of the effects of the cold work. In practice, however, oxygen contamination present either in the starting material or picked up during the heat treatment can be responsible for the development of α-phase precipitation (<10 vol %) even in alloys that have been cooled from the single-phase-β region.

In studies of the effect of heat treatment (300-1200°C) on Ti-Ta$_{50.9}$, prepared from Ti sponge containing the usual levels of oxygen contamination, decreases in both the *width* and *magnitude* of the superconducting transition have been noticed as the 1-h aging temperature increased within the range 300-1200°C . The *sharpening* of the transition could have been a result either of strain relief or an increase in alloy homogeneity. The decrease in its temperature was attributable to the T_c-depressive effect of α-phase precipitation occurring: *(b)* as a consequence of equilibration during heat treatment in the intermediate-temperature $\alpha+\beta$ region; *(b)* as a result of the α-stabilizing influence of dissolved O, picked up at some stage either during or after the

high-temperature anneal, in the case of alloys heat-treated in the β-region. By comparing the T_c of the not-heat-treated Ti-Ta$_{50(nom)}$-O$_{0.07}$ alloy, prepared from iodide Ti, with that of Ti-Ta$_{50(nom)}$-O$_{0.2}$, as prepared from sponge, COLLING et al [Col66[a]] concluded that dissolved O was responsible for an increase in the transition temperature of β-Ti-Ta alloys. The O-assisted *decrease* in T_c noted in the β-heat-treated Ti-Ta alloy must, therefore, have been due to the presence of α-precipitation stabilized by the O and in which it preferentially dissolved.

5.6 UPPER CRITICAL FIELDS OF Ti-Ta ALLOYS

5.6.1 Experimental Studies of the Resistive Upper Critical Field

Measurements of the upper critical fields of Ti-Ta alloys, together with those of an extensive series of other binary TM alloys were first conducted by BERLINCOURT and HAKE [Ber63[a]] in their classical study of the GLAG and CLOGSTON-CHANDRASEKHAR-paramagnetically-limited upper critical fields. Resistometrically determined (at $J = 10$ A cm^{-2}) values of $H_{r,1.2 K}$ were compared with the CLOGSTON fields ($H_p = 18.4$ $T_c(1-t^2)$ kOe, Eqn.(3-14)) computed with the aid of T_c values which had been measured on the same series of alloys by BLAUGHER and JOINER [Bla63], Fig.5-1. A comparison was not able to be made in this case with a computed H_{c2}^*, calculated for example using the GOR'KOV-GOODMAN formula, Eqn.(3-5). Had this been done it would no doubt have turned out, cf. Sect.3.6.1, Fig.3-3, that the experimentally determined upper critical field was less than the smaller of H_{c2}^* and H_p. As indicated in Fig.5-1, the upper critical field of Ti-Ta followed the general pattern that seems to have become established for (group IV)TM-(group V)TM alloys, viz: for group-IV-rich alloys (e.g. up to about 40 at.% Ta) H_r and H_p coincide (the upper critical field in this case is strongly Pauli limited) while at higher concentrations of the group-V constituent, H_r is significantly less than H_p, in which case it seems that the experimental upper critical field is controlled by the GLAG H_{c2}^* itself. In a summarizing paper in which the results for Ti-V, Ti-Nb, Ti-Ta, and Zr-Nb were intercompared [Ber63], it was shown that the peak (i.e. mid-composition) upper critical fields decreased in the sequence Ti-Nb > Ti-Ta > Zr-Nb > Ti-V.

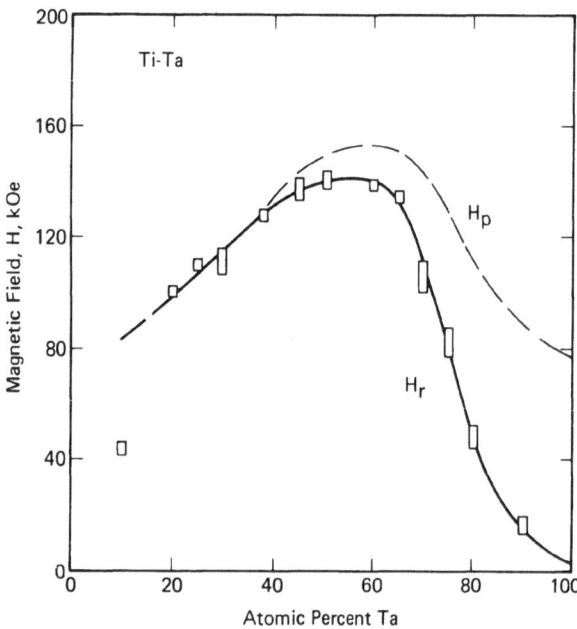

FIGURE 5-1. Resistive upper critical fields at 1.2 K for Ti-Ta alloys (H_r, from *threshold* to *full restoration* of resistivity at 10 A cm^{-2}); also shown is the paramagnetically limiting field (H_p, 1.2 K) calculated using the Clogston relationship, Eqn (3-14), and the BLAUGHER-JOINER [Zwi63] transition temperatures (cf. Table 5-4) — after BERLINCOURT and HAKE [Ber63[a]].

with Eqns.(3-7) and (3-11); viz with the aid of:

$$H_{c2}^* = 18.3 \, A(t) \, \rho_n \, \gamma \, T_c(1-t^2)(1+\Delta) \quad (kOe) \quad (5-1)$$

where Δ was a "small correction" related, presumably, to the BCS deviation function (as discussed, for example, in [Mon8.5]) and where $A(t)$ was the GOR'KOV temperature dependence, $1.77 - 0.43t^2 + 0.07t^4$, as it appears in Eqn.(3-5).

Unfortunately no direct experimental information on the low temperature specific heats of Ti-Ta alloys are available with which to compare the predictions of COLLING *et al*. As before, H_r was less than H_p over the entire composition range measured, but since Ta concentrations less than 30 at.% were not investigated, the equality of H_r and H_p observed earlier by BERLINCOURT and HAKE in the lower concentration members of the sequence was not able to be confirmed.

5.6.2 Influence of Spin-Orbit-Scattering Effects on the Paramagnetically Limited Upper Critical Fields of Ti-Ta Alloys

NEURINGER and SHAPIRA [Neu66] have measured the H_r temperature dependences of Ti-Ta$_{52}$ and two other related alloys, Ti-V$_{58}$ and Ti-Nb$_{44}$. The results obtained provided the opportunity for an experimental test of the full MAKI theory [Mak66] and the related theory of WERTHAMER, HELFAND, and HOHENBERG (WHH) [Wer66], both of which dealt with the partial relief of Pauli paramagnetic limitation by way of the development of superconductive-state paramagnetism through the intervening mechanism of spin-orbit scattering. A description of the mechanism has already been adequately provided in Sects.3.6.3 through 3.6.5 and need not be reiterated here. Paramagnetic upper critical field effects have also been fully discussed in [Mon15.00]. The importance of Ti-Ta$_{52}$ in this context is that its $H_r(t)$ was found to lie slightly *above* the MAKI-calculated curve for $\alpha = 0$, indicating an apparent *complete absence* of Pauli limitation. Since Ti-Ta, as with other such alloys, was expected to be significantly Pauli paramagnetic in the normal state, a very rapid spin-flip-scattering rate, λ_{so} in WHH theory, would have been needed to achieve the required full spin decoupling in the mixed state. Likewise, DOMB and JOHNSON [Dom78] later reported that the upper (and lower) critical field temperature dependences of a

BERLINCOURT and HAKE displayed their data as functions of both composition and e/a ratio (or average group number which, when the alloying elements are chosen from adjacent columns of the periodic table, bears a simple relationship to it). But using measured results for a sequence of binary, ternary, and quaternary TM alloys, including Ti-Ta$_{53}$, DeSORBO *et al* [Des67] found that an atomic-volume-weighted "effective electron/atom ratio, N_{eff}" was a more general descriptor of the positions of the maxima in both T_c and H_{c2}. In particular it was shown that for twenty of the twenty-six alloys considered, H_{c2} lay close to a curve which peaked at $N_{eff} \sim 4.4$. Critical-field experiments on a series of Ti-Ta alloys at 4.2 K were subsequently carried out by COLLING *et al* [Col66] who compared for each alloy the measured $H_{r,4.2 \, K}$ with an H_p computed from the CLOGSTON formula with the aid of their own measured values of T_c [Col66[a]].

The results of the H_r measurements were discussed in terms of the expected and observed microstructural properties of the alloys. Attempts were also made to calculate: *(a)* H_{c2}^*, using Eqn.(3-16b) with H_{c2} identified with H_r; *(b)* the electronic specific heat coefficient, γ, from a variant of Eqn.(3-5) in combination group of amorphous superconductors lay consistently above the MAKI curves. There the matter rested until: *(a)* ORLANDO *et al* [Orℓ79] pointed out that, as a consequence of electron-phonon interaction, the actual CLOGSTON field, H_p, is considerably higher by a factor $(1 + \lambda)^{1/2}$ than the bare density-of-states would suggest, where λ is the usual electron-phonon interaction parameter or coupling constant; *(b)* SCHOPAL and SCHARNBERG [Sch81] demonstrated that spin-orbit scattering could actually be the dominant scattering mechanism, in contrast to the restrictions that WHH had placed on their theory. The present status of upper-critical-field theory has been reviewed by BEASLEY [Bea82].

5.7 CRITICAL CURRENT DENSITIES OF Ti-Ta ALLOYS

5.7.1 Factors Which Influence Flux Pinning

Ti-Ta, together with the related systems Ti-V and Ti-Nb, had been classified as a potentially useful ductile alloy superconductor. Thus using Ti-Ta alloys as working substances, if not with actual applications in mind, the influences on the critical current density of several important mechanical, thermal, and physical properties have been considered. These were: *(a)* the effects of cold rolling on J_c and its anisotropy (i.e. the response of J_c to variation of the field direction relative to the rolling plane and the rolling direction) [Hak62^a, Hak63^a]; *(b)* the influence of field direction relative to the current direction in wire-drawn samples [Sek63] (cf. [Mon21.16] through [21.18] and [Mon22.00]); *(c)* the influence of heat treatment on J_c; *(d)* the effects on J_c of the addition of interstitial as well as other transition elements to the basic binary Ti-Ta alloy (as expressed in [Bet66] and in French Patent 1,517,216 [Ass68^a]). Also studied using Ti-Ta as a test material has been the discrepancy between measured short-sample and coil critical currents (i.e. so-called current degradation).

5.7.2 Influence of Cold Work on the Critical Current Density

The influences on J_c of anisotropic microstructures introduced by rolling and the subsequent

shearing-to-size of cold-rolled strips have been studied by HAKE and co-workers [Hak62^a, Hak63^a]. In the first of these papers it was noted that the "transverse" critical current density, i.e. that measured with a transverse magnetic field applied parallel to the rolling plane, $J_c(\|)$, was several times larger than $J_c(\perp)$. Also noted for the first time, with particular reference to a Nb-Sc alloy, was the appearance of a maximum in $J_c(H_a)$ as the applied field approached its critical value. A particularly interesting aspect of the "peak effect", as it is now called, is that it requires the existence, at intermediate-field values, of a region of $J_c(H_a)$ in which $\partial J_c/\partial H_a$ is *positive*. This implicit property of the peak effect offers interesting possibilities for intrinsic (flux-jump) stabilization, cf. [Mon25.9].

In a continuation of this work HAKE showed that maxima in J_c occurred whenever H_a was parallel to a set of texture planes [Hak63^a]. In a large sheet this of course implies that H_a is parallel to the rolling plane; but in a narrow sheared strip, secondary maxima can develop whenever the rolling texture is distorted, by the action of the cutting shears, away from that plane. In a study of the peak effect using rolled Ti-Ta$_{90}$ ribbon, it was noted that the peak appeared only in conjunction with the generally low values of J_c associated with a field direction perpendicular to the rolling plane, and disappeared as soon as H_a was rotated *into* the rolling plane so as to give maximal J_c for the alloy. During such rotations, the critical current density at fields close to H_{c2} tended to remain invariant, the major changes in J_c taking place within the "valley" region in the neighborhood of $H_r/2$. These were landmark papers in that they drew attention, for the first time, to the importance of metallurgical microstructure as a critical current determining variable at a time when the concept of the "anchoring of the transport supercurrents against the Lorentz force" was beginning to emerge. As a further contribution to these studies of generalized critical-current anisotropy SEKULA *et al* [Sek63], using commercially available superconducting alloy wires, investigated the influence on J_c of external fields applied parallel to the current direction. The sample materials used included commercial Ti-Ta$_{75}$ in the form of a 0.010 in$^\phi$ wire. Its critical current density as a function of longitudinal applied field passed through a broad maximum of height 3.8×10^5 A cm^{-2} symmetrically situated near 22 kOe, or mid-way between half-$J_{c,max}$

field values of 2 kOe and 56 kOe. Critical currents in a longitudinal applied field are the subject of [Mon22.00]. The significance of this maximum can be appreciated by comparing the $J_c(H_a)$ values, not with those obtained in the early work of HAKE *et al*, which were rather low in the practical sense, but with the transverse-field results of WOLGAST *et al* [Woℓ63] on Ti-Ta(70, 75 and 80 wt.%) wires (albeit, non-optimized) for which a critical-current plateau of about 1×10^4 A cm^{-2} existed at fields of between about 20 and 70 kOe.

5.7.3 Influence of Heat Treatment on the Critical Current Density

Critical current density is of course influenced not only by alloy composition and cold work but also by heat treatment. Chapter 7 deals fully with this subject. Detailed studies of the effects of 1-h heat treatments at temperatures between 300 and 1200°C (100°C intervals) were undertaken by COLLING *et al* [Coℓ66]. The subject of the investigation was Ti-Ta$_{50.9}$ prepared from sponge Ti and possessing an analyzed O impurity level of 249 wt. ppm. Heat treatments at between 400 and 600°C, which increased the intermediate-field J_c by a factor of about four, were accompanied by unidentified fine-scale precipitation which subdivided the cold-worked fiber structure. In that it maximized the intermediate-field J_c without reducing the upper critical field, the 400°C heat treatment was optimal. At temperatures above 700°C, H_r assumed a constant value. With increase in heat-treatment temperature, as a result of the elimination of the cold-worked fiber structure through recrystallization and grain growth, the intermediate-field ($\sim H_r/2$) critical current density decreased; at the same time, since $J_c(^2H_{c2})$ tended to remain invariant, a high-field peak (hence intermediate-field trough) developed in $J_c(H_a)$.

The influences of heat treatment on the J_c's of Ti-Ta$_{60}$ alloys both with and without the addition of small amounts of O (0.08 and 0.27 at.%) were considered briefly by BETTERTON *et al* [Bet66]. The heat treatment of course had the expected positive effect but no significant improvements accompanied the additions of oxygen in the levels mentioned above. As evidenced by French Patent 1,517,216 [Ass68[a]], Ti-Ta together with other binary and ternary alloys formed by transition elements of the IVth and V[th] groups of the periodic

table have been considered to be of value as possible candidates for the fabrication of commercial high-field superconducting magnet wire. An important ingredient of the patent which, to be sure, dealt chiefly with Ti-Nb and Ti-Nb-base alloys, was the specification of: *(a)* optional additions of the interstitial elements O, N and C; *(b)* heat treatments of 1h/400°C to be administered at certain stages during the wire-drawing process.

5.8 SPUTTERED Ti-Ta ALLOY FILMS

SPITZER [Spi74, Spi74[a]] has studied the superconducting transition temperatures and critical current densities of Ti-Ta alloy films prepared by DC triode co-sputtering onto a quartz substrate maintained at 700°C. Transition temperature, Fig.5-2, was determined by monitoring the sample resistivity as a continuous function of temperature, and critical current density, Fig.5-3, was measured by a pulsed-current pulsed-field technique. Although thick sputtered

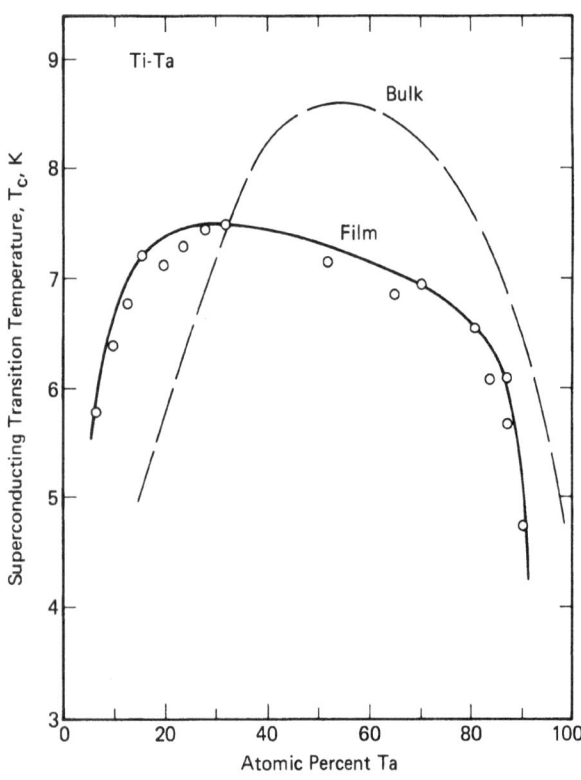

FIGURE 5-2. Intercomparison of the superconducting transition temperatures of sputter-deposited and bulk Ti-Ta alloys. Data sources: sputtered films (○) after SPITZER [Spi74]; bulk alloys (dashed line) from the Ti-Nb-Ta study of SUENAGA and RALLS [Sue69] — after SPITZER [Spi74].

films were expected to exhibit bulk properties, it turned out in practice that their transition temperatures and critical current densities differed considerably from those of bulk material. Examples of this have already been seen in sputtered Ti-V films.

While the transition temperatures of the Ti-Ta(>33 at.%) films were some 1 K lower than those of their bulk counterparts, the converse was true for the less concentrated members of the series, Fig.5-2. Although no structural investigations were undertaken, it is possible that sputtering permitted stabilization of the β-phase to lower Ta concentrations than is possible under conventional rapid solidification.

The $J_c(H_a)$ curves for sputtered Ti-Ta generally decreased monotonically with increasing applied field in a manner comparable to that encountered in contemporary conventionally-process material (cf. for example, Fig.7-44). It seems that sputtering as a process is immediately beneficial to the critical current density. Fig.5-3 compares the critical current density of >99.8% cold-worked and 1h/400°C aged Ti-Ta$_{50.9}$ with those for sputtered Ti-Ta(32, 52, and 70 at.%) alloys. The $J_c(H_a)$ characteristics of the sputtered material are suggestive of "higher degrees of optimization". The slopes, dJ_c/dH_a, are rather steep, a property unfavorable to flux-jump (intrinsic) stability. As pointed out in Sect.5.7.2, large negative dJ_c/dH_a's such as those depicted in Fig.5-3 generally accompany the "filling-in" of the mid-field ($\sim H_r/2$) trough in response to the appearance, as a result of thermomechanical processing in the case of bulk alloys, of a suitable density and arrangement of flux-pinning sites. The sources of flux pinning in the sputtered alloys were not discussed.

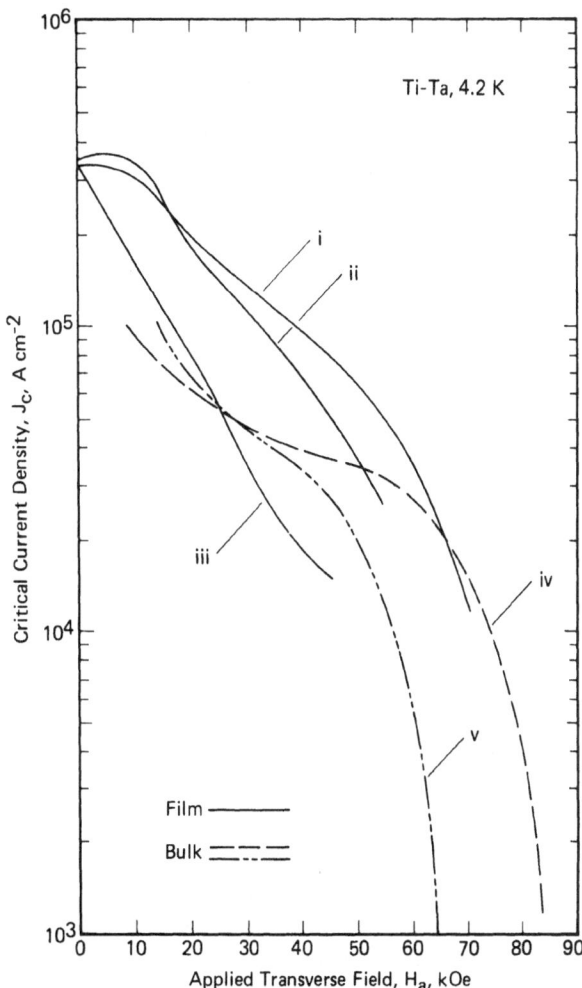

FIGURE 5-3. Intercomparison of the critical current densities of sputter-deposited and conventionally processed Ti-Ta alloys. Shown are the J_c *versus* H_a characteristics of sputtered: (i) Ti-Ta (32 at. %), (ii) Ti-Ta (52 at. %) and (iii) Ti-Ta (70 at. %) — after SPITZER [Spi74]; those of 99.8%-cold-worked bulk Ti-Ta (50.9 at. %) aged (iv) 1h/400°C and (v) 1h/500°C (criterion: 1/4 μV cm^{-1} at high fields) — after COLLING *et al* [Coℓ66].

5.9 SUPERCONDUCTIVITY IN Ti-Ta ALLOYS—TABULATED DATA

TABLE 5-4 TRANSITION TEMPERATURES (T_c) OF TITANIUM-TANTALUM ALLOYS

At.% Ta	T_c K	Data Sources[+]		At.% Ta	T_c K	Data Sources[+]	
2.5	1.3	[Rau65]	n	53.5	8.7	[Col66[a]]	g
5	2.9	"	"	56.5	8.7	"	"
26.4	6.1	[Zwi63]	g	59.2	8.6	[Zwi63]	g
30.1	7.1	[Col66[a]]	g	63.6	8.47	[Col66[a]]	ncw
34.2	7.6	"	"	"	8.07	"	na
34.8	6.9	[Zwi63]	g	68.5	8.28	"	ncw
39.2	7.4	"	"	"	8.02	"	na
42.3	8.4	[Col66[a]]	g	77.7	7.9	"	g
43.0	7.8	[Zwi63]	g	86.3	7.1	"	"
46.2	9.05	[Col66[a]]	ncw	97.7	4.9	"	"
"	8.26	"	na				
47.3	8.7	"	g	7	5.8	[Spi74[a]]	s
48.6	8.7	[Zwi63]	g	33	7.5	"	"
50.8	8.81	[Col66[a]]	ncw	53	7.2	"	"
"	8.11	"	na	71	7.0	"	"
53.0	8.4	[Zwi63]	g	90	4.8	"	"

+ n   = numerically supplied data
  na = numerically supplied -- annealed samples
  ncw = numerically supplied -- cold-worked samples

g = from plotted T_c vs composition data
s = selected data from T_c vs composition graph -- sputtered samples.

TABLE 5-5 TRANSITION TEMPERATURES OF TITANIUM-TANTALUM ALLOYS -- DATA SOURCES

Ta Concentration Range	Condition	Structures	Procedures	Literature
26-60 at.% (7 alloys)	Rapidly cooled.	β	Probably magnetic[+]	R.D. Blaugher and W.C.H. Joiner (1963) Referred to in U. Zwicker (1963) [Zwi63]
2.5 and 5 at.%	Annealed 650°C.	α	Not stated	C.J. Raub and U. Zwicker (1965) [Rau65]
30.1-97.7 at.% (13 alloys)	Homogenized and cold worked.	β	Resistometric	D.A. Colling et al. (1966) [Col66[a]]
7-90 at.% (16 alloys specified)	Rapidly cooled from 700°C.	β (sputtered films)	Resistometric	H.J. Spitzer (1974) [Spi74[a]]
17.5-97.5 at.% (10 alloys)	Probably quenched.[++]	β	Resistometric	A.F. Prekul et al. (1974) [Pre74]

+ Based on [Hul61]
++ Based on treatment of companion alloys Ti-V and Ti-Nb.

TABLE 5-6 UPPER CRITICAL FIELDS OF TITANIUM-TANTALUM ALLOYS

(a) Upper Critical Field (H_r) at 1.2 K [Ber63[a]]

At.% Ta	H_r[†] kOe	
10.0[††]	42.2	47.0
19.8[†††]	99.0	102.3
25.0[†††*]	108.0	112.0
29.5[†††*]	106.5	116.1
37.5	126.0	129.7
45.0	133.0	141.4
50.0	135.5	137.5
60.0	137.8	141.5
65.0	133.0	137.5
70.0	100.8	112.0
75.0	78.0	88.0
80.0	45.5	53.0
90.0	14.0	20.0

† Supplied as numerical data -- onset of detectable resistance and full restoration of resistance, respectively; measuring current density, 10 A cm^{-2}.

†† Mainly hcp but some bcc.

††† hcp and bcc in equal amounts.

†††* Mainly bcc but some hcp.

(b) Upper Critical Fields (H_r) at 4.2 K

At.% Ta	H_r kOe	Sources
30.1	67	[Col66][†]
34.2	78	"
42.3	87	"
46.2	93	"
47.3	92	"
50.8	92	"
52	86	[Neu66]
53	78	[Des67][††]
53.5	92	[Col66][†]
56.5	90	"
63.6	83	"
68.5	61	[Col66]
77.7	47	"
86.3	19	"
97.7	2	"

† H_r criterion: 1μV across 4 cm at 1 A cm^{-2}

†† Magnetically determined (H_u)

TABLE 5-7 UPPER CRITICAL FIELDS OF TITANIUM-TANTALUM ALLOYS -- DATA SOURCES

Ta Concentration Range	Condition	Procedures	Literature
10.0-90.0 at.% (13 alloys)	Arc melted and cold rolled.	Resistive (1.2 K, 10 A cm^{-2})	T.G. Berlincourt and R.R. Hake (1963) [Ber63[a]]
52 at.%	Vacuum annealed.	Resistive (1.3 K - T_c)	L.J. Neuringer and Y. Shapira (1966) [Neu66]
30.1-97.7 at.% (13 alloys)	Cold swaged and drawn to 0.010 in.[♀]	Resistive (4.2 K, 1 A cm^{-2})	D.A. Colling *et al.* (1966) [Col66]
53 at.%	Cold swaged and drawn to 0.010 in.[♀]	Magnetic (4.2 K)	W. DeSorbo *et al.* (1967) [Des67]

TABLE 5-8 CRITICAL CURRENT DENSITIES (J_c) OF TITANIUM-TANTALUM ALLOYS

(a) Cold-Rolled Strip -- Composition Dependence at 4.2 K [Hak62[a]].

At.% Ta	Thickness Reduction %	$J_c{}^+$(30kOe), 10^3 A cm^{-2}	
		$J_c(\perp)$	$J_c(\parallel)$
20	75		0.027
25	77	3.7	10
29	75	2.6	6.7
60	76	0.49	0.8
70	75	0.8	1.6
75	75	0.19	1.0

[+] Original data in tabular form
   J_c criterion: 0.25 µV across 7 mm.

(b) Cold-Rolled Ti-Ta(90 at.%) Strip (Thickness Reduction: 140:1, 99.3%) -- Anisotropy at 1.2 K [Hak63[a]].

Transverse Field kOe	$J_c{}^+$ vs Angle Between Field and Plane, 10^3 A cm^{-2}		
	9°[++]	90°	170°
4	40	0.63	5.1
8	29	0.41	5.2
11	--	----	3.4[+++]
11.5	--	0.63[+++]	---
12	9.5	----	---

[+] Data from published graphs,
   J_c criterion: 0.1 µV across sample.

[++] No peak-effect in $J_c(H_a)$ at this angle.

[+++] Peak in $J_c(H_a)$.

(c) Wire Samples (mostly "optimized")

At.% Ta	$J_c{}^+$(4.2K) in Various Transverse Applied Fields, 10^4 A cm^{-2}				Reference
	30kOe	40kOe	60kOe	80kOe	
38	---	1.1$_5$	1.0	---	[Woℓ63][++]
44	---	1.0	0.8$_9$	0.4$_3$	"
51	---	0.8$_1$	0.8$_5$	0.7$_1$	"
60	2.7	---	---	---	[Bet66]
60	3.4	---	---	---	"

[+] Data from published graphs.

[++] First appearance of voltage (system noise, $\overline{<}0.2$ µV), 1∿2 cm between potential leads.

(d) Wire Samples of Ti-Ta(50.8 at.%) Cold Worked[+] and Heat Treated 1 h [Coℓ66].

Heat-Treatment Temperature °C	J_c (4.2K, 50∿60 kOe)[++] A cm^{-2}
none	$1.0_0 \times 10^4$
200	$1.5_1 \times 10^4$
400	$3.5_5 \times 10^4$
500	$3.9_8 \times 10^4$
600	$3.0_2 \times 10^4$
800	$1.0_0 \times 10^3$
1000	$6.3_1 \times 10^2$

[+] Swaged 0.25 in.$^\phi$ to 0.14 in.$^\phi$ and drawn to 0.01 in.$^\phi$ (99.84% overall).

[++] Intermediate-field "plateau" current density. Data selected from published graphs, J_c criterion: 1 µV across 4 cm.

TABLE 5-8 CRITICAL CURRENT DENSITIES (J$_c$) OF TITANIUM-TANTALUM ALLOYS -- *continued*

(e) Longitudinal-Field[†] Critical Current Density of
    Cold-Drawn Ti-Ta(75 at.%) at 4.2 K [Sek63]

Wire Diameter in.	J$_c$[††] A cm^{-2}
0.005	4.3x10^5
0.010	3.8x10^5

† Field at current maxima, ∿22 kOe.
†† Data from published graph.

(f) Critical Current Densities of Sputter-Deposited
    Ti-Ta Films [Spi74, Spi74^a]

At.% Ta	J$_c$(4.2K), 10^4 A cm^{-2} at Various Fields[†]		
	30kOe	40kOe	60kOe
7	4.7	4.1	2.7
16	10.8	8.7	4.6
32	14.1	9.6	3.6
52	11.6	6.8	---
70	3.4	1.9	---
90	<1	---	---

† Values selected from plotted J$_c$(H$_a$) curves.

ALLOY GROUP 3: TITANIUM-MOLYBDENUM BINARY ALLOYS

5.10 SUPERCONDUCTIVITY IN Ti-Mo ALLOYS

Ti-Mo occupies a prominent place in the historical development of pure and applied type-II superconductivity. The earliest studies of the system, carried out in the late 1950's and early 1960's, dealt
with the response of T$_c$ to changes of solute content.
Ti-Mo was an important model system in that, as a member of the β-isomorphous class of Ti-TM alloys, its
electron/atom ratio could be varied over a wide range
(from 4 to 6) without danger of intermetallic-compound
formation as was the case with, for example, the
β-eutectoid alloy, Ti-Cr (cf. [Mon1.2] for a comparison of the equilibrium-phase diagrams). Ti-Mo has
been thermally aged in studies of the influences of
ω-phase and α-phase precipitation on T$_c$; the influence
of deformation on the calorimetrically measured T$_c$ has
also been examined. Ti-Mo has received considerable
attention as a short mean-free-path (i.e. "dirty")
superconductor and as such has played an important
role in experimental investigations of advanced theories of the mixed state. Extensions of the original
GLAG theory to include the effects of normal-state
Pauli paramagnetic limitation, and the establishment

of a Pauli paramagnetic mixed state aided by spin-orbit
scattering of the superconducting electrons, have been
examined using Ti-Mo as a "working substance".

Several studies relating to critical current
density have also been undertaken. Finally, again
within the context of current transport in superconductors, it is appropriate to mention that Ti-Mo has
contributed to studies of the mixed-state Hall effect,
anomalous normal-state resistivity, and fluctuation
superconductivity.

5.11 TRANSITION TEMPERATURES OF Ti-Mo ALLOYS

5.11.1 The Superconducting Transition Temperatures
         of bcc Ti-Mo Alloys

The first study of superconductivity in the Ti-Mo
system was carried out by BUCHER *et al* [Buc65] who
investigated a limited range of bcc alloys
(30-48.5 at.% Mo) using a magnetic technique and expressed the results, along with those for β-Ti-V and
V-Mo alloys, as functions of electron/atom ratio.

This work was quickly followed by a series of low-temperature calorimetric measurements by HAKE [Hak61] of the low-solute-concentration end of the same system (viz 6.25-8.60 at.% Mo). As a result of metallographic, x-ray, and hardness investigations conducted on these alloys it was claimed that they were free of ω-phase precipitation, and consequently were single-phase bcc; but by now it is recognized that a considerable volume fraction of ω-phase must in fact have been present in all of the alloys measured. At about the same time, BLAUGHER et al [Bla61] published the results of a magnetic study of an extensive series of Ti-Mo alloys covering the composition range 0 to 60 at.% Mo and, consequently, the entire range of possible structures: α^m, $\omega + \beta$, and β. The curve of T_c versus solute concentration, c, possessed two distinct features: (a) a steeply rising branch in the low-concentration α^m region, terminating near 6 at.% Mo and extrapolating back to 0.49 K for pure Ti (cf. Table 2-3); (b) for the remaining concentrations, a continuous arc with a well-defined maximum in the vicinity of 16 at.% Mo (e/a = 4.32). SINHA [Sin68] measured Ti-Mo(13-85 at.%) using low temperature specific heat, and in effect completed the work begun by HAKE [Hak61]. BLAUGHER et al [Bla61], SINHA [Sin68], and HAKE [Hak61], in an analysis of the latter's data, all discussed the calorimetric results in terms of the BCS-MOREL relationship, Eqn.(3-1), coming to the conclusion that the interaction parameter, V, was approximately constant for the entire series of Ti-Mo alloys, and consequently that T_c paralleled the variation of $n(E_F)$; cf. [Mon8.7] and [Mon12.1] for further discussion.

SINHA also failed to detect ω-phase in the two lower concentration alloys, Ti-Mo(13 and 15 at.%) -- and indeed it could well have been absent at room temperature -- and consequently was led to the conclusion that the T_c maximum, noted in this case at e/a = 4.3 ∿ 4.4, was a reflection of a maximum in the single-phase-bcc density-of-states. On the other hand, HO and COLLINGS [Ho69] noted as a result of transmission electron microscopy conducted by BOYD [Col71] [Boy78], the presence at room temperature of large volume fractions of ω-phase in quenched Ti-Mo($4\frac{1}{2}$, 5, and 10 at.%) alloys and claimed: (a) that the decreases in γ and T_c that took place with decreasing electron/atom ratio below about 4.4 were due to a continuously increasing density of ω-phase precipitation and, moreover, that in the absence of such precipitation, T_c would lie on an upper branch tending towards T_c ∿ 6.4 K

for bcc-Ti. Developed with the aid of high-temperature magnetic susceptibility data this lattice-stability-related effect was fully discussed in papers by COLLINGS, HO, and JAFFEE [Col72] and COLLINGS [Col72[a]], who at the same time were able to predict the superconducting transition temperatures of several dilute bcc-phase Ti-Mo alloys. In those papers were presented the results of a calorimetric study of a 12-member series of Ti-Mo(1-70 at.%) alloys. The structures of quenched Ti-Mo(1-3 at.%) were α^m; those of the remainder were $\omega + \beta$, tending to β as e/a increased beyond 4.15 ∿ 4.2. The structures generally to be expected in quenched alloys of this type are discussed in [Mon1.6] and [Mon1.9].

As indicated above, calorimetric data when applied to the BCS-MOREL expression for T_c, Eqn.(3-1), revealed an interesting conclusion regarding the constancy of V within an alloy series. In the approach used by BLAUGHER et al [Bla61], Eqn.(3-1) was supposed to have enabled $n(E_F)V$ to be directly computed. Then by assuming $n(E_F) \propto \gamma$, the quotient $n(E_F)V/\gamma$ was taken as being proportional to V. Thus if the quotient turned out to be constant, it was assumed that V, the electron-phonon pairing potential, was also constant. Such an approach, however, ignored the effect of the electron-phonon interaction on the density-of-states as a consequence of which, cf. [Mcm68], [Orl79]:

$$\gamma = \gamma_{bs}(1+\lambda) \qquad (5-2)$$

where γ_{bs} is the theoretical (band-structure) electronic specific heat coefficient (referred to above, but without the subscript), γ is now the measured value, and λ is the electron-phonon coupling constant, already encountered in Sect.5.6.2. Similarly the slope of the $\log(T_c/\theta_D)$ versus γ^{-1} plot of MORIN and MAITA [Mor63] (cf. Sect.3.2.2, also [Mon8.7], [Mon10.3], [Mon11.9]) is now seen to yield an apparent pairing potential, V_{app}, rather than an unmodified BCS value. This problem has been dealt with by COLLINGS and HO [Col71[a]] who have pointed out that the empirically obtained V_{app} is in fact just that needed to adjust the measured γ to the $n(E_F)$-proportional value, γ_{bs}, by means of the relationship:

$$\gamma_{bs} = \gamma(1 + 0.212 \gamma V_{app})^{-1} \qquad . \qquad (5-3)$$

This case study of the density-of-states and superconducting properties of the Ti-Mo system, cf. [Col72,

Col73], has been fully discussed in [Col72[a]] and [Mon10.4].

5.11.2 The Transition Temperatures of the Quenched Martensitic Alloys

Quenched martensitic alloys were studied using calorimetric and magnetic techniques by BUCHER et al [Buc65]. As with those of the other dilute Ti-TM alloys studied at that time, a rapid rise of T_c with composition was noted. COLLINGS and HO [Col69] carried out a calorimetric study of a series of six quenched α^m-Ti-Mo(1 to $4\frac{1}{2}$ at.%) alloys. The rapid increase in T_c with composition was not reflected in the behavior of the electronic specific heat coefficient, consequently it was claimed that T_c was the recipient of enhancement stemming from a "lattice-softness" effect deriving from the disordered state of the α^m-structure (cf., for example, [Str68]). Further such discussions of superconductivity in disordered structures have been advanced, within the context of the influences of deformation-induced martensitic and other structural transformations on the calorimetrically measured superconducting transitions, by WHITE and COLLINGS [Whi78] (Ti-V), [Whi76][Col78[a]] (Ti-Mo). The influence of deformation and the resulting transformation products on superconducting transitions in Ti-Mo alloys are considered below.

5.11.3 Influence of Deformation on the Superconducting Transition

In calorimetric measurements of Ti-Mo$_{4.5}$, an alloy whose composition placed it at the boundary of the α^m and $\omega+\beta$ regimes, COLLINGS and HO [Col70] noted that the microstructure, and associated T_c, could be controlled by suitable heat treatment and deformation (i.e. by thermomechanical processing). This was the first time in alloy research that a study had been carried out on the influence of structure on T_c at *fixed* average solute concentration. The height of the specific heat jump scaled qualitatively with the fraction of martensite visible in a metallographic section -- small in as-cast samples, and dense in alloys quenched from 1300°C. Working on the assumption that transformation in the latter alloys had been aided by quenching strains, a successful attempt was made to induce structural transformation -- and with it a

larger superconductive volume fraction -- by subjecting the as-cast Ti-Mo$_{4.5}$ sample to mechanical deformation by compression under an applied load of about 30 tons (roughly 200 ksi or 15×10^8 N m^{-2}). In response to this treatment, the transition temperature of Ti-Mo$_{4.5}$ was raised from an extrapolation-estimated value of about 0.5 K for the $\omega+\beta$-structure, to a measured 3.1 K at the threshold of the specific heat jump for the deformed alloy. To enable this to be claimed as an "enhancement effect" rather than simply the result of a deformation-structure-induced change in $n(E_F)$, calorimetric data with the associated γ-value are essential. For this alloy it was noted that, whereas $T_{c,\alpha^m}/T_{c,\omega+\beta} \sim 3.1/0.5 \sim 6$, the corresponding electronic specific heat coefficient ratio was $\gamma_{\alpha}m/\gamma_{\omega+\beta} = 4.65/3.85$, which itself would be responsible for a T_c-ratio of only about 2.2. The deformation can therefore be said to have led to a 3-fold enhancement in T_c. In a continuation of this work, HO and COLLINGS [Ho71] (see also COLLINGS et al [Col71] and HO and COLLINGS [Ho73]) studied the effect of deformation on the calorimetrically observed superconducting transition in Ti-Mo(5 and 7 at.%) alloys. As indicated in Fig.5-4, in which the calorimetric results for Ti-Mo$_5$ are depicted, very significant enhancements of T_c were again obtained in response to the deformation-induced structural change. It is also instructive to compare the ΔT_c obtained in response to this deformation-transformation with those which accompanied quench-transformations of the same base alloy following the additions of small amounts (viz 1 at.% and 3 at.%) of an α-stabilizing element such as Al, Fig.5-5. An interesting self-consistency between the transition temperature of the deformation-transformed binary alloy and those of the quench-transformed ternary alloys was noted: the latter when

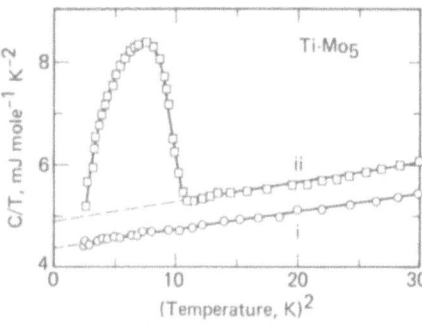

FIGURE 5-4. Low temperature specific heats of (i) quenched, and (ii) quenched-plus-deformed Ti-Mo (5 at. %). Martensitic transformation is responsible for the enhanced T_c — after HO and COLLINGS [Ho71].

plotted *versus* Al concentration back-extrapolated to the transition temperature of the deformed Ti-Mo$_5$ master alloy. Not unexpectedly, the calorimetric transition in the deformed alloy was broad and rounded, Fig.5-4. Some measurable widths were also associated with the transitions depicted in Fig.5-5. In order to extract quantitative information under such conditions it is necessary to mathematically fit the data to a suitable distribution of transition temperatures. The manner in which this can be accomplished is set out in [Mon10.3] and [Mon10.8]. Briefly, the procedure adopted consisted of *assuming* a transition temperature distribution function such as an asymmetric-exponential or an "asymmetric-Gaussian", combining it with an exponential or BCS sub-T_c electronic specific heat function, and fitting the result to the experimental data using the nonlinear least-squares-fitting code (NONLIN4) initially devised by WHITE [Whi74] for the analysis of calorimetrically detected magnetic transitions. The asymmetric-Gaussian/exponential-specific-heat option has been applied to the analysis of the calorimetric data of quenched-plus-deformed Ti-Mo$_5$ [Whi76], while the asymmetric-Gaussian/BCS-specific-heat approach has been applied to the same alloy, and also to quenched, and quenched-plus-deformed Ti-Mo$_7$ [Col78^a]. Such analyses enabled a surprising amount of information about the combined microstructural and physical states of the samples to be extracted. For example, when applied to the results presented in Fig.5-4, the Gaussian-BCS option yielded the following data: the observed superconducting transition was 68% complete on a BCS basis, the average value of γ was increased 12% by the deformation, while θ_D was slightly depressed; the distribution of T_c used to describe the rounding had a width at half-maximum of 0.47 K and was extremely skewed toward lower temperatures. The curve-fitting enabled the following conclusions to be drawn: the product of the deformation was 68% martensite with a T_c of 3.27 K, 32% $\omega+\beta$ with a T_c of 1 K, plus an internal proximity effect between the two; in addition, it was suggested that defects within the martensite might have played substantial roles in influencing the enhancement, completeness, and rounding of the transition.

5.11.4 The Structures of Quenched and Deformed Ti-Mo Alloys

A physical-property study of martensitic alloys is of course not complete without an accompanying structural investigation. In most cases this has not been possible to arrange. As indicated in a review of the subject by WILLIAMS [Wil73], the structures of quenched martensites may be either hexagonal, face-centered cubic, orthorhombic, or face-centered orthorhombic. The transformation-structure assumed by the pure elements Ti, Zr and Hf and the *dilute* Ti-TM alloys is hcp and has been assigned the symbol α' [Bag59][Wil73]. Although WILLIAMS [Wil73] has claimed an exception for Ti-V, it is generally recognized that quenched martensites in sufficiently concentrated Ti-TM alloys possess an orthorhombic structure; this is designated α'' [Bag59][Wil73]. The compositional boundaries between the quenched α' and α'' phases have been specified for a few Ti-TM alloys by BAGARIATSKII *et al* [Bag59] and are listed in [Mon1.7]. The α' variant is generally confined to the very low concentration range; thus, for example, in Ti-Mo the boundary between the α' and α'' regimes is given as 2 at.%. It would be convenient to be able to refer to the martensitic state without the necessity of specifying structure. To do this a separate symbol would need to be coined; α^m is suggested, and has in fact been adopted in this work.

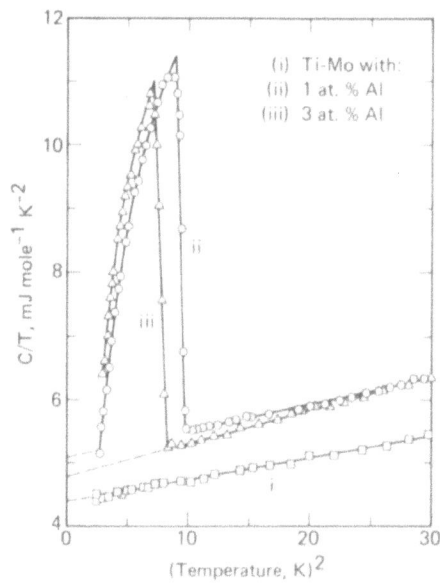

FIGURE 5-5. Low temperature specific heats of (i) quenched Ti-Mo (5 at. %) (as in Fig. 5-4, curve (i)); (ii) the same alloy base doped with 1 at. % Al (annealed-and-quenched); (iii) the same doped with 3 at. % Al (annealed-and-quenched). As in the previous figure the effect of martensitic transformation is responsible for the enhanced T_c — after COLLINGS *et al* [Col76].

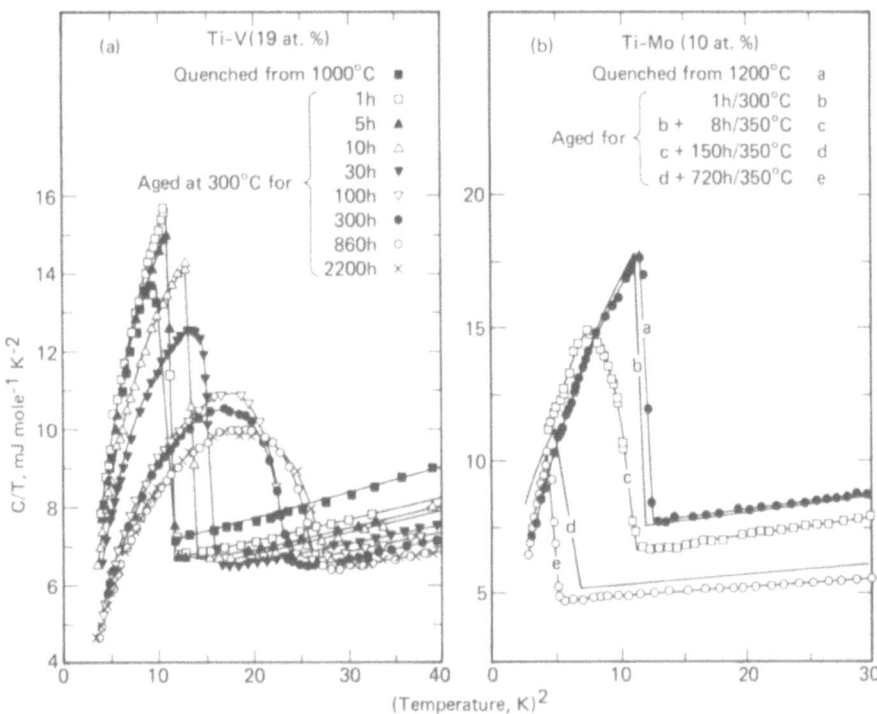

FIGURE 5-6. Influence of aging on the calorimetric superconducting transition in (a) Ti-V (19 at. %) and (b) Ti-Mo (10 at. %) — after COLLINGS [Col83].

Since stress-induced martensite can only occur in an alloy not already transformed by quenching it is confined to the higher concentration ranges. In the light of the foregoing observations on the structures of quenched martensites it is therefore not surprising to find that deformation martensite is invariably orthorhombic [Wil73]. Since both martensitic transformation and twinning are the results of bcc-shear instabilities it turns out that twins are to be found towards the solute-rich border of the deformation-martensite range. For example, deformed Ti-Mo$_7$ appears to be copiously twinned. Using light microscopy alone it is not possible to distinguish between deformation martensite and twins; a diffraction technique must be employed to determine the crystallographic structure of the deformation product.

5.11.5 Influence of Aging on the Superconducting Transition

The influence of aging on the superconducting transition temperature of a Ti-Mo alloy was first studied by RAUB and ZWICKER [Rau64[a]]; the observed decrease in T$_c$ which accompanied the aging of Ti-16Mo

(i.e. 8.7 at.% Mo) for up to 1000 min at 400°C was attributed to the growth of ω-phase precipitation. A detailed calorimetric study of the aging of Ti-Mo$_{10}$ at temperatures favorable for ω-phase precipitation, but unfavorable for α+β equilibration, was carried out by HO and COLLINGS [Ho73] (see also COLLINGS [Col72[a]] and HO et al [Ho70]). The results, as summarized in Fig.5-6, show that although T$_c$ decreased continuously with aging time from 3.45±0.10 K (for the as-quenched alloy) to 2.18±0.10 K (after 1h/300°C plus 878h/350°C) the relative height of the specific heat jump, $\Delta C_{es}/\gamma T_c$, remained close to the BCS value (viz 1.43) and well within the range of relative jump heights recorded for a series of quenched Ti-Mo(7-40 at.%) alloys. From this it can be concluded that a superconductive proximity effect is operative at all stages of precipitate development, Fig.5-7.

At the end of the aging experiment, the pre-quench temperature of the alloy was raised to 550°C thereby allowing it to attain thermodynamic equilibrium in the α+β field. Its T$_c$ then increased to 3.85±0.10 K, that of the equilibrium-β phase, and the relative height of the specific heat jump dropped to a value appropriate to the volume fraction of β-phase prescribed by the equilibrium phase diagram, Fig.5-8.

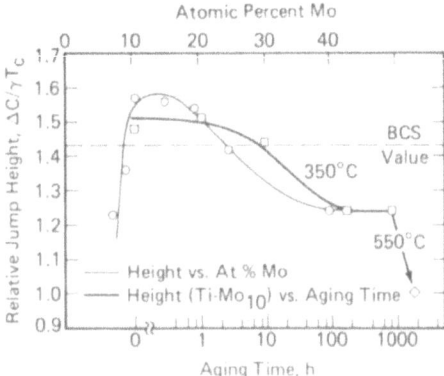

FIGURE 5-7. Relative height of the specific heat jump at T_c, $\Delta C/\gamma T_c$, versus aging time at 350°C in the meta-equilibrium $\omega + \beta$ field (cf. schedule, Fig. 5-6) followed by an aging of 300 h/500°C in order to equilibrate the $\alpha + \beta$-phase. The $\omega + \beta$-phase in this case yields a complete proximity-effect-controlled transition; that of the $\alpha + \beta$-phase is incomplete. Inserted for comparison is the jump-height versus composition curve for the entire quenched Ti-Mo series and the BCS-value of 1.43 – after COLLINGS [Coℓ83], see also HO and COLLINGS [Ho73].

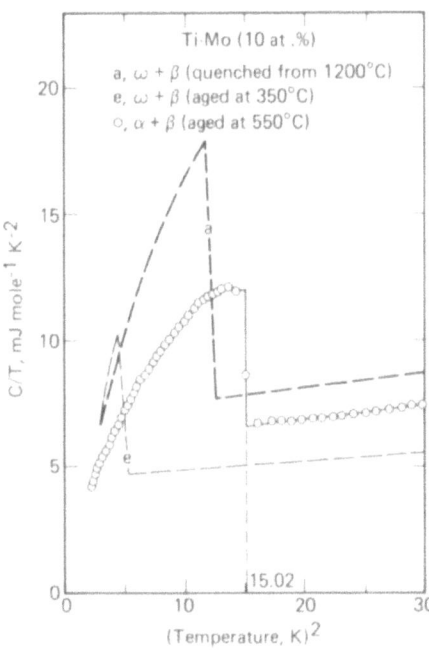

FIGURE 5-8. Low temperature specific heat in the usual format C/T versus T^2 for Ti-Mo (10 at. %) in the as-quenched condition (curve a), after aging for almost 1000 h at 350°C (curve e), and after aging for 300 h at 550°C in order to equilibrate the $\alpha + \beta$-phase (data points). Practically unrounded transitions are noted in all cases – after COLLINGS [Coℓ83], see also HO and COLLINGS [Ho73].

5.12 THE MIXED STATE OF Ti-Mo ALLOYS

5.12.1 The Development of Mixed State Theories

During the 1960's, the low-temperature magnetic properties of superconducting Ti-Mo alloys, particularly $Ti-Mo_{16}$, provided a forum for discussing the development of theories of the mixed state. If experimental values of the upper critical field are all that are sought, resistive measurements (yielding H_r) are adequate. However, an understanding of the mixed state itself demands the performance of magnetization measurements -- and these as functions of temperature -- while detailed analyses of the magnetic data are further assisted by the results of low temperature specific heat measurements preferably as a function of magnetic field. Such experiments have been undertaken by BARNES and HAKE [Bar66, Bar67].

In considering the thermodynamic order of the mixed-state/normal-state transition, HAKE [Hak67[b]] and CAPE [Cap66] have carefully followed the field dependence of the magnetization of $Ti-Mo_{16}$, noting by the manner in which it approached and merged into the normal-state $M(H_a)$ line at the magnetic upper critical field, H_u, that the transition was second-order, Fig. 5-9. The second-order nature was confirmed by the results of the specific heat measurements carried out contemporaneously by BARNES and HAKE [Bar66, Bar67] who showed that the calorimetric specific heat anomaly -- second-order in zero field (cf. for example Fig. 5-6) -- maintained its overall shape in the mixed state in the presence of applied magnetic fields of up to 29 kOe.

In a subsequent analysis of the same specific heat data, HASSING, HAKE, et al [Has73], after paying close attention to the width of the transition, have interpreted what appeared to be a field-induced transition-broadening in terms of a field-induced reduction in the dimensionality of the superconductivity.

Returning to the question of order of the transition into and out of the mixed state at H_{c2}, attention is drawn to an observation made by HAKE in connection with his discussion of Fig. 5-9(a), viz that as T dropped below $T_c/2$, $M(H_a)$ began to develop convexity, and thereby to approach the normal-state $M(H_a)$ limit at sharper and sharper angles, in a manner which suggested a trend towards first-order transitions at low temperatures where, as he put it, "the quasiparticle excitations are reduced".

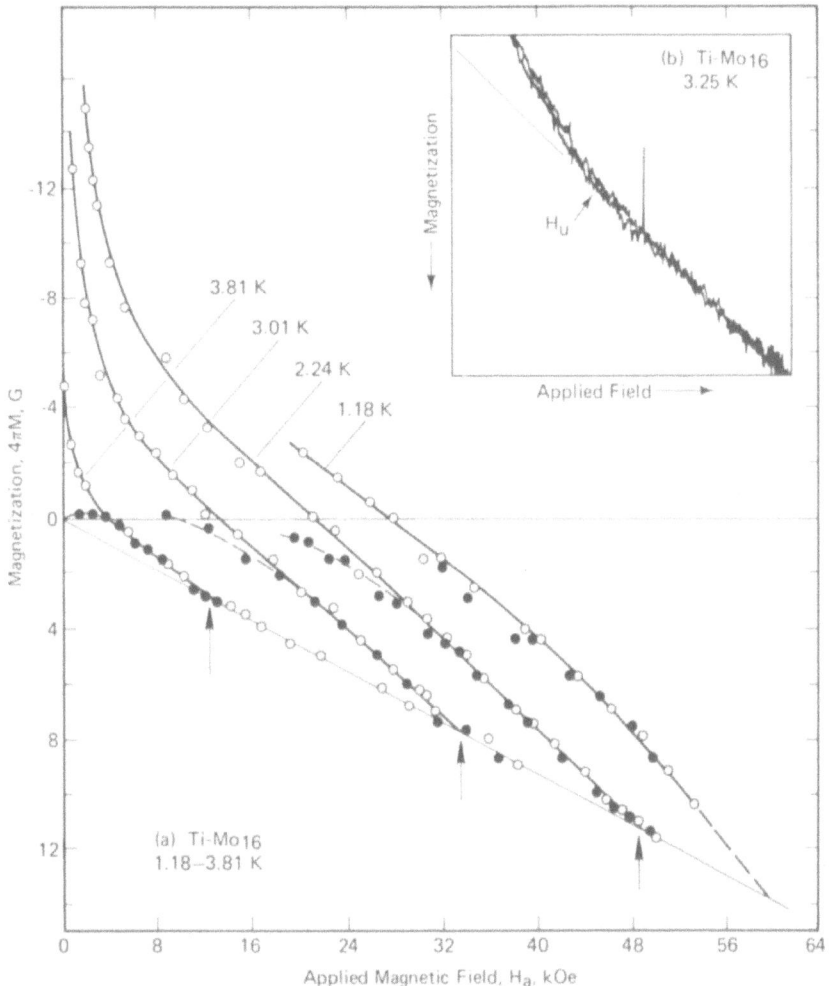

FIGURE 5-9. (a) Isothermal magnetization, $4\pi M$, *versus* applied magnetic field, H_a, for a Ti-Mo (16 at. %) alloy (T_c = 4.18 K, κ_{GL} = 68) (cf. Fig. 3-6). For clarity not all the data points are included. Arrows indicate the *s/n* transitions. The data taken with field increasing (○) followed by field decreasing (●) exhibit *complete reversibility* in a high-field *paramagnetic* mixed state — after HAKE [Hak67[b]]. (b) Continuous record of the reversible magnetization change with increasing and decreasing applied field across the second-order boundary of the mixed and normal states at H_u as measured using a vibrating-sample magnetometer — after CAPE [Cap66].

5.12.2 Early Studies of the Upper Critical Field

In a pioneering paper on alloy superconductivity at high fields, HAKE, BERLINCOURT, and LESLIE [Hak62], pointed out in discussing the critical current of Ti-Mo$_{16}$, that the "zero-resistance condition is noted in fields at least a factor 35 greater than the bulk material critical field as inferred from calorimetric data", i.e. as compared with the BCS-computed thermodynamic critical field, H_c. Subsequently HAKE and LESLIE [Hak63] recognized that GLAG theory should be applicable to alloys such as Ti-Mo(6.25, 9.05, and 16.3 at.%) yielding H_{c2}/H_c ratios of about 110, which

at the time was noted to be comparable to the experimentally obtained ratios of 84 < (H_r/H_c) ≲ 115, in which H_r is the resistive upper critical field (J = 1 A cm^{-2} in this case). In the companion paper, BERLINCOURT [Ber63] briefly considered the temperature dependence of H_r, thus initiating a series of studies of the relative merits of the GOR'KOV, SHAPOVAL, and other expressions, for the upper critical field temperature dependences. In a much more extensive report of this work, BERLINCOURT and HAKE [Ber63[a]] described the results of a study of Ti-Mo(6.25-50 at.%), and several other binary alloy systems, whose resistive upper critical fields had been measured (this time at a specimen current density of 10 A cm^{-2}) under pulsed-applied-field conditions. The measured H_r's were compared with: *(a)* GLAG-predicted values of $H_{c2}^*(t)$ (below) calculated, in the case of the Ti-Mo(6.25-8.60 at.%) alloys, from the calorimetric results of HAKE [Hak61]; *(b)* the CLOGSTON paramagnetic limit, H_p. Surprisingly H_r and H_p coincided for $4.1 < e/a < 4.6$, whereas according to the simple CLOGSTON-modified GLAG picture, the experimental upper critical field should lie below both H_{c2}^* and H_p. This and other evidence suggested that additional mechanisms, yet to be identified, were at work alleviating the effect of paramagnetic limitation. The question of upper critical field temperature dependence was also addressed, and it was in [Ber63^a] that the dirty-limit, GLAG-based, pre-MAKI, H_{c2}^*-temperature dependences were encapsulated in the form

$$H_{c2}^*(t) = A(t) \; \kappa_{GL}^d \; H_c(t), \qquad (5\text{-}4a)$$

where: κ_{GL}^d is the Ginzburg-Landau parameter in the dirty limit as defined in Eqn.(3-11), $H_c(t)$ is the thermodynamic critical field as presented in Eqn.(3-7), and the temperature-dependent pre-factor assumed the forms, or values:

$$A(t) = \sqrt{2} = \text{constant} \qquad \text{-- (ABRIKOSOV)} \quad (5\text{-}4b)$$

$$A(t) = 1.77 - 0.43t^2 + 0.07t^4 \qquad \text{-- (GOR'KOV)} \quad (5\text{-}4c)$$

$$A(t) = \begin{array}{l} 3.03, \; t=0 \\ \sqrt{2}, \; t=1 \end{array} \qquad \text{-- (SHAPOVAL)} \quad (5\text{-}4d)^{\dagger}$$

5.12.3 Pauli Paramagnetic Limitation and the Order of the Transition at H_{c2}

In the CLOGSTON model the paramagnetic lowering of free energy with increasing field, which resulted, according to Figs.3-3 and 3-6, in a lowering of the experimentally attainable upper critical field from H_{c2}^* (the GLAG-value) to $H_u(PP)$, was assumed to apply

---

† As pointed out subsequently by El BINDARI and LITVAK [Elb64], cf. Sect.3.6.1, the SHAPOVAL H_{c2} temperature dependence adopted the following forms between the end-points:

$$1-2t\ell n2 \qquad (t \ll 1)$$

$$\text{and } (1-t)8/\pi^2 \qquad (t \stackrel{\sim}{=} 1).$$

only to the normal state. If the paramagnetism of the superconducting state is now taken into account, two important effects follow: *(a)* the experimental upper critical field increases to $H_u(PP + SO)$; *(b)* the mixed-state/normal-state transition, first-order under the terms of the simple CLOGSTON model (as explained in connection with Fig.3-3), becomes second-order. These postulated effects have been examined in a series of some eight papers by HAKE and colleagues (e.g. [Hak65, Hak67, Hak67^a, Hak67^b]) who used Ti-Mo$_{16}$ and some related alloys such as Ti-V and Ti-V-Cr, as prototype systems.

5.12.4 Experimental Testing of the MAKI and WHH Theories of the Paramagnetic Mixed State

One step beyond Eqns.(5-4a through 5-4d) as descriptors of the upper critical field temperature dependence is the following pair of equations due to MAKI [Mak64, Mak64^a]:

$$H_{c2}^* = \sqrt{2} \; \kappa_1 \; H_c \qquad , \qquad (5\text{-}5)$$

$$4\pi(dM/dH)_{H_{c2}} = [1.16 \; (2\kappa_2^2-1)]^{-1} \qquad , \qquad (5\text{-}6)$$

in which part of the temperature dependence has been assigned to the generalized Ginzburg-Landau parameters $\kappa_1(t)$ and $\kappa_2(t)$. These have been evaluated by MAKI, first in the nonparamagnetic dirty limit as discussed in [Mak64] and subsequently, after the application of paramagnetic approaches of successively increasing levels of sophistication, in the often-quoted paper [Mak66].

The development, first of these nonparamagnetic, and then of these paramagnetic, theories of the mixed state have been traced in [Mon14.5] and [Mon15.4 through 15.6], respectively.

In testing the theories of the mixed state as they have developed from the initial GLAG model, appropriate points of contact between theory and experiment must be identified. It has been found adequate to deal in detail with the properties of a few representative alloys such as Ti-V(22.5 and 25 at.%), Ti$_{30}$-V$_{60}$-Cr$_{10}$, and Ti-Mo$_{16}$, which is to be considered in this section. Measurements which have contributed to the accumulation of microscopic physical property data are: normal-state electrical resistivity, mixed-state magnetization, and mixed-state calorimetry.

Results of these measurements: *(a)* expressed in terms of the temperature dependences of the reduced generalized Ginzburg-Landau parameters $\kappa_1(t,\beta)/\kappa_{GL}$ and $\kappa_2(t,\beta)/\kappa_{GL}$, as recommended by MAKI [Mak66]; *(b)* compared with the reduced upper critical fields $h^*(t,\beta)$ of MAKI [Mak66] or $h^*(t,\alpha,\lambda_{so})$ of WERTHAMER, HELFAND, and HOHENBERG (WHH) [Wer66], have enabled the necessary quantitative comparisons with theory to be made.

As discussed in detail in [Mon15.5], MAKI has developed analytical expressions for the temperature dependences of $\kappa_1(t,\beta)$ and $\kappa_2(t,\beta)$ upon which the behaviors of H_{c2} depended. These parameters are seen to be functions of β, which in turn is a mixed parameter involving both the paramagnetic-limitation parameter $\alpha \equiv \sqrt{2}\, H_{c20}^*/H_{p0}$ (a quantity which is calculable from either superconductive, Eqn.(3-21), or normal-state data, Eqn.(3-22)), and τ_{so}, the spin-orbit relaxation time, an *adjustable quantity* to which a value must be assigned in matching theory to experiment.

Next, as discussed in [Mon15.6], WHH have developed an expression suitable for numerical solution which enables a reduced upper critical field, $h^*(t,\alpha,\lambda_{so})$, to be evaluated as a function of temperature for assigned values of α and λ_{so}. Their α was the same paramagnetic-limitation parameter as before, while λ_{so}, an arbitrarily-chosen spin-orbit-scattering frequency, was reciprocally related (*via* some atomic constants) to τ_{so}.

Both theories were clearly dealing with the same physical phenomena. WHH have confined their calculations to the reduced critical field $h^*(t,\alpha,\lambda_{so})$. In that MAKI has extended his to the development of an expression for that same reduced field, symbolized $h^*(t,\beta)$ in his case, a common ground had been established upon which the two theories could be intercompared. HAKE [Hak67[b]] has presented such an intercomparison in the form of a plot of $h^*(t=0)$ *versus* α as in Fig.5-10. Reasonably close agreement between the results of the two approaches is evident. It is therefore adequate, in an experimental test of the now "MAKI-WHH" theory of the paramagnetic mixed state to intercompare the experimental and computed temperature dependences of *either* h^* *or* the κ-parameters. Using data acquired magnetically [Cap66][Hak67[b]] and calorimetrically [Bar67], HAKE [Hak67[b]] has in fact utilized *both* formats in his enquiry and obtained the results depicted in Figs.5-11 and 5-12.

It was concluded by HAKE [Hak67[b]] that, although effects interpretable in terms of Pauli limitation and

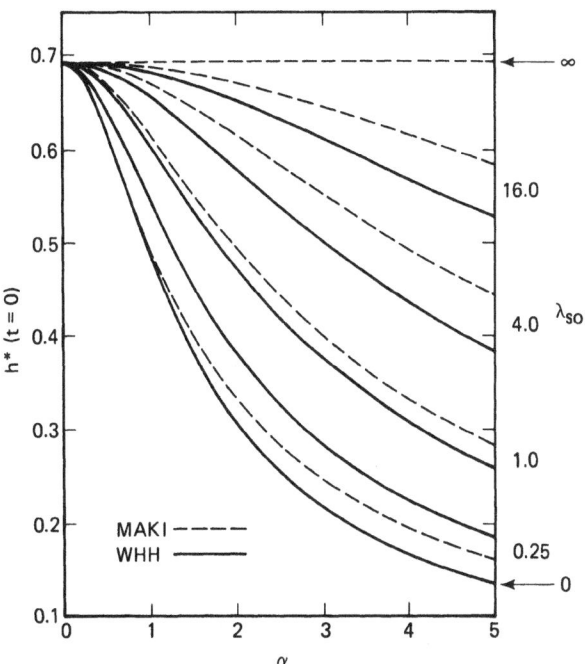

FIGURE 5-10. Reduced upper critical field at zero K in the dirty limit *versus* the Maki paramagnetic limitation parameter α (calculated according to Eqn (3-22)) for various values of the spin-orbit-scattering frequency parameter λ_{so} according to the theories of Maki and WHH — after HAKE [Hak67[b]].

an enhancement of mixed-state paramagnetism *via* the mechanism of spin-orbit-coupling-induced spin-flip scattering were undoubtedly being exhibited and were qualitatively explicable in terms of MAKI-WHH theory, the finer details of the H_u and κ_1 temperature dependences predicted by the theory were not corroborated. The inadequacy of the unmodified MAKI-WHH approach has been alluded to already in Sect.3.6.5; it is to be discussed in slightly more detail in Sect.7.11 in connection with the mixed-state properties of Ti-Nb, and the entire situation is reviewed in [Mon15.8].

5.12.5 The Mixed-State Hall Effect in Ti-Mo Alloys

This review of the mixed-state properties of Ti-Mo alloys concludes with a description of a flux-flow-related property, Hall effect. The usual $\vec{F} = \vec{J} \times \vec{B}$ relationship applied to a superconductor in the mixed state implies that the ABRIKOSOV flux lattice will experience a force, and will eventually flow, in a direction normal to that of the transport current density, $\vec{J}$. But if for some reason the fluxoids have a component of their velocity, $\vec{v}$, parallel

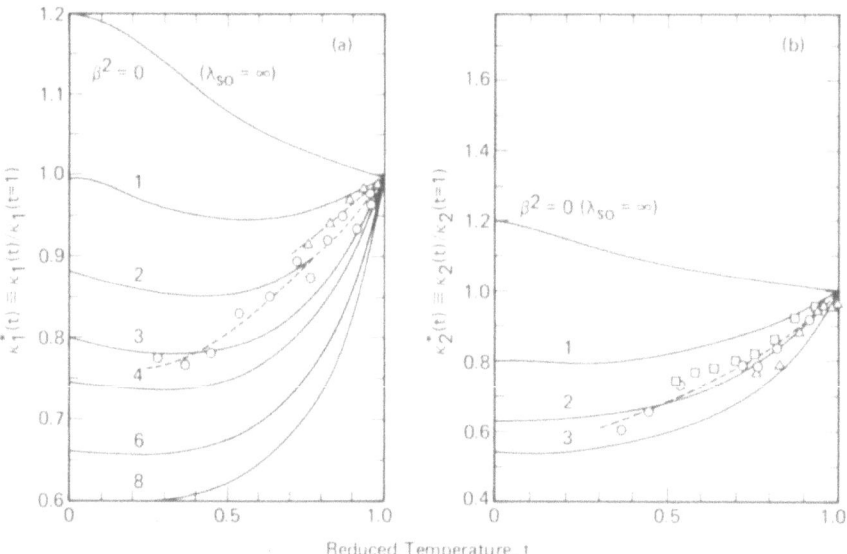

FIGURE 5-11. Temperature dependences of the Maki paramagnetic/spin-orbit parameters $\kappa_1(t, \beta)$ and $\kappa_2(t, \beta)$ plotted in the formats: (a) $\kappa_1^* \equiv \kappa_1(t)/\kappa_1(t = 1)$ *versus* t, and (b) $\kappa_2^* \equiv \kappa_2(t)/\kappa_2(t = 1)$ *versus* t for various values of the Maki β-parameter defined in this diagram as $\beta^2 \equiv \alpha^2/1.78\lambda_{SO}$. Comparison is made with some experimetal results for Ti-Mo (16 at. %). The data portrayed are: (i) the magnetic results of HAKE [Hak67[b]] (○); (ii) the magnetic results of CAPE [Cap66] (□); (iii) the calorimetric results of BARNES and HAKE [Bar66, Bar67] (△) — after HAKE [Hak67[b]].

to $\vec{J}$, the equation $\vec{E} = \vec{v} \times \vec{B}$ predicts the appearance of a *transverse* voltage. Provided that the longitudinal component of fluxoid motion has not been induced by metallurgical artifacts of the sample, i.e. by defect-guided "channelling", this transverse voltage is a Hall voltage, which arises in the following manner. According to KIM *et al* [Kim65], the dissipation associated with flux-flow resistance can be approximately accounted for by regarding it as being seated in the normally-resistive cores of the flux vortices. Retaining the spirit of this concept, BARDEEN and STEPHEN [Bar65] went on to propose that this normal transport current within the vortices, in the presence of the vortex field, gave rise to an incremental Hall voltage $\vec{v}_H$ within the fluxoid, perpendicular to $\vec{J}$, which then emerged as an aggregate Hall voltage, $\vec{V}_H$, across the sample. The reaction between $\vec{v}_H$ and $\vec{H}_{core}$ is a longitudinal force on the fluxoid -- hence the association of the Hall effect with a longitudinal or drift component of the usually transversely flowing flux.

HAKE [Hak68] has measured the transverse voltages, V_T, across a sample of annealed Ti-Mo$_{16}$ and has derived a Hall voltage from them with the aid of the following formula for V_H:

$$V_H = \frac{1}{4}[V_T(J_+,H_+) - V_T(J_+,H_-) - V_T(J_-,H_+) + V_T(J_-,H_-)] \quad (5-7)$$

(in which the '+' and '-' signs signify reversals of current and field directions). HAKE noted that, whereas the Hall component of V_T reverses with H and with J, the component of V_T due to defect-guided flux motion reverses with J but is insensitive to the sign of H. The quantitative results of HAKE's work appeared to exceed theoretical expectation, and to be larger than those obtained in prior experiments. Thus in spite of (or perhaps because of) a series of very careful measurements of annealed Ti-Mo$_{16}$, a dirty ($\kappa_{GL}^d = 68$) paramagnetic type-II superconductor, HAKE [Hak68] was unable to leave the theory of mixed-state Hall effect in a satisfactory state. The situation does not appear to have improved much since then. In a subsequent paper, however, WEIJSENFELD [Wei68] was able to match the mixed-state Hall angle to that appropriate to the normal-state core of the vortex simply by making suitable adjustments to the normal-state and core relaxation times. The physics of mixed-state Hall effect and the associated literature are reviewed in [Mon23.11].

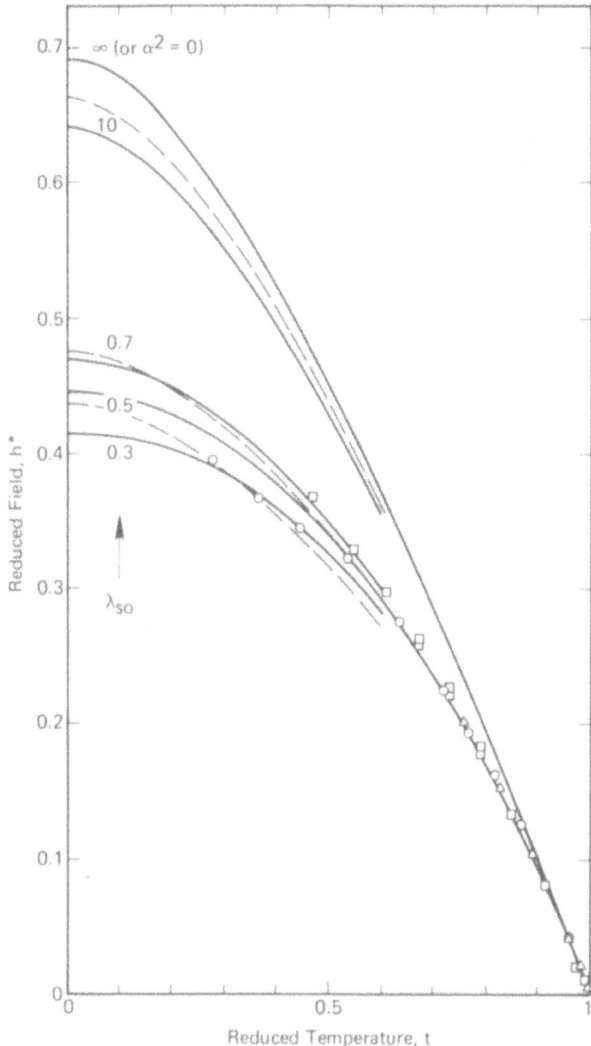

FIGURE 5-12. Temperature dependences of the reduced upper critical field $H^*(t) = H_u(t)/(-dH_u/dt)_{t=1}$ according to Maki theory (full line) and WHH theory (dashed line) for $\alpha^2 = 3.3$ and various values (as indicated) of λ_{so}, the spin-orbit-scattering frequency parameter. ($\lambda_{so} = \infty$, $\alpha^2 = 3.3$) is of course the "same" as (λ_{so} = finite, $\alpha^2 = 0$). Comparison is made with some experimental results for Ti-Mo (16 at. %). The data portrayed are: (i) the magnetic results of HAKE [Hak67b] (○); the magnetic results of CAPE [Cap66] (□); the calorimetric results of BARNES and HAKE [Bar66, Bar67] (△) — after HAKE [Hak67b].

5.13 CRITICAL CURRENT DENSITIES OF Ti-Mo ALLOYS

During a period of almost twenty years, current-carrying superconductivity has developed from a laboratory study of critical currents and critical-current anisotropy in small strips fabricated from numerous superconducting Ti-TM alloys towards the full-scale commercialization of the stabilized thermomechanically process-optimized multifilamentary composites, capable of carrying thousands of amperes, that are presently in use. This book, by implication, traces these developments. In that Ti-Mo was the "working substance" in many of the initial investigations of the subject, this section, which describes some of the current-transport experiments which have been performed on that alloy, augmented by Sects.5.7 and 5.9 (Ti-Ta) and Chapter 3 (Ti-V) Parts 3 and 4, encapsulate the early history. Subsequent chapters of this book, notably Chapter 7 (which deals with Ti-Nb), Chapter 12 (containing Ti-Nb-Ta) and Chapter 13 (containing Ti-Zr-Nb-Ta) report on the most recent developments.

The critical current densities of several Ti-Mo alloys (of concentrations 6.25, 9.05, and 16.3 at.% Mo) have been reported in a series of papers by BERLINCOURT, HAKE and LESLIE which appeared in 1962 and 1963 [Hak62, Hak62a, Hak63]. During that period, although the concepts of GINZBURG, LANDAU, ABRIKOSOV, and GOR'KOV were beginning to be introduced into discussions of the upper critical field, the principle of transport-current stabilization in the presence of a strong magnetic field in terms of the pinning of the flux lattice associated with that field, had yet to emerge. Instead, the stabilization of large transport currents, *via* the medium of metallurgical defects introduced by cold work, was thought of in terms of a filamentary model in which the current percolated along continuous paths established by dislocation networks and/or the effects of their strain fields.

HAKE and BERLINCOURT and LESLIE [Hak62] compared the transverse-field J_c's of cold-rolled strips of Ti-Mo$_{16}$ (in fields of up to 30 kOe and at temperatures of 1.2 and 2.4 K) with those of strips prepared by cutting and polishing (i.e. without mechanical deformation). It was noted immediately that when the transverse field was applied parallel to the plane of the cold-rolled ribbon, the critical current density, $J_c(||)$, was 12-24 times larger than $J_c(\perp)$, the critical current density for an applied field normal to that plane. That this had to do with microstructural anisotropy rather than specimen geometry, was demonstrated by the relative insensitivity of the J_c of the machined but unrolled strips to rotation of the transverse field about the specimen axis. In a subsequent paper, this work was extended by the same authors [Hak62a] to Ti-Mo$_9$ and other binary transition-metal alloys. A detailed study of critical current density anisotropy was conducted by HAKE and LESLIE [Hak63] on Ti-Mo(6, 9, and 16 at.%) again in fields of up to 30 kOe, as a result of which highly anisotropic polar diagrams were able to be constructed of J_c versus the angle between the transverse field direction and the rolling plane. If samples were sheared, rather than spark-cut, such that microstructural distortion occurred at the cut edges, minor J_c maxima would appear at intermediate angles of the applied field. It was noted at the time that with the development of a ribbon-like microstructure through cold rolling the high-field mixed-state structures were better anchored

against Lorentz forces perpendicular to the rolling plane than parallel to it.

Using a pulsed-field technique BERLINCOURT [Ber63] extended the field range of the critical current measurements up to about 60 kOe and went on to explore some interesting properties relating to the influence on J_c of the magnitude and direction of the applied field. (a) With H_a parallel to the conductor axis the critical current density was found to be an order of magnitude higher than was the case for H_a perpendicular to J and parallel to the rolling plane. This was the first indication of the existence of "force-free" critical-current transport, a phenomenon which was to be the subject of considerable study by subsequent authors, cf. [Mon21.16 et seq] and [Mon22.00]. (b) With H_a perpendicular to the current as in all the previous experiments, the anisotropy noted at low fields tended to diminish with increase in field -- in fact at fields just above 60 kOe, it was found that $J_c(||) \cong J_c(\perp)$. This equality came about, not through any general monotonic approach of $J_c(||)$ to J $(\perp)$ but through the existence of a high-field peak in $J_c(\perp)$, Fig.5-13. In other words it had been discovered that at fields close to H_r, J_c tended to lose its microstructural sensitivity and to become a function of H_r itself rather than a property of the pinning structure. The importance of upper critical field control in the optimization of high-field Ti-Nb superconductors is discussed in Sect.7.32.3.

5.14 ANOMALOUS TRANSPORT PROPERTIES OF Ti-Mo ALLOYS

5.14.1 Fluctuation Superconductivity

The electrical resistivities of strips of Ti-Mo$_{16}$, and numerous other binary TM alloys, were measured by HAKE [Hak69[a], Hak71] within a temperature range equivalent to 1.0<t<3.8 (T_c = 4.2 K). It was discovered that as the temperature of the normal state was reduced towards T_c, a noticeable decrease in the residual resistivity began to take place as early as t∿2.2. This was interpreted as a partial short-circuiting of the sample by superconductive fluctuations. The possibility of sample inhomogeneity as the cause of what is in effect a long superconductive tail must of course be considered, but may be immediately rejected on the grounds that the maximal T_c in the entire Ti-Mo system never exceeds 4.2 K, that reported for the subject alloy.

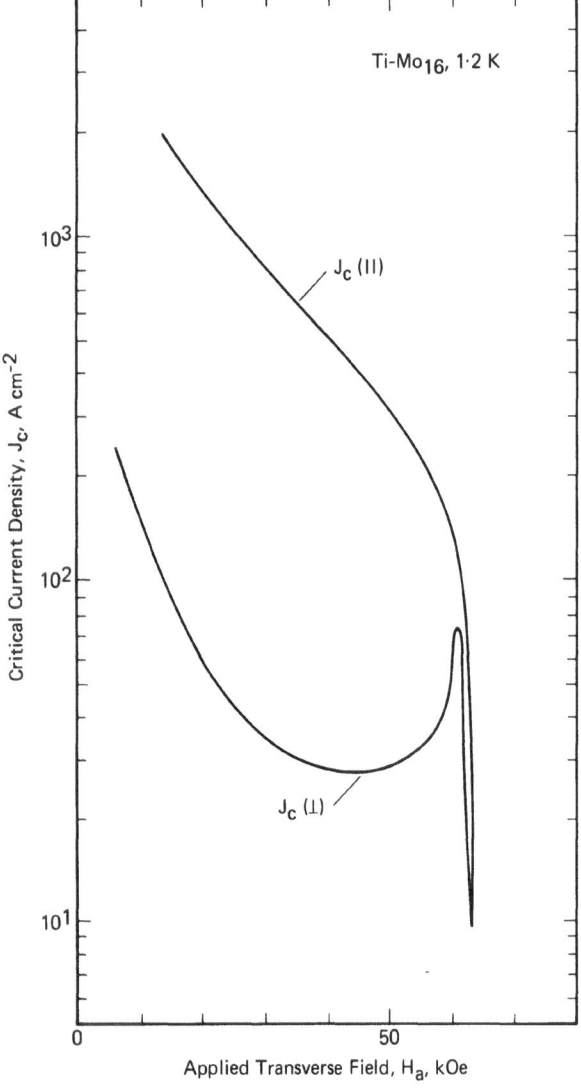

FIGURE 5-13. Critical current densities at 1.2 K of a 75%-cold-rolled ribbon of Ti-Mo (16.3 at. %) for two orientations of the transverse applied magnetic field: normal to the rolling plane (hence J_c ($\perp$)) and parallel to the rolling plane (hence J_c (||)) — after BERLINCOURT [Ber63], see also BERLINCOURT and HAKE [Ber63[a]].

Fluctuation superconductivity was also evident at temperatures below T_c but in fields well above both H_{c2} and H_{c3} (the third upper critical field [Hak69], see also Sect.3.7). It manifested itself as an increase in sample resistivity with increasing applied field as a consequence of the field-quenching of the superconductive fluctuations. At high fields and at high current densities (such as 300 A cm^{-2}) the incremental sample resistivity became field independent, in satisfactory confirmation that the quenching of fluctuations rather than magnetoresistance had in fact been observed.

5.14.2 Negative Normal-State Resistivity Temperature Dependence and Magnetoresistance

The resistivities of a series of quenched Ti-Mo(5-20 at.%) alloys, measured in the temperature range 4.2 to 620 K by HO and COLLINGS [Ho72], turned out to be anomalous in two respects: (i) as the Mo concentration was reduced into the range for which ω-phase precipitation was observed, the resistivity isothermals rose to extremely high values, e.g. 149×10^{-6} Ω cm for Ti-Mo$_7$ for all temperatures between 4.2 and 300 K; (ii) between the compositions 7 and 15 at.% Mo, $d\rho/dT$ was negative, Fig.5-14. It is claimed that the first effect is due to electron scattering from ω-phase precipitation present as a result of nucleation-and-growth during the quenching of the starting ingot, while the second is a consequence of reversible, so-called "athermal", ω-phase precipitation. The situation has been described in [Mon5.5] and [Mon5.6] in connection with a general discussion of anomalous electrical resistivity in Ti-TM alloys. Whether the athermal ω-phase in question is static (i.e. crystalline) or dynamic (i.e. a result of soft phonons) is open to speculation. Subsequently, HAKE and colleagues [Hak75] also addressed the question of negative $d\rho/dT$, expressed in terms of a parameter $\alpha^T \equiv \rho^{-1}(\partial\rho/\partial T)_{H=0}$, and introduced at the same time a companion parameter $\alpha^H \equiv \rho^{-1}(\partial\rho/\partial H)_T$ to describe the influence of an applied magnetic field on the resistivity near T_c. They made the suggestion that the negative α^T was the result of a localized spin fluctuation. The existence of such a mechanism, it was claimed, would also be consistent with a negative magnetoresistance, although the possibility of explaining both negative α^T's and negative α^H's in terms of soft-phonon modes was not excluded.

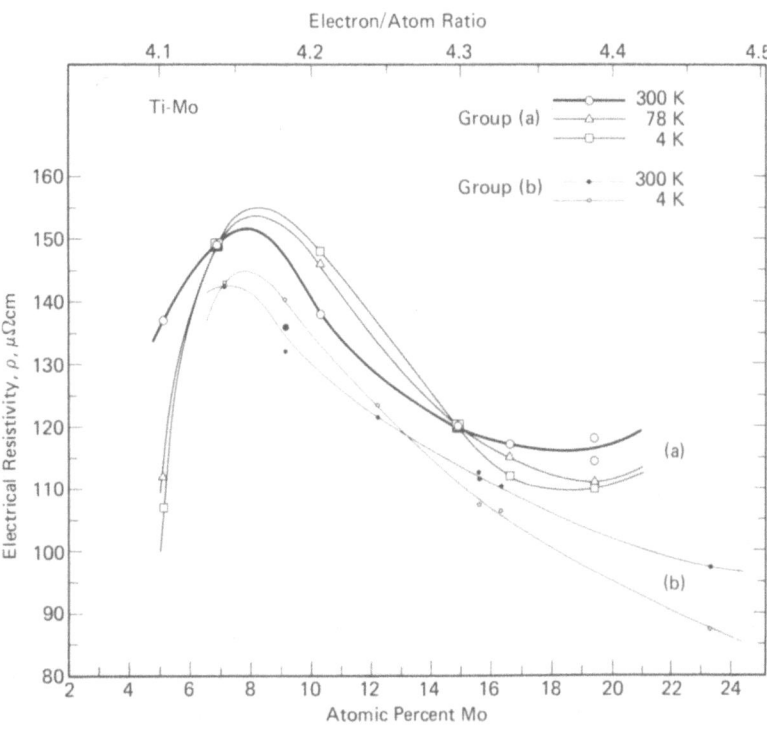

FIGURE 5-14. Resistivity-composition isothermals for Ti-Mo alloys. Depicted are the results of HO and COLLINGS [Ho72] (Group (a): ground from annealed-and-quenched material) and HAKE et al [Hak61a] (Group (b): ground from cast ingots). The occurrence and extents of the regimes of negative dρ/dT are in good agreement.

5.15 SUPERCONDUCTIVITY IN Ti-Mo ALLOYS—TABULATED DATA

TABLE 5-9   TRANSITION TEMPERATURES (T_c) OF TITANIUM-MOLYBDENUM ALLOYS

(a) Calorimetric Results

At.% Mo	T_c K	Data Sources (Tabulated Data)
0.9_9	<1.5	[Col72]
1.9_1	<2.0	"
2.8_7	2.1±0.3	"
3.4_8	2.6±0.2	"
3.9_4	2.9±0.2	"
4.5_1	3.0±0.2	"
5.0_0	<1.5	"
6.25	2.04±0.20	[Hak61]
6.50	2.23±0.20	"
6.9_4	2.1±0.1	[Col72]
7.5_4	2.60±0.25	[Hak61]
8.6_0	3.09±0.26	"
8.8_6	2.80±0.03	[Col72]
10.3_0	3.32±0.03	"
13	3.96±0.08	[Sin68]
14.9_2	3.85±0.10	[Col72]
15	4.06±0.09	[Sin68]
19.3_8	3.98±0.07	[Col72]
20	4.00±0.09	[Sin68]
24.7_3	3.70±0.14	[Col72]
39.8_1	2.79±0.11	"
40	<1.4	[Sin68]
71.0_2	<1.5	[Col72]
85	<1.4	[Sin68]

(b) Magnetic Results

At.% Mo	T_c K	Data Sources (Plotted Data)
1	2.2	[Bla61]
1.5	1.9_5	"
2	2.2	"
2.5	2.5	"
2.5	1.8_5	[Buc65]
3	2.4	[Bla61]
4	2.5	"
4	2.9_5	[Buc65]

(b) Magnetic Results -- *continued*

5.5	2.3	[Bla61]	
6.0	2.5	"	
8.5	3.0_2	"	
9.5	3.4	"	
10.0	3.3	"	
10.5	3.5	"	
21.2	3.3_2	"	
29.0	3.1	"	
30	2.8	[Buc59]	as-cast
"	2.4	"	annealed
33	3.2	[Bla61]	
35	2.4	[Buc59]	annealed
40	2.1	"	as-cast
	1.9	"	annealed
40	2.8_2	[Bla61]	
45	1.4	[Buc59]	annealed
$48._5$	1.6	"	as-cast
50.0	1.8	[Bla61]	
55.0	1.7	"	
60.0	1.4	"	

(c) Calorimetric Studies of Aging and Deformation

At.% Mo	Thermal and Mechanical Treatment	T_c K	Refs.
10.3	2h/1200°C/quench	3.45±0.10	[Ho73]
	above + 1h/300°C/quench	3.38±0.08	
	above + 8h/350°C/quench	3.05±0.30	
	above + 150h/350°C/quench	2.43±0.23	
	above + 720h/350°C/quench	2.18±0.10	
	above + 300h/550°C/quench	3.85±0.10	
4.5	As-cast	2.88±0.23	[Col70]
	above + compressive deformation	2.80±0.30	"
5.0	8h/1300°C/quench + compressive deformation	3.00±0.30	[Ho71]
6.9_4	" " "	3.00±0.70	"

TABLE 5-10 TRANSITION TEMPERATURES OF TITANIUM-MOLYBDENUM ALLOYS -- DATA SOURCES

Mo Concentration Range	Condition	Structures	Procedures	Literature
30-48.5 at.% (5 alloys)	As-cast and annealed.	β	Magnetic	E. Bucher *et al.* (1959) [Buc59]
1-60 at.% (19 alloys)	Quenched from 1100°C.	1-4 at.% Mo, α^m 5.5-60 at.% Mo, ω+β to β	Magnetic	R.D. Blaugher *et al.* (1961) [Bℓa61]
6.25-8.60 at.% (4 alloys)	As-cast.	ω+β	Calorimetric	R.R. Hake (1961) [Hak61]
2.5 and 4 at.%	Quenched from just below solidus.	2.5 at.% Mo, α^m 4 at.% Mo, α^m+β and possibly ω	Magnetic and calorimetric	E. Bucher *et al.* (1965) [Buc65]
13-85 at.% (7 alloys)	Quenched from 1400°C.	ω+β to β	Calorimetric	A.K. Sinha (1968) [Sin68]
0.99-71.02 at.% (15 alloys)	Quenched from 1300°C.	0.99-4.51 at.% Mo, α^m 5.00-71.02 at.% Mo, (ω+β)+β	Calorimetric	E.W. Collings *et al.* (1972) [Coℓ72]
10 at.%	Aged at 300, 350, 550°C.	300, 350°C aging, ω+β 550°C aging, α+β	Calorimetric	J.C.Ho and E.W.Collings (1973) [Ho73]
4-1/2 - 7 at.% (3 alloys)	Quenched and deformed.	Deformation martensite Deformation twins	Calorimetric	E.W.Collings and J.C.Ho (1970) [Coℓ70] See also [Ho71]

TABLE 5-11 CRITICAL FIELDS OF TITANIUM-MOLYBDENUM ALLOYS

(a) Lower Critical Fields (H_{c1}) of Ti-Mo(16 at.%) [Hak67[b]]

Temperature K	H_{c1}[+] Oe
1.25	72.5
1.95	67.5
2.41	59.5
2.79	53.4
3.08	46.0
3.36	37.0
3.61	26.6
3.80	19.0
4.00	10.4
4.11	3.2

+ Data extracted from plotted $M(H_a)$ curves.

(b) Resistometrically Determined Upper Critical Fields (H_r) at 1.2 K [Ber63[a]]

At.% Ta		H_r[+] kOe
6.25	27.2	32.4
9.07	45.0	51.6
11.1	49.4	52.2
16.3	63.3	64.4
25.0	58.8	59.8
37.5	31.0	----
50.0	14.0	----

+ Supplied as numerical data -- refers to onset of detectable resistance and full restoration of resistance, respectively; measuring current density, 10 A cm^{-2}.

TABLE 5-11 CRITICAL FIELDS OF TITANIUM-MOLYBDENUM ALLOYS -- *continued*

(c) Calorimetrically Determined Upper Critical
Fields (H_u) of Ti-Mo (16 at.%) [Bar67][†]

Temperature K	H_u[††] kOe
3.216	29.04
3.509	22.01
3.772	14.99
3.980	8.76
4.067	5.99
4.159	2.99
4.212	0.99
4.246	0.00

† Supplied as numerical data.

†† Data provides a least-squares fit (94.3%) to:

$$H_u = 72.3 - 71.2\ t^2\ .$$

(d) Magnetically Determined Upper Critical Fields
(H_u) of Ti-Mo(16 at.%) [Hak67[b]][†]

Temperature K	H_u[††] kOe
0.00	∿63[†††]
1.18	59.4
1.53	55.0
1.86	51.8
2.24	48.5
2.65	41.5
3.01	33.5
3.20	29.0
3.42	24.2
3.62	19.0
3.81	12.5
4.01	6.0

† Data extracted from plotted $M(H_a)$ curves.

†† With the exception of (†††) and with
T_c = 4.246 K, data provides a least-squares
fit (99.8%) to:

$$H_r = 65.0 - 64.1\ t^2\ .$$

††† Statement in text.

TABLE 5-12 CRITICAL FIELDS OF TITANIUM-MOLYBDENUM ALLOYS -- DATA SOURCES

Mo Concentration Range	Condition	Procedures	Literature
6.25-50.0 at.% (7 alloys)	As-cast, or as-cast plus cold rolled.	Resistometric (1.2 K, 10 A cm^{-2})	T.G. Berlincourt and R.R. Hake (1963) [Ber63[a]]
16 at.%	Homogenized 1h/∿1500°C and step cooled.	Magnetic (H_{c1}, 1.25 - 4.11 K) (H_u, 1.18 - 4.01 K)	R.R. Hake (1967) [Hak67[b]]
16 at.%	Homogenized 1h/∿1500°C.	Calorimetric (H_u, 3.2 - 4.2 K)	L.J. Barnes and R.R. Hake (1967) [Bar67]

TABLE 5-13   CRITICAL CURRENT DENSITIES (J_c) OF TITANIUM-MOLYBDENUM ALLOYS

(a) Transverse Applied Fields of 30 kOe

At.% Mo	Temp. K	Condition[+]	J_c, 10^2 A cm^{-2}[++]		$J_c(\|\|)/J_c(\perp)$	Literature	
			$J_c(\perp)$	$J_c(\|\|)$			
9	1.2	78% cr	2.9	3.1	1.1	[Hak62[a]]	n
16	1.2	80% cr	0.34	7.8	23	"	"
16	1.2	79% cr	0.36	7.9	22	[Hak62]	g
		ground	0.32	0.68	2.1	"	"
16	2.4	79% cr	0.084	2.6	31	"	"
		ground	0.058	0.11	1.9	"	"

(b) Longitudinal and Transverse Applied Fields of 80 kOe

At.% Mo	Temp. K	Condition	J_c, 10^2 A cm^{-2}[+++]			Literature	
			$J_c(\perp)$	$J_c(\|\|)$	J_c(long.)		
16	1.2	75% cr	0.5	1	10	[Ber63]	g

+ ground = reduction to ribbon by grinding and polishing
    cr = reduction to ribbon by cold rolling
    n = original data in tabular form
    g = data from published $J_c(H_a)$ graph
++ J_c criterion: 0.25μV across 7 mm
+++ J_c criterion: resistive-onset detection sensitivity: 100μV

ALLOY GROUP 4: TITANIUM-TECHNETIUM AND TITANIUM-RHENIUM BINARY ALLOYS

5.16 SUPERCONDUCTIVITY IN Ti-Tc ALLOYS

High transition temperatures are usually favored by the bcc structure. Thus technetium (Tc) is remarkable in being an hcp element whose transition temperature (∿7.8 K) [Sek67] is second only to that of Nb. Using an inductive technique, KOCH [Koc76[a]] has studied the superconducting transition temperatures of a series of ten high-purity arc-melted Ti-Tc alloys. Ti-rich as-cast bcc solid solutions in the electron/atom ratio range of 4.45-5.05 (15-35 at.% Tc) exhibited a local T_c maximum at 25 at.% Tc corresponding to

e/a = 4.75. Qualitatively this behavior parallels those of the related alloys Ti-Mn and Ti-Re for which, however, the T_c-peak occurs at e/a ∿ 4.4. In Ti-Mn and Ti-Re the T_c-maximum is claimed to be an artifact of ω-phase precipitation which takes place on the Ti-side of the peak and suppresses T_c there, rather than a feature of the single-phase electronic structure. In the absence of detailed electron microscopic study it is not possible to say whether Ti-Tc also conforms to this model.

The highest transition temperatures in the Ti-Tc system occur in the hcp Tc-rich region, and reflect the high T_c of hcp-Tc itself. Based on the only available data for the T_c of pure Tc [Sek67], it must be concluded that an absolute T_c-maximum for the system (10.6±0.3 K) exists at a Tc concentration of 97 at.% (e/a = 6.9). Heat treatment of the compound-forming and two-phase alloys at temperatures of from 700 to 1500°C, depending on the mean alloy composition, produced no significant changes in either the as-cast structures or the average transition temperatures, and only slight narrowings (by about 0.2 K) of the widths of the transitions.

5.17 SUPERCONDUCTIVITY IN Ti-Re ALLOYS

The superconducting transition temperatures of a series of as-cast Ti-Re(2-10 at.%) alloys were measured by MATTHIAS *et al* [Mat59] using a mutual inductance technique. The results of the T_c *versus* composition study of Ti-Re were juxtaposed against those for the Ti-Mn system in such a way that it appeared as if dT_c/dc for Ti-Mn was very much larger than that of the former system. Superconductivity in Ti-Mn has been discussed in Chapter 4. In presenting the results of the above study, in the context of a review of the superconducting transition temperatures of ten other Ti-TM alloys, ZWICKER [Zwi63] implied that the observed T_c-composition dependence was quite characteristic of the bcc regime (the importance of ω-phase precipitation being unrecognized in 1963) and that, had measurements been undertaken in the quenched α^m-regime, the dT_c/dc of Ti-Re would have been similar to that of Ti-Mn with which it is isoelectronic.

The complete set of data for Ti-Rh alloys which had already been acquired was qualitatively similar to those noted for the Ti-(3d)TM alloys, for which it was believed that the T_c-behavior was guided by structural as well as e/a considerations; it was assumed that Ti-Re (the subject of this section) and Ti-Ru (to be considered below) followed suit. According to MOLCHANOVA [Mol65, p.129], ω-phase is found in β-quenched Ti-Re alloys containing 10-15 wt.% (i.e. 2.8-4.3 at.%) Re. The lowest concentration alloy studied by MATTHIAS possessed a nominal Re content of 2 at.%, which is quite close to the lower limit for ω+β referred to above. After taking into consideration the possibility that compositional uncertainties were associated with both the structural and transition

temperature determinations, it is conceivable that all of the quenched alloys studied by MATTHIAS *et al* contained ω and β phases in various proportions, and that superconductivity in quenched α^m-Ti-Re alloys has in fact not yet been investigated.

5.18 SUPERCONDUCTIVITY IN Ti-Tc AND Ti-Re ALLOYS—TABULATED DATA

TABLE 5-14   TRANSITION TEMPERATURES (T_c) OF TITANIUM-TECHNETIUM ALLOYS [Koc76[a]][+]

(a) Cast Alloys

At.% Tc	Structures	T_c K
15	bcc	1.74 ± 0.04
25	bcc	2.70 ± 0.05
35	bcc	2.25 ± 0.40
50	bcc	<1.7
67	bcc	<1.7
75	bcc	5.1 ± 0.9
85	α-Mn	8.5 ± 1.3
87.5	α-Mn+hcp	9.9 ± 1.1
95	hcp+α-Mn	10.0 ± 0.6
97	hcp	10.56 ± 0.33
100	hcp	∿7.8[++]

(b) Annealed Alloys

At.% Tc	Anneal. Temp. °C	Structures	T_c K
50	700	CsCl	<1.70
67	1000	CsCl	<1.70
85	1500	α-Mn	7.9 ± 0.2
87.5	1500	α-Mn+hcp	8.7 ± 0.8
95	1500	hcp+α-Mn	9.8 ± 0.4

+ Magnetic determination; original data in tabular form.

++ Ref. [1] of [Koc76[a]]

TABLE 5-15   TRANSITION TEMPERATURES (T_c) OF
TITANIUM-RHENIUM ALLOYS[†] [Mat59]

At.% Re	T_c[††] K
2.0	1.2
4.0	1.6
5.5	1.8
10.0	2.8

† Condition: as-cast.

†† Magnetic determination; values obtained from
   plotted T_c *vs* composition data.

ALLOY GROUP 5: TITANIUM-RUTHENIUM, -OSMIUM, -RHODIUM, -IRIDIUM, -PALLADIUM, AND -PLATINUM BINARY ALLOYS

5.19 SUPERCONDUCTIVITY IN Ti-Ru ALLOYS

5.19.1 Composition Dependence of the Transition Temperature

In a discussion of the systematics of the successive appearances and disappearances of superconductivity in the periodic table of the elements, MATTHIAS [Mat57] pointed out that the superconducting transition temperature "is a periodic function in the long period with maxima near 3, 5 and 7 valence electrons per atom and minima in between and seeming to vanish at 2 and above 8 electrons". The increase in T_c with increasing e/a following the above-mentioned minimum at e/a = 4 was illustrated using binary alloys of Zr and Ti with elements selected from the nine members of the block which constitute group VIII of the periodic table. Combining a continuation of this systematic approach with an interest in examining the effects of ferromagnetic solute elements on the superconducting transition temperature, MATTHIAS *et al* [Mat59] made a direct comparison of the T_c-concentration dependences of the isoelectronic pair of alloys Ti-Fe and Ti-Ru, the results of which have already been depicted in Fig.4-4. Ti-Fe exhibited the rapid rise of T_c with composition, previously noted; but, in spite of the fact that data for the corresponding structure in

Ti-Ru were missing, it was claimed that the dT_c/dc for Ti-Ru was much less, thus elevating alloys such as Ti-Fe to a special position with regard to superconductive pairing interactions. This misleading juxtapositioning of the curves for Ti-Fe and Ti-Ru persisted for several subsequent publications (viz BARDEEN and SCHRIEFFER [Bar61], MATTHIAS [Mat63], MATTHIAS *et al* [Mat63[a]]) and with it a claim for the discovery, in the class of alloy exemplified by Ti-Fe, of a new magnetic pairing mechanism which could substitute for the Cooper-pair interaction. The situation has already been adequately discussed in Chapter 4, and in particular Sect.4.9.1 of that chapter.

In the meantime, other studies of dilute Ti-(group VIII)TM alloys were being undertaken, the results of which revealed a parallelism between the respective sets of T_c-data for hcp Ti-(3d)TM and Ti-(4d)TM alloys, thereby obviating the need for a new mechanism. The similarity among all Ti-TM alloys with regard to the variation of T_c with composition, and composition-controlled microstructure, was first recognized by ZWICKER [Zwi63] who predicted, by analogy with the then-available data for Ti-Rh, that in the hcp range the superconducting transition temperature of Ti-Ru should be similar to that previously found in the Ti-(3d)TM alloys. Unfortunately, no phase-equilibrium information on Ti-Ru is available in [Mol65], the most comprehensive compendium of Ti-alloy phase diagrams currently available, with which to complete the T_c-microstructure correlation.

5.19.2 Fluctuation Superconductivity in Ti-Ru Alloys

With reference to Ti-Ru$_8$, and the numerous other Ti-TM alloy superconductors mentioned in Table 1-24, HAKE and colleagues [Hak75][Lue75] have discussed the temperature and field dependences of electrical resistivity on the normal-state side of the mixed/normal second-order phase boundary, a regime where, on general grounds, fluctuations into the mixed state are to be expected. Using Ti-Ru$_8$ as a model system, HAKE [Hak69[a]] showed that as the reduced temperature ($t \equiv T/T_c$) decreased from $t > 3.5$ to 1.0, a significant reduction of resistivity began to take place at temperatures well above T_c. At $t \sim 2$, curvature in the resistivity temperature dependence became noticeable, and by $t = 1.2$ the resistivity of Ti-Ru$_8$ had dropped by 0.5% below its residual value. This was claimed to be a manifestation of fluctuations into the superconducting state, a phenomenon which increases the conductivity by an amount $\Delta\sigma_{fl}$, a quantity which is expressible in terms of the microscopic parameters of the GLAG theory. In an applied magnetic field such fluctuations, involving as they do the movement of flux and current, tend to be suppressed; this was, presumably, responsible for the observed increase in the resistivity of the sample with increasing field. That the effect was not due to normal magnetoresistance has been confirmed by its disappearance when, as a result of the application of high current densities, the sample was removed from the fluctuation regime [Hak69[a]]. The investigations of Ti-Ru$_8$ by HAKE [Hak69] have provided examples of both of the above effects as as parameterized by $\alpha^T \equiv (1/\rho)(\partial\rho/\partial T)_{H=0}$ and $\alpha^H \equiv (1/\rho)(\partial\rho/\partial H)_T$, respectively.

In recent work, HAKE and colleagues [Lue75] have extended these studies on Ti-Ru$_8$ to applied fields of up to 140 kOe with several interesting consequences. At high fields, conductivity fluctuations -- so-called paraconductivity[†] -- is completely quenched, thereby enabling a more accurate normal-state residual-resistivity reference-value to be acquired, and a better value of $\Delta\sigma_{fl}$ to be derived. At high fields (80-100 kOe), $\Delta\sigma_{fl}$ reached a maximum, its decrease in fields above 100 kOe being interpreted as a weak negative magnetoresistance. Negative magnetoresistance is usually associated with the influence of magnetic fields on s-d scattering, where s represents the spin of a localized magnetic moment (e.g. [Hed64]); but the usual Curie-Weiss-type of localized magnetic moment is not found in any bcc-Ti-TM alloy. Consequently, in the spirit of the above interrelationships, the observed weak negative magnetoresistance was postulated to be the consequence of a field-influenced conduction-electron interaction with "highly compensated or very rapidly fluctuating localized spins". Having gone this far, it was suggested by HAKE and colleagues [Hak75] [Lue75] that the negative $d\rho/dT$'s frequently encountered in those bcc Ti-TM alloys whose compositions were just sufficient to stabilize the bcc phase were also the result of localized spin fluctuations, although the operation of other possible mechanisms was not excluded [Hak75]. But since the negative $d\rho/dT$ effect is very strong in alloys of Ti with V, Nb and Ta for which the question of magnetic spin never arises, it is clearly much more likely to be associated with structural, rather than magnetic-spin, fluctuations. It would be interesting to make a detailed study of the resistivity of the Ti-Mn system with this phenomenon in mind. An ω-phase- or soft-phonon-related mechanism for negative-resistivity-temperature dependence in Ti-TM alloys has been considered in [Col74] and reviewed in [Cha74] and [Mon5.5].

5.20 SUPERCONDUCTIVITY IN Ti-Os ALLOYS

Apart from a reference to Ti-Os$_8$ by HAKE *et al* [Hak75] in connection with an investigation into the negative $\alpha^T \equiv \rho^{-1}(\partial\rho/\partial T)_{H=0}$ and negative $\alpha^H \equiv \rho^{-1}(\partial\rho/\partial H)_T$ effects, no information is available on superconductivity-related properties of Ti-Os alloys.

---

† According to HAKE [Hak69], R.A. FERRELL had suggested the term "paraconductivity" for $\Delta\sigma_{fl}$, the extra conductivity due to fluctuations into the superconductive state. The term was selected to be loosely analogous with the state of a ferromagnet above its Curie temperature.

5.21 SUPERCONDUCTIVITY IN Ti-Rh ALLOYS

5.21.1 Composition Dependence of the Transition
Temperature

Ti-Rh is a system whose superconducting transition temperature was the subject of considerable study
during the period between 1957 and 1969. The influences on T_c of composition and composition- and heat-
treatment-related microstructures have been investigated, and assessments were made within that context
of the manner in which the measuring technique (resistometric, magnetic, or calorimetric) can influence the
results obtainable when two- or three-phase alloys are
under examination.

The results of the first studies of Ti-Rh alloys
(5-15 at.%) were published by MATTHIAS in 1957 in a
paper which drew attention to the electron/atom ratio
systematics of T_c and the emergence of T_c from a low
minimum value at e/a $\sim$ 4 with increasing values of
that parameter [Mat57]. In a second paper MATTHIAS
and colleagues [Mat59] juxtaposed the data for Ti-Rh
with those for the related alloy system Ti-Co in a
demonstration of possible anomalous superconductive
interactions associated with "ferromagnetic" (i.e.
late-3d) solutes. Parallelism between the T_c *versus*
composition curves was not obtained. As had been the
case with the Ti-Ru system considered above, transition-temperature data for α^m-Ti-Rh were also missing.
Consequently T_c *versus* composition was forced, by
default, to rise much less rapidly than was the case
with Ti-Co, thereby substantiating the belief that a
special "magnetic interaction" was operative within
the Ti-(3d)TM class of alloys. Also using a magnetic
technique, RAUB and ANDERSON [Rau63] conducted a comparative study of the effects of Rh additions on the
mutually isoelectronic solvents Ti and Zr. In that
dilute Ti-Rh(0.5, 1 and 2 at.%) alloys were included,
this study provided the data that were missing from
the earlier work, and showed that T_c first increased
rapidly with Rh concentration, then dropped again to
low values at the boundary between the quenched-α^m
and quenched-β (actually $\omega+\beta$) regimes. This result
enabled ZWICKER [Zwi63] to suggest that similar behavior was to be expected in dilute Ti-Re and Ti-Ru,
not then measured, and to state by implication that
the three systems of alloys, Ti-(3d)TM, Ti-(4d)TM,
and Ti-(5d)TM, possessed similar structure-related
T_c *versus* composition characteristics.

The drop in T_c noted at the edge of the hexagonal
regime led RAUB and ANDERSON [Rau63], and subsequently
RAUB [Rau64], to the erroneous conclusion that for a
given solute concentration the T_c of bcc Ti-Rh was
lower than that of the α-phase. It was not until the
later 1960's that recognition of the occurrence and
properties of ω-phase precipitation was being transferred from the metallurgical to the physical literature. By now it is understood that the low T_c associated with the "bcc" alloys of threshold composition is
due to the presence of a high volume-fraction of low-
T_c ω-phase precipitation. A detailed discussion of
the T_c-suppressive influence of ω-phase precipitation
is provided in [Mon10.2].

Hypothetical low-concentration bcc Ti-TM alloys
containing no ω-phase should have significantly higher
T_c's than either $\omega+\beta$-phase or α-phase alloys [Mon10.4].
In a very early experiment, BUCKEL *et al* [Buc62] used
a resistive technique to measure the transition temperatures of a series of Ti-Rh alloys. It is usually
disadvantageous to study mixed-phase alloys using
electrical resistivity, since the presence of small
volume-fractions of high-T_c interconnected second
phases will give erroneously high T_c's if they are
attributed to the bulk. On the other hand, if the
transition temperature of the minority phase is actually being sought, the resistive technique offers an
excellent way of distinguishing its value from that of
the bulk, which can then best be extracted calorimetrically in a separate experiment. BUCKEL *et al* [Buc62]
measured the T_c's of Ti-Rh(0.063-15 at.%), paying
particular attention to the 5, 10, 12.5 and 15 at.%
members of the series which contained, according to
x-ray investigations, both α- and β-phases. The
superconducting transitions were attributed to the
β-phase component. It was, therefore, possible to
plot as in Fig.2-1 the transition temperature of
β-Ti-Rh as a function of composition, after which that
of unalloyed β-Ti, which turned out to be within the
range of 4 to 6 K, was obtainable by extrapolation to
zero solute content. As pointed out in Sect.2.2.4,
this value is in accord with the more recent estimate
of 6.4 K obtained by COLLINGS *et al* [Col72, Col72a]
[Ho73] who used an elaborate extrapolation argument
(explained fully in [Mon10.4]) based on the results of
a series of conjoint magnetic susceptibility and low
temperature specific heat studies of a number of
quenched Ti-Mo alloys. From a plot of the bcc-Ti-Rh
lattice parameter as function of composition, BUCKEL
et al were also able to obtain by extrapolation a

reasonable value[†] for the room-temperature lattice parameter of bcc-Ti. At 3.278 Å it was in good agreement with that derivable from the results obtained by HAKE et al [Hak61[a]] in a similar experiment on a series of Ti-Mo alloys.

In a detailed intercomparison of the superconducting properties of representative Ti-(3d)TM, -(4d)TM, and -(5d)TM alloys, RAUB and HULL [Rau64[c]] measured the T_c composition dependences of Ti-Co, -Rh, and -Ir. The rapid increase in T_c with composition exhibited by hcp-Ti-Rh, and claimed not to be associated with β-filaments, was again shown to be as large as that encountered in the hcp-Ti-Co alloys. A maximum in T_c near e/a = 4.5 was noted.

5.21.2 Calorimetric Studies of Superconductivity in Ti-Rh Alloys

When studying superconductivity in quenched α^m-Ti-TM alloys, in order to resolve uncertainties associated with possible electrical short-circuiting or magnetic shielding by minority-phase precipitates, the results of calorimetric measurements are usually needed. DUMMER and OFTEDAL [Dum68] carried out a calorimetric (>0.9 K) and magnetic (>1.2 K) investigation of the superconducting transition temperature in a series of as-cast Ti-Rh(1-10 at.%) alloys, while DANNER and DUMMER [Dan69] focused attention on α^m alloys in the concentration range below 3 at.% Rh, using He[3] and He[4] calorimetry (0.4-4 K) and resistive measuring techniques. The results of these studies may be compared with those of the magnetic measurements of RAUB and ANDERSON [Rau63] who also investigated as-cast alloys, although prepared from different starting materials. Although the T_c's observed by DANNER and DUMMER [Dan69] were generally higher than those obtained by DUMMER and OFTEDAL, they were generally lower than the results of the magnetic measurements of RAUB and HULL [Rau64[c]]. In showing a rapid rise in T_c with Rh content in the α^m-regime, followed by a discontinuous drop at the foot of the ω+β-curve,

---

† The lattice constant of bcc-Ti at 900°C is 3.312 Å. An expansion of from 3.278 Å (at say 20°C) to this value represents a mean coefficient of thermal expansion of 11.7×10^{-6} °C^{-1}; that of hcp-Ti within the temperature range 500-600°C is 11.5×10^{-6} °C^{-1} [Moℓ65, p.xvii].

the results were all qualitatively similar. According to DUMMER et al [Dum68] [Dan69], T_c scaled with γ throughout the composition range studied, suggesting that T_c was controlled by $n(E_F)$ throughout all the structural regimes. In order to illustrate the joint influences of structural state and measuring technique on the transition temperature, DUMMER and OFTEDAL plotted T_c versus composition both for the calorimetrically measured as-cast alloys as well as for smaller samples which had been measured magnetically and resistively after heat treatment at elevated temperatures followed by: (a) He-gas quenching (≳300°C sec^{-1}); (b) furnace cooling (2-3 min); (c) slow cooling (∼10 min to 600°C). The range of transition temperatures obtained can be appreciated by noting that for Ti-Rh$_3$, T_c varied from 0.9 K (calorimetric) to 4.1 K (slow-cooled, resistometric). The comparison illustrated the utility of combining resistive with calorimetric measurements when studying superconductivity in systems of this kind. In a mixed-phase alloy the resistive result may be the transition temperature of a minority high-T_c phase while calorimetry, which is a bulk measurement, will respond most strongly to the major superconductive component and tend to ignore in most cases the presence of low volume-fractions (a few percent) of thread-like precipitates no matter what their disposition within the sample or how high their transition temperature. On the other hand, if evenly proportioned amounts of two superconducting phases are present, calorimetry may yield a double transition as exemplified by the results of the measurements by DANNER and DUMMER [Dan69] on slow-cooled Ti-Rh(1 at.%).

5.22 SUPERCONDUCTIVITY IN Ti-Ir ALLOYS

5.22.1 Magnetic Measurements

Ti-Ir, a companion to Ti-Co and Ti-Rh, has been studied by RAUB and HULL [Rau64[c]], who used magnetic techniques in investigating the superconducting transition temperatures of 14 as-cast alloys in the composition range 0.4 to 13.5 at.%. In the α^m regime, with regard to which this is the only study, the rate of increase of T_c with composition was not as pronounced as for the other two alloys of the group, Ti-Co and Ti-Rh. It was noted that extrapolation of the α^m-Ti-Ir data led to an estimated 0.4 K for the T_c of pure α-Ti in reasonable agreement, according to Table 2-3, with the results of direct measurement.

5.22.2 Calorimetric Measurements

As part of a larger investigation of Ti-Ir-Pt pseudobinary A15 compounds, JUNOD et al [Jun76] measured the bcc solid solutions: Ti-Ir(5, 10, 12.5, 15, and 17.5 at.%). As a result of this work it was noted that the boundary of the solid-solution regime was about 18 at.% Ir. In agreement with the previous work it was noted that T_c passed through a maximum of 4.0 K at 10 at.% Ir; i.e. at an electron/atom ratio of 4.5 (assuming an e/a of 9 for Ir itself).

5.23 SUPERCONDUCTIVITY IN Ti-Pt ALLOYS

JUNOD et al [Jun76] have measured the low-temperature specific heats of a series of as-cast Ti-Pt alloys as part of the above-mentioned investigation of superconducting transition temperature in the Ti-Ir-Pt pseudobinary system. As a result of this work they suggested that the quenched-bcc regime extended out to about 10-15 at.% Pt. The results for the few solid-solution or near-solid-solution alloys measured indicated that the maximal T_c occurred at an e/a of about 4.5, as before, assuming an e/a of 10 for Pt.

5.24 SUPERCONDUCTIVITY IN BINARY ALLOYS OF Ti WITH THE 4d- AND 5d-GROUP-VIII TRANSITION ELEMENTS Ru, Rh, Ir, AND Pt— TABULATED DATA

TABLE 5-16   TRANSITION TEMPERATURES (T_c) OF TITANIUM-RUTHENIUM ALLOYS[+] [Mat57][++]

At.% Ru	T_c[+++] K
5	2.5
10	3.4
15	1.9

+   Condition: as-cast
++ See also [Mat59, Mat63][Bar61]
+++ Magnetic determination; values obtained from plotted T_c vs composition data (cf. Fig. 4-4).

TABLE 5-17   TRANSITION TEMPERATURES (T_c) OF TITANIUM-RHODIUM ALLOYS

(a) Magnetic Results

At.% Rh	T_c K	Data[+] Sources	
0.5	1.8	[Rau63, Rau64[C]]	n
1	2.4	" "	"
2	3.2	" "	"
3	1.9	" "	"
4	2.0	" "	"
5	2.3	[Mat57, Mat59]	g
5.5	2.6_5	[Rau64[C]]	g
7	3.6_5	"	"
8	4.0	"	"
9	4.3	"	"
10	3.2_5	[Mat57, Mat59]	g
10.5	4.3_5	[Rau64[C]]	g
12	4.0	[Rau63, Rau64[C]]	n
13.5	3.7_5	[Rau64[C]]	g
15	3.9_5	[Mat57, Mat59]	g

+ g = from plotted T_c vs composition data
  n = original data in tabular form

(b) Calorimetric (cal.) and Resistometric (res.) Results of Dummer et al.

At.% Rh	T_c(cal.)[++] K	T_c(res.)[+++] K	Data Sources
0.5	0.79	2.8	[Dan69]
1	1.16	3.0	"
1	∼0.9		[Dum68]
1.5	1.6	2.9	[Dan69]
2	1.73	2.9	"
2	1.7		[Dum68]
2.5	1.79	2.4	[Dan69]
3	1.34	1.9	"
3	∼1.0		[Dum68]
6	2.6		[Dum68]
8	3.5		"
10	4.0		"

++ Original data in tabular form.
+++ From plotted T_c vs composition data.

TABLE 5-18   TRANSITION TEMPERATURES (T_c) OF
TITANIUM-IRIDIUM ALLOYS[+]

At.% Ir	T_c[++] K (1)	(2)	Literature
0.4	0.6		[Rau64[C]]
0.5$_5$	0.7		"
0.8	0.7		"
1.0	1.0		"
1.5	1.2$_5$		"
1.9	1.4		"
3.0	1.4$_5$		"
4.0	1.6$_5$		"
5.0	2.5		"
5.0	3.0	2.7	[Jun76]
6.0	3.0		[Rau64[C]]
7.0	3.5$_5$		"
8.5	4.0		"
10.0	4.0		"
10.0	4.0	4.0	[Jun76]
12.5	3.2	3.0	"
13.5	3.1$_5$		[Rau64[C]]
15.0	2.0$_5$	1.7	[Jun76]
17.5	1.2[+++]	---	"

+   Condition: as-cast -------------- column (1)
                 homogenized 1400°C --- column (2)
++ From plotted T_c vs composition data.
+++ Transition commenced at 1.2 K.

TABLE 5-19   TRANSITION TEMPERATURES (T_c) OF
TITANIUM-PLATINUM ALLOYS[+] [Jun76]

At.% Pt	T_c[++] K
5	2.1
10	4.4
15	3.0

+   According to the T_c measurements the limit of
    Pt-solid-solubility in Ti is between 10 and
    15 at.% Pt.
++ From plotted T_c vs composition data.

Concentration Range	Condition	Structures	Procedures	Literature
(a) TITANIUM-RHODIUM ALLOYS				
5, 10 and 15 at.% Rh	As-cast.	Probably $\omega+\beta$ to β	Magnetic	B.T. Matthias *et al.* (1957,1959) [Mat57, Mat59]
0.5-12 at.% Rh (6 alloys)	As-cast and annealed 100h/700°C, and quenched.	0.5, 1, 2 at.% Rh, α^m 3 at.% Rh, mixed 4 and 12 at.% Rh, $\omega+\beta$ to β	Magnetic	C.J. Raub and C.A. Anderson (1963) [Rau63]
0.5-13.5 at.% Rh (12 alloys)	As-cast.	0.5, 1, 2 at.% Rh, α^m 3 at.% Rh, mixed 4-13.5 at.% Rh, $\omega+\beta$ to β	Magnetic	C.J. Raub and G.W. Hull (1964) [Rau64[c]]
1-10 at.% Rh (6 alloys)	As-cast.	1, 2 at.% Rh, α^m 3-10 at.% Rh, $\omega+\beta$ to β	Calorimetric and resistometric	G. Dummer and E. Oftedal (1968) [Dum68]
	As-cast plus heat treated just below m.p. and: (a) gas-quenched, (b) furnace cooled (2-3 min), (c) slowly cooled ($\gtrsim$10 min to 600°C).	Various phases	Magnetic and resistometric	
0.5-3 at.% Rh (6 alloys)	As-cast.	1-2.5 at.% Rh, α^m 3 at.% Rh, mixed	Calorimetric and resistometric	S. Danner and G. Dummer (1969) [Dan69]
(b) TITANIUM-IRIDIUM ALLOYS				
0.4-13.5 at.% Ir (14 alloys)	As-cast.	0.4-1.9 at.% Ir, α^m 3.0 at.% Ir, mixed 3.0-13.5 at.% Ir, $\omega+\beta$ to β	Magnetic	C.J. Raub and G.W. Hull (1964) [Rau64[c]]
5-17.5 at.% Ir plus compounds	As-cast, also homogenized at 1400°C.	bcc solid solutions A15 compounds	Calorimetric	A. Junod *et al.* (1976) [Jun76]
(c) TITANIUM-PLATINUM ALLOYS				
5 and 10 at.% Pt plus compounds	As-cast.	bcc solid solutions A15 compounds	Calorimetric	A. Junod *et al.* (1976) [Jun76]

6

TERNARY ALLOYS OF TITANIUM WITH SIMPLE METALS AND TRANSITION METALS (EXCEPT NIOBIUM)

Alloys considered in this chapter include pseudo-binary research alloys based on $Ti-Zr_{50}$, a metal "like" Ti or Zr, but one whose transition temperature of 1.6 K is more easily accessible than that of either of the pure elements Ti ($T_c = 0.36 \pm 0.03$ K, Table 2-3) or Zr ($T_c = 0.6 \sim 0.7$ K). Also included are those alloys based on Ti-V and Ti-Ta which have been regarded as competitors to Ti-Nb as current-carrying high-field (hard) superconductors, and in connection with which the influences of simple-metal and interstitial-element additions on flux pinning have been investigated. Developments in the latter direction culminated in French Patent No. 1,517,216 [Ass68[a]] which laid claim to the above-mentioned Ti-(group V)TM alloys doped with one or all of the so-called "interstitial elements" C, N, and O.

On the purely fundamental side, the superconducting transition temperatures of β-Ti-V-Cr and alloys based on Ti-Rh have been examined, as have the mixed-state and "paraconductive state" properties of β-Ti-V-Cr. In this chapter, as in Table 1-26, the ternary Ti-base alloys to be treated have been organized into the groupings: (a) Ti-TM-SM, (b) Ti-Zr-TM, (c) Ti-V-TM and Ti-V-Cr, (d) Ti-Rh-TM, and (e) Ti-V/Ta-Interstitial. The number of alloying species investigated has been quite small, for instance, the only simple metals to appear under item '(a)' above are Al and Sn.

6.1 SUPERCONDUCTIVITY IN Ti-TM-SM TERNARY ALLOYS

Studies of the alloys $Ti_{40}-V_{50}-Al_{10}$ and $Ti_{40}-V_{50}-Sn_{10}$ by EFIMOV et al [Efi70[a]] showed that additions of Al and Sn brought about rapid reductions in the transition temperature of the erstwhile equiatomic Ti-V alloy. Similarly COLLINGS and GEGEL [Coℓ75] noted a rapid reduction of the resistively measured T_c (0.38 K per at.%) following additions of 1, 2, 4, and 6 at.% of Al to a β-$Ti-Mo_{25}$ starting alloy. Using low temperature specific heat, COLLINGS and HO [Coℓ76] studied the effect of adding 1 and 3 at.% Al to $Ti-Mo_5$ and $Ti-Mo_{10}$ both of which possessed the ω+β structure in the quenched state. The addition of the 1 and 3 at.% Al to the latter alloy brought about no change in the optically-observed microstructure, but caused T_c to decrease at the rate of 0.10 K per at.% Al. On the other hand, the addition of 1 and 3 at.% Al to the initially ω+β-phase $Ti-Mo_5$ alloy caused: (a) the quenched structure to be martensitic; (b) the calorimetrically measured transition temperature to jump from an undetectably low value (<1.5 K) up to 3.1 K (with the addition of the 1 at.% Al), as illustrated in Fig.5-5; (c) the T_c of the ternary $Ti-Mo_5-Al_x$ alloys then to decrease at the rate of 0.16 K per at.% Al. These results are summarized in Fig.6-1 which, as explained in the caption,

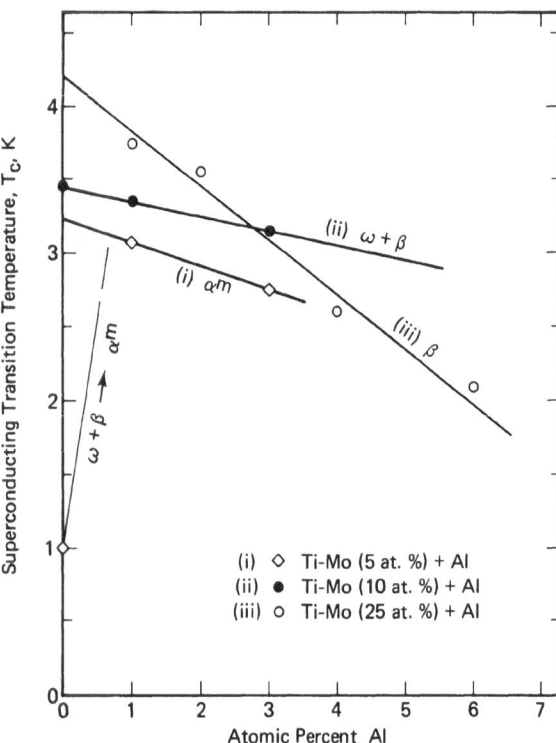

FIGURE 6-1. Influence of Al additions on the superconducting transition temperatures of three Ti-Mo alloys. Solution strengthening rapidly lowers T_c in the β-alloy Ti-Mo (25 at. %); but is less effective in Ti-Mo (10 at. %), already partially stabilized by ω-phase formation. The addition of only 1 at. % of Al to Ti-Mo (5 at. %) results in a martensitic quenched structure and an enhanced value of T_c. Further addition of Al in the now martensitic regime again tends to lower T_c. Note: Curve (i) intersects the axis at 3.2 K — within the range of 2.7–3.3 K measured for mechanically deformed Ti-Mo (5 at. %). Curve (iii), determined resistometrically, intersects the axis at T_c = 4.2 K, higher than that obtained calorimetrically (3.6–3.8 K) in Ti-Mo (25 at. %); however, the slope itself is expected to be similar under both measuring conditions — after COLLINGS and HO [Coℓ76].

shows the transition temperatures of the ternary Ti-Mo-Al alloys to be in agreement with those of the similarly structured low-concentration binaries.

In one of the first applications of superconducting transition temperature measurement to metallurgical diagnostics, RAUB and ZWICKER [Rau64[a]] studied the influence of isothermal aging at temperatures of 250, 350, 400, 450, 500, 550, and 650°C, for times of up to 1000 min, on the T_c of the commercially interesting alloy Ti-16V-3Al and interpreted the changes that took place in terms of the anticipated development of ω- and α-phase precipitation in the initially 900°C-quenched β-phase alloy.

More recently WOLFF and colleagues [Lep72][Woℓ73] conducted a similar and detailed study of transition-temperature/microstructure relationships in the aged

commercial alloy Ti-6Al-4V. Using an AC-susceptibility measuring technique, the effects of solution-treatment temperature, delay prior to quench, cooling rate, aging time and temperature, and slight compositional differences, on the mechanical properties, microstructures, and transition temperatures were examined. In general, the magnetically determined T_c's of alloys of this kind depend on: (a) special features of the microstructure such as the degree of connectedness of the β-phase and, in particular, whether or not it surrounds and shields the major α-phase component; (b) the transition temperatures of the various component phases, viz: (i) the α^m-phase having the alloy's average composition, (ii) the low-T_c V-lean α-phase, and (iii) the high-T_c V-rich β-phase. The superconducting transition temperatures of binary Ti-V alloys have been given in Chapter 3. Although not studied in the paper under discussion, β-quenched Ti-6Al-4V is martensitic and can be expected to have a transition temperature slightly lower (cf. the Ti-Mo-Al work, above) than that of binary Ti-4V (i.e. α'-phase Ti-$V_{3.8}$) which has a T_c, according to Fig.3-2, of 3.6 K. During aging in the $\alpha+\beta$ field, the Al dissolves preferentially in the α-phase and lowers an already low T_c (viz 1.9 K for α-phase Ti-$V_{3.8}$). On the other hand, V dissolves preferentially in the β-phase, which is able in this way to support more than 30 at.% V, and raise its T_c. In the experiments by WOLFF et al, Ti-6Al-4V was given a heat treatment of $\frac{1}{2}$h/940°C/water quench plus 4h/538°C/air cool. At the final aging temperature the equilibrium β-phase contained 23 at.% V, whose Al-reduced T_c would be expected to be slightly lower than the 4.8 K of the binary alloy (cf. Fig.3-1). In practice, a T_c of 4.14 K was obtained after the heat treatment.

6.2 SUPERCONDUCTIVITY IN Ti-Zr-TM TERNARY ALLOYS

In basic studies of the effects of transition-metal additions to a group-IV base, Ti-Zr_{50} (T_c, $\sim$1.6 K) has occasionally been used as substitute for either pure Ti or pure Zr with their less readily accessible transition temperatures of 0.36 $\pm$ 0.03 and $\sim$0.6$_5$ K, respectively. Thus HAKE and CAPE [Hak64], in an investigation of the influence of the possible local-moment character of Mn on the superconductivity of group-IV elements, studied the low temperature specific heat and magnetic susceptibility of $Ti-Zr_{49.9}-Mn_{0.2}$. In contrast to that of the binary Ti-Zr base, the magnetic susceptibility of the Mn-containing alloy was strongly temperature dependent within $10 < T < 100$ K, and when fitted to the Curie-Weiss equation

$$\chi = \chi_0 + C/(T - \theta_W) \qquad (6-1)$$

yielded a $\theta_W \cong 0$ K and, via the Curie constant C, a Mn-moment of 1.1 Bohr magnetons.

Whereas calorimetric measurements on the binary Ti-Zr alloy revealed the usual kind of bulk superconducting transition (mid-point of the C/T versus T^2 "jump", cf. Fig.5-5, at $T_c = 1.60$ K) the results for the ternary alloys revealed no such evidence of a superconducting transition within the temperature range of the measurement, the lower limit of which was 1.24 K. Indeed the C/T versus T^2 plot appeared subjectively to possess a T^{-1} type of temperature dependence at low temperatures, suggestive of the existence of a magnetic-clustering-induced constant term in the low temperature specific heat. In experiments conducted at about the same time, BUCHER et al [Buc65] also used magnetic and calorimetric techniques to study the superconducting transitions of an extensive series of $Ti-Zr_{50}-TM_{0.2}$ alloys, including $Ti-Zr_{49.9}-Mn_{0.2}$, an alloy similar to that just referred to. BUCHER et al noted: (a) that trace additions of TM = V, Fe, Nb, Mo, Ru, W and Re all resulted in slight increases in T_c, (b) that the addition of Cr slightly reduced T_c, while (c) Mn brought about a drastic reduction in T_c, in agreement with the observations of HAKE and CAPE [Hak64]. Measurements on as-cast and annealed (72h/300°C) alloys yielded qualitatively similar results.

This section concludes by noting that the upper critical field of $Ti_{30}-Zr_7-Ta_{63}$, along with those of a large family of related ternary and quaternary alloys, was measured using a pulsed-field resistometric technique by DeSORBO et al [Des67] in a study of H_r versus atomic-volume-adjusted effective electron/atom ratio systematics.

6.3 SUPERCONDUCTIVITY IN Ti-V-TM TERNARY ALLOYS

6.3.1 General Discussion

Extensive studies of the influence of dissolved transition elements on the superconductive properties (generally T_c) of Ti-V alloys have been undertaken by SAVITSKII and colleagues. In particular, it has been noted by EFIMOV et al [Efi70[a]] that the substitution of 10 at.% of the elements Zr, Hf, Nb, Mo, W, and Re for Ti in the alloy $Ti-V_{50}$ invariably decreased the superconducting transition temperature. The amount of this decrease was least for Zr and greatest for Re; i.e., it scaled with the "distance" of the alloying element from Ti and V in the periodic table. Although the raising of the average electron concentration by alloying to the "right" reduced T_c, this cannot be interpreted as implying the existence of an optimal average electron concentration at $Ti-V_{50}$ since T_c is also reduced by the isoelectronic substitution of Zr or Hf for Ti in that alloy.

Further details concerning the influence of transition-metal solutes on the T_c of a $Ti-V_{50}$ alloy base, originating in the work of EFIMOV, BARON, and SAVITSKII, have been presented by SAVITSKII [Sav73, p.337]. The particular solute elements referred to were Zr, Ta, and Mo. Within the context of a discussion of the ternary equilibrium phase diagrams, transition temperature data were reported for the β-phase alloys: $Ti-V_{50}-Zr_{0,1,5,10}$, $Ti-V_{50}-Ta_{0,0.5,1,5,10}$, and $Ti-V_{50}-Mo_{0,1,5,10}$. The suppressions of T_c which were found to accompany the additions of Zr, Ta, and Mo to $Ti-V_{50}$ (which took place at rates of about 0.2 K per at.%, 0.1_5 K per at.%, and 0.1_5 K per at.%, respectively) were about one-half as severe as that which accompanied the addition of Al to $β-Ti-Mo_{25}$, Sect.6.1.

The entire Ti-V-Ta alloy system has been subjected to extensive investigation. SAVITSKII has offered ternary diagrams of phase equilibria and superconducting transition temperatures [Sav73, p.339] and has also reported in considerable detail, with the aid of ternary diagrams, the results of: (a) microhardness studies; (b) electrical resistivity measurements at 12 and 77 K; (c) T_c measurements. All of these were made on alloys which had been quenched from 1500°C [Sav71[a]]. Finally, attention is drawn to Soviet Patent, No. 223357, filed by SAVITSKII, BARON, and EFIMOV [Sav68], which claimed, with brief reference to metallurgical and superconductive data, the alloy group: (60.5-40.5)Ti-(0.01-0.95)W-(balance)V.

6.3.2 Superconductivity in Ti-V-Cr Alloys

Ti-V-Cr has been the subject of several basic investigations of transition temperature and mixed-state properties. Both dilute (α^m-phase) and concentrated (β-phase) members of the family have been studied. BUCHER et al [Buc65], using magnetic and calorimetric techniques, have investigated the superconducting transition in predominantly-α^m Ti-$V_{2.2}$-$Cr_{1.1}$ as well as in the hcp binary alloys Ti-V, -Cr, -Fe, -Nb, and -Mo. In the ternary alloy, the transition which took place at a temperature of 3.6 K (magnetic method) was extremely broad, presumably on account of a complicated metallurgical structure.

The annealed β-phase alloy Ti_{30}-V_{60}-Cr_{10} has been referred to in five publications. In the first group of papers, HAKE [Hak65, Hak67, Hak67[b]] has studied this alloy, together with Ti-V and Ti-Mo, from the standpoint of mixed-state paramagnetism. At a given reduced temperature, Ti_{30}-V_{60}-Cr_{10} had about the same upper critical field[†] as Ti-Mo_{16}; but in the annealed state, unlike annealed Ti-Mo_{16}, it exhibited considerable magnetic irreversibility except at fields very close to H_u. According to HAKE [Hak67[b]], the critical field temperature dependence of Ti_{30}-V_{60}-Cr_{10} was in qualitative, and partial quantitative, agreement with the WHH and MAKI theories as delineated in Sect.5.12.4 and elsewhere in this book.

Again with the aid of the same alloy, HAKE [Hak69[a]] studied pre-transition reductions in electrical resistivity which began to set in as the temperature was lowered below $\sim 3T_c$, citing the effect as evidence for superconductive fluctuations. Ti_{30}-V_{60}-Cr_{10} was also a member of a group of alloys

(other members being Ti-Mn, -Fe, -Mo, -Ru, and -Os) whose normal-state electrical resistivity temperature dependences, $\alpha^T \equiv \rho^{-1}(\partial\rho/\partial T)_H$, were negative. For all except Ti_{30}-V_{60}-Cr_{10} itself, $\alpha^H \equiv \rho^{-1}(\partial\rho/\partial H)_T$ was also negative. The negativities of α^T and α^H, least in the present alloy and greatest in Ti-Mn and Ti-Fe, were tentatively attributed either to: electron scattering from magnetic ions, details of the energy-band structure, or electron scattering from soft phonons. The subject has been briefly reviewed in Sect.3.12 during a discussion of anomalous normal-state transport properties in Ti-V itself.

6.4 Ti-Rh-TM (INCLUDING NOBLE-METAL) TERNARY ALLOYS

Using a magnetic technique, RAUB [Rau64, Rau64[b]] has measured the superconducting transition temperatures of a series of bcc alloys designated Ti_{91}-Rh_4-TM_5. Here, TM represents twenty-five transition metal and noble-metal members of the three long rows of the periodic table, and with a few exceptions includes the elements: (a) Sc through Cu, (b) Y through Ag, (c) Hf through Au. The alloys were all bcc in the as-cast condition and needed no further heat treatment, except

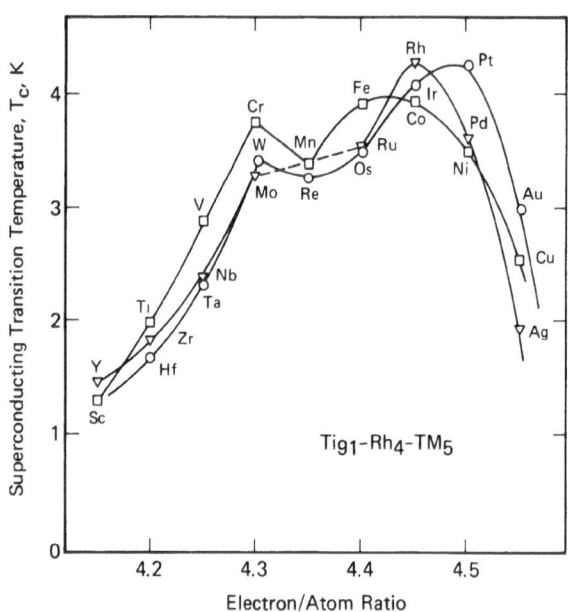

FIGURE 6-2. Superconducting transition temperatures of Ti_{91}-Rh_4-TM_5 alloys as a function of (s+d) electron/atom ratio[†] — after RAUB [Rau64[b]].

[†] $(e/a)_{Rh} = 9$, $(e/a)_{Pd} = 10$, etc.

[†] Ti-Mo_{16}:   at t = 0.72 (3.01 K), H_u = 34.0 kOe

Ti_{30}-V_{60}-Cr_{10}:   at t = 0.68 (3.81 K), H_u = 34.5 kOe those defined by TM ≡ Hf, V and Ta, whose martensitic minor components were removed by homogenization for 2h/950°C. The results, which are summarized in Fig.6-2, display two important features: *(i)* the transition temperature of Ti-Rh was always reduced by the addition of the third element, the amount of reduction increasing roughly with the magnitude of the group-number difference, just as was reported in Sect.6.3.1 to be the case for the Ti-V-TM alloys studied by EFIMOV *et al* [Efi70[a]]; *(ii)* the three curves of T_c *versus* e/a were approximately coincident, suggesting that T_c was responding to common electron-density-related properties.

6.5 TERNARY Ti-V-BASE AND Ti-Ta-BASE ALLOYS WITH C, N, OR O

The influences of C, N and O in interstitial solid solution, on the superconductivity of Ti-V and Ti-Ta alloys, have been considered from the standpoints of possible improvements to the critical current density. It is reasonable to assume that the solute-solvent interaction that promotes the observed solution-strengthening effects should also create centers conducive to flux-pinning and consequent stabilization of the current-carrying state. However, in spite of the existence of three references to the subject, nothing definite can be stated about the effects of C, N and O on Ti-V or Ti-Ta alloys.

COLLING *et al* [Col66] have intercompared, with null result, the 4.2-K "plateau" critical current densities (i.e. J_c in fields of 30-40 kOe) of a pair of Ti-Ta(∼50 at.%) alloys which, having been prepared from iodide-Ti and sponge-Ti, contained about 114 ppm and 249 ppm of oxygen, respectively. Likewise, BETTERTON *et al* [Bet66] failed to detect any significant differences between the 4.2-K, 30-kOe, J_c's of "optimally-heat-treated" Ti-Ta$_{60}$ containing zero, 0.08, and 0.27 at.% O.

With the aid of qualitatively illustrative curves for Ti-Nb, Ti-Nb-Ta, and Ti-Hf-Nb alloys, French Patent No. 1,517,216 [Ass68[a]] claimed that beneficial effects derived from the addition of one or more of the elements O, N, and C (to the extents of 500-4000 ppm, 500-2000 ppm, and 500-1500 wt. ppm, respectively) to

Ti-V, Ti-Ta, Ti-Nb, Ti-Nb-Ta, and Ti-Hf-Nb alloys. In addition to claiming improvements resulting from the use of commercial-purity, rather than high-purity, starting materials, the patent recommended the inclusion of further controlled amounts of C, N, and O through the addition of appropriate carbides (Nb_2C, TiC), nitrides (TiN, Nb_2N), and oxides (TiO_2, Nb_2O_5) to the melt, due attention of course being paid to the *total* impurity level (starting-plus-added) and its effect on workability.

The influence of interstitial elements on the superconductivity of Ti-Nb and a few related alloys is the subject of Chapter 8, while a general discussion of the flux-pinning aspects of B, C, N, and O additions to Ti-Nb and Ti-Nb-base alloys is to be found in [Mon21.13].

6.6 SUPERCONDUCTIVITY IN TERNARY ALLOYS OF Ti WITH SIMPLE METALS AND TRANSITION METALS (EXCEPT Nb)—TABULATED DATA

TABLE 6-2   TRANSITION TEMPERATURES (T_c) OF TITANIUM-BASE TERNARY ALLOYS (NO NIOBIUM)

(a) Transition Temperatures of Ti-Mo-Al [Col75, Col76]

Alloy	T_c K
(Ti-Mo$_5$)-Al$_0$	<1.5
-Al$_1$	$3.0_8 \pm 0.0_8$
-Al$_3$	$2.7_5 \pm 0.1_0$
(Ti-Mo$_{10}$)-Al$_0$	$3.4_5 \pm 0.1_0$
-Al$_1$	$3.3_5 \pm 0.0_5$
-Al$_3$	$3.1_5 \pm 0.0_5$
(Ti-Mo$_{25}$)-Al$_0$	$3.7_0 \pm 0.1_4$
-Al$_1$	3.7_0
-Al$_2$	3.5_5
-Al$_4$	2.6_0
-Al$_6$	2.1_0

(b) Transition Temperatures of Ti-V-Zr [Sav73, p.338]

Composition (at.%)			T_c
Ti	V	Zr	K
50	50	--	7.8
49	50	1	7.7
45	50	5	7
40	50	10	5.5

(c) Transition Temperatures of Ti-V-Ta [Sav73, p.340]

Composition (at.%)			T_c
Ti	V	Ta	K
50	50	--	7.8
49.5	50	0.5	7.8

(c) Transition Temperatures of Ti-V-Ta - *continued*

49	50	1	7.0
45	50	5	7.0
40	50	10	6.7

(d) Transition Temperatures of Ti-V-Mo [Sav73, p.341]

Composition (at.%)			T_c
Ti	V	Mo	K
50	50	--	7.8
49	50	1	6.9
45	50	5	6.9
40	50	10	6.2

TABLE 6-1   TRANSITION TEMPERATURES OF TITANIUM-BASE TERNARY ALLOYS (NO NIOBIUM) -- DATA AND DATA SOURCES

Compositions	Condition	Structures	Procedures	T_c	Literature
Ti-16V-3Al	Quenched from 1h/900°C.	α^m		4.2	C.J. Raub and U. Zwicker (1964) [Rau64[a]]
Ti-4V-6Al	1h/941°C (1725 F) quench.	$\alpha + \alpha^m$ + some retained β	a.c. susceptibility	1.96	R. Lepper *et al.*
	2h/871°C (1600 F) furnace cool.	α+β (equal fractions)		<1.1	(1972) [Lep72]
	1h/732°C (1350 F) furnace cool.	α+β (mostly α)		1.73	
(Ti-Mo$_{5,10}$)-Al$_{1,3}$	Quenched from 1 week at 1000°C.	(Ti-Mo$_5$)-Al$_{1,3}$, α^m (Ti-Mo$_{10}$)-Al$_{1,3}$, ω+β	Calorimetric	See below	E.W. Collings and J.C. Ho (1976) [Col76]
(Ti-Mo$_{25}$)-Al$_{1,2,4,6}$		β	Resistometric	"	E.W. Collings and H.L. Gegel (1975) [Col75]
Ti-V-Ta (ternary system)	Cold worked and recrystallized 2h/1500°C, cooled 50°C/sec.	β		"	E.M. Savitskii *et al.* (1971) [Sav71[a]]
Ti-V$_{50}$-Zr$_{0,1,5,10}$	As-cut from cast alloys.	β		"	E.M. Savitskii *et al.* (1973) [Sav73, p.338]
Ti-V$_{50}$-Ta$_{0,0.5,1,5,10}$	As-cut from cast alloys.	β		"	E.M. Savitskii *et al.* (1973) [Sav73, p.340]
Ti-V$_{50}$-Mo$_{0,1,5,10}$	As-cut from cast alloys.	β		"	E.M. Savitskii *et al.* (1973) [Sav73, p.341]
Ti-Rh$_4$-TM$_5$	As-cast except for TM ≡ Hf, V, Ta, in which case quenched from 950°C.	β	Magnetic	Fig. 6-2	C.J. Raub (1964) [Rau64[b]]

7

TITANIUM-NIOBIUM BINARY ALLOYS

This chapter offers complete coverage of the early literature of Ti-Nb superconductivity, but as a result of the burgeoning applicability of Ti-Nb alloys during the mid 1960's, full coverage of the literature published after 1965-1966 is no longer possible or desirable. With regard to the post-1965 period, only key publications dealing with fundamental physical phenomena, metallurgical optimization of critical current density, and the most recent (through 1980-82) literature, have been selected for consideration. Many important developments in Ti-Nb superconductivity are also treated in a separate monograph intended as a companion to this book [Mon00.00] and Tables 1-27, 1-28, and 1-29 are further guides to the associated literature. In preparation for the writing, the literature was sorted into five categories which then gave rise to the following five parts of this chapter:

PART 1, *THE SUPERCONDUCTING TRANSITION*, deals primarily with T_c but also considers other effects related to the transition, such as fluctuation superconductivity. PART 2, *THE MIXED STATE*, deals primarily with the critical fields (particularly the upper critical field) and also with other magnetic properties such as the paramagnetic mixed state, magnetic hysteresis, the critical state, and the motion of magnetic flux. PART 3, *THE CRITICAL CURRENT DENSITY*, deals primarily with the influence of metallurgical variables on flux pinning and process optimization, then goes on to review some commercial conductor development programs.

PART 4, *RECENT ADVANCES*, deals with the relative flux-pinning potencies of subbands and composition-controlled precipitation and some recent advances in the optimization of high-field Ti-Nb superconductors. PART 5, *CRITICAL CURRENT DATA*, provides data describing the high-field current-carrying capacities of a group of commercial Cu/Ti-Nb composite conductors representing the products of eight companies worldwide.

A subdivision of the relevant literature into the first three of these categories has been presented in Chapter 1, Tables 1-27, 1-28, and 1-29. Although many papers, especially the more comprehensive ones dealing with the influence of metallurgical variables on the critical current, could properly have been listed there under more than a single category, they are usually included only once in those tables, due mention being made if "other properties of Ti-Nb" have in fact been studied.

USAGES AND CONVERSIONS

Alloy Compositions

Following a convention adopted throughout this book, alloy compositions are specified either in weight percent (in the format Ti-Nb(n wt.%) or Ti-nNb for brevity) or in atomic percent (in the format Ti-Nb(n at.%) or Ti-Nb$_n$ for brevity).

A composition expressed in at.% carries more scientific information than does the same composition expressed in wt.%, which is the format preferred by the engineer. Expressed in at.%, Ti is the dominant constituent of a typical commercial alloy and as such would normally be listed first. Expressed in wt.%, the level of Nb may in a commercial alloy be a little more *or* a little less than 50%; accordingly it is not uncommon, in wt.% format, to see the Nb component listed first. To suite the convenience of all readers whether they are used to seeing compositions expressed in wt.% Nb, wt.% Ti, or at.% Nb, or are interested in the electron/atom ratio of the alloy concerned, Table 7-1, a conversion chart, is included in this introduction.

Magnetic Fields

The system of cgs-practical units is generally used throughout this book. In it, magnetic field strength is given in oersteds (Oe) and the corresponding induction in gauss (G). Thus applied magnetic fields are expressed in oersteds (or kOe) towards the beginning of the chapter particularly since it is there, in describing magnetization experiments, that a distinction has to be clearly made between the applied field, H_a, and the induction it produces. The SI unit of magnetic field is $A\ m^{-1}$. Modern writers, even when purporting to employ SI units, tend to eschew the use of $A\ m^{-1}$, presumably on account of its awkwardness, and in so doing are forced to abandon reference to magnetic field strength. In its place the tesla, the induction unit, is often found. Thus towards the latter part of this chapter, in conformity with the contemporary literature under discussion (especially that dealing with the field-dependence of J_c rather than magnetization phenomena), the tesla (T) is often used as a replacement for 10 kOe, in full implicit recognition of course that it refers not to H_a itself but to $\mu_0 H_a$, the coefficient being the permeability of vacuum.

TABLE 7-1 WEIGHT-PERCENT $\rightleftharpoons$ ATOMIC-PERCENT CONVERSION TABLE FOR Ti-Nb ALLOYS[†]

Wt.% Nb	Wt.% Ti	At.% Nb	At.% Nb	e/a	Wt.% Nb
10	90	5.42	10	4.1	17.73
15		8.34	15		25.50
20		11.42	17		28.43
25		14.67	19		31.27
30		18.10	20	4.2	32.66
35		21.73	23		36.68
40		25.58	25		39.27
45	55	29.67	30	4.3	45.39
46	54	30.52			
47	53	31.38	31		46.56
48	52	32.25	32		47.72
49	51	33.13	33		48.86
50	50	34.02	34		49.98
51	49	34.92	35		51.09
52	48	35.84	36		52.18
53	47	36.76	37		53.25
			38		54.31
53.5	46.5	37.23	39		55.36
54	46	37.70	40	4.4	56.39
55	45	38.66			
56	44	39.62	45		61.34
			50	4.5	65.98
60		43.61	55		70.33
65		48.91	60	4.6	74.42
70		54.61	65		78.27
75		60.73	70	4.7	81.90
80		67.34	75		85.33
85		74.50	80	4.8	88.58
90		82.27	85		91.66
95	5	90.74	90	4.9	94.58

† Atomic weight, Ti: 47.90
  Atomic weight, Nb: 92.906

PART 1: THE SUPERCONDUCTING TRANSITION IN TITANIUM-NIOBIUM ALLOYS

7.1 THE SUPERCONDUCTING TRANSITION TEMPERATURE

Although adequate information is available concerning the responses of the superconducting transition temperatures of Ti-Nb alloys to changes of composition (quenched alloys) and microstructure (quenched-and-aged alloys) it is a surprising fact that, in spite of its emerging utility as a practical superconducting alloy, Ti-Nb has played a less active role than Ti-V during the early development of type-II superconductivity. Nevertheless Ti-Nb alloys have over the years been the subjects of many fundamental investigations: The results of systematic studies of the transition temperature over a wide composition range have been correlated with electronic property and microstructural changes; low-concentration Ti-Nb alloys have been studied within the context of the effects of dilute additions of transition metals to Ti; the results of low-temperature specific heat, electrical resistivity temperature dependence, and low-temperature thermal conductivity measurements have been interpreted in terms of microscopic superconductive mechanisms.

7.2 SYSTEMATICS OF THE TRANSITION TEMPERATURE

In an important and comprehensive study of superconducting transition temperature in binary TM alloys selected from among members of the three long rows of the periodic table, HULM and BLAUGHER [Huℓ61] have investigated a series of homogenized Ti-Nb(2-97 at.%) alloys. Samples were measured in the as-quenched as well as slow-cooled conditions and the responses of T_c to variation of composition and thermal treatment were interpreted in terms of microstructural states estimated from the equilibrium phase diagrams of HANSEN [Han51], Fig.7-1, in conjunction with a martensitic transformation curve developed by DUWEZ [Duw53], Fig.7-2. The existence of ω-phase which was to play an important role in the interpretation of aging effects in β-phase alloys was not recognized at the time; from a heuristic standpoint, however, the simplified assumption of β → α+β decomposition is perfectly adequate. A T_c-composition diagram for Ti-Nb

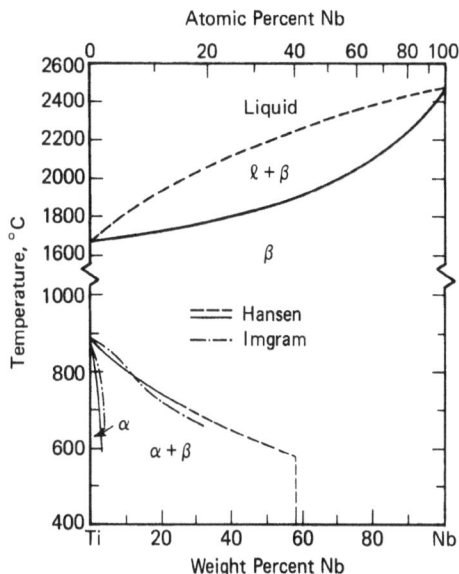

FIGURE 7-1. The Ti-Nb equilibrium phase diagram due to HANSEN et al [Han51] and IMGRAM et al [Img61] as modified by the observation that no appreciable α-phase precipitation takes place during the aging at 400°C of alloys with more than about 40 at. % Nb [Nea71, Wes80, Wes81, Wes82].

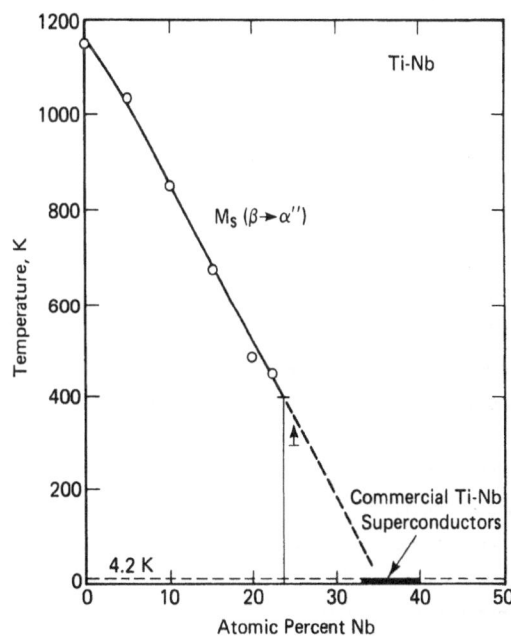

FIGURE 7-2. The martensite $\beta \rightarrow \alpha^m$ transformation curve for Ti-Nb based on the results of DUWEZ [Duw53] and the more recent data of BROWN et al [Bro64] and BAKER [Bak71] — after KOCH and EASTON [Koc77].

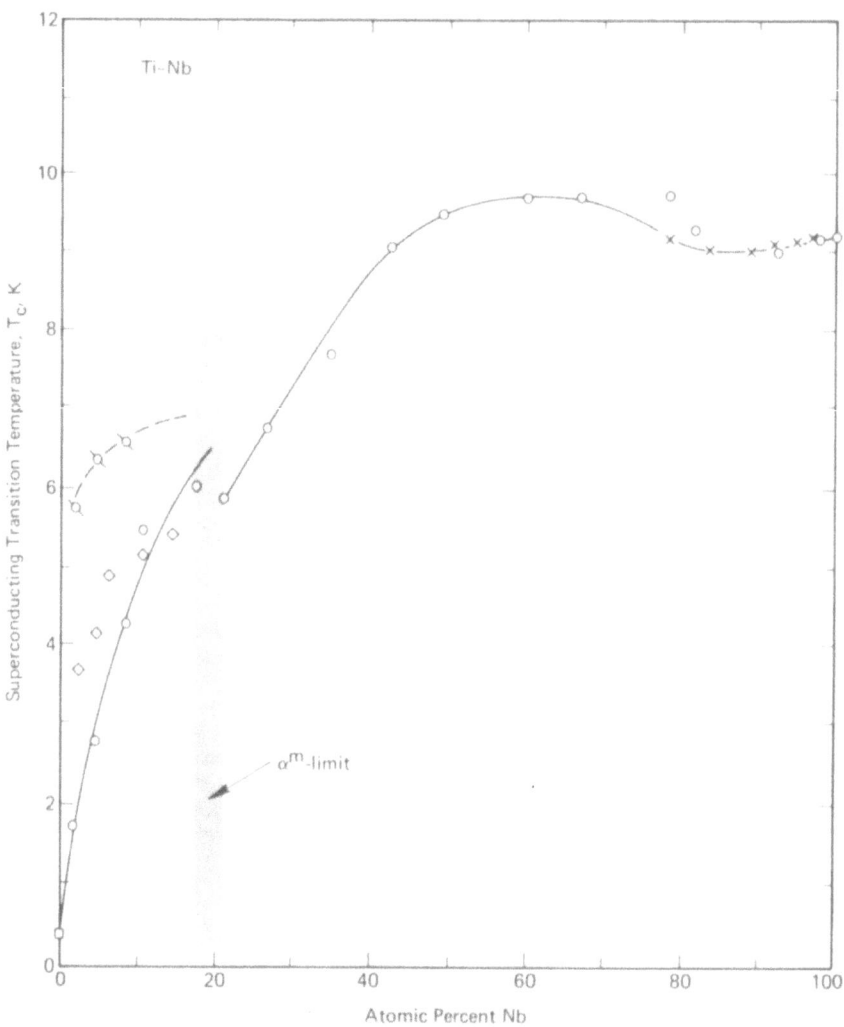

FIGURE 7-3. Superconducting transition temperature as function of composition for Ti-Nb alloys. Data sources are: HULM and BLAUGHER [Huℓ61], water-quenched (○) and slow-cooled (ℚ) alloys; BUCHER and MÜLLER [Buc61] (◇); DeSORBO [Des63] (x). The unalloyed-Ti value is from Table 2-3.

alloys, assembled from the results of HULM and BLAUGHER [Huℓ61] and some others is given in Fig.7-3. A comparable plot of the calorimetrically-measured T_c-composition dependence is to be found in [Mon10.3].

The transition temperature of Ti-Nb, unlike those of other Ti-TM alloys, varies fairly continuously with composition throughout the entire concentration range, consequently a clear subdivision of the data into two regimes would not be possible without accompanying microstructural evidence. Within the "quenched-β" (but equilibrium α+β) field the development during slow cooling of structural and compositional inhomogeneity was responsible for broad transitions with relatively high upper-temperature limits [Huℓ61]. The

low-concentration alloys were regarded as α^m-martensite (plus a possible admixture of α-phase) when quenched, and α plus a β-phase (of almost constant composition, independent of parent composition) when slow cooled. The occurrence of α^m-martensite and other phases found in quenched Ti-Nb alloys are described in Sect.7.20. HULM and BLAUGHER's early investigation of superconductivity and microstructure was extended by ZWICKER whose results were described in a review article dealing with the T_c of Ti-base alloys and intermetallic compounds [Zwi63][Zwi74, p.81] and in other papers [Zwi65][Loh70, Loh73]. Other studies of T_c in β-Ti-Nb alloys have been undertaken as components of more comprehensive investigations of superconductivity in binary and ternary transition metal alloys. For example, DeSORBO [Des65] explored the validity of an "effective electron/atom ratio" defined as:

$$N_{eff} = \frac{\Sigma_0^n f_i \zeta_i}{f_0 + \Sigma_0^n f_i (v_i/v_0)} \quad , \qquad (7\text{-}1)$$

where f represents atomic fraction, ζ is valence, v is unit cell volume (subscript-o designating the solvent, and subscript-i the solute) as a general index of the variation of both T_c and H_{c2} with composition. This quantity, which takes into account both electron/atom ratio (e/a) and atomic volume, is a measure of electron density. It turned out that maxima in T_c versus N_{eff} occurred at $N_{eff} \sim 4.7$ and ~ 6.4 independent of whether the solvent or solute atom came from the 3d, 4d, or 5d series. Furthermore it was claimed that the method could also account for the variations in T_c noted among certain pairs of elements from the same column in the periodic table.

BELLIN et al [Bel69], using a magnetic technique, measured the transition temperatures of a series of Ti-Nb alloys in constructing one of the three boundaries of the ternary Ti-Nb-V composition triangle. The results, although not correlated with any accompanying physical-property information, were interesting in their own right. Referring to them, COLLINGS and GEGEL offered a lattice-stability-based interpretation of the variation of T_c with composition [Col75]. It is significant that the substitution for Nb of V, one of its electronic isomers, always produced a lowering of the transition temperature, and that at ~ 10 K the T_c of Ti-Nb(50-60 at.%) is an absolute maximum in the ternary Ti-Nb-V system.

Resistive measurements of T_c, peripheral in the sense that they were undertaken as part of a larger investigation of Ti-Nb-base ternary alloys, were made by HAWKSWORTH and LARBALESTIER [Haw80]. Their major goal at the time was to determine the influence of Hf and Ta additions on the resistively measured upper critical fields, H_r, of Ti-Nb alloys whose composition range included members of potential commercial interest. The results of such measurements of T_c, upper critical field, and the residual resistivity, ρ_n Ω cm, itself are coupled through $H_{c20} = 3.06 \times 10^4 \rho_n \gamma T_c$ (Oe), a relationship which was introduced in Chapter 3 (Eqn.(3-13)) and which is fully discussed in [Mon14.6].

7.3 TRANSITION TEMPERATURES OF LOW-CONCENTRATION Ti-Nb ALLOYS

Early studies of the superconducting transition temperatures of low-concentration Ti-TM alloys were stimulated by a suggestion by MATTHIAS et al [Mat59, Mat63, Mat63a] that the presence of small amounts of "magnetic" 3d transition elements might be responsible for unexpectedly high T_c's in the resulting alloys, through the operation of a new magnetic superconductive-interaction mechanism. In an investigation of this suggestion, HAKE [Hak62b] studied the resistive superconducting transition in Ti-Nb(1 at.%) in association with similar measurements upon dilute Ti-Cr, -Mn, -Fe, -Co and -Ni alloys. Where comparison was possible the resistive transitions, appropriately defined, agreed with those measured magnetically by MATTHIAS et al [Mat59], however, some important questions were raised concerning the existence of localized magnetism in Ti-TM alloys. For further discussions of the magnetic and related physical properties of such alloys, [Mon7.10] is recommended. At about the same time as HAKE's measurements were being carried out, BUCHER and MÜLLER [Buc61] conducted a series of magnetic measurements of the superconducting transition temperatures of a number of Ti-V and Ti-Nb alloys. The results showed that Nb and V elevated T_c just about as strongly as did Cr, Mn, Fe, and Co, indicating that magnetic moments, if they did exist on those atoms, were not contributing to the superconductivity. Although the controversy might very well have ended there, confirmatory studies on selected alloys were carried out subsequently by BUCHER, HEINIGER and MÜLLER [Buc65] and by HEINIGER and MÜLLER [Hei64], who augmented magnetic measurements with low temperature specific heat measurements through the superconducting transition. After fitting the results to a distributed specific-heat function, it was determined that, although broad, the superconducting transition in quenched Ti-Nb$_4$, which took place between about 3 and 2.4_5 K, was more than 95% complete. But the anomalous behavior of Ti-Fe which yielded a broad transition only about 30% complete, although the subject of considerable subsequent study (cf. AGARWAL [Aga74] for the most recent account) did not receive a completely satisfactory explanation. Problems associated with the calorimetrically-observed superconducting transition in dilute Ti-TM alloys are discussed in [Mon10.3],

7.4 CALORIMETRIC MEASUREMENTS OF THE TRANSITION TEMPERATURE

methods for dealing with rounded calorimetric transitions are described in [Mon10.3] and [Mon10.8], while a reason for the anomalous breadth of the transition in quenched dilute Ti-Fe and related alloys has been offered in Sect.4.14.

Just referred to are the results of some low temperature specific heat measurements of the superconducting transition in relatively dilute Ti-Nb alloys. The concentrated members of the Ti-Nb system have been the subjects of extensive calorimetric study by workers at the Physico-Technical Institute for Low Temperatures (Kharkov) [Suk68, Suk71] [Shc73] [Ver80]. In the experiment on Ti-Nb$_{25}$ conducted by SUKHAREVSKII et al [Suk68], since no special precautions had been taken to ensure sample homogeneity, the inevitable presence of precipitates resulted in severe smearing of the transition. This result provided SUKHAREVSKII et al with an opportunity to devise a technique for dealing with calorimetric rounding; a method was developed which they claimed could be extended to a study of second-order phase-transition rounding in general. Later, in conjunction with a different pair of co-authors, SUKHAREVSKII [Suk71] conducted a series of low temperature specific heat measurements on the alloys Ti-Nb(10, 25, 50, 75, and 80 at.%) over the temperature range 2.5-20 K. Unfortunately, of the wealth of useful information that derives from such measurements, only the T_c data were published. Nevertheless a complete set of calorimetric data, for the alloys Ti-Nb(20, 25, 50, 75, and 90 at.%) including T_c, γ, and θ_D, has been courteously provided by VERKIN [Ver80], and these are to be found in [Mon8.2].

SUKHAREVSKII et al [Suk71] and subsequently SHCHETKIN and KHARCHENKO [Shc73], who measured Ti-Nb$_{92.8}$, decided to analyze their results in terms of a two-band model. According to BCS [Bar57], the temperature dependence of the superconductive electronic specific heat, C_{es}, is expressible in the form:

$$C_{es}/\gamma T_c = a \exp(-b/T) \qquad (7-2)$$

in which according to RICKAYZEN [Ric65, p.205], for the temperature range $0.4 > t > 0.17$, $a = 8.5$ and $b = 0.82 \, \Delta_{00} \, k_B^{-1}$, Δ_{00} being the zero-K BCS half-gap,

cf. [Mon8.4]. It follows that a plot of $\ln(C_{es}/\gamma T_c)$ versus t^{-1} should be linear, with a negative slope containing the BCS gap. This has been frequently demonstrated. The experimental results for Ti-Nb$_{50}$ were plotted in this format by SUKHAREVSKII et al [Suk71]. A distinct break in slope at $t \sim 0.4$ separated the data-plot into segments with two very different slopes, suggesting to those authors and subsequently to SHCHETKIN and KHARCHENKO [Shc73], that a description in terms of two energy gaps -- one corresponding to d-electrons and the other to s-electrons would be valid for Ti-Nb. But according to PREKUL et al [Pre74], not to mention the condition $0.4 > t > 0.17$ imposed on Eqn.(7-2) itself, other interpretations are possible. In concluding this paragraph on calorimetric properties it is interesting to note, with regard to the above work on Ti-Nb$_{92.8}$ [Shc73], that a deviation-function plot placed that alloy near Pb, thereby classifying it as a strong-coupled superconductor (cf. [Mon8.6] for further discussions of this subject).

7.5 FLUCTUATION EFFECTS—TRANSPORT PROPERTIES

7.5.1 Electrical Resistivity

PREKUL et al [Pre74] have measured the electrical resistivity temperature dependences of Ti-Nb (also Ti-V and Ti-Ta) alloys from the ice point to T_c. Apart from a somewhat gradual transition from the normal to the superconducting state as T_c is approached (not explicitly referred to by PREKUL et al) the principal anomaly noted was a reduction in the slope, $d\rho/dT$, with decreasing solute concentration. This effect, which increased in magnitude in the sequence V < Nb < Ta, and which led to a negative $d\rho/dT$ in low-concentration Ti-V alloys, was tentatively construed as evidence for the existence of some form of localized spin fluctuation -- a mechanism which in normal-metal alloys is related to the resistance-minimum effect. This author prefers to interpret the resistance anomalies -- at least for the particular case of Ti-V alloys -- in terms of the precipitation of ω-phase and/or the lattice fluctuations from which the ω-phase derives [Coℓ74]. That such fluctuations can be expected in quite concentrated Ti-Nb alloys has been demonstrated by BALCERZAK and SASS [Baℓ72] as a result of electron diffractometry and by IKEBE and co-workers [Ike77] on the basis of thermal and electrical conductivity measurements. The possible existence of fluctuation-related resistive anomalies in Ti-TM alloys with particular reference to Ti-V$_{25}$ was considered by HAKE [Hak70] and more generally by HAKE et al [Hak75]. The microstructural or lattice dynamical aspects of anomalous $d\rho/dT$ are considered in detail in [Mon5.5].

7.5.2 Thermal Conductivity

Samples of Ti-Nb$_{45}$, cold worked 99%, and heat treated for 1, 5 and 25h/500°C and for 25h/1600°C (i.e. recrystallized) were the subjects of thermal conductivity measurements by IKEBE et al [Ike77]. The thermal resistivities corresponding to all the heat treatment conditions were anomalously large. The dislocation density needed to explain the effect in the recrystallized sample was two orders of magnitude too large. Precipitation was suspected in the 500°C-aged samples, but no clear evidence of α-phase precipitation had been observed in Ti-Nb alloys with more than 40 at.% Nb. On the other hand, electron diffractograms of the Ti-Nb$_{45}$ alloys in the cold-rolled (cr), cr plus 25h/500°C, and cr plus 25h/1600°C conditions exhibited the circular diffraction streaks characteristic of precurser ω-phase precipitation. There seemed to be little doubt that the anomalously high thermal resistivity was associated with this effect. Several microscopic mechanisms were postulated for the phonon scattering, the essential ingredients of which were: (a) fine ω-like particles -- more or less dynamical in nature according to aging conditions; (b) a relationship between enhanced J$_c$ and enhanced phonon scattering; (c) a modulated electronic structure as a result of the presence of fine normal precipitates in a superconducting matrix (coherence length $\xi \sim 50$ Å). The thermal conductivities of transition-metal alloy superconductors are discussed in detail in [Mon6.9] and data for some ten Ti-Nb alloys are presented in [Mon6.10].

7.6 INFLUENCE OF AGING ON THE TRANSITION TEMPERATURE

According to the results of HULM and BLAUGHER [Hul61] referred to above, the slow cooling of low-concentration Ti-Nb alloys resulted in a sharp, almost parent-composition-independent, transition temperature. Slow cooling is of course short-term aging. The influence of aging for times of up to 15,000 min at temperatures of 400 and 500°C on Ti-Nb(4, 9, 20, 25 at.%) and Ti-Nb$_{33}$ were investigated by NISHIMURA and ZWICKER [Nis68] and RAUB et al [Rau64^a]. The results of those studies, which included microstructural and hardness measurements, and electrical resistivity measurements in the case of the former work, were easily interpretable in terms of the effects of solute diffusion and precipitation, and with regard to the low-concentration alloys, in terms of the relationship between the martensitic transformation temperature, M$_s$, and the aging temperature. As part of a comprehensive study of the influence of aging on the superconductive, normal-state, and microstructural properties, SALTER [Sal65, Sal66] investigated the influence heat treatment for up to 50h at 200-650°C on the T$_c$ of Ti-Nb$_{22}$. Maximum transition temperatures of 8.8 to 9.8 K were obtained after extended aging at temperatures of 350 to 500°C. Again the results could be reasonably well interpreted in terms of phase-decomposition-driven Nb-enrichment of the β-phase. The variation of T$_c$ with aging, compared against the Nb concentration of the β-phase as determined by x-ray diffraction, agreed reasonably well with the T$_c$ versus composition curve of HULM and BLAUGHER. A slight deviation towards the low side was thought to be due to a depressive effect resulting from precipitate proximity. Proximity effects of this type are discussed in detail in [Mon11.9]. Further detailed studies of the influence of aging on the transition temperature of Ti-Nb$_{22}$, and an alloy of comparable properties, Ti-Nb$_{25}$, were undertaken by McINTURFF and CHASE [Mci73]. Although no longer in favor on account of their poor workability, the Ti-rich alloys investigated by those authors, and SALTER [Sal65, Sal66] before them, had in the early 1960's been placed in commercial production by Atomics International (subsequently to become a division of the North American Rockwell Corporation). The influence of aging, i.e. precipitation heat treatment (PHT), on the transition temperature of Cu-clad and >99.9% cold worked Ti-Nb$_{22}$ and Ti-Nb$_{25}$ alloys is depicted in Fig.7-4. The cold-worked wires quickly developed a high density of α-phase precipitation at the dislocation-cell boundaries, cf. Sect.7.20; the accompanying enrichment of the β-matrix elevated the resistively measured T$_c$ to about 10 K.

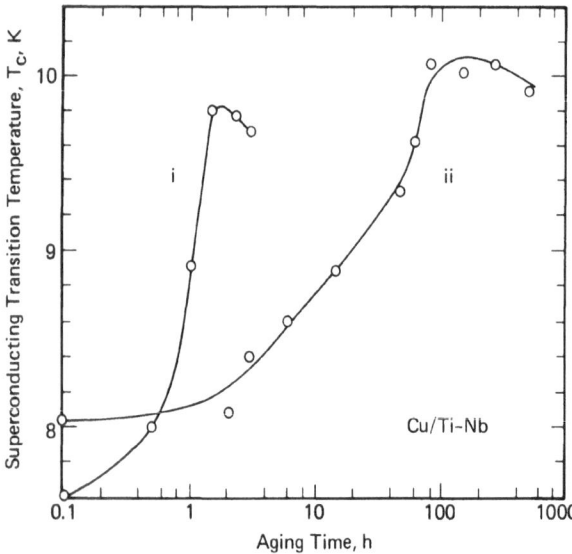

FIGURE 7-4. Critical temperature (T_c,K), measured resistometrically at a current density of ~ 20 A cm^{-2}, *versus* aging time for cold-worked (> 99.9%) Cu-clad low-concentration Ti-Nb monofilaments. (i) Ti-Nb (22 at. %) aged at 450°C; (ii) Ti-Nb (25.2 at. %) aged at 350°C — after McINTURFF and CHASE [Mci73].

7.7 COMMERCIAL ALLOYS

The superconducting transition temperatures of Ti-Nb research alloys are well known as a result of the work summarized in Fig.7-3. Consequently in describing the T_c's of composite superconductors a few representative data are adequate. The transition temperatures of Ti-Nb alloys of commercially useful compositions have been studied in the as-cold-worked and the optimized conditions, respectively, by HAWKSWORTH and LARBALESTIER [Haw80] and HILLMANN *et al* [Hil79]. Some of their results are given in Table 7-2.

TABLE 7-2   SUPERCONDUCTING TRANSITION TEMPERATURES OF SOME COMMERCIAL Ti-Nb ALLOYS

Niobium Content		T_c		
Wt.%	At.%	K	Condition	Reference
52	36	9.0	Monofilaments, cold-worked>97%	[Haw80]
53.5	37	9.25	" "	"
53.5	37	9.25	Commercial multifilament	"
50.1	34	8.5	Optimized monofilament	[Hil79]
52.1	36	8.8	"	"
61	44.5	9.1	"	"

PART 2: THE MIXED STATE IN TITANIUM-NIOBIUM ALLOYS

Ti-Nb has shared the center of the stage with Ti-V during the development and experimental testing of theories of the mixed state. Advantage has been taken of the wide solubility range of Nb in Ti in order to study the magnetic properties of dilute as well as concentrated Ti-Nb alloys. FIETZ [Fie67], for example, using small additions of Ti to Nb has been able to control the electronic mean free path within wide limits while keeping T_c and H_c fairly constant. On the other hand, numerous workers have focused their attentions on Ti-rich alloys with short electronic mean free paths in studies of "dirty" and extremely dirty (i.e. "extreme") type-II superconductors. Following what might be referred to as the "experimental rediscovery" of type-II superconductivity in 1962 (see [Ber63^a] for further references), the microscopic picture of the mixed state quickly developed in sophistication on several fronts. A review of the early studies of magnetic effects in Ti-Nb automatically traces this development. Of considerable interest initially were the lower and upper boundaries (H_{c1} and H_{c2}) of the mixed state, the manner in which H_{c2} varied with alloying (and with the electronic implications of alloying), and the influence on H_{c2} of the metallurgical state of the alloy.

7.8 THE MAGNETIC PROPERTIES OF TYPE-II SUPERCONDUCTORS

A parameter of central importance in superconductivity is the κ_{GL} of the 1950 GINZBURG-LANDAU theory which in type-II superconductors is greater than $1/\sqrt{2}$ and which, as a result of the work of GOR'KOV (cf. [Ric65, p.315]) aided by the calculations of GOODMAN [Goo62], is expressible in terms of atomic constants and physical measurables. Thus, as has already been pointed out in Sect.3.5 (with reference to [Mon13.5] and [Mon14.4]), κ_{GL} may be expressed as the sum of intrinsic (or "clean") and extrinsic (or "dirty") components according to:

$$\kappa_{GL} = \kappa_{GL}^c + \kappa_{GL}^d \qquad \text{[3-9] (7-3)}$$

with $\kappa_{GL}^c = 1.60\text{x}10^{24}\, T_c\, \gamma^{3/2} (n^{2/3} S/S_F)^2 \qquad \text{[3-10] (7-4)}$

and $\kappa_{Gl}^d = 7.49\text{x}10^3\, \gamma^{1/2}\, \rho_n \qquad \text{[3-11] (7-5)}$

Since the GINZBURG-LANDAU equations were valid only in a narrow temperature range below T_c and close to H_{c2} where the gap function is small and forms a convenient expansion parameter, theorists following GOR'KOV attempted to extend the temperature range of the microscopic theory and to further generalize it in terms of macroscopic materials properties. Taken into consideration has been the effect of a wide range of electronic mean free path lengths, ranging from the "clean" ($\ell \gg \xi_0$) to the "dirty" ($\ell < \xi_0$) limits, where ℓ is the electronic mean-free-path and ξ_0 represents the pure solvent coherence length (cf. [Mon13.5]). The theoretical extensions to the GLAG theory soon resulted in the defining of three generalized parameters: κ_1, κ_2, and κ_3, all of which converged on κ_{GL} as $t \equiv T/T_c$ increased towards 1.0. The dirty limits of the MAKI [Mak64] parameters, $\kappa_1(t)$ and $\kappa_2(t)$, are defined by:

$$H_{c2}^*(t) = \sqrt{2}\,\kappa_1(t)\, H_c(t), \qquad \text{[5-5] (7-6a)}$$

$$4\pi (dM/dH)|_{H_{c2}} = [\beta_A(2\kappa_2^2(t)-1)]^{-1} \qquad \text{[5-6] (7-6b)}$$

(where $\beta_A = 1.16$ for a triangular flux lattice)

while the dirty limit of $\kappa_3(t)$, according to MAKI [Mak64[a]], is given by:

$$H_{c1}(t) = \frac{\ell n\, \kappa_3(t)}{\sqrt{2}\,\kappa_3(t)}\, H_c(t) \qquad (7-6c)$$

With the upsurge of interest in GLAG theory, beginning in about 1961, it was immediately recognized that normal-state Pauli paramagnetism if it existed must play an important role in perturbing the superconducting-to-normal free-energy balance. As pointed out in [Mon15.00], theoretical studies of the effects of Pauli paramagnetism on the upper critical field were initiated independently by CLOGSTON [Clo62] and CHANDRASEKHAR [Cha62] and extended by MAKI [Mak64[a], Mak66] and WERTHAMER, HELFAND and HOHENBURG (WHH) [Wer66]. The sequence of events which led to the contemporary picture of the mixed state are summarized in Table 7-3 and developed further in [Mon15.00]. The goals of the numerous studies of the field and temperature dependences of mixed-state magnetization undertaken since about 1962 were to put these new extensions and generalizations of GLAG theory ("neo-GLAG" theory according to HAKE [Hak67[b]]) to the experimental test. Over the years, theory and experiment have been coupled in studies of the temperature dependences of H_{c2} and of the discrepancies between the experimentally determined and GLAG-predicted upper critical fields. This has led to a series of papers dealing with the effects on H_{c2} of Pauli paramagnetic limitation, and subsequently its partial cancellation as a consequence of a spin-orbit-scattering induced decoupling of Cooper pairs and the resulting superconductive-state Pauli paramagnetism.

TABLE 7-3 EVOLUTION OF CONTEMPORARY THEORIES OF THE PARAMAGNETIC MIXED STATE

_____

Literature and Contribution

_____

B.S. Chandrasekhar (1962) [Cha62] and A.M. Clogston (1962) [Clo62]: *Recognized the existence of a limitation on H_{c2} imposed by normal-state Pauli paramagnetism. The surviving equation is the Clogston paramagnetically limiting field given by*
$$H_{po} = 1.84\text{x}10^4\, T_c \ (Oe).$$

K. Maki (1964) [Mak64[a]]: *Expressed H_{c2} and $4\pi M$ as functions of t and the effect of normal-state Pauli paramagnetism in terms of $\kappa_1(t)$ and $\kappa_2(t)$ in the dirty limit.*

K. Maki (1966) [Mak66]: *Combined the influence of normal-state Pauli paramagnetism with strong spin-flip-scattering induced mixed-state paramagnetism into a theory for H_{c2} and $4\pi M$ in the dirty limit expressed in terms of $\kappa_1(t)$ and $\kappa_2(t)$.*

TABLE 7-3 (Continued)

N.R. Werthamer, E. Helfand, and P.C. Hohenburg (1966) [Wer66]: *Also studied the compensatory influence on Pauli limitation of mixed state spin-flip-scattering aided mixed state paramagnetism. Calculated the temperature dependence of H_{c2} in terms of a parameter $h^*(t)$ for unrestricted electron mfp specialized to the dirty limit.*

T.P. Orlando, E.J. McNiff, S. Foner, and M.R. Beasley (1979) [Orℓ79]: *Recognized that the Maki and WHH theories were incomplete in that by neglecting the effect of electron-phonon renormalization of normal-state parameters they frequently required an anomalously high spin-orbit-scattering frequency ($\lambda_{so} \gg \lambda$) to explain observed cancellations of Pauli paramagnetic limitation.*

N. Schopohl and K. Scharnberg (1981) [Sch81]: *Re-examined WHH theory itself and showed as a result that spin-orbit scattering could actually be the dominant scattering mechanism, in contrast to the restrictions that WHH had placed on their theory.*

M.R. Beasley (1982) [Bea82]: *Has offered an up-to-date review of the current status of upper-critical-field theory bringing into play such factors as the usual Pauli limitation and spin-orbit scattering as well as the electron-phonon interaction and the effect of spin fluctuations.*

Flux pinning in the paramagnetic mixed state has been studied in considerable detail by BRAND [Bra72]. Related to flux pinning is of course magnetic hysteresis, flux creep, flux jumping (hence flux-jump instability) and the use of cylinder or tube magnetization as a means of measuring critical current density; related to the *absence* of strong pinning is flux flow. All these phenomena have been considered within the context of Ti-Nb alloys.

7.9 THE UPPER CRITICAL FIELD, H_{c2}, AS A FUNCTION OF METALLURGICAL VARIABLES

7.9.1 Alloying

According to ABRIKOSOV [Abr57], the upper critical field H_{c2} is given by:

$$H_{c2}^* = \sqrt{2}\,\kappa_{GL}\,H_c \quad , \tag{7-7}$$

valid only near $t \sim 1$; where the asterisk admits, as usual, that limitation of the upper critical field through normal-state Pauli paramagnetism has not been taken into consideration. MAKI theory in the dirty limit of course gave:

$$H_{c2}^* = \sqrt{2}\,\kappa_1\,H_c \quad , \tag{7-6a}$$

valid over the entire temperature range provided the appropriate temperature dependence of κ_1 was taken into account. The thermodynamic critical field is also temperature dependent according to $H_c = H_{c0}(1-t^2)$, where H_{c0} from BCS theory [Bar57] is given by:

$$H_{c0} = 2.43\,\gamma^{1/2}\,T_c \qquad (Oe) \qquad [3-7] \tag{7-8}$$

But again, according to BCS theory and also to experiment (HULM and BLAUGHER [Huℓ61]), T_c itself scales with γ through the Fermi density-of-states, $n(E_F)$. Consequently it is to be expected that H_{c2} for a series of binary TM alloys will also mimic the common compositional variation of either γ or T_c.

The first, and most comprehensive, set of measurements of the resistive upper critical field, H_r, of binary TM alloys, comparable to the transition temperature study of 1961 by HULM and BLAUGHER [Huℓ61], was reported in 1963 by BERLINCOURT and HAKE [Ber63^a]. Fourteen cold-rolled Ti-Nb alloy specimens of compositions 10-95 at.% Nb were measured at 1.2 K using a current density, J, of 10 A cm^{-2} in a superconducting magnet capable of being pulsed to fields of up to 160 kOe. As expected from the T_c data of HULM and BLAUGHER [Huℓ61], H_r described an arch-like curve, with a pronounced maximum situated between 34 and 39 at.% Nb.

Comparable results were obtained subsequently by RALLS [Raℓ64] (see also RICKETTS [Ric69]) during the study of a series of eleven cold worked (99.94%) Ti-Nb alloys measured resistively at 4.2 K, again using $J = 10$ A cm^{-2}, as part of a critical current density investigation. In this series the maximum in H_r fell between 34 and 49 at.% Nb. The data of BERLINCOURT and HAKE [Ber63[a]] have been plotted as a function of electron/atom ratio (e/a) which, of course, is the same as the average group number. DeSORBO [Des67] on the other hand had preferred to use an "effective electron/atom ratio", N_{eff}, as defined in Sect.7.2 which refers to his work on the transition temperature, in order to display the results of critical field measurements on two Ti-Nb and numerous other binary, as well as ternary and quaternary alloys. As a size-effect-corrected quantity, N_{eff} may be regarded as an electron concentration and consequently of greater physical significance than the simple e/a. The purpose of the change from e/a to N_{eff} was to obtain a parameter with which H_{c2} (and T_c -- see above) would be more closely and generally correlated. Its use did not facilitate the addressing of any of the more important topics of the day which had to do with the temperature dependence of H_{c2} and paramagnetic effects.

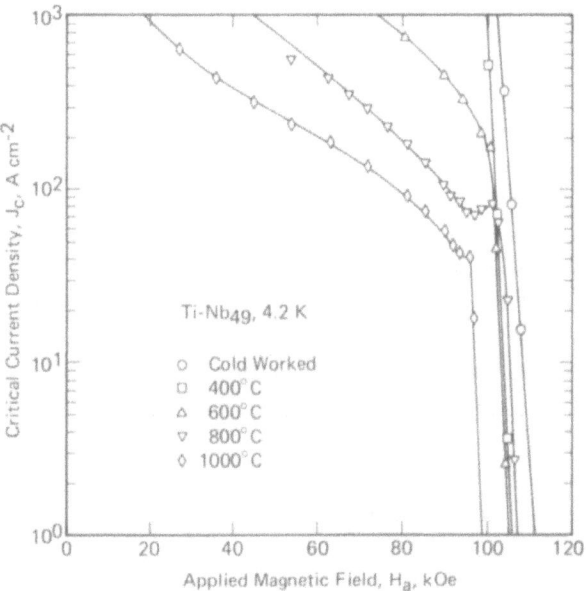

FIGURE 7-5. Critical current density at 4.2 K *versus* applied field for a Ti-Nb (49 at. %) alloy wire in the cold-swaged-and-drawn condition (99.94% reduction in area) and after subsequent final heat treatments for 1 h at the temperatures indicated (criterion: 1/4 μV cm^{-1}) — after RALLS [Raℓ64].

7.9.2 Deformation and Heat Treatment

RALLS [Raℓ64] has investigated the influence of cold work (cw) and intermediate heat treatment on H_r. The data were obtained as part of a J_c *versus* H_a experiment by increasing the applied field strength to such an extent that J_c dropped below 10 A cm^{-2}. A typical set of curves for the single-phase-β alloy Ti-65Nb, after cw and heat treatment according to 1h/(400, 600, 800 or 1000°C), Fig.7-5, supports the general conclusion that a high level of residual cold work must be present, or restored following heat treatment, if H_r is to attain its maximal value. But H_r is not H_{c2}. The possible artifactual nature of the mechanically induced change in H_r was recognized by RALLS [Raℓ64], who drew attention to the fact that H_r anisotropy has been observed in cold-rolled strip. It may be advisable in each case to examine the possibility of interpreting these effects in terms of changes in critical current density (since finite transport currents must be used) brought about as a result of changes in flux pinning. Assuming this is not the case and that $H_r \stackrel{\sim}{=} H_{c2}$ it is interesting to examine RALL's result (Fig.7-5) in the light of the above-mentioned theories of the mixed state commencing with

$$H_{c2}^* = \sqrt{2} \; \kappa_1 \; H_c \quad , \qquad (7\text{-}6a)$$

due attention being paid to the temperature dependence of κ_1 (as discussed in [Mon14.5]), the end-points of which are:

$$\kappa_1(t=0) = 1.195 \; \kappa_{GL}^d \qquad (7\text{-}9a)$$

$$\kappa_1(t=1) = \qquad \kappa_{GL}^d \qquad (7\text{-}9b)$$

A very important result arising from a combination of Eqns.(7-6a), (7-9a), (7-5) and (7-8), is the well-known relationship:

$$H_{c20}^* = 3.06 \times 10^4 \; \rho_n \; \gamma \; T_c \quad , \quad (Oe) \quad [3\text{-}13] \; (7\text{-}10)$$

and the useful conclusion that dislocations and other defects which lower the electronic mean free path and raise the normal-state resistivity will increase κ_1 and with it the upper critical field.

BERLINCOURT and HAKE [Ber63[a]] had also examined the critical current density question in connection with a discussion of the validity of the resistometric method of upper critical field measurement for Ti-TM alloys, and noted that values measured at $J < 10$ A cm^{-2} should be "approximately identifiable with H_{c2} or H_p" and "appear to represent the best compromise between available detection sensitivity and allowable perturbation of the high-field superconducting state by measuring current". The acceptability of this comment has endured rather well. In subsequent measurements of H_r in Ti-Nb and related alloys current densities of about 10 A cm^{-2} have been widely used, for example by CHANDRASEKHAR *et al* [Cha63], EL BINDARI and LITVAK [Elb63], JONES *et al* [Jon64], and SALTER [Sal66]. More recently, however, the tendency has been to employ even lower measuring-current densities such as 5 A cm^{-2} [Haw80], 1.3 A cm^{-2} [Hel71], and 1 A cm^{-2} [Wad81]. As early as 1964, RALLS [Ral64] had intercompared three H_r *versus* composition curves for Ti-Nb prepared from data taken at current densities of 10^3, 10, and 1 A cm^{-2}, respectively. At the optimum composition ($\sim$45 at.% Nb), the resistive upper critical fields $H_{r, 10 \text{ A cm}^{-2}}$ and $H_{r, 1 \text{ A cm}^{-2}}$ (of which the latter was the higher) differed by only 1 kOe.

It is interesting to note that BERLINCOURT and HAKE's studies on Ti-Nb$_{35}$ (and Ti-Mo$_{16}$) revealed that H_r was nearly independent of the amount or type of cold work introduced by rolling, and the relative orientation of the applied field with respect to the rolling-induced texture. For these alloys, presumably, cold work introduced negligible changes into already high residual resistivities. But when metallurgical changes do effect increases in the resistivity, it does not necessarily follow that corresponding increases in H_r will take place. For example, as RICKETTS has demonstrated [Ric69], it is possible for the *increases* in H_r which occur in precipitate-forming Ti-Nb alloys during 1-h heat treatments at temperatures between 250 and 600°C to be accompanied by *decreases* in ρ_n, Fig.7-6. The following explanation serves to emphasize that in relying on Eqn.(7-10) for guidance when attempting to predict the influence of mechanical or thermomechanical processing on H_r, the presence of the *composition-dependent* factors, γ and T_c should not be overlooked. During aging at temperatures near 500°C, dissolved O and dislocations migrate to the deformation-cell walls (as discussed in [Mon2.11, Mon 2.12]) and as a result the resistivity decreases; at the same time the O-assisted α-phase precipitation, which then takes place in the cell walls, raises the Nb content of the matrix, and increases both γ and T_c. Whether H_{c2} then increases or

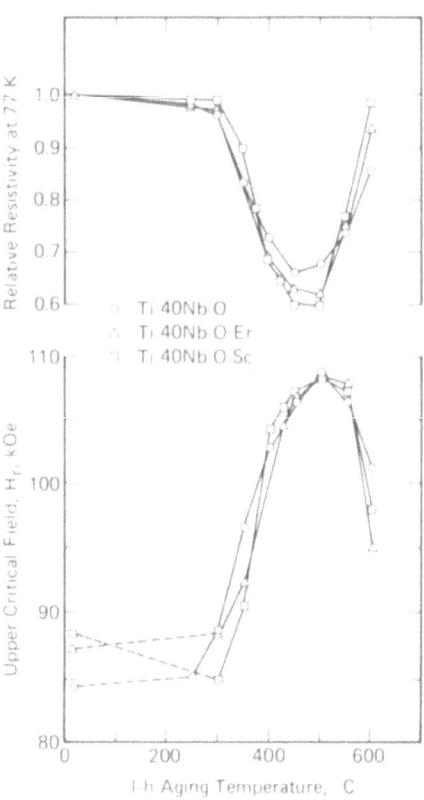

FIGURE 7-6. Variation of relative resistivity (measured at 77 K) and resistive upper critical field (H_r, kOe) with temperature of 1-h aging for a nominal Ti-40Nb alloy containing various small amounts of O, Er and Sc, viz: 2400 ppm O (○); 1.67 vol. % Er$_2$O$_3$ plus 60 ppm O (△); 1.20 vol. % Sc$_2$O$_3$ plus 1140 ppm O (□). (H_r criterion: 1/2 μV cm^{-1} at 1 mA in 10-mil wire; i.e., 4×10^{-7} Ω cm) — after RICKETTS [Ric69].

decreases depends on the outcome of the competition between ρ_n and the γT_c product; obviously in the example cited the latter is dominant. Complementary results have been obtained by SALTER [Sal65[a]] during a comprehensive study of the influence of aging on the properties of Ti-Nb$_{22}$. In accordance with the above observations, pronounced changes in H_r were encountered during heat treatment of that alloy at temperatures between 200°C and 650°C for times of between 10 min and about 50 h. From an initial 4.2 K value of 70 kOe for the β-annealed sample, H_r(4.2 K) rose to between 106 and 110 kOe after extended aging at temperatures between 350 and 500°C. As had been the case in the transition temperature study referred to in Sect.7-6, the maximum in H_r could be attributed to precipitation-induced Nb enrichment of the matrix. The variation of H_r with aging correlated positively with the Nb concentration in the β-phase, as determined by x-ray diffraction, and was compared with an

H_r *versus* composition diagram constructed from the data of BERLINCOURT and HAKE [Ber63^a] (corrected for the change in temperature from 1.2 K to 4.2 K). Excellent qualitative agreement was obtained. Nevertheless the new data points lay significantly ($\sim$10 kOe) below the projected curve, a deviation that was interpreted as the result of a proximity-effect-induced depression of the superconductive properties. The existence of such a mechanism, if operative at all, was never conclusively demonstrated; indeed the spacing between the α-phase precipitate particles was estimated to be about 10^{-4} cm, some two orders of magnitude larger than the coherence length in a concentrated transition-metal alloy -- typically about 300 Å in, for example, Ti-Nb$_{32}$ according to [Mon13.5].

The results reported in this section have confirmed that the *composition-* and *thermomechanical-processing-dependences* of H_{c2} depend on the net variation of the product of the three variables ρ_n, γ, and T_c as prescribed by Eqn.(7-10).

7.10 THE UPPER CRITICAL FIELD, H$_{c2}$, AS A FUNCTION OF TEMPERATURE

7.10.1 Early Studies

Following the appearance of the ABRIKOSOV formulation for H_{c2}^*, Eqn.(7-7), restricted to t$\sim$1, interest centered upon developing prescriptions for H_{c2} that would be valid over the entire sub-T_c temperature range. A full discussion of the topic is provided in [Mon14.1]. In the early 1960's, the results of measurements of the temperature dependences of H_{c2} in Ti-Nb and other alloys were interpreted in terms of simple extensions to GLAG theory. As pointed out by CHANDRASEKHAR *et al* [Cha63][Jon64], one such equation can be obtained by starting with Eqn.(7-7), inserting the usual $H_c(t) = H_{c0}(1 - t^2)$ for the classical temperature dependence of the thermodynamic critical field, and associating with κ_{GL} a reasonable but arbitrary temperature function of the form:†

$$\kappa_{GL}(t) = 2\ \kappa_{GL1}(1 + t^2)^{-1} \quad . \qquad (7-11)$$

The result was an upper critical field temperature dependence of the form:

$$H_{c2}^*(t) = 2\sqrt{2}\ \kappa_{GL1}\ H_{c0}\ \frac{1-t^2}{1+t^2} \quad . \qquad [3\text{-}6]\ (7\text{-}12)$$

Secondly, according to GOR'KOV:

$$H_{c2}^*(t) = \kappa_{GL1}\ H_{c0}(1.77 - 0.43t^2 + 0.07t^4)(1-t^2),$$

$$[5\text{-}4c]\ (7\text{-}13)$$

which of course reduces to:

$$H_{c20}^* = 1.77\ \kappa_{GL1}\ H_{c0} \quad \text{at t = 0} \qquad (7\text{-}14)$$

and tends, as it must, to Eqn.(7-7) when $t \cong 1$. Finally, SHAPOVAL concluded that in the dirty limit

$$H_{c20}^* = 3.03\ \kappa_{GL1}\ H_{c0} \quad , \qquad [3\text{-}15c]\ (7\text{-}15)$$

and associated this with a nearly linear temperature dependence described, according to EL BINDARI and LITVAK [Eℓb64], by:

$$H_{c2}^*(t)/H_{c20}^* \begin{cases} \cong 1-2t\ell n2, & t\ll1 \quad [3\text{-}15a]\ (7\text{-}16a) \\ \cong (1-t)8/\pi^2, & t \cong 1 \quad [3\text{-}15b]\ (7\text{-}16b) \end{cases}$$

The validity of these relationships was explored by CHANDRASEKHAR *et al* [Cha63] and also by JONES *et al* [Jon64] using the results of critical field measurements of a series of Ti-Nb(44, 68, 83, and 96.3 at.%) and other alloys. When plotted in the format $H_{c2}(t)/H_{c20}$ *versus* t as in the latter paper it became clear that: *(a)* alloys with lower critical fields lay close to the ABRIKOSOV-GINZBURG curve, *(b)* the behavior of alloys with higher H_{c2}'s approached the GOR'KOV curve, *(c)* the SHAPOVAL prediction was completely out of the range of the experimental data.† The latter

---

$\dagger$ GINZBURG-LANDAU theory was restricted to t near 1. The subscript-1 is used here to indicate the value of κ_{GL} at t = 1. In modern theory, κ_{GL} is again regarded as constant, and expressible in terms of Eqns.(7-3) through (7-5).

$\dagger$ CHANDRASEKHAR, JONES, *et al* did not specify the SHAPOVAL functional form as given in Eqns.(7-16a,b) but merely referred to it as being "nearly linear with positive second derivative".

observation refuted the conclusion reached by EL BINDARI and LITVAK [Elb64] who, on the basis of low temperature electrical resistivity and low temperature heat capacity measurements, and a single result at 4.2 K for the H_r of Ti-Nb(90 at.%), had claimed that the SHAPOVAL expression adequately described both the magnitude and temperature dependence of H_{c2}.

This experimentally-oriented discussion of early extensions to GLAG theory is concluded with reference to the work of FIETZ [Fie67, Fie67[a]] who studied the applicabilities of the MAKI [Mak64, Mak64[a], Mak65], HELFAND and WERTHAMER [Hel64], and EILENBERGER [Eil67] spin-ignoring theories of the mixed state to the critical field temperature dependences of Ti-Nb(87.5, 91, 95.5, 98.5 and 99.5 at.%) -- essentially Nb alloys doped with Ti. These nonparamagnetic theories of mixed-state temperature dependence are considered in detail in [Mon14.5] *et seq*. The results of nonparamagnetic MAKI theory may be expressed in terms of the generalized GINZBURG-LANDAU parameters $\kappa_1(t)$ and $\kappa_2(t)$ as defined in Eqns.(7-6a) and (7-6b). HELFAND and WERTHAMER expressed H_{c2} as a function of t and electron-lattice-scattering frequency parameterized by λ in terms of a $\kappa_1(t,\lambda)$ [Hel64], and subsequently in terms of a reduced parameter
$h^*(t,\lambda) \equiv H_{c2}(t,\lambda)/(-dH_{c2}/dt)_{t=1}$ [Hel66]. EILENBERGER introduced p-wave (hence anisotropic) scattering, in addition to s-wave scattering, into the description of κ_1 (hence H_{c2}) temperature dependence, which was then parameterized by ℓ_{tr}/ℓ where ℓ_{tr} then became the average mfp with ℓ the s-wave scattering component of it. FIETZ *et al* [Fie67] undertook an experimental evaluation of the predictions of HELFAND-WERTHAMER theory expressed in terms of the $h^*(t,\lambda)$, and the predictions of EILENBERGER [Eil67] expressed in terms of $\kappa_i[t,(\ell_{tr}/\ell),(\xi_0/\ell_{tr})]/\kappa_{GL}$ *versus* t. In all cases significant quantitative discrepancies between theory and experiment were detected. The possible sources of these, which although small in the dirty limit ($\lambda > 10$) were large in the clean limit ($\lambda < 1$), were discussed in terms of nonsphericity of the Fermi surface and the influence of unaccountable details of the electron-scattering process. It seemed more likely, however, that the discrepancies were a result of serious inadequacies in the theories themselves; consequently FIETZ's conclusions set the stage for the introduction of paramagnetic effects to be considered below.

7.10.2 Paramagnetic Theories of Mixed-State Temperature Dependence

The years between 1963 and 1966 saw a rapid evolution in theories of the paramagnetic mixed state during which the influence of Pauli spin paramagnetism, and its moderation by the introduction of mixed-state paramagnetism created as a consequence of the spin-orbit-scattering-induced spin-decoupling of Cooper pairs, were introduced. A review of the composition dependences and temperature dependences of H_{c2} in Ti-Nb, traces the interplay of theory and experiment during the development of models for the paramagnetic mixed state. For a detailed review of the subject [Mon15.00] is recommended. Before proceeding, the reader is reminded that an asterisk is used to denote the theoretical spin-ignoring upper critical field, while the experimental values are designated by H_r (resistive) and H_u (magnetic or calorimetric); the subscript 0 refers to zero-K.

With reference to Fig.3-3, in the spin-ignoring case, an increasing applied magnetic field increases the superconductive free-energy density, g_s, with respect to a constant but higher g_n to which it becomes equal at H_{c2}^*. Then if normal-state Pauli paramagnetism is to be taken into account, g_n is no longer constant but decreases with increasing field, reaching $g_s(H_a=0)$ at the CLOGSTON field, H_p. As the field corresponding to the intersection of an increasing $g_s(H_a)$ and a decreasing $g_n(H_a)$, H_{c2} is less than both H_{c2}^* and H_p. That either of these fields may be the higher can be seen from an inspection of the (1972) results of BRAND [Bra72] on Ti-Nb(32, 55, 62 and 83 at.%), as reported in Table 7-4.

BERLINCOURT and HAKE [Ber63[a]] were the first to undertake a comprehensive study of the resistive upper critical fields of binary TM alloys. Although in the case of Ti-V it was possible to compare the composition dependence of H_r with those of H_{c2}^* (from GLAG theory) and H_p (from CLOGSTON's formula), for Ti-Nb insufficient data were available at the time to enable H_{c2}^* to be evaluated. H_r was of course expected to fall below the lower of H_{c2}^* and H_p, and in these initial measurements on Ti-Nb, H_r certainly never exceeded H_p. In fact as shown in Fig.7-7, $H_r(1.2 K)$ appeared to be equal to H_{p0} for bcc alloys of about 30 at.% Nb, but after passing through a broad maximum

TABLE 7-4   COMPARISON OF H_{c20}^{*} (TWO METHODS[+]) AND H_{po} FOR FOUR Ti-Nb ALLOYS -- Brand [Bra72]

At.%Nb	32	55	62	83
H_{po} (kOe)	150	179	178	170
H_{c20}^{*n} (kOe)	254	192	128	68
H_{c20}^{*s} (kOe)	166	143	134	68

+ The superscripts n and s refer to H_{c2}^{*}'s being calculated from normal-state and superconducting-state parameters, respectively. For further details, see [Mon14.6].

of 145 kOe in the vicinity of 34-45 at.% Nb it rapidly dropped below H_{p0}, which tended to retain its high value in response to the rather small compositional variation of T_c exhibited by Ti-Nb in the range above 50 at.% (cf. Fig.7-3).

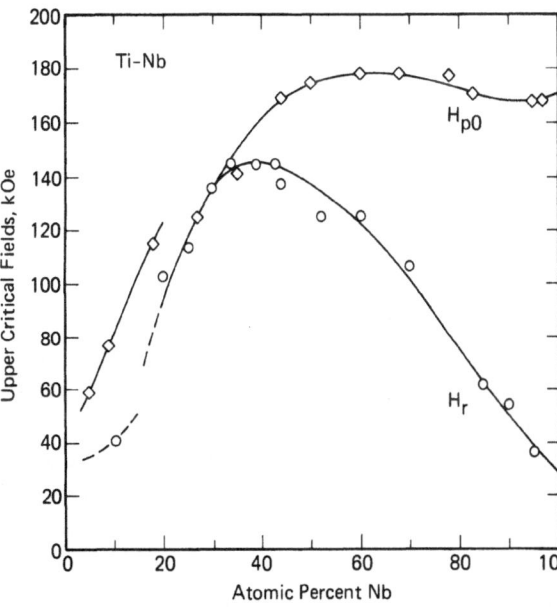

FIGURE 7-7. Resistive upper critical fields at 1.2 K for Ti-Nb alloys (H_r, transition midpoints in most cases, J = 10 A cm^{-2}) after BERLINCOURT and HAKE [Ber63^a] compared with the zero-K Clogston limiting field H_{po} = 18.4 T_c kOe (Eqn (3-14)) computed using the transition temperature results of HULM and BLAUGHER [Hul61] (cf. Fig. 7-3).

As indicated in [Mon15.2], which traces the evolution of paramagnetic theories of the mixed state, MAKI [Mak64^a] was the next to deal with the influence of normal-state paramagnetism on H_{c2} and the mixed-state magnetization, $4\pi M$, again in terms of the modified Ginzburg-Landau parameters $\kappa_1(t)$ and $\kappa_2(t)$. In that paper, which was the first generalization of GLAG theory to take into account a Pauli paramagnetic energy term, the relative importance of the paramagnetic limitation of H_{c2} was gauged by a parameter, α, given by

$$\alpha = \sqrt{2}\, H_{c20}^{*}/H_{p0} \quad , \qquad [3-17]\ (7-17)$$

associated with which was the following relationship between the paramagnetically limited (H_{c20}) and the GLAG (H_{c20}^{*}) upper critical fields:

$$H_{c20}/H_{c20}^{*} = 1/\sqrt{1 + \alpha^2} \quad , \qquad (7-18)$$

but cf. [Mon17.2]. As pointed out in Sect.3.6.3, α may be calculated from superconductive parameters (hence α_s, Eqn.(3-21)) or normal-state parameters (hence α_n, Eqn.(3-22)). To obtain it from Eqn.(7-17), a knowledge of H_{c20}^{*} is required; this may be obtained from flux-flow resistivity measurements -- to be considered in Sect.7.16.2 -- or from the relationship:

$$H_{c20}^{*} = 0.68\, T_c(-dH_u/dT)_{T_c} \quad , \qquad (7-19)$$

cf. [Mon14.6]. H_{p0} is of course given by:

$$H_{p0} = 18.4\, T_c \ \text{(kOe)} \quad . \qquad [3-14]\ (7-20)$$

In one of the first experimental tests of the new theory, SHAPIRA and NEURINGER [Sha65] obtained qualitative agreement between the resistively measured upper critical field temperature dependences of Ti-Nb(44 and 63 at.%) and the predictions of MAKI theory. But when H_r was compared with the predicted H_{c2} from Eqn.(7-18), α having been determined from Eqn.(7-17) with both calculations relying on a flux-flow resistive determination of H_{c20}^{*}, it was concluded that the simple paramagnetically-limited MAKI theory overestimated the effect of normal-state Pauli paramagnetism.

Further developments of the theory of paramagnetic limitation were aimed at improving the agreement between the experimental and calculated values of H_{c2} -- in other words, at investigating the accumulating evidence that applications of the effects of uncompensated normal-state Pauli paramagnetism were overestimating the critical-field depression. One possible source of the discrepancy had already been identified by CLOGSTON [Clo62] but was subsequently forgotten, not to be rediscovered until the late 1970's when ORLANDO *et al* [Orl79] looked into the influence on the Fermi density-of-states, hence the normal-state Pauli paramagnetic susceptibility, of electron-phonon (and electron-electron) interactions. In the meantime, attention was being focused on attempts to deal with ways of increasing the *mixed-state* paramagnetism in order to decrease the difference between the magnetic free energies of the *s* and *n* states (which in the original GLAG theory was zero). The mechanism selected by MAKI and TSUNETO [Mak64b], MAKI [Mak66], and WERTHAMER, HELFAND, and HOHENBERG (WHH) [Wer66] was a spin decoupling (but not momentum decoupling) of the Cooper pairs brought about by way of a spin-orbit-coupling-induced spin-flip-scattering mechanism characterized by a frequency-parameter, λ_{so}.

Experimental case studies of the upper critical field temperature dependences of Ti-Nb alloys, with reference to the newer theory of MAKI and the theory of WHH, were conducted by NEURINGER and SHAPIRA [Neu66] using Ti-Nb$_{44}$, and by BRAND [Bra72] using Ti-Nb(32, 55, 62 and 83 at.%). In examining the new theory of WHH, NEURINGER and SHAPIRA took the results for Ti-Nb$_{44}$ just referred to, added to them those for Ti-V$_{58}$ and Ti-Ta$_{52}$, and compared the measured upper critical field temperature dependences with predictions from WHH theory and the old (spin-orbiting excluding) theory of MAKI [Mak64a]. In fitting the WHH theory, α was calculated from Eqn.(7-17) with inputs from Eqns.(7-19) and (7-20), while λ_{so} was regarded as an adjustable parameter. In the case of Ti-Nb$_{44}$, α was found equal to 1.34, and a good fit to the data was obtained with λ_{so} = 4.5. The corresponding spin-orbit relaxation time was $\tau_{so} = 2\hbar/3\pi k_B T_c \lambda_{so} \cong 2\times10^{-14}$ sec, cf. [Mon15.6]. This value seemed reasonable in comparison with a typical transport-scattering relaxation time of about 3×10^{-15} sec for that class of alloy. Qualitatively at least, there seemed to be no doubt that a reduction of the unrestrained influence of Pauli paramagnetism was required, and that the spin-orbit-scattering mechanism was operating in the direction to accomplish this. The form of the temperature dependence of the WHH theory was satisfactory, so that

a suitable adjustment of λ_{so} always guaranteed a good fit to the experimental points. All that remained was then to pass a value judgment upon whether the value of λ_{so} so obtained was physically acceptable.

WHH themselves fitted the critical field data for Ti-Nb$_{44}$ [Sha65] to their theory. With the help of an *estimated* value of γ for substitution in Eqn.(3-22) they obtained α_n = 1.22. For fitting purposes H_r was converted to $h^*(t,\lambda_{so})$ using the normalizing parameter $(-dH_r/dt)_{t=1}$, cf. Eqn.(3-23),[+] in the manner outlined in [Mon15.4]. Best visual fit was obtained with λ_{so} = 1.5 for the adjustable parameter, also a physically acceptable value. No direct experimental verification of the full MAKI (spin + spin-orbit) theory has been undertaken using Ti-Nb as a working substance.

BRAND [Bra72] has discussed the paramagnetic mixed state of Ti-Nb(32, 55, 62 and 84 at.%) alloys with the aid of the MAKI [Mak66] and WHH [Wer66] theories. Guided by Eqns.(7-10) and (7-20), BRAND had selected this particular set of alloys in order to enable the entire limitation range, from $H_{p0} < H_{c20}^*$ (strong limitation) to $H_{p0} > H_{c20}^*$ (weak limitation), to be covered (cf. Table 7-4). Using atomic constants and physical measurables H_{p0}, H_{c20}^* and the α's were calculated, after which λ_{so} was determined by trial-and-error fitting of the WHH function to the experimental $H_u(t)$ data. A MAKI [Mak66] β-function was next computed from

$$\beta^2 = \alpha^2/1.78\lambda_{so} \qquad (7-21)$$

(which is now known to be in error by a factor of 2 [Mon15.7]) in order to determine $\kappa_2(t)/\kappa_{GL}$ for use in paramagnetic-mixed-state flux-pinning studies. The results of these calculations, and applications of the WHH fitting procedure, are exemplified by Fig.7-8 and summarized in Table 7-5 and Fig.7-9. They show that: α, whose variation is dominated by that of ρ_n in the composition range concerned [Mon5.3], increases monotonically between 83 and 32 at.% Nb following a decrease in the electronic mean-free-path; the ratio $H_{c20}^{*n}/H_{c20}^{*s}$ departs from unity as the alloy becomes dirtier; λ_{so} has no smooth variation, and with decreasing Nb content must be assigned the values ∞, 5, ∞, 6 in order that $H_{c2}^{WHH}(t)$ fit the experimental $H_u(t)$ curve. Thus two of the alloys treated in this theory

---

+ It should be mentioned that an incorrect value was
  chosen for the slope.

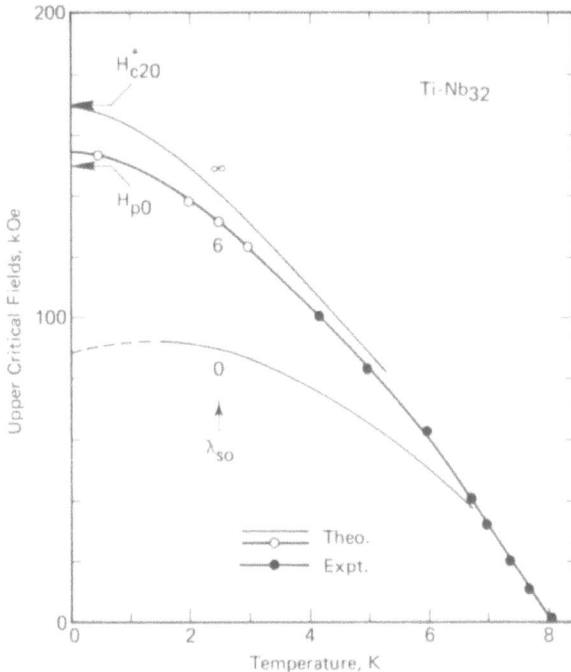

FIGURE 7-8. Theoretical and experimental upper critical fields for Ti-Nb (32 at. %) measured in the cast-and-cold-rolled (~99.5%) condition. The normal-state-calculated Maki paramagnetic limitation parameter was $\alpha_n = 1.56$; a best WHH-fit to the experimental critical fields was obtained with $\lambda_{so} = 6$, enabling them to be extrapolated to inaccessibly low temperatures. The two other theoretical curves were WHH-calculated using the λ_{so}-values indicated; the $\lambda_{so} = \infty$ calculations yield H_{c2}^* (hence H_{c20}^*) — after BRAND [Bra72].

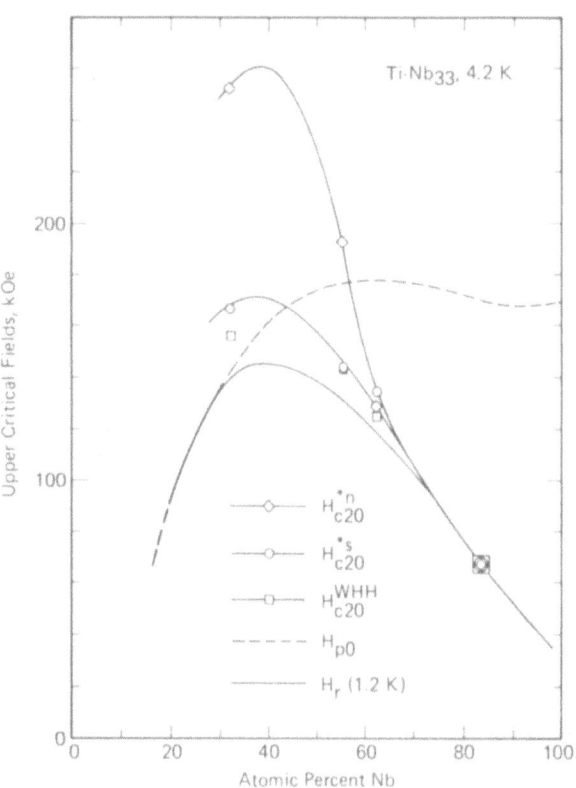

FIGURE 7-9. Theoretical (zero-K) and experimental (H_r, 1.2 K) upper critical fields as functions of composition in Ti-Nb alloys — after BRAND [Bra72] (cf. Table 7-4).

TABLE 7-5   INTERCOMPARISON OF H_{c20}^{*n}, H_{c20}^{*s}, (kOe) AND THE RESULTS OF THE WHH FITTING PROCEDURE -- Brand [Bra72]

At.%Nb	32	55	62	83
H_{c20}^{*n}	254	192	128	68
H_{c20}^{*s}	166	143	134	68
H_{c20}^{WHH}	155	143	123	68
α	1.56	1.13	1.06	0.565
λ_{so}	6	∞	5	∞
mean-free-path (Å)	4.1	6.8	10	16

(i.e. those not paramagnetically limited -- at least with respect to H_{c20}^{*s}) demanded that $\lambda_{so} = \infty$, which violated the restrictions placed on WHH, and corresponded to an unphysically rapid spin-orbit-scattering frequency. It would now be appropriate to re-examine this result in the light of the recalculation of WHH theory by SCHOPOHL and SCHARNBERG [Sch81][†] who showed, not only that spin-orbit could be the dominant scattering mechanism, but that the value of λ_{so} needed to explain an experimental critical field was significantly smaller than that required under the older theory.

7.11 CONCLUSIONS FROM MAKI-WHH THEORY

Due to the lack of spin-orbit-scattering-frequency data, it is not possible to perform *ab-initio* calculations of H_{c2} using the MAKI or WHH theories. Nevertheless they do have certain practical

---

† I am indebted to M.R. BEASLEY for drawing attention to this paper.

utility: *(i)* Having fitted $H_u(t)$ using a calculable α and adjustable λ_{so}, the critical field can be extrapolated to inaccessibly low temperatures as in Fig.7-8; in addition by allowing $\lambda_{so} \to \infty$ a value for $H_{c2}^*(t)$ can be obtained. *(ii)* Such theoretically extrapolated values of $H_{c2}(t)$ and $H_{c2}^*(t)$ are needed for a full understanding of the paramagnetic mixed state. In particular, in order to study *flux pinning* in that state, it is necessary to calculate a paramagnetic Ginzburg-Landau parameter. This can be accomplished by combining some of the results of the MAKI [Mak66] and WHH [Wer66] theories in the manner of [Bra72] as summarized in [Mon21.8].

For a given degree of Pauli limitation, the higher that H_{c2}^* is to begin with, the higher will be the attainable upper critical field, H_u. According to Eqn.(3-13), increasing ρ_n is a step in the right direction. Assuming that all formula remain valid as the electronic mfp in the normal state becomes comparable to the interatomic spacing, and with the additional assumption that the spin-orbit-scattering frequency is one half the transport scattering frequency (in terms of mfp's $\ell_{so} = 2\ell_0$), HAKE [Hak67[a]] has predicted for Ti-Nb$_{30}$ a GLAG limit of H_{c20}^* = 361 kOe and a WHH limit $H_{c2}(0, \alpha_{lim}, \lambda_{lim})$ of 310 kOe (with α_{lim} = 3.9, λ_{lim} = 24). With reference to Fig.7-8, as the spin-orbit-scattering frequency increases, and with it the superconducting state Pauli paramagnetism, H_u approaches H_{c2}^*. It was this conclusion that suggested to NEURINGER and SHAPIRA [Neu66] that the use of high-atomic-number (Z) alloys, in which spin-orbit-scattering is strong, should favor the cancellation of the influence of Pauli paramagnetic limitation. Unswerving belief in this principle has led numerous workers to seek improvements in the upper critical fields of Ti-Nb alloys through the substitution in them of heavy transition-element solutes. The influence of Hf, as a heavy-element substitute for Ti, has been investigated by HAWKSWORTH and LARBALESTIER [Haw80] and by WADA *et al* [Wad80]. Ta as a substitute for Nb was first investigated in this context by SUENAGA and RALLS [Sue68, Sue69] and subsequently by HAWKSWORTH and LARBALESTIER [Haw80]. HORIUCHI *et al* [Hor73] have also quoted the MAKI-WHH-based high-Z principle of NEURINGER and SHAPIRA [Neu66] as having provided the guidance for their choice of both Zr and Ta for substitution into the basic Ti-Nb alloy, the result of which was a Zr,Ta-doubly substituted quaternary alloy based on Ti-Nb$_{33}$, cf. Chapter 13.

But some clues as to the incompleteness of this approach have already been dropped: In experiments on Ti-Ta$_{52}$, NEURINGER and SHAPIRA found that the experimental h(t) data lay *above* the MAKI curve for α = 0 (or finite-α,λ_{so} = ∞) suggesting an unphysically high level of superconductive spin depairing, hence the operation of an additional "paramagnetic delimitation" process. Comparable examples are to be found in the critical-field results for Ti-Nb$_{55}$ and Ti-Nb$_{83}$ presented in Table 7-5.

According to ORLANDO *et al* [Orℓ79] an electron-phonon renormalization of all the relevant density-of-states related quantities must be taken into consideration. In particular, if λ is the usual electron-phonon coupling constant, it turns out that

$$H_{p0} = \eta \, (1+\lambda)^n \, 18.4 \, T_c \qquad (\text{kOe}), \qquad (7\text{-}22)$$

where $n = \frac{1}{2}$ for a first-order transition and n = 1 for the more usual second-order transition, and in which η is the strong-coupling correction to H_c (equal to unity for a weak-coupled superconductor). Comparing Eqns.(7-20) and (7-22), it can be appreciated that strong-coupled superconductors with high electron-phonon coupling strengths do not suffer as much from Pauli limitation as had previously been believed, and as a consequence call for less spin depairing to account for the observed critical fields. To this conclusion should be coupled the observation made by SCHOPOHL and SCHARNBERG [Sch81], following a recalculation of WHH theory, Sect.7.10, to the effect that in any case less spin-orbit scattering than the WHH-predicted amount seems now to be needed to explain a given measured H_{c2}.

7.12 THE LOWER CRITICAL FIELD, H_{c1}

Eqn.(7-6c) expresses the MAKI dirty-limit lower critical field in terms of κ_3 and H_c. Next by substituting κ_{30} = 1.53 κ_{GL}^d, as explained in [Mon14.8], and after converting κ_{GL}^d to atomic constants according to Eqn.(7-5), followed by the use of Eqn.(7-10) to eliminate γ, and Eqn.(7-8) to remove H_{c0}, it is possible to express H_{c10} (dirty limit) in the form:

$$H_{c10} = 7.50 \times 10^{-5} \, (T_c/\rho_n) \ell n (4.30 \times 10^3 \, H_{c20}^* \rho_n / T_c) \qquad (7\text{-}23)$$

where ρ_n is in Ω cm, and the fields are in Oe. This

TABLE 7-6 MEASURED AND CALCULATED LOWER CRITICAL
FIELDS AND OTHER PROPERTIES OF SOME Ti-Nb
ALLOYS -- Lubell and Kernohan [Lub71].

At.% Nb	T_c K	ρ_n 10^{-6} Ω cm	κ_{GL}	H_{c20}^* kOe	H_{c10} (expt.) Oe	H_{c10} (theo.) Oe
75	9.93	35.5	25	90.5	350	151
90	9.61	12.1	9	35	500	310
95	9.41	6.4	5	18	675	433

equation was developed by LUBELL and KERNOHAN [Lub71] in order to permit values of H_{c1} to be obtained from atomic quantities and simple measurables, rather than by direct measurement with its attendant uncertainty as to the exact location of the transition from the MEISSNER to the mixed state. In order to test it, they conducted a series of studies on the alloys Ti-Nb(75, 90, 95 at.%) whose Ginzburg-Landau parameters, κ_{GL}, went from 24 to 5. In carrying out the investigation, ρ_n was measured at 4.2 K in the presence of a field $H_a > H_r$, T_c and $H_{c1}(T)$ were measured magnetically and H_{c10} was obtained by extrapolation; finally H_{c20}^*, the non-limited GLAG upper critical field was determined from flux-flow resistivity as discussed in Sect.7.16.2. The results of the study, as summarized in Table 7-6, showed the MAKI-predicted values of H_{c10} to be some 36~57% below the measured values. This discrepancy between theory and experiment could perhaps have been ascribed to the inadequate dirtiness of the alloys measured ($\kappa_{GL} < 24$) except that, as pointed out by LUBELL and KERNOHAN [Lub71], other authors had noted discrepancies even in high-kappa ($\kappa_{GL} \sim 60$) material. The result led to the conclusion that additional refinements to GLAG lower-critical-field theory, including perhaps paramagnetic effects, were needed if non-adjustable-parameter predictions of the experimental results were to be obtained.

7.13 THE ROLE OF Ti-Nb ALLOYS IN THE FORMULATION OF MACROSCOPIC MODELS OF THE MIXED STATE

In introducing a discussion of flux pinning, based on the results of critical current density measurements, BRAND [Bra72] has offered a detailed description of the microscopic and macroscopic properties of the paramagnetic mixed state. Investigated was a series of Ti-V alloys, and a series of Ti-Nb alloys with 32, 55, 62, and 83 at.% Nb. It is well known by now that the ABRIKOSOV flux lattice is restrained in place by pinning forces, against possible dislodgement by transport-current or flux-gradient-instigated Lorentz forces. In a magnetized superconductor the flux lattice is always just on the point of breaking away from its pinning sites -- i.e. it exists in a "critical state". The properties of this state are discussed in detail in [Mon16.00] and elsewhere. By medium of the common pinning mechanism the critical-state magnetization is directly related to the critical current density, the consequences of which are explored in [Mon19.00]. With time, or under the influence of a small Lorentz force, it is believed that flux in the form of clusters or bundles will "creep" out of the sample. The status of flux creep and the theories that describe it are discussed in [Mon 18.5] through [18.10]. If the pinning is especially weak, as in a well-annealed sample, or alternatively if the flux-lattice itself is "soft" as a consequence of the applied field's being close to H_{c2}, the flux will "flow" under the influence of a Lorentz force. Finally if H_a is moderate, and the flux-lattice strongly pinned, the application of a sufficiently large Lorentz force will cause a catastrophic disintegration of the lattice -- i.e. a "flux jump". The analogy between these properties of a flux lattice and the creep, plastic flow, and brittle failure, respectively, of crystal lattices is discussed in [Mon18.2]. Ti-Nb alloys have played roles in the study of all of these properties as indicated by the following catalog of early investigations:

Critical State: DIETRICH and WEYL [Die64], Ti-Nb$_{20}$; HECHT and HARPER [Hec65] and GANDOLFO and HARPER [Gan66[a]], Ti-Nb$_{33}$; WITCOMB and NARLIKAR [Wit72], Ti-Nb$_{60}$.

Flux Creep: GANDOLFO [Gan67], Ti-Nb$_{33}$.

Flux Flow: SHAPIRA and NEURINGER [Sha65], Ti-Nb$_{44,63}$; KROEGER [Kro66], Ti-Nb$_{75}$; KIM *et al* [Kim65], Ti-Nb$_{90}$.

Flux Jumping: HECHT and HARPER [Hec65] and GANDOLFO *et al* [Gan66^a, Gan69], Ti-Nb$_{33}$; KROEGER [Kro66], Ti-Nb$_{75}$.

7.14 STATIC MAGNETIZATION AND THE CRITICAL STATE

7.14.1 Magnetization in the Mixed State

Prior to the common acceptance of the ABRIKOSOV model, what is now referred to as the mixed state was regarded as a spongy (MENDELSSOHN [Men35, Men64]), filamentary (BERLINCOURT and HAKE [Ber63^a]), or laminar (GOODMAN [Goo62]) arrangement of superconducting and non-superconducting components. Magnetization of a superconductor was envisioned in terms of the establishment of macroscopic circulating currents, and in phenomenological theory it may still be viewed in this light, e.g. [Mon22.1]. Indeed, in the early studies of cold-worked Ti-Nb$_{20}$ by DIETRICH and WEYL [Die64], magnetic irreversibility and residual magnetization were regarded as manifestations of such circulatory currents, trapped in the multiply-connected superconducting mesh of a sponge-like structure.

In several classical papers, e.g. [Hak67^b], HAKE has studied the microscopic and macroscopic magnetic properties of the mixed state, but largely with reference to the alloys Ti-V, Ti-V-Cr and Ti-Mo. It is important to note at the outset, that magnetic irreversibility in such systems is unrelated to the size of the Ginzburg-Landau parameter κ_{GL} in spite of the relationship between this and solute content which, under suitable conditions, may lead to precipitation and its attendant flux-pinning properties. Indeed as pointed out by HAKE [Hak67^b], magnetic *reversibility* is quite possible in extremely dirty (so-called "extreme") type-II superconductors with κ_{GL} values in the range 30-100.

Today, of course, magnetic hysteresis is recognized to be the result of the presence of flux-pinning microstructures. Since these also favor high critical current densities, the study of magnetization made a useful adjunct to the usual potentiometric measurements in investigations of the metallurgical and superconductive properties of what, by analogy with ferromagnetic materials, began to be referred to as "hard"

superconductors. For example, BAKER and SUTTON [Bak69] conducted an extensive series of magnetization measurements on Ti-Nb$_{20}$ in conjunction with a detailed study of the effects of thermomechanical processing on its flux-pinning microstructure, hence the critical current density. Subsequently, WITCOMB and NARLIKAR [Wit72] for similar reasons (although without reporting the results of any actual J_c measurements) investigated the effects of various heat treatments on magnetic irreversibility in 90%-cold-worked Ti-Nb$_{60}$. During this research it was noted that annealing at 300°C produced a marked increase in the magnetic hysteresis. This became even more pronounced when the annealing temperature was raised to 500°C, the microstructural result of which was to produce, from heavily cold-worked tangles, a well-defined dislocation cell structure, as evidenced by the electron micrographs (at 76,000X) which accompanied the discussion of the results. The optimal annealing temperature, from the standpoints of hysteresis and remanent magnetization, was ascertained to be 500°C; the optimal annealing time at that temperature was about 1 h.

7.14.2 Magnetization and Critical Current

An important model of type-II magnetization is that authored by BEAN [Bea64]. A phenomenological theory, it regarded the superconductor as continuous and homogenous at all levels. Yet, conceptually, a material with a filamentary structure was implied, such that during the penetration of magnetic flux, successive layers of filaments carried supercurrents of critical density. In the original model, flux entering the superconductor, in response to a field applied parallel to its surface, did so with a linearly decreasing flux density and generated a field-independent current of density J_c wherever there was flux. The BEAN model, which associated the assumption J_c = constant with the usual Maxwell relation:

$$\nabla \times \vec{H}_a = (4\pi/10)\vec{J}_c \quad , \qquad (7\text{-}24)$$

is discussed in [Mon16.4] and the standard texts. A special one-dimensional case of Eqn.(7-24) predicts that a field H^* applied parallel to the surface will just penetrate the wall of a superconducting slab of thickness, w, provided that

$$H^*/w = -(4\pi/10)\vec{J}_c \quad . \qquad (7\text{-}24a)$$

It also follows from considerations of this kind that by measuring the actual field gradient across a superconductor immersed in a magnetic field, its J_c can, in principle, be directly obtained [Mon19.00].

KIM, HEMPSTEAD, and STRNAD [Kim62] introduced the long hollow cylinder as an experimental sample configuration well suited for the study of superconductive field gradients and their associated critical currents. Eqns.(7-24) and (7-24a) predict a field-gradient in the wall given by:

$$dH(B_z)/dr = -(4\pi/10)J_{c\theta} \qquad (7\text{-}25a)$$

$$\text{i.e.} \quad dB_z/dr = -(4\pi/10)\mu_{eq}J_{c\theta} \quad , \qquad (7\text{-}25b)$$

where $H(B_z)$ is the field needed to produce an induction B_z reversibly within the sample, and μ_{eq} is the differential permeability.[†] But in order to take into account the fact that J_c *increases* with decreasing field, KIM *et al* [Kim62] proposed that an average circulating current, $\langle J_c \rangle$, in the cylinder of wall-thickness w, should be related to an average internal field, $\langle B \rangle$, cf. [Mon16.6], according to:

$$\langle J_c \rangle = \frac{\alpha(T)}{b_0 + \langle B \rangle} \qquad (7\text{-}26)$$

where $\alpha(T)$ and b_0 are empirical constants. Combining this with the Maxwell relation in the form of Eqn.(7-25) yields:

$$(H'-H_a)/w = \pm k|\langle J_c \rangle| \qquad (7\text{-}27)$$

where k replaces $4\pi/10$ and the '-' sign is used when $H_a > H' > 0$ and the '+' sign when $H' > H_a > 0$, H' being the field at the surface of the bore of the cylinder. The coupling of Eqns.(7-26) and (7-27) then gives:

$$(H'-H_a)^{-1} = \pm \frac{b_0 + \langle B \rangle}{k\,w\,\alpha(T)} \qquad (7\text{-}28)$$

showing that a plot of $(H'=H_a)^{-1}$ *versus* $\langle B \rangle$ is linear with slope proportional to $1/\alpha(T)$ and with intercept proportional to $b_0/\alpha(T)$. Provided with such information, it is possible with the aid of Eqn.(7-26) to compute J_c as a function of the average field $\langle B \rangle$.

GANDOLFO and HARPER [Gan66[a], Gan67] have employed the KIM static tube-magnetization technique (the "type-A" procedure of [Mon18.4]) to study the critical current density of a commercial Ti-Nb$_{33}$ alloy in two metallurgical conditions: warm forged and cold rolled, and that plus aged 2h/400°C. The results are given in Fig.7-10. GANDOLFO and HARPER [Gan66[a]] also studied the influence of interior- and exterior-surface Cu plating (thickness $\sim 2\times10^{-3}$ cm) on the magnitude and stability of the critical current densities of the two Ti-Nb$_{33}$ alloy samples referred to above. Tube-magnetization experiments were performed under the static ("type-A") conditions just described, and also with flux jumping (to be considered below). KROEGER [Kro66] has performed a static tube-magnetization experiment on annealed Ti-Nb$_{75}$ as well as some Zr-Nb alloys in various conditions.

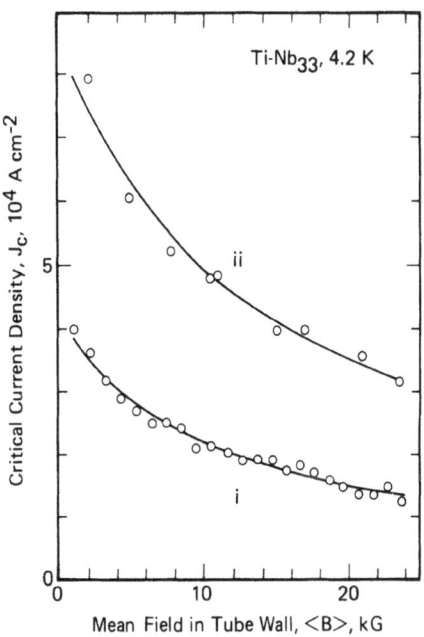

FIGURE 7-10. Results of a static tube magnetization measurements of critical current densities as functions of the mean field $\langle B \rangle$ in the sample wall for Ti-Nb (33 at. %) in two conditions, viz: (i) cast, 300°C forged, and cold rolled ($\sim$ 70%); (ii) above plus aged 2h/400°C. J_c (4.2 K) is seen to be well fitted in each case to $\alpha/(b_0 + \langle B \rangle)$ with: (i) $\alpha = 0.47 \times 10^6$ kG A cm^{-2}; $b_0 = 11.5$ kG; (ii) $\alpha = 1.10 \times 10^6$ kG A cm^{-2}; $b_0 = 12.0$ kG — after GANDOLFO [Gan67].

7.15 FLUX CREEP

Having defined the critical state in terms of a flux lattice, an array of pinning sites, and an equilibrium between Lorentz and pinning forces, the question of long-term stability inevitably arises. The first experimental exploration of the durability of

---

† In this context, μ_{eq} can be put equal to 1.

the critical state was also conducted by KIM, HEMP-STEAD, and STRNAD [Kim62]. Although the time-dependence of critical-state decay has been subjected to more detailed study in recent years by BEASLEY *et al* [Bea69], the essential features of the original observations have remained valid, viz: *(a)* The trapped flux gradient, or "persistent current", decreases logarithmically with time. Fig.7-11, for example, a typical set of results representative of the relaxation of the shielding critical state, shows a trapped field, initially of 4 kG, decaying at the rate of 1.5 G per decade of time. *(b)* Although the rate of decay is measurable near the critical point, the decay from *subcritical* states takes place immeasurably slowly. According to KIM *et al* [Kim62], from a starting trapped current density of 90% of the critical value, seven years are required for a decay of one part in 10^4.

A model of thermally activated creep was devised by ANDERSEN [And62] to explain these two related experimental observations of KIM, HEMPSTEAD, and STRNAD [Kim62, Kim63], viz: *(a)* the temperature dependence of the pinning-force parameter $\alpha(T)$; *(b)* the measured logarithmic decay of the critical state at constant temperature. Recent concerns over the validity of the ANDERSON model are expressed in [Mon18.1]. Underlying this picture of flux creep as the thermally activated negotiation of pinning barriers, is the idea of creep as a diffusional process. With this in mind, LUBELL and WIPF [Lub66] have applied the usual equations of cylindrical heat diffusivity to a description of field change during the transition between successive critical states following a stepwise perturbation of the

external field (an experiment referred to in [Mon18.4] as "type-B" tube magnetization).

Potentiometric evidence for flux creep has also been claimed as a result of four-terminal measurements of the foot of the current-voltage characteristic where sample voltage is just beginning to be detected (cf. [Mon18.7], [Mon23.6]). In investigating this regime using Zr-Nb$_{75}$ wire, KIM *et al* [Kim63, Kim64] found creep resistivities of typically 5×10^{-12} to 5×10^{-13} Ω cm. GANDOLFO *et al* [Gan66^a, Gan67], as a result of step-pulse tube magnetization studies of Ti-Nb$_{33}$, estimated them to lie within the range of 10^{-10} to 10^{-12} Ω cm. Thus by all accounts creep resistivities are seen to span the detectability threshold (10^{-12} Ω cm) inherent in the usual four-terminal critical-current-density experiment.

7.16 FLUX FLOW AND FLUX JUMPING

Flux flow and jumping under the influence of the "magnetic pressure" associated with a field gradient are discussed in [Mon18.12, Mon18.14] and in the presence of a current-induced Lorentz force in [Mon23.7, Mon23.14].

7.16.1 Magnetic Studies of Flux Flow

Under the conditions outlined in Sect.7.13, if the individual-fluxoid driving force, f_ϕ, exceeds the elementary pinning force, f_p, flux is able to flow at a viscosity-controlled velocity, v_f, given by:

$$\eta \vec{v}_f = \vec{f}_p + \vec{f}_\phi \quad . \quad (7\text{-}29)$$

With the aid of a suitably designed tube-magnetization experiment (in particular the "type-C" procedure of [Mon18.4]) direct measurements of $\vec{v}_f$ and η are possible. In the arrangement selected by KROEGER [Kro66], a pickup coil wound on the outer surface of the tube and another positioned inside it were able to detect during the steady increase of the applied field, H_a, the first penetration of flux into the cylinder (at $H_a = H_{c1}$) followed by its arrival at the inner surface, by which time H_a had risen to a value H^{**}. Signals from these coils applied to an oscilloscope enabled the transit time, hence velocity, of the flux front to be measured.

The equations of force and flux motion, when solved in the manner described in [Mon18.12], lead to

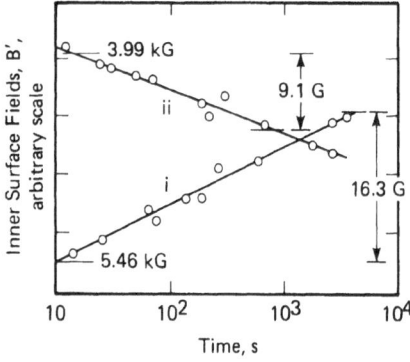

FIGURE 7-11. Decay of flux density gradients within the wall of a magnetized superconducting Zr-Nb (75 at. %) tube: (i) represents the decay of the "shielding" magnetization; (ii) represents that of "trapped" magnetization. A few representative data points have been inserted to indicate that the curves are experimental — after KIM *et al* [Kim62].

the relationship

$$\frac{H^{**}-H_{c1}}{w} = \frac{4\pi}{\Phi_0}(\eta v_f + f_p) \qquad (\text{Oe cm}^{-1}) \qquad (7\text{-}30)$$

where Φ_0 is the flux quantum (2.07×10^{-7} G cm^2).
Eqn.(7-30) predicts that a plot of H^{**} versus v_f, both
of which are measurable, will be linear with slope
$4\pi w\eta/\Phi_0$ and intercept $(4\pi wf_p/\Phi_0) + H_{c1}$. An example of
such a plot for a weakly pinned (i.e. 2h/1250°C/
quenched) sample of Ti-Nb$_{75}$ is to be seen in Fig.7-12.
Finally, according to the argument advanced in
[Mon18.13], a magnetically-derived value of flux-flow
resistivity can be obtained from the equation:

$$\rho_f = 10^{-9}\frac{B_z\Phi_0}{\eta} \qquad (\Omega\ \text{cm}) \qquad (7\text{-}31)$$

where B_z is some mean longitudinal field.

Example: From the tube of weakly pinned Ti-Nb$_{75}$ referred to above the following data were acquired:

H_{c1}	570 ± 20	Oe
w	3.2	mm
slope	9.2	Oe s cm^{-1}
η	4.7×10^{-7}	Oe G s
Field selected for ρ_f estimation, B_z	35	kG
$\rho_f = 10^{-9} B_z\Phi_0/\eta$	15	$\mu\Omega$ cm

Corollary: From the lower critical field, H_{c1}, and the
expression for the intercept given above, KROEGER
[Kro66] was also able to extract values for the elementary pinning strength, f_p.

7.16.2 Transport Studies of Flux Flow

Flux may also be induced to flow in response to
the transverse Lorentz force which appears in the presence of a longitudinal transport current. Recognizing
that this combination of flux-motion and transport-
current vectors equates to an electric field in the
current direction, it is apparent that a resistance-
like dissipation accompanies flux flow. Flux-flow
resistivity, as a transport property, was first
studied and discussed in a series of papers by KIM
et al [Kim65, Kim67, Kim69] during which the macroscopic expression for the flow resistivity, presented

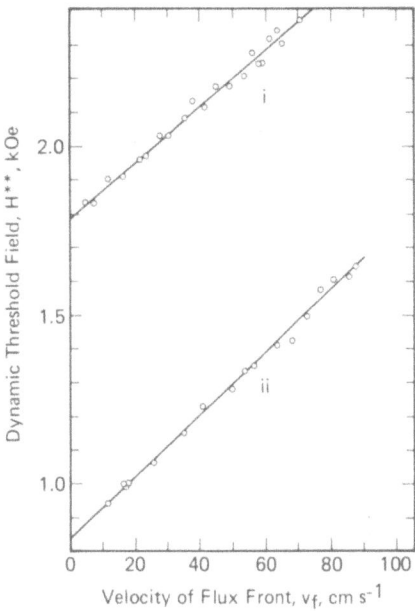

FIGURE 7-12. Plot of the dynamic full-penetration-threshold applied field H^{**} *versus* flux-front velocity, v_f, for two annealed-
and-quenched alloy samples: (i) Zr-Nb (75 at. %); (ii) Ti-Nb (75
at. %) — after KROEGER [Kro66].

above as Eqn.(7-31) was developed. Their second
approach was couched in terms of the microscopic
properties of the mixed state. According to this
argument, the fluxoids in the act of moving across the
specimen, compel the transport current to encounter
the normal regions presented by the fluxoid cores.
The flux-flow resistivity, ρ_f, is thus ρ_n modified by
a factor representing the relative volumes of the
normal and superconducting components of the mixed
state. This simple picture, considered in more detail
in [Mon23.8], led to the following expression for the
flux-flow resistivity:

$$\rho_f/\rho_n = H_a/H_{c20}^* \qquad . \qquad [3\text{-}28b]\ (7\text{-}32)$$

Eqn.(7-32) was first discovered experimentally by KIM,
HEMPSTEAD, and STRNAD [Kim65]; its interpretation in
terms of current-sharing between a superconducting
matrix and the normal fluxoid cores was substantiated
by the subsequent theoretical work of BARDEEN and
STEPHEN [Bar65].

Transport-current induced flux flow in Ti-Nb
alloys was extensively studied by GAUSTER, KROEGER,
LUBELL, and others at the Oak Ridge National Laboratory. A series of published reports provides a detailed description of that work, which was directed

towards an understanding of superconductor stabilization. In tracing the progress of the Oak Ridge flux-flow investigations, the final report for the year 1969 [Gau70] makes a useful point of entry. If in performing a critical-current experiment, the circuit current is increased beyond the critical value, I_c, it is possible for a weakly pinned sample to enter a flux-flux state and begin to dissipate energy. Upon further increase of circuit current, the sample will eventually quench at some current value, I_t, the take-off current, and enter the normal state. Provided it remains intact, the sample will return to the normal state after the circuit current has been reduced to I_r, the recovery current. In order to investigate the properties of current take-off and recovery, and the influence of heat transfer to a liquid He bath on them, GAUSTER *et al* [Gau67] conducted a series of model ex-experiments using a number of heat treated ($2\frac{1}{2}$h/1200°C/quenched) and cold-rolled bare Ti-Nb alloy ribbons whose high Nb contents (60-95 at.% Nb) guaranteed extensive flux-flow regimes. A set of I-V characteristics for the bare Ti-Nb$_{75}$ ribbon, 3.3 x 0.117 mm^2,

measured with the applied field normal to the rolling plane, is presented in Fig.7-13. The absence of any Cu stabilizer eliminated any possible confusion between flux flow and current sharing which, if present in a composite conductor, could have yielded similar I-V characteristics. Data extracted from these curves in the form of take-off and recovery powers, $P_t = I_t V_t$ and $P_n = I_n V_r$, respectively, explicable in terms of nucleate and film boiling at the conductor surface, the transition between them and the stability properties of the surface-cooled superconductor, are discussed in [Mon23.15].

Returning to flux-flow resistivity itself, a typical result -- for a sample of bare annealed (2h/1400°C) Ti-Nb$_{75}$ which yields an experimental value of H_{c20}^* = 90.5 kOe -- is given in Fig.7-14.

7.16.3 Flux Jumping

The pinning of the flux lattice by defects and precipitates of appropriate size is analogous to the

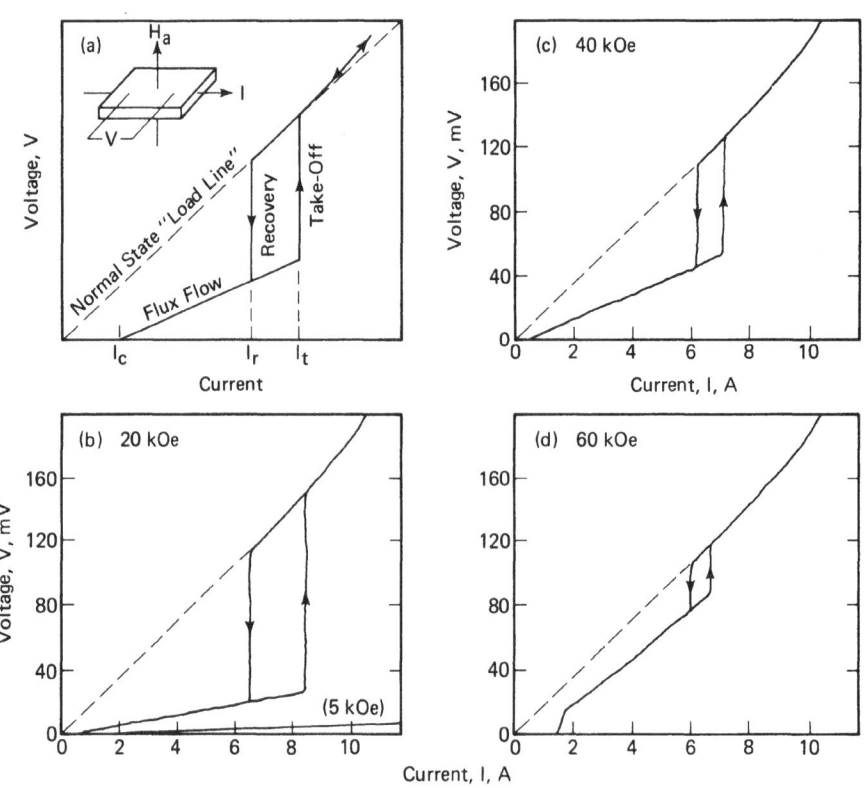

FIGURE 7-13. Current-voltage characteristics of a bare cold-rolled Ti-Nb (75 at. %) ribbon in various transverse fields applied normal to the rolling plane (the weakest pinning attitude). The absence of Cu cladding eliminates possible confusion with current sharing. The extensive flux-flow regimes are terminated by discontinuous switching to the normal state followed, upon reduction of the circuit current, by a hysteretic discontinuous return to the superconducting state — after GAUSTER *et al* [Gau67].

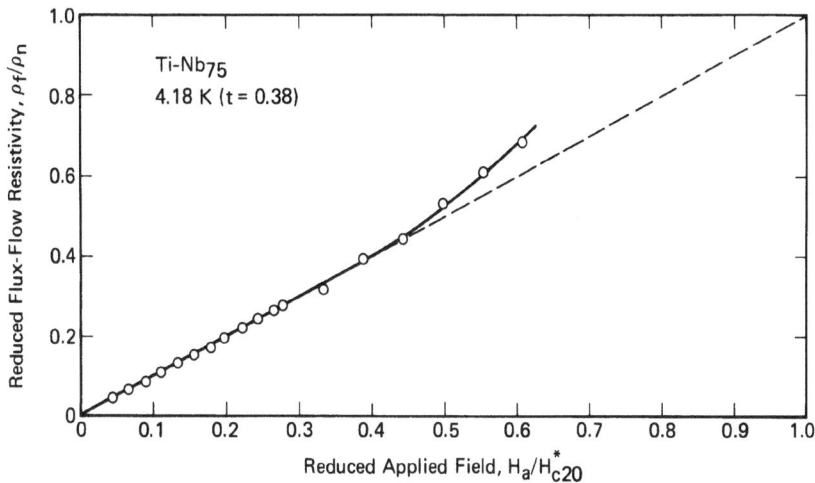

FIGURE 7-14. Reduced flux-flow resistivity ρ_f/ρ_n *versus* applied field normalized to H^*_{c20} = 90.5 kOe for a vacuum-annealed (2h/1400°C) 0.77 mmϕ Ti-Nb (75 at. %) wire drawn from a 2h/1250°C/quenched ingot. H^*_{c20} had been computed from Eqn (7-32) after ρ_f = dV/dI had been measured from the slope of the linear portion of the I-V curve taken at constant H_a and bath temperature (cf. Fig. 7-13) — after LUBELL and KERNOHAN [Lub71].

strengthening of a metal through the pinning of dislocations by comparable departures from regular crystallinity. In response to extreme additions of such strengthening agents, a formerly ductile alloy becomes brittle, and instead of extending plastically under tensile stress, undergoes brittle fracture with little elongation. Such a catastrophic change of state also occurs in the flux lattice as a result of extreme flux pinning. Strong bulk pinning will resist flux creep, but as the Lorentz force builds up, a flux jump will eventually take place, perhaps liberating sufficient heat to drive the superconductor locally into the normal state. Early studies of flux jumping in Ti-Nb$_{33}$ were carried out by HECHT and HARPER [Hec65] using a tubular sample (length, 31.8 mm; O.D., 6.4 mm; wall thickness, 1.3 mm) machined from the commercial (Westinghouse) alloy HI-120. The results obtained on the as-received (∿75% cold-worked) sample at 6.2 K at an applied-field sweep rate of 100 Oe s^{-1} are shown in Fig.7-15. As the applied field, H_a, directed parallel to the axis of the tube, is slowly increased from zero, the tube interior remains shielded up to a value H^*, corresponding to flux penetration of the entire wall. H_a and B' (the inner-surface field) then increase together in the usual way up to the value H_{fj} where the tube, or perhaps a portion of it, becomes transparent to the applied field. HECHT and HARPER [Hec65] noted that no flux jumping would take place at sweep rates of 1 Oe s^{-1} even at temperatures near 4.2 K. After the sample had been precipitation heat treated ($2\frac{1}{2}$h/400°C), although its J_c "increased five-fold", it

showed a much greater tendency to flux jump as a consequence of the development of a dense array of flux pinning centers and their support of a steeper critical vortex-density gradient. An increase in critical current density had thereby been achieved at the cost of a reduction in stability. Flux-jump instability of course ceases to be a problem if the diameter of the superconductor, no matter how well pinned, is made sufficiently small [Mon25.7].

Again, referring to Fig.7-15, as soon as the tube recovers from the "quench" or "flux-jump", it again shields the bore from the applied field until the quench repeats itself. If the effect is represented

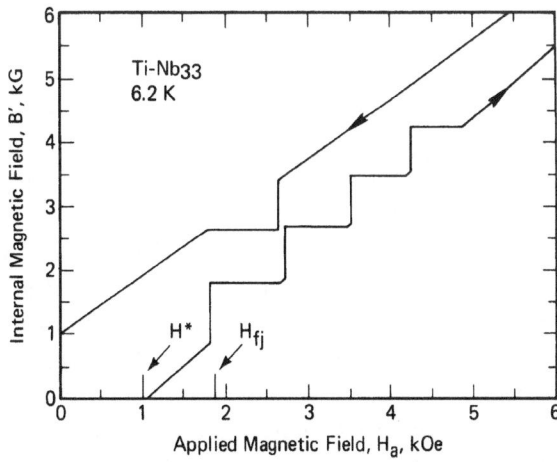

FIGURE 7-15. Slow-rate magnetization (100 Oe sec^{-1}) of a cold-worked (∼ 75%) Ti-Nb (33 at. %) tube (1.25 in. long; 0.25 in. O.D.; 0.050 in. wall) at 6.2 K — after HECHT and HARPER [Hec65].

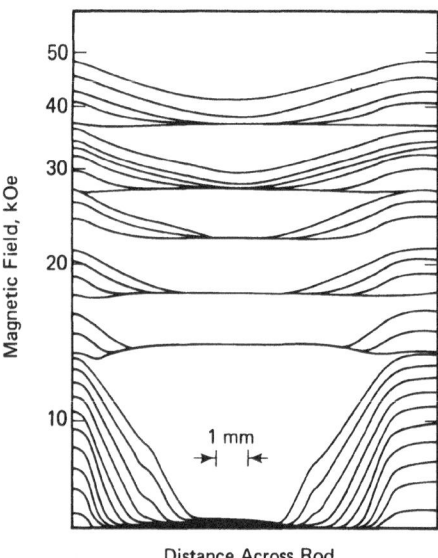

FIGURE 7-16. Recorder tracings of the field distribution across a Ti-40Nb cylinder obtained using moving Hall probes. The applied field was increased at the rate of about 5 Oe s^{-1} between successive traces. Five flux jumps are indicated. Note the deeper penetration of the field after each jump, and flux flow after the last two jumps — after COFFEY [Cof67].

as a succession of critical states as in Fig.7-16, a field-penetration diagram for Ti-40Nb due to COFFEY [Cof67], it appears as if quenching occurs periodically with increasing applied field strength. Indeed as numerous authors have discovered (see [Swa68] for references) the spacing between successive flux jumps has turned out to be approximately constant at a given temperature and applied-field sweep rate. Upon variation of these parameters, however, it has been noticed that this spacing is no longer constant but maximizes at temperatures intermediate between zero and T_c, and at constant temperature decreases with increasing sweep rate.

Tube-magnetization experiments were the historical starting points of studies of flux-jump stability. The classical papers of WIPF and LUBELL [Wip65], WIPF [Wip67], and SWARTZ and BEAN [Swa68] have been considered in detail in [Mon18.15] *et seq*; therein the idea of a stability cycle was introduced and an equation, expressing the conditions under which a small thermal disturbance, ∆T, would be self-extinguishing, was developed. As a consequence of such an analysis, it became apparent that a superconducting rod would be "intrinsically" flux-jump stable provided it was sufficiently thin. Flux-jump stability in all its ramifications has been fully discussed in [Mon25.6] through [Mon25.20]. Intrinsic stability is achieved by fila-

mentary subdivision, plus twisting, in order to mutually decouple the resulting filaments when they are imbedded in a conducting matrix; "dynamic stability" is achieved by the presence of such a matrix, usually OFHC Cu.

This brief review of flux jumping in Ti-Nb is concluded by recalling (cf. Sect.7.14.2) that GANDOLFO and HARPER [Gan66[a]] have investigated the influence of Cu plating on the magnetization of a Ti-Nb$_{33}$ alloy tube, and CHIKABA [Chi70] has studied the effects of thermal insulation on flux jumping in rods of Ti-50Nb. IWASA [Iwa69] has measured the magnetizations, as functions of applied field, of three classes of Cu/Ti-Nb (3:1) composite wire, *(a)* single filament, *(b)* 131-filament, *(c)* 131-filament, twisted to a pitch of 8.5 mm. From observations of the relative sizes of the various magnetization loops and their responses (or otherwise) to sharp applied field pulses, Fig.7-17, it was possible to deduce that not only was the twisted multifilamentary wire the most stable, but it also possessed the smallest hysteresis and consequently was the most suitable for AC applications.

7.17 PHENOMENOLOGICAL STUDIES OF H_{c2}

7.17.1 The Significance of H_{c2} in Technical Superconductivity

The need for high magnetic fields, in order to achieve dense plasma confinement in fusion reactor

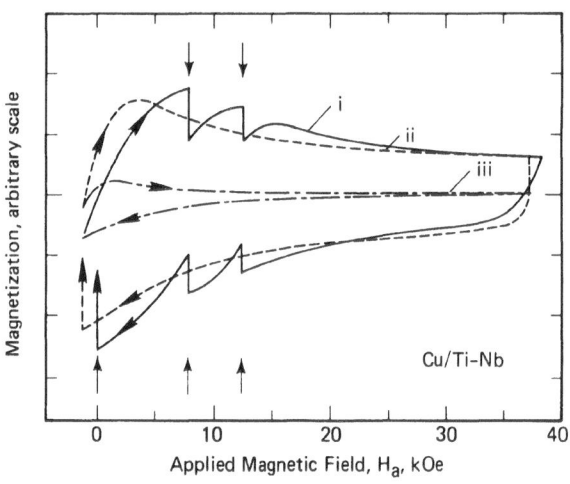

FIGURE 7-17. Magnetization *versus* applied field for the three Cu/Ti-Nb composites referred to in the text. Sample (i) is not field-sweep-rate dependent, but is unstable against magnetic field pulses (60 µs, 150 Oe) applied at the arrows; sample (ii) is pulse-stable but rate dependent — the loop shown is for 320 Oe s^{-1}; sample (iii) is pulse-stable and rate independent — after IWASA [Iwa69].

vessels and to increase the efficiency of MHD generators, has aroused fresh interest in the development of
superconductors with high upper critical fields. The
stimulating but short-lived "Twelve-Tesla" program
instigated by the U.S. Department of Energy is an example of this. A high H_{c2} in this context is necessary
for two overlapping reasons. In order to guarantee
the *existence* of the mixed state to high fields it is
an obvious requirement. The second reason is more
subtle: at high fields, even in densely pinned material, the flux lattice eventually softens as $H_a \to H_{c2}$;
flux is then able to flow, thereby ending the utility
of the state. The goal of increasing H_{c2} is to defer
this effect to as high an applied field as possible.
Thus in attempting to increase the high-field J_c,
efforts directed towards the optimization of a flux-
pinning microstructure are futile unless they are
accompanied by improvements to H_{c2} itself. The microscopic features of the elastic-lattice flux-pinning
model just alluded to are described in [Mon21.4]
et seq.

7.17.2 Composition Dependence of H_{c2}

The influence of multicomponent alloying on H_{c2}
is treated in subsequent chapters of this book. With
regard to binary Ti-Nb, the composition dependences of
H_r, based on the early measurements of BERLINCOURT and
HAKE [Ber63[a]] (1.2 K, $J \stackrel{\simeq}{=} 10$ A cm^{-2}), RALLS [Ral64]
(4.2 K, $J = 1$ A cm^{-2}) and other workers, are presented
in Fig.7-18. The microscopic basis for the general
shape of that curve in terms of Eqn.(7-10), for example, has been considered in Sect.7.9.2. According to
Fig.7-18, H_r(1.2 K) appears to possess a broad maximum
occupying the composition range 34∿42 at.% Nb (i.e.
50∿58 wt.% Nb). H_r(1.2 K) drops off rather rapidly
for [Nb]$\stackrel{\sim}{<}$ 34 at.%, while at 4.2 K the drop-off appears
to begin at a slightly higher Nb content ($\stackrel{\sim}{<}$36 at.%).
Further work would be needed to verify this statement,
since results to be considered below suggest that the
H_r's of all alloys in this composition range have similar temperature dependences. The technically important Ti-Nb superconductors occupy the peak of the H_r
versus composition curve. Detailed information relating to the structure of this peak is given in Fig.7-19,
assembled from the work of HILLMANN *et al* [Hil79] and
HAWKSWORTH and LARBALESTIER [Haw80]. No predictions
are offered with regard to the durability of the local
maximum at 35 at.% Nb in the face of further measure-

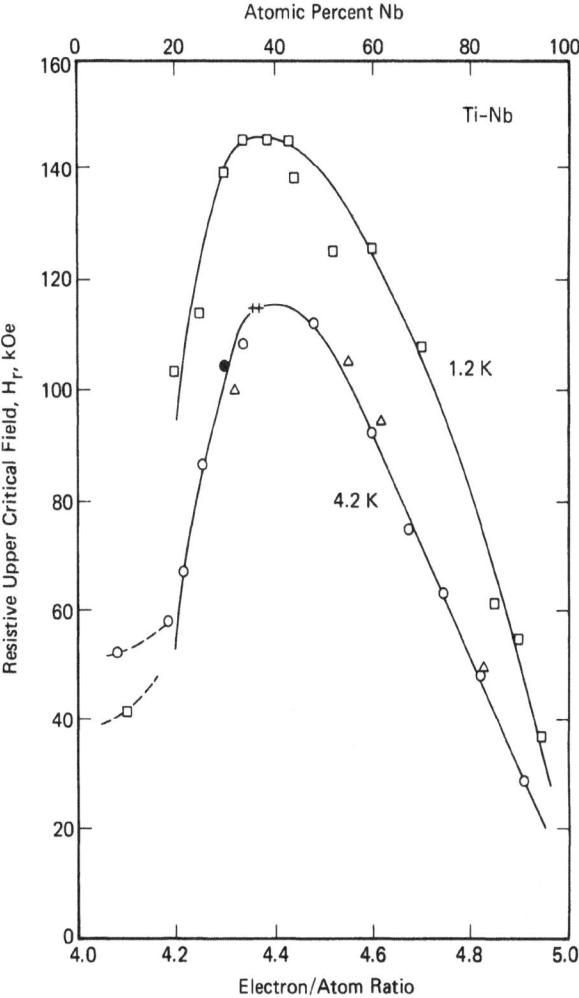

FIGURE 7-18. Resistive upper critical fields, H_r, as functions of
Nb content in Ti-Nb alloys. Represented are: the early 1.2-K results of BERLINCOURT and HAKE [Ber63[a]] (□) (cf. Fig. 7-7),
the subsequent results of RALLS [Ral64] (○) (1/4 μV cm^{-1} at
1 or 10 A cm^{-2}), and the more recent results of BRAND [Bra72]
(△) (from $J_c \to 0$; ∼ 10^{-8} V cm^{-1} criterion), HAWKSWORTH and
LARBALESTIER [Haw80] (++) (resistive onset at 5 A cm^{-2}), and
WADA *et al* [Wad80] (●) (resistive midpoint at 1 A cm^{-2}).

ment. The figure does reiterate the existence of the
abrupt critical-field drop-off already observed for
alloys with Nb contents less than about 35 at.% (i.e.
52 wt.%) and in so doing emphasizes the importance of
maintaining the Nb concentration at as high a level as
metallurgical and economic considerations will permit.
If the Nb concentration were to the right of the
shaded band in Fig.7-19 appreciable, α-phase precipitation, essential for J_c maximization, would be absent
[Nea71]. However, as several authors (e.g. SEGAL *et al*
[Seg80] and HILLMANN *et al* [Hil79]) have proclaimed,
the best of both worlds can to some extent be achieved
by appropriately processing less-Nb-rich alloys.
HILLMANN *et al* [Hil79], for example, have mentioned

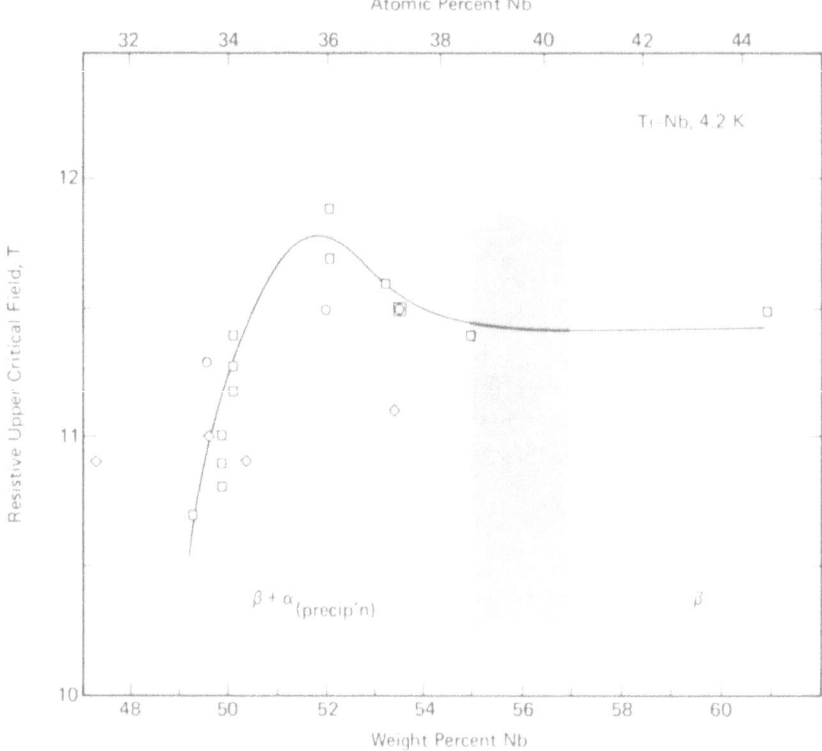

FIGURE 7-19. Resistive upper critical field H_r (4.2 K) as a function of Nb content within a narrow concentration range covered by some Cu/Ti-Nb composites: measurements by HAWKSWORTH and LARBALESTIER [Haw80] on laboratory monofilaments (○) (at $J = 5$ A cm^{-2}) and commercial multifilaments (◇) (at $J = 500 \sim 900$ A cm^{-2}); measurements by HILLMANN *et al* (Hiℓ79] on optimized wires (□). The smooth curve through the data was recommended by the latter authors. The shaded band subdivides the alloys into two classes depending on whether or not significant amounts of α-phase precipitation are present; precipitation is expected (after thermomechanical processing) on the Ti-rich side of the division.

that the deformation and aging of a *ca* 50-wt.% Nb alloy can deposit flux-pinning α-phase precipitates in the cell walls (cf. [Mon21.15]) accompanied by a rejection of Nb into the β-matrix thereby moving its composition into a more H_{c2}-favorable region. According to those authors, an observed α-phase precipitate density[†] of 6×10^9 particles cm^{-2} combined with an assumed particle volume of 8×10^{-18} cm^3 leads to an increase of the matrix Nb level of 0.2 wt.%. SEGAL *et al* [Seg80], in discussing the relative merits of Ti-53.5Nb and Ti-50Nb for use in high-field conductors, stated a preference for the latter for several reasons, one of which was the possibility that heat treatment of it would result in a high density of flux-pinning precipitates, whose presence would enhance J_c and whose formation and growth would release additional Nb into the β-matrix thereby increasing its critical field. The experimental outcome of this prediction is dis-

cussed in the original article. Such a solute-redistributive approach to superconductive-property improvement (with particular reference to the merits of $\alpha+\beta$-aging) had been suggested as early as 1965 by VETRANO and BOOM [Vet65] in connection with the development of the Ti-Nb$_{22}$ (i.e. Ti-35Nb) class of superconductor.

7.17.3 Temperature Dependence of H_{c2}

The temperature dependence of H_r has already been alluded to in connection with Fig.7-18, according to which a lowering of the temperature from 4.2 K to 1.2 K generally increases H_r by about 30 kOe (i.e. 3 tesla (T) in terms of the induction below the sample's surface). Thus to a first approximation at least, the effect of temperature seems able to be accounted for simply in terms of a composition-independent scale-factor. The upper critical field temperature dependences of optimized commercial Cu/Ti-Nb composite superconductors have been studied by HILLMANN

[†] Two dimensions were being referred to.

et al [Hiℓ79]. A set of results for some monofilamentary composites with Nb concentrations within the range 49.4-61 wt.% is presented in Fig.7-20. According to the figure:

$$H_r \stackrel{\sim}{=} H_{r0}(1-t^2) \quad , \qquad (7-33)$$

which is comparable to the thermodynamic critical field temperature dependence of classical theory (cf. the equations of Sect.7.10.1, Eqn.(3-7), and the equations of [Mon13.5]). If Eqn.(7-33) were strictly obeyed, with $T_c \stackrel{\sim}{=} 9$ K for the technical alloys concerned (cf. Table 7-2), a reduction in the temperature of from 4.2 K to 1.2 K should increase H_r by about 25%, i.e. from $H_r(4.2 \text{ K}) = 11.5$ T to $H_r(1.2 \text{ K}) = 14.4$ T. Indeed, a close a examination of the results for a number of technical alloys over the narrow temperature range of interest confirms this expectation. According to Fig.7-21, from a recent review paper by LARBA-LESTIER [Lar80], the upper critical fields of a group of Ti-Nb(47.3-53.5 wt.%) alloys increased by 3 T, starting at $H_r(4.2 \text{ K}) = 11$ T, as the temperature was lowered from 4.2 to 1.7 K.

7.17.4 The Status of Resistive Upper Critical Field Determination -- Experimental Artifacts

As mentioned in Sect.7.9.2, it is generally preferred that resistive upper critical fields be measured at sample current densities of $1\sim10$ A cm^{-2}. The results recorded in Fig.7-18 were obtained within that range, but those of Fig.7-21 were obtained at current densities of between 500 and 900 A cm^{-2}. It is important to inquire into the influence of measurement current density on H_r. RALLS in an early study of superconductivity in the Zr-Nb and Ti-Nb systems has already done this [Raℓ64]. According to Fig.7-22, taken from the results of his investigation, an increase in J of from 1 to 10^3 A cm^{-2} caused an apparent decrease in the H_r of mid-composition-range alloys of about 6.5 kOe (i.e. 0.65 T). Similarly, in recent studies of a laboratory-prepared Cu/Ti-52Nb monofilament, HAWKSWORTH and LARBALESTER [Haw80] have demonstrated that an increase in J of from 5 to 10^3 A cm^{-2} produced a reduction in the 4.2-K H_r of 0.65 T. The existence of this effect, whose magnitude is related to the slope of $J_c(H_a)$ at $J_c \to 0$, calls for the establishment of both measuring-current-density and threshold-resistivity standards for the performance of the resistive-critical-field test.

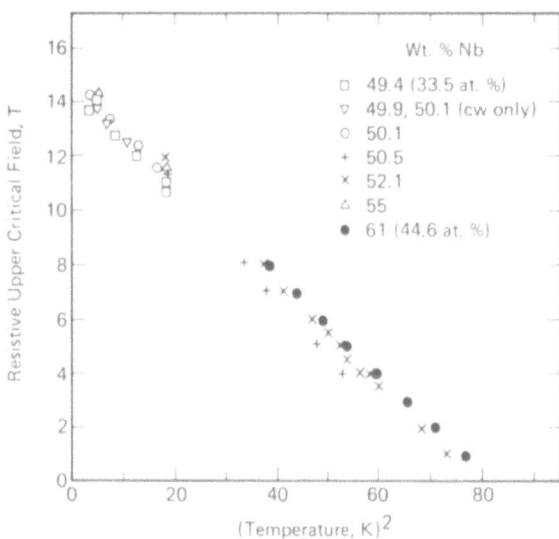

FIGURE 7-20. Temperature dependence of resistive upper critical field for seven technical monofilamentary or 60-filamentary Cu/Ti-Nb composite conductors. All seven were measured in the current-optimized condition; for comparison, one wire was measured after cold work (cw) only. H_r can be seen to follow a $(1-t^2)$ temperature dependence — after HILLMANN *et al* [Hiℓ79].

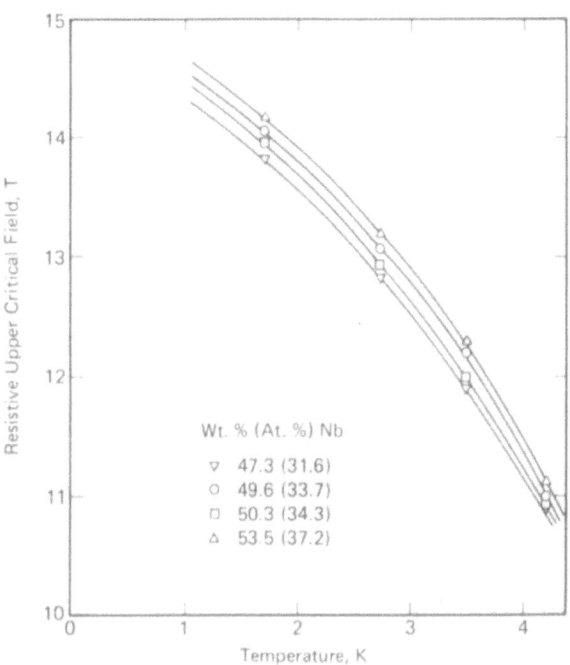

FIGURE 7-21. Temperature dependences of resistive upper critical fields corresponding to the "first appearance of resistance" (at J = 500 ~ 900 A cm^{-2}; onset criterion probably ~ 10^{-12} Ωcm [Haw80ª]) for four optimized multifilamentary Cu/Ti-Nb composites — after LARBALESTIER [Lar80].

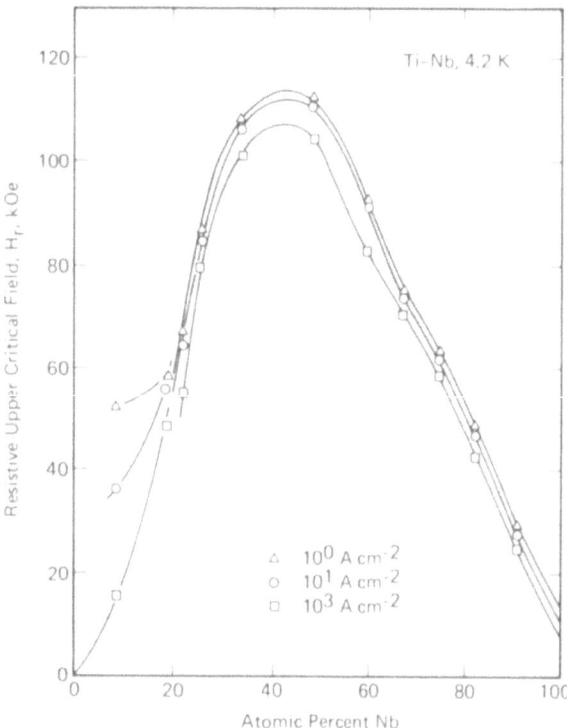

FIGURE 7-22. Influence of measuring current density, in association with a resistive-onset criterion of 1/4 μV cm^{-1}, on the 4.2-K resistive upper critical fields of a series of bare heavily-cold-worked 0.25 mm$^\phi$ Ti-Nb alloy wires — after RALLS [Ral64].

Another source of H_r-variability has to do with the metallurgical condition of the sample over and above that which influences H_{c2}^* via Eqn.(7-10). Since the measuring current density is actually a *critical* current density when $H_a = H_r$, factors which influence J_c without altering the bulk properties of the alloy can bring about unpredictable changes in the apparent H_r. The operation of this effect, which seems to have been considered first by BERLINCOURT and HAKE [Ber63^a] in their classical 1963 paper, and discussed further

by RALLS in his thesis [Ral64] and again in reference [Ral66], can to some extent be appreciated from a glance at Fig.7-5. RALLS has shown in studies of numerous Zr-Nb and Ti-Nb alloys [Ral64], that metallurgical factors which enhance J_c also increase H_r, giving rise to a limiting value of that quantity for each alloy measured and a smooth H_r-composition dependence as in Fig.7-22 for the alloy series. As pointed out in Sect.7.15, when working at a threshold of about 10^{-12} Ω cm, the quantity being detected in the four-terminal potentiometric measurement is the *creep resistivity* of the flux, thus all variables which effect creep resistance -- particularly the strength and distribution of the pinning sites -- will have an influence on H_r.

Finally, it should be noted that when following the course of a superconducting-to-normal state transition, using either magnetic or resistive techniques, the reported value of critical field will depend on which part of the transition is accepted as fiduciary. SUENAGA and RALLS [Sue69] gave serious consideration to this problem. After suggesting that either the onset resistivity (yielding H_{cs}) or the half-normal resistivity could be accepted as criteria, they decided themselves to use the threshold of the normal state (yielding H_{cn}). The reason advanced was that, although H_{cn} might be higher than the "true" critical field, it was less sensitive to pinning effects (of the type referred to above) and consequently would lead to more self-consistent results when mapping the critical fields of an extensive matrix of ternary alloys. Other authors still prefer to use the onset resistivity as critical-field criterion, especially when dealing with technical alloys [Lar80], since the result so obtained is the most conservative, and consequently the one of greatest value to the design engineer.

PART 3: CRITICAL CURRENT DENSITY IN TITANIUM-NIOBIUM ALLOYS

7.18 INTRODUCTION

7.18.1 Early Literature and Patents (pre-1966) Relating to Technical Ti-Nb Superconductors

It was recognized more than twenty years ago by AUTLER [Aut60], that from the standpoints of both economy and stability considerable advantage was to be

gained from the use of superconductive coils rather than conventional magnets in large-scale devices such as plasma-containment magnets in thermonuclear confinement, and that "hard" type-II superconductors suitable for their construction were in existence. This suggestion was followed rapidly by the demonstration at the Bell Telephone Laboratories [Kun61] of a 0.3-cm bore superconducting magnet constructed from 30,000 turns

of 0.007 cm$^\phi$ gold-plated Mo-Re$_{25}$ wire, the discovery
of the high-field current-carrying potential of the
brittle intermetallic compound Nb$_3$Sn, and the development of the high-field ductile-alloy superconductors
Zr-Nb and Ti-Nb. The commercial importances of Ti-Nb
wire, and of coils made from it, were immediately
recognized; and even though the discipline was in its
scientific infancy, a series of patents were applied
for in 1961 and the following four years which, although primitive in detail, were conceptually predic-
tive of future commercial superconductor development.

The patent literature up to 1967 (year of issu-
ance) is reviewed below. Numerous subsequent patents
on the subject of Ti-Nb alloy superconductors have of
course been granted, the more recent ones dealing with
the commercial development of superconductive wires.

In 1961 MATTHIAS [Mat65] applied for a patent
(granted, 1965) to cover a superconducting solenoid,
or other unspecified devices, to be constructed using
cold-worked Ti-Nb. ZWICKER [Zwi63^a] in applying for a
patent for a system of Ti-Nb alloys (granted, 1963) to
be cold deformed into either wire or ribbon, precipitation heat treated, and provided with a final (optimal) stage of cold deformation (≲60%), laid out at
least qualitatively a processing scheme that is followed to this day. Although more technical details
were included in the application, REYNOLDS [Rey66]
(see also [Wes66, Wes67]) assigned to the Westinghouse
Electric Corporation (1963-1966) a somewhat similar
idea for the preparation of cold-rolled-and-aged Ti-Nb
ribbon, and in a general way, a superconducting coil
constructed from it. The first patent for a composite
superconductor was applied for by CHESTER in 1964 and
1965 and assigned to the Central Electricity Generating Board [Che65]. It dealt with the fabrication of
laminated strip and laminated multifilamentary conductors in which Ti-Nb foils or filaments, ∼25 μm in
thickness or diameter, respectively, were to be in
intimate contact with a normal metal, specifically Cu
but permissibly Ag, Al, In or Cd; large-scale applications were envisioned such as magnets for fusion devices. A patent for the preparation of clad cylindrical Ti-Nb superconductor was granted to the Compagnie
Francaise Thomson-Houston in 1967 [Com67^a]. In this,
the cladding of Ti-Nb in annealed Cu was specified,
but allowance was also made for the possible addition
of elements such as V, Zr, Hf, and Ta, to the basic
Ti-Nb alloy, as well as for the cladding of the wire-
rod during production in Al, Ag, Au, and stainless
steel, and the wrapping of the final product with

mylar tape. Although with today's alloys, large un-
interruped cold reductions are possible, the Thomson-
Houston patent called for frequent intermediate
annealings of (2-5h)/(500-700°C) presumably for metalworking reasons. Precipitation heat treatment and
final cold work did not form part of the claim.

7.18.2 Early Studies of Pulse and AC Effects and Long-Sample (Coil) J$_c$-Measurements in Ti-Nb Superconductors

The study of Ti-Nb under pulse and AC conditions,
the construction of coils, and investigations of coil
stability did not await the final optimization of the
wire fabrication process. In fact, comparisons of
short-sample and in-coil properties of Ti-Nb wires
were already taking place in 1962. For example,
WOLGAST et al [Wol63] intercompared the short-sample
critical current densities of some eight alloy wires
including a sample of non-optimized Ti-Nb$_{43}$, which was
recognized at the time to be a "readily fabricable
alloy", and went on to test the performances of some
small coils,[†] for example a Ti-Nb-wound coil conforming to the specifications:

Length	8.1	cm
O.D.	5.8	cm
I.D.	1.3	cm
Number of Turns	18,429	
Length of Wire	2,045	m

Comparisons of short-sample and coil properties
were also made by WEYL and DIETRICH [Wey65] using cold-
worked 0.25 mm$^\phi$ Ti-Nb$_{20}$ wire. In the short-sample
tests, both DC and pulse-current measurements of critical current were made on material that had been: (a)
cooled from the normal state in constant field (yielding I$_{c,H}$); (b) cooled in the absence of field and the
field subsequently applied (I$_{c,0}$). In the DC measurements I$_{c,H}$ ≡ I$_{c,0}$, but during the pulse measurements
it was found that I$_{c,H}$ > I$_{c,0}$. The difference was interpreted at the time as evidence for the existence of
sponge-trapped circulating currents induced in the
superconducting component by the changing field.

Early examples of pulse-current, pulse-field, and
AC effects in Ti-Nb samples are to be found in the

---

† A general review of the small-coil testing of Ti-Nb
  superconductors is given in [Mon24.00].

papers of BERKL and WEYL [Ber64], YOUNG and SCHENK [You64] and VALLIER [Val65].

BERKL and WEYL have discussed the experimental techniques of critical current measurement, especially as they relate to sample heating either by thermal conduction *via* the current leads, or as a result of the measuring current itself. Although in the case of DC measurements, the electrical circuitry is simpler, the measurement of very large current densities using heavy current leads is accompanied by severe refrigeration difficulties. It was claimed that these could be avoided if pulse currents were applied. BERKL and WEYL [Ber64] concluded, as a result of their experiments with 0.2-mm$^\phi$ Ti-Nb$_{33}$ wires, that not only could reliable values of critical current density be obtained using the pulse method but that intrinsic heating of the sample could be distinguished from heating due to the current leads. By contrast, it was believed that spurious heating effects could invalidate the results of DC measurements.

YOUNG and SCHENK [You64] have studied AC critical current density at frequencies of up to about 3300 Hz in five superconducting wires including 0.13- and 0.23-mm$^\phi$ Ti-Nb. In measurements made at 4.2 K in the short-sample self-field, J_c was found to decrease from 2.90 to 1.70×10^5 A cm^{-2} as the frequency increased from 0.01 to 4.5 kHz. In short-sample tests using Nb-Zr wire, it is interesting to note that the critical AC current density decreased with the logarithm of the wire diameter, thus anticipating the results of subsequent stability calculations.

In an early investigation of the stability of commercial superconducting wires, VALLIER [Val65] studied the influence of a pulsed longitudinal magnetic field (0 to 110 kOe in 40 ms) on the re-establishment in fields less than critical of the normal state in current-carrying conductors. The effect considered had to do with the thermal propagation of normally conducting regions. The present status of normal-zone propagation is considered in [Mon25.37, Mon25.38] within the context of a general discussion of conductor stability.

7.18.3 Scope of the Discussion of Critical Current Density

As in the foregoing treatments of other aspects of Ti-Nb superconductivity, an attempt has been made to give complete coverage of the early literature up to that of 1965-1966, except perhaps in cases where Ti-Nb has been considered merely as a control alloy in studies of multicomponent Ti-Nb base alloys. With regard to the period between about 1966 and the present (i.e. through 1982), reason required only the most important contributions relating to critical current transport to be taken into consideration.

In the development of Ti-Nb as a practical superconductor, attention has been divided between the Ti-Nb(20-22 at.%) and Ti-Nb(34-42 at.%) classes of alloy. In the former, a rich variety of microstructures (including the martensitic, ω, and α phases, and mixtures of them) can be developed by thermomechanical treatment. In Ti-Nb(34~37 at.%), some of whose properties are superior to those of the lower concentration alloys, appreciable α-phase precipitation is possible. Ti-Nb(20-22 at.%) alloys were being developed as commercial superconductors in the early to mid 1960's. Development of the Ti-Nb(34-42 at.%) alloys had its roots in that same time period. They were eventually to supersede the lower concentration alloys and to be subjected to intense research and commercial development through the 1970's.

7.19 METALLURGICAL INTRODUCTION

7.19.1 Microstructure and Macrostructure in Ti-Nb Alloys

Essential to an understanding of the current carrying state in Ti-Nb is a knowledge of its microstructural properties as a function of composition and in response to deformation and heat treatment. The first of these topics is discussed in detail in [Mon1.00] and the second in [Mon2.00]. Numerous authors have also conducted detailed studies of Ti-Nb microstructures with direct reference to their flux-pinning potentialities. Flux pinning by precipitates, precipitate-free subband structures, and precipitation-decorated subbands, is considered in [Mon2.13] through [Mon21.15].

NARLIKAR and DEW-HUGHES [Nar66] were among the first to recognize, with reference to Ti-Nb$_{47}$ and other alloys, the importance of flux pinning in the stabilization of the current-carrying mixed state. They noted that magnetic hysteresis (a measure of flux pinning) was related to current-carrying capacity and that flux could be pinned by individual particles as well as by dislocation tangles, particularly those forming the walls of dislocation cells. The mechanistic model offered was based on the proposition that the "free energy of a super-current vortex is reduced if it lies in a region of higher κ_{GL}". Since κ_{GL} is related to resistivity, hence lattice disorder, through Eqn.(7-5), the dislocation-cell walls provided flux-pinning sites by what was to become known as the "κ_{GL}-modulation" mechanism.

COURTNEY and WULFF [Cou69], recognizing on the basis of their own work and that of others the importance of ω-phase as a flux pinner, have considered the nucleation and growth of ω-phase particles and the associated solute/solvent interdiffusion. It was claimed that a high J_c was favored by: *(a)* a large volume density of fine ω-phase precipitates; *(b)* a high concentration of Ti within them, c_{Ti}^{ω}, a quantity which they described by the equation:

$$c_{Ti}^{\omega} = \frac{\phi_0 c_0}{\phi} + \frac{\delta}{\delta-1}(1-\frac{\phi_0}{\phi}) \qquad (7-34)$$

where ϕ represented the linear dimensions of the ω-phase precipitate particle of initial size ϕ_0, c_0 was the initial concentration of Ti in the ω-phase, and δ (>1) was the diffusivity ratio, D_{Ti}/D_{Nb}. In order to simultaneously satisfy requirements *(a)* and *(b)* above, a substantial Ti-enrichment of the precipitates must occur before significant growth takes place. This requirement is met by carefully adjusting the aging time/temperature, and is assisted by a small value of δ.

RAUCH *et al* [Rau68[a]] suggested that the presence of O, which gives rise to Ti-O groups associated with vacancies, should lead to an increase in D_{Ti} and consequently δ, hence for a given amount of particle growth to less Ti-enrichment of the ω-phase. Although O may of course be deliberately introduced into Ti-Nb it is frequently already present as a contaminant of one or other of the starting elements. The superior current-carrying capacities of U. S. alloys after optimization, as compared with those produced by other countries, has been attributed to the beneficial influence, on the former, of O contamination [Cur81]. Similarly, RASSMANN and ILLGEN [Ras72] have noted that the impurities present in Kroll-process Ti led to higher critical current densities in the Ti-Nb(20-70 at.%) alloys prepared from it than in alloys prepared from iodide Ti. The influence of interstitial-element additions to Ti-Nb is the subject of an entire chapter in this book.

A discussion of hard superconductors, in metallurgical terms, was presented in 1964 by ZWICKER [Zwi64]. Of particular importance at the time was the investigation of the quenched, cold-deformed, and cold-deformed-plus-heat-treated microstructures of Ti-26Nb and Ti-33Nb. It was recognized that a laminated structure resulted from cold work, and that a lamellar width of about 0.1 μm was particularly favorable from a current-carrying standpoint. By now it is well known that the existence of flux-pinning centers, i.e. suitable products of structural and compositional micro-inhomogeneity, are necessary for the support of large transport currents. The question then naturally arises as to what extent the starting ingot should be homogenized before it is converted, *via* extrusion and drawing, into superconducting wire. PAKHOMOV *et al* [Pak66] examined the situation experimentally. In short-sample comparisons of Ti-Nb$_{54}$ wires, cold-reduced 99.96% either from the as-cast ingot, or from one that had been annealed for various times of up to 5 h at 1500°C, the critical current density was found to *decrease* with increasing degrees of ingot homogeneity.

From a different standpoint the inhomogeneity which arises through the segregation of alloying elements during ingot solidification (coring) has been found to be deleterious. According to MOON [Moo68] it may be conducive to flux-jump instability. His argument proceeded along the following lines. Although the motion of flux between *closely spaced* pinning sites may be jerky, no detectable effects other than perhaps noise were produced; on the other hand in the presence of macro-inhomogeneities, hence *long-range* variations in κ_{GL}, he claimed that the rate of flux motion in the wake of a flux front would be large compared with that of the front itself, giving rise to a situation which under adiabatic conditions could trigger a flux jump.

Coring in the starting ingot also translates into longitudinal compositional modulation in the finished wire and to a practical critical current density no larger than that of the poorest sections.

7.19.2 Equilibrium and Nonequilibrium Phases and the Effects of Deformation and Aging

Necessary to an understanding of the thermal and compositional ranges of the equilibrium and nonequilibrium phases in Ti-Nb alloys are the appropriate phase diagrams. Fig.7-1 is the equilibrium phase diagram for Ti-Nb, taken primarily from standard sources [Han51][Img61], but modified by the information that, although practically no α-phase precipitation results from the prolonged aging at ∿400°C of alloys with Nb contents greater than about 40 at.% [Nea71], it certainly contributes to flux pinning in alloys with Nb contents as high as 37 at.% [Wes80, Wes82]. Equilibrium phases in Ti-TM alloys in general are discussed in [Mon1.1] through [Mon1.5].

The nonequilibrium phases which occur when Ti-TM alloys are rapidly quenched from the bcc field (β-quenched), and which depend on composition, quench rate, and the final temperature, are: the martensitic phases α' and α" (herein, collectively,[†] α^m) and the athermal and isothermal varieties of ω-phase. Fig.7-23 is a generalized nonequilibrium phase diagram for Ti-TM alloys indicating, as explained in [Mon1.6], the martensite-start transus (M_s) and the location of the narrow zone of athermal ω-phase which forms the solute-lean boundary of the α+β field within which isothermal ω-phase of terminal composition will form during moderate-temperature aging. Fig.7-24, which combines the equilibrium and nonequilibrium diagrams for Ti-Nb,

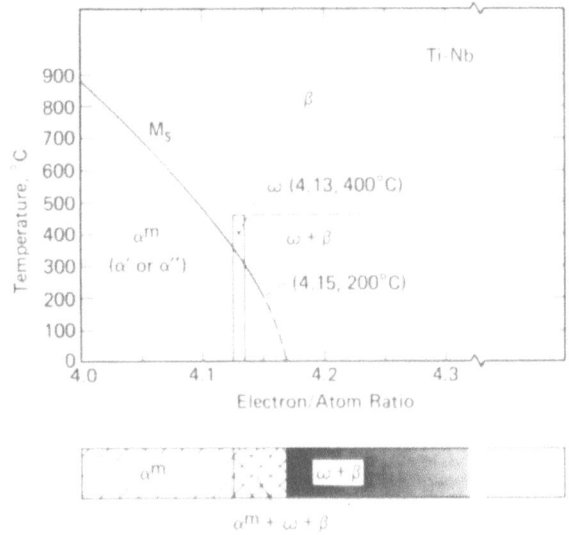

FIGURE 7-23. Generalized metastable equilibrium phase diagram for Ti-TM alloys indicating the average α^m-threshold and the zone of occurrence of athermal ω-phase.

---

† According to BAGARIATSKII *et al* [Bag59], the compositional boundary between the hexagonal (α') and orthorhombic (α") variants of the quenched martensite in Ti-Nb alloys is 6∿8 at.% Nb.

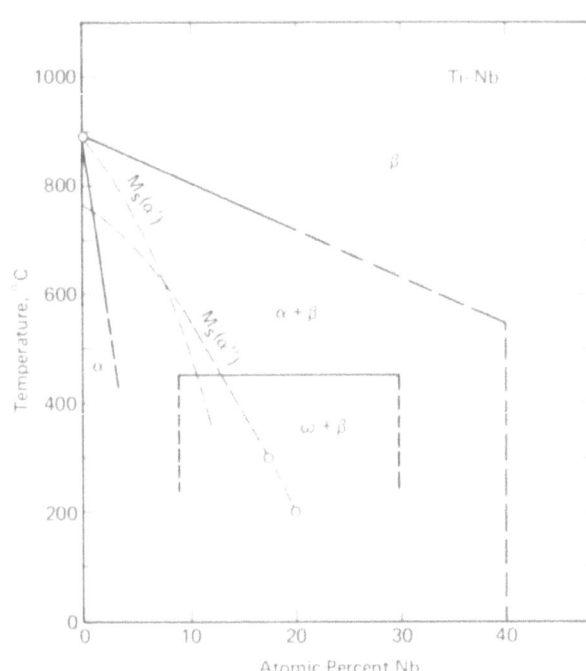

FIGURE 7-24. A combined equilibrium and metastable equilibrium phase diagram for Ti-Nb indicating, in particular, the compositional ranges of the ω + β and α + β regimes; cf. Fig. 7-1.

assists in describing the decomposition of intermediate-concentration Ti-Nb alloys during aging at temperatures below about 400°C.

Complete understanding of the processing and metallurgical properties of technical superconductive wire requires knowledge of: (i) the influence of cold deformation such as rolling, swaging, flattening, and wire drawing on the microstructures of quenched Ti-Nb alloys; (ii) the influences of stress, strain, and interstitial-element additions on the transformation kinetics of the quenched alloys; (iii) the formation during heavy plastic deformation of a fine highly directional dislocation-cell structure (subband structure) and, during subsequent aging, the motion of dislocations to the cell walls and the growth of the cells. These topics are fully discussed in [Mon2.8] and [Mon2.11].

7.20 QUENCHED-AND-AGED MICROSTRUCTURES OF Ti-Nb ALLOYS

7.20.1 The Occurrence of the Martensitic and Omega Phases in Quenched Ti-Nb Alloys

For a useful discussion of martensitic transformation kinetics in Ti-Nb alloys the work of JEPSON,

BROWN and GRAY is recommended [Jep70]. For Ti-Nb, an upper concentration limit for α'-martensite of $6\sim8$ at.% has been reported by BAGARIATSKII et al [Bag59]. The upper limit of 3 at.% referred to by HATT and RIVLIN [Hat68] and perpetuated by some subsequent authors, may have originated in what seems to have been an error in Fig.1(d) of [Bag59] but which had been corrected, by implication, later on in that paper.

According to BAGARIATSKII et al [Bag59], the solute concentration in Ti-Nb alloys for which ω-phase is formed on quenching is 17-18 at.%. This should not be confused with the composition of the ω-phase which forms during prolonged aging of higher concentration alloys at temperatures below about 450°C (the "saturation composition of ω-phase") which, according to HICKMAN [Hic69[a]], lies within the range $6\sim11$ at.% Nb.[†] The picture with regard to alloys in the concentration range 18-22 at.% Nb is complicated, although there are no significant points of disagreement among the several contributors to the subject. It has been claimed that the quenched microstructures depend not only on the quench rate (as is usually the case), but also on the temperature at which the sample is held immediately prior to the quench [Bal72]. In Ti-Nb$_{18.4}$, BALCERZAK and SASS [Bal72] showed that the oil-quenched microstructure was essentially either α'' or $\omega+\beta$ depending on whether the prior annealing temperature had been 900 or 1000°C, respectively. In general, the transformation to α'' seems to preempt the formation of ω-phase, although a *trace* of α'' can coexist with $\omega+\beta$. For Ti-Nb$_{20.7}$, HATT and RIVLIN [Hat68] showed that water quenching from 900°C yielded α'' plus a trace of β (whereas slow cooling, of course, led to $\omega+\beta$). In a similar alloy, BAKER and SUTTON [Bak69] obtained the same result. In Ti-Nb$_{22.6}$, BALCERZAK and SASS [Bal72] obtained essentially $\alpha''+\beta$ during a quench from 800°C, and $\omega+\beta$ in quenching from either 900 or 1000°C. From a heat-transfer standpoint it is reasonable that a quench from a lower annealing temperature should be more rapid than one from a higher temperature. Thus, in controlling quenched microstructures, there may not, in fact, be a real distinction between "quench rate" and "prior annealing temperature". HICKMAN [Hic69[a]] achieved relatively rapid cooling

rates (10^3 °C s^{-1}) during the He-gas quenching of resistively heated strips of alloy. In so doing, he was able to suppress ω-phase in all the alloys examined, viz Ti-Nb($\stackrel{>}{\sim}22$ at.%). In particular, Ti-Nb$_{22}$ yielded $\alpha''+\beta$, in agreement with the results of the 800°C quench of BALCERZAK and SASS [Bal72]; gas-quenched Ti-Nb$_{25}$ was also $\alpha''+\beta$, whereas the structure of Ti-Nb$_{25.6}$ oil-quenched from 900°C by BALCERZAK and SASS [Bal72] was essentially $\omega+\beta$. HICKMAN's Ti-Nb$_{27}$ was found to be "all β" [Hic69[a]], while for Ti-Nb$_{34}$ the quenched structure obtained by BALCERZAK and SASS [Bal72], although designated "β", yielded a selected area diffraction (SAD) pattern exhibiting so-called "lines of intensity". These lines, which are also present in association with ω-reflections in the lower concentration alloys, persisted in decreasing intensity through Ti-Nb$_{57}$ and have actually been noted in pure Nb [Sas72]. It is permissible to regard the lines of intensity as arising from "diffuse-ω". The designation 'ω' is retained, since diffuse-ω is believed to originate from the influence on the bcc lattice of that same vibrational state [Sas72] which gives rise, in the lower concentration alloys, to the athermal ω-phase itself. The occurrence of this effect as a function of composition in binary Ti-TM alloys has been summarized in tabular form (with particular reference to Ti-V) by COLLINGS [Col74]. All aspects of the ω-phase phenomenon have been fully reviewed by SIKKA, VOHRA, and CHIDAMBARAM [Sik82].

7.20.2 The Occurrence of the Isothermal-ω, Separated-β, and Equilibrium-α Phases in Aged Ti-Nb

(a) The Isothermal ω-Phase. As depicted in Fig.7-24, the maximum Nb concentration for isothermal ω-phase precipitation at 450°C is 30 at.%. This is probably also true for lower aging temperatures, according to the results of isothermal aging studies of the comparable alloy systems Ti-V [Hic68] and Ti-Cr [Hic69[a]] at temperatures between 300 and 400°C, which revealed practically vertical ($\omega+\beta$)/β transi.

This estimate of a compositional limit of 30 at.% for ω-phase precipitation in Ti-Nb alloys derives from two independent observations by HICKMAN [Hic69[a]], viz (i) his failure to detect ω-phase precipitation in Ti-Nb$_{30}$ and Ti-Nb$_{35}$; (ii) his finding, by means of x-ray lattice parameter measurement, that the Nb content of the β-component of $\omega+\beta$-aged Ti-Nb$_{22}$ and

---

† During quenching, the precipitation of ω-phase within the composition range between $6\sim11$ at.% Nb and 17-18 at.% Nb is preempted by martensitic transformation.

Ti-Nb$_{25}$ (for up to ∿30h/450°C) was 30 ± 1 at.%. More recently, OSAMURA *et al* [Osa80] have conducted a detailed study of precipitation in cold-worked and aged Ti-Nb$_{36}$ foils and wires. With regard to isothermal ω-phase precipitation, they noted, in agreement with HICKMAN, that no precipitation appeared during aging at 380°C for times of less than 10^4 min (∿167 h). On the other hand, after aging for longer times than this, a small amount of ω-phase was observed. Apparently, ω-phase precipitation is generally quite sluggish in the higher-concentration Ti-TM alloys. Since, however, extremely long heat-treatment times are not contemplated for technical Ti-Nb superconductors, a practical limit for ω-phase precipitation of 30 at.% Nb is adopted in this book.

(b) β-Phase Separation in Ti-Nb Alloys. When the temperature is too high [Luh70][Wiℓ71] or the alloys too concentrated [Wiℓ78] to support ω-phase precipitation a solute-lean bcc precipitate, designated β', separates out during moderate-temperature aging. Thus a metastable region designated β'+β, overlapping and extending beyond the upper-temperature and upper-concentration boundaries of the ω+β region, is formed. This relationship between the locations of the phases is depicted in Fig.9-1 for use in connection with a discussion of β→β'+β decomposition in Ti-Nb-SM alloys. As with ω+β, the β'+β mixed phase is metastable, but unlike ω, the β' precipitate stems from the chemical (or electronic) differences between solute and solvent atoms. Thus the β→β'+β phase-separation reaction is a clustering reaction characteristic of alloy systems which show positive heats of mixing or equivalent manifestations of a tendency for the constituents to unmix. Accordingly it would be expected to form within a restricted temperature-composition range in Ti-Nb alloys which, according to RUDMAN's x-ray diffuse scattering study of three such alloys [Rud64], exhibit clustering over a wide concentration range. Indeed unmistakable evidence for β' precipitation in aged Ti-Nb alloys -- Ti-35Nb (aged 9h/365°C), Ti-41Nb (aged 100h/375°C), and Ti-45Nb (aged 100h/400°C) -- has been obtained by MENDIRATTA *et al* [Men71]. Partly on account of the similarity between their respective aging requirements, the β' precipitate (or at least the precursor stage of it) was for a time confused with ω-phase. This is no longer the case, the status of β' as a metastable *bcc* precipitate now being well established [Wiℓ78]. Unfortunately, even as recently as 1981, some authors [Fon81, p.157] are still confusing β-phase separation with β-phase immiscibility which,

although it leads to a pair of bcc phases according to a reaction designated β→β'+β" [Kit70^c, Kit70^d, Kit70^e] (cf. Sect.11.6) characteristic in this context of alloys based on Nb-Zr, does so for entirely different thermodynamic reasons. The distinctions are clarified in [Mon2.3] (β-phase separation) and [Mon2.7] (β-phase immiscibility).

(c) The Equilibrium α-Phase. Fig.7-23 implies that, as a consequence of the sluggishness of the β to α+β decomposition reaction, it has not been possible using conventional methods to extend the equilibrium (α+β)/β transus to temperatures much below about 700°C. Thus information regarding phase equilibria in the interesting temperature range around 400-500°C has up until recently been scarce or lacking. If for no other reason than to inhibit excessive cell growth (cf. [Mon2.11]) heat treatment of Ti-Nb superconductor is restricted to temperatures below about 500°C. Nevertheless the same deformation-cell structure that "moderate-temperature" (∿500°C) heat treatment is designed to protect provides, in the form of lattice strain, sufficient energy to enable the alloy to proceed towards phase equilibrium within moderate periods of time at those temperatures. An illuminating series of moderate-temperature phase equilibrium (or approaches to equilibrium) investigations have been conducted by PFEIFFER and HILLMANN [Pfe68], NEAL *et al* [Nea71], OSAMURA *et al* [Osa80], and WEST and LARBA-LESTIER [Wes80, Wes82], on heavily-cold-worked and aged Ti-Nb alloys, whose compositions were not only technically important but which also, among themselves, appeared to span the boundary between the α+β and β fields at about 400°C.

According to PFEIFFER and HILLMANN [Pfe68], Ti-35Nb (22 at.% Nb) yielded α-phase precipitation during heat treatment either before or after deformation; when heavily deformed, ω-phase which appeared after 10min/380°C (cf. Fig.7-24) gave way to α-phase after 2 h at that temperature. Such results were of course entirely to be expected from existing equilibrium phase diagrams (cf. Fig.7-1). Although occurring only after cold deformation, α-phase precipitation also took place in Ti-50Nb (34 at.% Nb) during heat treatment [Pfe68]. These results have been amply confirmed in subsequent investigations. WILLBRAND and SCHLUMP [Wiℓ75], for example, noted the presence of α-phase precipitation in Ti-50Nb wires after moderate aging (e.g. several h/380°C) and were able to measure the sizes and abundances of the particles. On the other hand it is well known by now that NEAL *et al*

[Nea71] in studies of Ti-58Nb (i.e. 42 at. % Nb, analyzed composition), although able to detect the presence of some α-phase after aging usually for 1 h at temperatures of 350 to 500°C, noted that its density was "far too small to provide any significant degree of fluxoid pinning".

The situation with regard to alloys in the composition range beyond the 50 wt.% studied by [Pfe68], but below the 58 wt.% Nb of [Nea71], remained unclear until WEST and LARBALESTIER conducted a series of careful TEM/SAD [Wes80] and STEM/EDAX [Wes82] studies of some Ti-Nb alloys (both bare, and extracted from commercial multifilamentary composites) with 47.3, 49.6, 50.3 and 53.5 wt.% Nb. All of them were found to contain α-phase precipitates. Ti-53.5Nb (37 at.% Nb) was of particular interest, being the most concentrated Ti-Nb alloy known to support significant levels of precipitation.

Accordingly, on the basis of the combined results reported to date, the composition 56 ± 2 wt.% Nb (40 at.% Nb) is tentatively taken as demarcating the practically attainable "equilibrium" boundary between the $\alpha+\beta$ and β phases at about 400°C. This conclusion is reflected in Figs. 7-1, 7-19, and 7-24.

7.21 METALLOGRAPHY OF DEFORMED-AND-AGED Ti-Nb ALLOYS

In the processing of superconductor, the primary purpose of the metalworking is simply to produce wire or ribbon from a massive starting billet. It was soon discovered, however, that the fibrous substructure resulting from cold drawing had a most favorable influence on the current-carrying capacity, and that although the structure deteriorated or disappeared completely under excessive heat treatment, aging within a restricted temperature range could not only improve the properties of the subbands themselves but, under suitable conditions, decorate them with α-phase precipitation thereby enhancing still further the critical current density.

Fig. 7-25(a) shows a typical subband or dislocation-cell structure as produced in Ti-53.5Nb by cold drawing to an area-reduction of 97.5%. Annealing for $1\frac{1}{2}$ h/800°C was sufficient to completely recrystallize the material and remove all traces of the cold-worked structure, Fig. 7-25(b). The J_c-optimal heat-treatment temperature for superconductive Ti-Nb alloys, whether or not α-phase precipitation is compositionally possible, is about 400°C.

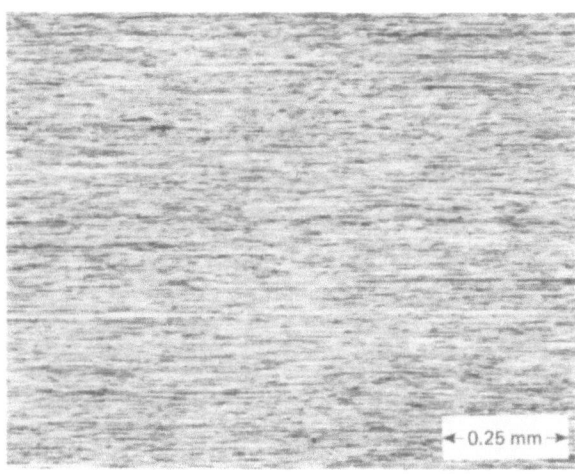

(a)

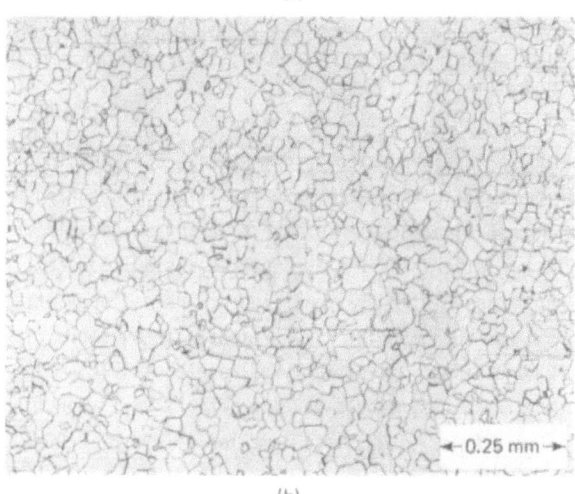

(b)

FIGURE 7-25. Ti-53.5Nb 2.1 mm$^\phi$ wire: longitudinal section, in the (a) as-drawn 97.5% and (b) drawn 97.5%-and-recrystallized 1.5 h/800°C conditions. Optical micrographs, original magnification 100X, etchant H_2O_2-lactic-HNO_3-HF. Photographs courtesy of Teledyne Wah Chang Albany — Mike Shields, photographer.

The influences of Cu and Ge additions on the superconducting properties of Ti-Nb alloys have been investigated by LÖHBERG [Loh71] and HELLER [Heℓ71], respectively. Naturally those of unalloyed Ti-50Nb, prepared under similar conditions as a control, needed to be measured as part of the investigation. Alloys were prepared by multiple arc melting and pouring into a 6-mm$^\phi$ rod-mold. X-ray microfluorescence analysis of the product indicated the presence of dendritic segregation (coring) in the form of compositional undulations of amplitude about 5% Nb and wavelength some 5-10 μm. The rods were cold-swaged 97.2% to 1-mm$^\phi$ wire, a portion of which was given a homogenizing anneal of 3h/1000°C in an evacuated quartz tube.

FIGURE 7-26. Ti-50Nb 1 mm$^\phi$ wire: longitudinal section in the as-cast plus cold-swaged-97.2% condition. Optical micrograph, original magnification 1000X, etchant 30% HNO_3, 10% HF, 60% H_2O, — after LÖHBERG [Loh71]; photograph courtesy of U. Zwicker.

Studies were conducted on both the as-cast and the homogenized wires. Fig.7-26 is a light micrograph of a longitudinal section of the cast-and-swaged wire. The striations present are not the dislocation-cell boundaries of the previous figure; instead they arise from the coring originally present. This compositional modulation, if initially of wavelength ∿10 μm in all directions, should develop after a 36:1 reduction in area (i.e. 97.2%) into lamellae of width ∿3000 Å, equivalent in size to over-aged dislocation cells. Cross-sectional views of the cast-plus-deformed 1-mm$^\phi$ wire, after aging for 16h/400°C, are presented at two magnifications in Fig.7-27(a) and Fig.7-27(b). As is well known by now, precipitation processes are accelerated by the presence of deformation strains,

partly as a result of which a fine distribution of precipitate particles is to be found between the lamellae. According to LÖHBERG [Loh71], the presence of segregation striations also stimulated the formation, during such aging (actually for 17h/400°C in his case), of copious amounts of α-phase precipitation, Fig.7-28, not visible under similar conditions (and, presumably, under light microscopy) in the previously-homogenized wire.

ARNDT and EBELING [Arn74] (see also WILLBRAND et al [Wiℓ75^a]) have also investigated the influences of Cu and Ge additions on the superconductivity of Ti-50Nb, correlations between critical current densities and microstructures being aided by the extensive use of transmission electron microscopy (TEM). Rods of Ti-50Nb the reference alloy, 3 mm in diameter, were

FIGURE 7-28. Ti-50Nb 1 mm$^\phi$ wire: longitudinal section as for Fig. 7-26 but final heat treated 17 h/400°C — photograph courtesy of U. Zwicker.

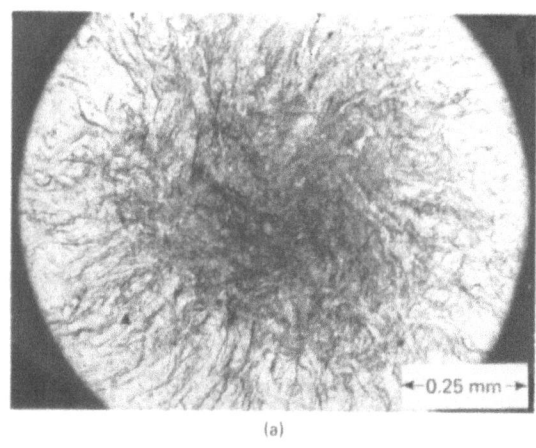

(a)

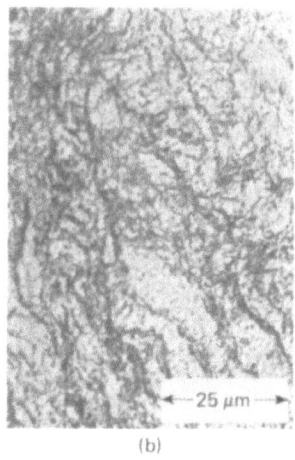

(b)

FIGURE 7-27. Ti-50Nb 1 mm$^\phi$ wire: transverse section in the cast, cold-swaged-97.2%, and aged 16 h/400°C, condition. Optical micrographs, original magnification (a) 100X, (b) 1000X — after HELLER [Heℓ71]; photographs courtesy of U. Zwicker.

produced by extrusion and drawing, clad with OFHC Cu
and further drawn to wire of diameter 0.25 mm (core
diam., 0.13 mm). The *total* cold reduction was estimated to have been more than 99.98%. Important structural features examined using TEM were: the subband
diameter, the sharpness of the subband boundaries, and
the sizes and abundances of any precipitates. WILL-
BRAND *et al* [Wiℓ75[a]] investigated the influences of
multiple intermediate heat treatment and final heat
treatment temperature on the current-carrying capacity.
A plot of the 5-T J_c *versus* subband diameter, d, for
wires which had received only final heat treatment
(from 3h/300°C to 100h/500°C depending on the subband
diameter desired) rose steeply to a maximum in the
vicinity of $450 < d < 850$ Å and fell away slowly as d
increased to large values. With multiple intermediate
heat treatment on the other hand, no such systematic
relationship between J_c and d could be recognized.
Thus although the presence of subbands, with diameters
of about 300-500 Å, was a necessary condition for the
attainment of high J_c's, it was apparently an insufficient one in that other products of the heat treatment,
presumably precipitates, seemed to be playing very
important roles.

Fig.7-29(a), a transmission-electron micrograph
of as-cold-worked (>99.98%) Ti-50Nb, exhibits a precipitate-free, sharply defined, set of subbands of
diameters about 375 Å. The 5-T J_c of this wire was
only 2×10^3 A cm^{-2}. All such samples which had been
annealed for times longer than 3 h at temperatures
above 300°C gave rise to α-phase precipitation visible
under the electron microscope. Thus for example
Fig.7-29(b), taken from an intermediate-heat-treated
sample of Ti-50Nb, exhibits according to ARNDT and
EBELING [Arn74] and WILLBRAND *et al* [Wiℓ75[a]] long rows
of particles, or stripes, lying parallel to the subbands. Table 7-7 which correlates the 5-T J_c's of
Ti-50Nb wires with their microstructural states and
the processing conditions which gave rise to them,
appears to confirm the above statement that precipitation, in addition to optimal subband width, is needed
for high J_c. But as indicated above, this conclusion
is incomplete; whether α-phase precipitation is the
principal J_c-enhancing product of the thermomechanical
process, or only complementary or subsidiary to other
microstructural factors, will be considered in Part 4
of this chapter.

In pursuing this investigation into the realm of
higher degrees of cold deformation WILLBRAND and
SCHLUMP [Wiℓ75] prepared 61-filamentary Cu/Ti-50Nb

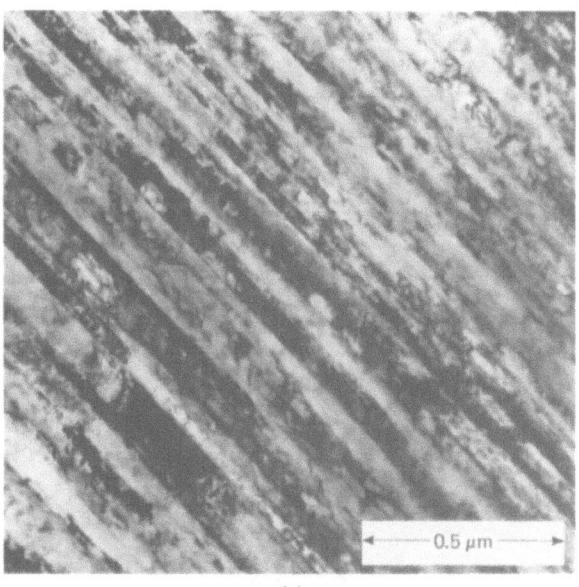

(a)

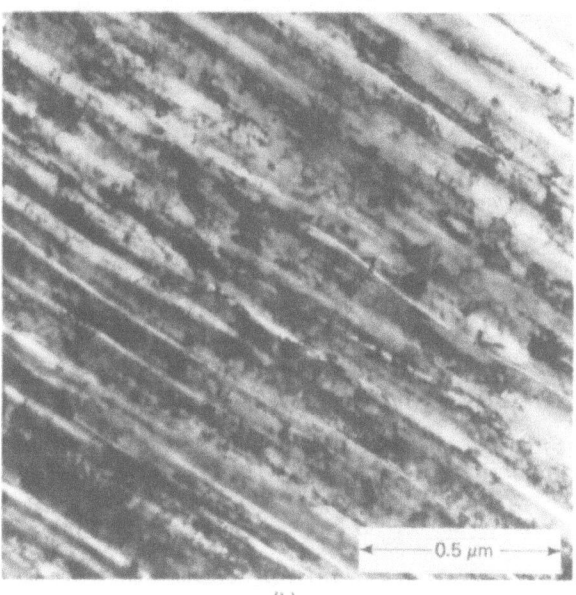

(b)

FIGURE 7-29. Ti-50Nb wire: samples prepared for TEM from a
single-filament (core, 0.13 mm$^\phi$) Cu/Ti-Nb composite in the (a)
as-cold-worked (> 99.98%) condition and (b) as-processed with
intermediate heat treatment; original magnifications 80,000X —
after WILLBRAND *et al* [Wiℓ75[a]] (see also ARNDT and
EBELING [Arn74]). Micrographs courtesy of J. Willbrand; reproduced by permission of Zeitschrift für Metallkunde.

composites again drawing down to a final wire diameter
of 0.25 mm but this time to a final core diameter of
21 µm. As mentioned in the original article and again
in [Mon21.19] in connection with a general discussion
of pinning-force optimization, the wire was subjected
to several intermediate heat treatments prior to final
cold deformations (to a fixed final wire diameter) of

TABLE 7-7 THE EFFECTS OF THERMOMECHANICAL PROCESSING
ON CRITICAL CURRENT DENSITY (J_c), SUBBAND
STRUCTURE, AND PRECIPITATION IN Ti-50Nb,
LISTED IN ORDER OF INCREASING J_c --
Arndt and Ebeling [Arn74].

Heat Treatment	$J_c(5T)^†$ 10^5A cm^{-2}	Subband Size Å	Precipitate Size		
			Not Visible	<500Å	>500Å
--	0.02	375	X		
100h/500°C	0.13	2680			X
3h/500°C	0.20	1950			X
3h/300°C	0.59	425	X		
100h/390°C	0.88	860			X
3h/390°C	0.94	500		X	
100h/300°C	1.12	475		X	
10x3h/390°C	1.49	495		X	
6x3h/390°C	1.55	385		X	
4x3h/390°C	1.60	385		X	

† J_c criterion: 5 µV across 1 cm.

from 80 to 95%. According to WILLBRAND [Wiℓ80] the
equivalent total diameter at the first stage of heat
treatment (which took place before the bundling) was
31 mm. By slight variation of the deformation and
heat treatment cycles it was found possible to produce
two "series" of wire products: *Series A* in which both
subband diameter and precipitate size remained fixed
at 250 Å and 35 Å, respectively, but in which the precipitate number density ranged between about 18×10^{16}
and 24×10^{16} cm^{-3}; *Series B* in which the subband diameter varied from 350 to 400 Å, the precipitate size
from 50 to 130 Å, and the particle abundance was with-
in the range $4 \times 10^{16} \sim 8 \times 10^{16}$ cm^{-3}. Three typical electron micrographs representative of this work are pre-
sented in Fig.7-30. Fig.7-30(a) represents a typical
Series-A sample with subband width of 250 Å and precipitate diameter 35 Å. Figs.7-30(b) and (c) are for
two members of the B Series: in Fig.7-30(b) the subband width is 390 Å and the precipitate diameter 60 Å;
in Fig.7-30(c) the precipitate diameter is 130 Å. In
both cases the precipitates were spherical and strung

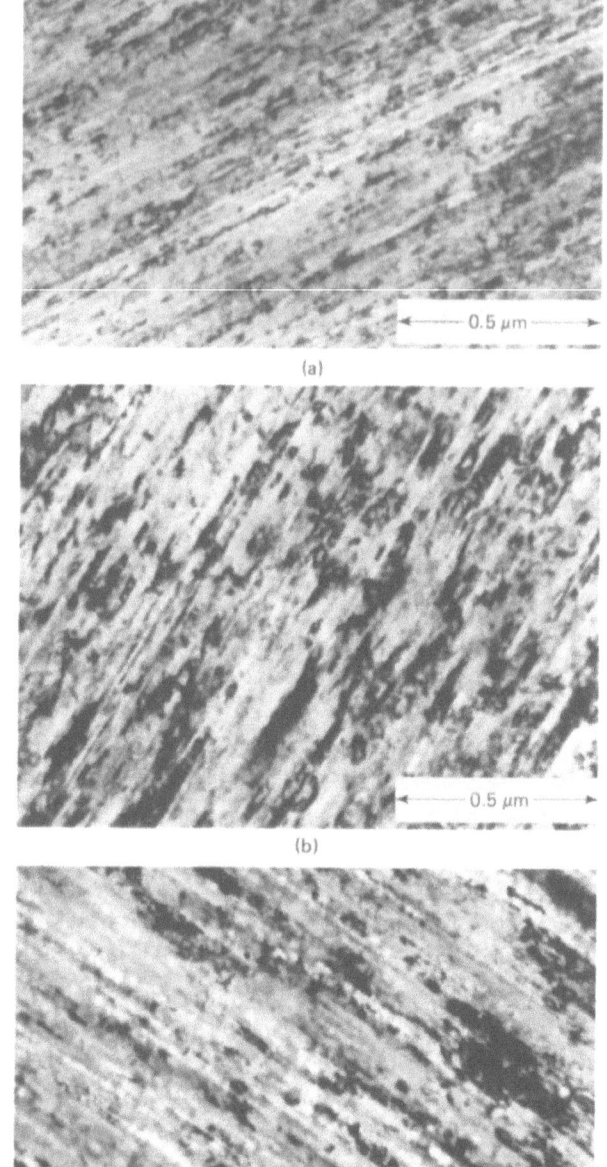

(a)

(b)

(c)

FIGURE 7-30. Ti-50Nb wire: samples prepared for TEM from a
61-filament (fil, 21 µm$^\phi$) Cu/Ti-Nb composite under slight variations of the processing conditions referred to in the text. (a) Series-A
sample: subband ϕ 250 Å; precipitate ϕ, 35 Å. (b) Series-B
sample: subband ϕ, 390 Å; precipitate ϕ, 60 Å. (c) Series-B
sample: precipitate ϕ, 130 Å. Original magnifications 80,000X —
after WILLBRAND and SCHLUMP [Wiℓ75]. Micrographs
courtesy of J. Willbrand; reproduced by permission of Zeitschrift
für Metallkunde.

out in rows along the subband boundaries; the wire with the larger particle diameter possessed the higher current-carrying capacity.

In a continuous series of investigations spanning the last 10-15 years, and still under way, HILLMANN and co-workers have studied the effects of thermomechanical processing on the superconductive properties of Ti-Nb alloys, chiefly Ti-50Nb. Special attention was paid to an effect not considered by other workers, viz precipitate redistribution during final cold work. As part of this superconductor development program TEM was performed on cold-drawn and precipitation-heat-treated Ti-50Nb [Hil73]. Fig.7-31, a dark-field micrograph taken during this investigation shows as bright striations the resulting α-phase precipitates aligned along the deformation-cell boundaries. The subband widths are roughly 200 Å, and the precipitate distribution is seen to be extremely anisotropic. The purpose of *final* cold deformation according to HILLMANN [Hil73] is to redistribute the precipitates in such a way as to increase their efficiency as a flux-pinning system. This goal is achieved as post-heat-treatment cold deformation reduces still further the distance between the precipitate-decorated deformation-cell boundaries and at the same time increases the separation of the precipitates in the drawing direction. Optimization, according to the HILLMANN model, is achieved when the precipitate distribution is approximately isotropic -- thus overoptimization in this model consists of the development of an excess precipitate density in the *radial* direction. The influence of precipitate redistribution on critical current optimization has been investigated by BEST *et al* [Bes79, Bes79[a]] (see also [Mon21.19]) who studied the effects of wire flattening. The influence of final

cold deformation on J_c-optimization has been considered in [Mon21.20].

In studies of Ti-50Nb process optimization, PFEIFFER and HILLMANN [Pfe68] prepared wire specimens by cold rolling an electron-beam-melted starting ingot from 70 mm$^\phi$ to 7 mm$^\phi$ with heat treatments at 65 mm$^\phi$ and 32 mm$^\phi$ of 1h/600°C; cold deformation by wire drawing with intermediate heat treatment then reduced the specimen to diameters of from 1.5 to 0.25 mm. Fig.7-32, a transmission electron micrograph of optimized 0.25-mm$^\phi$ wire (at 20,000X), exhibits a subband "density" of 4.9×10^{11} cm^{-2}. Fig.7-33 shows the same wire after having been rolled to a thickness of 0.1 mm. In this view, taken perpendicular to the plane of the ribbon, roll flattening is seen to have extensively damaged the subband structure, the dislocations previously aligned along the subband boundaries being

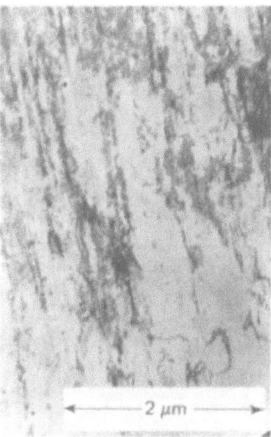

FIGURE 7-32. Ti-50Nb 0.25 mm$^\phi$ wire prepared by optimized thermomechanical processing. TEM original magnification 20,000X — after PFEIFFER and HILLMANN [Pfe68]; micrograph courtesy of H. Hillmann.

FIGURE 7-31. Ti-50Nb wire after deformation and precipitation heat treatment. The α-phase precipitates appear as light striations in dark-field TEM — after HILLMANN [Hil73]; micrograph courtesy of H. Hillmann.

FIGURE 7-33. Process-optimized Ti-50Nb 0.25 mm$^\phi$ wire after roll-flattening to a thickness of 0.1 mm. TEM original magnification 20,000X — after PFEIFFER and HILLMANN [Pfe68], HILLMANN [Hil69]; micrograph courtesy of H. Hillmann.

smeared out to such an extent that they partially ob-literate the strong directionality observable in the previous figure.

In conclusion, attention is drawn to an important metallographic study of a Ti-Nb alloy whose Nb content placed it outside the practical range of α-phase pre-cipitation. NEAL *et al* [Nea71] have examined the microstructures and superconducting properties of Ti-Nb(42 at.%)(analyzed composition, 58 wt.% Nb; nomi-nal composition, 56 wt.% Nb) as a function of deforma-tion and final heat treatment. The results of trans-mission electron microscopy (original magnification 110,000X) performed on selected members of a matrix of thermomechanically processed wires are shown in

Figs.7-34 and 7-35. As mentioned above, the Nb con-tent was such that no appreciable precipitation took place, consequently there was no call for *final* cold deformation which, following HILLMANN (e.g. [Hiℓ73]), is only beneficial if α-phase precipitation is present and in need of redistribution. Two views of the as-cold-drawn cellular structure are shown in Figs.7-34(a) and 7-34(b). The sample depicted was taken from the core of a Cu/Ti-Nb monofilament after core-diameter reduction by cold drawing of from 57 mm$^\phi$ to 0.25 mm$^\phi$ (i.e. 5x10^4:1). In the latter figure especially, a high density of dislocations is visible within the cells. During aging, these dislocations migrate to the cell walls leaving the interiors fairly clear of them. The result of applying an optimal heat treat-ment (unspecified by NEAL [Nea80] but probably 5h/385°C or thereabouts) is shown in Fig.7-35. Unfor-tunately a longitudinal view is not available, but nevertheless the dislocation-migration and wall-sharpening results of optimal heat treatment can be visualized by comparing Figs. 7-34(a) and 7-35.

(a)

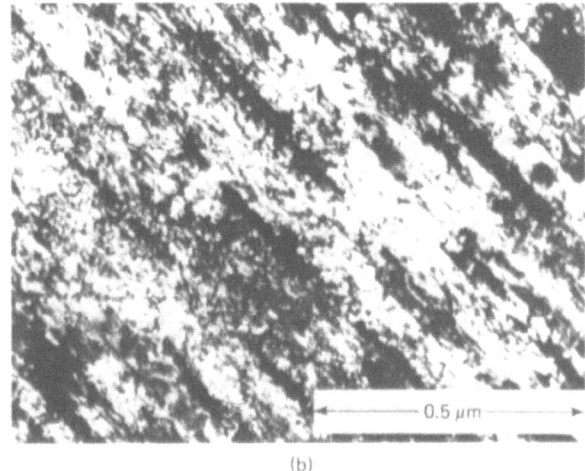

(b)

FIGURE 7-34. Ti-58Nb wire sample prepared for TEM from a single-filament (core, 0.25 mm$^\phi$) Cu/Ti-Nb composite in the as-cold-drawn (5 x 10^4:1) condition: (a) transverse section, (b) longitudinal section. Original magnification, 110,000X — after NEAL *et al* [Nea71]; micrograph courtesy of D. F. Neal.

FIGURE 7-35. Ti-58Nb wire in current-optimized condition; transverse section. Original magnification, 110,000X — hitherto unpublished; micrograph courtesy of D. F. Neal.

7.22 INFLUENCE OF METALLURGICAL VARIABLES ON THE CRITICAL CURRENT DENSITY

Immediately following the experimental renais-sance of type-II superconductivity in 1961 as a result of work performed on Ti-Nb, Mo-Re, and other materials by KUNZLER, MATTHIAS, and others (e.g. [Kun61][Mat65]) at the Bell Telephone Laboratories, the current-carry-ing mixed state in binary TM alloys became the subject of intense investigation. Inquiry into the properties of Ti-base alloys was expanded by BERLINCOURT and HAKE at what was then the Atomics International Division of North American Aviation. Their initial studies centered upon cold-worked alloys, but it soon became apparent that precipitation-aging led to substantial improvements in the high-field current-carrying capacity. A great deal of attention was devoted to an understanding of the metallurgical processes that took place during aging, and in this regard the study of superconductivity both aided, and was aided by, a knowledge of deformation and precipitation mechanisms. Table 7-8 is an index, arranged in order of increasing Nb content, to the research and development literature of Ti-Nb alloys as current-carrying superconductors. In that table the term "aging" is used as an abbreviation for both "precipitation heat treatment" (valid for at.% Nb $\lesssim$ 40) and "major heat treatment" (which does not specify precipitation as a product); in either case, heat treatment for several hours at 380∿390°C is usually implied. Other heat treatments, when administered, serve the following purposes: (a) the promotion of ductility prior to twisting [Col77]; (b) in the form of a 2h/300°C vacuum anneal at final wire size, to increase the conductivity of the Cu stabilizer without affecting the properties of the Ti-Nb.

TABLE 7-8 THE TECHNICAL DEVELOPMENT OF TITANIUM-NIOBIUM SUPERCONDUCTORS

(A) LOW CONCENTRATION (at.% Nb $\lesssim$ 30) TITANIUM-NIOBIUM ALLOYS

At.% Nb	Condition	Reference
8	99.94% cold deformed to wire // aged	[Ral64]
13	97.2% cold deformed to wire // aged	[Hel71]
18	99.94% cold deformed to wire // aged	[Ral64]
20	89% cold rolled	[Hak62[a]]
20	β-quenched // aged	[Sut66]
20	(a) 95.3% cold deformed to wire // aged // 97.1% cold deformed (1967/70) (b) 99.86% cold deformed to wire // aged	[Byc70]
20	(a) quenched from 800°C // 78% cold deformed (b) quenched // 78% cold deformed // aged (c) recrystallized, annealed, quenched (d) recrystallized, annealed, quenched // aged	[Bak69]
20	97.2% cold deformed to wire // aged	[Hel71]
20	(a) recrystallized 15min/1000°C, quenched (b) cold deformed 99.9% (c) cold deformed 99.9% // aged	[Ras72]
21	80% cold rolled // aged	[Vet65]
22	99.94% cold deformed to wire // aged	[Ral64]
22	61% cold drawn // aged to 30 specifications	[Sal65, Sal65[a], Sal66]
22	quenched from 800°C // aged	[Ato66, Ato67][Kra67]
22	(a) 99.8 to 99.99% deformed (b) deformed // aged (c) deformed // multiple aged // deformed	[Pfe68]
22	(a) 83 to 99.98% cold deformed (b) cold deformed // aged (c) recrystallized // aged	[Byc69]

TABLE 7-8 (Continued)

22 { (a) 99.95% cold drawn // aged
{ (b) cold drawn // aged // 90-98% cold drawn (same overall)   [Mci73]

25   99.5% cold work, with and without intermediate heat treatment   [Die62]

25 { (a) 95.3% cold deformed // aged // 97.1% cold deformed   (1967/70)   [Byc70]
{ (b) 99.86% cold deformed // aged

25   97.2% cold deformed to wire // aged   [Heℓ71]

25 { (a) 99.95% cold drawn // aged
{ (b) cold drawn // aged // 90-98% cold drawn (same overall)   [Mci73]

26   99.94% cold deformed to wire // aged   [Raℓ64]

26   99.95% cold deformed to wire // aged   [Ric69, Ric70]

28 { (a) 99.95% cold drawn // aged
{ (b) cold drawn // aged // 90-98% cold drawn (same overall)   [Mci73]

30   cold rolled   [Ber63]

30   99.9% cold deformed // with and without 1h/400°C   [Bet66]

30 { (a) 95.3% cold deformed // aged // 97.1% cold deformed   (1967/70)   [Byc70]
{ (b) 99.86% cold deformed // aged

      { (a) recrystallized 15min/1000°C, quenched
      { (b) cold deformed 99.9%
30   { (c) cold deformed 99.9% // aged   [Ras72]
      { (d) 98.8% cold work // 1h/400°C // 91.7% cold work // 1h/(RT-425°C)
      { (e) above with two stages of intermediate 1h/400°C

30 { (a) 99.95% cold drawn // aged
{ (b) cold drawn // aged // 90-98% cold drawn (same overall)   [Mci73]

30   cold deformed // aged // 75-98% cold drawn   [Ste78][Seg79]

(B) INTERMEDIATE CONCENTRATION (30 ≲ at.% Nb ≲ 40) TITANIUM-NIOBIUM ALLOYS

---

31 { (a) quenched from 1150°C // aged
{ (b) 98% cold rolled // aged   [Voz68]

31   97.2% cold deformed // aged   [Heℓ71]

32   99.95% (2000:1) cold rolled   [Bra72]

33 { (a) 97.2% cold deformed // aged
{ (b) homogenized and quenched // aged   [Loh71]

      { (a) 97.2% cold deformed // aged
33   { (b) 93-99% cold deformed // aged   [Heℓ71]
      { (c) homogenized and quenched // aged

34   86-99% cold rolled   [Laz64]

34   99.94% cold deformed to wire // aged   [Raℓ64]

      { (a) 99.8 to 99.99% deformed
34   { (b) deformed // aged   [Pfe68]
      { (c) deformed // multiple aged // deformed

34 { (a) 99.9% cold deformed // aged
{ (b) recrystallized // aged   [Rau68, Rau68[a]]

34   cold deformed // multiple aged // 75-97.2% cold deformed (multifil.)   [Hiℓ72]
                                                                                 [Hiℓ73[a], Hiℓ74]

TABLE 7-8 (Continued)

34	99.98% cold deformed // multiple aged (mono- and multifil.)		[Arn74][Wiℓ75[a]]
34	cold deformed // multiple aged // 80-95% cold drawn (multifil.)		[Wiℓ75]
34	cold deformed // multiple aged // final cold drawn		[Seg80]
35	90% cold rolled		[Hak62[a]]
35	9:1, 31:1 and 200:1 (99.5%) cold rolled		[Ber63]
35	{ (a) 95.3% cold deformed // aged // 97.1% cold deformed { (b) 99.86% cold deformed // aged	(1967/70)	[Byc70]
35	97.2% cold deformed // aged		[Heℓ71]
36	55-99.95% cold deformed // aged 1h/(350-525°C)		[Ras72]
37	{ (a) 99.95% cold drawn // aged { (b) cold drawn // aged // 90-98% cold drawn (same overall)		[Mci73]
37	cold deformed // aged // 90-97% cold drawn (multifil.)		[Coℓ77]
37	cold deformed // aged // 95% cold drawn (monofil.)		[Cur79]
37	cold deformed // aged // 87-99.5% cold drawn (multifil.)		[Ste78][Seg79]
39	cold drawn		[Ber63]
39	97.2% cold deformed // aged		[Heℓ71]
39	deformed // multiple aged // 75-97.2% cold drawn (multifil.)		[Hiℓ73[a]]
39	cold deformed // aged // 90-97% cold drawn (multifil.)		[Coℓ77]
40	{ (a) 95.3% cold deformed to wire // aged // 97.1% cold drawn { (b) 99.86% cold deformed to wire // aged	(1967/70)	[Byc70]
40	{ (a) recrystallized 15min/1000°C, quenched { (b) cold deformed 99.9% { (c) cold deformed 99.9% // aged { (d) 98.97% cold work // 1h/400°C // 90.5% cold work // 1h/(RT-425°C) { (e) above with two stages of intermediate 1h/400°C		[Ras72]
40	Soviet alloy 60 T		[Laz73]

(C) HIGH CONCENTRATION (at.% Nb ≥ 40) TITANIUM-NIOBIUM ALLOYS

42	{ (a) 33 to 99.999% cold deformed // fixed aging { (b) 99.998% cold deformed // various agings		[Nea71]
42	99.998% cold deformed // aged		[Ham72]
43	"nonoptimized wire"		[Woℓ63]
44	97.2% cold deformed // aged		[Heℓ71]
45	99.87% cold deformed to wire // aged		[Cha70]
49	99.94% cold deformed to wire // aged		[Raℓ64]
50	99.5% cold deformed with and without intermediate heat treatment		[Die62]
50	92% cold rolled		[Hak62[a]]
50	275:1 (99.6%) cold rolled		[Ber63]
50	{ (a) 95.3% cold deformed to wire // aged // 97.1% cold deformed { (b) 99.86% cold deformed to wire // aged	(1967/70)	[Byc70]

TABLE 7-8 (Continued)

50	(a) recrystallized 15min/1000°C, quenched (b) cold deformed 99.9% (c) cold deformed 99.9% // aged	[Ras72]
54	(a) as-cast (b) homogenized 1500°C 99.96% cold deformed	[Pak66]
55	99.95% (2000:1) cold rolled	[Bra72]
60	24 and 240:1 (99.6%) cold rolled	[Hak63[a]]
60	99.94% cold deformed to wire // aged	[Raℓ64]
60	(a) recrystallized 15min/1000°C, quenched (b) cold deformed 99.9% (c) cold deformed 99.9% // aged	[Ras72]
62	99.95% (2000:1) cold rolled	[Bra72, Bra75]
67	99.94% cold deformed to wire // aged	[Raℓ64]
67	(a) 99.9% cold deformed // aged (b) recrystallized // aged	[Rau68, Rau68[a]]
70	(a) recrystallized 15min/1000°C, quenched (b) cold deformed 99.9% (c) cold deformed 99.9% // aged	[Ras72]
72	90% cold rolled	[Hak62[a]]
75	99.5% cold work with and without intermediate heat treatment	[Die62]
75	99.94% cold deformed to wire // aged	[Raℓ64]
82	99.94% cold deformed to wire // aged	[Raℓ64]
83	99.95% (2000:1) cold rolled	[Bra72]
90	98.4% (62:1) cold rolled	[Hak63[a]]
91	99.94% cold deformed to wire // aged	[Raℓ64]

The extreme lattice distortion that accompanies heavy cold work has always hindered the interpretation of transmission electron micrographs. Consequently there was initially a tendency for physical metallurgists to study the relationships between J_c (flux pinning) and microstructure on materials which had not been subjected to cold deformation. In one sense this could be regarded as a temporary removal of one of the variables. But on the other hand: (a) since alternate heat treatment and deformation is empirically necessary for the optimization of J_c; (b) since deformation is an inseparable parameter in that dislocation-cell boundaries, either with or without precipitation, are needed for flux pinning, the results of such studies made, at best, very indirect contributions to the literature of superconductive-alloy process optimization. Finally it has been determined, more-or-less empirically, that the results of applying the processing sequence cold work/heat treatment/cold work are necessary for the support of large transport currents in strong applied magnetic fields. This procedure, whose metallurgical and superconductive consequences were first discussed by HILLMANN and colleagues at Vacuumschmelze GmbH (e.g. [Pfe68], [Hiℓ73, Hiℓ79] and so on) and later by SEGAL et al at Magnetic Corporation of America (e.g. [Coℓ77][Ste78][Seg79, Seg80]) forms the basis of general commercial practice. Thus the development of Ti-Nb into a commercially useful superconducting material can be traced with reference to Table 7-9 which contains the same entries as the previous table but rearranged according to thermo-mechanical process.

The contents of Table 7-9 are discussed under the appropriate headings in Sects.7.23 through 7.28. Of these, Sect.7.26 is subdivided into three compositional ranges according to previously explained schema (cf. for example Sect.7.20.2) partly for convenience, and partly as a scientific aid to an understanding of the products of the deformation-plus-aging process. As discussed in [Mon2.00], moderate-temperature (~400°C) aging has three distinct consequences: *(i)* the precipitation of isothermal ω-phase provided at.% Nb $\stackrel{\sim}{<}$ 30; *(ii)* α-phase precipitation provided at.% Nb $\stackrel{\sim}{<}$ 40; *(iii)* an acceleration of the precipitation processes in the appropriate composition ranges, and a refinement of the subband structures of all the deformed alloys.

TABLE 7-9 DEVELOPMENT OF THERMOMECHANICAL
PROCESSING PROCEDURES FOR Ti-Nb ALLOYS

Condition[†] and At.%Nb	Literature	
R,Q		
20	Sutton and Baker	[Sut66]
20	Baker and Sutton	[Bak69]
20	Rassmann and Illgen	[Ras72]
C,Q,C // D		
20 - 72	Hake *et al.*	[Hak62[a]]
25 - 75	Dietrich *et al.*	[Die62]
30 - 50	Berlincourt	[Ber63]
60, 90	Hake *et al.*	[Hak63[a]]
34	Lazarev *et al.*	[Laz64]
54	Pakhomov *et al.*	[Pak66]
30, 35	Betterton	[Bet66]
22	Bychkov *et al.*	[Byc69]
20	Baker and Sutton	[Bak69]
32 - 83	Brand	[Bra72]
20 - 70	Rassmann and Illgen	[Ras72]
R,Q,C // A		
20	Sutton and Baker	[Sut66]
22	Kramer and Rhodes	[Kra67]
31	Vozilkin *et al.*	[Voz68]
34- 67	Rauch	[Rau68]
22	Bychkov *et al.*	[Byc69]
20	Baker and Sutton	[Bak69]
33	Heller	[Heℓ71]

D // A		
49	Ralls	[Raℓ64]
21	Vetrano *et al.*	[Vet68]
22	Salter	[Saℓ66]
30, 35	Betterton	[Bet66]
20 - 50	Bychkova *et al.*	[Byc70]
31	Vozilkin *et al.*	[Voz68]
22, 34	Pfeiffer and Hillmann	[Pfe68]
34 - 67	Rauch	[Rau68]
26	Ricketts	[Ric69]
20	Baker and Sutton	[Bak69]
22	Bychkov *et al.*	[Byc69]
45	Charlesworth and Madsen	[Cha70]
42	Neal *et al.*	[Nea71]
13 - 44	Heller	[Heℓ71]
42	Hampshire and Taylor	[Ham72]
20 - 70	Rassmann and Illgen	[Ras72]
22 - 37	McInturff and Chase	[Mci73]
D // A // D		
20 - 50	Bychkova *et al.*	[Byc70]
30 - 40	Rassmann and Illgen	[Ras72]
22, 25	McInturff and Chase	[Mci73]
37, 39	Colling *et al.*	[Coℓ77]
37	Curtis and McDonald	[Cur79]
30, 37	Segal *et al.*	[Seg79]
D // A-D-A // D		
25 - 75	Dietrich *et al.*	[Die62]
22, 34	Pfeiffer and Hillmann	[Pfe68]
30, 40	Rassmann and Illgen	[Ras72]
34, 39	Hillmann	[Hiℓ73[a]]
34	Hillmann	[Hiℓ74]
34	Arndt and Ebeling	[Arn74]
34	Willbrand and Schlump	[Wiℓ75]
34	Segal *et al.*	[Seg80]

† R,Q	Recrystallized or β-quenched.
C,Q,C // D	Cast, β-quenched or β-cooled, and cold deformed.
R,Q,C // A	Recrystallized, β-quenched or β-cooled, and aged.
D // A	Cold deformed and aged.
D // A // D	Cold deformed, aged, and final deformed.
D // A-D-A // D	Cold deformed, multiple intermediate aged, and final deformed.

7.23 R,Q—RECRYSTALLIZED OR β-QUENCHED Ti-Nb ALLOYS

Critical current density in recrystallized and quenched rolled sheets of Ti-Nb$_{20}$ was first studied by SUTTON and BAKER [Sut66] as a preliminary to an investigation of the effects of aging at 334°C. The relatively high J_c of the initial as-quenched condition was attributed to flux pinning by a very fine internal structure ($\stackrel{\sim}{<}$1000 Å) of α"-martensite platelets. In a subsequent and related paper, BAKER and SUTTON [Bak69] reported on a multiple-peaked critical-current-anisotropy effect peculiar to a monocrystalline short sample in which <110> was parallel to the pre-recrystallization rolling direction of the sheets. It was suggested that the anisotropy was associated with preferred orientations of the dislocation planes associated with the orthorhombic α"-martensite, which has twelve allowed orientations with respect to the original bcc structure. In the meantime, PFEIFFER and HILLMANN [Pfe68] had noted that Ti-Nb$_{22}$ after recrystallization and air cooling exhibited practically no martensite, while unfavorable (presumably rapid) cooling conditions could lead to dense martensitic transformation which, of course, may be redissolved by annealing at 300°C. In an intercomparison of the effects of: *(a)* no deformation; *(b)* 99.9% cold work; *(c)* 99.9% cold work followed by 1h/400°C aging, RASSMANN and ILLGEN [Ras72] measured the 45-kOe J_c as a function Nb content between 20 and 70 at.%. The critical current density of the as-quenched wire was of course almost two orders of magnitude lower than that after cold working and aging ($\sim$1.5x10^3 as compared to $\sim$7.5x10^4 A cm^{-2}).

7.24 C,Q,C//D—CAST, β-QUENCHED OR β-COOLED AND COLD-DEFORMED Ti-Nb ALLOYS

HAKE *et al* [Hak62^a, Hak63^a] studied the transverse-field critical current anisotropies of cold-rolled ($\sim$90%) Ti-Nb alloys in the composition range 20-72 at.%. It was noted that sharp peaks in J_c occurred whenever the direction of the applied field was parallel to the rolling plane (i.e. parallel to the planes of the lamellate fiber structure) even to the extent that subsidiary peaks were encountered whenever the field was parallel to locally tilted regions such as those at the edges of a sheared strip. A fila-

mentary current-carrying mechanism was suggested. The results of comparable studies of cold-rolled Ti-Nb$_{34}$ undertaken in the Soviet Union by LAZAREV *et al* [Laz64] were also interpreted in terms of a superconductive filamentary model in which the conduction of high current densities in strong magnetic fields was supposed to rely on the existence of a well-developed system of very thin layers, lying approximately parallel to the rolling plane. Thus although it was early recognized that the effects of cold work were extremely beneficial, it was thought that current was being transported along filamentary paths associated in some way with the work-developed fibrous microstructures. In a related paper, BERLINCOURT [Ber63] reported some results for both cold-rolled ribbon and moderately-cold-worked Ti-Nb wire, interpreting the results, with reference to the newly recognized ABRIKOSOV flux-lattice model, in terms of a favorable match between the vortex lattice and an array of structural defects such as a dislocation network. This principle of bulk flux pinning by "matching", which was subsequently advocated on several occasions by HILLMANN [Hiℓ72, Hiℓ74], is reviewed in [Mon21.11].

In later studies, BAKER and SUTTON [Bak69] and BYCHKOV *et al* [Byc69] investigated cold-worked Ti-Nb$_{20}$ and Ti-Nb$_{22}$, respectively, prepared as first steps in comprehensive investigations of the effects on J_c of thermomechanical processing. The former authors found that β-quenching, applied to Ti-Nb$_{20}$, resulted in an orthorhombic α"-martensitic structure which then persisted (although the x-ray diffraction lines became broad and diffuse) during cold deformation. PFEIFFER and HILLMANN [Pfe68] noted that in Ti-Nb$_{22}$ α-phase may be present in the starting alloy depending on the cooling conditions and that, if so, it also would remain during subsequent deformation. In the case of Ti-Nb$_{34}$ (i.e. Ti-50Nb) α-phase, although not initially present, may appear in the heavily deformed wire after suitable aging. For the reasons given above, RASSMANN and ILLGEN [Ras72] had measured the 45-kOe J_c's of quenched-plus-deformed (99.9%) members of the entire Ti-Nb(20 through 70 at.%) alloy series.

7.25 R,Q,C//A—RECRYSTALLIZED, β-QUENCHED OR β-COOLED AND AGED Ti-Nb ALLOYS

Alloys which have not received mechanical deformation are more amenable to TEM investigation than they are after heavy deformation. With undeformed materials it is a relatively straightforward matter to correlate critical current with microstructural changes taking place during intermediate-temperature (~400°C) aging. All alloys included in this section (cf. Table 7-9) are now known to be amenable to isothermal ω-phase precipitation except those investigated by RAUCH [Rau68] (34-67 at.% Nb) and HELLER [Hel71] (33 at.%), but cf. Sect.7.20.2(a). The aging of β-Ti-TM alloys is reviewed in [Mon2.1].

The earliest investigation of the influence of aging (at 334°C) on the critical current density of β-quenched Ti-Nb (in particular, $Ti-Nb_{20}$) was conducted by SUTTON and BAKER [Sut66]. In spite of the development of a fine precipitate dispersion (platelets some $\gtrsim 100$ Å$^\phi$ by $\gtrsim 25$ Å^t) incorrectly identified as α-phase particles, aging was found to be detrimental to J_c. Flux pinning by the initial as-quenched α″-structure appeared to be the more effective; for example whereas J_c(4.22 K, 30 kOe) for the as-quenched α″-alloy was $\gtrsim 4 \times 10^3$ A cm^{-2}, at no time during the aging did it exceed 2×10^2 A cm^{-2}. In a subsequent paper, BAKER and SUTTON [Bak69] correctly identified the precipitation product of 330°C aging as ω-phase developed according to the reaction:

$$\alpha'' \longrightarrow \beta \longrightarrow \omega + \beta$$

$$\text{or} \longrightarrow \omega + \beta' + \beta \quad \text{(cf. [Mon2.3]),}$$

and went on to discuss in detail the flux-pinning characteristics of ω-phase with reference to the results of both critical current density and magnetization measurements.

If a 22 at.% Nb alloy is cooled from the β-field, not only is a higher critical current density obtained, but the metallographic features are simpler to interpret than those of the quenched 20 at.% Nb alloy. KRAMER and RHODES [Kra67] noted that $Ti-Nb_{22}$, rapidly cooled from a 1h/800°C anneal in the β-field, possessed an ω-phase precipitate density of at least 10^{17} cm^{-3}, and that the particle size and spacing were about 50 Å. During aging, J_c increased as the precipitates became enriched with Ti such that a maximal J_c(4.2 K, 30 kOe) of 1×10^5 A cm^{-2} was obtained after 1h/450°C -- higher than that quoted above for the α″-$Ti-Nb_{20}$ alloy. Longer anneals at 400 and 450°C yielded considerably lower values of J_c.

In a companion TEM study of $Ti-Nb_{22}$, BRAMMER and RHODES [Bra67] noted that samples that were vacuum cooled from 800°C contained an "unresolvable ω-phase",

but that aging at 450°C for several tens of h resulted in well developed ω precipitates. After 72h/450°C α-phase began to nucleate at the β grain boundaries. In general studies of the α-phase precipitation process, no evidence is found for its *direct* conversion from ω-phase precipitates. Two common modes of α-phase formation in ω+β-phase alloys are encountered. In one of these, cells of α+β lamellae grow out from the grain boundaries to consume the entire prior ω+β structure; in the other, the α-phase nucleates at the ω/β interfaces [Bla68].

Critical-current results comparable to those referred to above were obtained by BYCHKOV et al [Byc69] with 1h/800°C-recrystallized $Ti-Nb_{22}$ wires after aging at 390 and 425°C. The optimal heat treatment time at 425°C was 3 h, after which J_c(4.2 K, 40 kOe) was about 4.5×10^4 A cm^{-2}.

In a detailed metallographic study of quenched $Ti-Nb_{31}$, an alloy clearly of threshold composition, VOZILKIN et al [Voz68] claimed evidence for ω-phase precipitation, present in equivalent amounts after aging for either 10h/400°C or (2-4h)/500°C. In the former case the precipitate size was some 200-250 Å while after 28h/500°C it had grown to about 500-600 Å. Its J_c(4.2 K, 23 kOe) after heat treatment for (8-15h)/500°C was 2.8×10^4 A cm^{-2}.

RAUCH's studies of recrystallized-and-aged Ti-Nb(34, 44, and 67 at.%) alloys were unaided by metallographic analysis [Rau68]. His interpretations of the effects of heat treatment on $Ti-Nb_{34}$ (viz that during 1-h agings within the temperature range 200-700°C, a temperature of 375°C was optimal for ω-phase precipitation, while α-phase precipitation took place between 450 and 500°C) were based solely on the results of hardness measurements. The accompanying critical current tests revealed that J_c(4.2 K, 55 kOe) attained a maximum of ~3.5×10^4 A cm^{-2} after 1h/400°C, while that in response to 1h/500°C had dropped to 1.2×10^4 A cm^{-2}. As RAUCH [Rau68] correctly pointed out, ω-phase precipitation from the Nb-rich members of the group was not possible. (Indeed, more recent studies indicate that ω-phase is unlikely to form, during aging for reasonable periods of time, in any pure binary Ti-Nb alloys with Nb contents greater than 30 at.%, Sect.7.20.2(a)). In the 1-h aging of $Ti-Nb_{67}$ at temperatures between 250 and 750°C, a broad maximal J_c(4.2 K, 40 kOe) of $1.8-2.2 \times 10^4$ A cm^2 was attained within the temperature range 500-600°C. In casting about for a precipitation mechanism to which this peak in J_c could be attributed, RAUCH suggested the decomposition of the alloy into a pair of β-phases as a possibility. Phase separation in β-Ti-TM alloys is discussed in Sect.7.20.2(b) and [Mon2.3] through [Mon2.6].

As part of a study of critical current density in Ti-Nb-Cu and Ti-Nb-Ge alloys, LÖHBERG [Loh71] and HELLER [Heℓ71] investigated the influence of deformation and subsequent aging on the J_c of the binary alloy bases. But in order to separate the influence of aging from that of deformation, experiments were carried out on a homogenized (3h/1000°C) and quenched Ti-Nb$_{33}$ alloy for comparison with the properties of the deformed-and-aged alloys to be considered in the next section. The most favorable 400°C heat treatment time was 200 h, after which the 40-kOe and 60-kOe J_c's were 5.0×10^4 and 4.3×10^4 A cm^{-2}, respectively. These values may be compared with the 5.5×10^4 and 4.0×10^4 A cm^{-2} obtained after 97.2% cold deformation and an optimal 250h/400°C heat treatment. Anticipating the results of the "deformation//aging" work to follow, it can be seen that 97% deformation is *quite inadequate* if high J_c's are to be developed.

7.26 D//A—COLD-DEFORMED AND AGED Ti-Nb ALLOYS

If strain energy is introduced into the lattice through cold work in the form of rolling, swaging, or wire drawing, an alloy is able to attain thermodynamic equilibrium, during moderate-temperature aging, much faster than would otherwise be the case. Thus not only do the cold-worked-and-aged alloys possess well-developed fibrous dislocation-cell types of microstructure, but the precipitates which form under intermediate-temperate aging conditions are often different from those which appear during the aging of a recrystallized alloy for the same period of time. The situation is summarized in [Mon2.9].

Modern superconductive-alloy wire processing methods may be regarded as extensions of the elementary deformation-plus-aging sequence. It is therefore instructive to review the effects of such a simple two-stage process on J_c and the alloy microstructure. This is done after the relevant literature has been subdivided into three concentration ranges in the manner of Fig.7-36. A limited amount of work has been done on high-Nb-content alloys (>40 at.%) most of which, for fundamental reasons, turn out to be unattractive from a practical standpoint. In contrast, a

considerable amount of attention has been devoted to the low-Nb-content alloys (<30 at.%) especially those of compositions near 20-22 at.% Nb which, based on the early discovery of their potential as practical conductors, had been put into commercial production as early as 1964 by the Atomics International Division of North American Aviation. Unfortunately the Nb-lean alloys suffer from poor fabricability. Ti-Nb alloys in the intermediate-concentration range, 30-40 at.% Nb (45-56 wt.%) are more ductile. This, and the fact that they can support large current densities as a result of appropriate thermomechanical processing, has led to their being the choice for use in present-day commercial superconducting alloy wires. In what follows we trace, with reference to Fig.7-36, the development of cold-worked-plus-heat-treated Ti-Nb superconductors within three broad categories. Metallurgically, the range-terminating compositions of 30 and 40 at.% Nb refer to the limit of the isothermal ω+β field, and the upper "practical" limit of the ∿400°C α+β-phase field, respectively.

7.26.1 Low-Concentration (<30 at.% Nb) Ti-Nb Alloys

The results of several important studies on the influence of moderate temperature, or "warm", aging on the critical current density of cold-worked Ti-Nb(21∿28 at.%) alloys were reported during the period 1965-1973. A comparative survey of the effects of deformation and near-optimal heat treatment on their 30- and 50-kOe critical current densities is presented in Table 7-10. The extent and nature of the cold deformation has varied from 80% cold rolling (following recrystallization), as in the work of VETRANO and BOOM [Vet65] with Ti-Nb$_{21}$, to the more than 99.9% cold deformation administered by RASSMANN and ILLGEN [Ras72] and McINTURFF and CHASE [Mci73]. The numerous attempts which have been made to study the effects of the cold work and heat treatment, regarded as separate variables, will be reviewed below.

Two important studies, those due to SALTER [Saℓ64, Saℓ65, Saℓ65^a, Saℓ66] and HELLER [Heℓ71], some of whose results make up Table 7-11, are included in this discussion. The former work, together with additional studies by RASSMANN and ILLGEN [Ras72] and McINTURFF and CHASE [Mci73] will, however, also be included in Sects.7.27 and 7.28 which deal with final deformed (D//A//D) and intermediate-heat-treated alloys (D//A-D-A//D), respectively. SALTER had administered

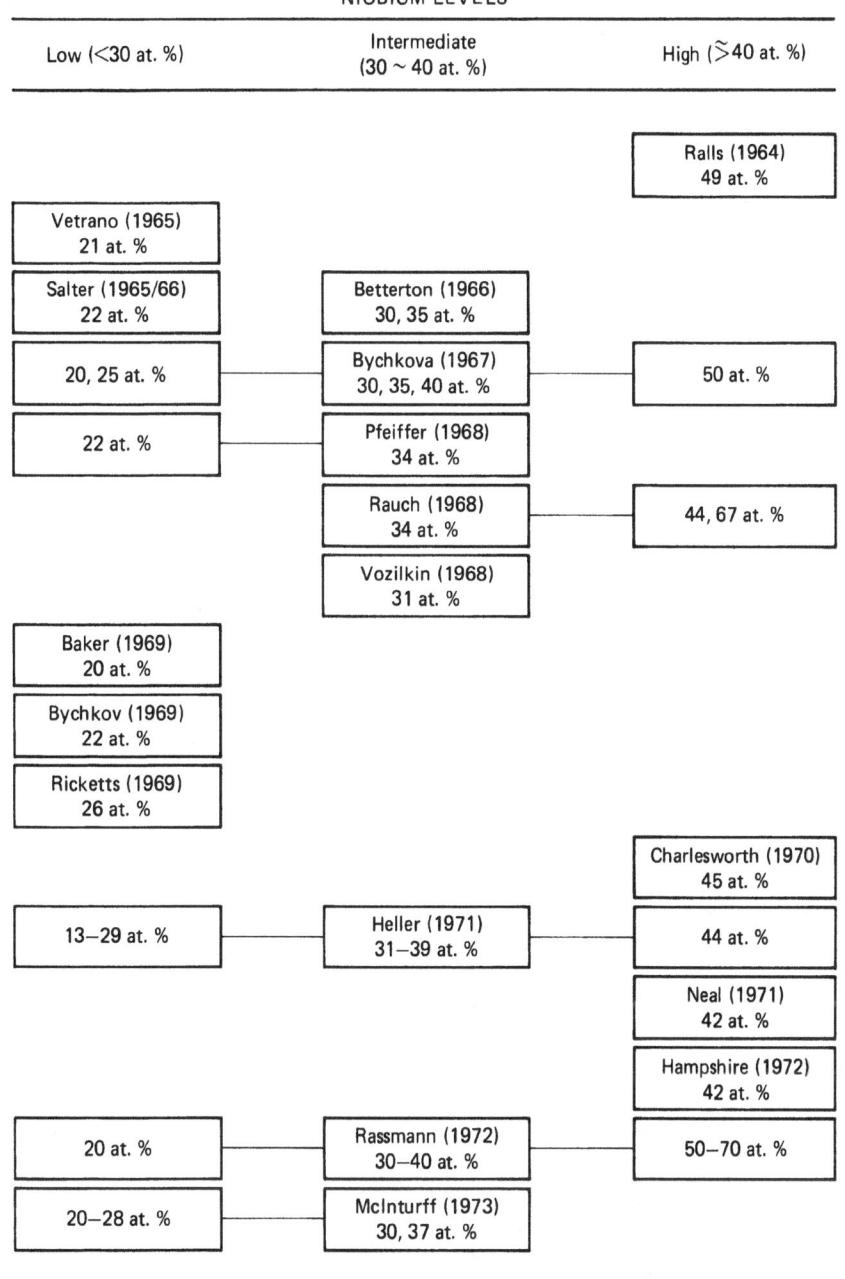

NIOBIUM LEVELS

Low (<30 at. %)	Intermediate (30 ~ 40 at. %)	High (≳40 at. %)
		Ralls (1964) 49 at. %
Vetrano (1965) 21 at. %		
Salter (1965/66) 22 at. %	Betterton (1966) 30, 35 at. %	
20, 25 at. %	Bychkova (1967) 30, 35, 40 at. %	50 at. %
22 at. %	Pfeiffer (1968) 34 at. %	
	Rauch (1968) 34 at. %	44, 67 at. %
	Vozilkin (1968) 31 at. %	
Baker (1969) 20 at. %		
Bychkov (1969) 22 at. %		
Ricketts (1969) 26 at. %		
		Charlesworth (1970) 45 at. %
13–29 at. %	Heller (1971) 31–39 at. %	44 at. %
		Neal (1971) 42 at. %
		Hampshire (1972) 42 at. %
20 at. %	Rassmann (1972) 30–40 at. %	50–70 at. %
20–28 at. %	McInturff (1973) 30, 37 at. %	

FIGURE 7-36. List of pioneering studies of cold-deformed-and-aged Ti-Nb alloys presented in three concentration ranges.

an intermediate anneal of 4h/550°C to his Ti-Nb$_{22}$ in order to facilitate metalworking. Although not specifically recognized at the time, it was presumably this that was instrumental in significantly enhancing J_c to the extent indicated in Table 7-11. The work of HELLER [Heℓ71] was intended to be a comparative evaluation of: *(a)* the variation of the 50-kOe and 100-kOe J_c's of 97%-cold-worked plus 1h/400°C-aged wires with Nb content (in which maxima were discovered at 50 kOe,

33 wt.% Nb and 100 kOe, 50 wt.% Nb, respectively); *(b)* the influence of Ge additions on the critical current density of Ti-49Nb. In it, thermomechanical processes were standardized rather than optimized, for which reason it was inappropriate to include the numerical results in Table 7-10. A comparison of the 30-kOe data of HELLER with those of the other workers suggests that the critical current values are anomalously low. This could conceivably have been due to a

TABLE 7-10 COMPARATIVE CRITICAL CURRENT DENSITIES AFTER DEFORMATION AND OPTIMAL OR NEAR-OPTIMAL
HEAT TREATMENT -- LOW-CONCENTRATION-RANGE TITANIUM-NIOBIUM ALLOYS

First Author	Date	At.% Nb	Percent Cold Deformation	Heat Treatment	Critical Current Densities[*] Field(kOe)/J_c(A cm^{-2})		Reference
Vetrano	(1965)	21	80	3h/412°C	30/1.9x10^5		[Vet65]
Bychkova	(1967/70)	25	99.86	1h/450°C	26/1.0x10^5		[Byc70]
Ricketts	(1969)	26	99.95	1h/400°C		50/7.4x10^4	[Ric69]
Bychkov	(1969)	22	99.64	3h/390°C		50/4.1x10^4	[Byc69]
Rassmann	(1972)	20[+]	99.97	1h/400°C		50/6.7x10^4	[Ras72]
McInturff	(1973)	22 25 27.6	99.95	3h/350°C 400h/325-350°C 400h/325-350°C	30/1.7x10^5 30/2.9x10^5 30/2.9x10^5		[Mci73]

+ Kroll-process Ti

* J_c criteria:

[Byc70] : 50 mV across 12 cm.
[Ric69] : 1 µV across 4 cm.

[Byc69]: 10^{-8}-10^{-7} V across sample.
[Ras72]: 40 µV across sample.
[Mci73]: 10^{-9} Ω cm.

TABLE 7-11 COMPARATIVE CRITICAL CURRENT DENSITIES OF INTERMEDIATE-WARM-ANNEALED (SALTER) AND
STANDARD-HEAT-TREATED (HELLER) LOW-CONCENTRATION-RANGE TITANIUM-NIOBIUM ALLOYS

First Author	Date	At.% Nb	Condition Prior to Heat Treatment	Heat Treatment	Critical Current Densities[+] Field(kOe)/J_c(A cm^{-2})		Reference
Salter	(1965/1966)	22	swaged to 0.025 in.$^\phi$ annealed 4h/550°C cold drawn to 0.010 in.$^\phi$ (final cold work: 84%)	1.7h/400°C	30/2.5x10^5	50/1.0x10^5	[Sal66]
Heller	(1971)	13 20 25	cold drawn 97.2% from cast ingot	1h/400°C	30/1.0x10^4 30/5.5x10^4 30/4.7x10^4	50/2.0x10^3 50/4.0x10^4 50/2.7x10^4	[Hel71]

+ J_c criteria: [Sal66]: current at quench
[Hel71]: 100µV across 3.5 cm

current-lead (ohmic-heating) induced premature transition to the normal state, brought about by the need to supply much larger currents to HELLER's 1-mm$^\phi$ wires than were required by the 0.25-mm$^\phi$ and 0.38-mm$^\phi$ samples measured by SALTER and McINTURFF, respectively.

(a) Cold Work. As indicated in Table 7-10, cold work has been varied from 80% by rolling to 99.95% by cold drawing with rather a small impact on the final critical current density. This suggests that in the low-Nb-concentration range the critical current density is controlled by precipitation effects rather than by details of the cold-worked microstructure. In order to investigate the influence of pre-heat-treatment cold work in a controlled manner, BYCHKOV et al [Byc69] focused attention on only a single alloy (Ti-Nb$_{22}$) and varied the area reduction ratio from 6.3:1 (84%) to 5300:1 (99.98%) prior to aging for 1h/425°C. As part of the same investigation, various aging times at 390°C and 425°C were also studied. As shown in the Fig.7-37, the effect of prior cold work was slight. This conclusion is substantiated, at least with respect to the 50-kOe data, by the results of HELLER [Hel71], Table 7-11, who obtained for Ti-Nb$_{20}$ (97.2% cold worked plus 1h/400°C) a 50-kOe J_c of 4.0×10^4 A cm^{-2}, which may be compared with, for example, BYCHKOV's 50-kOe J_c for Ti-Nb$_{22}$ (99.98% cold worked plus 1h/425°C) of 3.4×10^4 A cm^{-2} [Byc69].

(b) Aging Temperature. In the initial experiments of VETRANO and BOOM [Vet65] with Ti-Nb$_{21}$, the temperature of 3-h heat treatments were varied up to 460°C during which J_c(4.2 K, 30 kOe) passed through a maximum at 412°C. SALTER [Sal66] obtained comparable results with Ti-Nb$_{22}$ for which, during aging for an unspecified time at temperatures between 200 and 650°C, J_c passed through a sharp maximum (width ±19°C at

0.7J_c) centered on 407°C. BYCHKOV et al [Byc69] in studies of Ti-Nb$_{22}$ focused on only the two temperatures, 390 and 425°C. The results of SALTER's work [Sal65] on a similar alloy indicated that of the three variables -- degree of cold work, aging time, and aging temperature -- the latter was the most sensitive. Thus J_c(4.2 K, 30 kOe), when plotted vertically against aging time and aging temperature on a three-dimensional isometric diagram, peaked *broadly* within the aging-time range of 5∿15 h and *sharply* at a temperature of about 350°C. The results of McINTURFF and CHASE [Mci73] were essentially in agreement with this. With reference to their data, if a fixed aging time of 10 h is selected, maximum J_c(4.2 K, 30 kOe) occur at temperatures of 325°C (Ti-Nb$_{22}$), 375°C (Ti-Nb$_{25}$), 375-350°C (Ti-Nb$_{27.6}$), and 350°C (Ti-Nb$_{30}$). RICKETTS [Ric69], in studies of 99.95% cold-deformed Ti-Nb$_{26}$, selected 1 h as a fixed aging time and 50 kOe as the fixed field for J_c determination. Within these constraints, as the aging temperature was varied up to 600°C, J_c(4.2 K) passed through a sharp maximum at 400°C. Similar results were reported by RASSMANN and ILLGEN [Ras72] for the 45-kOe J_c of Ti-Nb$_{30}$ during 1-h aging at temperatures of up to 800°C. Just how these maxima might shift in temperature if (i) the fixed aging time were extended to 5 or 10 h or (ii) if appreciably different fixed fields were selected, was not explored.

(c) Aging Time. Aging times have been varied from about 1 to 400 h. In the initial experiments of VETRANO and BOOM [Vet65], 3 h was selected as a fixed aging time. RICKETTS [Ric69] used 1 h. Several workers have investigated time as a parameter in isothermal annealing experiments. For example, BYCHKOV et al [Byc69] varied the aging time from 1 to 10 h, and as the result of one series of experiments on Ti-Nb$_{22}$, claimed that at 390°C an aging time of 3 h was optimal. In studies of Ti-Nb$_{20}$, RASSMANN and ILLGEN [Ras72] intercompared the effects on J_c(4.2 K, 45 kOe) of aging for times of up to 24 h at 200 and 500°C. In experiments on the somewhat higher concentration alloys Ti-Nb$_{25}$ and Ti-Nb$_{27.6}$, McINTURFF and CHASE [Mci73] noted that the 30-kOe J_c increased monotonically as the aging time, at temperatures of about 325-350°C, increased beyond 400 h. As a result of SALTER's comprehensive study of Ti-Nb$_{22}$ mentioned above [Sal65], in which the influences of both aging time and temperature on J_c were investigated, it can be concluded that J_c(4.2 K, 30 kOe), in passing "isothermally" through a

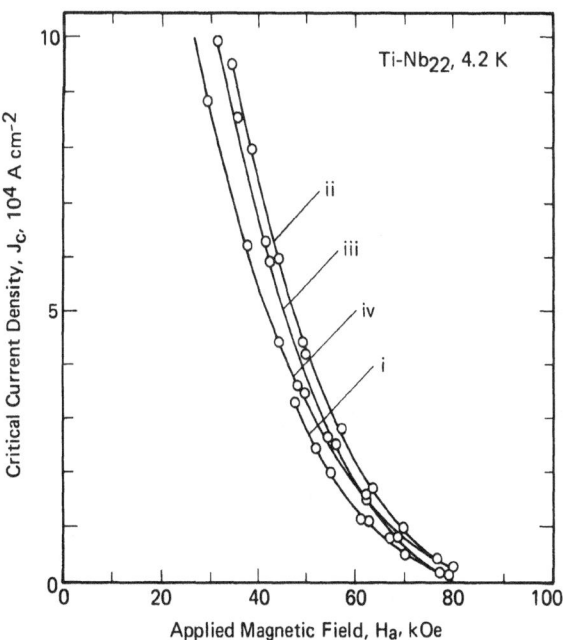

FIGURE 7-37. Critical current densities at 4.2 K of Ti-Nb (22 at. %) aged 1 h/425°C after cold deformations of: (i) 84.13%, (ii) 98.55%, (iii) 99.88% and (iv) 99.98% (transition criterion: 10^{-8} to 10^{-7} V) — after BYCHKOV et al [Byc69].

broad maximum, was rather insensitive to aging time at 350°C.

(d) Precipitation. As was pointed out above, as well as in [Mon1.6] and [Mon1.8], ω-phase is the precipitate which forms during the β-quenching or β-cooling of low concentration Ti-Nb alloys. If, as BRAMMER and RHODES [Bra67] have claimed, it is diffuse or unresolvable in the immediately cooled alloy, it soon develops into crystalline precipitates upon aging at temperatures of up to about 450°C. The situation with regard to deformed-plus-aged alloys is distinctly different. In fact, in the literature being reviewed under this heading there is very little reference to ω-phase precipitation. Basing his conclusions on the results of x-ray diffractometry, RICKETTS [Ric69] stands alone in his confident diagnosis of ω-phase precipitation as a product of the 1h/400°C aging of $\sim$99.95% cold-deformed $Ti-Nb_{26}$ wire. The x-ray diffraction studies of aged $Ti-Nb_{21}$ by VETRANO and BOOM [Vet65] indicated the presence of 2000-Å size α-phase precipitates. SALTER [Sal66], using $Ti-Nb_{22}$, confirmed that α-phase was present in the thirty deformed-and-aged members of his heat-treatment matrix. But since the approach used was to measure the relative intensities, I_α/I_β, of the strongest diffraction peaks -- (011) for α and (110) for β -- the results do not *exclude* the possible presence of ω-phase. For a detailed discussion of phase decomposition in cold-worked $Ti-Nb_{22}$ further reading of [Sal66] is recommended. BYCHKOVA *et al* [Byc70] observed that α-phase precipitation had taken place after 1h/450°C in $Ti-Nb_{20}$ and after 5h/450°C in $Ti-Nb_{25}$; they did not record the presence of any ω-phase but suspected that some might have been present as a result of aging at temperatures below 450°C. According to BAKER and SUTTON [Bak69], in β-quenched $Ti-Nb_{20}$ the as-quenched orthorhombic martensite α'' is retained during cold working, and on aging passes through a strained-β stage before precipitating α-phase. BYCHKOV *et al* [Byc69] claimed that although ω-phase was to be found in recrystallized-and-aged $Ti-Nb_{22}$, the precipitation product of deformation-and-aging was α-phase, although the methods used for arriving at this conclusion were not discussed. The extremely useful work of McINTURFF and CHASE [Mci73] on cold-worked-and-aged $Ti-Nb_{22-30}$ was not, unfortunately, accompanied by any microstructural examinations. In what was essentially an extension of the procedure used by SALTER [Sal66] in examining the precipitates in aged $Ti-Nb_{22}$, RICKETTS

[Ric69]·used relative-intensity x-ray diffractometry in a search for both α *and* ω precipitation during the 1-h aging of $Ti-Nb_{26}$. The amount of α-phase present was followed using the $(11\bar{2}0)$ α-peak, while that of ω-phase was followed using the $(2\bar{1}\bar{1}1)$ ω-peak. After normalization to the intensity of $(211)_\beta$ the resulting ratios I_α/I_β and I_ω/I_β were plotted as functions of 1-h aging temperature as in Fig.7-38. The results showed: *(i)* that ω- and α-phase precipitation can be expected to coexist at some stage during the aging; *(ii)* that ω-phase reaches peak abundance in 1-h anneals at about 400°C; *(iii)* α-phase precipitation begins to dominate at about 550°C. With reference to the equilibrium phase diagram, Fig.7-1, we note that all precipitates in $Ti-Nb_{26}$ must dissolve at $\sim$650°C; i.e. at the $(\alpha+\beta)/\beta$ transus. In studies of $Ti-Nb_{22}$ (as well as the intermediate-concentration alloy $Ti-Nb_{34}$) using light and electron microscopy and electron diffraction, PFEIFFER and HILLMANN [Pfe68] noted that, after strong deformation, aging caused α-phase to precipitate out at the boundaries of the subbands. Since these boundaries or cell walls act as nucleation sites for α-phase, the stronger the deformation the greater the density, fineness, and uniformity of distribution of the subsequent precipitation. Heavy

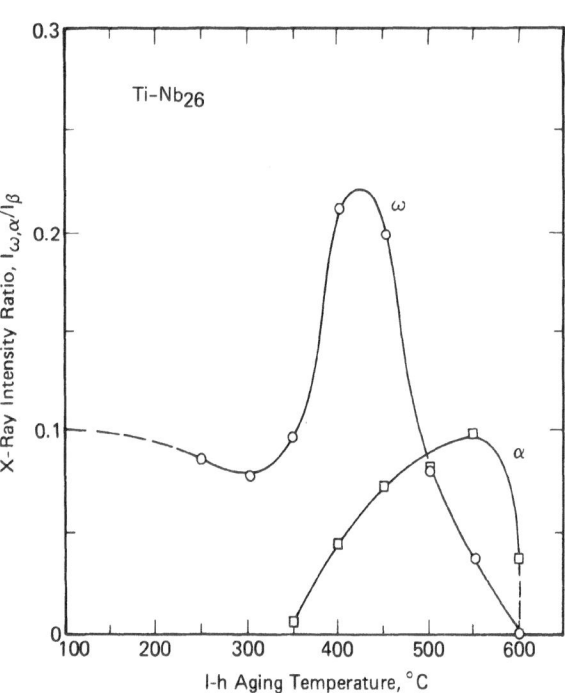

FIGURE 7-38. Precipitation of ω and α phases during the 1-h aging of previously $\sim$ 99.95% cold-deformed Ti-Nb (26 at. %) wires (Ti-40Nb plus 440 ppm oxygen) as determined by relative-intensity X-ray diffractometry — after RICKETTS [Ric69].

deformation also increases the speed of the precipitation process which depends very strongly on the number per unit area of the subbands. PFEIFFER and HILLMANN [Pfe68] also discovered that the aging of strongly deformed Ti-Nb$_{22}$ for 10 min/380°C yielded ω-phase precipitation, which after only 2 h at that temperature was converted to α-phase. Although the extensive lattice damage, in the form of subbands and dislocations, prevented the ω-phase from being imaged in the electron microscope its presence could be confirmed by electron diffraction.

7.26.2 Intermediate-Concentration (30-40 at.% Nb) Ti-Nb Alloys

A brief report of the influence of cold work followed by heat treatment on the critical current densities of intermediate-concentration Ti-Nb alloys (viz Ti-Nb$_{30}$ and Ti-Nb$_{35}$) was published in 1966 by BETTERTON et al [Bet66]. The research described, which was based on the results of an earlier study of Zr-Nb alloys, was part of an investigation into the effects of interstitial elements on flux pinning. This preliminary study was followed by a succession of investigations of cold-worked and heat-treated alloys covering the "intermediate" composition range (and beyond) as summarized in Table 7-12. Towards the end of this sequence of investigations of two-stage (D//A) thermomechanical processing, Cu-clad monofilamentary wires were introduced as sample materials by McINTURFF and CHASE [Mci73]. At the same time heavy deformation was applied, and high critical current densities began to be achieved. Table 7-12 indicates that although wide ranges of cold-work area reduction and heat-treatment time have been used, heat-treatment *temperatures* over the period have tended to be centered near 400°C.

As a prelude to an investigation of the influence of Ge additions on the J_c's of Ti-Nb alloys, HELLER [Hel71] measured a number of binary Ti-Nb control alloys after 97.2% cold work and 1h/400°C heat treatment. In addition, the response of a selected member of the

TABLE 7-12 COMPARATIVE CRITICAL CURRENT DENSITIES OF DEFORMED AND OPTIMALLY, OR NEAR-OPTIMALLY, HEAT TREATED INTERMEDIATE-CONCENTRATION-RANGE TITANIUM-NIOBIUM ALLOYS

First Author	Date	At.% Nb	Percent Cold Deformation	Heat Treatment	Critical Current Densities[+] Field(kOe)/J_c(A cm^{-2})		Reference
Betterton	(1966)	30	99.8	1h/400°C	30/3.0x10^4		[Bet66]
Bychkova	(1967/70)	30	99.86	1h/450°C	26/6.3x10^4		[Byc70]
		35	99.86	10h/450°C	26/3.2x10^4		
Rauch	(1968)	34	99.9	1h/400°C	30/7.0x10^4	50/4.0x10^4	[Rau68]
Vozilkin	(1968)	31	98	28h/500°C	23/4x10^4		[Voz68]
Heller	(1971)	33	97.2	250h/400°C	30/6.0x10^4	50/5.1x10^4	[Hel71]
		39	97.2	250h/400°C	30/5.4x10^4	50/2.5x10^4	
Rassmann	(1972)	30	99.9	1h/400°C		45/7.2x10^4	[Ras72]
		36	99.95	1h/375°C		45/9.2x10^4	
		40	99.9	1h/400°C		45/2.6x10^4	
McInturff	(1973)	30	99.95	400h/350°C	30/2.7x10^5		[Mci73]
		38	99.95	400h/350°C	30/4.7x10^4		

[+] J_c criteria: [Bet66]: probably 1μV across 4 cm. [Hel71]: 100μV across 3.5 cm.
[Byc70]: 50 mV across 12 cm. [Ras72]: 40μV across sample.
[Rau68]: 1μV across 4 cm. [Mci73]: 10^{-9} Ω cm.

series, viz Ti-49Nb, to a wide range of thermomechanical processing variables was investigated. Some representative data from this very extensive investigation are included among other early "optimal" results in Table 7-12, and also in Table 7-13 wherein J_c at fixed "standard" heat treatment is presented as a function of composition within the range under consideration. A study of HELLER's results, as they appear in Tables 7-11, 7-12, and 7-13, indicates that if the deformation is light (area reduction $\sim$97%) a modest degree of heat treatment (1h/400°C) is adequate to permit the *low-concentration* alloys to support relatively high critical current densities; but that under the same deformation conditions, the development of adequate precipitation in the *intermediate-concentration* range alloys requires much longer heat treatment times ($\sim$250 h) at the standard temperature of 400°C.

The effects of thermomechanical processing on the critical current density of intermediate concentration Ti-Nb alloys will be considered further under the headings: *(a)* starting-ingot condition; *(b)* cold work; *(c)* aging temperature; *(d)* aging time.

TABLE 7-13 CRITICAL CURRENT DENSITIES (J_c) OF 97.2% COLD WORKED Ti-Nb ALLOYS IN RESPONSE TO AGING AT 400°C FOR THE TIMES SPECIFIED -- after Heller [Heℓ71].

At.% Nb	Heat Treatment Time	J_c, 10^3A cm^{-2}, at Fields Specified		
		30 kOe	50 kOe	100 kOe
31	1 h	11	7.8	2.6
33	"	11	6.8	3.2
35	"	7.2	4.4	3.1
39	"	4.6	2.8	2.0
[44[††]	"	4.4	2.8	1.4]
33	250 h	60	51	
39	"	54	25	
[44[††]	"	24	12]	

† J_c criterion: 100 μV across 3.5 cm of 1-mm$^\phi$ bare wire.

†† A "high-concentration" result included for later reference.

(a) Starting Ingot Condition. In studies of J_c optimization the questions of starting material purity and homogeneity eventually arise. HELLER [Heℓ71] has evaluated the consequences of using Kroll-process Ti sponge, as compared to iodide Ti, in the preparation of Ti-Nb$_{33}$. The improvement in critical current density obtained upon substituting sponge for iodide Ti as one of the starting materials, was discussed in terms of precipitation kinetics. It was considered that the higher impurity content of the sponge led to an increase in the reaction kinetics and consequently to a higher J_c than was obtainable with iodide Ti under similar processing conditions. Typical impurity contents of these two grades of Ti are given in Table 7-14; see also Table 8-12.

TABLE 7-14 IMPURITY CONTENTS (WEIGHT %) OF KROLL-PROCESS AND IODIDE-PROCESS TITANIUM -- Rassmann and Illgen [Ras72].

Element	Iodide Ti	Kroll Ti
Mg	0.01	0.13
Si	0.01	0.05
Al	0.02	
Fe	0.01	0.20
Ni	0.01	
Co		0.02
Cr	0.01	
Mn	0.005	0.02
C	0.01	0.08
N	0.02	0.04
O	0.02	0.11

The question of starting-Ti purity was also addressed by RASSMANN and ILLGEN [Ras72]. Fig.7-39, taken from their study, shows that in response to $\sim$99.97% cold work followed by 1h/400°C heat treatment, Ti-Nb$_{40}$ prepared from Kroll-process Ti had almost one order of magnitude advantage over the same alloy prepared from an iodide-Ti base.

HELLER [Heℓ71] also intercompared, again using Ti-Nb$_{33}$, the critical current densities of wires prepared from as-cast and homogenized, respectively, starting ingots. Measurements were made of the J_c field dependences of 1-mm$^\phi$ bare wires, aged at 400°C for times of up to 500 h, after swaging-and-drawing

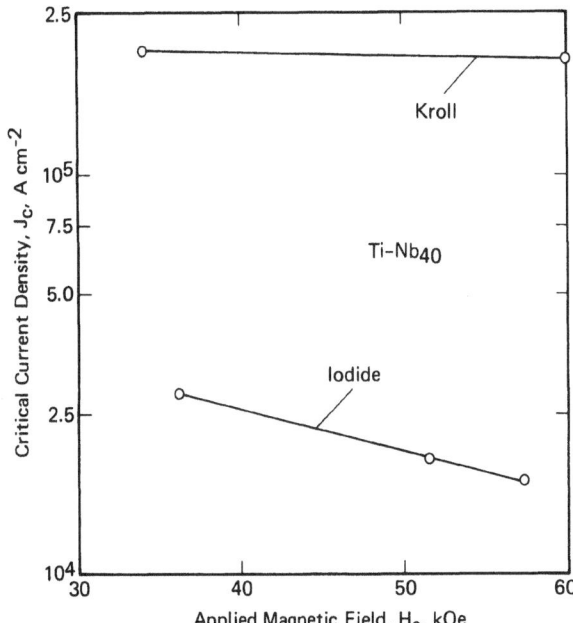

FIGURE 7-39. Influence of starting material purity on J_c: an intercomparison of the critical current densities of 99.97% cold deformed and 1 h/400°C aged Ti-Nb (40 at. %) alloys prepared from either Kroll-process or iodide-process Ti (transition criterion: 40 μV) — after RASSMANN and ILLGEN [Ras72].

from 6-mm$^\phi$ rods whose conditions were: *(i)* as chill cast; *(ii)* homogenized for 3h/1200°C plus water quenched. After 8h/400°C, the wire prepared from the as-cast ingot possessed a higher 50-kOe J_c (viz 2.5x10^4 A cm^{-2}) than that of the 10h/400°C-aged homogeneous wire (viz 6.1x10^3 A cm^{-2}). This finding was in qualitative agreement with the results of a similar experiment performed much earlier by PAKHOMOV *et al* [Pak66], cf. Sect.7.19.1. To be sure, HELLER [Heℓ71] went on to show that subsequent aging for 500h/400°C raised the critical current densities of both wires to comparable values: viz 3.2x10^4 A cm^{-2} for the "as-cast" wire, and 4.0x10^4 A cm^{-2} for the "homogenized" wire, and erased any initial advantage held by the as-cast material.

(b) Cold Work. Preliminary studies of the influence of varying degrees of prior cold deformation on the critical current density of aged Ti-Nb$_{33}$ were carried out by HELLER [Heℓ71]. Unfortunately, both the average degree of deformation (∿97%) and its range (93-99%) were too small to yield satisfactory data. At about the same time RASSMANN and ILLGEN [Ras72] were investigating the effect of cold work on the 45-kOe J_c of Ti-Nb$_{36}$. Measurements were made on samples that had experienced 55, 92.5, 96.5, 99.3, and

99.95% deformation followed by heat treatments of 1h/(350-525°C). The results showed that J_c did not respond much to the pre-heat-treatment deformation until it had exceeded the 92.5% level, after which a significant increase began to take place, especially between 99.3 and 99.95% (the limit of the test). This conforms exactly to the experience of NEAL *et al* [Nea71] during the processing of Ti-Nb$_{42}$ (a nonprecipitating alloy), namely that in samples which were cold worked from 33 to 99.999% and heat treated 1h/385°C, the critical current density increased sharply as soon as the cold work exceeded the 99% level. These results, as well as those of RASSMANN and ILLGEN [Ras72], were explained in terms of the development of a suitably fine subband texture. The relative influences on J_c of subbands and precipitates, when they occur together in the same alloy, are to be considered below.

(c) Aging Temperature. The influence of aging temperature on J_c has been studied by BYCHKOVA, RAUCH, HELLER, RASSMANN and ILLGEN, McINTURFF and CHASE, and many others. It is interesting to note that in spite of the different degrees of prior cold work administered (cf. Table 7-12), and differing claims with regard to the principal flux-pinning mechanisms (and the observed presence or absence, as the case may be, of precipitation), maximum J_c was always achieved for aging temperatures close to about 400°C, provided the measuring field was equal to or less than 50 kOe. BYCHKOVA [Byc70], using a 26-kOe measuring field and 99.86% cold-worked alloys of compositions between 25 and 40 at.% Nb, noted that optimal J_c was obtained for 1-h aging temperatures of between 400 and 450°C. In studies by RAUCH [Rau68] of 99.9% cold-worked Ti-50Nb, J_c as a function of 1-h aging temperature rose to a sharp maximum at 400°C with a small shoulder at 500°C. In experiments by HELLER [Heℓ71] on 97.2% cold-worked Ti-49Nb, the influences on J_c of aging times extending beyond 100 h, and aging temperatures of 300, 400, 500, and 600°C, were explored. Employing a common aging temperature of 16 h, which was meaningful for these materials in the temperature range 300-500°C,[†] HELLER [Heℓ71] noted that: *(i) at 50 kOe,* J_c peaked sharply at 400°C, in agreement with the other work; *(ii) at 100 kOe,* J_c decreased continuously with 16-h aging

---

[†] J_c did not pass through a maximum during 0-16 h aging at these temperatures, although at 600°C, J_c was already decreasing with time beyond about 10 min.

temperature from the as-worked condition. Both RAUCH and HELLER attributed the very favorable flux pinning noted during the 400°C aging of the above-mentioned alloys to the formation of ω-phase -- but cf. Sect.7.20.2(a) for a discussion of the conditions under which this precipitate is expected. The experiments of RASSMANN and ILLGEN [Ras72] also laid claim to some attention under this category; for each level of deformation between 55 and 99.95%, the 45-kOe J_c was measured as a function of 1-h aging temperature. The results, aided by some license as a substitute for the missing data, are plotted in Fig.7-40. Portrayed is an optimal 1-h aging temperature which, not unexpectedly, decreases with increasing levels of prior cold work. In experiments by McINTURFF and CHASE [Mci73], the 30-kOe J_c of Ti-Nb$_{30}$ was measured as function of aging temperature between 275 and 520°C and aging times of up to 400-500 h. The favored temperature for all aging times was 350°C; in all cases J_c increased monotonically with aging time and at 100 h, J_c(30 kOe)

lay between 1.2×10^5 and 2.0×10^5 A cm^{-2}. The 30-kOe J_c of Ti-Nb$_{37.7}$, the other intermediate-concentration alloy investigated, was relatively low ($\sim 0.4 \times 10^5$ A cm^{-2}) and rather insensitive to aging times of more than a few tens of hours.

(d) Aging Time. In maximizing J_c, a wide range of aging times have been required depending, it seems, on the degrees of prior cold work that have been administered. In the studies by BYCHKOVA *et al* [Byc70] of 99.86% cold-worked Ti-Nb(30 and 35 at.%), maximal J_c was not achieved during aging for up to 10 h at 450°C. In HELLER's measurements of 97.2% cold-worked Ti-49Nb [Heℓ71], the generally highest member of a family of J_c *versus* H$_a$ curves representing $\frac{1}{4}$- to 500-h agings at 400°C was that for 250 h. This result is consistent with the observation that in plots of J_c(50 kOe) *versus* aging time up to 180 h, the 300°C and 400°C J_c-isothermals increased monotonically with time. At higher temperatures, J_c-maxima were encountered within that time span -- that for J_c(50 kOe, 500°C) occurring at about 20 h and for J_c(50 kOe, 600°C) at about 10 min. The optimal aging time at 400°C for J_c(100 kOe) is about 17 h. With regard to precipitation, HELLER claimed that the heat treatment of these samples (i.e. the 97.2% cold-worked Ti-49Nb) resulted in ω- and α-phase precipitation.[†] This experience then stimulated an intercomparison of the J_c *versus* H$_a$ characteristics of 97%-cold-worked Ti-Nb(33, 39, and 44 at.%) alloys, aged *(i)* for 40h/400°C, to favor the high-field behavior, and *(ii)* for 250h/400°C, to optimize the 50-kOe current-carrying capacity. The results of this work showed that, within a group of similarly treated alloys, J_c decreased as the Nb content increased beyond 33 at.%. Whereas in the latter alloy the longer aging time produced only a slight improvement in J_c throughout most of the applied-field range, for Ti-Nb$_{39}$ and Ti-Nb$_{44}$, increasing the 400°C aging time from 40 to 250 h had the effect of doubling the mid-range critical current densities; the respective pairs of J_c(H$_a$) curves did,

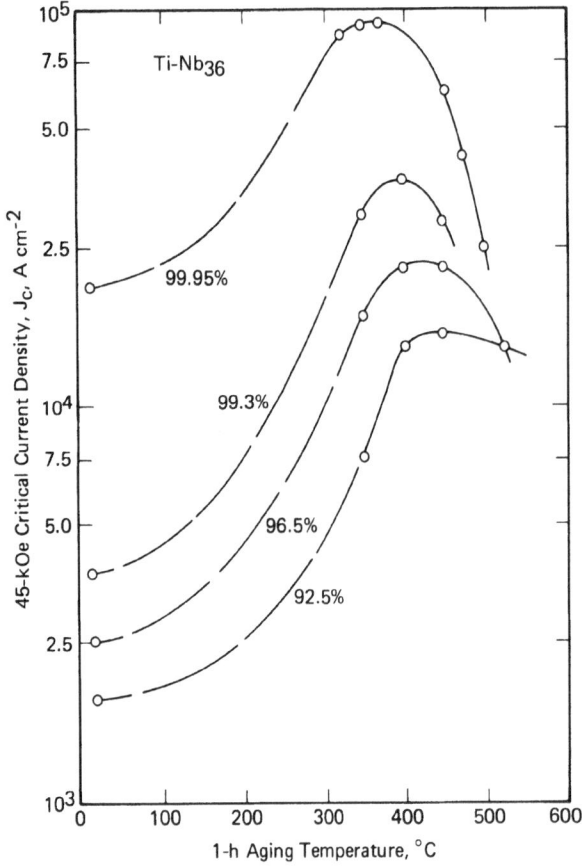

FIGURE 7-40. Influence of deformation level and 1-h aging temperature on the 45-kOe J_c of an intermediate-concentration Ti-Nb (36 at. %) alloy (transition criterion: 40 μV) — after RASSMANN and ILLGEN [Ras72].

---

† Cf. the above comments, and those of Sect.7.20.2(a), on the non-appearance of ω-phase during moderate-time aging of an intermediate-concentration Ti-Nb alloy [Hic69[a]] but the appearance of small amounts [Osa80] after extremely long-time aging. Cf. also Sect.7.26.4 which mentions the possibility of solute-segregation artifacts.

however, converge at very high fields, to meet at about 100 kOe.

In studies already referred to, McINTURFF and CHASE [Mci73] also employed long aging times at temperatures of 275-520°C and 300-375°C in investigating the 30-kOe J_c's of Ti-Nb$_{30}$ and Ti-Nb$_{37.7}$, respectively, both alloys having experienced deformations of 99.95% (i.e. 2000:1 area reduction). It is interesting to note that in neither case was a maximal J_c encountered. For example, during the aging of Ti-Nb$_{30}$ at temperatures of 275-520°C, the 30-kOe J_c increased monotonically with time for up to about 500 h, at which point the test was terminated.

7.26.3 High-Concentration (>40 at.% Nb) Ti-Nb Alloys

One of the earliest studies of the influence of aging on the critical current density of Ti-Nb was that by RALLS [Ral64] on 99.94% cold-deformed Ti-Nb$_{49}$, aged for 1 h at 400, 600, 800 and 1000°C. Within this range 400°C was the optimal heat-treatment temperature. Typical 4.2-K J_c's in applied fields of 30, 40 and 60 kOe were 2.3, 1.9, and 1.1x10^4 A cm^{-2}, respectively. In excellent agreement were the subsequent results of BYCHKOVA et al [Byc70], who noted that the 26-kOe J_c of cold-deformed (99.86%) and aged 1h/450°C Ti-Nb$_{50}$ was 2.4x10^4 A cm^{-2}. Optimal annealing temperatures to achieve this were found to be in the range 500-700°C, although the heat treatment did not result in any x-ray-detectable precipitation. RAUCH [Rau68] has studied the J_c of Ti-Nb$_{67}$ during aging for 1 h at 300, 450, 600, 750 and 1000°C, after: (i) cold deformation of about 99.9%; (ii) recrystallization -- as mentioned in the previous section. In both cases the optimal 1-h heat-treatment temperature was 600°C. The 1h/600°C aging was essentially a recrystallization heat treatment during which, it was suspected, the high-concentration Ti-Nb (a clustering system) decomposed into two bcc phases of differing compositions, cf. Sect.7.20.2(b) and [Mon2.3]. Under the circumstances it is not surprising that both cold-worked and recrystallized alloys, after heat treatment, yielded the same optimal J_c(4.2 K, 40 kOe), viz 2.1x10^4 A cm^{-2}.

Subsequently, NEAL et al [Nea71] reported the results of a very important case study of the deformation-plus-aging processing of a Cu/Ti-Nb monofilamentary 0.5-mm$^\phi$ wire. Cu-clad samples of Ti-Nb$_{42}$ (nominally Ti-58Nb, a commercial superconductor manufactured by IMI Ltd.) was subjected to area reductions by

cold work of between 33% and 99.999% and final heat treatments, mostly for 1 h, at temperatures of 300 to 600°C. Complementary to that study, in the sense that although less extensive processing was explored a wider range of alloy compositions was investigated, was the work of HELLER [Hel71] on Ti-Nb(13-44 at.%) and that of RASSMANN and ILLGEN [Ras72] on Ti-Nb(20-70 at.%). In moving out of the intermediate-concentration range the advantage of α-phase precipitation as a flux pinner was relinquished. This represents a serious sacrifice if only modest cold work is to be administered; for example, RASSMANN and ILLGEN showed that for 99.9%-cold-deformed wires, aged 1h/400°C, the 45-kOe J_c continuously decreased from 7.2x10^4 A cm^{-2} through 2.6x10^4 A cm^{-2} to 1.5x10^4 A cm^{-2} in going from 30 through 40 to 70 at.% Nb. Thus in high-concentration-range alloys, heavy cold work is essential for the development of microstructures suitable for supporting high critical current densities. Such a philosophy, as we shall see, *also* paves the way for the optimization of precipitation-range alloys.

(a) Cold Work. NEAL et al [Nea71] investigated the influence on the 50-kOe J_c of 1h/385°C-aged Ti-Nb$_{42}$ of cold deformation ranging from 33 to 99.999% (1.5:1 to 1.4x10^5:1 area reduction ratio). Particularly interesting was the manner in which critical current density increased sharply as the cold work exceeded 99% (100:1 area reduction ratio, ARR) and the perhaps surprising insensitivity of J_c to cold work in excess of that value, Table 7-15. The results were discussed in terms of the development by cold work and heat treatment of a well-defined dislocation-cell structure.

(b) Aging Temperature. A systematic investigation of the influence of aging temperature on the J_c of very heavily deformed (99.998%, i.e. 5x10^4:1 ARR) Ti-Nb$_{42}$ was also undertaken by NEAL et al [Nea71], whose results are summarized in Table 7-16. Although in this case flux pinning was regarded as being the result of a favorable dislocation-cell structure rather than precipitation, the optimal aging temperature was again within the range 385-425°C.

(c) Aging Time. In contrast to the observations by McINTURFF and CHASE [Mci73] on the aging responses

TABLE 7-15 50-kOe CRITICAL CURRENT DENSITIES (J_c)
OF COLD-DEFORMED-AND-AGED 1h/385°C
Ti-Nb(42 at.%)-- Neal *et al.* [Nea71].

Area Reduction Ratio	Percent Reduction in Area	J_c[+] A cm^{-2}
1.4×10^5	99.999	1.22×10^5[++]
5×10^4	99.998	1.20×10^5
1×10^3	99.9	1.13×10^5
1×10^2	99.0	1.09×10^5
2	50	2.7×10^4
1.5	33	1.8×10^4

[+] J_c criterion: 100 μV across sample, Ti-Nb-core diameter generally 0.25 mm.

[++] Core diameter, 0.15 mm.

TABLE 7-16 50-kOe CRITICAL CURRENT DENSITIES (J_c)
OF COLD-DEFORMED 99.998% AND 1-h AGED
Ti-Nb(42 at.%)-- Neal *et al.* [Nea71].

1-h Aging Temperature °C	J_c[+] 10^5 A cm^{-2}
None	0.41
300	0.97
325	1.00
350	1.10
385	1.20
425	1.20
450	1.19
500	0.77

[+] J_c criterion: 100 μV across sample; Ti-Nb core diameter, 0.25 mm.

of Ti-Nb$_{30}$ and Ti-Nb$_{37.7}$, as outlined above, were those of NEAL *et al* [Nea71] concerning Ti-Nb$_{42}$. In studies of very heavily cold-worked (99.998%; 5×10^4:1, ARR) composites they found that an aging time of 5 h at 385°C was sufficient to *maximize* the 50 kOe J_c.

7.26.4 Flux-Pinning Microstructures of Cold-Worked-and-Aged Ti-Nb Alloys

Flux pinning in cold-worked-and-aged Ti-Nb alloys is generally associated with structural features such as dislocations, dislocation tangles, cell boundaries, and precipitation. Earlier studies tended always to focus on the precipitational aspects, while more recent work placed the emphasis on the formation by cold work and heat treatment of optimal dislocation-cell (i.e. "subband") structures either with or without precipitation. In evaluating the effects of heat treatment, evidence for the occurrence of precipitation and the identification of precipitated species has often been incomplete or unsatisfactory. Since phase kinetics and equilibria are controlled both by alloy constitution (both major and minor) and the level and nature of the deformation, discrepancies between the reported results of ostensibly similar experiments are to be expected as a consequence of unavoidably small differences in compositions and degrees of cold work. It is difficult to examine, by either optical or electron microscopy, the structures of very heavily cold-worked alloys.

The most satisfactory transmission electron microscopy[+] has been carried out on annealed-and-quenched alloys, the results of which were alluded to in Sect.7.25 which dealt with quenched-and-aged (R,Q,C//A) alloys. BALCERZAK [Baℓ72] has noted the presence of weak *diffuse* ω-reflections in quenched Ti-Nb$_{34}$. HICKMAN [Hic69[a]], in studies of Ti-Nb(22, 25, 27, 30, 35 at.%) noted that although Ti-Nb$_{27}$ (which was all-β in the as-quenched condition) yielded ω-phase precipitation during aging at temperatures above 200°C, no ω-phase precipitation could be detected in the 30 and 35 at.% alloys either after quenching or as a result of aging. On the other hand, OSAMURA *et al* [Osa80] have detected ω-phase in very long-time aged Ti-Nb$_{36}$, cf. Sects.7.20.1 and 7.20.2(a).

In discussions of the relationship between precipitation and J_c-enhancement in heat-treated Ti-Nb alloys, confusion has frequently arisen for the following reasons: *(a)* metallographic data obtained from as-quenched alloys has often been used to aid in the

---

[+] Important TEM studies of Ti-Nb alloys are those of BRAMMER and RHODES [Bra67], BALCERZAK and SASS [Baℓ72], and WEST and LARBALESTIER [Wes80], see also [Wes82].

interpretation of results obtained from cold-deformed samples; *(b)* precipitation effects confirmed for alloys in low-concentration ranges have been assumed to be valid in nearby higher concentration ranges; *(c)* distinctions between precipitation effects in different composition ranges have tended to become obscured by the interesting fact that during the isochronal aging of *all* Ti-Nb alloys, J_c as a function of aging temperature maximizes at about 400°C. The observations of various workers in this controversial area are considered below.

In a series of studies of cold-worked Ti-Nb(30 and 35 at.%) alloys, BYCHKOVA and co-workers [Byc70] reported on the absence of precipitation after intermediate-temperature annealing. The x-ray measurements by CHARLESWORTH and MADSEN [Cha70] of Ti-Nb$_{45}$ (cold reduced 99.94%) failed to reveal after aging any evidence for precipitation -- not unexpectedly for such a high-concentration alloy. VOZILKIN [Voz68] in studies of 98%-cold-deformed Ti-Nb$_{31}$ noted a tendency for Ti enrichment at dislocation tangles, a possible precursor to α-phase precipitation, and the actual appearance of α-phase particles as determined by electron microscopy after prolonged aging, for example 150h/500°C.

Strong proponents of precipitation as the principal result of post-cold-work aging were RAUCH [Rau68] and HELLER [Heℓ71]; see also LÖHBERG, HELLER and ZWICKER [Loh73]. Both ω- and α-phase precipitation was claimed, although the evidence was not conclusive. In 99.9%-cold-worked Ti-50Nb, RAUCH [Rau68] noted: *(a)* a pronounced peak at 400°C in the hardness *versus* 1-h aging temperature curve (necessary, but not sufficient evidence for the precipitation of ω-phase -- a known hardener); *(b)* a shoulder at 450-500°C, the temperature range within which α-phase would be anticipated for this system. A comparable hardness maximum was also noted by HELLER [Heℓ71] during the 16-h aging at various temperatures of 97.2%-cold-worked Ti-49Nb, an alloy now known to lie outside the isothermal $\omega + \beta$-phase field. As a result of transmission electron microscopy HELLER detected the presence of precipitates, 1000 Å in length and 100 Å in diameter, whose orientations at 45° to the wire axis were thought to be controlled by wire texture. At the time, the precipitates were believed to be ω-phase, although direct attempts to confirm their identify using TEM were not successful. Nevertheless, HELLER [Heℓ71] and co--workers (see for example LÖHBERG, HELLER and ZWICKER [Loh73]) were led to the conclusion that alloys such

as Ti-49Nb were able to yield both ω-phase and α-phase precipitation, in addition to dislocation networks, after sufficient cold deformation followed by a 400°C heat treatment. It is interesting, and perhaps significant, to note that HELLER [Heℓ71] suspected that the ω-precipitates were formed in the Ti-rich portions of segregation zones. Certainly this interpretation averts conflict with HICKMAN's observations, which would otherwise occur, and might explain why the J_c of previously homogenized wire can be significantly less than that produced from as-cast (and presumably segregated) starting material.

As mentioned above, PFEIFFER and HILLMANN [Pfe68] who studied Ti-Nb(22 and 34 at.%) with the aid of light and electron microscopy and electron diffraction, were able to claim that after strong deformation α-phase precipitates out at the boundaries of the subbands. Since these act as nucleation sites for the α-phase, the stronger the deformation the greater the density, fineness and uniformity of distribution of the subsequent precipitation. Heavy deformation also increases the speed of the precipitation process which depends very strongly on the number per unit area of the subbands.

NEAL *et al* [Nea71] using a Ti-Nb alloy of considerably higher Nb content (42 at.%) attributed all the cold-work and heat-treatment-induced J_c enhancement to the establishment of an optimal substructure of dislocation cells. No evidence of useful precipitation either within cells or cell walls was found. Hardness is a useful indicator of the formation of fine precipitate dispersions. But during heat treatment at various temperatures no noticeable hardness anomaly, of the kind observed earlier in lower concentration alloys by both RAUCH [Rau68] and HELLER [Heℓ71], was detected near the favorable 400°C temperature. In studies of material with higher interstitial content than that used in the principal work, no evidence of accumulation in the cell walls was found, and moreover the bulk pinning force continued to fit the previously determined universal linear relationship with inverse cell size (to be considered below, cf. Fig.7-45). Such a result would still of course admit a possible, and beneficial, interstitial-element induced refinement of cell size. According to NEAL *et al*, flux pinning in heavily cold-worked-and-aged Ti-Nb$_{42}$ stems from a "pencil-shaped" sub-cell structure which arises during an early stage of cold work, and which becomes refined during further increases in the level of cold work which, in their research, was varied from less than 10^2:1 to more that 10^5:1. During heat treatment, dislocation annihilation and rearrangement leaves the cell-interiors free of dislocations; the optimal heat treatment is that which achieves this important objective while maintaining constant cell size.

HAMPSHIRE and TAYLOR [Ham72] attempted a mechanistic interpretation of NEAL *et al*'s phenomenological study of flux-pinning effects in Ti-Nb$_{42}$. After reviewing, and finding inapplicable to this alloy, the approaches of COFFEY [Cof68], BYCHKOV *et al* [Byc69], FIETZ and WEBB [Fie69], FREYHARDT [Fre69], LABUSCH [Lab67, Lab68, Lab69^a], and CAMPBELL *et al* [Cam68], HAMPSHIRE and TAYLOR revived an early suggestion by DEW-HUGHES [Nar66] and proposed that flux pinning could be interpreted in terms of a "modulated-κ_{GL} model". Further details of cell-wall pinning, as it is also termed, are to be found in [Mon21.3].

7.26.5 Section Summary -- Characteristics of Deformed-and-Aged (D//A) Ti-Nb Alloys

Sect.7.26 in its entirety has been dealing with the manner in which the critical current densities of Ti-Nb alloys respond to sequential applications of cold work and aging heat treatment, usually in the temperature range near 400°C, a procedure designated D//A. In order to clarify the discussion, the alloys had been divided into the three concentration regimes:

low concentration: <30 at.% Nb

intermediate concentration: 30-40 at.% Nb

high concentration: >40 at.% Nb,

the selection of which was guided by the possible appearance or nonappearance of the ω- and α-phases in the thermomechanically processed wire. It may be recalled that LARBALESTIER, in a recent review article [Lar80], has employed a narrower subdivision of Ti-Nb alloys into the ranges:

high Ti: 55∿65 wt.% Ti (22∿30 at.% Nb)

intermediate: 50 wt.% Ti (34 at.% Nb)

high Nb: 40∿48 wt.% Ti (36∿44 at.% Nb),

a scheme in which the "high-Nb" range spans the practical boundary between the regimes of appearance and nonappearance of α-phase precipitation, Fig.7-1.

Whereas in low-concentration alloys, precipitation plays an important role in flux pinning, it seems to be absent from the high-concentration alloys (>40 at.%) for which dislocation cell size is the critical parameter. An interesting property common to all deformed-and-aged alloys, and one which could confuse the interpretation of flux-pinning mechanisms is the existence, for the entire range of alloys, of a *common optimal aging temperature*, viz ∿400°C. Towards the lower end of the Nb-concentration scale, precipitates are easily formed at 400°C after deformation, while for *all* alloys this temperature readily activates dislocations and aids in the formation of fine dislocation-free cell structures. Optimization arises in both cases for different reasons both of which, however, are related to atomic diffusion. An increase in temperature of some 100°C leads in the low-concentration case to a dissolving of the ω-phase and its replacement by α-phase, a less effective flux pinner, and in all alloys to an increase in cell size as a precursor to recrystallization.

Table 7-17 lists the 30-kOe and 50-kOe critical current densities of deformed-plus-aged Ti-Nb(20-50 at.%) alloys. These data, assembled from Tables 7-10 through 7-13 and elsewhere may be regarded as "semi-optimized" in the sense that, although only the highest J_c values have been selected (when aging time and/or temperature has been varied in the studies concerned), even these have been subject to severe limitation, in some cases by fixed and occasionally low levels of deformation, and in others by short aging times. The current densities listed are of course very much lower than those to be reported below in sections dealing with full optimization in the form of multiple intermediate heat treatment and final cold deformation. Nevertheless, when the data are plotted as in Fig.7-41 an instructive picture of the results of the early research emerges and some useful trends become discernible: *(a)* the highest low-field (30 kOe) critical current densities were obtained for Nb concentrations less than about 30 at.%; *(b)* the development of high J_c's in the low-Nb range is favored by heavy cold work (McINTURFF and CHASE [Mci73]) followed by protracted aging (HELLER [Heℓ71], McINTURFF and CHASE [Mci73]); *(c)* if high-concentration alloys are selected, advantage can be taken of their greater workability -- in such alloys, high current-carrying capacities may be developed by means of extreme cold deformation (NEAL *et al* [Nea71]) after which a short aging time is adequate.

TABLE 7-17 CRITICAL CURRENT DENSITIES OF DEFORMED-PLUS-AGED TITANIUM-NIOBIUM ALLOYS AT APPLIED FIELDS OF ABOUT 30 AND 50 kOe

At.% Nb	Condition % Cold Deformation	Condition Heat Treatment	Serial Number[†]	J_c, 10^5 A cm^{-2} ~30 kOe	J_c, 10^5 A cm^{-2} ~50 kOe	First Author	Reference
20	99.9	1h/400°C	1		0.67	Rassmann	[Ras72]
21	80	3h/412°C	2	1.9		Vetrano	[Vet68]
22		1.7h/400°C	3	2.5	1.0	Salter	[Sal66]
	99.95	3h/350°C	4	1.7		McInturff	[Mci73]
	99.64	3h/390°C	5		0.41	Bychkov	[Byc69]
25	99.86	1h/450°C	6	1.0		Bychkova	[Byc70]
	99.95	400h/325-350°C	7	2.9		McInturff	[Mci73]
26	99.95	1h/400°C	8		0.74	Ricketts	[Ric69]
28	99.95	400h/325-350°C	9	2.9		McInturff	[Mci73]
30	99.8	1h/400°C	10	0.30		Betterton	[Bet66]
	99.86	1h/450°C	11	0.63		Bychkova	[Byc70]
	99.9	1h/400°C	12		0.72	Rassmann	[Ras72]
	99.95	400h/350°C	13	2.7		McInturff	[Mci73]
31	98	28h/500°C	14	0.4		Vozilkin	[Voz68]
33	97.2	250h/400°C	15	0.60	0.51	Heller	[Hel71]
34	99.9	1h/400°C	16	0.70	0.40	Rauch	[Rau68]
35	99.86	10h/450°C	17	0.32		Bychkova	[Byc70]
36	99.95	1h/375°C	18		0.92	Rassmann	[Ras72]
38	99.95	400h/350°C	19	0.47		McInturff	[Mci73]
39	97.2	250h/400°C	20	0.54	0.25	Heller	[Hel71]
40	99.9	1h/400°C	21		0.26	Rassmann	[Ras72]
42	99.998	5h/385°C	22	1.87	1.25	Neal	[Nea71]
44	97.2	250h/400°C	23	0.24	0.12	Heller	[Hel71]
49	99.94	1h/400°C	24	0.23	0.14	Ralls	[Ral64]
50	99.86	1h/450°C	25	0.24		Bychkova	[Byc70]
	99.9	1h/400°C	26		0.20	Rassmann	[Ras72]

† Numbers refer to Figure 7-41.

By way of contrast it is instructive to review the results of the detailed investigation of J_c composition dependence conducted by HELLER [Hel71] as summarized in Tables 7-11 and 7-13 and plotted in Fig.7-42. Since only a low level of cold deformation was employed, the critical current densities recorded were also low. The pair of curves corresponding to the 30-kOe and 50-kOe J_c's for 1h/400°C aging maximize for compositions between about 20 and 25 at.% Nb. After heat treatment for an optimal (for these samples)

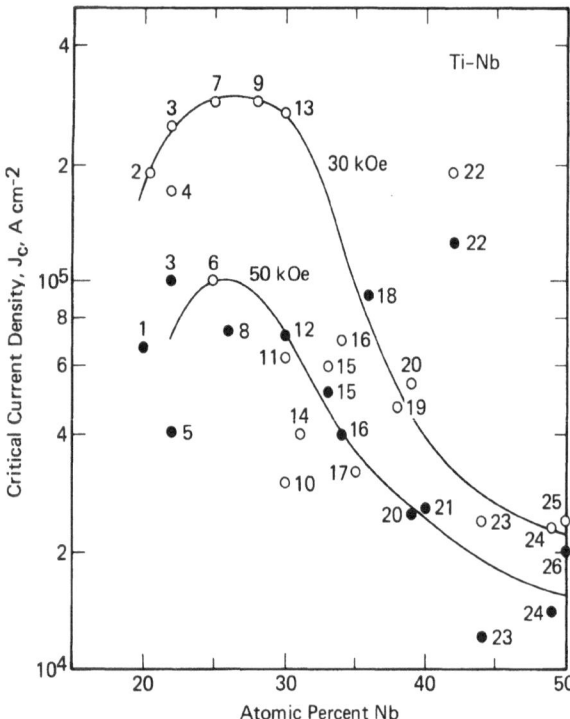

FIGURE 7-41. Critical current density *versus* composition for the alloys listed in Table 7-17. The numbers against the points correspond to the serial numbers of the table. The curves are intended to guide the eye through data corresponding to fields of ~ 30 kOe (○, upper curve) and ~ 50 kOe (●, lower curve). The 30-kOe data for the poorly processed samples also cluster about the lower line. The maximum in the low-concentration regime signifies precipitate-dominated flux pinning; on the other hand, with deformation-plus-aging optimization, high J_c's can be achieved outside the precipitation regime as indicated by the points labelled "22".

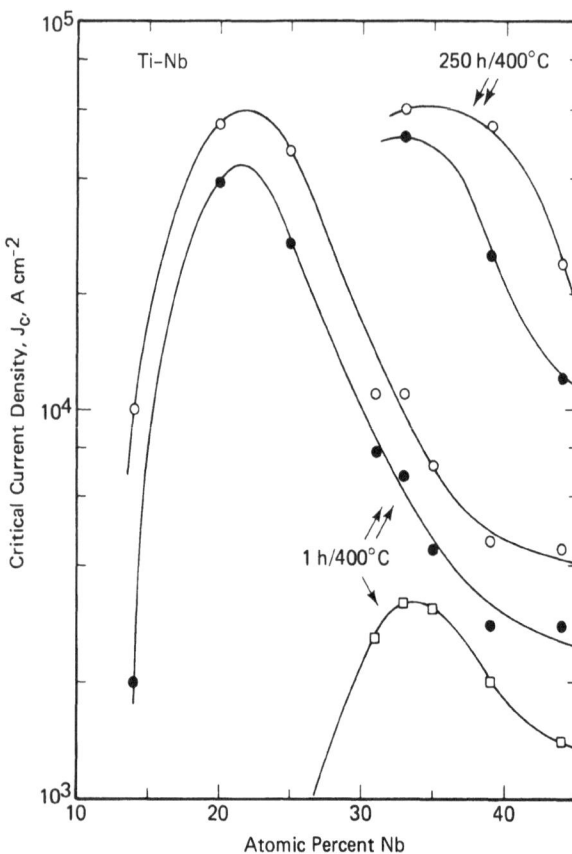

FIGURE 7-42. Critical current density *versus* composition after the results of HELLER [Heℓ71] as listed in Tables 7-11, 7-12, and 7-13 (transition criterion: 100 μV across 3.5 cm). With regard to the 30-kOe (○) and 50-kOe (●) data, the low-concentration-range peak for low aging times (and presumably attributable at least in part to ω-phase precipitation (at. % Nb < 30)) shifts or extends to the right with longer aging times as α-phase precipitation (at. % Nb < 40) contributes to the flux pinning. The 100-kOe data (□) are discussed in the text.

250h/400°C, the critical current densities are an order of magnitude higher for alloys of compositions about 30-35 at.% Nb, and it appears as if the J_c-maximum with respect to composition is also being shifted into that region. These results, judged in the light of others to be presented below, suggest that the "separation of the variables concept" prevalent during alloy and process development in the 1960's, in which the variables controlling J_c were examined individually without going to full process optimization in each case, must be treated with caution.

Finally, attention is drawn to another important result due to HELLER [Heℓ71], viz that the 100-kOe J_c peaks at between 33 and 35 at.% Nb. This composition range also corresponds to a peak in H_{c2}, Fig.7-18, suggesting that at very high fields, J_c responds to those same atomic properties of the alloy that control H_{c2}. Indeed the intrinsic nature of the high-field critical current density is suggested by the

manner in which, for a given alloy, the family of $J_c(H_a)$ curves which can be generated in response to variation of the thermomechanical processing conditions, tends to crowd into a common curve, independent of alloy condition, at fields near H_{c2} (cf. Fig.11-7). Upper critical field control of high-field J_c is to be considered in Sect.7.32.3.

7.27 D//A//D—COLD-DEFORMED, AGED AND FINAL-DEFORMED Ti-Nb CONDUCTORS—AN INTRODUCTION TO TECHNICAL PROCESS DEVELOPMENT

Intermediate heat treatment during the cold deformation of superconductive wire as a means of altering its structure and the precipitate distribution was

introduced as early as 1962 by DIETRICH *et al* [Die62].
Subsequently, SALTER [Saℓ65] employed a 4h/550°C
anneal (cf. Table 7-11) solely for the purpose of
stress relief during the otherwise uninterrupted cold
reduction of Ti-Nb$_{22}$; the exceptionally high critical
current density obtained (cf. Tables 7-10 and 7-11)
may be attributable in part to the influence of this
intermediate heat treatment. BYCHKOVA *et al* [Byc70]
(1967/70) in an early investigation of
Ti-Nb(20-50 at.%) have intercompared, at the same
overall level of deformation, the effects of intermediate and final heat treatment. Their thermomechanical process can be represented in general by:

$$\text{as-cast, 36 mm}^2 \xrightarrow{\text{cold roll}} 1.69 \text{ mm}^2 \,|\text{HT}_1|$$

$$\xrightarrow{\text{cold draw}} 0.25 \text{ mm}^\phi \,|\text{HT}_2|$$

where HT$_1$ and HT$_2$ are alternative 1h/(300∿600°C) heat
treatments. Thus, for example, a comparison was made
between:

$$\text{as-cast} \xrightarrow[95.3\%]{\text{cw}} |\text{HT}| \xrightarrow[97.1\%]{\text{cw}} 0.25 \text{ mm}^\phi$$

$$\text{and} \quad \text{as-cast} \xrightarrow[99.85\%]{\text{cw}} 0.25 \text{ mm}^\phi \,|\text{HT}|$$

In these trials it turned out that final heat treat-
ment was preferable to intermediate heat treatment.
Of the entire series of 1h/(300∿600°C)-heat-treated
Ti-Nb(20-50 at.%) alloys, 1h/450°C-aged Ti-Nb$_{25}$, with
J$_c$(26 kOe) = 1.0x10^5 A cm^{-2}, possessed the highest
critical current density.

RASSMANN and ILLGEN [Ras72] explored the effects
of both single and dual (separated by cold work) heat
treatments, each of 1h/400°C, in association with
final heat treatments of 1h/(RT-425°C) and an overall
reduction in area of 99.9% on the critical current
densities of Ti-Nb$_{30}$ and Ti-Nb$_{40}$ prepared from Kroll-
process Ti. The results are summarized in Fig.7-43.
The single-intermediate-heat-treatment processing option, the subject of this section, consisted of the
following stages:

Ti-Nb$_{30}$: 8.0 mm$^\phi$ ⟶ 0.87 mm$^\phi$(1h/400°C)
　　　　　　　　　⟶ 0.25 mm$^\phi$(1h/RT-425°C)

Ti-Nb$_{40}$: 8.0 mm$^\phi$ ⟶ 0.81 mm$^\phi$(1h/400°C)
　　　　　　　　　⟶ 0.25 mm$^\phi$(1h/RT-425°C)

Again we note, with reference to Fig.7-43, that with

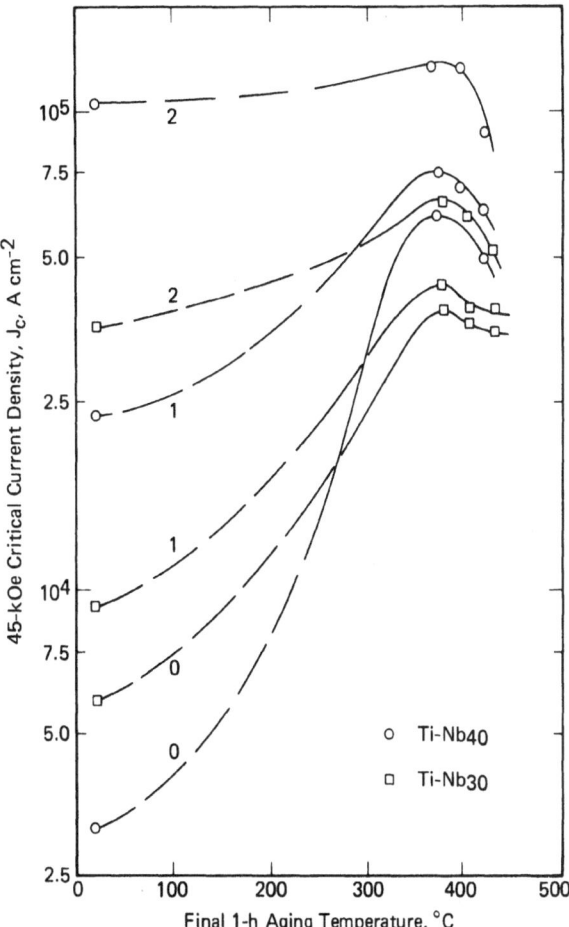

FIGURE 7-43. Critical current densities of Ti-Nb (30 at. %) and
Ti-Nb (40 at. %) as functions of 1-h final aging temperature and
for 0, 1 and 2 stages of 1 h/400°C intermediate heat treatment as
indicated. The comparison specified in the text is achieved by
combining 1 h/400°C data from curves-0 with the respective 1 h/
RT data from curves-1 – after RASSMANN and ILLGEN [Ras72].

99.9%-overall-deformed wire the shifting of a 1h/400°C
heat treatment from the end to some intermediate position in the processing sequence (in this case such
that some 90% cold work still remained to be done) resulted in the following *reductions* in 45-kOe J$_c$:

Ti-Nb$_{30}$: from 5.7x10^4 to 2.3x10^4 A cm^{-2},
Ti-Nb$_{40}$: from 3.6x10^4 to 9.2x10^3 A cm^{-2}.

Subsequent work very quickly began to show that, in
order for intermediate heat treatment to be fully
effective, the process must be designed such that
heavy deformation is able to be administered *prior to*
the first aging step.

Present-day methods for the processing of technical superconductors stem from the studies of McINTURFF

TABLE 7-18   INTERMEDIATE *VERSUS* FINAL HEAT TREATMENT IN THE FIXED-OVERALL-DEFORMATION PROCESSING
OF TITANIUM-NIOBIUM ALLOYS -- McInturff and Chase [Mci73]

| | Maximum J_c and Corresponding Processing Conditions | | | | |
| | Final Aging | | Aging and Final Cold Work | | |
At.% Nb	Time/Temperature h/°C	3-T $J_c^\dagger$ 10^5 A cm^{-2}	Time/Temperature h/°C	Area Reduction %	3-T $J_c^\dagger$ 10^5 A cm^{-2}
22	50/350	1.7	1.5/480	94	3.4
	18/480	0.9			
25	500/350	2.9	6/400	91	3.1
	25/400	2.3			

$\dagger$ J_c criterion: sample resistivity, 10^{-9} Ω cm.

and CHASE [Mci73] on the influence of final cold work, within the context of fixed overall deformation, on the 30-kOe J_c's of what would then become intermediate-heat-treated Ti-Nb$_{22}$ and Ti-Nb$_{25.2}$ alloys. Quite spectacular increases in critical current density were obtained upon exercising the final-cold-work option. Solution-treated 17-mm$^\phi$ bars were Cu-clad, reduced to 0.38-mm$^\phi$ wire (hence overall alloy deformation >99.95%) and heat treated. The results of this were reported above. Alternatively the drawing was interrupted, and the wires subjected to (73-1.5h)/(300-380°C) aging, before cold drawing to the same final wire diameter was resumed. The effectiveness of the intermediate-heat-treatment (or final-cold-work) technique, when properly applied, can be appreciated by scrutinizing Table 7-18, a data summary.

As a result of the work of McINTURFF and CHASE [Mci73] on the Ti-rich alloys Ti-Nb(22 and 25 at.%), it appears that the administering of final cold work -- at constant overall area reduction -- has two important advantages: *(a)* not only is J_c increased, but *(b)* the aging time needed to maximize J_c is significantly reduced.

In studies aided by transmission electron microscopy, the beneficial effects of heat treatment followed by final cold work were being independently considered by PFEIFFER and HILLMANN [Pfe68], who used as subjects the alloys Ti-Nb$_{22}$ and Ti-Nb$_{34}$, taken from the low- and intermediate-concentration ranges (as

defined above), respectively. Heavy cold work was found to lead to a fine subband structure, and a high density of nucleation sites for possible α-phase precipitation during subsequent heat treatment. But PFEIFFER and HILLMANN [Pfe68] also found that precipitation heat treatment was responsible for a partial recovery of the deformed structure, a so-called "softening" of the subband boundaries, counteracting to some extent the beneficial effect of precipitate formation. Accordingly they recommended that best results are obtained when the available amount of overall cold deformation is sufficient to produce the desired density of fine subbands well before the final wire diameter is reached, so that the deformation process can be continued after a pause for precipitation heat treatment. The "final cold work" on one hand improved the subband structure -- compensating for the recovery -- and on the other hand led to a "refinement", recognized subsequently as a redistribution of the α-phase precipitates, cf. [Mon21.19]. Some recent advances in process optimization achieved by STEKLY and colleagues [Ste78][Seg79] during the heavy-reduction (up to 10^6:1) plus single-intermediate-heat-treatment processing of Ti-Nb$_{30}$ (i.e. Nb-55%Ti) and Ti-Nb$_{37}$ (i.e. Nb-46.5%Ti) are to be discussed in Sect.7.32.

The characteristics of more elaborate thermomechanical processing programs, those involving several steps of alternate heat treatment and cold work, are to be considered next.

7.28 D//A-D-A//D—COLD-DEFORMED, MULTIPLE-INTERMEDIATE-AGED AND FINAL-DEFORMED Ti-Nb CONDUCTORS—A FURTHER INTRODUCTION TO TECHNICAL PROCESS DEVELOPMENT

The effect of multiple intermediate heat treatments on the critical current density of Ti-50Nb was first studied in detail by PFEIFFER and HILLMANN [Pfe68]. In that work, up to three stages of heat treatment were associated with various degrees of final cold deformation, qualitatively described as "average" and "strong". Application of these three stages of thermomechanical processing yielded 30-kOe and 50-kOe J_c's of 3.2×10^5 A cm^{-2} and 2.0×10^5 A cm^{-2}, respectively. RASSMANN and ILLGEN [Ras72] were also responsible for some early studies of the effect of multiple intermediate heat treatment on critical current densities, in their case of Ti-Nb$_{30}$ and Ti-Nb$_{40}$. This contribution, already mentioned in the previous section, consisted (referring to the Ti-Nb$_{30}$) of two stages of 1h/400°C aging separated by 52% cold work, administered *after* 98.8% cold work and *followed by* 83.2% deformation to a final wire diameter of 0.25 mm. The overall deformation was 99.90%, and the processing concluded with 1h/(RT-425°C) heat treatments. The Ti-Nb$_{40}$ was subjected to comparable thermomechanical processing. Within the present context, the experiments must be regarded as preliminary since relatively small deformations were involved at all stages, and the stated primary purpose of the heat treatment was to relieve work-hardening in the Cu cladding rather than to enhance J_c. Nevertheless, from a critical current standpoint some promising results were obtained: in 99.90%-total-deformed Ti-Nb$_{30}$ the replacement of a final 1h/400°C aging by two stages of intermediate treatment administered according to the above prescription, raised the 45-kOe J_c from 5.7×10^4 to 1.05×10^5 A cm^{-2}.

In subsequent studies by HILLMANN and colleagues [Hiℓ72, Hiℓ73[a], Hiℓ74], the influence of multiple intermediate heat treatments of 6h/380°C [Hiℓ74] administered at regular intervals during the cold deformation process, plus final cold work, was discussed in terms of the development of a suitable flux-pinning microstructure. Since these quantitative studies, as well as dealing with single-core Cu-clad conductors, also included detailed information on the preparation and properties of multifilamentary (∼60-filament) Cu-matrix composites they, together with the work of

ARNDT, WILLBRAND, *et al* [Arn74][Wiℓ75[a]], formed a link between research conducted in the mid-1960's, mostly on bare wire, and modern commercial conductors. These two investigations complemented each other in that, whereas HILLMANN and colleagues [Hiℓ72, Hiℓ73[a], Hiℓ74] directed their attention towards J_c-optimization primarily through variation of find-cold-reduction-induced precipitate distribution, ARNDT and co-workers [Arn74][Wiℓ75[a]] studied the influences on J_c of subband and precipitate morphology, as influenced by the number and duration of the 390°C intermediate heat treatments -- both with and without final cold deformation. This latter work was actually a comparative investigation of the influence of thermomechanical process variation (the microstructural results of which were determined using TEM) on the critical current densities of Ti-50Nb and that alloy doped with Cu and Ge, respectively. The Ti-50Nb component of the study is described immediately below.

7.28.1 Fundamental Contributions by WILLBRAND, ARNDT, *et al*, Krupp Forschungsinstitut, Essen, BRD

WILLBRAND *et al* studied the critical current density of Ti-50Nb (nominal) as a function of multiple heat treatment. With the aid of electron and optical microscopy, and assisted by the results of a parallel investigation into the effects of Cu and Ge additions on the microstructures and critical current densities, bulk flux pinning was discussed in terms of subband and precipitate morphology.

(a) Sample Preparation. The work was carried out using monofilamentary [Arn74] as well as multifilamentary [Wiℓ75[a]] Cu-matrix composite conductors. Extruded 3-mm$^\phi$ Ti-50Nb wire rod was Cu-clad and reduced to 0.25 mm$^\phi$, its filament or core diameter then being 0.13 mm. Prior to cladding, the wire rod had been heat treated 3h/390°C. After cladding, heat treatments when they were administered, took place at wire O.D.'s of 2.0, 1.5, 1.2, 1.0, 0.82, 0.68, 0.60, 0.50 and 0.39 mm; there was thus an opportunity for up to ten heat treatments. If less than ten were given, they were distributed throughout the range of reduction in such a way that the final cold deformation varied from 75% (the usual level) to 97%, as indicated in Table 7-19. A discussion of final-cold-work optimization, including the results of these studies, is given in Sect. 7.31.2.

TABLE 7-19   THERMOMECHANICAL PROCESSING SCHEDULE FOR A MULTIPLY-INTERMEDIATE-HEAT-TREATED Cu/Ti-Nb MONOFILAMENT -- Arndt and Ebeling [Arn74], see also [Wiℓ75[a]].

Property Group	Serial Number	Outside Diameter of Wire, mm											% Final Cold Deformation[††]	$5T\ J_c$[†††] $10^5\ A\ cm^{-2}$
		3.0[†]	2.0	1.5	1.2	1.0	0.82	0.68	0.60	0.50	0.39	0.25		
I	1	H	H	H	—	H	H	H	—	H			75	1.6_5
	2	H	—	H	—	H	H	H	—	H			75	1.5_5
II	3	H					H			H			75	1.4_0
	4	H					H						91	1.2_5
	5	H	—	H									97	1.1_5
	6	H								H			75	0.9_1
III	7	HT	—	HT	—	H	H	H	—	H			75	0.8_5
	8	HT	—	HT	—	HT	H	H	—	H			75	0.6_3
	9	HT	—	HT	—	HT	HT	H	—	H			75	0.4_5

H = 3h/390°C

HT = 3h/500°C

†   Diameter of unclad wire rod; others are overall diameters.

†† Assuming that the O.D. and fil. diams. reduce at the same rate.

††† J_c criterion: 5 μV across 1 cm.

(b) Flux Pinning -- Experimental. The results of the critical current density studies performed on the samples listed in Table 7-19 are given in Fig.7-44, scrutiny of which yields some very interesting systematics. Table 7-19 shows that when the data are listed in decreasing order of 50-kOe J_c, they fall naturally into three groups: Group I -- six or seven heat treatments of 3h/390°C; Group II -- two or three heat treatments of 3h/390°C; Group III -- two, three, or four stages of 3h/500°C heat treatment (which seems to have a cumulative detrimental effect) followed by several 3h/390°C heat treatments. Quite clearly the multiple 3h/390°C heat treatment option is the most effective in establishing the desired flux-pinning microstructure.

WILLBRAND et al [Arn74][Wiℓ75[a]] have also conducted detailed optical and electron microscopic studies of the relationship between J_c, the deformation structure, and the nature of the precipitation. Microstructural data corresponding to ten heat-treatment conditions have already been listed in Table 7-7. As a result of this work, and a companion study of Ti-Nb-Cu and Ti-Nb-Ge alloys, they were led to the conclusion that "a subband diameter of 300-500 Å is a necessary but not sufficient condition for a high current carrying capacity. A further condition for high J_c is the presence of precipitates with axes at right angles to the wire and lengths of the order of the band diameter. Thus it has been ascertained that the interplay between subband structure and precipitates plays an essential role in that uninterrupted lines of particles parallel to the wire axis possess smaller pinning force than do short particles perpendicular to the wire axis".

Critical currents were usually relatively low in those wires whose subbands, although fine, contained no precipitation, or when precipitation was accompanied by a coarse subband structure. Thus high J_c's required both fine subbands and small precipitates.

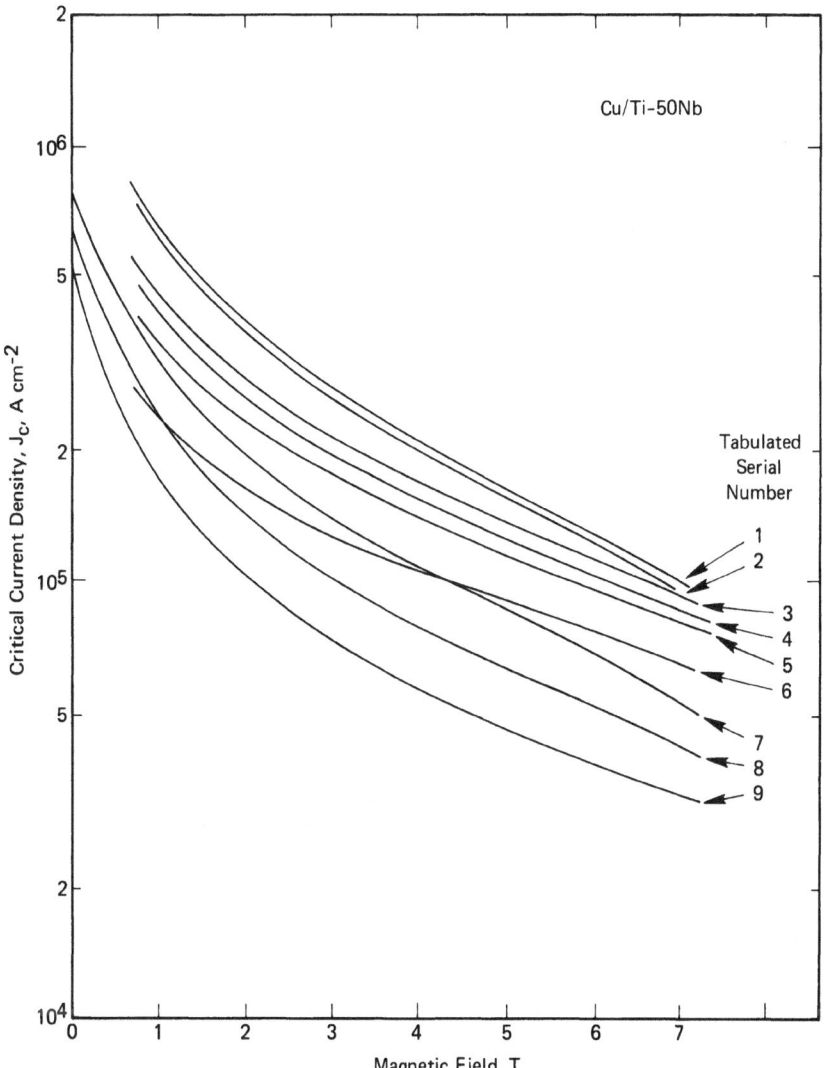

FIGURE 7-44. Critical current density *versus* applied magnetic field for Cu/Ti-50Nb monofilaments processed according to the schedule of Table 7-19 (transition criterion: 5 μV across 1 cm). Guided by the table a separation of the curves into three property-groups, depending on the kind of processing administered, is discernible — after WILLBRAND *et al* [Wil75^a].

7.28.2 Fundamental Contributions by HILLMANN *et al*, Entwicklungsabteilung, Vacuumschmelze GmbH, Hanau, BRD

Using Cu-clad multifilamentary conductors prepared from Ti-50Nb and Ti-55Nb, HILLMANN and colleagues have investigated the final cold-work requirements needed to maximize J_c in various applied fields, and have interpreted their results in terms of a pinning-site/flux-lattice matching theory (critically discussed in [Mon21.11]).

(a) Sample Preparation. Cast rods of Ti-50Nb, 10 cm in diameter, were cold deformed to 12 mm$^\phi$, clad in Cu and cold rolled to 10 mm$^\phi$, diffusion bonded at 600°C, bundled as a group of sixty-one wire rods into a Cu tube, and again diffusion bonded at some stage during the drawing to an outside diameter of 7.8 mm. The resulting rod was further reduced to diameters of from 1.02 to 0.2 mm with heat treatments of 6h/380°C at diameters of 7.8, 5.4, 3.7, 2.6 and 1.2 mm. During this process, subbands with diameters of about 2×10^{-6} cm developed, and α-precipitates were formed [Hil74]. In a like manner the 60-filament composites referred to below and in Table 7-20 were prepared for measurement [Hil73^a].

Ti-55Nb, with its slightly higher 4.2-K H_{c2} than that of Ti-50Nb (Fig.7-18), was also regarded as a

TABLE 7-20   CRITICAL CURRENT DENSITIES (J_c) OF
              OPTIMIZED 60-FILAMENT Ti-50Nb AND Ti-55Nb
              COMPOSITES -- [Hiℓ73[a]]

| Magnetic Field T | J_c, 10^5 A cm^{-2}[†] | | |
| | Ti-50Nb[††] | | Ti-55Nb |
	Series A	Series B	
3	3.56		2.96
4	3.06	3.22	2.45
5	2.53	2.58	2.08
6	2.03	2.10	1.76
7	1.57	1.59	1.43
8	1.06	1.1	1.04

† J_c criterion: 1 μV cm^{-1}

†† Heat treatments of 1h/380°C (Series A) and
    2h/380°C (Series B) were administered prior to
    final cold work -- cf. Figs. 7-49 and 7-50.

likely candidate for the development of high-field
superconductors [Hiℓ73[a]]. Accordingly, multifilamentary composites based on it were prepared for measurement using procedures similar to those just outlined.
As it turned out, the optimal critical current densities obtained were uniformly lower throughout the magnetic field range of measurement (3-8T) than those of
the Ti-50Nb-base conductor, a disappointing fact which
was attributed to the "sluggishness" of the α-phase
precipitation. Ti-55Nb (39 at.% Nb) it will be recalled, Fig.7-1, lies close to the upper "concentration limit" of the <600°C α+β-phase field.

(b) Optimization Studies. HILLMANN et al [Hiℓ73[a]]
have studied the influence of final cold deformation
on the critical current density at various applied
fields. In preparing material for measurement, wire
drawing was terminated at 0.6, 0.5, 0.4, 0.3, 0.25,
and 0.2 mm$^\phi$, yielding samples with final cold reductions from 1.2 mm$^\phi$ of 75, 83, 89, 94, 95.6 and 97.2%,
respectively. It was noted that: (i) as a function of
final cold work, each constant-field J_c passed through
a maximum; (ii) the position of the maximum was a function of the applied field. It follows from this, that
proper conductor design must take into account the expected operating field strength.

Some numerical results from HILLMANN's 1972/73
study of Ti-50Nb and Ti-55Nb multifilamentary composites [Hiℓ73[a]] are given in Table 7-20. Depicted are
the critical current densities of conductors which had
been cold reduced to 7.8 mm$^\phi$ (Ti-50Nb) or 8 mm$^\phi$
(Ti-55Nb) then alternately cold reduced and heat
treated to 1.2 mm$^\phi$ (the diameter at which the last
380°C heat treatment took place) and finally cold reduced by from 75 to 97.2%, as mentioned above. The
maximum critical current densities are recorded in the
table. A discussion of find-cold-work optimization,
which includes the results of these studies, along
with those of WILLBRAND et al mentioned above, is
given in Sect.7.31.2

(c) Flux Pinning -- Experimental. As a result of
the wire drawing, a dense structure of subbands parallel to the wire axis is developed [Hiℓ72] (cf. also
[Wes81]). The subbands, of diameter typically ∿10^{-5} cm
[Hiℓ72], act as sites for α-phase precipitation in
response to appropriate heat treatment (cf. also
[Wes80, Wes82]). The cold-work optimization process,
according to HILLMANN, consists of the establishment
of the most favorable (from a flux-pinning standpoint)
spacial arrangement of these precipitates in the manner
discussed in [Mon21.15] and outlined below. As justification for the precipitate-arrangement sensitivity
of the bulk pinning force, a geometrical precipitate-
spacing/fluxoid-spacing resonance mechanism known as
"matching" was proposed (cf. [Mon21.11] for a complete
discussion).

7.29 SPUTTERED Ti-Nb ALLOY FILMS

Early and unconventional processing techniques
for the fabrication of Ti-Nb superconductors are reviewed in [Mon28.5]. Sputtering (which, in contrast
to the processing methods just described, is "non-
mechanical") might also be designated as an "unconventional process". Its application to Ti-V superconductors has been considered in Sect.3.11 and to
Ti-Ta in Sect.5.8 (in particular, Figs.5-2 and 5-3).
SPITZER, the author of that work has also investigated
the properties of sputtered Ti-Nb alloy films [Spi71,
Spi74[a]]. The J_c versus H_a characteristics of more
than twenty different sputter-deposited Ti-Nb alloy
films were measured with the applied field parallel to
the film surface. The alloys had been prepared by DC
triode sputtering at an average deposition rate of

100 Å min^{-1} onto fused quartz substrates kept at 700°C. After deposition they were annealed for up to 3 h and rapidly cooled. Critical current densities, very high by comparison both with bulk data and the results of an earlier experiment on a sputtered Ti-Nb alloy [Edg64], were obtained. For example, at 3.7×10^5 A cm^{-2}, the 40-kOe critical current density of SPITZER's Ti-Nb$_{43}$ sample was much greater than the 6.3×10^4 A cm^{-2} reported by EDGECUMBE et al [Edg64] for sputter-deposited Ti-Nb$_{40}$, and compares favorably with the 3.23×10^5 A cm^{-2} obtained subsequently by LARBALESTIER [Lar80] upon optimized Ti-50Nb.

Resistive measurements of the superconducting transition temperatures were also undertaken. The transitions, which were sharp across the entire composition range, generally took place at temperatures higher than those reported for the bulk material. In a comparison with the classical work of HULM and BLAUGHER as summarized in Fig.7-3, the T_c composition dependence lay parallel to, but some 1 to 2 K higher than, the bulk results -- for example at its peak near 43 at.% Nb, T_c reached 10.6 K.

In the absence of any information having to do with the structures or impurity contents of the films, one can only speculate upon the causes of these enhancements.

PART 4: RECENT ADVANCES IN TITANIUM-NIOBIUM SUPERCONDUCTORS

7.30 FLUX-PINNING MICROSTRUCTURES IN Ti-Nb ALLOYS

7.30.1 Precipitate-Free Subbands

The seminal study of this topic has been conducted by NEAL et al [Nea71] using β-Ti-Nb. The composition of the alloy selected (an analyzed 42 at.% Nb) placed it so close to the practical boundary of the $\alpha + \beta$-phase field at, say, 400°C (Fig.7-1) that its precipitate-free status needed to be confirmed by TEM. This was carried out, and although some traces of α-phase were detected, its abundance was claimed to be inadequate to provide any significant degree of flux pinning. Sample preparation took place in the manner outlined in Sect.7.26.3, after which critical currents were measured at 4.2 K, using a 100 μV criterion in fields of up to 60 kOe, and converted to J_c's using measured wire cross-sectional areas. The electrical results were correlated with fiber diameters (i.e. subband diameters, or cell sizes) determined by the intercept method from transverse-section electron micrographs a typical example of which has been given in Fig.7-34. From the original article, and from other quantitative descriptions of the effects of cold drawing and heat treatment on the subband structure (cf. [Mon2.8]) and the subband-related J_c (cf. [Mon21.14]), the following conclusions can be drawn: (i) J_c appears to be extremely sensitive to the initially administered cold work which in NEAL's experiments produced its greatest effect during the first stage of area reduction (viz 10^2:1), further cold work in the range of 10^2:1 to 10^5:1 resulting in only moderate increases in J_c; (ii) in 1-h aging experiments, J_c increases rapidly as the temperature is raised above ∼300°C. This was interpreted as being due to a migration of dislocations from the interiors of cells to their walls, with little increase in cell diameter, a process which takes place all the more actively above about 400°C but which must compete with the deleterious effect of cell growth which begins to accelerate in the temperature range 400-500°C, [Mon2.11]; (iii) after 5×10^4:1 area reduction by cold work, the optimal 1-h aging temperature is 385 ∼ 425°C; (iv) during aging at 385°C, although the cell diameter appears to be quite stable within any reasonable estimate of experimental uncertainty, J_c increases 5% to a maximum after 5 h; (v) cell growth data, [Mon2.11], and the results of NEAL et al point to ∼400°C as an optimal heat treatment temperature. The fact that aging at this temperature can also yield ω-phase and α-phase precipitation in the appropriate concentration ranges has been a source of confusion in the literature; (vi) according to NEAL et al [Nea71], and BAKER [Bak70] before them, the bulk pinning force, $J_c H_a$, is proportional to the reciprocal of the cell diameter, d. As shown in Fig.7-45, for heat-treated Cu/Ti-Nb$_{42}$ monofilaments, J_c (at 50 kOe) is linear with slope 6.06×10^7 A cm^{-2} Å. The curve depicted represents a norm for optimized unprecipitated Ti-Nb superconductors, against which the effects of additional intrinsic α-phase precipitation, characteristic of less Nb-rich alloys, can be compared.

The effect of heat treatment on dislocation cell structure, within this context, has been discussed in

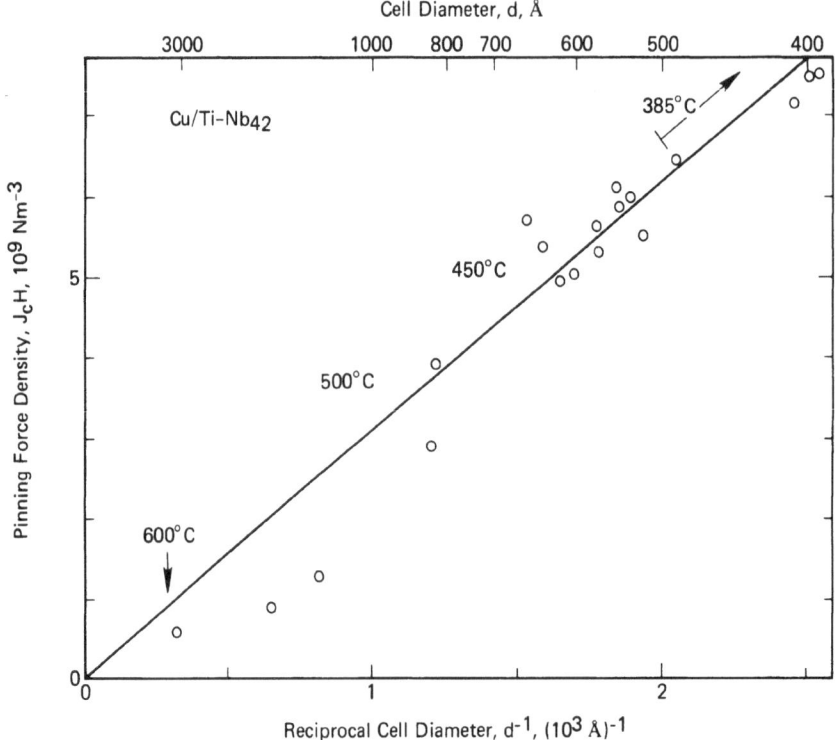

FIGURE 7-45. Bulk pinning force density at 4.2 K, 50 kOe *versus* reciprocal cell diameter for Cu/Ti-Nb (42 at. %), cold worked and heat treated 1 h at the temperatures indicated — after NEAL *et al* [Nea71].

detail by CHARLESWORTH and MADSEN [Cha70] with reference to the classical work of FRIEDEL [Fri64, p.276] and the contemporary studies of BAKER [Bak70] referred to above. All workers agree that critical current density is favored by a fine dislocation cell structure and that an optimal heat treatment is one that clarifies the interiors of the cells without contributing to their growth. The case study of NEAL *et al* [Nea71], augmented by the results of some recent studies of α-phase precipitation (e.g. [Wes80]) and "interstitial"-element effects and intermetallic-compound precipitation as discussed in other chapters, all seemed to point to subband structure as the dominant controller of J_c; but cf. Sect.7.28.1(b). NEAL's studies of precipitate-free alloys indicated that an optimal sub-cell structure must always be achieved as a basis for further improvements in flux-pinning strength, whether it be by cell-wall precipitation, or more fundamentally *via* an alloying-effect increase in the upper critical field.

7.30.2 Subbands and Precipitates

From the very beginning when it was first recognized that flux penetrated a type-II superconductor in

the form of an ABRIKOSOV lattice, it was realized that in order to stabilize transport current, flux drift had to be prevented by the establishment of a matrix of pinning centers of one kind or another, in general, by departures from crystalline perfection. The metallurgical options available were *intrinsic* defects such as dislocations and grain boundaries, or *extrinsic* defects such as precipitates. Process optimization then consisted of reducing the scale of these defects to that of the flux lattice to be pinned. Deformation and precipitation, as independent events, were treated early in the development of hard superconductors and as such have been considered in some of the preceding sections. But as indicated above in association with a discussion of Figs.7-41 and 7-42, it has often been difficult to separate improvements in J_c resulting from subband refinement from extra improvements which may result from precipitation. Fig.7-46 is an example of this. Fig.7-46(a) displays the composition dependences of J_c in response *only* to cold work [Haa75, p.237]. The interesting peak in J_c located at about 53.5 wt.% Nb (37 at.% Nb) is presumed to be traceable to the operation of a pronounced "peak effect" (cf. [Mon21.12]) in association with a Nb-concentration-controlled rapid decrease in H_{c2} below 34 $\sim$ 36 at.% Nb

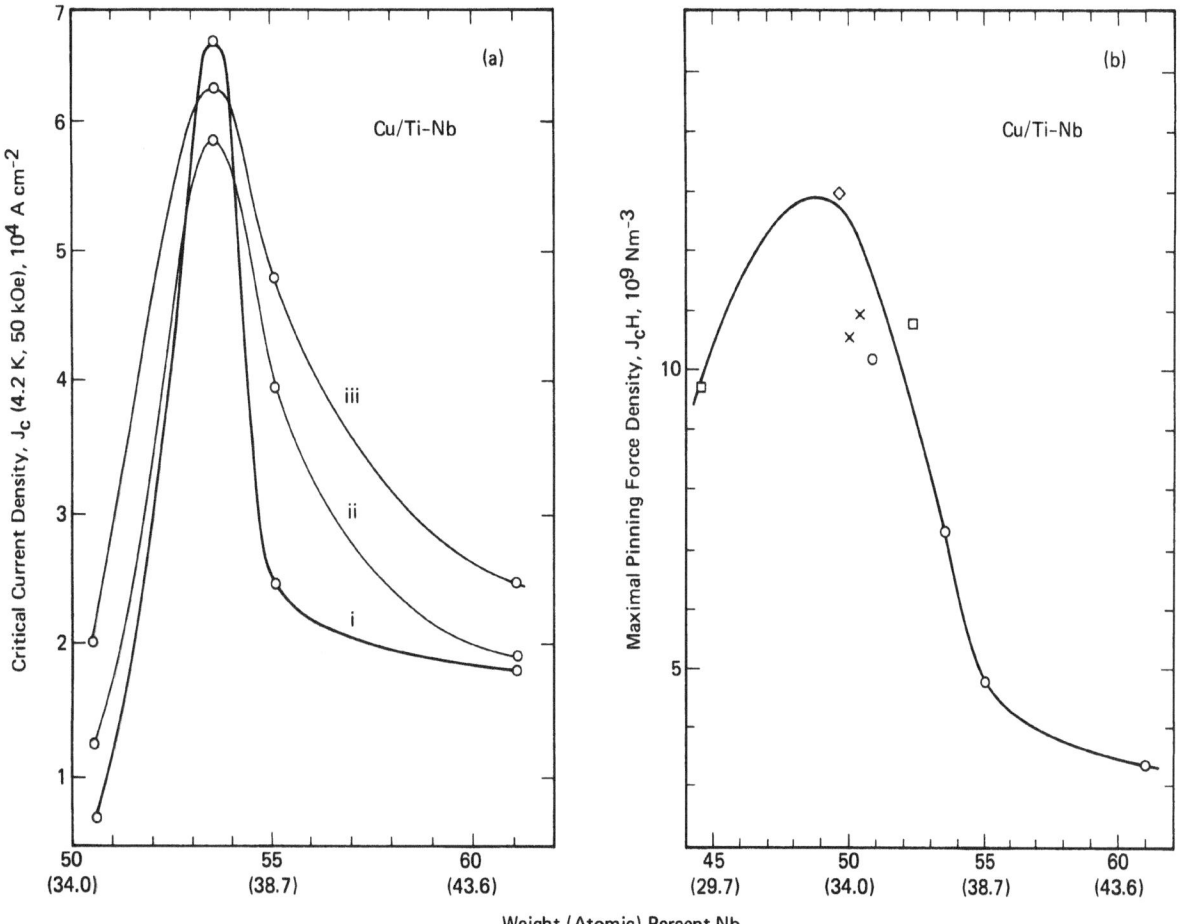

FIGURE 7-46. (a) Critical current density, J_c (4.2 K, 50 kOe) *versus* Nb content for Cu/Ti-Nb monofilaments cold reduced from a 30 min/600°C anneal at 10 mm$^\phi$ to: (i) 1.02 mm$^\phi$ (98.96%); (ii) 0.5 mm$^\phi$ (99.75%); (iii) 0.2 mm$^\phi$ (99.96%) — after HILLMANN [Hiℓ74]. (b) Maximal pinning force, $F_{p,max}$ (4.2 K) *versus* Nb content for process-optimized multifilamentary Cu/Ti-Nb composites (x) — HILLMANN *et al* [Hiℓ79], (◇) — LARBALESTIER [Lar80], (□) — STEKLY *et al* [Ste78], and Cu/Ti-Nb monofilaments, (○) — HILLMANN *et al* [Hiℓ79].

(Fig.7-19). It is important to note that J_c is generally low, and particularly so near Ti-Nb$_{35}$. On the other hand, an optimization heat treatment applied to the same series of alloys [Hiℓ79] results in a curve of the form given in Fig.7-46(b) which demonstrates how rapidly the maximum-pinning-force increases as decreasing Nb content causes the alloy to enter the precipitation regime. The pair of figures demonstrates, for example, that the 50-kOe J_c of Ti-53.5Nb rises from about 6.6x10^4 A cm^{-2} to about 1.5x10^5 A cm^{-2} as a result of process optimization. But in comparisons of this type it is not possible to separate out the contributions that precipitation -- as distinct from cell optimization -- has made to the increase in J_c. To accomplish this successfully, a different approach is called for. A brilliant solution to the problem was offered by WEST and LARBALESTIER [Wes80],

Fig.7-47. Instead of comparing heat-treated with as-cold-worked material, a comparison was made between the optimized critical current densities of the precipitate-free alloys of NEAL *et al* [Nea71], Ti-Nb$_{42}$, with those of a series of conventional α-precipitated alloys of neighboring compositions, Ti-Nb$_{32,37}$, using cell diameter as the common variable. In Fig.7-47 it is seen that precipitation does little to increase flux pinning in large subband material, but that provided the wire is already structurally optimized (*heat-treated* cell size $\stackrel{\sim}{<}$550 Å), precipitation may contribute an additional ∿0.5x10^5 A cm^{-2} to the critical current density. These results added strength to a body of opinion that had by then accumulated [Arn74] [Wiℓ75] (and with which the results of some recent work [Wes81] are in accord) to the effect that: although a fine heat-treated subband structure is a

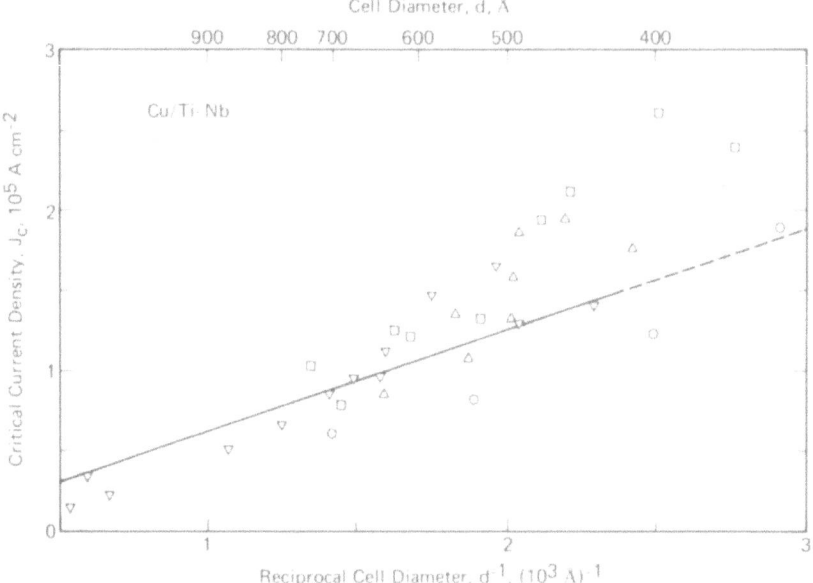

FIGURE 7-47. Influence of precipitation on critical current density. Critical current density, J_c (4.2 K, 50 kOe), of optimized commercial multifilamentary conductors *versus* inverse cell (subband) diameter. Alloys represented are: Ti-53.5Nb ($\triangle$); Ti-50.3Nb ($\circ$); Ti-49.6Nb ($\square$); Ti-47.3Nb ($\triangledown$). The straight line is a repetition of the results of NEAL *et al* on precipitate-free Ti-Nb (42 at. %) (Ti-58Nb) from Fig. 7-45 — after WEST and LARBALESTIER [Wes80].

necessary prerequisite for strong bulk flux-pinning, full optimization demands the presence of a suitable dispersion of suitably dimensioned precipitates.

A quantitative study of precipitation in Ti-50Nb conductors was carried out by WILLBRAND and SCHLUMP [Wil75] according to which it appeared that for precipitate particles of constant diameter, J_c was proportional to their number density, n_p; and that normalized to constant n_p, J_c was proportional to the square of the particle diameter, ϕ^2 ($\propto$ the area presented to a moving flux line). It is indeed fortunate from the standpoint of wire fabrication that J_c does not, however, depend on a unique combination of n_p and ϕ^2. In fact, as indicated in Fig. 7-48, J_c increased almost linearly with $n_p\phi^2$, values of which between about 3 and 6 were associated with 50-kOe J_c's as high as 3.5×10^5 A cm^{-2} [Wil75].

The most recent study of subband and precipitational effects in Ti-Nb alloys is that of WEST and LARBALESTIER [Wes81]. The starting material for the investigation was a sample of production Fermilab conductor (cf. [Mon27.16][Mon28.5]), 3.66 mm$^\phi$, taken from an intermediate stage in the manufacturing process after having received post-extrusion cold-work area-reduction of ~255:1. TEM measurements of subband diameter were performed on Ti-Nb samples in the following six conditions: *(a)* at "3.6 mm$^\phi$" in the as-

received condition; *(b)* at "3.66 mm$^\phi$" after 80h/375°C and 160h/375°C; *(c)* at "0.66 mm$^\phi$" after cold drawing from conditions *(a)* and *(b)*, respectively. Critical

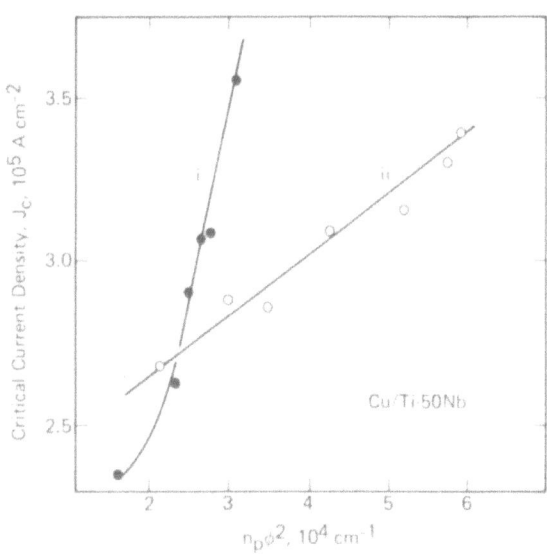

FIGURE 7-48. Influence of distribution and size of α-phase precipitate particles on flux pinning. Critical current density (4.2 K, 50 kOe, criterion: 5 μV cm^{-1}) of a 61-filament Cu/Ti-50Nb composite *versus* the product $n_p\phi^2$ where n_p is the number density of particles and ϕ is the mean diameter. Two series of alloys designated (i) and (ii) (distinguished by slightly different heat-treatment and deformation cycles within similar deformation regimes) are represented — after WILLBRAND and SCHLUMP [Wil75].

current density measurements were performed on long samples of the 0.66-mm$^\phi$ Cu/Ti-Nb strand. The results of the work, as summarized in Table 7-21, show that as a consequence of heat-treatment-induced supposed increase in α-phase precipitate-particle number density, very large increases in J_c may take place in the presence of only slight reductions of subband diameter.

TABLE 7-21 INFLUENCE OF HEAT TREATMENT ON THE SUBBAND DIAMETER, IMPLIED INCREASE IN α-PHASE PRECIPITATE PARTICLE NUMBER-DENSITY, AND CRITICAL CURRENT DENSITY[†] OF Ti-53.5Nb COMPOSITE STRANDS[††] -- after West and Larbalestier [Wes81].

| Condition | Subband Diameter $\overset{\circ}{A}$, at Strand Diameters Specified | | 4.2 K, 5 T, J_c of 0.66 mm$^\phi$ Strand, 10^5 A cm^{-2} |
	3.66 mm	0.66 mm	
As received (a.r.)	690	510	0.37
a.r. plus 80h/375°C	1,160	370	1.49
a.r. plus 160h/375°C	1,230	440	1.54

† J_c criterion: I_c determined at a sample resistivity of 10^{-12} Ω cm.

†† As-received samples were incompletely processed Fermilab basic strands. For the fully processed strand: strand diam. $\sim$ 0.68 mm; fil. No. $\sim$ 2046; Cu/SC ratio $\sim$ 1.8:1

7.31 PROCESS OPTIMIZATION OF Ti-Nb SUPERCONDUCTORS

It has been shown how the critical current density of a heavily deformed Nb-rich alloy may be optimized by the application of a moderate temperature ($\sim$385°C) heat treatment, and that further improvements can be obtained when the composition is such that precipitation, also as a result of moderate temperature heat treatment (e.g. 390°C -- WILLBRAND et al, Sect.7.28.1; 380°C -- HILLMANN et al, Sect.7.28.2) takes place within the cells or cell walls. Optimization is the process of further increasing J_c, through

fine adjustments of deformation structure and precipitate size and distribution. This is achieved: (a) by administering heat treatments in stages during the wire-drawing process (so-called "intermediate heat treatment"); (b) by administering, following intermediate heat treatment, a rather precise degree of final cold deformation.

7.31.1 Intermediate Heat Treatment

Fig.7-44 is an example of the steady increase in critical current density which accompanies the addition of more and more stages of intermediate heat treatment followed in most cases by a fixed 75% final cold reduction [Wiℓ75^a]. The experimental processing schedule was listed in Table 7-19. Numerous other cases are to be found in the literature: a classical example, appearing in the early work of PFEIFFER and HILLMANN [Pfe68] and cited several times subsequently [Hiℓ73][Lar80], made reference to the beneficial effects of properly administered final cold deformation.

7.31.2 Final Cold Deformation

It has been found empirically that it is better to terminate the alternating cold-work/anneal cycle of the intermediate heat treatment sequence with the cold-work option. Some of the first systematic studies of the effect of final cold deformation on the critical current density of intermediate-heat-treated alloys, those of PFEIFFER and HILLMANN [Pfe68] on Ti-50Nb, have been mentioned in Sect.7.28.2. Since even the *detection* of α-phase precipitates in heavily deformed wire is difficult, it has usually been possible only to speculate from indirect evidence, on the manner in which precipitation is *affected* (i.e. refined or redistributed) by the final cold work. Mechanical-property evidence, for example, has suggested that the initially formed large α-phase particles were reduced in size and distributed more uniformly during the post-anneal deformation [Aℓb76]. But very recently, as a result of some very careful STEM investigations by WEST and LARBALESTIER [Wes82] it has been possible to portray not only the sizes and shapes of the individual precipitate particles (cf. [Mon1.4]) but the manner in which they became deformed during wire drawing.

According to some earlier work by HILLMANN and HAUCK [Hiℓ72], heat treatment of the fiber structure yielded a very anisotropic distribution of precipitate particles -- a number-density greater *in* the drawing direction than perpendicular to it. During final cold reduction by wire drawing, it was supposed that the density of particles naturally increased in the transverse direction at the same time as they became more spread out in the longitudinal direction -- leading eventually to an isotropic distribution. Based on such a model it was predicted that J_c should increase with increasing final cold work, eventually to pass through a maximum at some optimal level of it. Indeed that was observed experimentally. Further evidence in support of this hypothesis has been acquired by BEST *et al* [Bes79, Bes79a][Hiℓ79] in studies of the J_c anisotropy of wires that had been removed from various stages of the final cold deformation sequence and flattened by rolling.

Optimized final cold deformation in the processing of superconducting wire has been examined using

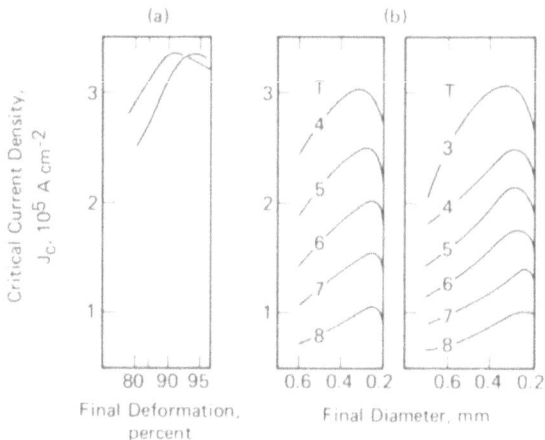

FIGURE 7-50. Process optimization of 61-filament Cu/Ti-50Nb composites — influence of cold deformation after the final heat treatment. (a) Critical current density *versus* final cold deformation,[†] according to process-schedule "a" of the previous figure, for two specimens of the "series-(ii)" alloy referred to in Fig. 7-48 — after WILLBRAND and SCHLUMP [Wiℓ75]. (b) Critical current density *versus* final wire diameter, according to process-schedule "b" of the previous figure, for two specimens of wire — after HILLMANN *et al* [Hiℓ73a, Hiℓ74].

[†]According to WILLBRAND [Wiℓ80] the horizontal scale is proportional to the logarithm of the square of the (cold-reduction) wire-diameter ratios.

two slightly differing experimental formats: *(a)* fixed overall reduction and fixed final wire diameter, with variable reduction within the intermediate-heat-treatment segment [Wiℓ75]; *(b)* fixed reduction within the heat-treatment segment and variable final wire diameter [Hiℓ72, Hiℓ73, Hiℓ73a, Hiℓ74]. This pair of experimental modes, the first due to WILLBRAND and SCHLUMP and the second to HILLMANN *et al* are defined in Fig.7-49. The results, typified by Figs.7-50(a) and 7-50(b), indicate that at 50 kOe, for example, optimal final cold reduction in Ti-50Nb lies somewhere within the range: 91∿94% [Wiℓ75] or ∿94% [Hiℓ74]. According to HILLMANN *et al* [Hiℓ72, Hiℓ73, Hiℓ73a, Hiℓ74], the source of Fig.7-50(b), optimization for higher fields calls for higher levels of cold work -- up to about 95% at 80 kOe. The significance of *total* reduction is considered below. The existence of the maximum implies the possibility of "overoptimization" through excessive final cold work. These effects, which were claimed to result from precipitate-particle redistribution, have been examined in the experiments on flattened wires [Bes79, Bes79a][Hiℓ79] mentioned above and dealt with fully in [Mon21.19].

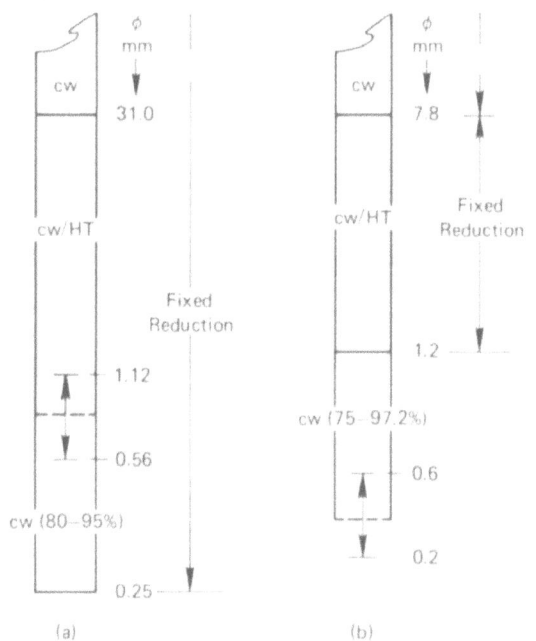

FIGURE 7-49. Thermomechanical processing by alternate cold work (cw) and heat treatment (HT) of 61-filament Cu/Ti-50Nb composite wires. Schema for two investigations of final cold deformation optimization. (a) Variation of the final cold formation within 80-95% at constant overall area reduction hence fixed final wire diameter — after WILLBRAND and SCHLUMP [Wiℓ75]. (b) Variation of the final cold deformation within 75-97.2% at fixed prior wire diameter necessitating variable final wire diameter — after HILLMANN *et al* [Hiℓ73a].

7.32 RECENT ADVANCES IN PROCESS OPTIMIZATION

7.32.1 Total Area Reduction and Final Cold Deformation

Considered above were the effects of multiple intermediate heat treatment and variation of the level of final cold deformation on the critical current density of Ti-50Nb, the alloy favored by European groups. In conductor optimization, composition is of course another important variable whose influence on the pinning-force density has already been summarized in Fig. 7-46. Accordingly, it is instructive to introduce into a continuation of the discussion of cold-work effects a pair of alloys, Ti-45Nb and Ti-53.5Nb, whose compositions bracket that of the alloy just treated. In the recent work of STEKLY et al [Ste78][Seg79], commercial-size multifilamentary billets (Ti-45Nb: 3 and 8 in.$^{\phi}$, with 156 and 1272 cores, respectively; Ti-53.5Nb: 8 and 10 in.$^{\phi}$, with 132 and 2046 cores, respectively) were reduced to wires of diameters ranging from $\sim$0.1 to 0.025 cm, during which various single-stage and multiple-stage heat treatments were administered. An optimal single-stage heat treatment was discovered for each composition (Ti-45Nb: 75h/350°C; Ti-53.5Nb: 75h/375°C) and in a two-part investigation an attempt was made to separate the influence on J_c of the $total$ area reduction from that of the post-heat-treatment $final$ cold work.

The mechanical processing, in conjunction with the single-stage "optimal" heat treatments specified above, took place according to the schema of Fig.7-51. It will be recalled that a somewhat similar comparison

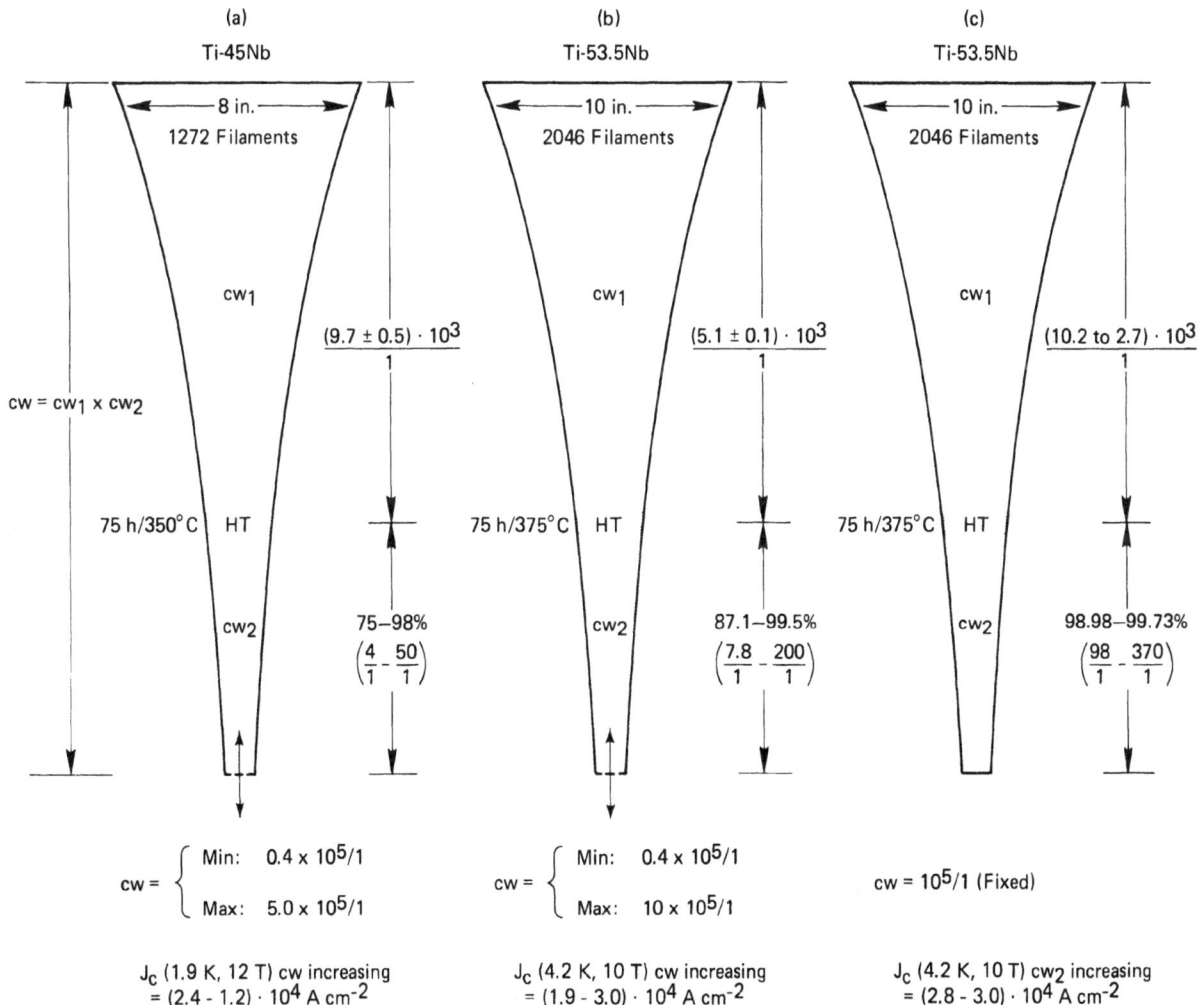

FIGURE 7-51. Schema for comparative studies of the influences of total or final cold deformation on the critical current densities of singly-heat-treated Ti-45Nb and Ti-53.5Nb multifilamentary composites prepared from commercial-size billets. Trends are indicated by a listing of the cold-work extrema and the corresponding high-field critical current densities — after STEKLY et al [Ste78, Seg79].

has already been synthesized by assembling, as in Fig.7-49 and 7-50, the independently acquired results of WILLBRAND and SCHLUMP [Wiℓ75] and HILLMANN et al [Hiℓ73^a] who, however, had employed multistage intermediate heat treatment schedules. In their case the results, according to Fig.7-50, turned out to be similar irrespective of whether the final cold reductions were carried out in association with (a) fixed, or (b) variable, total area reduction.

Critical current density versus area-reduction data have been acquired by STEKLY et al [Ste78] for processing schedules represented by Figs.7-51(a), (b), and (c). As indicated in (a) and (b) the influence of final cold reduction, cw_2, on the critical current densities of Ti-45Nb and Ti-53.5Nb were intercompared. In either case the wires had been subjected to practically constant reduction, cw_1, (viz 9.7x10^3 and 5.1x10^3:1, respectively) prior to the heat treatment. According to (b) and (c) of the same figure, a study was also made of the influence on J_c of variation of final cold reduction at fixed cw_1, as compared to variation of final cold reduction at fixed overall reduction, cw, (implying a compensatory variation in cw_1). The results, presented in detail in the original articles, and in summary form in the above figures themselves, are as follows: (i) with the Ti-45Nb composite, the critical current density decreased monotonically with final cold work suggesting that, for the magnetic field range under consideration, this alloy was already overoptimized at about 75% final cold deformation; (ii) with Ti-53.5Nb, J_c(1.9 K, 13 T) and J_c(4.2 K, 10 T) were found to be close to each other and to increase monotonically with increasing final cold work. Such a result is not inconsistent with the high-field response of Ti-50Nb to cold work as suggested in Fig.7-50(b) in which the peak of the curve shifts in the direction of smaller wire diameter as the applied field increases. In the present case it was noted that J_c increased insignificantly with final reductions greater than about 98%, which from a practical standpoint, must be regarded as optimal; (iii) the purpose of the comparison represented by Figs.7-51(b) and (c) was to determine whether total reduction or final reduction had the greater influence on critical current density. Unfortunately the design of the experiment prevented the question from being definitively answered. With regard to final cold reduction it is interesting to note that whereas in (b) a variation in final cold reduction ratio of 26:1 increased the critical current density by 58%, in (c)

a variation of final cold reduction ratio of 3.8:1 increased J_c by 7% (in spite of the fact that in this case cw_1 was undergoing a large decrease). This remarkable proportionality of increase in J_c to increase in final cold reduction ratio serves to emphasize the latter's importance as a determiner of current-carrying capacity in the magnetic field range under investigation. To STEKLY et al [Ste78], the results depicted in the figure suggested that total area reduction, from pre-extrusion to final wire size, was the determining factor. SEGAL, in commenting privately on these results also took the view that, at least for Ti-53.5Nb, total reduction is more important than final reduction and went on to point out that, although it would always be possible to design a large conductor with a relatively large final reduction, it would not always be possible in such a conductor, if monofilamentary, to fit an adequate amount of cold work between the billet diameter and the final strand size. This is one reason why large conductors are cabled.

7.32.2 Thermomechanical Process Optimization

Studies of the heat treatment variables were undertaken by STEKLY et al [Ste78][Seg79] on wire prepared from billets somewhat smaller than those referred to in Fig.7-51. With Ti-53.5Nb, heat treated for times of 40 to 150 h at temperatures between 350 and 400°C, it was found that J_c(1.9 K and 4.2 K, 12 T) passed through a maximum in the vicinity of 50-75h/375°C. As a result of this work, a single heat treatment of 75h/375°C was selected for the deformation studies just outlined. Similarly after subjecting Ti-45Nb to heat treatments definable by (50-150h)/(325-375°C), and observing the behavior of J_c(1.1 K, 12 T), an optimal single heat treatment of 75h/350°C was chosen. Some attempts at investigating the effects of multiple heat treatments were also made. It was noted that the J_c(1.9 K, 12 T) of singly-heat-treated Ti-45Nb could be raised 8% with one additional heat treatment or 25% with two more.

The optimized critical current densities of the two alloys referred to here are juxtaposed against that of the European alloy Ti-49.5Nb, as measured by LARBALESTIER [Lar80], in Fig.7-52. On this evidence, and with reference to a J_c(intermediate-field) versus composition format, Fig.7-46, the alloys Ti-45Nb and Ti-53.5Nb lie on either side of a maximum centered near Ti-50Nb. It is not known whether this maximum

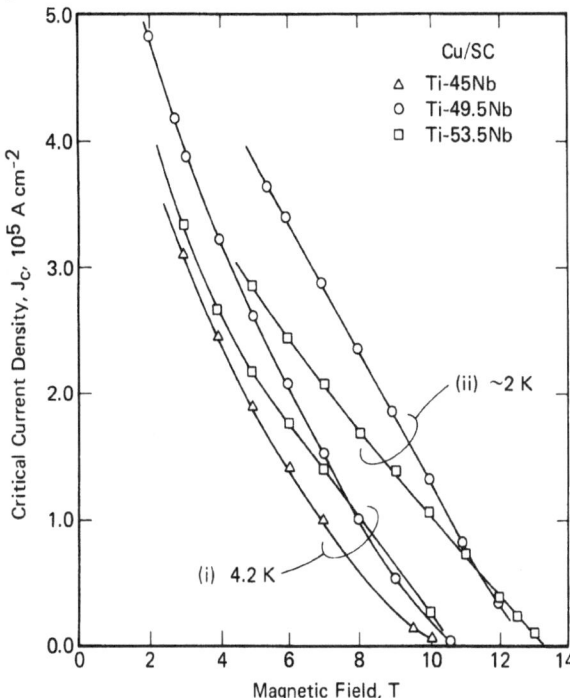

FIGURE 7-52. Group (i) — comparison of the J_c (4.2 K) field dependences of optimized Ti-45Nb and Ti-53.5Nb as measured by STEKLY et al [Ste78, Seg79] with that of Ti-50Nb from the work of LARBALESTIER et al [Lar80, Haw80ª]. Group (ii) — J_c (~ 2 K) field dependences of Ti-49.5Nb (at 2 K) and Ti-53.5Nb (at 2.05 K) conductors — after HAWKSWORTH and LARBALESTIER [Haw80ª].

will survive further attempts at the optimization of Ti-53.5Nb by means of multistage intermediate heat treatment, such as is administered by the German groups to Ti-50Nb (cf. Sects.7.28.1 and 7.28.2). All evidence points to the technical desirability of multiple, rather than single-stage, cold work, thus it is suspected that further improvements to the J_c of Ti-53.5Nb are possible. Multiple heat treatment is more expensive to carry out than a single-stage process, but if successfully applied to Ti-53.5Nb the reward would be a conductor that is more workable than Ti-50Nb and one that, as already suggested in Fig.7-52, has a greater high-field critical current density.

7.32.3 Critical Field Limitation

Having been carried as far as possible by empiricism in the form of thermomechanical process optimization, J_c is subject to further increase only if closer attention is paid to some of the fundamental aspects of flux pinning such as those reviewed in

numerous standard works on the subject (e.g. [Cam72] [Haa75]). A review of flux pinning useful in the present context is offered in [Mon21.00] (cf., in particular, [Mon21.4] through [Mon21.9]). According to [Mon21.5], ABRIKOSOV theory provides for a $(1-H_a/H_{c2})$ factor in the bulk (or "summed") pinning-force expression. This has nothing to do with the elementary pinning force, but with the *existence* of the flux lattice. In this sense then there is an *existence cutoff* for flux pinning such that, as LARBALESTIER [Lar80] has reminded us, at high fields J_c tends to become dominated by H_{c2}. The control that H_{c2} exerts over the optimized bulk pinning force is exemplified by a comparison of the composition dependences of these quantities as in Figs.7-19 and 7-46(b). The implications of this are twofold:

(a) For a given class of superconducting alloy, e.g. Ti-50Nb, the $J_c(H_a)$ curves corresponding to various thermomechanical processing conditions tend to coalesce at very high fields, and converge linearly upon an almost fixed value of H_{c2}. Numerous examples of this are to be found in the literature, although very high field performance generally tends to be obscured through the conventional use of the semilogarithmic format for the display of the J_c *versus* H_a data; but the point we wish to make here is adequately illustrated in Fig.7-52, a linear plot.

(b) A small shift in H_{c2} to higher fields, carrying with it the entire $J_c(H_a)$ curve whose typical slope, according to published results for Ti-53.5Nb [Lar80][Haw80ª], is close to

$$dJ_c/dH_a = -3.2 \qquad (kA\ cm^{-2}\ kOe^{-1}) \qquad (7-35)$$

would reflect a significant improvement upon J_c at any field. Thus, given that existing superconductors have been successfully optimized, any further increases in high-field J_c can be won only by increasing the upper critical field. There are two ways of doing this: *(i)* by reducing the temperature; *(ii)* through alloying.

(a) Temperature. According to the HELFAND and WERTHAMER nonparamagnetic theory of upper critical field temperature dependence [Heℓ64, Heℓ66], as discussed in [Mon14.5]:

$$H_{c20}^{*} = 0.680(-dH_u/dt)_{t=1} \qquad (7-19)$$

It follows, after replacing H_{c20}^* with the right-hand-side of Eqn.(7-10), that:

$$-dH_u/dT = 4.50 \times 10^4 \, \rho_n \, \gamma \quad (Oe \; K^{-1}) \qquad (7-36)$$

with ρ_n in Ω cm and γ in erg cm^{-3} K^{-2}. Next, after substituting some numerical values: e.g.

$$\rho_n = 78 \times 10^{-6} \, \Omega \; cm \; [Ber63^a], \; and$$

$$\gamma = 1.04 \times 10^4 \; erg \; cm^{-3} \; K^{-2},$$

$$-dH_u/dT = 36.5 \; kOe \; K^{-1} \qquad (7-37a)$$

in satisfactory agreement with the

$$-dH_u/dT = 26.4 \; kOe \; K^{-1} \qquad (7-37b)$$

obtained experimentally by HAWKSWORTH and LARBALESTIER upon a Ti-52Nb monofilamentary conductor [Haw80].

Combining these data with the J_c field dependence quoted in Eqn.(7-35) gives for the H_u-controlled high-field-J_c temperature dependence the values:

$$-dJ_c/dT = -(dJ_c/dH_u)(dH_u/dT)$$
$$= 84.5 \sim 116.8 \; kA \; cm^{-2} \; K^{-1} \qquad (7-38)$$

in semiquantitative, but nevertheless useful, agreement with the 46.5 kA cm^{-2} K^{-1} obtained by intercomparing the 4.2-K and 2.05-K, 8-T, J_c's of Ti-53.5Nb as reported by HAWKSWORTH and LARBALESTIER [Haw80^a] and depicted in Fig.7-52. The improvement in J_c which can be expected when the temperature of operation of a Ti-Nb conductor is reduced from 4.2 to 2 K is adequately illustrated in that figure.

(b) Alloy Design. The importance of the H_{c2}-control of critical current density having been established within the context of change of operating temperature, attention can next be directed towards the second alternative, alloying, as a means of increasing H_{c2}, hence the high-field J_c. The fundamental electronic and magnetic factors which influence H_{c2} have already been adequately discussed in Part 2 of this chapter (in particular Sect.7.11, see also [Mon15.9]) wherein it was stated that a reduction of the effect of Pauli paramagnetic limitation by means of heavy-element substitutional alloying should increase the experimental upper critical field. With this in mind, the following substitutions have been made in Ti-Nb alloys:

Hf for Ti -- HAWKSWORTH and LARBALESTIER [Haw80], WADA et al [Wad80],

Ta for Nb -- SUENAGA and RALLS [Sue68, Sue69], HAWKSWORTH and LARBALESTIER [Haw80],

Zr for Ti, and Ta for Nb -- HORIUCHI et al [Hor73].

These deliberate attempts to increase H_{c2} through heavy-element substitutions for the basic components of Ti-Nb alloys were intended, at least in recent years, to form the bases for improved high-field current-carrying alloy superconductors; the results of so doing are discussed in the appropriate chapters of this book.

PART 5: CRITICAL CURRENT DATA—SOME GRAPHICAL REPRESENTATIONS

7.33 COMPARATIVE SURVEY OF SOME CONTEMPORARY HIGH-FIELD Cu-STABILIZED Ti-Nb MONOLITHIC COMPOSITE CONDUCTORS

In support of the Japanese Large Coil Task, the critical currents and critical current densities (Ti-Nb component) of seventeen different Cu/Ti-Nb monolithic composite conductors from Japanese, British, and American suppliers were evaluated by TADA, ANDO, OKA, and SHIMAMOTO of the Tokai Research Establishment, Division of Thermonuclear Fusion Research, JAERI [Tad80]. This unique study is particularly valuable for the following reasons: (a) it has presented for intercomparison the properties of optimized representatives of the widest range of Ti-Nb compositions that could conceivably be of practical importance -- viz alloys with Nb contents of from 30 to 50 at.%; (b) it has enabled this comparison to be based on the results of tests conducted under similar conditions in association with identical, explicitly defined, transition criteria. Technical descriptions of sixteen of the conductors tested are listed in Table 7-22, and the geometries of some of them are depicted in Fig.7-53.

The samples which were spool wound, with voltage taps separated by 25 to 190 mm depending on the requirements of the test, were mounted under liquid He within the bore of a 13-T Nb_3Sn solenoid. Tests were made in fields of up to 11 T, the transition to the normal state being specified as having taken place when the sample resistance rose to either 10^{-10} or 10^{-11} Ω cm. The resulting critical current properties are presented in Figs.7-57(a) through (p).

TABLE 7-22   SPECIFICATIONS OF SIXTEEN DIFFERENT Cu/Ti-Nb SUPERCONDUCTING COMPOSITES TESTED BY TADA *et al.* AT THE LABORATORIES OF THE JAPAN ATOMIC ENERGY RESEARCH INSTITUTE [Tad80].

Sample Code (cf. Figs.7-53,54)	At.% Nb	Overall Diameter or Dimensions mm	Filament Diameter μm	Number of Filaments	Cu/SC Ratio	Twist Pitch mm	Manufacturer[+]
(a)	37.5	1.005	46.16	132	2.6	25.4	MCA
(b)	38.85	1.00	34.02	144	5.0	12.7	IGC
(c)	41.69	0.8	24.31	361	2.0		IMI
(d)	34.02	1.9x3.8	36	2,302	2.08	27	FEC
(e)	34.02	2.2	11.5	11,000	3.25		FEC
(f)	34.29	7.0x7.0	96	750	8.0	150	SEI
(g)	36.48	0.87x2.65	25.87	1,566	1.8	22	SEI
(h)	37.5	1.53	47.2	331	2.17	20	HC
(i)	50	0.998	38.16	114	5.0	12.7	VM
(j)	30	0.775	25.6	568	2.1		SW
(k)	30	1.0	33	568	2.1		SW
(l)	37	0.95	31.3	568	2.1		SW
(m)	37	1.0	33	568	2.1		SW
(n)	40	0.775	25.6	568	2.1		SW
(o)	40	1.0	34	514	1.9		SW
(p)	50	0.69	20	169	3.0		SW

+ MCA: Magnetic Corporation of America
  IGC: Intermagnetics General Corporation
  IMI: Imperial Metal Industries, Ltd.
  FEC: Furukawa Electric Company, Ltd.
  SEI: Sumitomo Electric Industries, Ltd.
  HC : Hitachi Cable, Ltd.
  VM : Vacuum Metallurgical Company, Ltd.
  SW : Showa Electric Wire and Cable Company, Ltd.

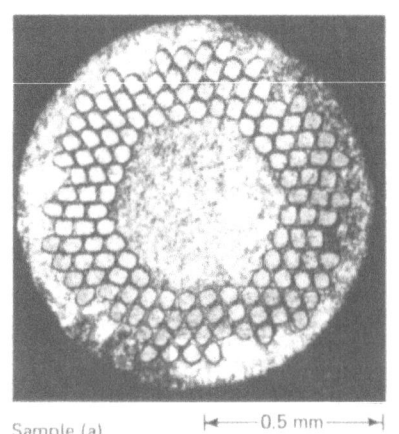

Sample (a) |◄——— 0.5 mm ———►|

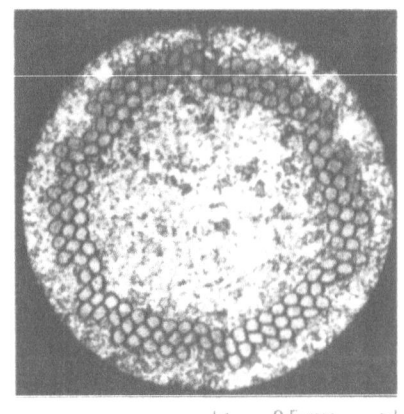

Sample (b) |◄——— 0.5 mm ———►|

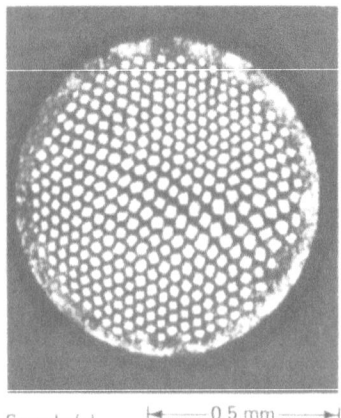

Sample (c) |◄——— 0.5 mm ———►|

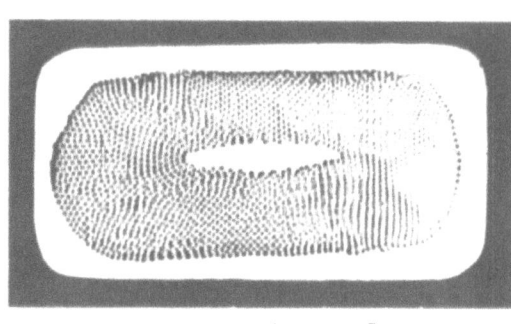

Sample (d) |◄——————— 2 mm ———————►|

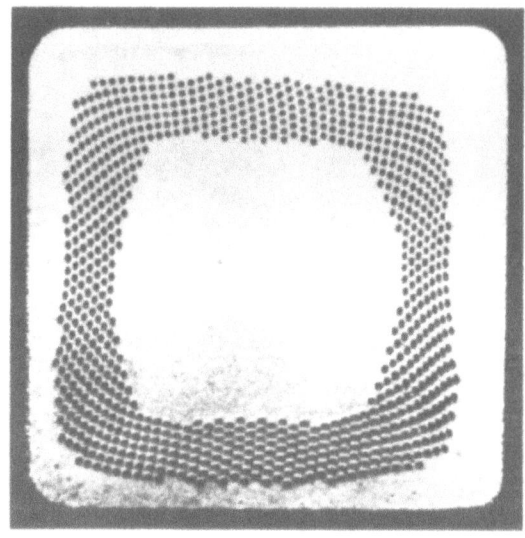

Sample (f) |◄——— 3 mm ———►|

FIGURE 7-53. Cross-sectional photomicrographs of some representative Cu/Ti-Nb small and large conductors selected from Table 7-22. Depicted are Samples: (a) 1.0 mm$^\phi$, (b) 1.0 mm$^\phi$, (c) 0.8 mm$^\phi$, (d) 1.9 x 3.8 mm^2, (f) 7.0 x 7.0 mm^2, (g) 0.87 x 2.65 mm^2, and (i) 1.0 mm$^\phi$ — after TADA *et al* [Tad80] — micrographs courtesy of S. Shimamoto; reproduced by permission of H. Nakamoto, JAERI.

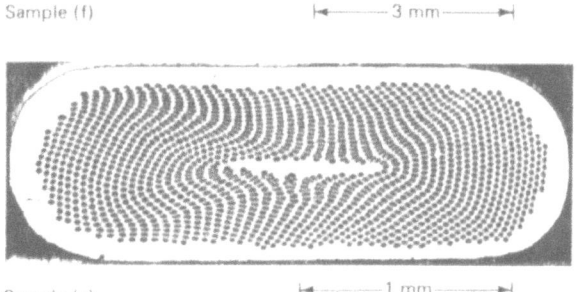

Sample (g) |◄——— 1 mm ———►|

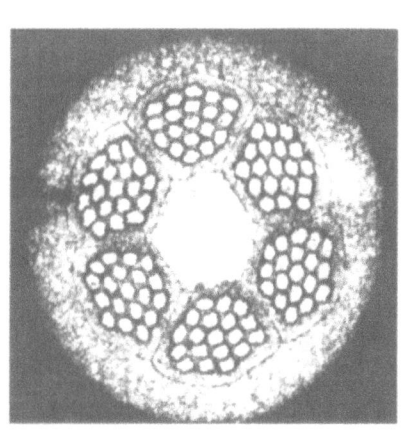

Sample (i) |◄——— 0.5 mm ———►|

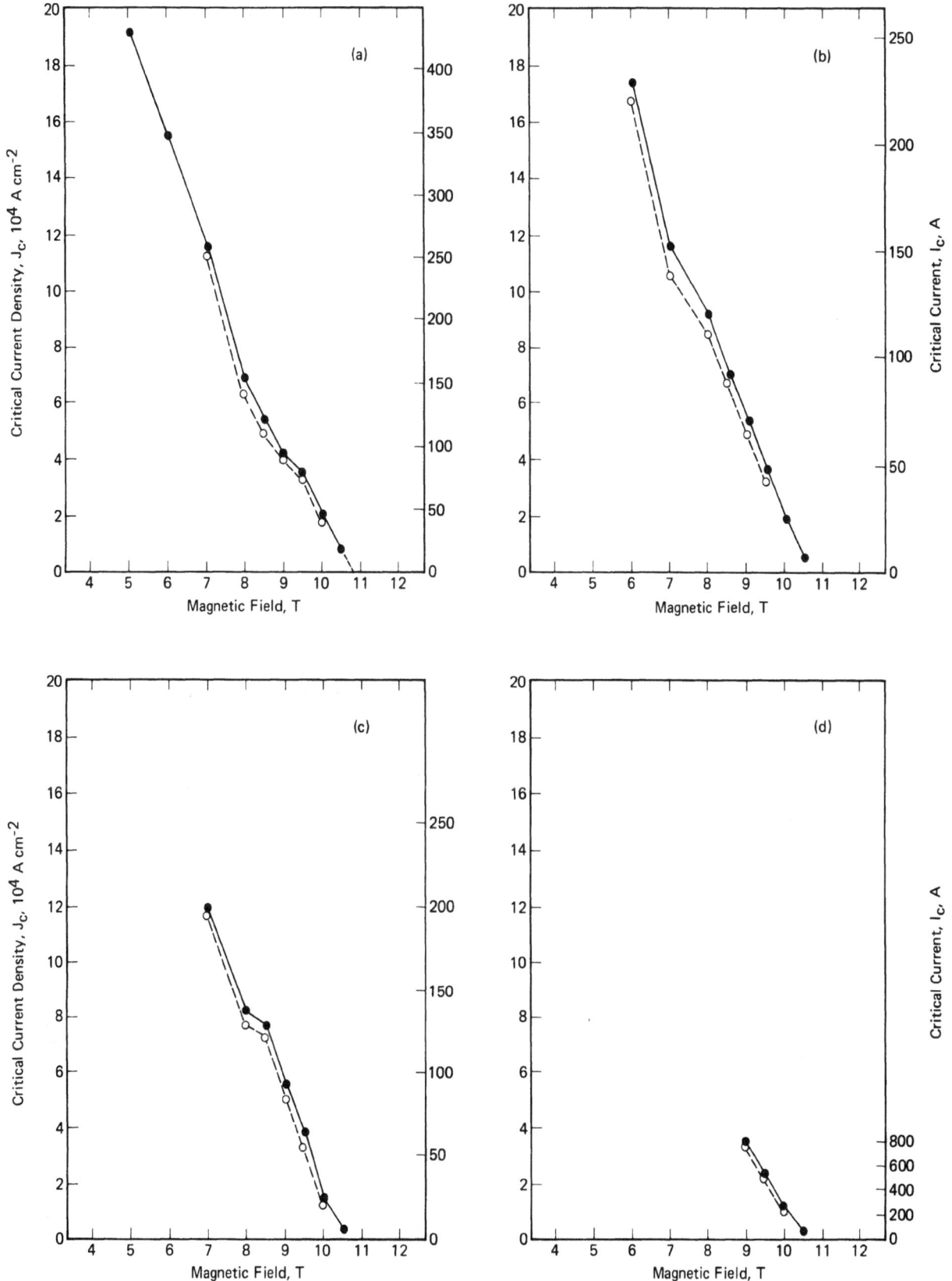

FIGURE 7-54. Short-sample characteristics of the conductors listed in Table 7-22. Critical criteria: 10^{-10} Ω cm ($\bullet$), and 10^{-11} Ω cm ($\circ$), respectively — TADA *et al* [Tad80]; reproduced by permission of H. Nakamoto, JAERI.

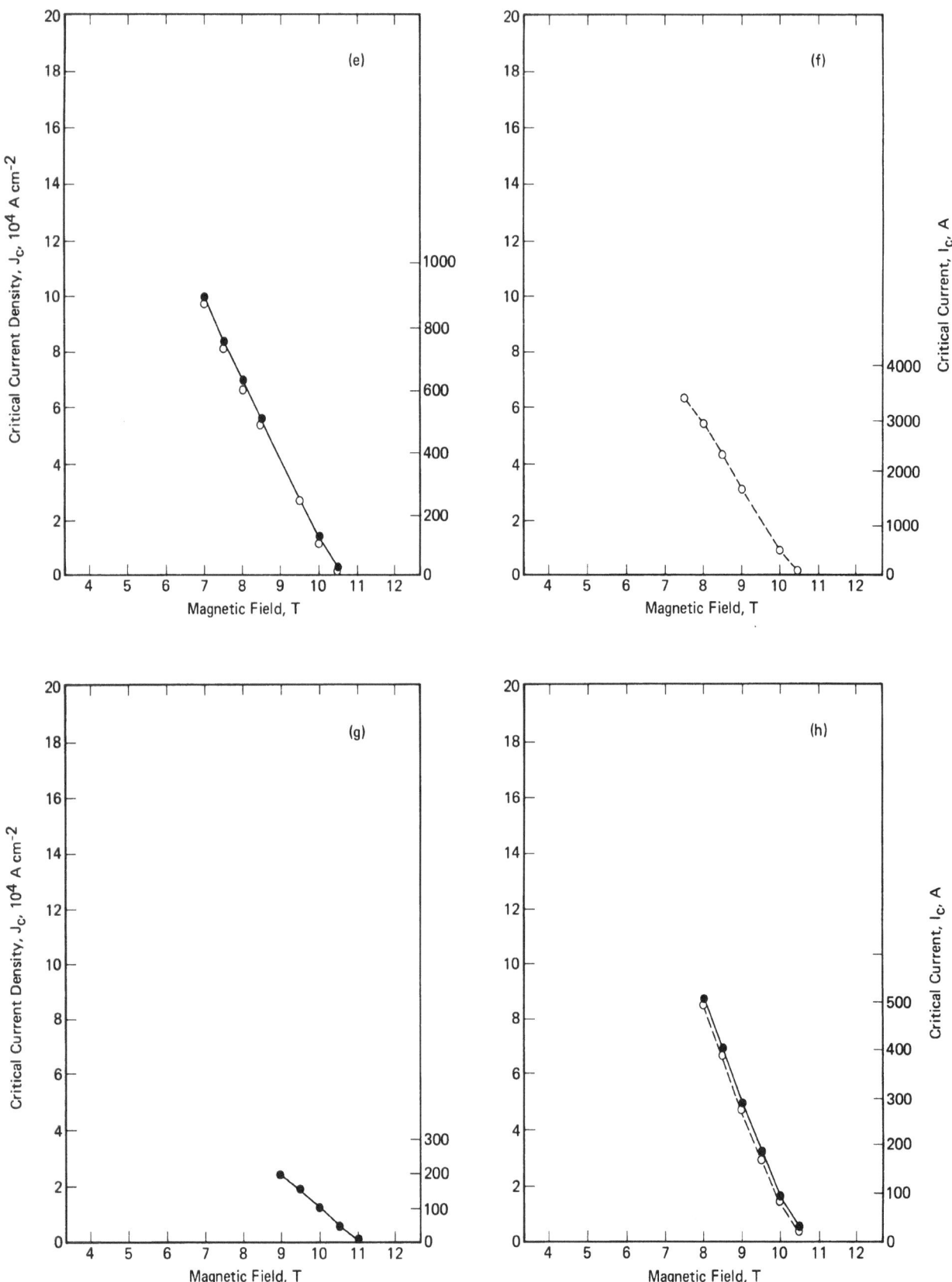

FIGURE 7-54. — (continued)

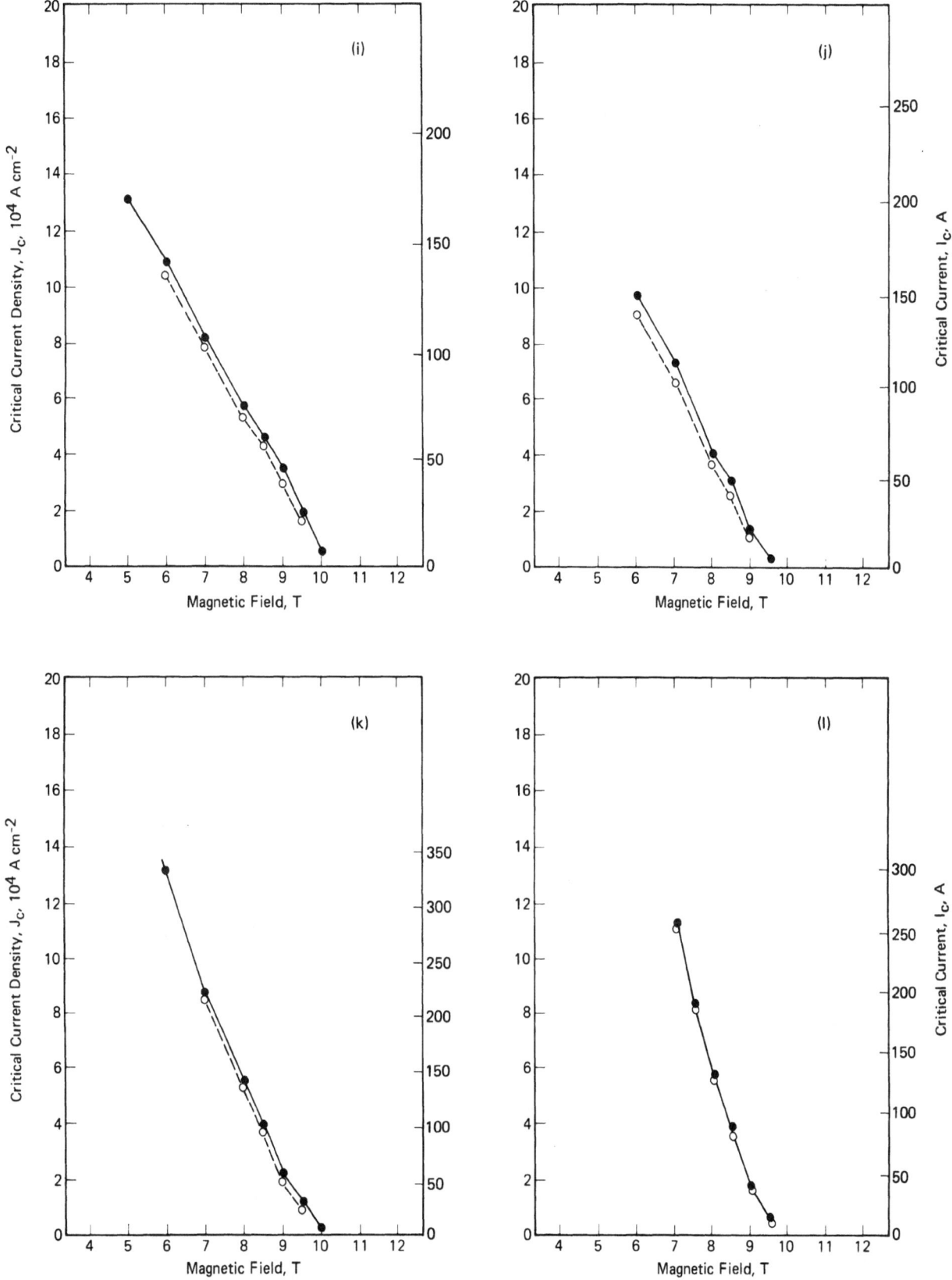

FIGURE 7-54. — *(continued)*

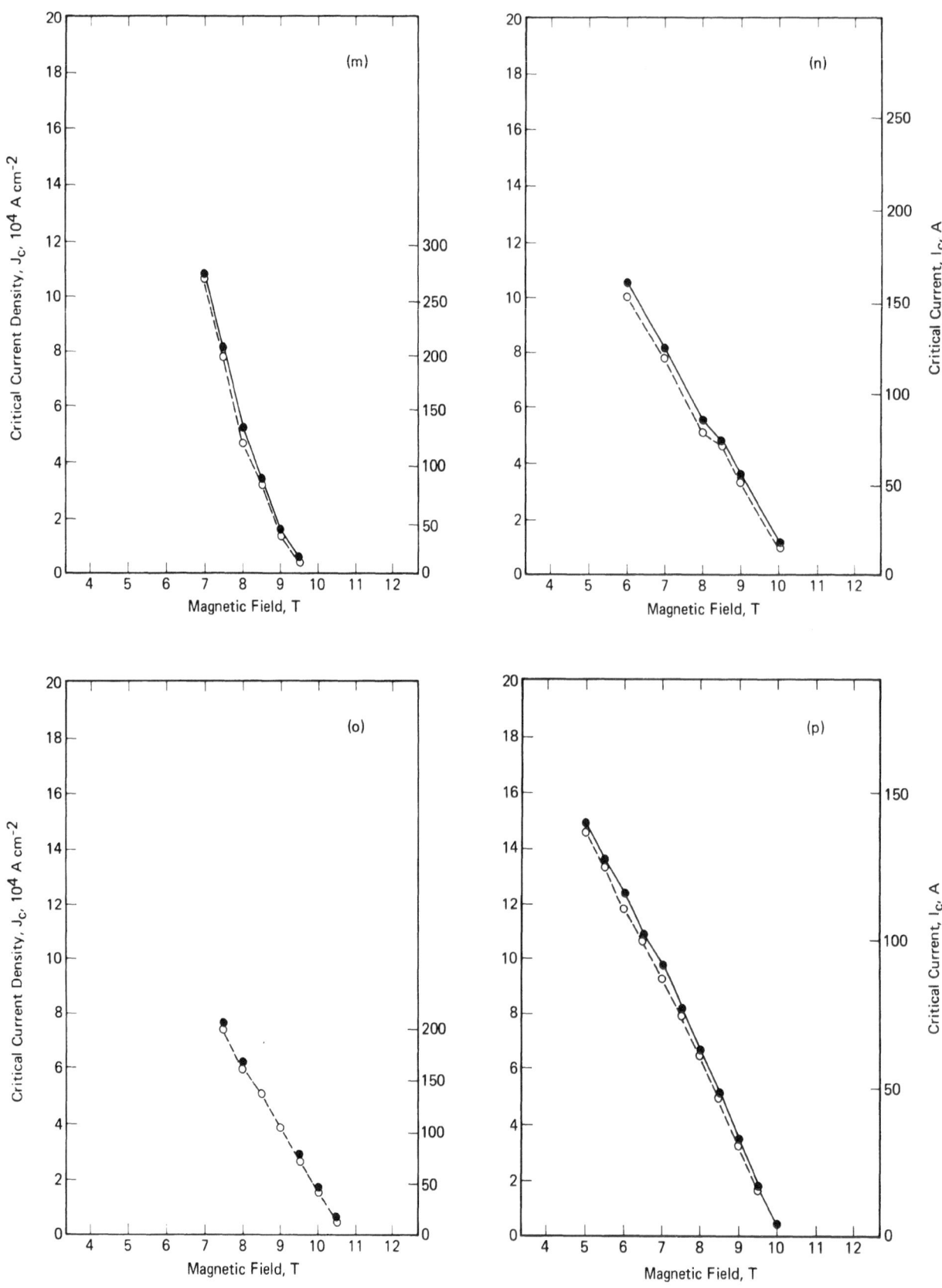

FIGURE 7-54. — *(continued)*

8

TITANIUM-NIOBIUM AND TITANIUM-NIOBIUM-BASE ALLOYS CONTAINING SMALL ADDITIONS OF BORON, CARBON, NITROGEN, OR OXYGEN

This chapter deals with the effects of the so-called "interstitial elements", B, C, N, and O (as listed in order of increasing atomic number) on the superconducting properties of Ti-Nb and related alloys. C, N, and O are always present to some extent in the commercial grades of starting material. Typical interstitial levels in commercial Ti are: C, 100 ppm; N, 100 ppm; and O, 800 ppm; commercial Nb may contain: C, 20 ppm; N, 20 ppm; and O, 100 ppm; while commercial Ti-52Nb superconductor may contain at least: C, 70 ppm; N, 50 ppm; and O, 500 ppm (see also Table 8-12, p.316).

The effects of B, C, N, and O on the superconducting properties of Ti-Nb have been studied for several reasons: (a) It is important to know whether their presence is directly beneficial, or whether their influences are sufficiently detrimental to the superconducting and mechanical properties as to warrant the expense of additional purification. (b) If the former turns out to be the case, then consideration can be given to deliberately augmenting the indigenous interstitial content. (c) The presence of elements in solid solution, and the interstitial-elements in particular, can be expected to alter the kinetics of precipitation, lead more quickly to the attainment of phase equilibrium at a given temperature, and while directly influencing processing procedures exert an indirect influence on superconducting properties.

The term "interstitial" is understood within the context of early transition metals to represent the elements B, C, N, and O, recognizing that although when dissolved they occupy interstitial positions in the lattice, they may in practice also become incorporated into compound phases.

ALLOY GROUP 1: BORON AND CARBON ADDITIONS TO TITANIUM-NIOBIUM

8.1 BORON ADDITIONS TO Ti-Nb

In spite of its high melting point (2800°C) small amounts of TiB_2 dissolve readily in Ti-Nb during arc melting. In this way, RASSMANN and ILLGEN [Ras72b] have added excess TiB_2 to Ti-56Nb such that, according to optical metallography, precipitated dendritic needles of TiB_2 were present in the as-cast alloy, leaving some small residual amount of B ($\lesssim$0.1 wt.%) in solid solution. The hardnesses of both the as-cast and the annealed (24h/500°C/water quench) alloys were investigated. Also measured, for comparison with that of the binary starting material, was the critical current density at 4.2 K and 50 kOe of 99.9% cold-worked and 1-h-heat-treated wire. That J_c was maximized by the 1h/400°C anneal was due principally to an optimization of the microstructure of the binary solvent rather than to the presence of the B itself. At high

B levels, Ti-Nb alloys become difficult to work. In the experiments cited it was found that the increases in hardness and J_c were both slight, presumably on account of the low solubility of B in Ti-Nb.

8.2 CARBON ADDITIONS TO Ti-Nb

Some useful discussions of the influence of C additions on the workability and superconducting properties of Ti-Nb alloys appeared in the research and patent literature during 1967-1968.

KUNAKOV *et al* [Kun70] reported on the effect of adding up to 0.4 wt.% C to Ti-Nb$_{50}$. The specimens were prepared by arc melting, C levels of 0.1, 0.15 0.4 wt.% being introduced in the form of Nb$_2$C. Ingots were hot forged, then cold drawn to 0.25-mm$^\phi$ wire, during which it was noted that, although the alloys with 0.1 and 0.15 wt.% C were quite ductile, that with the highest amount of C was brittle. In critical current density measurements it was found that although the J_c($\sim$25 kOe) of the cold-drawn wires increased rapidly as the C content was increased up to 0.15 wt.% (at the rate of about 2×10^4 A cm^{-2} per 0.1 wt.% C), no further increase in J_c above 4.5×10^4 A cm^{-2} was noted in going on to 0.4 wt.% C. In the absence of metallographic information or the results of electrical resistivity measurement, it is possibly only to speculate that: *(a)* in these alloys, 0.15 wt.% C represented the limit of solid solubility (although in subsequent work, to be mentioned below, precipitated carbide needles $\sim$25 mm in length were present in Ti-52Nb-0.15C); *(b)* precipitated carbides are much more effective work-hardeners than dissolved C, which is, on the other hand, a more effective flux pinner. Critical current density measurements of Ti-Nb$_{50}$ doped with 0.15 wt.% C, after heat treatment for probably 1 h at temperatures of up to 800°C, revealed a pronounced maximum in the range 350-450°C wherein the critical current density was claimed to exceed about 1×10^5 A cm^{-2}. In the absence of binary Ti-Nb$_{50}$ control data, the actual contribution that the C itself made to this increase cannot be assessed. Also using Ti-Nb$_{50}$ as a solvent, BARANOV *et al* [Bar67[a]] studied the effect of C additions of up to about 0.29 wt.% on the transition temperature. Corresponding to a rapid increase in hardness at about 0.08 wt.% C was a jump of about 0.3 K in the transition temperature, after which T_c continued to increase slightly with further additions of C. These results, considered in the light of the critical

current density results of KUNAKOV *et al* [Kun70], suggest that the solubility of C in Ti-Nb alloys is within the range 0.08-0.15 wt.% depending on thermomechanical processing. Bearing this work in mind, RASSMANN and ILLGEN [Ras72[a]] studied the effect of 0.15 wt.% C on the superconductivity of Ti-52Nb. Optical metallography performed on the as-cast alloys revealed the presence of star-like arrangements of precipitated carbide needles, probably (Ti,Nb)C. Although the hardness of the alloy remained at $\sim$165 kg mm^{-2} during heat treatment, the critical current density of the 99.9% cold-worked wire increased to a maximum of 4.2×10^4 A cm^{-2}. The optimal heat treatment temperature for the C-containing alloy was about 350°C. That of the Ti-Nb$_{40}$ (i.e. 56.4 wt.% Nb) "reference" sample was 400°C, aging at which temperature produced a critical current density of 2.2×10^4 A cm^{-2}. The reference alloy, Ti-56Nb, had been prepared from Kroll-process Ti, the doped specimen of Ti-52Nb-C from iodide-process Ti. As part of the same study the alloy Ti-32Nb-0.11C, whose hardness prevented its being drawn into wire, was briefly considered from a metallurgical standpoint. During a 24h/500°C heat treatment, the carbide needles became surrounded with α-phase precipitates, which were believed to be responsible for the observed increase in hardness of from 265 kg mm^{-2} (as-cast) to 340 kg mm^{-2}.

At least three important early patents have dealt with the influences of C on the critical current densities of Ti-Nb, Ti-Hf-Nb, Ti-Zr-Nb, Ti-Nb-Ta and other alloys.[†] French Patent No. 1,517,216 [Ass68[a]] (see also FP No. 1,512,971 [Ass68]), issued to Associated Electrical Industries Ltd. of Great Britain, claimed 500-1500 ppm C (also N, 500-2000ppm; O, 500-4000 ppm, separately or conjointly) as beneficial additions to Ti-V, Ti-Ta, Ti-Nb, Ti-Nb-Ta, and Ti-Hf-Nb alloys. Supporting their claim was data for J_c *versus* applied-field dependence (up to 70 kOe) of the alloy Ti-Nb$_{40}$ which, in addition to its starting interstitials, contained 1000 ppm of C introduced in the form of TiC. The experimental materials had been prepared from commercial grade Ti (80 ppm C, 50 ppm N, 800 ppm O) and commercial grade Nb (300 ppm C, 60 ppm N, 5000 ppm O), melted and cast into 19-mm$^\phi$ ingots, hot forged at

---

† Other solvent alloys considered in British Patent 1,089,786 were Nb-25Zr, Nb-50Zr, and Nb-70Zr, while French Patent 1,517,216 (see also FP 1,512,971) cited Ti-V and Ti-Ta.

650°C to 5 mm$^\phi$, cold swaged to 1.8 mm$^\phi$, wire drawn to 0.25 mm$^\phi$, and final heat treated 1h/400°C both with and without a similar intermediate heat treatment administered at a wire diameter of 0.5 mm. Critical current measurements made on these and, for comparison, the cold-worked and final-heat-treated commercial binary starting material, indicated that the added C and the intermediate heat treatment were doubly beneficial. At fields of 30 and 70 kOe, for example, the 4.2-K J_c's were 1.3×10^5 and 6.7×10^4 A cm^{-2}, respectively.

In British Patent No. 1,089,786, issued in 1967 to the Westinghouse Electric Corporation, the preparation and properties of C-doped Zr-Nb and Ti-Nb wires were discussed in considerable detail [Wes67]. Since the effects of C on the superconductive properties of Ti-Nb differ according to C concentration, solute composition, and the temperature of the final heat treatment it was necessary, in order to obtain a valid estimate of the effects of C under optimal conditions, to carry out a broad matrix of tests involving the above-mentioned variables. The results of such an evaluation program, applied to a series of Ti-Nb alloys of compositions 20, 40, 60 and 80 wt.% Nb, each of which was doped with nominally 0.01, 0.02, 0.04 and 0.08 wt.% C, and heat treated for 1 h at temperatures of up to 500°C, were presented in the Westinghouse patent. Commercial grade starting materials might be expected to yield an alloy of composition typically Ti~50Nb-0.015C, which then serves as the "zero-C" control. Unfortunately in the absence of a true zero-C alloy the effect of ultimate purification can only be deduced by extrapolation. Sample materials were prepared by consumably arc melting the appropriate amounts of 99.2% Ti and 99.7% Nb, with 99.9% C in

the form of carbon cloth. The cast ingots were induction annealed using a Mo susceptor at 16h/1550±20°C in 4×10^{-5} torr, machined to 0.423-in^t slabs, hot rolled to 0.253 in^t at 880°C with reductions of about 5-20% per pass, grit blasted and pickled, cold rolled in a two-high rolling mill with reductions of about 2-10% per pass to 0.031 in^t, and finally cold rolled in a four-high rolling mill with reductions of about 10-25% per pass to a final thickness of 0.003 in. The results of the critical current density investigations were plotted in the format J_c(20 kOe) versus the annealing temperatures (up to 500°C). Although in many cases J_c increased monotonically with temperature, suggesting that optimal conditions were not being achieved, critical current densities as high as 1.5×10^5 A cm^{-2} were frequently obtained. It can be concluded that additions of C in the range considered generally resulted in significant increases in J_c. For example: (a) in Ti-60Nb-C, cold worked about 99.3% and heat treated 1h/400°C, increasing the C content from 0.007 to 0.043 wt.% raised J_c(4.2 K, 20 kOe) from 0.62×10^5 to 1.61×10^5 A cm^{-2} -- almost by a factor of 3; (b) the addition of 0.077 wt.% C to Ti-80Nb-0.006C, cited as an example of a cold-worked Nb-rich alloy, raised J_c(4.2 K, 20 kOe) from 0.39×10^5 to 1.06×10^5 A cm^{-2}, representing a rate of increase of 8.7×10^4 A cm^{-2} per 0.1 wt.% C (which may be compared with the ΔJ_c(25 kOe)$/\Delta c = 2 \times 10^4$ A cm^{-2} per 0.1 wt.% C reported by KUNAKOV et al [Kun70]). It was also briefly noted that: (a) although O and N also had positive influences on J_c, it has proved difficult to control the concentrations of interstitially introduced additions of these elements; (b) the total level of O and N to be expected in commercial alloys such as those referred to above was about 1500 ppm (0.15 wt.%).

ALLOY GROUP 2: NITROGEN ADDITIONS TO TITANIUM-NIOBIUM, TITANIUM-HAFNIUM-NIOBIUM, AND TITANIUM-NIOBIUM-TANTALUM

8.3 NITROGEN ADDITIONS TO Ti-Nb

The results of some studies dealing with the structural, electrical, and superconductive properties (both T_c and J_c) of Ti-33Nb and Ti-40Nb alloys containing various amounts of N, O and numerous other elements have been published by BACHMANN et al [Bac68]. Ti-Nb$_{20-70}$ and other alloys with 500-2000 ppm N,

and/or other interstitial elements, were claimed in French Patent No. 1,571,216 (see also FP No. 1,512,971) which was issued in 1968 to Associated Electrical Industries Ltd. of Great Britain [Ass68, Ass68^a]. In it, particular reference was made to the properties of Ti-56Nb containing residual interstitial impurities augmented by 1000 ppm N. In subsequent research articles, the properties of the same basic alloy, Ti-56Nb, with 1100 ppm N were discussed by BAKER [Bak70] and with 500 ppm N by RASSMANN and ILLGEN [Ras72[b]]. The variation of T_c with N content in Ti-Nb$_{50}$ was studied by BARANOV et al [Bar67[a]]; finally, with regard to quaternary alloys, to be discussed under a subsequent heading, French Patent No. 1,517,216 [Ass68[a]] also considered the effects on J_c of N additions to Ti-Hf-Nb and Ti-Nb-Ta.

In what follows, the influence of dissolved N on the superconducting properties of a wide range of Ti-Nb compositions (viz 20-50 at.%) is discussed. The report is subdivided into sections according to the compositions of the host alloys, viz:

    (a) Ti-33Nb (20 at.% Nb)

    (b) Ti-40Nb (25.5 at.% Nb)

    (c) Ti-56Nb (39.5 at.% Nb)

and (d) Ti-66Nb (50 at.% Nb).

It is noted with reference to Chapter 7 (Ti-Nb) that these alloys fall into categories which are meaningful in terms of the precipitation which takes place in response to moderate-temperature (i.e. ∿400°C) aging: During the aging of quenched and heavily-cold-worked Ti-33Nb and Ti-40Nb, ω-phase should quickly yield to α-phase precipitation. Ti-56Nb lies at the practical limit of the α+β-phase regime (cf. Fig.7-1), and as a binary alloy should support practically no α-phase precipitation. Ti-66Nb is in the single-phase-bcc regime. The effects of third element, and in particular interstitial-element, additions on transformation processes in Ti-TM alloys have been outlined in [Mon2.6, Mon2.9, Mon2.12].

8.3.1 Nitrogen Additions to Ti-33Nb (20 at.% Nb)

The influence of N additions of up to 0.4 at.% on the T_c (defined as the $\frac{1}{2}$-point of the induction change) of (a) deformed and (b) recrystallized (at 950°C) Ti-33Nb alloys has been studied by BACHMANN et al [Bac68]. With the deformed samples, no significant changes in the temperatures of the rather broad super-conducting transitions could be detected. The transition temperatures of the recrystallized alloys decreased at the rate of about 2.0 K per at.% N. Critical current density studies were not undertaken.

8.3.2 Nitrogen (and Occasionally Nitrogen plus Oxygen) Additions to Ti-40Nb (25.5 at.% Nb)

BACHMANN et al [Bac68] also investigated the influence of N on the transition temperatures of (a) deformed, and (b) partly recrystallized (at 950°C) Ti-40Nb alloys. As before, the T_c of the deformed material did not change significantly with interstitial alloying, whereas that of the recrystallized alloy decreased much more rapidly than was the case for Ti-33Nb-N (at the rate, in fact, of about 4.2 K per at.% N). The influence of aging time at 500°C on the T_c of deformed Ti-40Nb-0.05N-0.05 wt.% O[+] was also studied. Control samples were Ti-40Nb in the conditions: (a) deformed, and (b) deformed plus annealed ($\frac{1}{2}$h/950°C/water-quench) plus deformed. Increases in T_c with aging time were noted in all cases, with a leveling-off after about 15 h. The average transition temperature of Ti-40Nb-N-O, with 0.1 wt.% total interstitial content, was at all times comparable to those of the addition-free controls. Since, as indicated above, no significant changes accompanied the presence of interstitial elements in deformed Ti-33Nb-N and Ti-40Nb-N alloys this experiment might be regarded as a test of reaction kinetics rather than solute-induced transition-temperature change.

The 4.2-K critical current densities of Ti-40Nb-0.05N and Ti-40Nb-0.05N-0.05 wt.% O in both the deformed, and annealed-plus-deformed conditions have also been measured. Alloys were prepared by arc melting and casting into the form of 6-mm$^\phi$ rods. Samples for measurement were prepared by cold swaging about 96% to 1.75 mm$^\phi$, sometimes annealed $\frac{1}{2}$h/950°C (2×10^{-5} torr)/water quenched, then drawn to wire, 1-0.2 mm$^\phi$, without further annealing. Comparative studies against a Ti-40Nb control alloy, of the self-field critical current densities of Ti-40Nb with O, N and O+N, and a commercial alloy were undertaken, and a comparison was made between the J_c(4.2 K) versus H_a($\lesssim$100 kOe) curve for Ti-40Nb-0.05N-0.05 wt.% O and that for the unalloyed control. Slight increases as well as decreases of the critical current density can

---

+ When referring to the weight-percent *oxygen* content of an alloy, the usual short-hand format (which in this case would be written "-nO") is abandoned in favor of the full "-n wt.% O".

be discerned in the results for the deformed but not-heat-treated samples. Samples that were annealed-and-deformed generally showed decreases of J_c as compared to the unalloyed control.

In studies of the critical current densities of two cold-worked alloys, Ti-40Nb and Ti-40Nb-0.05N-0.05 wt.%O, in fields of up to 100 kOe it was noted that the J_c of the interstitial-containing alloy was generally lower than that of the addition-free control. However, the existence of a pronounced peak-effect in the former indicated that considerable process optimization was in order. The critical current densities at the peaks were:

Ti-40Nb (50 kOe): 1.5×10^4 A cm^{-2},

Ti-40Nb-0.05N-0.05 wt.% O (70 kOe): 4.1×10^3 A cm^{-2}.

8.3.3 Nitrogen (and Occasionally Nitrogen plus Oxygen) Additions to Ti-56Nb (39.5 at.% Nb)

The influence of N on the critical current density of Ti-56Nb has been thoroughly investigated by RASSMANN and ILLGEN [Ras72[a]], by BAKER [Bak70], and in French Patent No. 1,517,216 [Ass68[a]] (see also FP No. 1,512,971 [Ass68]) which also includes in its list of claims the addition of one or more of the interstitial elements C, N and O to Ti-V, Ti-Ta, Ti-Hf-Nb, and Ti-Nb-Ta, as well as two alloy melting practices.

The situation with regard to Ti-56Nb-N can be briefly represented in Table 8-1 which shows that at sufficiently high interstitial concentrations (including starting-impurity levels) current densities in the high 10^4 A cm^{-2} range can be expected. Having experienced moderately large area reductions by cold work, and a final heat treatment of 1h/400°C, but no *final* cold work, the alloys can be characterized as "partially optimized". In all cases a N-free control alloy was studied, thus enabling a valid and -- on account of the partial optimization -- probably reasonably general evaluation of the effect of N-additions to be made.

Each of the papers cited made its individual contribution to the subject. Thus: *(a)* RASSMANN and ILLGEN [Ras72[a]], while studying the effect on J_c of 1-h annealing at temperatures of up to 500°C, compared the properties of Ti-Nb$_{40}$($\sim$56 wt.%), prepared from high-purity starting materials of low initial interstitial content, with those of the same binary alloy

TABLE 8-1   CRITICAL CURRENT DENSITIES (J_c) OF COLD-DEFORMED AND FINAL-HEAT-TREATED (1h/400°C) Ti-56Nb-N ALLOYS AT 4.2 K, 50 kOe.

Nitrogen Content, ppm	Area Reduction, %	J_c[†], 10^4 A cm^{-2}		Literature
		Without N	With N	
1000[††]	99.98	4.1	7.7	[Ass68[a]]
1100[†††]	99.93	4.3	8.0	[Bak70]
500[†††*]	99.9	2.1	2.6	[Ras72[a]]
500 ppm N +500 ppm O[†††*]	99.9	2.1	6.0	"

[†] J_c criteria: 5 µV across 0.5 cm [Bak70]; 40 µV across sample [Ras72[a]].

[††]   In addition to 3169 ppm O, 56 ppm N, and 204 ppm C, starting impurity.

[†††]   In addition to 645 ppm O, 50 ppm N, and 405 ppm C, starting impurity.

[†††*]   In addition to 200-300 ppm O, and 50-100 ppm N, starting impurity -- both alloys were based on iodide Ti.

doped with N, O, and N+O; *(b)* BAKER [Bak70] accompanied studies of the influence of heat treatment on the J_c of Ti-56Nb-0.11N, prepared from commercial-grade starting materials, with optical- and electron-microscopic studies of the quenched, cold-worked, and worked-plus-heat-treated microstructures in an attempt to determine the metallurgical sources of the flux pinning; *(c)* French Patent No. 1,517,216 [Ass68[a]] intercompared the effect on the J_c of a given alloy of *(i)* final and intermediate-plus-final heat treatments (of 1h/400°C), and *(ii)* the influence, on N-doped Ti-Nb-base alloys, of the addition of a third solvent component in the form of Hf and Ta. Before going on to consider in more detail some relevant features of the individual papers we note, with reference to Table 8-1, that bearing in mind the slight differences in *(i)* prior cold work, *(ii)* added solute level, and *(iii)* *starting-material* interstitial content, the results are in most satisfactory agreement with each other. Furthermore it is clear that within the context of "partially optimized" alloys, cold worked and heat treated to the extents specified, the addition of 1000 ppm N to Ti-56Nb can be expected to result in a doubling of the 50-kOe J_c.

As shown in Table 7-14, Kroll-process Ti is rich in interstitial impurities. In order to examine the influence of these on a Ti-Nb$_{40}$ alloy prepared from it, after aging for 1 h at temperatures of 200, 300, 400 and 500°C, RASSMANN and ILLGEN [Ras72[a]] compared the corresponding 50-kOe critical current densities with those of Ti-Nb$_{40}$ doped with 500 ppm O, 500 ppm N, and 500 ppm O + 500 ppm N. These alloys had been prepared from Nb and pre-alloyed iodide Ti in which oxygen had been introduced by oxidation for 3h/800°C, and nitrogen by nitriding for 3h/1350°C. For the Kroll-Ti-Nb$_{40}$, and all the iodide-Ti-Nb$_{40}$-Intl alloys, J_c peaked at 400°C indicating that, although the addition of N and other interstitials were responsible for some increases in J_c (e.g. a 24% increase with 500 ppm N), even more important were the microstructural properties of the host alloy.

BAKER [Bak70] has studied the microstructures and critical current densities of cold-worked and heat-treated sheet and wire specimens of a commercial Ti-56Nb alloy both without and with the addition of 1100 ppm of N. The nitriding in this case was carried out by arc melting the binary alloy with the appropriate amount of TiN. Following a solution treatment of 1h/1100°C and quenching, only 480 ppm of the N was retained in solution, the rest precipitating out in the form of a nitride phase. The cold work administered to the wire specimens was 99.93% (area reduction ratio, ~1400:1) and the heat treatments were for 1h at 300, 400, 500, 600 and 700°C. Studied in detail under the electron microscope were the precipitation, the cell size, and the fiber size, from which tentative conclusions were able to be drawn as to the roles played by the various microstructural features. It was asserted that the fiber cross-sectional diameter of the drawn structure (some 1000 Å) was related to the spacing between precipitates or other such inhomogeneities rather than to the original grain structure or substructure which would have drawn down under the same conditions to diameters of 50,000 Å or 50 Å, respectively, neither of which matched the observed structure. It was able to be concluded from the study that: *(a) precipitated* interstitials favor an increase in J_c; *(b)* interstitials in *solution* call

for a higher (>100°C) optimizing heat treatment temperature; *(c)* maximum J_c occurs *(i)* when the cell or fiber wall is clearly differentiated from the fiber interior and *(ii)* when the result of the heat treatment is a "fine dispersion of precipitates and a high percentage of interstitials in supersaturated solid solution". Flux pinning by precipitates is discussed in [Mon21.13].

French Patent No. 1,517,216 [Ass68[a]] refers to alloys prepared either by consumable or nonconsumable arc melting. In the examples given, commercial grades of Ti and Nb, with the addition of TiN, were argon arc melted and cast into 19-mm$^\phi$ ingots. These were then cold swaged to 1.8 mm$^\phi$ (hot forging as a preliminary step was not employed) and drawn to a final wire diameter of 0.25 mm, after which a final heat treatment of 1h/400°C was administered. Also, as a process variant, an intermediate heat treatment of 1h/400°C at a diameter of 0.5 mm was occasionally administered. As was the case with Ti-Nb-C, the application of this additional step was beneficial and resulted in a further increase of J_c(4.2 K, 50 kOe) which then attained the value 1.1×10^5 A cm^{-2}.

8.3.4 Nitrogen Additions to Ti-66Nb (50 at.% Nb)

BARANOV *et al* [Bar67[a]], in studies of Ti-Nb$_{50}$ alloyed with up to about 0.44 at.% N, reported a continuous decline of the superconducting transition temperature from 9.3_3 to 8.9_1 K.

8.4 NITROGEN ADDITIONS TO Ti-Hf-Nb AND Ti-Nb-Ta ALLOYS

As examples of the general class of Ti-Hf-Nb and Ti-Nb-Ta alloys to be doped with one or more of the interstitial elements C, N and O, French Patent No. 1,517,216 [Ass68[a]] has cited the alloys Ti$_{65}$-Hf$_5$-Nb$_{30}$ and Ti$_{65}$-Nb$_{30}$-Ta$_5$ whose critical current densities in response to the addition of 500 ppm N were investigated.

Sample preparation was similar to that mentioned above. Only final heat treatments of 1h/400°C were administered, and the critical current densities of alloys based on commercial-grade starting materials, both without and with the extra N (which was added to the melt in the form of TiN), were intercompared.

Table 8-2, which presents data for an arbitrarily chosen field of 50 kOe, suffices to indicate that the J_c's of $Ti_{65}-Hf_5-Nb_{30}$ and $Ti_{65}-Nb_{30}-Ta_5$, which were slightly larger than, and equal to, respectively, that of $Ti-Nb_{40}$ were both increased by the addition of N.

TABLE 8-2   INFLUENCE OF 500 ppm OF NITROGEN ON THE CRITICAL CURRENT DENSITIES (J_c) OF TWO SIMILARLY PROCESSED[†] TERNARY ALLOYS [Ass68[a]].

Solvent Alloy	J_c(4.2K, 50 kOe) 10^4 A cm^{-2}	
	without N	with N
$Ti_{65}-Hf_5-Nb_{30}$	5.1	9.4
$Ti_{65}-Nb_{30}-Ta_5$	4.1	6.6

† Cold-swaged and drawn 99.98% plus final-heat-treated 1h/400°C.

ALLOY GROUP 3: OXYGEN ADDITIONS TO TITANIUM-NIOBIUM AND SOME TITANIUM-NIOBIUM-BASE TERNARY AND QUATERNARY ALLOYS

8.5 TRANSITION TEMPERATURES OF Ti-Nb-O ALLOYS

Using neutron-activation analysis as an analytical tool to measure O concentration in a series of $Ti-Nb_{60}$ alloys containing 0.2 to 3.1 at.% O, RODRIGUEZ-GONZALES [Rod70] has studied the rate of change of T_c with interstitial content. It was discovered that T_c decreased linearly with O content at the rate of 0.56 K per at.% O, equivalent in this alloy to 2.5_5 K per wt.% O. In an earlier investigation, BACHMANN et al [Bac68] had noted that O additions of up to about 0.3 at.% decreased the T_c of annealed (at 950°C) Ti-33Nb and Ti-40Nb alloys at rates of about 2.2 K per at.% O and 1.5 K per at.% O, respectively, but left those of the deformed alloys practically unchanged. In the results of studies published at about the same time, BARANOV et al [Bar67[a]] showed that O levels of up to about 0.44 at.% reduced the superconducting transition temperature of $Ti-Nb_{50}$ (66 wt.% Nb) from 9.3_3 to 8.8_8 K, equivalent to an average rate of decrease of about 1.0 K per at.% O. A similar rate of decrease was noted with respect to N additions. Thus, although it is clear that O reduces the T_c of Ti-Nb alloys there is, according to

Table 8-3, considerable disagreement as to the rate at which it does so.

The influence of aging at temperatures of 400 and 500°C, for times up to about 19 h, on the T_c of deformed Ti-33Nb-0.09 wt.% O (i.e. $Ti-Nb_{20}-O_{0.32}$) were

TABLE 8-3   INFLUENCE OF OXYGEN ON THE SUPERCONDUCTING TRANSITION TEMPERATURES OF Ti-Nb ALLOYS

At.% Nb	Rate of Change of Transition Temperature	Reference
26	-2.2K per at.%	[Bac68]
50	-1.0K per %[†]	[Bar67[a]]
56	-1.5K per at.%	[Bac68]
60	-0.5_4K per at.%	[Rod70]

† It is uncertain as to whether the interstitial levels were given in wt.% or at.%. We interpret BARANOV's paper [Bar67[a]] to indicate that his C additions were given in wt.%, while N and O were in at.%.

studied by BACHMANN *et al* [Bac68] and compared with that of an O-free control sample. Even during the first hour, rapid and comparable *increases* of T_c took place. During heat treatments at 400 and 500°C, the transition temperatures of the test alloy increased from 6.9 K to 7.9 K and 8.4 K, respectively. At the termination of the test, after 19 h of aging had been administered at each of the temperatures 400 and 500°C, the transition temperatures had climbed to 8.9 K and 8.7_5 K, respectively. Most of the features of the T_c *versus* time curve for the test sample were reflected by that of the O-free control (although shifted in temperature) indicating once more, that indigenous properties of the solvent were dominating the changes in T_c. The initial and final transition temperatures of the O-free alloy were 6.7 and 8.7 K, respectively.

KITADA and DOI [Kit70[b]] have measured the superconducting transition temperatures of $Ti_{10}-Zr_{40}-Nb_{50}$ with 0.1, 0.3 and 0.95 wt.% O. As indicated in Table 8-4, although aging for 3h/700°C following solution heat treatment for 3h/1100°C increased T_c by about 1 K, as the O level increased, T_c decreased by a few tenths of a degree.

TABLE 8-4   SUPERCONDUCTING TRANSITION TEMPERATURES (T_c) OF OXYGEN-CONTAINING $Ti_{10}-Zr_{40}-Nb_{50}$ ALLOYS IN RESPONSE TO SOLUTION HEAT TREATMENT (ST) FOLLOWED BY PRECIPITATION HEAT TREATMENT [Kit70[b]].

	Transition Temperature, K	
Wt.% Oxygen	ST (3h/1100°C)	ST plus 3h/700°C
0.1	9.3	10.3
0.3	9.2	10.1
0.95	9.2	10.0

8.6 CRITICAL CURRENT DENSITIES OF Ti-Nb-O ALLOYS

Listed in Table 8-5 are the binary Ti-Nb alloys whose critical current densities, in response to the addition of O, have been reported during the period 1966 to 1972. A glance at the table indicates the

TABLE 8-5   CATALOG OF Ti-Nb ALLOYS WHOSE CRITICAL CURRENT DENSITIES IN RESPONSE TO OXYGEN ADDITIONS HAVE BEEN INVESTIGATED, AND A "ppm" TO "At.% O" CONVERSION TABLE.

Nb Concentration in Solvent Alloy			Molar Weight of Solvent	At.% O equiv. to 0.1 wt.% O (1,000 ppm)
wt.%[+]	at.%[++]	Literature		
33	20	[Bac68] [Ras72[a]]	57.01	0.36
39	25	[Com67]	59.06	0.37
40	26	[Bac68] [Rau68[a]] [Ric70]	59.41	0.37
45	30	[Bet66]	61.25	0.38
50	34	[Rau68[a]] [Zwi70]	63.21	0.39
51	35	[Wit73]	63.62	0.40
56	40	[Ras72[a]]	65.73	0.41
60	44	[Reu66] [Rau68[a]] [Bid70] [Zwi70]	67.53	0.42
70	55	[Bet66]	72.48	0.45
80	67	[Rau68[a]]	78.21	0.49

+   Starting point for conversions (basis of calculation).

++ Rounded to nearest integer, cf. Table 7-1.

substantial contributions that have been made by COMEY, REUTER, RAUCH, and RICKETTS, who studied this rather complex topic at MIT during the mid-1960's under the direction of J. WULFF.

One of the earliest studies of the present topic was undertaken by BETTERTON *et al* [Bet66], who investigated the effects of 0.08 and 0.2 at.% O on the 4.2-K critical current densities of $Ti-Nb_{30}$ and $Ti-Nb_{55}$ in the form of cold-drawn (about 99.9%) and "optimally heat-treated" wires. In response to the addition of 0.08 at.% O, for example, the

J_c(4.2 K, 30 kOe)'s of the cold-worked and cold-worked-plus-1h/400°C-aged samples underwent the following increases:

4.5 to 8.5×10^3 A cm^{-2} (cw only)

3.0 to 5.0×10^4 A cm^{-2} (cw + age)

This preliminary study of the effect of interstitial elements on the J_c's of Ti-Nb alloys was followed by an extensive series of investigations of a wide range of alloy compositions. To be considered below are the effects of 0 on the critical current densities of:

    (a) Ti-40Nb (25.5 at.% Nb),
    (b) Ti-50Nb (34 at.% Nb),
    (c) Ti-56Nb (39.5 at.% Nb),
and (d) Ti-60Nb (43.5 at.% Nb),

representative of four important metallurgical regimes, viz: (a) the region between about 10 and 30 at.% Nb which is common to both the meta-equilibrium (or isothermal) ω+β-phase and the equilibrium α+β-phase (cf. Sect.7.20); (b) the region of equilibrium α+β-phase between 30 and 40 at.% Nb from which isothermal ω+β decomposition is excluded; (c) the "practical" boundary, at ∼40 at.% Nb, between the equilibrium α+β and β fields; (d) the equilibrium single-phase-bcc regime beyond ∼40 at.% Nb, cf. Fig.7-1.

8.6.1 Oxygen Additions to Ti-40Nb (25.5 at.% Nb)

In the mid 1960's COMEY [Com67] conducted an investigation into the current-carrying capacity of Ti-39Nb alloyed with what were termed "low" and "high" levels of 0. The report of the results of this study would have been extremely valuable had it not been marred by the absence of any information on the actual 0 levels employed. From an intercomparison of the J_c versus H_a curves for the "low-" and "high-oxygen" alloys it can be concluded that, although J_c(low-0) was greater than J_c(high-0) for $H_a <$ 70 kOe (the "peak" field), in that the high-0 alloy had the stronger peak effect it possessed the greater potential for J_c increase in response to heat treatment. In spite of this, heat treatments of 1h/425°C ("low"-0) and 1h/525°C ("high"-0) still left the former alloy with the higher 40-kOe J_c. Ti-39Nb is of course in a different class from the Nb-rich Ti-60Nb to be considered

below. In COMEY's case its selection was guided by a desire to study an alloy of higher Nb content, and therefore better workability, than the Ti-Nb$_{20.7}$ which had been investigated earlier by VETRANO and BOOM [Vet65], but yet sufficiently rich in Ti as to yield a deformation martensite (as indeed was observed) during plastic deformation.

In a subsequent study of Ti-40Nb, alloyed with 0.044 and 0.24 wt.% 0, cold rolled and aged for 1 h at temperatures of up to 600°C, RICKETTS et al [Ric70] noted as a result of x-ray observations that maximum α-phase precipitation occurred near 500°C, and more abundantly in the 0-rich alloy, while an annealing temperature of 400°C favored ω-phase precipitation which was influenced only slightly by variation of the 0 content. In this alloy, J_c(4.2 K, 50 kOe) appeared to be dominated by metallurgical effects other than interstitial content, in that the height and position (i.e. temperature) of a sharp peak in the plot of J_c versus 1-h aging temperature, Fig.8-1, was independent

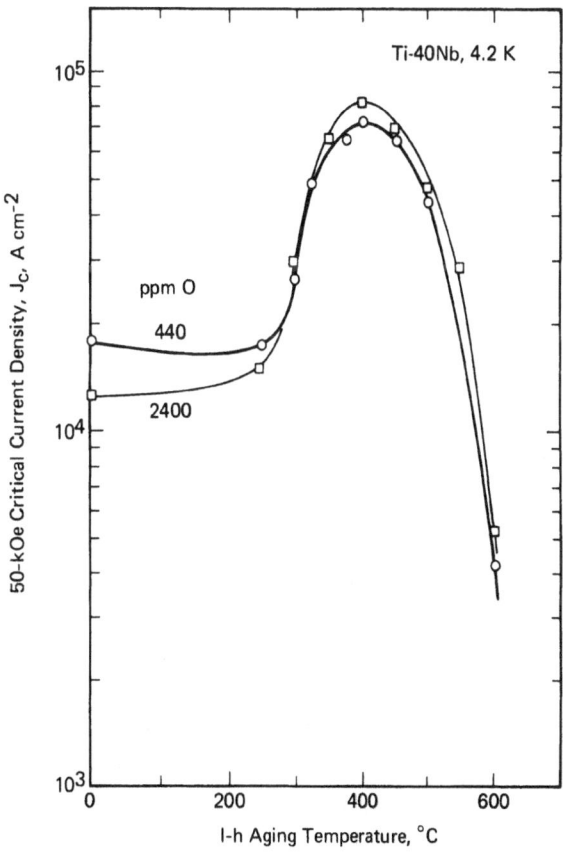

FIGURE 8-1. Critical current density at 4.2 K, 50 kOe ($\cong H_r$/2) as a function of 1-h aging temperature for a Ti-40Nb (nominal) alloy doped with 440 ppm of oxygen (○) and 2400 ppm of oxygen (□) (criterion: 1/4 μV cm^{-1}) — after RICKETTS et al [Ric69, Ric70].

of the O concentration. That the peak occurred at 400°C was taken as evidence for the flux-pinning efficacy of ω-phase precipitation in these alloys. However, as pointed out elsewhere (e.g. Sect.7.30.2) the deformation-cell structure itself is also a beneficiary of the 1h/400°C heat treatment. It is interesting and important to note that, whereas RAUCH et al [Rau68[a]] working with Ti-60Nb-O (see below) found that an increase in the O level of from 0.052 to 0.27 wt.%, accompanied by an increase in the 1-h aging temperature of from 400 to 500°C, was associated with an improvement in $J_c(H_r/2)$ of from 2.6×10^4 to 5.5×10^4 A cm^{-2}, RICKETTS et al [Ric70] using Ti-40Nb-O saw no such changes in the peak value of J_c(4.2 K, 50 kOe) versus 1-h aging temperature as the O content was increased from 0.044 to 0.24 wt.%.

Whereas it is certain that precipitation will not take place in an alloy as rich in Nb as Ti-60Nb, direct x-ray evidence for the presence of precipitates in Ti-40Nb-O alloys, and the relative abundances of them after 1 h at temperatures between 200 and 600°C has been presented by RICKETTS et al [Ric70]. Both ω-phase and α-phase precipitation were investigated. Fig.8-2 (cf. Fig.7-38) shows that annealing at 400°C maximizes the density of ω-phase precipitation to about the same extent in both Ti-40Nb-0.04 wt.%O and Ti-40Nb-0.24 wt.% O, whereas 550°C is the optimal

temperature for α-phase precipitation from Ti-40Nb-0.04 wt.% O, and the threshold of a relatively much more abundant precipitation of α-phase from Ti-40Nb-0.24 wt.% O. These results, coupled with the J_c data, suggested that ω-phase precipitation when it is permitted on compositional grounds, is the more effective of the two types of flux-pinning precipitate. In the "intermediate-concentration" alloys (as defined in Sect.7.26) beyond the range of ω-phase precipitation, or perhaps even in some "high-concentration" alloys, if the possibility of abundant O-assisted α-phase precipitation exists, a heat treatment temperature near 500°C must be employed. But although, as the figure shows, a greater abundance of α-phase can be expected to accompany the use of an even higher heat-treatment temperature, the resulting beneficial influence of an increasing density of flux-pinning precipitates would have to compete with the deleterious effect of microstructural recovery which acts to reduce J_c.

According to the results of RAUCH [Rau68[a]] and RICKETTS (cf. [Rau68]), Ti-40Nb-O when cold worked and 1-h aged responded with a J_c that maximized at 400°C for both O levels (cf. Fig.8-1), but when recrystallized and 1-h aged, it gave rise to a J_c-maximum which increased both in magnitude and temperature with increase in the O content, a situation comparable to that encountered in the high-concentration (in particular Ti-60Nb) precipitate-free range. An explanation of these results contains as essential ingredients the following known facts relating to the deformation-and-aging of precipitation-prone Ti-TM alloys as discussed in [Mon2.11], and flux pinning by deformation structures and precipitates as summarized in Sect.7.30. Briefly, J_c is enhanced both by thermally-induced "refinement" of the cold-worked structure, and by precipitation; and α-precipitation itself is encouraged both by deformation and the presence of dissolved O. A set of representative critical current density data acquired by the MIT group on O-bearing Ti-40Nb is presented in Table 8-6, inspection of which reveals the quite promising values that have been achieved by the cold-worked + aged members of the set.

Data of the kind presented are the result of a complicated interplay between precipitational effects and the influence of heat treatment on the cold-worked microstructure. It can be appreciated that no serious attempts had been made during the 1960's to separate out the flux-pinning contributions of the precipitates from those of the deformation

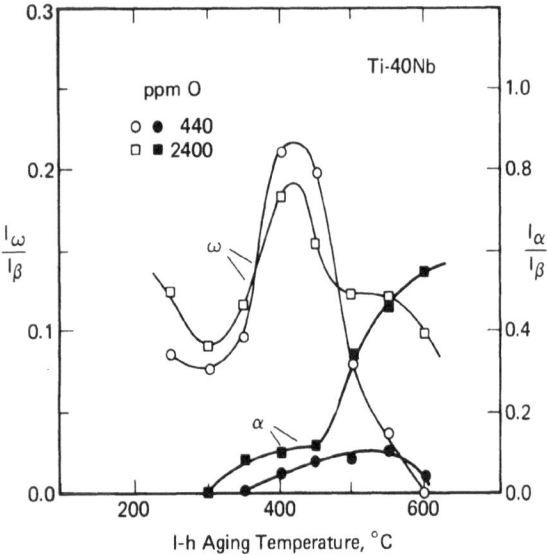

FIGURE 8-2. Precipitation of ω and α phases during the 1-h aging of previously ~ 99.95%-cold-deformed Ti-40Nb (nominal) alloys doped with 440 ppm (○, ●) and 2400 ppm (□, ■) oxygen as determined by relative-intensity X-ray diffractometry. Note the relative compression (X4) of the I_α/I_β scale — after RICKETTS et al [Ric69, Ric70].

TABLE 8-6 OXYGEN-DOPED Ti-40Nb ALLOYS. TEMPERATURES
OF 1-h AGINGS REQUIRED TO MAXIMIZE J_c
AND THE CORRESPONDING VALUES OF $J_{c,max}$.

Oxygen Level, ppm	Condition	Applied Field, kOe	Temperature at $J_{c,max}$, °C	$J_{c,max}$, 10^4 A cm^{-2}	References
1600	cw_1+age	50	400	7.8	+
1600	rc+age	40	450	1.7	++
1400	cw_2+age	40	400	12	++
2400	rc+age	40	500	2.0	+++
2400	cw_2+age	50	400	8.3	+++*
2400	cw_2+age	50	400	7.3	+++*

cw_1 = 0.42 in.$^\phi$ →0.010 in.$^\phi$ (99.94%)

cw_2 = 0.45 in.$^\phi$ →0.010 in.$^\phi$ (99.95%)

rc = recrystallized 1h/1000°C

 +   [Rau68^a]

 ++ [Rau68], see also appendix to [Rau68^a]

 +++ [Rau68], not otherwise published

 +++* [Ric70]

structures; in fact it seems as if, in many of the
early studies, the role of precipitation may have been
overemphasized.

BACHMANN *et al* [Bac68], contemporaneously with
the MIT work just discussed, conducted a limited study
of the extent to which the critical current density of
Ti-40Nb responded to additions of O and N+O. The
alloys Ti-40Nb-0.045 wt.% O, Ti-40Nb-0.09 wt.%O, and
Ti-40Nb itself as a control, were measured in the deformed and deformed-plus-annealed (950°C) condition.
In zero-field measurements, rather low J_c's were obtained ($\sim$5x10^3 A cm^{-2}), and for the annealed alloys
the critical current density *decreased* when Ti-40Nb
was alloyed with 0.045 wt.% O, then decreased still
further when the O level was increased to 0.09 wt.%.

8.6.2 Oxygen Additions to Ti-50Nb (34 at.% Nb)

Using as background a general discussion of
deformation-induced-microstructures in bcc metals and
alloys, WITCOMB and DEW-HUGHES [Wit73] studied the
development by cold work and heat treatment of flux-
pinning microstructures in Ti-Nb$_{35}$-O alloys, and went
on to assess the influence of those structures on the
superconducting properties. The interstitial content
of the supplied commercial-grade material was approximately: N, 400 ppm; and O, 3000 ppm. TEM studies and
critical current density measurements were made on the
as-received 90%-cold-worked 0.2-mm^t strip after heat
treatment in vacuum ($\sim$10^{-4} torr) for: (1 and 2h)/400°C,
5h/450°C, (1,2 and 5h)/550°C, and (1 and 2h)/750°C;
and in Ar for: 1h/300°C, and (0.5,1,10 and 50h)/550°C.
Compared to vacuum annealing, heat treatment performed
under Ar was stated to result in a "complex" microstructure due, it was thought, to the presence of O$_2$
as an impurity.† Of course, only the vacuum-anneal
results were of practical importance, since in clad
multifilamentary composites, the Ti-Nb filamentary
surface is well protected from oxidation. For reference in a discussion of flux-pinning microstructures
the critical current density, data were plotted in the
format $J_c H_a$ (i.e. the bulk pinning force) *versus* H_a,
the applied field. The resulting curves were characterized by two significant features: *(a)* a peak near
H_r attributed to pinning by dislocations, hence referred to as the "D-peak"; *(b)* an additional (or
alternative) peak located towards to the middle of the
applied magnetic field range which, being attributed
to precipitates, was referred to as the "P-peak".
Prior to vacuum heat treatment, J_c and consequently
$J_c H_a$, was small and only a D-peak was observable. The
largest values of $J_c H_a$ occurred after aging 1h/550°C
as a result of the enlargement of the D-peak and the
appearance of a large broad P-peak. Metallographi-
cally this could be correlated with the growth into
lenticular platelets (2 μm x $\sim$1000 Å), with internal
twinning, of δ-TiO$_x$ precipitate particles. The growth
of the D-peak in the early stages of aging was attributed to an increasing improvement in the definition of

---

+ It is estimated that the partial pressure of O$_2$ in
  the usual clean industrial welding gases
  (99.995% He or Ar) is within the range 4x10^{-3} to
  5x10^{-4} torr.

the cell structure as dislocations migrated from cell-interiors to cell-walls -- no doubt assisted by the diffusion of interstitials to the cell-walls.

Having offered a mechanistic description of pinning-force optimization in physical-metallurgical terms, the paper of WITCOMB and DEW-HUGHES [Wit73] concluded with a brief discussion and evaluation of so-called "cell-wall" and other pinning mechanisms. For further descriptions of these, [Mon21.3] and the standard literature should be consulted.

8.6.3 Oxygen Additions to Ti-56Nb (39.5 at.% Nb)

As mentioned in the introduction to this section Ti-56Nb, at moderate temperatures, lies at the boundary of the equilibrium $\alpha+\beta$ and β-phase fields. As part of an extensive series of experiments, RASSMANN and ILLGEN [Ras72[a]] studied the influence of O on the J_c of that alloy, as well as on the J_c of Ti-Nb$_{20}$, an alloy previously investigated (along with the Ti-40Nb of Sect.8.6.1) by BACHMANN et al [Bac68]. In the first of a group of studies Ti-Nb$_{20}$ and Ti-Nb$_{40}$, prepared from iodide-process Ti, were alloyed with O, added in the form of Nb$_2$O$_3$. The effect of dissolved O, introduced in this way, in association with precipitated Y$_2$O$_3$ particles (produced by the addition of metallic Y to the melt) was also explored. Alloys were studied in the 99.9% cold-worked condition, and also after cold work followed by (1 and 5h)/500°C heat treatments. Some results are summarized in Table 8-7. Changes in microstructure brought about by cold work, heat treatment at 500°C, solid-solution alloying, and precipitation, all influenced J_c at a given field. With Ti-Nb$_{20}$, the addition of O decreased the amount of martensite present in the as-cast alloy. With the all-β Ti-Nb$_{40}$-base alloys, J_c increased in all cases by an order of magnitude following the post-cold-work heat treatment; and within the cold-worked and heat-treated groups, respectively, J_c increased as the interstitial solute and precipitate content increased. It was noted, however, with reference to other work by the same authors [Ras72], that the critical current densities were all less than those obtained using Kroll-process Ti as starting material, suggesting, it was claimed, that an O level of between 600 and 800 ppm is desirable.

These observations formed the basis for a second series of measurements in which an attempt was made to identify the impurities present in Kroll-process Ti

which were responsible for its acclaimed superior performance as a starting material for the preparation of Ti-Nb superconductors, cf. Sect.7.26.2.

TABLE 8-7 CRITICAL CURRENT DENSITIES, J_c(4.2K,50kOe) OF 99.9%-COLD-WORKED-AND-AGED TITANIUM-NIOBIUM ALLOYS CONTAINING INTERSTITIAL IMPURITIES [Ras72[a]]

Impurity Content	Heat Treatment	Critical Current Density[†] 10^3 A cm^{-2}
(a) Ti-Nb$_{20}$		
3500 ppm O introduced as Nb$_2$O$_5$ of which 1500 ppm is combined as Y$_2$O$_3$ -- Alloy "c"	1h/500°C	47
1500 ppm O introduced as Nb$_2$O$_5$ -- Alloy "b"	1h/500°C	25
300 ppm O -- Alloy "a"	1h/500°C	19
Alloy "a"	5h/500°C	15
Alloy "c"	None	
Alloy "a"	None	6.0
Alloy "b"	None	3.2
Alloy "b"	5h/500°C	<1
(b) Ti-Nb$_{40}$		
3000 ppm O introduced as Nb$_2$O$_5$ of which 1500 ppm is combined as Y$_2$O$_3$ -- Alloy "c"	1h/500°C	54
1500 ppm O introduced as Nb$_2$O$_5$ -- Alloy "b"	1h/500°C	29
225 ppm O -- Alloy "a"	1h/500°C	8.2
Alloy "c"	None	3.8
Alloy "b"	None	∿2.5
Alloy "a"	None	1.4

† J_c criterion: 40 μV across sample.

8.6.4 Identification of the Active Impurities in Kroll-Process Ti

In a continuation of the experiments just outlined, RASSMANN and ILLGEN [Ras72[a]] prepared a Ti-Nb$_{40}$ alloy containing 500 ppm O introduced by the oxidation (for 3h/800°C) of iodide-Ti-base starting material. J_c(4.2 K, 50 kOe) was measured and plotted *versus* 1-h annealing temperature up to 500°C. The juxtapositioning of a pair of such curves for Ti(Kroll)-Nb$_{40}$ and Ti(iodide)-Nb$_{40}$, respectively, revealed a remarkable degree of coincidence indicating that: *(a)* the use of Kroll-process Ti as a starting material could be expected to lead to a Ti-Nb$_{40}$ alloy containing some 500 ppm O; *(b)* J_c maximization during the 1-h aging of such alloys is again able to take place at, or near, the usual 400°C.

8.6.5 Oxygen Additions to Ti-60Nb (43.5 at.% Nb)

In the first of a series of investigations to be conducted at MIT, by a group headed by J. WULFF [Reu66] [Com67][Rau68, Rau68[a]][Ric70], on the effects of additional solutes on the superconductivity of Ti-Nb and related alloys, REUTER *et al* [Reu66] carried out a detailed study of the microstructure and critical current density of Ti-60Nb alloyed with 0.239 wt.% O. In the completeness of its metallurgical detail this research set the stage for the numerous investigations that were to follow. Although the effect of O was addressed in microstructural terms, the absence of a control alloy prevented the making of a quantitative and unequivocal assessment of its influence on J_c. Comparative studies were undertaken with wire samples which had been subjected to the processing schedule: cold swaged and drawn from 0.415 in$^\phi$ to a final wire size of 0.010 in$^\phi$ (i.e. 99.94% cw) plus aged/annealed 1h/(290, 370, 500, 600, 700, 800, 900 and 1000°C). Also explored were the effects of the following variants of that schedule: *(a)* cold worked 99.94% plus aged ($\frac{1}{2}$, 1, 2, 8, 32h)/500°C; *(b)* cold worked 99.94% plus recrystallized 1h/1000°C plus aged 1h/500°C; *(c)* cold swaged from 0.415 in$^\phi$ to 0.113 in$^\phi$ (92.6%), 0.078 in$^\phi$ (96.5%), and 0.030 in$^\phi$ (99.5%) plus aged 1h/500°C and cold drawn 99%, 98% and 89%, respectively, to the final wire diameter of 0.010 in.

Of the 1-h heat treatments, 500°C was optimal. Reducing the time of this heat treatment to $\frac{1}{2}$ h was slightly detrimental; increasing it to 2 h made little

different to the J_c at $H_r/2$, but had the beneficial effect of slightly increasing H_r itself. All intermediate heat treatments in which cold work was traded between post- and pre-heat treatment operations reduced J_c indicating that for a given amount of cold work, as dictated by starting and finishing materials constraints, it was better to administer it before the final heat treatment. This of course is contrary to present-day process-optimization experience as discussed, for example, in Sect.7.31; nevertheless at 5.5×10^4 A cm^{-2} the J_c(4.2 K, 60 kOe) attained under the best processing conditions was quite substantial even by contemporary standards. REUTER *et al* [Reu66] attempted to identify the microstructural features of the processed wires using transmission electron microscopy. The essential results of the observations were interpreted as follows: *(a)* heat treatment at 370°C and below permitted the development of a fine (0.03 µm$^\phi$) precipitate-free fibrous structure; *(b)* at 500°C the fibers apparently developed into elongated "grains" (grain diameter 0.01-0.03 µm) with precipitates of ω-phase and O-saturated α-Ti-Nb; *(c)* at 600°C, a 1-h heat treatment led to the precipitation of TiO and possibly some α-Ti-Nb.

In view of its Nb-richness, the alloy in question, if homogeneous, could not support ω-phase precipitation; otherwise the observations reported are in reasonable accord with the currently accepted picture of the influence of aging on deformation structures as summarized in [Mon2.11]. In an alloy of this composition, that precipitation takes place at all is due to the so-called "α-stabilizing" influence of the dissolved oxygen as discussed, for example, in [Mon2.6].

In the final study of this series conducted by the MIT group, RAUCH *et al* [Rau68, Rau68[a]] investigated the superconducting properties of Ti-40Nb alloyed with 0.14 and 0.16 wt.% O, Ti-50Nb with 0.051 wt.% O, Ti-60Nb with 0.052, 0.11, and 0.27 wt.% O, and Ti-80Nb with 0.034 wt.% O. Although the O had been deliberately introduced into the melt (in the form of Nb$_2$O$_5$ powder) this was primarily an aging study in that the results were displayed in most cases using the heat-treatment parameters, *aging time* and *aging temperature*, as variables. In fact, as stated above, only for Ti-60Nb were three levels of oxygen introduced. Aging studies of Ti-60Nb-O alloys with 520 and 2700 ppm O were carried out at temperatures of up to 700°C. In the resulting plots of J_c(55 kOe, 520 ppm O) and J_c(50 kOe, 2700 ppm O) *versus* 1-h aging temperature, that corresponding to 520 ppm O peaked at 450°C

($J_{c,max}$ = 2.6x10^4 A cm^{-2}) while that for 2700 ppm O peaked at about 525°C ($J_{c,max}$ = 5.5x10^4 A cm^{-2}). Without recourse to the use of TEM or x-ray techniques these results were interpreted in terms of a supposed greater flux-pinning effectiveness of α-phase as compared to ω-phase precipitates. It is of course unlikely that any precipitation at all was taking place in these alloys under the conditions described.

High-Nb alloys are definitely out of the range of ω-phase precipitation, but the possibilities of: (a) separation into two β-phases as a result either of thermodynamic clustering or some preparational artifact such as residual coring, (b) moderate-temperature precipitation of α-phase in the non-equilibrium environment of a cell wall, cannot be ignored. On the basis of the results for Ti-60Nb-O, the only alloy which had been provided with a wide range of O levels (viz 0.052-0.27 wt.%), it must be concluded that the presence of O is generally beneficial to J_c provided that the heat treatments are administered at the appropriate temperatures. Peak values of

J_c(4.2 K, ∿50 kOe) when plotted against the l-h aging temperatures were: 2.6x10^4 (520 ppm), 2.0x10^4 (1100 ppm) and 5.5x10^4 A cm^{-2} (2700 ppm).

As part of a commercial-alloy development program, BIDAULT and DOSDAT [Bid70] investigated the influence of heat treatment and dissolved O and N on the microstructure (as determined by TEM at 60,000X) and critical current density Ti-60Nb. In experiments on cold-worked and, presumably, optimally heat-treated wire (internal evidence suggested a protracted anneal at 380°C), they found that the 50-kOe J_c increased from 1.0x10^5 to 1.5x10^5 A cm^{-2} as the O concentration was raised from 490 to 2400 ppm, and to 1.7$_5$x10^5 A cm^{-2} for material with an interstitial element content of 996 ppm N plus 1870 ppm O. The microstructures developed are listed in Table 8-8, and a typical set of results is presented in Fig.8-3. Clearly the interstitial elements, through their influences on the cold-worked and heat-treated microstructures, had strong indirect influences on the critical current densities. This conclusion relates closely to the results of (but not the conclusions arrived at by) RASSMANN and ILLGEN [Ras72^a].

8.7 CRITICAL CURRENT DENSITIES OF QUATERNARY ALLOYS CONTAINING OXYGEN

8.7.1 Oxygen Additions to Ti-Nb-TM Alloys

BETTERTON et al [Bet66] have reported briefly on the results of a study of critical current density in cold-worked Ti-Nb$_{30}$-Fe$_{0.1}$-O$_{0.16}$ and Ti-Nb$_{55}$-Fe$_{0.52}$-O$_{0.13}$ both before and after "optimal" heat treatment (i.e. one whose temperature and duration gave the best J_c). With regard to the second alloy, it was shown that the addition of O to the Ti-Nb$_{55}$-Fe$_{0.52}$ base brought about, in the presence of cold work and heat treatment, a slight increase in the J_c(4.2 K, 30 kOe) of from 1.5x10^4 to 1.7x10^4 A cm^{-2}.

Increases in J_c as a result of O additions of 0.1, 0.3, and 0.95 wt.% to Ti$_{10}$-Zr$_{40}$-Nb$_{50}$ were noted by KITADA and DOI [Kit70^b], who studied microstructures, mechanical properties, and J_c(4.2 K) versus H_a. For all the conditions investigated: solution-treated 3h/1100°C (ST), ST and aged 4h/700°C, ST and aged 100h/700°C, J_c increased with O concentration. At all fields up to 60 kOe, the J_c of the ST-and-aged-4h/700°C alloy was greater when the aging took place in a vacuum of 5x10^{-5} torr than when a vacuum of 8x10^{-7} torr

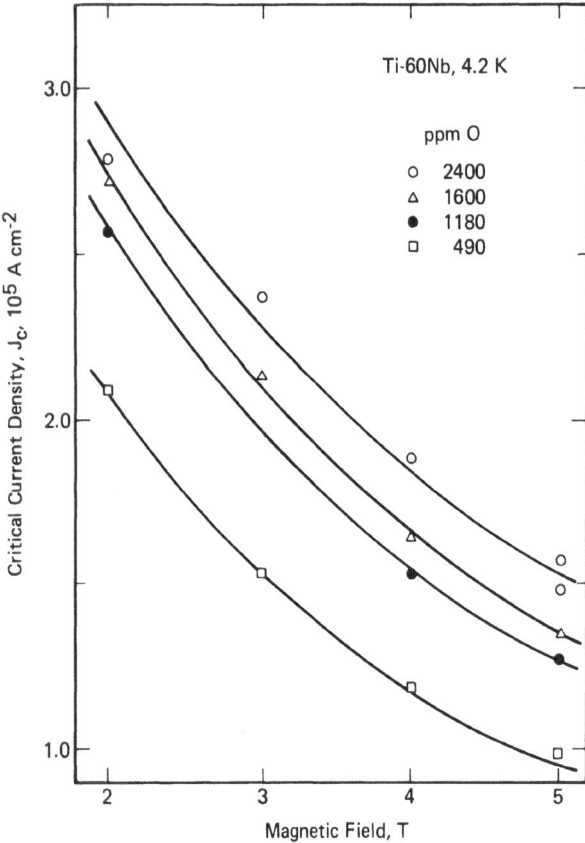

FIGURE 8-3. Influence of oxygen on the critical current density of Ti-60Nb. The oxygen levels depicted are: 2400 ppm (○); 1600 ppm (△); 1180 ppm (●); and 490 ppm (□) — after BIDAULT and DOSDAT [Bid70].

Dopant:	490 ppm O	2,400 ppm O	1,870 ppm O plus 966 ppm N
Microstructure:	(1) Bands 0.1∿0.2μm in thickness -- very little internal structure.	(1) Filamentary bands 0.03∿0.1μm in thickness -- fine granular internal structure.	(1) Fine lenticular bands 0.05∿0.1μm in thickness -- fine cellular internal structure of mesh size less than 0.01μm.
	(2) No precipitation visible.	(2) Globular precipitates, 0.01∿0.03μm$^{\phi}$.	(2) No precipitation visible.
J_c at 5 T:	1.0×10^5 A cm^{-2}	1.5×10^5 A cm^{-2}	1.7×10^5 A cm^{-2}

was used. The reason for this had to do with O pickup. The positive influences of O were claimed to be: *(i)* when in solution, a *direct* influence on the pinning strength; *(ii)* an *indirect* influence on pinning, in that the kinetics of formation and degree of refinement during 700°C-aging of the β'+β" lamellar structure (cf. [Mon2.7]) which is responsible for flux pinning in the ternary alloy under investigation is influenced by the presence of O.

8.7.2 Oxygen Additions to Ti-Nb-Rare-Earth Alloys

During the search for possible improvement to the critical current density of Ti-Nb, consistent with the preservation of its workability, several studies have been undertaken of the combined influences on J_c of additions of O *and* rare-earth elements (including Sc -- which is not a rare earth, and Th -- which is an actinide). Systems studied are listed in Table 8-9.

Oxygen when added to Ti-Nb alloys in the form of Nb_2O_5 goes into solid solution. The elements Sc, Y, Gd, Er, and Th when added in metallic form react with the dissolved O to form M_2O_3 oxides (except Th, which forms a dioxide). The elements themselves all have undetectable or very small ($\lesssim 2\%$) solid solubilities in α- and β-Ti. The element Y is claimed to be insoluble in Nb. Although Th has no detectable solid solubility in both α- and β-Ti it has unlimited solubility in the melt, as have Sc and Gd. Consequently these elements are able to dissolve in the melt, become internally oxidized, and precipitate out upon solidification into finely divided oxide particles and occasionally metallic ones, as in the case of Th, Gd, and Y [Cou67]. The experimental plan of COURTNEY and WULFF [Cou67] did not provide for the inclusion of a ternary Ti-Nb-O control alloy. This deficiency was remedied by RICKETTS *et al* [Ric70] in another contribution from the MIT school. By combining the two sets of data it is possible to evaluate the relative merits of adding either Sc, Y, Gd, Er, Th, or nothing at all to O-containing Ti-Nb$_{25(or\ 26)}$. In comparing data from rows 2, 3, and 8 of Table 8-9 with the J_c of the high-O control, we note that precipitates of the oxides of Y, Gd, or Sc are beneficial; while in comparing rows 9 and 10 with the low-O control we see that the presence of finely divided Er oxide is also slightly beneficial. RICKETTS *et al* [Ric70] studied as a function of 1-h aging temperature, the relative volume fractions of ω-phase and α-phase precipitation, as illustrated in Fig.8-2 for the O-only members of the group, correlating the results with those of electrical resistivity, critical field, and critical current density studies. It appeared that although O, Er, and Sc stabilized the α-phase (which was favored by heat treatment at 500°C) they had little effect on ω-phase precipitation (which was maximized during heat treatment at 400°C). It is, however, necessary to issue the usual cautionary reminder that these metallurgical effects and electrical properties, which have been ascribed to the presence of interstitial solutes or oxide precipitates, can easily be masked by small variations in the heat treatment conditions which control the deformation microstructures, as evidenced in the trend in average J_c noticeable in proceeding from column to column in Table 8-9.

Row No.	At.% Nb	"Rare Earth" Content, Wt.%					Oxygen Content		Critical Current Density[++] J_c (4.2K, 50kOe), 10^4 A cm^{-2} Temperature of 1-h Heat Treatment, °C			Literature
		Sc	Y	Gd	Er	Th	Added as Nb_2O_5 ppm	Dissolved O ppm	400	425	500	
1	20	--	0.56	--	--	--	3,500	2,000			4.7	[Ras72[a]]
2	25	--	1	--	--	--	4,500	1,800	9.6[+++]	8.0		⎫
3		--	--	1	--	--	4,500	2,980	9.6[+++]	8.0		⎬ [Cou67]
4		--	--	--	--	1	4,500	3,120	7.9[+++]	6.6		⎭
5	26			control			440	440	7.2		4.5	⎫
6		--	--	--	--	--	2,400	2,400	8.2		4.8	⎪
7		--	0.49	--	--	--	3,800	1,140	8.0		3.5	⎬
8		--	0.62	--	--	--	4,500	1,020	9.0		4.3	[Ric70]
9		--	--	--	1.83	--	2,900	200	8.0		4.5	⎪
10		--	--	--	1.99	--	3,000	60	9.8		4.6	⎭
11	40	--	0.56	--	--	--	3,000	1,500			5.4	[Ras72[a]]

[+] *Condition*: 99.9% cold work plus heat treatment. The Gd- and Th-containing alloys needed two separate 2h/1000°C intermediate vacuum anneals.

[++] J_c criteria: 40 μV across sample [Ras72[a]]; $\frac{1}{4}$ μV cm^{-1} [Cou67][Ric70].

[+++] Increased 20% from published 425°C data.

TABULATED DATA—INFLUENCES OF C, N, AND O ON THE T_c OF Ti-Nb

TABLE 8-10   TRANSITION TEMPERATURES OF TITANIUM-NIOBIUM-INTERSTITIAL-ELEMENT ALLOYS -- DATA SOURCES

Host Alloy	Solute Additions	Condition	Procedures	Literature
	CARBON ADDITIONS			
Ti-66Nb	0.07, 0.08, 0.12, 0.21, 0.23 and 0.28 wt.% C			I.A. Baranov *et al.* (1967) [Bar67[a]]
	NITROGEN ADDITIONS			
Ti-66Nb	0.05, 0.11, 0.29, 0.35 and 0.44 at.% N			I.A. Baranov *et al.* (1967) [Bar67[a]]
Ti-33Nb	0.2, 0.41 at.% N	Deformed by swaging, and recrystallized by vacuum annealing, 1/2h/950°C.	Inductive	D. Bachmann *et al.* (1968) [Bac68]
Ti-40Nb	0.21 at.% N	Same as above, except 1/2h/950°C resulted in only partial recrystallization.	"	" "
	OXYGEN ADDITIONS			
Ti-33Nb	0.32 at.% O	Deformed by swaging, and recrystallized by vacuum annealing, 1/2h/950°C.	"	" "
Ti-40Nb	0.17 and 0.34 at.% O	Same as above, except 1/2h/950°C resulted in only partial recrystallization.	"	" "
Ti-66Nb	0.05, 0.12, 0.29 and 0.44 at.% O			I.A. Baranov *et al.* (1967) [Bar67[a]]
Ti-74Nb	0.17, 0.68, 1.17, 1.66, 2.17, 2.60 and 3.13 at.% O	Arc melted with Nb_2O_5 to provide O. Ingot homogenized *in situ* by "fire polishing" which melted only a thin upper surface layer.	Schawlow-Devlin technique	F.A. Rodriguez-Gonzalez (1970) [Rod70]

TABLE 8-11   TRANSITION TEMPERATURES OF TITANIUM-NIOBIUM-INTERSTITIAL ALLOYS (from plotted data)

(a) CARBON Additions to Ti-66Nb [Bar67[a]]

Wt.% C	T_c K
0	9.29
0.07	9.18
0.08	9.35
0.12	9.54
0.21	9.54
0.23	9.68
0.28	9.65

(b) NITROGEN Additions to Ti-33Nb [Bac68]

At.% N	T_c, K	
	Deformed	Fully Recrystallized
0	6.7	6.9
0.2	6.6	6.7
0.41	6.7	6.1

(c) NITROGEN Additions to Ti-40Nb [Bac68]

At.% N	T_c, K	
	Deformed	Partly Recrystallized
0	7.6	7.9
0.21	7.3	6.8

(d) NITROGEN Additions to Ti-66Nb [Bar67[a]]

At.% N	T_c K
0	9.29
0.05	9.32
0.11	9.28
0.29	9.17
0.35	9.08
0.44	8.91

(e) OXYGEN Additions to Ti-33Nb [Bac68]

At.% O	J_c, K	
	Deformed	Fully Recrystallized
0	6.7	6.9
0.32	6.9	6.2

(f) OXYGEN Additions to Ti-40Nb [Bac68]

At.% O	J_c, K	
	Deformed	Partly Recrystallized
0	7.6	7.9
0.17	7.6	7.6
0.34	7.6	7.4

(g) OXYGEN Additions to Ti-66Nb [Bar67[a]]

At.% O	T_c K
0	9.29
0.05	9.29
0.12	9.24
0.29	9.09
0.44	8.87

(h) OXYGEN Additions to Ti-74Nb [Rod70]

At.% O (nominal)	T_c[+] K
0.17	9.45
0.68	9.18
1.17	8.82
1.66	8.62
2.17	8.25
2.60	8.09
3.13	7.88

[+]   Intercept/slope/(correlation coefficient) = 9.51 K/-0.54 K per at.% oxygen/(99.6%).

TYPICAL INTERSTITIAL-ELEMENT LEVELS IN Ti-50Nb AND ITS CONSTITUENTS

TABLE 8-12 CARBON, NITROGEN, AND OXYGEN CONTENTS OF SEVERAL GRADES OF Ti, Nb, AND Ti-50Nb SUPERCONDUCTOR

Solvent	Interstitial Content, ppm			Source
	C	N	O	
Ti:				
MRC (MARZ-grade)	78	6	63	a
MRC (VP-grade)	150	40	350	b
TMC electrorefined sponge (grade ELXX)	---	40	370	c
Kroll-process (Toho sponge)	---	110	860	d
Kroll-process	800	400	1100	e
Iodide-process	100	200	200	e
Nb:				
MRC (MARZ-grade)	8	4	23	a
MRC (VP-grade)	25	15	100	b
KBI-A	20	20	115	f
KBI-B	30	25	165	g
Fansteel, beam-melted	36	10	32	h
Ti-Nb Superconductor:				
Nb-46.94Ti, KBI	70	50	470	i
Nb-48.48Ti, KBI	70	50	470	j
Nb-45.7Ti, TWCA	80	49	930	k

a Materials Research Corporation: Zone-refined; supplied typical analysis.

b Materials Research Corporation: Vacuum-melted; supplied typical analysis.

c Titanium Metals Corporation: See also E.W. Collings and J.C. Ho, Phys. Rev. 2 235-44 (1970).

d See E.W. Collings and J.C. Ho, Phys. Rev. 2 235-44 (1970).

e See Table 7-14, p.268.

f Kaweki-Berylco Industries: Metallurgical-grade Nb; Lot No. 507-313-347.

g Kaweki-Berylco Industries: Metallurgical-grade Nb; Lot No. 9350-1.

h Fansteel Corporation: Supplied analysis; see also E.W. Collings and R.D. Smith, J. Less-Common Metals 27 389-401 (1972).

i Kaweki-Berylco Industries: Superconductor-grade NbTi; Lot. No. 8794.

j Kaweki-Berylco Industries: Superconductor-grade NbTi; Lot. No. 8792.

k Teledyne Wah Chang, Albany: As supplied to the Fermi National Accelerator Laboratory, September 1974.

TERNARY ALLOYS OF TITANIUM-NIOBIUM WITH SIMPLE METALS

This chapter deals with the effects of the so-called "simple metals" on the superconducting properties of Ti-Nb alloys. Simple metals to be considered, listed in order of increasing atomic number, are: Al, Si, Cu, Ga, Ge, Y, Ag, In, Sn, Sb, Au, Pb, and U. The treatment proceeds in that order, excepting that the effects of Cu and Ge additions which have been dealt with quite extensively in the literature are considered separately and sequentially in the latter part of the chapter.

The effects of simple metals on the superconductivity of Ti-Nb alloys, unlike those of transition-element substitutions, tend to be dominated by metallurgical rather than electronic considerations. In other words, although substitutions of elements such as Zr, Ta, and so on, influence the Fermi density of states (hence T_c) and both normal-electron and super-electron scattering (hence H_{c2}), simple metals exert their strongest influences on the deformed-and-aged microstructures. Deformation-subbands and precipitates play crucial roles in the optimization of Ti-Nb-alloy superconductors, Sect.7.30. It is within such a context that it is appropriate to discuss the properties of simple-metal additions to them, and the manner in which they may: *(a)* influence the deformation process, hence the deformation microstructure; *(b)* enhance the intrinsic tendency of the host alloy to form α-phase precipitates; *(c)* deposit intermetallic-compound precipitates in their own right, which may, either directly or indirectly, have some value in flux pinning.

9.1 METALLURGICAL CONSIDERATIONS

The structures of quenched Ti-Nb alloys in the precipitate-rich low- and intermediate-concentration ranges (as defined in Sect.7.26) have been described in detail in Sect.7.20. There it was pointed out that the composition of athermal ω-phase, the product which appears spontaneously during brine quenching, is 18 at.% Nb [Bag59]. As a result of moderate-temperature aging (at 450°C), the saturation composition of the resulting "isothermal" ω-phase was claimed to be $6 \sim 11$ at.% Nb [Hic69[a]]. In Fig.9-1, which is a composite metastable (quenched), meta-equilibrium

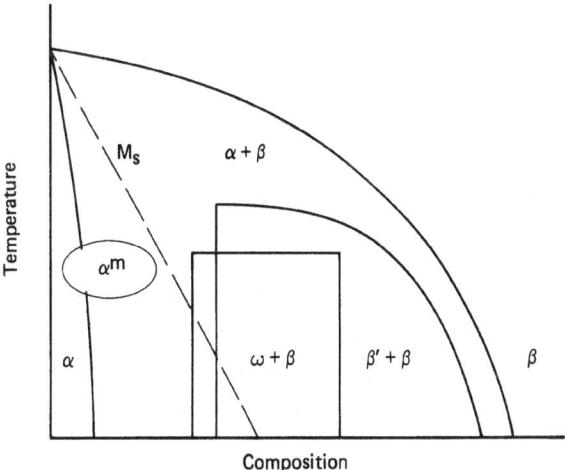

FIGURE 9-1. Schematic representation of the locations of the two metastable regimes $\omega + \beta$ and $\beta' + \beta$ within the equilibrium $\alpha + \beta$ field in a typical Ti-TM alloy — after WILLIAMS *et al* [Wil71].

(moderately aged), and equilibrium phase diagram for Ti-TM alloys, the athermal-ω-decomposition process is depicted as being contained within the rectangle labelled ω+β. But, as described in Sect.7.20.2 and indicated in Fig.9-1, if the temperature is too high or the alloy too concentrated for ω-phase precipitation to take place, a solute-lean bcc phase, designated β', separates out instead. Unlike the ω-phase which can be generated by lattice-displacement waves in a virtual crystal if need be, β' precipitation stems from the chemical dissimilarity between solute and solvent atoms that leads eventually to α+β decomposition. At least in its precursor stages, β' was for a time [Nar71] confused with ω-phase precipitation; this is no longer the case, its status as a metastable bcc precipitate being now well established [Wiℓ73, Wiℓ78].

If an alloy, previously aged in the ω+β field, is elevated to the β'+β field the isothermal ω-phase "reverts" at the higher temperature to a bcc phase, leaner in solute than the matrix. The reaction is not reversible, the bcc phase remaining untransformed when the alloy is returned to room temperature. The new phase could also be thought of as β', since a contributor to its stability is the inherent tendency for Ti-TM alloys to cluster. With regard to Ti-Nb alloys, clustering has been discussed by RUDMAN [Rud64], and β' precipitation by MENDIRATTA et al [Men71].

In some systems, notably Ti-Cr, prolonged aging in the β'+β field (i.e. >9min/450°C) has resulted in the nucleation of α-phase at the β'/β interfaces [Luh70]; in others, such as Ti-Mo-Al, which is more closely related to the systems to be considered in this chapter, α-phase precipitation has been shown to take place within the β' grains [Wiℓ71]. Again with regard to α-phase precipitation, it should be noted that although during the overaging of the ω+β phase several routes towards the attainment of the equilibrium α+β structure have been identified, none of them involve the *direct* conversion of ω to α [Bra67].

Our current understanding of the manner in which third-element (and in particular, simple-metal) alloying influences the decomposition of the ω+β and β'+β phases, stems from an investigation by WILLIAMS et al [Wiℓ71] of the effects of O, Al, Sn, and Zr additions to Ti-V$_{20}$ and Ti-Mo$_6$. The essential results of that are briefly as follows:

(i) α-Phase Precipitation from ω+β. Ternary additions of the so-called "α-stabilizers", O and Al,

reduce the time of stability of the ω-phase and promote an early precipitation of α-phase.

(ii) α-Phase Precipitation from β'+β. "Neutral" solutes such as Sn and Zr tend to preserve the stability of the bcc structure when dissolved in it. In doing so they are not, however, "β-stabilizers" in the usual sense, but strengtheners of the bcc lattice against the ω-instability. In the presence of these and similar solutes, the ω+β region shrinks to such an extent that a given host alloy previously within it may find itself by default in the β'+β region, hence prone to α-phase precipitation at the β'-sites.

Thus simple-metal additions, like the interstitial-element additions of the previous chapter, have profound effects on the "intrinsic" α-phase precipitation process in Ti-Nb alloys and in so doing play roles in the development of flux-pinning microstructures. In addition, if present in sufficient quantities, solutes such as Cu and Ge can lead to the formation of what might then be termed "extrinsic" precipitates of Ti$_2$Cu and Ti$_5$Ge$_3$, respectively, which either themselves, or in association with the deformation subband structure, may contribute to the flux pinning.

9.2 SUPERCONDUCTIVITY IN Ti-Nb-SM ALLOYS— A COMPARATIVE SURVEY

9.2.1 The Superconducting Transition Temperature

BACHMANN et al [Bac68] have studied the superconducting transition temperatures of ternary alloys of Ti-Nb(10-20 at.%) with some 2 at.% additions of the simple metals Al, Cu, Ga, Ge, Ag, In, Sn, Sb, Au, Pb, and U (as well as the N and O, referred to in Chapter 8) either singly or in the combinations: Al + Sb, Ga + Sb, In + Sb, and Sn + U. The actual compositions investigated and some of the results obtained are listed in Table 9-1. In this work various base-alloy compositions were selected according to the valence of the solute element, the goal being to preserve in the final alloy a constant average electron/atom ratio of 4.20. It is debatable whether this was achieved in a meaningful way.

Transition temperatures were measured on 1.75-mm$^\phi$ wires cold-worked from as-cast material. As indicated in the table, that of the basic unalloyed binary alloy, Ti-33Nb, 6.7$_5$ K, could be either raised or lowered by

TABLE 9-1 SUPERCONDUCTING TRANSITION TEMPERATURES (T_c) OF SOME Ti-Nb-SM ALLOYS -- Bachmann *et al.* [Bac68].

Compositions, at.%			T_c, K
Nb	SM		
20.3	Control		6.7_5
22.0	Al	2.0	7.5
23.0	Cu	1.0	7.6_5
22.0	Ga	2.0	7.0_5
20.0	Ge	2.0	7.3
23.0	Ag	1.0	7.6_5
22.0	In	2.0	6.9
20.0	Sn	2.0	6.6
16.0	Sb	2.0, 4.0, 10.0	6.6_5, 5.9, 4.3
20.0	Pb	2.0	5.1-5.7
10.0	U	5.0	3.8_5

the presence of the simple metals. The highest T_c, 7.6_5 K, was obtained with the noble-metal additions, while U brought about the largest "depression" (presumably because this alloy had experienced the most severe reduction of Nb content). If the transition temperatures of the ternary alloys are compared with those of binary Ti-Nb as presented in Fig.7-3, it will be noted that most of the Nb-levels fall in the "gap" between the α'' and $\omega+\beta$ structures. It could well be, therefore, that the enhanced T_c's were a manifestation of a solute-induced extension of the α''-branch of the curve to higher Nb levels, in the same way that low-concentration quenched Ti-Mo alloys responded to small additions of Al, cf. Fig.5-5.

9.2.2 The Upper Critical Field

The upper critical fields of Ti-Nb(33, 39, and 44 at.%), alloyed with small amounts of Cu and Ge, as well as those of the alloys Ti_{65}-Nb_{33}-SM_2, where SM represents Al, Cu, Ga, Ge, Ag, In and Sn (also Ti_{65}-Nb_{33}-TM_2, where TM represents Cr, Mn, Fe and Zr), have been studied by HELLER, LÖHBERG and BABISKIN [Heℓ71^a]. Their samples in the form of 97% cold-deformed 1-mm$^\phi$ wires were generally measured in the cold-worked-and-aged condition, although in some cases for comparison purposes, the as-worked structure was erased prior to aging by a homogenization anneal of 3h/1000°C (or 3h/1080°C, for the sample with 2 at.% Ge).

The results of the resistive upper critical field, H_r, measurements on the Ti_{65}-Nb_{33}-SM_2 alloys, which had been carried out using a sample current density of 1:3 A cm^{-2} and a 1 μV transition criterion in a water-cooled Bitter solenoid (maximum field 150 kOe) at the U.S. Naval Research Laboratory, are presented in Table 9-2. The observed variation of H_r with alloying was thought to have been due primarily to a variation of the Nb-content of the β-phase, which in turn depended on the amount of (Ti-rich) ω-phase and α-phase precipitation which took place.

TABLE 9-2 RESISTIVE UPPER CRITICAL FIELDS (H_r) OF Ti_{65}-Nb_{33}-SM_2 ALLOYS -- Heller *et al.* [Heℓ71^a]

Simple Metal	H_r(4.2K),[†] kOe
--	114.0
Al	110.0
Cu	114.0
Ga	103.5
Ge	111.5
Ag	111.0
In	105.0
Sn	95.8

† Criterion: 1 μV across 1 cm at 10 mA (1.3 A cm^{-2}).

ALLOY GROUP 1: TITANIUM-NIOBIUM-SIMPLE-METAL TERNARY ALLOYS

9.3 Al ADDITIONS TO Ti-Nb

9.3.1 The Transition Temperature

As part of a study of the influence of thermo-mechanical-processing-induced microstructural variation on the superconducting properties of alloys of interest in the construction of superconducting magnets, ZWICKER, LÖHBERG and HELLER [Zwi70] conducted a brief investigation of the effect of Al on the mechanical and superconducting properties of Ti-Nb. Although the metallurgical conditions of the samples were not explicitly specified, indirect evidence suggests that they had been heat treated 34h/400°C after deformation. It was found that, although additions of 1, 1.52, and 2.5 at.% Al in solid solution increased both the hardness and the elastic modulus of Ti-Nb$_{33}$, they had detrimental effects on all the superconducting properties. For example, it was shown that T$_c$ *decreased* with Al concentration at approximately 0.35 K per at.% Al.

Detailed studies of the influence of heat treatment on the superconducting transition temperatures of a large number of Ti-Nb-Al alloys with Nb concentrations in the range 5-46 at.%, and Al concentrations between 5 and 25 at.%, have been undertaken by COTTON, TAGGART and POLONIS [Cot74]. Also studied were Ti$_{74}$-Nb$_{12}$-Mo$_4$-Al$_{10}$ and Ti$_{64}$-Nb$_{12}$-Mo$_4$-Al$_{20}$. Ingots 0.5 cm in diameter and 5 cm in length, prepared by levitation melting and chill casting, were solution treated at 1050°C under gettered static-vacuum conditions and either quenched or slow cooled. The water-quenched (WQ) specimens were subjected to the following heat-treatment schemes: *(a)* WQ plus salt-bath aged at 350°C in the ω+β regime (WQ + ωAge); *(b)* WQ + ωAge plus salt-bath reversion heat treated at 500°C to produce solute fluctuations in the β-phase; *(c)* furnace cooled at 100-150°C per h (FC); *(d)* FC followed by aging for 40 h at each of the three temperatures 550, 650 and 750°C, respectively, in order to produce isothermal transformation to the α+β structure. The influences on the transition temperatures of these heat treatments designed to promote the formation of the ω+β, β'+β, and α+β phases, respectively, were discussed in detail in [Cot74].

The rationale for the work was the possibility of developing high transition temperatures in alloys with relatively low average concentrations of β-stabilizer. In binary Ti-TM alloys, T$_c$ generally maximizes in the e/a-ratio range 4.5 ∿4.8. This requires the addition to Ti of some 50 to 80 at.% of a group-V transition element. In the particular case of Ti-Nb, Fig.7-3 indeed shows T$_c$ passing through a maximum between 60 and 70 at.% Nb. However, it has been postulated that by taking advantage of one or other of the three-phase decomposition reactions cited above, solute-rich zones of relatively high T$_c$ could be produced in alloys of low *average* compositions. A summary of the results obtained is presented in Table 9-3. It was noted that when a mixture of the α and β phases was produced in Ti-Nb-Al, the Al separated to the α-phase leaving the β-phase enriched with Nb and in possession of a T$_c$ higher than that of the quenched alloy. In general, thermal treatments designed to enrich the Nb content of the β-component make it possible for the T$_c$ to approach the binary-phase limit of ∿9.7 K. This is illustrated by the results for Ti$_{30}$-Nb$_{45}$-Al$_{25}$, whose transition temperature was increased from 4.1 K (as-quenched) to 9.3 K by an application of the "furnace-cooling" process.

In a recently published study, WADA *et al* [Wad80] substituted 2, 6, and 10 at.% Al into each of the binary master alloys Ti-Nb$_{30}$, Ti-Nb$_{50}$, and Ti-Nb$_{75}$, thus extending the Nb-concentration range of the earlier research. Alloys were measured in the 50% cold-rolled condition and then after aging for 24 h at temperatures between 300 and 800°C. The general conclusion reached was that, although in all cases the transition temperature was reduced by the third-element addition, it was partially recovered again by aging, but that even in the aged alloys the T$_c$ was lower than that of the starting binary alloy.

9.3.2 The Critical Field

According to ZWICKER, LÖHBERG, and HELLER [Zwi70], the upper critical field of Ti-Nb$_{33}$ (temperature unspecified, but presumably 4.2 K) was *diminished* by addition of Al at the rate of 6.7 kOe per at.%. Resistive measurements of the upper critical field of

TABLE 9-3 SUPERCONDUCTING TRANSITION TEMPERATURE AS FUNCTION OF COMPOSITION AND
HEAT TREATMENT IN TITANIUM-NIOBIUM-ALUMINUM ALLOYS -- Cotton *et al*. [Cot73, Cot74].

Composition		Transition Temperature After Heat Treatment						
At.% Nb	At.% Al	A	B	C	D	E	F	G
5.00	15.01	<1.5			6.89	4.80		6.98
5.00	25.03	1.83						6.81
5.02	5.01	3.09			7.42	4.93		7.04
5.62	22.47	1.67						
6.57	9.36	2.98	2.73	2.90	6.65			7.17
10.23	9.65	3.51	3.25	3.60	6.55			7.10
12.97	6.82	4.08	3.92	5.05	7.38			
14.98	9.99	3.81	3.47	4.19	6.73			7.17
18.02	6.07	4.80	4.55	5.46	6.98			7.15
19.55	9.75	4.70		5.34	6.24	5.29	6.93[+]	7.65
19.59	9.78	4.70	4.32	5.09				
22.71	6.47	5.15	4.87	5.15	6.75			
24.97	10.03	5.00	4.95	5.20	6.36	6.04	6.85[+]	7.89
34.94	10.00	6.20			8.82	6.68	7.48	7.90
42.52	19.64	3.65			7.89	7.00		7.22
44.98	25.02	4.05			9.30			
45.00	10.03	6.96			9.10	7.09		7.89
45.07	4.98	7.12			8.60	8.12		7.14
46.11	19.58	4.36						7.36

A - Annealed 1050°C/water quench (ST+WQ)

B - Above plus 40h/350°C/water quench (WQ+ωAge)

C - Above plus 2min/550°C/water quench (ωAge+Revert)

D - Annealed 1050°C/furnace cooled (ST+FC)

E - (ST+FC) to 750°C; 40h/750°C/water quench

F - (ST+FC) to 650°C; 40h/650°C/water quench

G - (ST+FC) to 550°C; 150h/550°C/water quench

+ Furnace cooled from 650°C.

cold-worked (97%) and aged (10h/400°C/quenched) 1-mm$^\phi$ wires of Ti$_{65}$-Nb$_{33}$-Al$_2$ by HELLER *et al* [Hel71[a]] yielded a value of 110.0 kOe which, referred to the the 114.0 kOe of the unalloyed Ti-Nb$_{33}$ control, represented a rate of *decrease* of 2 kOe per at.% Al. In the measurements by WADA *et al* [Wad80] of the Ti-Nb-Al alloys specified in Sect.9.3.1, the 4.2-K H$_r$ of the cold-rolled Ti-Nb$_{30}$-base ternaries *decreased* at the rate of 9.8 kOe per at.% Al. The rates of H_r-decrease exhibited by the Ti-Nb$_{50}$- and Ti-Nb$_{75}$-base alloys were less rapid, but in all cases, as expected from the T_c results, the H_r's of the ternary alloys increased with aging.

9.3.3 Critical Current Density

The influences of Al substitutions on the critical current densities of Ti-Nb alloys have been investigated by ZWICKER et al [Zwi70], RASSMANN and ILLGEN [Ras72a], and WADA et al [Wad80].

In the first of these experiments the addition of 1 at.% Al to Ti-Nb$_{33}$ reduced the J_c (probably 4.2 K, 100 kOe) of deformed-only wire from about 2.5×10^3 to 1.3×10^3 A cm^{-2}. RASSMANN and ILLGEN also demonstrated that Al had a detrimental effect on the critical current density of Ti-50.6Nb, the addition of 2.8 wt.% Al reducing the 40-kOe J_c of the ~99.9% cold-worked alloy from 2.5×10^3 A cm^{-2} to about 3×10^2 A cm^{-2}. Annealing for 1h/500°C then increased the 50-kOe J_c of Ti-50.6Nb-2.8Al to about 5×10^3 A cm^{-2}. No control

study was undertaken; nevertheless the results of an independent test showed that the 50-kOe J_c of cold-worked and 1h/500°C-aged Ti-56Nb was 5×10^3 A cm^{-2}.

The description of the earlier results makes a useful introduction to the relatively comprehensive investigations recently completed by WADA et al [Wad80]. Intercompared were the critical current densities at 80 kOe of Ti-Nb$_{30}$, and two ternary alloys prepared by dissolving 2 at.% Al in Ti-Nb$_{30}$ and Ti-Nb$_{50}$, respectively. Alloys were measured in the cold-rolled (50%) condition and after 24-h agings at 400, 600, and 800°C. The results are best summarized with the aid of a figure. As shown in Fig.9-2, J_c passed through the usual maximum as the fixed-time aging temperature increased through ~400°C (cf. Fig.7-40). Although Al was detrimental to the J_c of cold-worked or over-aged Ti-Nb$_{33}$, the converse was true in the presence of appropriate heat treatment. Thus the 80-kOe J_c of cold-worked-and-aged 24h/400°C (Ti-Nb$_{30}$)$_{98}$-Al$_2$ exceeded that of the binary base alloy by an appreciable margin (1.5×10^4 as compared to 4.3×10^3 A cm^{-2}). It was anticipated that this ternary-solute-induced improvement would survive further stages of process optimization.

9.4 Si ADDITIONS TO Ti-Nb

RASSMANN and ILLGEN [Ras73] studied some metallurgical properties, including hardness, of the family:

$$Ti_{78.9}\text{-}Nb_{19.0}\text{-}Si_{2.1}$$
$$Ti_{79.2}\text{-}Nb_{20.5}\text{-}Si_{0.3}$$
$$Ti_{60.2}\text{-}Nb_{33.6}\text{-}Si_{6.2}$$
$$Ti_{62.4}\text{-}Nb_{35.2}\text{-}Si_{2.4}$$

But since the alloys selected were unable to be drawn into wire, a critical-current investigation had to be abandoned.

ISHIDA et al [Ish70] noted that the 40-kOe J_c of Ti-Nb$_{40}$, aged at temperatures between 350 and 500°C, generally increased rapidly with the addition of 0.5~2 at.% Si. For wire processed according to the schedule:

$$8 \text{ mm}^\phi \longrightarrow 2 \text{ mm}^\phi (5h/1100°C) \longrightarrow 0.25 \text{ mm}^\phi (3h/500°C)$$

J_c (40 kOe) rose by almost a factor of 10 to 7×10^4 A cm^{-2}

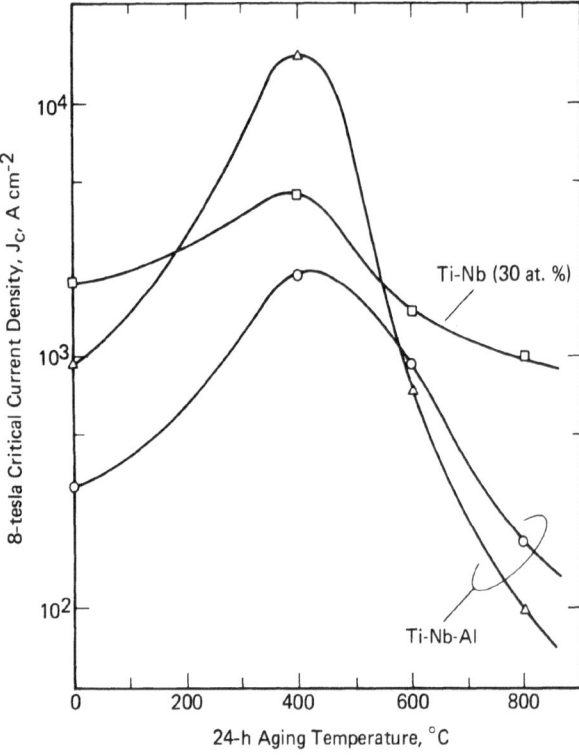

FIGURE 9-2. Critical current density as a function of 24-h aging temperature for previously cold-rolled (50%) strips of (Ti-Nb$_{50}$)-Al$_2$ (○), (Ti-Nb$_{30}$)-Al$_2$ (△) and a Ti-Nb (30 at. %) control (□) (criterion: 5 µV cm^{-1}) — after WADA et al [Wad80].

during the addition of the first 0.5 at.% Si and there-after remained constant. Bearing in mind the manner in which Si degrades the mechanical properties of the host alloy, 0.5 at.% Si must be regarded as optimal in this case.

WATANABE [Wat74] has also studied the critical current densities of heat-treated Ti-Nb$_{40}$ with 2 at.% Si as function of heat-treatment time (up to 1.8×10^3 h) and temperature (350, 400, 500°C), comparing the re-sults with those of a similarly heat-treated binary control. During aging at 350°C, J_c(70 kOe) continu-ously increased with time in both Ti-Nb-Si and Ti-Nb. The rate of increase in Ti-Nb was the greater, so that at the end of the aging (after 1.8×10^3 h) the critical current densities were almost the same in both alloys ($\sim 5.6 \times 10^4$ A cm^{-2}). At higher temperatures, shorter times were of course adequate to attain a given J_c; thus during aging at 500°C only 3 h was needed to reach a J_c(70 kOe) of 6.8×10^4 A cm^{-2}.

The mechanism underlying the improved J_c, when it occurred in alloys containing Si, was attributed to finer and more uniformly dispersed α-phase precipita-tion, in association with a refinement of the β-phase grains induced by the presence of α-phase and γ-phase (an intermetallic compound) precipitates. At the same time, Si did have a deleterious influence on the fab-ricability [Ish70]. According to the metallurgical introduction to this chapter, the presence of an ele-ment such as Si is expected to increase the kinetics of precipitation. Thus, as WATANABE has discovered [Wat74], whereas less than 3 h were needed at 500°C to enable the ternary Ti-Nb-Si alloy to reach the J_c(70 kOe) of 6.8×10^4 A cm^{-2} referred to above, it took as long as 20 h at that temperature for the binary alloy to attain 5×10^3 A cm^{-2}.

9.5 Ga ADDITIONS TO Ti-Nb

In a pair of overview papers in which the effects of numerous simple-metal additions to Ti-Nb were dis-cussed, members of the Erlangen-University group (U. ZWICKER *et al*) have considered the influence of 2 at.% Ga on the transition temperature of Ti-35Nb [Bac68] and the upper critical field of Ti-49Nb [Hel71^a].

In measurements on as-cast-plus-cold-deformed wire, the transition temperature of Ti-35.1Nb-2.5Ga was 7.0_5 K, apparently a little higher than the 6.7_5 K of the Ti-33Nb binary "control" (an attempt having

been made to preserve a constant electron/atom ratio of 4.20 throughout the alloying process). It will be recalled, however, that on account of the variability of the microstructure that accompanies this particular e/a (cf. the shaded region in Fig.7-3) some degree of uncertainty is associated with the transition tempera-ture of an alloy such as Ti-33Nb (i.e. 20.3 at.% Nb) when specified simply by composition.

Resistive measurements of the upper critical field of cold-worked (97%) and aged (10h/400°C/quenched) 1-mm$^\phi$ wires of Ti$_{65}$-Nb$_{33}$-Ga$_2$ yielded a value of 103.5 kOe, somewhat lower than the 114.0 kOe of the binary Ti-Nb$_{33}$ control after a similar heat treatment [Hel71^a].

9.6 Y ADDITIONS TO Ti-Nb

KOCH and SCARBROUGH [Koc76] have studied the in-fluence of 2 at.% Y on the J_c's of the alloys Ti-39.3Nb and Ti-55Nb. Yttrium is practically insoluble in Ti-Nb, coming out as a ductile precipitate which de-forms into stringers during wire drawing. It may contribute to flux pinning in its own right, or *directly* and *indirectly* (i.e. as a getter of O and N if they are present) alter the ω-phase and α-phase precipitation kinetics.

9.6.1 Y Additions to Ti-39Nb

The addition of Y to Ti-39Nb was found to accel-erate the ω-phase and α-phase precipitation kinetics, such that precipitation would appear sooner at a given temperature, or coarser after a given reaction time; the coarsening of precipitate morphology degraded J_c. Electron microscopy failed to reveal the presence of Y in association with the precipitates themselves. It was believed at the time (erroneously, in the light of the discussions of the previous chapter) to be a get-ter of O and N and thereby to exert an indirect influ-ence on the precipitation kinetics.

9.6.2 Y Additions to Ti-55Nb

Yttrium had a beneficial influence on the criti-cal current density of suitably processed Ti-55Nb. The 60-kOe J_c of 94%-cold-worked and 6h/350°C-aged Ti-55Nb increased from 8×10^3 A cm^{-2} to 1.7×10^4 A cm^{-2}

with the addition of 2 at.% Y. Since electron microscopy had failed to reveal any change in dislocation-
cell size, dislocation density, or cell-wall width,
it was believed that the Y was in this case contributing to the flux pinning in its own right. Unfortunately at the higher reduction ratios encountered in
practice the improvement in J_c diminished. For further comment on the effects of insoluble, internally
oxidizable, additions to Ti-Nb alloys the reader is
referred to previous discussions (Sect.8.7) of the
systems:

$$\text{Ti-Nb-}\begin{pmatrix} \text{Y} \\ \text{Gd} \\ \text{Th} \end{pmatrix}\text{-O}$$

$$\text{and}\quad \text{Ti-Nb-}\begin{pmatrix} \text{Er} \\ \text{Sc} \end{pmatrix}\text{-O}.$$

9.7 Ag ADDITIONS TO Ti-Nb

The influences of small amounts of Ag on the T_c,
H_{c2}, and 100-kOe J_c of Ti-Nb alloys has been studied
by ZWICKER and members of the Erlangen group.

The transition temperature of Ti_{76}-Nb_{23}-Ag_1, in
the form of cold-deformed wire, was found by BACHMANN
et al [Bac68] to be 7.6_5 K. Since the alloy selected
lay within a metallurgically complicated composition
range (cf. Sect.9.5) in which various amounts of
martensite, deformation-martensite, and ω-phase, may
coexist depending on the details of the thermomechanical processing, a useful comparison of its T_c with
that of an unalloyed control alloy is not possible.

Upper critical field data, such as those presen-
ted by HELLER et al [Heℓ71[a]] for Ti_{65}-Nb_{33}-Ag_2 and
ZWICKER et al [Zwi70] for Ti-Nb_{33}-$Ag_{1,2}$, indicated
that Ag lowered the upper critical field of Ti-Nb_{33}.
The reduction rates noted by those authors were
1.5 kOe per at.% Ag and 4 kOe per at.% Ag, respectively.

According to ZWICKER et al [Zwi70], small additions of Ag also had a deleterious effect on the
100-kOe critical current density, which dropped from
2.5×10^3 A cm^{-2} to 2.0×10^3 A cm^{-2} with 1 at.% Ag and
then to 0.5×10^3 A cm^{-2} with 2 at.% Ag.

9.8 In ADDITIONS TO Ti-Nb

The influences of 2 at.% of In on the critical
temperature of cold-deformed Ti-Nb_{22} wire and on the

critical field of cold-deformed (97%) and aged
(10h/400°C) Ti-Nb_{33} wire have been studied by BACHMANN
et al [Bac68] and HELLER et al [Heℓ71[a]], respectively.
The critical temperature of Ti_{76}-Nb_{22}-In_2 was 6.9 K,
again a little higher than the 6.7_5 K of the Ti-Nb_{20}
(i.e. Ti-33Nb) reference sample (cf. Sects.9.5 and
9.7). The upper critical field of Ti_{65}-Nb_{33}-In_2, at
105.0 kOe, was also lower than the 114.0 kOe claimed
for the binary Ti-Nb_{33} control.

9.9 Sn ADDITIONS TO Ti-Nb

According to BACHMANN et al [Bac68], the critical
temperature of cold-worked Ti_{78}-Nb_{20}-Sn_2 was 6.6 K;
that of the binary Ti-Nb_{20} control was 6.7_5 K. The
upper critical field of cold-worked (97%) and aged
(10h/400°C/quenched) Ti_{65}-Nb_{33}-Sn_2 1-mm$^\phi$ wire, according to HELLER et al [Heℓ71[a]], was 95.8 kOe, much lower
than the 114.0 kOe determined by them for the binary
Ti-Nb_{33} control.

The influences of 1-h aging temperature on the
critical current densities of ∿99.9%-cold-worked
Ti-31.5Nb-0.78Sn and Ti-51.8Nb-0.79Sn wires were compared by RASSMAN and ILLGEN [Ras72[b]] with those of
the corresponding binary Ti-Nb alloys. As usual the
Ti-Nb alloys exhibited maximal critical current densities in response to the 400°C aging temperature. Of
the two higher-Nb alloys, that containing 0.79 wt.% Sn
possessed, with 3.4×10^4 A cm^{-2}, a slightly higher
"optimized" 50-kOe J_c than the 2.6×10^4 A cm^{-2} of the
binary control. The critical current densities of the
binary and ternary Ti∿30Nb-base alloys varied errati-
cally with increase of the 1-h aging temperature; in
some temperature regimes $J_{c,binary}$ was above, in
others below, $J_{c,Ti-Nb-Sn}$.

9.10 ADDITIONS OF Sb, Au, Pb, U, AND PAIRS OF SIMPLE METALS TO Ti-Nb

As mentioned in Sect.9.2.1, studies of the influence on the T_c of Ti-Nb of various simple metals,
either singly or in pairs, have been made by BACHMANN
et al [Bac68]. The elements Sb, Au, Pb, U, and various pairs of simple metals, as well as those mentioned
in the preceding paragraphs, were added to Ti-Nb in
such a way as to preserve (according to the authors)

a constant electron/atom ratio of 4.20. The superconducting transition temperatures of the cold-deformed
wires are listed in Table 9-4.

TABLE 9-4 SUPERCONDUCTING TRANSITION TEMPERATURES
(T_c) OF TITANIUM-NIOBIUM-SIMPLE-METAL
ALLOYS[†] -- Bachmann *et al.* [Bac68]

Niobium Concentration		Solute Species (Concentration, at.%)		T_c, K
wt.%	at.%			
33.0	20.3	Control		6.75
15.5	10.0	Sb	10.0	4.3
25.6	16.0		4.0	5.9
29.1	18.0		2.0	6.65
35.8	23.0	Au	1.0	7.65
30.9	20.0	Pb	2.0	5.1 - 5.7
15.0	10.0	U	5.0	3.85
32.4	20.0	Al(1.0) + Sb(1.0)		7.0
32.1	20.0	Ga(1.0) + Sb(1.0)		5.8
31.9	20.0	In(1.0) + Sb(1.0)		7.0
14.6	10.0	Sn(2.5) + U (5.0)		3.75

† Condition: Swaged 92%, 30 min/950°C/quenched,
plus 96.0 - 98.7% final cold work.

ALLOY GROUP 2: TITANIUM-NIOBIUM-COPPER TERNARY ALLOYS

9.11 SUPERCONDUCTIVITY IN Ti-Nb-Cu ALLOYS

The influence of Cu on the microstructural properties of deformed and heat-treated Ti-Nb alloys and on
their related critical current densities have been the
subjects of extensive investigation. Major studies
were conducted by the Erlangen group headed by ZWICKER
[Zwi70][Loh71, Loh73][Heℓ71[a], Heℓ72], and by a team of
workers including WILLBRAND, ARNDT, EBELING, and MOHS
at the Krupp Forschungsinstitut, Essen [Arn74][Wiℓ75[a]],
who investigated and intercompared the effects of both
Cu and Ge additions on the superconducting properties
of Ti-Nb. The most recent investigations under this
heading are those of WADA *et al* [Wad80], who measured
the responses to thermomechanical processing of the
superconducting properties, T_c, H_{c2}, and J_c, of five
ternary Ti-Nb-Cu alloys having Nb contents of 5, 10,
15, and 20 at.%, and Cu concentrations in each case of

about 5 at.%. Finally, within the context of alloys
involving Ti, Nb, and Cu, mention must be made of an
interesting study of an entirely different nature to
those referred to above, viz an unusual series of
investigations by OLIVEI [Oℓi74], whose goal was the
development of an *in situ* process for the preparation
directly from the melt of a composite conductor consisting of Ti-Nb filaments imbedded in a Cu-base
matrix.

9.12 TRANSITION TEMPERATURES OF Ti-Nb-Cu ALLOYS

9.12.1 Low-Concentration Ti-Nb-Cu Alloys

The influences of small additions of Cu (up to
about 5 at.%) in association with thermomechanical

processing on the T_c's of alloys based on Ti-Nb(5, 10, 15, 20, and 33 at.%) have been studied. The Nb-lean members of this group have been the subjects of rather extensive investigations by WADA et al [Wad80]. In preparation for measurement, the alloys:

$$Ti_{89.8}-Nb_5-Cu_{5.2},$$
$$Ti_{85.0}-Nb_{10}-Cu_{5.0},$$
$$Ti_{80.3}-Nb_{15}-Cu_{4.7},$$
$$\text{and} \quad Ti_{75.6}-Nb_{20}-Cu_{4.4},$$

were homogenized 2h/1200°C plus 5h/1000°C and aged for various times and temperatures (occasionally after cold working or hot working at 500 or 600°C). T_c was measured as functions of: (a) 48-h aging temperature to 600°C during which T_c maximized at 500°C; (b) aging time to 240 h at 500°C, possibly one of the more interesting results of which was that the T_c's of the Ti-Nb$_{15}$-Cu and Ti-Nb$_{20}$-Cu alloys reached 8.5 K, the transition temperature of binary Ti-Nb with more than $\sim$35 at.% Nb (cf. Fig.7-3); (c) aging time to 21 h at 500°C after hot working at 600°C.

As part of an overview of the influence of alloying and microstructures on a wide range of Ti-Nb-base alloys, BACHMANN et al [Bac68] have measured the transition temperatures of eleven cast, swaged, and drawn Ti-Nb-SM ternary alloy wires (cf. Sects.9.3, 9.5, and 9.7 through 9.10). Ti$_{76}$-Nb$_{23}$-Cu$_1$, a member of this group, was found to have a T_c of 7.6_5 K, slightly higher than the 6.7_5 K of Ti-Nb$_{20.2}$, the reference sample.

9.12.2 Intermediate-Concentration Ti-Nb-Cu Alloys

In a study related to that just described, ZWICKER et al [Zwi70] have measured the T_c's of Ti-Nb$_{33}$-Cu$_{0.5,1,3}$ wires (in what appears to be the as-worked plus 33h/400°C-aged condition) finding that T_c $versus$ Cu-concentration passed through a maximum at 1 at.% Cu. The maximum T_c attained was 8.8 K, lower than the 9.2_5 K recorded for the Ti-Nb$_{33}$ master alloy.

LÖHBERG [Loh71] has investigated the effects of aging at 400°C for times of up to 150 h on the T_c's of: (a) cast, homogenized (3h/1000°C), and quenched and (b) cast-plus-deformed (97.2%) Ti-Nb$_{33}$-Cu$_{0,3,5}$ alloys. The T_c results were expressed as functions of aging time, but could equally well have been re-cast in the form of isochronal curves of T_c $versus$ composition

(0, 3, 5 at.% Cu) for the two starting conditions referred to above and for various aging times. The following are some typical results:

(a) Homogenized (3h/1000°C) and Quenched Alloys. For the two ternary alloys Ti$_{64}$-Nb$_{33}$-Cu$_3$ and Ti$_{62}$-Nb$_{33}$-Cu$_5$, T_c first passed through a minimum during aging for a few minutes at 400°C after which it steadily increased with aging time at the rate of about 0.5 K per 100 h. The width of the transition was about 0.2 K in the early stages, increasing to 0.4-0.6 K for annealing times of more than $\frac{1}{2}$ h. The addition of 5 at.% Cu decreased T_c by 1.1 K in the as-quenched alloys and by <0.1_5 K in those which had been annealed for 100 h or more.

(b) Cast and Deformed (97.2%) Alloys. Aging experiments were also carried out on Ti-Nb$_{33}$-Cu$_{0,3,5}$ alloys which had experienced 97.2% deformation from the cast condition. In the as-deformed alloys T_c decreased monotonically with Cu content at the rate of about 0.2 K per at.% Cu. For aging times longer than about 5 min, T_c $versus$ composition passed through a minimum such that for aging times longer than 1 h, the T_c of Ti$_{62}$-Nb$_{33}$-Cu$_5$ was always higher than those of the other two alloys (Ti-Nb$_{33}$ and Ti$_{64}$-Nb$_{33}$-Cu$_3$). The T_c's of Ti-Nb$_{33}$-Cu$_{3,5}$ increased monotonically with aging time, presumably due to a partitioning of Nb to the β-phase; for example that of Ti$_{62}$-Nb$_{33}$-Cu$_5$ increased from 8.0 to 9.9 K during the 100 h aging treatment. LÖHBERG discussed in considerable detail the precipitation and solute redistribution effects which were thought to be responsible for the observed changes in T_c.

9.13 UPPER CRITICAL FIELDS OF Ti-Nb-Cu ALLOYS

In the investigations by WADA et al [Wad80] introduced above, the upper critical fields of four low-Nb-concentration alloys (i.e. 5, 10, 15, and 20 at.% Nb) were measured in the conditions: (a) as-quenched, (b) 600°C hot worked, and (c) 600°C hot worked plus 2h/500°C aged. Under the first two heat treatment conditions, H_{c2} for each of the alloys except Ti-Nb$_{20}$-Cu was predictably low (i.e. $\overset{\sim}{<}$50 kOe); on the other hand, as a result of the metallurgical changes brought about by hot working followed by aging, H_{c2} took on uniformly "high" values (in particular, 91$\sim$103 kOe) throughout the entire Nb-concentration range of 5 through 20 at.%. The reason for this again had to do with Nb-enrichment of the β-phase component of the multiphase system in accordance with the suggestion offered earlier by COTTON *et al* [Cot74] in connection with their studies of the transition temperatures of Ti-Nb-Al, Sect.9.3.1.

The critical fields of Ti-Nb$_{33}$-Cu$_{0.5,1,3}$ have been considered briefly by ZWICKER *et al* [Zwi70] and in greater detail by HELLER *et al* [Hel71[a]], who also compared the results with those of a comprehensive study of a matrix of Ti-Nb-Ge and numerous Ti$_{65}$-Nb$_{33}$-SM$_2$ alloys. The Ti-Nb-Cu component of this survey stemmed from the work of LÖHBERG [Loh71], who studied under this heading the metallurgical and superconductive properties of nine alloys defined by Ti-Nb$_{33,39,44}$-Cu$_{0,3,5}$. Critical fields of homogenized (3h/1000°C) and quenched Ti-Nb$_{33}$-Cu$_{3,5}$ were studied as functions of aging time up to 200 h at 400°C, while that of Ti$_{64}$-Nb$_{33}$-Cu$_3$ was studied as a function of aging time to 15 h at 700°C. The critical fields of Ti-Nb$_{33,39,44}$-Cu$_{0,3,5}$, deformed 97.2% from the as-cast condition, were studied as functions of aging for times up to 200 h at 400°C. The results are briefly as follows:

(a) Homogenized (3h/1000°C) and Quenched Alloys. The critical fields of the as-quenched alloys are reduced by Cu additions at the rate of about 4 kOe per at.% Cu, paralleling the behavior of the critical temperature. As was the case with the response of T_c to aging, the H_r of Ti$_{64}$-Nb$_{33}$-Cu$_3$ passed through a minimum before increasing to a value higher than the as-quenched starting critical field. For example, the values of H_r (quenched), H_r (minimum), and H_r (100h/400°C) for that alloy were, respectively: 102 kOe (10 kOe below that measured upon Ti-Nb$_{33}$ itself), 97 kOe, and 111 kOe. In those studies it turned out that the critical fields of the Cu-containing alloys were generally lower than those of binary Ti-Nb$_{33}$ throughout the aging.

(b) Cast and Deformed (97.2%) Alloys. Upper critical fields as functions of aging time at 400°C were determined for all six alloys defined by Ti-Nb$_{33,39,44}$-Cu$_{3,5}$ and compared with the Ti-Nb$_{33,39,44}$ control values. The deformed alloys had higher critical fields than those of the corresponding homogenized samples. In all as-deformed alloys, H_r decreased with Cu content, the rate being greatest in the case of the Ti-Nb$_{44}$-Cu alloys whose H_r's assumed the values

115.$_5$, 103, and 97 kOe at 4.2 K for Cu contents of 0, 3 and 5 at.%, respectively. Although the H_r of Ti-Nb changed rather slightly with aging time, those of the Ti-Nb-Cu alloys increased rapidly with aging time at 400°C, reflecting the increase in reaction kinetics that occasionally accompanies third-element additions to Ti-Nb. In Ti$_{62}$-Nb$_{33}$-Cu$_5$, for example, H_r increased at the rate of 1 kOe per h. With one exception, the H_r's always remained below those of the binary alloys. The microstructural changes bearing on these effects have been discussed in detail by LÖHBERG [Loh71].

9.14 CRITICAL CURRENT DENSITIES OF Ti-Nb-Cu ALLOYS

Preliminary studies of critical current density in Ti-Nb-Cu alloys have been reported by ZWICKER *et al* [Zwi70] who compared the current-carrying capacities of Ti-Nb$_{33}$-Cu$_{0.5,1,3}$ with those of Ti-Nb$_{33}$-Al and Ti-Nb$_{33}$-Au. Of the elements Al, Au, and Cu, only the latter was found to increase the 100-kOe J_c. The critical current density of Ti$_{66}$-Nb$_{33}$-Cu$_1$ was also measured at 100 kOe as a function of aging for times of up to about 85 h at 400°C following cold deformation. As a result, it was concluded that additions of Cu (and, as will be mentioned below, Ge) could lead to an increase of the critical current density of technically interesting Ti-Nb alloys. Since Cu is expected to contaminate the outer layers of Ti-Nb filaments during the processing of Cu-clad multifilamentary composite conductors, it was important to know that its presence is not necessarily detrimental to the superconducting properties.

9.14.1 Critical Current Densities of Research Alloys

The five alloys:

Ti$_{79.7}$-Nb$_{18.5}$-Cu$_{1.8}$,
Ti$_{78.5}$-Nb$_{21.3}$-Cu$_{0.2}$,
Ti$_{56.9}$-Nb$_{32.1}$-Cu$_{11.0}$,
Ti$_{63.0}$-Nb$_{35.2}$-Cu$_{1.8}$,
and Ti$_{62.1}$-Nb$_{36.1}$-Cu$_{1.8}$,

some of which were subjected to metallographic, hardness, and superconductive-property testing, were prepared by RASSMANN and ILLGEN [Ras73] during a study of the relationship between structure and J_c in alloys of Ti-Nb with β-stabilizing elements such as Ta, V, and Mo (yielding "β-isomorphous" structures) and Si and Cu (the "β-eutectoid" formers, [Mon1.2]). In measurements of J_c as a function of 1-h aging temperature it was noted that, although the addition of 0.2 at.% Cu to Ti-Nb$_{21.3}$ produced only small increases or decreases in J_c (depending on aging temperature), the addition of 1.8 at.% Cu to Ti-Nb$_{35.2}$ resulted in significant increases in the 45-kOe J_c's of cold-worked (~99%) and appropriately aged wires, viz:

(Ti-Nb)   (Ti-Nb-Cu)

2.5×10^4 to 6.0×10^4 A cm^{-2}, at 1h/400°C

and 7.5×10^3 to 6.9×10^4 A cm^{-2}, at 1h/500°C.

As reported in the doctoral thesis of LÖHBERG [Loh71] and in subsequent papers, extensive studies of the influence of Cu additions on the critical current densities of Ti-Nb$_{33}$, Ti-Nb$_{39}$, and Ti-Nb$_{44}$, have been undertaken at the University of Erlangen-Nürnberg under the direction of U. ZWICKER (e.g. [Loh73]). LÖHBERG has studied the microstructures and critical current densities of homogenized (3h/1000°C)-and-quenched Ti-Nb$_{33}$ and Ti$_{66}$-Nb$_{33}$-Cu$_3$ after aging them for 0, 10, 20, 40, and 80 h at 400°C and $\frac{1}{2}$, 1, 2, 6, and 15 min at 700°C. Also investigated were the J_c field dependences of 97.2%-cold-worked binary- and ternary-alloy wires after heat treatment at 400°C for the times listed in Table 9-5. As pointed out by LÖHBERG et al [Loh73], the various kinds of heat treatment that had been administered were designed to produce the following specific classes of inhomogeneity whose effects on J_c could, it was hoped, be independently evaluated. These were:

(i) *Vacancies* -- produced by homogenization of the binary alloys for 3h/1000°C followed by water quenching. It was thought that vacancy aglomeration might occur during subsequent annealing at 700°C (in the β-phase field).

(ii) *Dislocation Networks* -- produced by following the homogenization heat treatment with a 97% area-reduction by swaging.

(iii) *Intermetallic Compound Precipitates* -- produced by annealing the homogenized ternary alloys at 700°C during which small particles of Ti$_2$Cu were precipitated.

TABLE 9-5   THE AGING OF Ti-Nb-Cu ALLOYS -- An Index of Aging Studies Reported by Löhberg [Loh71]

Alloy	Aging Times at 400°C, h
Ti-Nb$_{33}$	4, 16, 32, 64, 250
Ti-Nb$_{39}$	20, 40, 250
Ti-Nb$_{44}$	40, 250
Ti$_{64}$-Nb$_{33}$-Cu$_3$	$\frac{1}{4}$, 1, 12, 20, 40, 85
Ti$_{62}$-Nb$_{33}$-Cu$_5$	$\frac{1}{4}$, 8, 20, 40
Ti$_{58}$-Nb$_{39}$-Cu$_3$	20, 40
Ti$_{56}$-Nb$_{39}$-Cu$_5$	40
Ti$_{53}$-Nb$_{44}$-Cu$_3$	40

(iv) *ω-Phase*[†] -- formed by heat treating the homogenized speciments at 400°C.

(v) *Dislocation Networks, ω-Phase,*[†] *and α-Phase Precipitation* -- formed by heat treating at 400°C alloys which had received sufficient cold deformation (cf. Sect.9.1).

For both the homogenized//quenched//aged alloys, and the deformed//aged alloys, the results were summarized in the form of curves of J_c(50 and 100 kOe) *versus* aging time at 400°C; some representative data are presented in Table 9-6. It was concluded that the J_c of Ti-Nb was increased by the addition of Cu. The results of the Erlangen studies were discussed in terms of: *(a)* flux pinning by ω-phase,[†] α-phase, and Ti$_2$Cu precipitates; *(b)* the influence of dissolved Cu on precipitation kinetics, and consequently on the size of the precipitates; *(c)* solute partitioning between phases, and the consequent Nb-enrichment of the β-matrix.

WADA et al [Wad80] have measured the 4.2-K critical current densities, in fields of up to 90 kOe, of

---

† The occurrence of ω-phase is questionable; for a relevant comment on this subject see Sect.7.20.2(a).

TABLE 9-6 MAXIMUM CRITICAL CURRENT DENSITIES (J_c)
ACHIEVED DURING AGING FOR TIMES OF UP TO
600h AT 400°C -- Löhberg [Loh71].

Alloy	t_{max}[†] h	Applied Field, kOe	J_c, 4.2K[††] 10^4 A cm^{-2}
(a) Homogenized-and-Quenched Alloys			
Ti-Nb$_{33}$	200	50	4.6
	50	100	0.063
Ti$_{64}$-Nb$_{33}$-Cu$_3$	60	50	4.5
	30	100	1.1
(b) Cast-plus-Deformed (97.2%) Alloys			
Ti-Nb$_{33}$	200	50	4.6
	14	100	0.42
Ti$_{65}$-Nb$_{33}$-Cu$_2$	83	50	4.4
	12	100	0.90
Ti$_{62}$-Nb$_{33}$-Cu$_5$	20	50	6.0
	20	100	1.1

[†] Aging time at 400°C to achieve maximum J_c.

[††] J_c criterion: 100 µV across 3.5 cm [Hel71].

the four low-Nb-concentration alloys whose heat treatments (three conditions) and upper critical fields
were referred to in Sect.9.13. Useful critical currents were obtained in all alloys with 10 or more
at.% Nb after hot working; and the application of the
final 2h/500°C aging resulted in a pronounced improvement to J_c. The highest high-field J_c was achieved
by the double-heat-treated Ti-Nb$_{20}$-base alloy. At
50-kOe the critical current density of
Ti$_{80.3}$-Nb$_{15}$-Cu$_{4.7}$ was 1.5x10^4 A cm^{-2}; it is anticipated, on the basis of its superior high-field perfor-
mance, that the 50-kOe J_c of Ti$_{75.6}$-Nb$_{20}$-Cu$_{4.4}$ would
be higher. Although the current-carrying capacities
were low by comparison with those listed in Table 9-5,
let alone contemporary practical standards, it was
anticipated that the ternary alloys would respond
favorably to attempts to "optimize" their properties
by suitable adjustment of the fabrication and heat-

treatment variables. It was suggested that if the
pre-optimization promise of the ternary alloys did in
fact become realized, further consideration would then
need to be given to the inclusion of small amounts of
Cu in the formulations of technical superconducting
alloys.

9.14.2 Critical Current Densities of Technical Alloys

ARNDT and EBELING [Arn74] and WILLBRAND et al
[Wil75[a]] have studied the influence of thermomechanical process variation on the critical current densities of Cu- and Ge-doped alloys based on Ti-50Nb,
interpreting their results in terms of subband-struc-
tures and associated precipitation in the spirit of
the early and contemporary work of HILLMANN et al (cf.
for example [Hil75] as discussed in Chapter 7 and, in
particular, Sect.7.28). A master alloy containing
about 50 wt.% Nb was prepared by arc melting. With a
portion being retained as a control, this was used as
the basis for a series of Ti-Nb-Cu alloys having the
compositions listed in Table 9-7. Alloy wires prepared by sequential applications of extrusion, swaging,
Cu-cladding, and wire drawing, experienced overall
reductions in area of more than 99.98%. Heat treatments under flowing Ar were carried out at temperatures of 200, 300, 390 and 500°C, of which that at
390°C was optimal. Subsequent studies compared the
effects of up to 6 intermediate heat treatments for
3 h at that temperature. During the final processing
of Cu/Ti-Nb monofilaments, Cu-clad composite rods of
core diameter 3 mm, were reduced to wire of core diameter about 0.13 mm by drawing, interspersed by heat
treatments of 3h/390°C administered at diameters of

TABLE 9-7 COMPOSITIONS OF THE Ti-Nb-Cu ALLOYS (wt.%)
STUDIED BY Willbrand et al.[Arn74][Wil75[a]]

Symbol	Ti	Nb	Cu	C + N + O
Cu-05	53.2	46.1	0.48	0.121
Cu-1	51.7	47.1	0.90	0.140
Cu-2	51.2	46.8	1.85	0.180
Cu-3	51.2	45.7	2.90	0.165
Cu-4	49.4	46.3	4.05	0.179

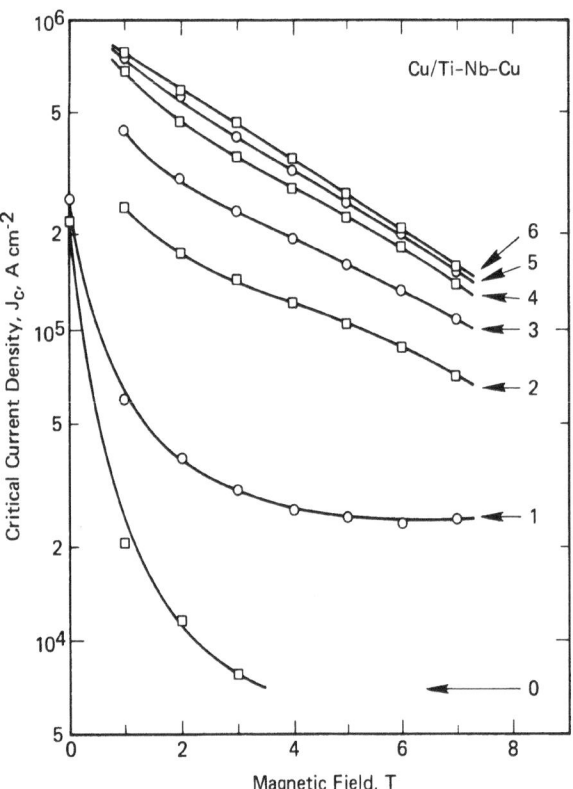

FIGURE 9-3. Critical current density (criterion: 5 μV cm^{-1}) *versus* applied field for 0.25 mm$^\phi$ Cu/Ti-Nb-Ge wire (alloy Ge-1 of Table 9-7) in seven final thermomechanical process conditions. The numbers against the curves correspond to the number of intermediate heat treatments administered according to Table 9-8 — after WILLBRAND *et al* [Wiℓ75^a].

TABLE 9-8 FINAL-STAGE THERMOMECHANICAL PROCESSING OF THE REPRESENTATIVE Cu/Ti-Nb-Cu MONOFILAMENT Cu-2 (cf. Table 9-7) WHOSE PERFORMANCE IS DEPICTED IN Fig. 9-3 -- Willbrand *et al.* [Arn74][Wiℓ75^a]

Number of Heat Treatments	Outside Diameter (mm) at which Treatments of 3h/390°C were Administered					
	3.0†	1.5	1.0	0.82	0.68	0.5
0						
1	H					
2	H	H				
3	H	H	H			
4	H	H	H	H		
5	H	H	H	H	H	
6	H	H	H	H	H	H

† Administered <u>before</u> applying the Cu sheath.

TABLE 9-9 CRITICAL CURRENT DENSITIES (J$_c$) OF SOME REPRESENTATIVE Ti-Nb-Cu ALLOYS -- Willbrand *et al.* [Arn74][Wiℓ75^a]

Alloy	Heat Treatment Code†	J$_c$ (4.2K) at the Fields Specified,†† 10^5 A cm^{-2}	
		5T	7T
Ti-50Nb	F	0.9$_8$	0.6$_3$
	6	1.5$_8$	1.0$_2$
Ti-46.1Nb-0.48Cu	6	2.1$_4$	1.2$_3$
Ti-46.8Nb-1.85Cu	1	0.2$_5$	0.2$_5$
	6	2.6$_3$	1.5$_5$

† F = Final heat treatment of 3h/390°C.

  1 = Single intermediate heat treatment of 3h/390°C.

  6 = 6-fold intermediate heat treatment of 3h/390°C.

†† J$_c$ criterion: 5 μV across 1 cm.

3.0, 1.5, 1.0, 0.82, 0.68 and 0.5 mm. Critical current densities were then measured at 4.2 K in fields of up to 70 kOe. Results for the Cu-clad monofilaments were summarized in [Arn74] where they were intercompared with those of a similar study of Ti-Nb-Ge to be considered below. A wealth of detailed information is to be found in [Wiℓ75^a] which also deals with the properties of a 61-filament prototype conductor. Some typical results only can be given here. Fig.9-3 presents a set of J$_c$ *versus* applied field characteristics for the alloy referred to as Cu-2 in Table 9-7, final-processed according to the schedules of Table 9-8. Some selected data are also given in Table 9-9 wherein it can be seen, by comparison with Table 9-6, that the current densities were generally appreciably higher than those obtained in LÖHBERG's laboratory-scale study. For example, using the 50-kOe J$_c$'s of the cold-worked-and-final-heat-treated binary alloys as bases for comparison, 9.8x10^4 A cm^{-2} (Table 9-9) is to be compared with 4.6x10^4 A cm^{-2} (Table 9-6). These current densities reflect, presumably, the differing amounts of prior cold work that had been administered in each of the two examples -- 99.98% as compared with 97.2%. Also, as indicated in Table 9-9, multiple thermomechanical processing (intermediate heat treatment) resulted in a 60% improvement to the 50-kOe critical current density. The results of Table 9-6, judged in the light of an imagined *fixed* heat treatment time, are consistent with those of Table 9-9 which shows J_c(6-fold heat treatment) increasing with the addition of Cu. Taken together, these are important results which suggest that improvements in the critical current densities of nonoptimized wires, occurring as a result of precipitation, tend to be reflected in those of the optimized technical conductors. Thus ZWICKER *et al* [Zwi70] and WILLBRAND *et al* [Arn74][Wil75^a] agree that small additions of Cu are able to improve (*via* a mechanism to be discussed below)

the current-carrying capacities of Ti-Nb alloys in strong magnetic fields.

Returning to the technical superconductors, we enquire into the beneficial effects of successive stages of cold deformation which accumulate during the processing of a 61-filament conductor. In so doing it is noted that at 50-kOe the critical current density of the multifilamentary wire was a factor of 2.5 higher than that of the best monofilament. Now the 50-kOe current-carrying capacity of Cu-2 (6 heat treatments), at 2.62×10^5 A cm^{-2}, is a factor of 1.7 higher than that of the corresponding Ti-50Nb (1.55×10^5 A cm^{-2}); if these ratios are maintained, it would follow that multifilamentary Ti-50Nb-2Cu would have a 50-kOe J_c of $1.55 \times 1.7 \times 2.5 = 6.6 \times 10^5$ A cm^{-2}. Further discussion is postponed to the concluding sections wherein the properties of Ti-Nb, Ti-Nb-Cu, and Ti-Nb-Ge are intercompared.

ALLOY GROUP 3: TITANIUM-NIOBIUM-GERMANIUM TERNARY ALLOYS

9.15 TRANSITION TEMPERATURES OF Ti-Nb-Ge ALLOYS

In a survey of the superconducting transition temperatures of a series of Ti-Nb-SM ($\sim$2 at.%) alloys with Nb contents selected so as to preserve a constant value of 4.2 for the electron/atom ratio, BACHMANN *et al* [Bac68] have placed the T_c of cast plus cold-drawn Ti$_{78}$-Nb$_{20}$-Ge$_2$ at 7.3 K. In a related study by the Erlangen group, ZWICKER *et al* [Zwi70] have reported on the influence of 400°C-aging for times of up to 34 h on the T_c of Ti-Nb$_{33.3}$-Ge$_{0.5,1,1.5,2}$, during which small increases in T_c were frequently noted. The transition temperatures of the cold-worked and heat-treated (34h/400°C) members of both the "[Ti] = const." and "[Nb] = const." families, see below, decreased by about 0.5 K after the addition of the first 0.5 at.% Ge and continued to decrease slightly with further additions of that solute. More details were presented by HELLER [Hel71] who studied 18 ternary Ti-Nb-Ge alloys conforming to the formulae:

Ti$_{bal.}$-Nb$_{33.3}$-Ge$_{0.5-5}$, a "[Nb] = const." series,

Ti$_{66.7}$-Nb$_{bal.}$-Ge$_{0.5-5}$, a "[Ti] = const." series,

Ti-Nb$_{39}$-Ge$_{2,4}$, and Ti-Nb$_{44}$-Ge$_{2,4}$.

In a comprehensive report of his investigations, HELLER [Hel71] presented his results in the following formats:

(a) With regard to the [Nb] = const. and [Ti] = const. series of Ti-Nb$_{\sim 33}$-Ge$_{0.5-5.0}$ alloys, in the *(i)* cast plus 97% cold worked and *(ii)* homogenized plus 97% cold worked conditions, T_c was plotted *versus* Ge content.

(b) T_c *versus* aging times of up to 140 h at 400°C for the [Ti] = const. series of Ti-Nb$_{33.3}$-Ge$_{0.5-4}$ alloys.

(c) J_c *versus* 16-h annealing temperature for 97% cold-worked Ti-Nb$_{33.3}$-Ge$_{2,4}$ wires and a binary Ti-Nb$_{33}$ control.

The alloying and aging results were generally discussed in terms of the partitioning of Nb and Ge between the host alloy and precipitated phases. For example: *(a)* When Ge was added to a homogenized-plus-cold-deformed alloy, T_c first of all decreased with *small* additions of Ge before increasing again as the Ge content was further raised. The decrease was attributed to the dissolving of Ge in the β-phase, and the subsequent increase, to an increase in the Nb content of it as a result of the loss of Ti to the precipitate, Ti_5Ge_3. (b) The behaviors of homogenized-and-aged samples were discussed in terms of the dissolving of Ge in the β-phase at 1080°C followed by precipitation of Ti-rich phases at 400°C, leading to Nb-enrichment of the β-matrix (cf. Sect.9.3.1) and a T_c higher than that of the corresponding binary alloy. Then, after aging 8h/400°C it was thought that further Nb enrichment of the matrix, and a concomitant increase of T_c, stemmed from further precipitation of Ti_5Ge_3 as well as (Ti-rich) ω- and α-phases, whose formation also rejects Nb to the matrix.

KITADA [Kit73], as part of an investigation of the influence of heat treatment on the J_c(4.2 K) field dependences of Ti-Nb-Ge alloys, has measured the T_c of Ti_{60}-Nb_{39}-Ge_1 (condition unstated) finding a value of 9.6 K, suggestive in their case, of the domination of the measured T_c by the presence of Nb-rich zones.

9.16 UPPER CRITICAL FIELDS OF Ti-Nb-Ge ALLOYS

The upper critical fields of numerous Ti-Nb-Ge alloys have been studied in detail by the Erlangen group. A preliminary survey of the results, in particular those for cold-worked-plus-aged (34h/400°C) members of the Ti-$Nb_{\sim33}$-Ge, [Ti] = const. and [Nb] = const. series, has been presented in a review paper by ZWICKER et al [Zwi70]. Details of the critical field behaviors, as functions of thermomechanical processing, of the three [Nb] = const. sequences: Ti-$Nb_{33.3}$-Ge, Ti-Nb_{39}-Ge, and Ti-Nb_{44}-Ge, were considered fully in HELLER's thesis [Hel71], and again in an article by HELLER, LÖHBERG and ZWICKER [Hel71[a]], where comparisons were made between the critical field properties of Ti-Nb-Ge and Ti-Nb-Cu. Samples which had been cast into 6-mm$^\phi$ rods were cold reduced by swaging to 1-mm$^\phi$ wires. The cold work, which represented an area reduction of 97%, yielded the usual fibrous subband structure favorable to good current-carrying capacity. Critical fields were measured resistively at 4.2 K using a current density of 1.3 A cm^{-2} and a 1 μV transition criterion. The annealing conditions selected for the critical field study corresponded in many cases to that used in an accompanying critical current density investigation. In an attempt to separate the influence of deformation from that of aging, the effects of cold work prior to aging were, in some samples, erased by a heat treat-

ment of 3h/1000°C/quench[+] in order to homogenize the structures.

The influences of both isochronal (16 h) aging temperature and isothermal (300, 400, 500, and 600°C) aging time on the critical fields of 97%-cold deformed Ti-$Nb_{33.3}$, $Ti_{64.7}$-$Nb_{33.3}$-Ge_2, and $Ti_{62.7}$-$Nb_{33.3}$-Ge_4 were intercompared. Fig.9-4 presents the results of the first experiment. Fig.9-5, a typical set of results, represents the influence of aging at 400°C and 500°C on the critical fields of the same set of alloys. The maximum values of H_r attained during aging by the initially 97%-cold-worked alloys are given (together with the optimal heat-treatment conditions) in Table 9-10.

Summary

The addition of Ge to Ti-$Nb_{33.3}$ reduced the upper critical field of the homogenized-and-quenched samples at the rate of about 8 kOe per at.% Ge. With the 97%-cold-worked samples, an increase of 1 kOe accompanied the addition of 0.5 at.% Ge to Ti-$Nb_{33.3}$ (113.3 kOe), after which H_r dropped to 112.4 kOe with

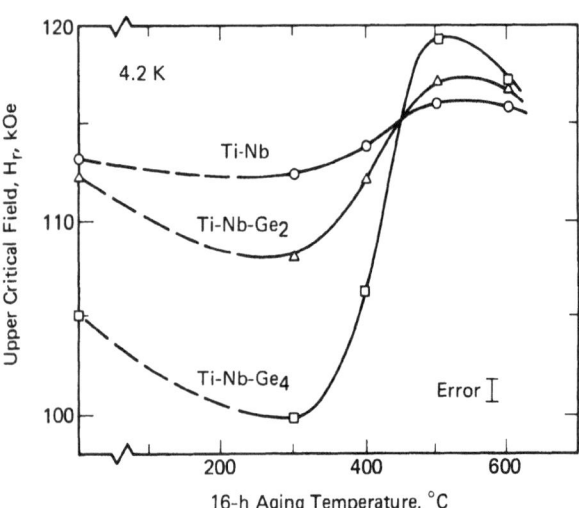

FIGURE 9-4. Resistive upper critical fields (4.2 K) of 1-mm$^\phi$ wires as functions of 16-h aging temperature in previously cold-worked (97%) wires of Ti-$Nb_{33.3}$-Ge_4 (□), Ti-$Nb_{33.3}$-Ge_2 (△) and a Ti-Nb (33.3 at. %) control (○) (criterion: 1 μV cm^{-1} at 10 mA; i.e., 8×10^{-7} Ω cm) — after HELLER [Hel71].

[+] An exception was the 2%-Ge sample, which required 3h/1080°C for homogenization.

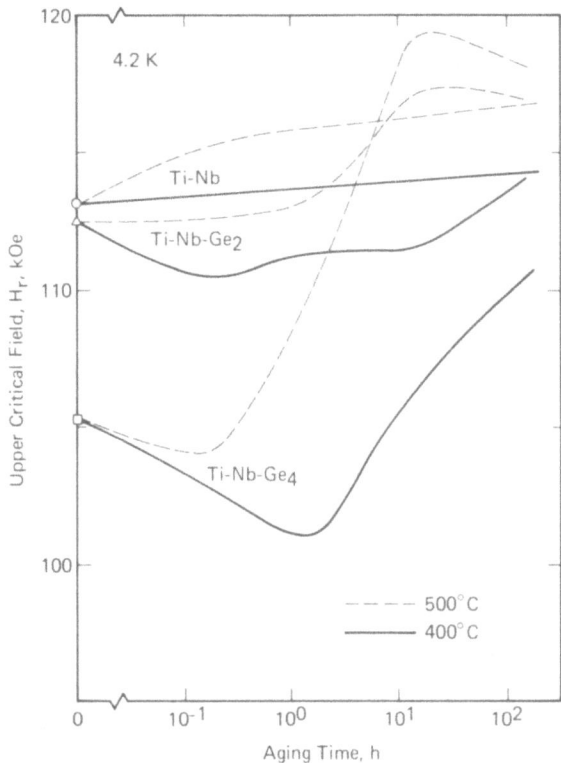

FIGURE 9-5. Resistive upper critical fields (4.2 K) of 1-mmϕ wires as functions of aging time at 400°C and 500°C of the three alloys referred to in Fig. 9-4 (criterion: 1 μV cm^{-1} at 10 mA; i.e., 8 x 10^{-7} Ω cm) — after HELLER [Heℓ71].

TABLE 9-10 MAXIMUM VALUES OF H$_r$(4.2K) ATTAINED DURING THE AGING OF INITIALLY 97%-COLD-DEFORMED Ti-Nb-Ge ALLOYS -- Heller [Heℓ71]

Alloy	Aging Condition	Resistive Upper Critical Field, H$_r$† kOe
Ti-Nb$_{33}$	160h/500°C	117
Ti$_{65}$-Nb$_{33}$-Ge$_2$	16h/500°C	117.5
Ti$_{63}$-Nb$_{33}$-Ge$_4$	16h/500°C	119.4

† Criterion: 1 μV across 1 cm at 10 mA (1.3 A cm^{-2}).

2 at.% Ge and to 105.2 kOe with 4 at.% Ge. Improvements in H$_r$ could be achieved through aging. As shown in Fig.9-4, the H$_r$ of initially cold-worked Ti$_{62.6}$-Nb$_{33.3}$-Ge$_4$ rose to 119.4 kOe when 500°C was selected as the 16-h aging temperature. At 119.4 kOe, H$_r$ exceeded by 2 kOe the highest-then-obtained binary-alloy value, that of cold-worked Ti-Nb$_{39}$. Thus although as pointed out above the presence of Ge in solid solution is detrimental to H$_r$, it is *indirectly beneficial during aging* in that by precipitating out as Ti$_5$Ge$_3$ it induces a Nb enrichment of the matrix. In indirect effects of this kind a competition exists between the beneficial effects of Nb enrichment, and the generally deleterious influence of the simple metal in solid solution.

9.17 CRITICAL CURRENT DENSITIES OF Ti-Nb-Ge ALLOYS

Considerable attention has been given to the possibility of increasing the critical current density of Ti-Nb alloys through the addition of Ge, in amounts which exceed equilibrium solid solubility, and which during heat treatment precipitates out in the form of Ti$_5$Ge$_2$ particles. Major contributors to the subject were ZWICKER *et al* (Universität Erlangen-Nürnberg), WILLBRAND *et al* (Krupp Forschungsinstitut) and KITADA and DOI (Central Research Laboratory, Hitachi Ltd.). RASSMANN and ILLGEN, whose work on other Ti-Nb-SM systems has already been discussed, also undertook a brief examination of the properties of three Ti-Nb-Ge alloys. The results of the Erlangen group, stemming from the dissertational research of HELLER [Heℓ71], have been summarized in a review article by ZWICKER *et al* [Zwi70] and papers by HELLER *et al* [Heℓ72] and LÖHBERG *et al* [Loh73]; those of the Krupp group have been published in a research paper by ARNDT and EBELING [Arn74] and discussed in detail in a report authored by WILLBRAND *et al* [Wiℓ75^a], while the work of the Hitachi group appears in two papers by KITADA and DOI [Kit72, Kit73].

9.17.1 Critical Current Densities of Research Alloys

RASSMANN and ILLGEN [Ras72^b] prepared for examination three Ti-Nb-Ge alloys of compositions:

Ti-31.4Nb-0.63Ge

Ti-32.8Nb-0.45Ge

and Ti-51.3Nb-1.49Ge.

The Nb-lean pair were martensitic, could not be drawn into wire, and were unable to be further examined. The alloy Ti-51.3Nb-1.49Ge, although homogeneously bcc in the as-cast condition, yielded small amounts of grain-boundary precipitation after heat treatment for 24h/500°C. In studying the current-carrying capacity of the cold-deformed (∿99.9%) and aged alloy, pronounced current instabilities were noted in wire which had been heat treated 1h/500°C and quenched. These instabilities, which were claimed to be characteristic of other precipitated alloys, were no longer present after the heat-treatment times had been extended to 5 to 10 h, and at the same time no deterioration in current-carrying capacity was noted. The critical current density of Ti-51.3Nb-1.49Ge was compared with that of the equivalent binary alloy as function of 1-h heat-treatment temperature. The presence of Ge shifted the optimal heat-treatment temperature from 400 to 500°C, and increased the J_c of the ∿99.9% cold-worked-and-aged wires from 2.2×10^4 A cm^{-2} to 5.0×10^4 A cm^{-2} (unstable). Further increases in J_c would of course be anticipated following optimization, however, the poor workability[†] of the Ge-containing alloy was regarded as a serious disadvantage from the standpoint of applied superconductivity.

HELLER [Heℓ71] and LÖHBERG et al [Loh73] have studied the metallurgical and microstructural properties of the Ti-Nb-Ge alloys specified in Sect.9.15. Critical current densities at 4.2 K in fields of up to 100 kOe were measured on 1-mm$^\phi$ wires which had been cold worked 97% and heat treated at 400°C for times of up to 500 h (12 steps). Studies were also made on alloy wires which had been annealed 3h/1000°C (1080°C in the case of the 2 at.% Ge sample), quenched, and aged for various times at 400°C, and samples prepared from an ingot which had been annealed prior to the 97% cold deformation. The results were compared with those for the corresponding binary control alloys. It was noted, for example, that in the case of $Ti_{64.7}$-$Nb_{33.3}$-Ge_2, cold worked 97% to 1 mm$^\phi$ from the

as-cast condition and aged 250h/400°C, that the 100-kOe J_c was 1.3×10^4 A cm^{-2}, significantly higher than the 2.6×10^3 A cm^{-2} of the corresponding binary alloy.[†] By aging for 16h/500°C, J_c could be further raised to 1.7×10^4 A cm^{-2}. The attainment of the usual 97% cold deformation was prevented by prior grain-boundary precipitation of Ti_5Ge_3 which, however, could be made to form as a flux-pinning precipitate in the already-cold-worked wire by heat treatment at 700°C. Precipitates formed during aging were identified as ω-phase, but see Sect.7.20.2(a). Augmenting this postulated precipitate as flux-pinning agents were α-phase precipitates and the dislocation network (i.e. the subband structure).

9.17.2 Critical Current Densities of Technical Alloys

KITADA [Kit73] prepared for critical current density study wire samples of the fifteen Ti-Nb-Ge alloys defined in Table 9-11. Three of these alloys, viz Ti_{55}-Nb_{44}-Ge_1, Ti_{60}-Nb_{39}-Ge_1, and Ti_{65}-Nb_{34}-Ge_1 were also taken by KITADA and DOI [Kit72], annealed 3h/1000°C, cold reduced 99% to 0.05-mm^t ribbon and aged from 3 to 10 h at 500°C prior to J_c versus H_a measurement. With the rolling direction parallel to the current direction and the applied field perpendicular to the current direction (both usual arrangements) critical currents were measured with the applied field:

TABLE 9-11 COMPOSITIONS OF THE Ti-Nb-Ge ALLOYS
STUDIED BY KITADA AND DOI [Kit73]

Ti(at.%)	Nb	Ge(at.%)	O (ppm)[†]
55	bal.	0.5, 1, 2	882, 305, 496
60	"	" " "	468, 607, 427
65	"	" " "	510, 601, 437
70	"	" " "	636, 616, 479
75	"	" " "	632, 769, 531

[†] The 882 ppm corresponds to the 0.5 at.% Ge, the 305 ppm to the 1 at.% Ge, and so on.

[†] Hardnesses: as-cast ------------ 190 kg mm^{-2},
∿99.9% cold worked -- 230 kg mm^{-2},
cw plus 1h/500°C ---- 275 kg mm^{-2},
after 2h/500°C, "decreasing hardness".

[†] The 100-kOe J_c of the binary alloy, aged 16h/400°C, was about 4×10^3 A cm^{-2}.

(a) parallel to the rolling plane: *(b)* perpendicular to the rolling plane.

In a study conducted concurrently with that of Ti-Nb-Cu considered above, ARNDT, WILLBRAND, *et al* [Arn74][Wiℓ75[a]] investigated the critical current densities of composite wires fabricated from the Ti-Nb-Ge alloys listed in Table 9-12.

(a) Properties of Ti-Nb-Ge Ribbon. Three interesting observations emerged from the results of the rolling study referred to above: *(i)* Although to a first approximation the $J_c(H_a)$ characteristics of the three subject alloys *when in the form of wire* were fairly insensitive to composition within the range considered, those of the ribbons varied markedly from alloy to alloy. *(ii)* In all cases J_c (||) was very much higher than both J_c (wire) and, of course J_c ($\perp$). Consider for example the 80-kOe J_c's of the Ti_{60}-Nb_{39}-Ge_1: for the 6h/500°C-aged ribbon, J_c (||) = 1.1×10^5 A cm^{-2} and J_c ($\perp$) = 3.6×10^4 A cm^{-2}; for the corresponding wire sample (of unspecified, but probably 99.69%, prior cold work) J_c = 5.4×10^4 A cm^{-2}, part-way between the two, suggesting that some advantage may attend the use of ribbon conductor, provided that a field orientation parallel to the rolling plane can always be ensured. *(iii)* An extreme minimum, associated with a pronounced "peak-effect" in the J_c ($\perp$) of 6h/500°C-aged Ti_{55}-Nb_{34}-Ge_1, "filled in" as the Ti content moved towards 65 at.%. The peak effect, as discussed in [Mon21.12], is an indication of inadequate intermediate-field pinning strength; it is a common feature of the J_c ($\perp$) *versus* H_a characteristics of rolled ribbons and reflects the coarseness of the subband structure when viewed in the direction normal to the rolling plane. This deficiency can be partially cured by precipitation, and for

TABLE 9-12   COMPOSITIONS OF THE Ti-Nb-Ge ALLOYS (wt.%)
STUDIED BY Willbrand *et al.* [Arn74][Wiℓ75[a]]

Symbol	Ti	Nb	Ge	C + N + O
Ge-05	51.5	47.8	0.50	0.070
Ge-1	52.4	46.2	0.98	0.100
Ge-2	50.9	46.9	2.10	0.150
Ge-3	50.9	46.0	2.90	0.074
Ge-4	50.2	45.7	3.90	0.096

that reason the peak effect washed out in the 500°C-aged alloys as the Nb content was reduced from 44 at.%, outside the α-phase precipitation regime, to 34 at.%, well within it (cf. Sect.7.20.2 and [Mon2.13]). Results for: *(i)* Ti_{55}-Nb_{44}-Ge_1, aged 6h/500°C; *(ii)* Ti_{60}-Nb_{39}-Ge_1, aged 3h/500°C; and *(iii)* Ti_{65}-Nb_{34}-Ge_1, aged 6h/500°C, are offered in Fig.9-6 as an example of this process.

(b) Properties of Ti-Nb-Ge Wire. KITADA [Kit73] has studied in detail the critical current densities of the alloys listed in Table 9-11 in fields of up to about 90 kOe under a broad matrix of final heat treatment conditions. Sample fabrication consisted of 74% reduction in area followed by a 3h/1000°C anneal and 99.69% cold reduction to 0.25-mm$^\phi$ wire, prior to final aging heat treatments at 350, 400, 450, and 500°C for times of typically 1, 3, 5, 10, 15, and 20 h. The results were expressed in several useful formats:

   (i) J_c *versus* H_a (for various aging temperatures/ times)
   (ii) J_c *versus* aging temperature (80 kOe)
   (iii) J_c *versus* Ge concentration (various fields and aging temperatures)
   (iv) J_c *versus* Ti concentration (80 kOe, various aging temperatures) etc.

The largest 80-kOe J_c, viz 7×10^4 A cm^{-2}, was attained with Ti_{60}-Nb_{39}-Ge_1 aged 10h/450°C. A typical set of data, those for Ti_{65}-Nb_{34}-Ge_1, against which the results of the Krupp experiments to be discussed below may be compared, are presented in Fig.9-7.

ARNDT, WILLBRAND, *et al* [Arn74][Wiℓ75[a]] have conducted detailed studies of the influence of thermomechanical processing on the critical current densities of Ti-Nb-Ge alloys, accompanying their measurements with a TEM investigation of the deformation-induced bands and the precipitates which develop therein. Extruded-and-drawn 3-mm$^\phi$ wire rod was clad in Cu and reduced 99.8% to a monofilamentary composite of O.D. 0.25 mm and superconducting core diameter about 0.13 mm. Including the preparation of the starting wire rod, the overall reduction amounted to more than 99.98%. Up to six intermediate heat treatments were administered at outside diameters of 3.0, 1.5, 1.0, 0.82, 0.68 and 0.5 mm, usually for 3h/390°C but occasionally for 5h/350°C. J_c ($\lesssim$70 kOe) increased with the number of intermediate heat treatments. According to a plot of J_c (70 kOe) *versus* Ge content for various heat treatments, the favored Ge content was 1 wt.%. The highest 70-kOe J_c attained in this

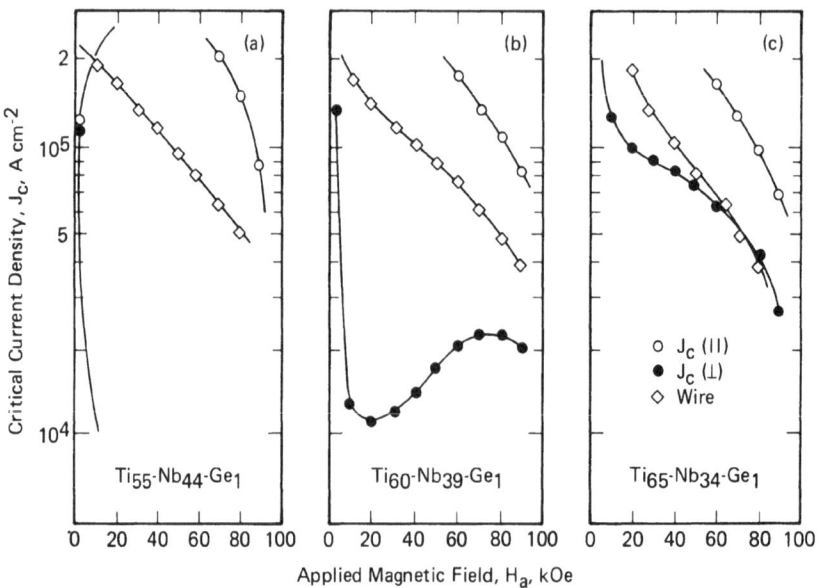

FIGURE 9-6. Critical current density anisotropy and peak effect in cold-rolled (99%)-plus-aged Ti-Nb-Ge ribbons measured with the field directed parallel, J_c (‖), and perpendicular, J_c (⊥), respectively, to the rolling plane. The three figures are arranged to demonstrate the washing-out of the peak effect and the reduction in J_c-anisotropy which accompany α-phase precipitation as the Nb level passes across the $\beta/(\beta + \alpha)$ phase boundary at about 40 at. % Nb. Included for comparison are data for the corresponding alloys in wire form (◇, cold worked, probably 99.69%, prior to the heat treatments) — after KITADA and DOI [Kit72].

study, that for Ti-46.9Nb-2.1Ge after six intermediate heat treatments of 3h/390°C each, was 2.3×10^4 A cm^{-2}. This may be compared with the 1.6×10^5 A cm^{-2} for Ti-46.8Nb-1.85Cu and the 1.1×10^5 A cm^{-2} for Ti-53.5Nb itself, again after six-fold 3h/490°C heat treatment. Some typical results only can be given here. Fig.9-8 presents a set of J_c *versus* applied field characteristics for Ge-1 (cf. Table 9-12), final processed according to the schedules of Table 9-13. Particularly noteworthy are three related observations:
(i) In Fig.9-8, $J_c(H_a)$ continues to increase with increase in the number of 3h/390°C heat treatments, the lack of a maximum indicating that process optimization had not been achieved; *(ii)* the $J_c(H_a)$ curves display a barely submerged peak effect (i.e. a rather broad "plateau" region); *(iii)* the critical current densities depicted are markedly inferior to those obtained by KITADA [Kit73] with a comparable alloy, cf. Fig.9-7. These results combine to suggest that the Krupp Ti-Nb-Ge alloys had been insufficiently heat treated. Indeed, according to Fig.9-7, heat treatments for 2 or 5 h at 400°C are quite inadequate; heat inputs to these classes of alloy equivalent to more than 5h/450°C, or many hundreds of hours at 350°C, seem to be needed to develop the best flux-pinning structures.

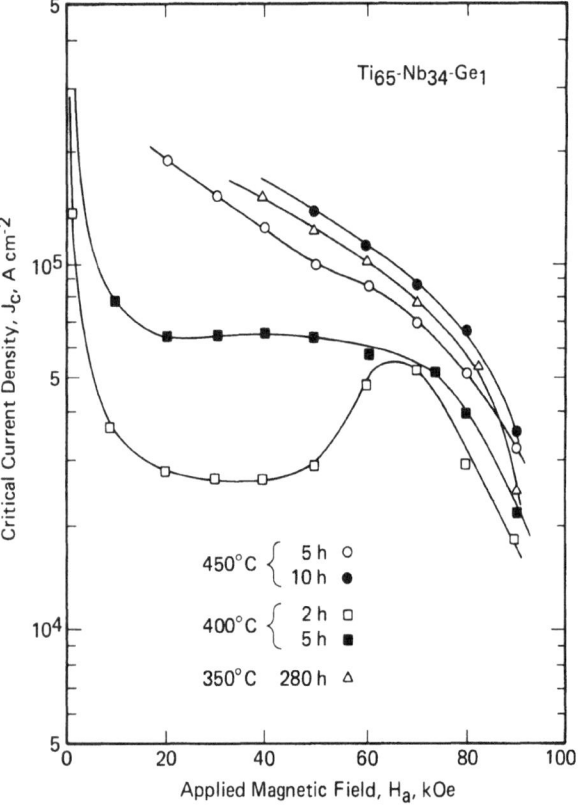

FIGURE 9-7. Critical current density *versus* applied magnetic field for previously 99.69% cold reduced 0.25 mm$^\phi$ Ti$_{65}$-Nb$_{34}$-Ge$_1$ wire under the five final-heat-treatment conditions specified (criterion: 5 μV across 3.5 ~ 4 cm [Doi68ª, 68ᵇ]) — after KITADA [Kit73].

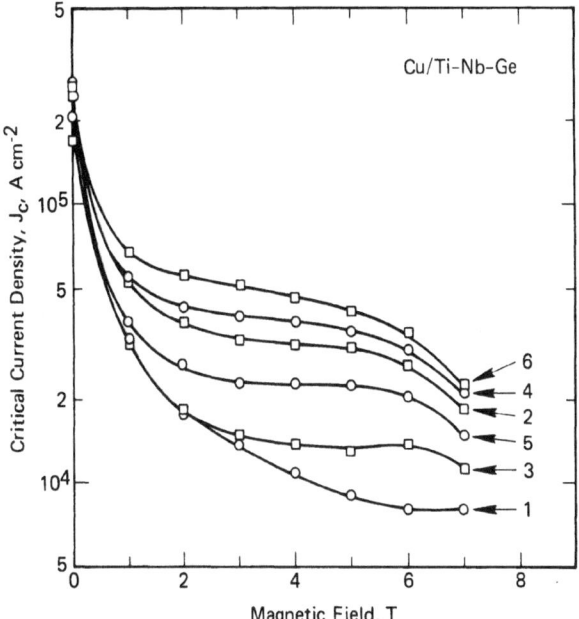

FIGURE 9-8. Critical current density (criterion: 5 μV cm^{-1}) *versus* applied field for 0.25 mm$^\phi$ Cu/Ti-Nb-Ge wire (alloy Ge-1 of Table 9-12) in six final thermomechanical process conditions. The numbers against the curves correspond to the numbers of intermediate heat treatments administered according to Table 9-13 — after WILLBRAND *et al* [Wiℓ75^a].

TABLE 9-13 FINAL-STAGE THERMOMECHANICAL PROCESSING OF THE REPRESENTATIVE Cu/Ti-Nb-Ge MONOFILAMENT Ge-1 (cf. Table 9-12) WHOSE PERFORMANCE IS DEPICTED IN FIG. 9-8 -- Willbrand *et al*. [Arn74][Wiℓ75^a]

Number of Heat Treatments	Outside Diameter (mm) at which Treatments of 3h/390°C were Administered					
	3.0†	1.5	1.0	0.82	0.68	0.5
1	H	—————————————————				
2	H	————————————				H
3	H	H	H	—————————		
4	H	H	H	—————————		H
5	H	H	H	H	H	——
6	H	H	H	H	H	H

† Administered *before* applying the Cu sheath.

CONCLUDING DISCUSSION

9.18 FLUX PINNING IN Ti-Nb-Cu AND Ti-Nb-Ge ALLOYS

Fig.9-9 from the work of WILLBRAND and colleagues [Arn74][Wiℓ75^a] juxtaposes the J_c's of six-fold-3h/390°C intermediate-heat-treated Ti-Nb, Ti-Nb-Cu, and Ti-Nb-Ge alloys, and enables an intercomparison to be made, useful not only in its own right, but also as a vehicle for the discussion of optimal current-carrying microstructures. Important microstructural features from the standpoint of flux pinning are: the dimensions of the drawing-induced subbands, the crystallographic textures associated with them, and the morphology and disposition of the precipitation occurring within them.

It is generally agreed that drawing produces a fibrous structure. The fibers, or subbands as they have also been called, have diameters typically lying between 500 and 3000 Å. It is natural to conclude that such fibers are the elongated effect of heavy area reduction (typically 99.95%) on the grain or sub-grain structure of the forged or cast starting ingot. However, BAKER [Bak70] has claimed that this approach would yield fibers of diameter 5 μm (50,000 Å) or 50 Å, respectively, neither of which matches the diameters observed in practice. Accordingly he concluded that the fiber size was related to the "spacing between precipitates or some other inhomogeneity". This was not a satisfactory conclusion, leading as it does to the possibility of an extremely wide variability in fiber diameter -- which is not observed. Various researchers, studying Ti-Nb alloys of varying impurity content, and doped with various solute elements, have encountered fiber diameters which tend to cluster within a common range of values. Such demonstrations of the lack of dependence of the fiber diameter on the composition of the starting material serve to emphasize that subband formation is a rather general response to a metalworking process, and that the as-worked subband diameters will tend to be invariant among the various members of a given class of alloy under comparable metalworking conditions. ARNDT and

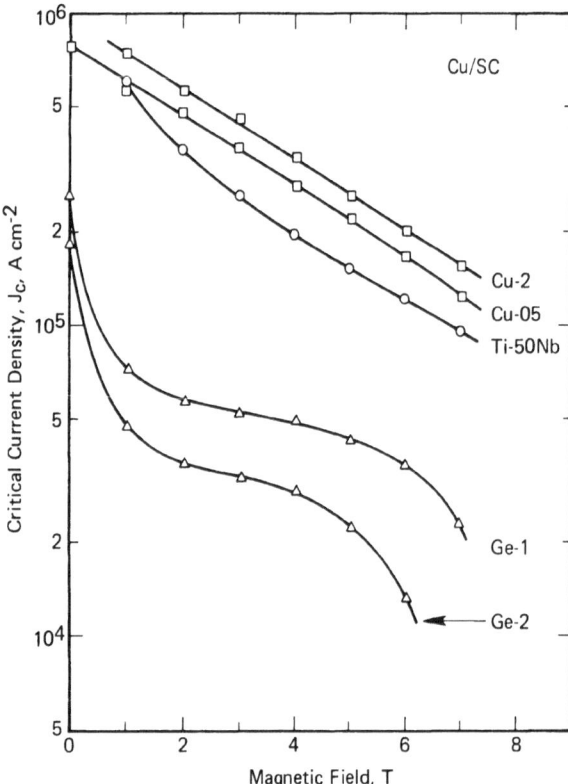

FIGURE 9-9. Intercomparisons of the critical current densities (criterion: 5 μV cm^{-1}) of best-processed (i.e., 6-fold 3 h/390°C heat treated) Ti-Nb-Cu and Ti-Nb-Ge composite conductors (viz., Cu-05 and Cu-2 of Tables 9-7 and 9-8; Ge-1 and Ge-2 of Tables 9-12 and 9-13). Inserted for comparison are data for similarly processed Cu/Ti-50Nb (for which 7-fold final processing yielded a slightly higher $J_c(H_a)$) — after WILLBRAND et al [Wil75a].

EBELING [Arn74] correlated fiber diameter with critical current density in Ti-53.5Nb, final annealed for various times at temperatures of 300, 390 and 500°C, in response to which the fibers acquired diameters within the range 375 to 2680 Å. Optimal diameters were shown to be in the vicinity of 500 Å. On the other hand, when the comparison was extended to Ti-46.8Nb-1.85Cu and Ti-46.9Nb-2.1Ge, it was noted that for fiber-diameters of about 300 Å the 50-kOe critical current density of the Cu-containing alloy was about ten times greater than that containing Ge (cf. Fig.9-9). Likewise, BAKER [Bak70] noted in studies of Ti-56Nb, both with and without the addition of N, that in wires with fiber diameters of, say 800 Å, the critical current density was lower by an order of magnitude in those which contained a "high uniform density of dislocations". Evidence was accumulating to support the conclusion that a fiber or subband diameter of 300-500 Å was a necessary but not sufficient condition for critical-current-density optimiza-

tion. Other factors to be considered were: the initial stages of fiber disintegration during heat treatment, the crystallographic texture of the fibrous alloy, the dislocation density within the fibers as distinct from that in their boundaries, and precipitation. ARNDT and EBELING [Arn74] noted that a high J_c was not necessarily associated with a unique value of the subband diameter. Thus whereas all wires with high J_c's had subband diameters of 300~500 Å, some conductors were found that, although possessing subband diameters within this range, exhibited current-carrying capacities as low, it was claimed, as those of subband-free recrystallized material. They concluded, therefore, that since no one-to-one relationship between subband diameter and J_c came out of their measurements, other factors must also have been influencing the critical current.

Subband diameters in six-fold heat-treated Ge-1 (Table 9-12) and Cu-2 (Table 9-7) were found to be 205 Å and 370 Å, respectively. Studies of Ti-Nb$_{42}$ by NEAL et al [Nea71] have shown that the 50-kOe J_c varied inversely with subband diameter in this range (see Fig.7-45, Sect.7.30.1). But according to Fig.9-9, Ge-1 and Cu-2 taken together are certainly exceptions to this rule; a clue to this apparently anomalous behavior is contained in Figs.9-10(a) and (b). Whereas the subbands in ~99.98%-deformed and intermediate-aged Ge-1 are narrow, linear, and sharply defined, those of similarly processed Cu-2 are disintegrated and weakly textured; furthermore the latter alloy, according to TEM, revealed evidence for round or elliptical precipitate particles, about 500 Å in diameter.

In the as-drawn condition the alloy wire is strongly textured in the drawing direction: <110> according to ARNDT and EBELING [Arn74] who studied Ti-53.5Nb and Ti-Nb-Cu,Ge by electron diffraction after about 99.98% reduction, and <100> according to BAKER [Bak70] as a result of measurements on 99.93% reduced Ti-56Nb with or without the addition of N. But as-drawn fiber structures, even if optimally dimensioned, are usually associated with relatively low J_c's. The microstructural result of aging is first to clear dislocations from cell interiors and deposit them or annihilate them at the walls [Nea71], then to reduce the texture and commence the coarsening [Bak70] and initial disintegration [Arn74] of the fiber structure. Detexturization and coarsening, both of which tend to interrupt the continuity of the fibers, have beneficial influences on J_c. Elongated precipitates lying perpendicular to the wire axis and

FIGURE 9-10. Longitudinal alloy sections prepared for TEM from single-filament (core, 1.13 mmϕ) cold-worked and thermomechanically processed Cu/SC composites. (a) Ti-Nb-Ge (alloy Ge-1) after 6-fold 3 h/390°C intermediate heat treatment retains sharply delineated subbands (diam., ~ 205 Å) but exhibits no precipitation. (b) Ti-Nb-Cu (alloy Cu-2) after intermediate heat treatment (4- or 6-fold 3 h/390°C) exhibits elliptical precipitate particles, maximum diameter ~ 500 Å, oriented at right angles to the wire axis. TEM original magnification, 80,000X — after WILLBRAND et al [Wiℓ75a] (see also ARNDT and EBELING [Arn74]). Micrograph courtesy of J. Willbrand; reproduced by permission of Zeitschrift für Metallkunde.

of length commensurate with the fiber diameter, as distinct from particles aligned parallel to the wire axis, are also claimed to be beneficial.

Intermetallic-compound precipitation does not result from the moderate-temperature aging of Ti-Nb-Cu and Ti-Nb-Ge alloys [Heℓ71][Loh73][Arn74]. The effect of the Cu and Ge additions is to modify the precipitation processes normally expected to take place in the binary host alloy during 300-600°C heat treatment. At temperatures above 300°C, fine coherent precipitation commences (reminiscent of ω-phase -- but cf. Sect.9.1). Because of its fineness, this precipitate is often only indirectly detectable by means, for example, of a "coherency strain". Between 390 and 500°C incoherent precipitation takes place. This is generally

regarded as α-phase, the morphology and orientation of which is controllable by ternary doping. Thus, according to ARNDT and EBELING [Arn74], whereas in Ti-Nb the α-phase precipitation took the form of long narrow strips parallel to the fiber axis, the addition of Cu resulted in the growth of round or elliptical particles with diameters perpendicular to the wire axis of about 500 Å, an arrangement especially favorable for high J_c. The addition of Ge, on the other hand, was found to retard the precipitation that would otherwise take place in the binary Ti-Nb alloy. Thus during the heat treatment of Ti-Nb-Ge, the strong drawing texture and fiber structure were maintained [Arn74], thereby inhibiting the optimization of J_c that would otherwise be expected.

10

SOVIET TECHNICAL ALLOYS

Soviet technical alloys include the Ti-Nb types T 60 and NT 50, and the Ti-Zr-Nb types SS 2, 35 BT, 50 BT, and 65 BT, whose typical compositions are listed in Table 10-1. The positions of the ternary alloy types on the Ti-Zr-Nb composition triangle are indicated in Fig.10-1. A cable designated KSMI-6 consists of six Cu-plated 65 BT wires, and one Cu wire, twisted together and impregnated with In.

In that the Soviet alloys include both binary Ti-Nb and ternary Ti-Zr-Nb types, the logical place

for a discussion of them is somewhere between Chapters 7 (Ti-Nb) and 11 (Ti-Zr-Nb). Otherwise this review could have been included within Chapter 11 itself, in order to place it in closer proximity to the ternary-alloy metallurgical information, and the discussion of the Japanese Ti-Zr-Nb technical alloys, contained therein.

Reviewed below are a group of representative papers from the Soviet literature of the period 1967-1976 dealing with the superconductive and microstructural properties, small-scale applications, and other topics involving the above-mentioned technical alloys. The microstructural properties are dealt with quite briefly, since the small Zr levels used (up to 10 wt.%) are not expected to bring about pronounced departures from those of the nearby Ti-Nb binary alloys of Chapter 7 to which the reader is referred for further information. In the same vein, it is helpful to consider the heat treatment, microstructures, and properties of the Soviet alloys in the light of the general discussions of the properties of Ti-Zr-Nb alloys which follow in Chapter 11.

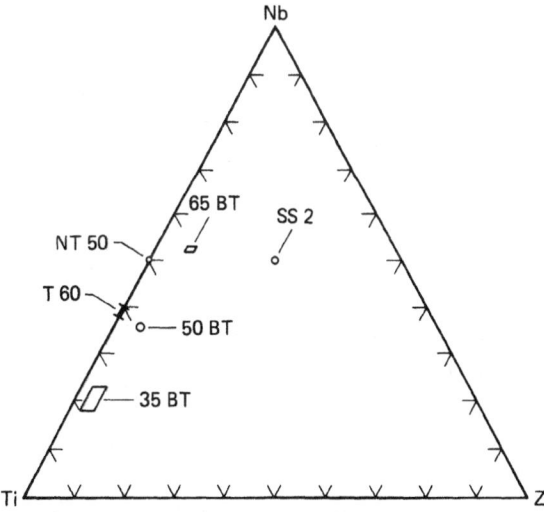

FIGURE 10-1. Composition triangle (at. % linear) indicating the locations of the Soviet alloys T 60, SS 2, 35 BT, 50 BT and 65 BT, according to the compositions and compositional ranges listed in Table 10-1.

10.1 PROCESSING OF SOVIET ALLOYS— HOMOGENEITY OF THE STARTING BILLET

In the processing of Ti-alloy superconducting wire the condition of the starting billet, as well as the subsequent thermomechanical processing parameters, must be taken into consideration. KUNAKOV *et al*

TABLE 10-1 COMPOSITIONS OF SOVIET ALLOY TYPES

Soviet Alloy Type	Composition, Wt.%			Composition, At.%			Literature
	Ti	Zr	Nb	Ti	Zr	Nb	
T 60	46-44	--	54-56	62-60	--	38-40	Lazarev *et al.* [Laz68, Laz72, Laz73, Laz74] Krainski *et al.* [Kra71] Kadykova [Kad73] Verkin *et al.* [Ver76]
NT 50	33	--	67	50	--	50	Verkin *et al.* [Ver76]
SS 2				25	25	50	Kunakov *et al.* [Kun70] Lazarev *et al.* [Laz68, Laz72, Laz73]
35 BT	67-57.5	3.1-7.7	30-35	80-72	2-5	18-23	Kadykova *et al.* [Kad72, Kad73, Kad74]
50 BT	42.5	7.5	50	59	5.5	35.5	Kadykova *et al.* [Kad72, Kad73]
65 BT	27.5-25	7.5-10	65	42-39	6-8	52-53	Alekseevskii *et al.* [Ale67, Ale68] Prekul *et al.* [Pre67] Gorina *et al.* [Gor69] Sychev *et al.* [Syc69] Buynov *et al.* [Buy70] Krainski *et al.* [Kra71] Kadykova *et al.* [Kad72, Kad73]

[Kun70] have investigated the effect that coring[†] in the starting billet has on the critical current density of the final cold-worked wire product. In agreement with the results of earlier studies by KNEIP *et al* [Kne62] on Nb-Zr alloys, which indicated that billet homogenization resulted in a decrease in the critical current density of the wire produced from it, KUNAKOV *et al* showed that the 16-kOe J_c's of cold-worked Ti-Nb$_{54}$, Zr-Nb$_{25}$, Ti$_{39}$-Zr$_8$-Nb$_{53}$, and Ti$_{25}$-Zr$_{25}$-Nb$_{50}$ (SS 2) wires were all decreasing functions of the 1500°C-vacuum-annealing time of the starting billets. Carrying the experiment one logical step further, however, they noted (in tests on Ti-Nb$_{35}$) that final heat treatment at an optimal temperature of 350-400°C not only increased by an order of magnitude the J_c's of both prior-as-cast and prior-homogenized wire samples, but slightly reversed their order. It was therefore

---

† The term "coring" refers to the dendritic-structured compositional modulation produced as a result of solute-solvent segregation which takes place as the ingot cools through the liquid + solid region of the phase diagram.

regarded as advantageous to *homogenize the starting billet* since this led to greater longitudinal uniformity in the properties of the final wire, the critical current density of which is after all dominated not by starting-billet inhomogeneities, but by the thermomechanical processing to which it has been subjected.

10.2 PROCESSING AND STRUCTURES OF 35 BT

The precipitation which takes place during the aging of cold-worked 35 BT naturally depends strongly on the Nb concentration which in this alloy type, according to KADYKOVA and FEDOTOV [Kad72] and KADYKOVA [Kad73], is permitted to vary from 30 to 35 wt.%. As indicated in Table 10-1, the Zr level may also be varied, values of 3.1 to 7.7 wt.% having been cited by KADYKOVA *et al* [Kad74]. According to the x-ray diffraction and TEM results of KADYKOVA and FEDOTOV [Kad72], traces of diffuse ω-phase already present in cold-deformed 35 BT (30 wt.% Nb) developed into dense ω-phase precipitation after 4h/400°C. In addition α-phase precipitation which started to form at 350°C coexisted with the ω-phase in the temperature range 350-400°C. At 450°C, according to KADYKOVA and FEDOTOV [Kad72], the structure was $\alpha+\beta$. In 35 BT (35 wt.% Nb), an increase in the Zr level of from 4.4 to 7.4 wt.%, or the addition of 0.7 wt.% Fe, hindered the formation of the ω- and α-phases during annealing at 400°C. Cold-deformed 35 BT (35 wt.% Nb) contained traces of diffuse ω-phase, but decomposition of this alloy during heat treatment in the temperature range 350-550°C took place mostly through α-phase precipitation [Kad72, Kad74]. According to KADYKOVA *et al* [Kad73] the highest critical current density was obtained following an α-phase-precipitation heat treatment of 4h/450°C.

10.3 PROCESSING AND STRUCTURES OF 50 BT

The type 50 BT alloy, with a typical composition Ti-7.5Zr-50Nb, is metallurgically comparable to the European Ti-50Nb class of alloy. Information on Soviet-type 50 BT is scarce. KADYKOVA [Kad73] has quoted a post-cold-work heat treatment of 4h/450°C for this alloy. In the paper coauthored with FEDOTOV [Kad72], in which the results of TEM were discussed, it was noted that an unidentified "intermetallic phase" appeared during cold deformation. Formed as a result of annealing for 100h/450°C were α-phase platelets, and in addition to the unidentified phase, a fine uniformly distributed dispersion of precipitates about 50 Å in size. Further insight into the properties of 50 BT is provided by the work of KITADA and DOI [Kit69] who have conducted a detailed study of the superconductive and microstructural properties of cold-worked-plus-aged and quenched-plus-aged $Ti_{60}-Zr_5-Nb_{35}$ (i.e. Ti-6.9Zr-49Nb), an alloy whose composition assigns it to the Soviet 50 BT group. Anticipating a detailed discussion of Ti-Nb-rich Ti-Zr-Nb alloys to be presented in Chapter 11, some of the essential metallurgical results obtained by KITADA and DOI are outlined below:

10.3.1 Quenched-Plus-Aged 50 BT-Type Alloys

According to [Kit69]: *(i)* during aging at temperatures near 350°C (e.g. 100h/350°C) "G.P. zone" precipitation (size, 50-200 Å) took place; *(ii)* more extensive heat treatment (e.g. 1000h/400°C) transformed this into an "intermediate phase" (size, 500-3000 Å) identified as being hexagonal in structure (with c/a = 9.67 Å/2.80 Å); *(iii)* after aging at temperatures above 400°C (e.g. 1000h/450°C) α-phase formed as a disc-shaped precipitate.

10.3.2 Deformed-Plus-Aged 50 BT-Type Alloys

Although cold deformation prior to aging accelerated the precipitation processes they were similar to those which took place during the aging of the quenched alloys. It was noted that best superconducting properties were obtained in cold-worked-plus-aged (250h/350°C or 50h/400°C) material in which what were referred to as "G.P.-zone precipitates" about 25-200 Å in size were responsible for the flux pinning. Complete identification of this precipitate was not made. Nevertheless the precipitation temperature ranges and sequences [Kit69], viz:

$$\beta \longrightarrow \beta + G.P. \longrightarrow \beta + int. \ phase \longrightarrow \beta+\alpha$$

were similar to those with which the occurrence of the β' precipitate is associated, Sect.9.1.

In view of the fact that the heat treatments were all taking place within the equilibrium $\alpha+\beta$ field, and that the composition range for ω-phase precipitation dictated by at.% Nb$\lesssim$30 was exceeded, it is most likely that a conventional phase-separation reaction $\beta\rightarrow\beta'+\beta$ was being observed by KITADA and DOI.

10.4 PROCESSING AND STRUCTURES OF 65 BT

65 BT with typical compositions of Ti-(7.5-10)Zr-65Nb (i.e. Ti$_{39-42}$-Zr$_{8-6}$-Nb$_{53-52}$) occupies a position in what appears to be the single-phase-bcc field of the equilibrium composition triangle. In this light it is difficult to interpret some of the precipitation effects reported in the Soviet literature.

10.4.1   Quenched-Plus-Aged 65 BT

The highest J_c noted in quenched-plus-aged 65 BT was achieved in response to an aging of 2h/700°C which according to KADYKOVA [Kad73] is responsible for the precipitation of a very distorted α-phase, referred to as an "intermediate phase" (cf. 50 BT), of particle size 250-300 Å and particle spacing 400 Å. Dislocation loops associated with precipitate-matrix incompatibility have been photographed by BUYNOV et al [Buy70] in 65 BT quenched and aged 45h/600°C, 2h/700°C, and 6h/800°C, after which heat treatments the precipitates had grown to 700-800 Å. From an electron micrograph of 65 BT, quenched from 1250°C and aged 2h/700°C, the presence of β' precipitates in a β matrix can be positively identified; cf. [Wil73], also Sect.9.1.

10.4.2   Deformed-Plus-Aged 65 BT

In deformed 65 BT the "intermediate phase" (presumably related to the "distorted α-phase" referred to above) appeared as a fine precipitate ($\sim$50 Å in size) in places of high dislocation density. According to GORINA et al [Gor69], who studied the influence of final heat treatment on J_c, much less severe heat treatments than the 2h/700°C referred to above are needed to maximize the critical current density of 65 BT. Experiments were performed on 99.9% drawn (8 mm$^\phi$ to 0.25 mm$^\phi$) and intermediate-heat-treated (at 3, 1.5 and 0.8 mm$^\phi$) wire. The variations in critical current density in response to final heat treatments at temperatures up to 600°C for times of 20 sec and 2, 4 and 10 h were recorded, particular

attention being paid to the effects of aging for times of up to 10 h at temperatures of 250°C and 400°C. The best 4-h and 10-h heat-treatment temperatures were about 500 and 475°C, respectively. As function of time at 400°C the critical current density appeared to reach a broad maximum after 4 h, with a 20-kOe value of 7.9x10^4 A cm^{-2}.

The moderate temperature thermal stability of the heat-treated wire was also investigated in a search for possible degradation of properties during the baking of an insulating enamel coating, following the procedures used in the electric-cable industry. No deterioration of critical current density was noted.

10.5 CRITICAL CURRENT DENSITIES OF THE SOVIET ALLOYS

10.5.1   The Critical Current Density of 65 BT

The critical current density of 65 BT, in the form of 0.25-mm$^\phi$ wire has been measured by ALEKSEEVSKII et al [Ale67] at 4.2 K using a steady-field/pulse-current method in fields of up to 50 kOe, followed by the pulse-field/steady-current method, in fields of up to about 90 kOe. It is interesting to note that these authors, in discussing mechanisms for the support of high current densities in heavily drawn wire, chose to regard the fibrous banded structure as a superconducting composite (comparable to a "synthetic superconductor", or the heterogeneous superconductor referred to in an early paper by BEAN [Bea64]) consisting of superconducting filaments imbedded in normal metal,[+] rather than a bulk superconductor whose current-carrying ability relied on flux pinning by line and point defects. A J_c $versus$ applied field curve, from the work of ALEKSEEVSKII et al [Ale67], is presented in Fig.10-2.

10.5.2   Critical Current Densities of Other Alloys

LAZAREV et al [Laz72] have studied the critical current densities of short samples of T 60 (Ti-Nb$_{40}$)

[+] An interesting discussion of the filamentary-mesh model, presented within the context of current transport in cold-worked alloy superconductors, has been presented by BERLINCOURT and HAKE [Ber63[a]].

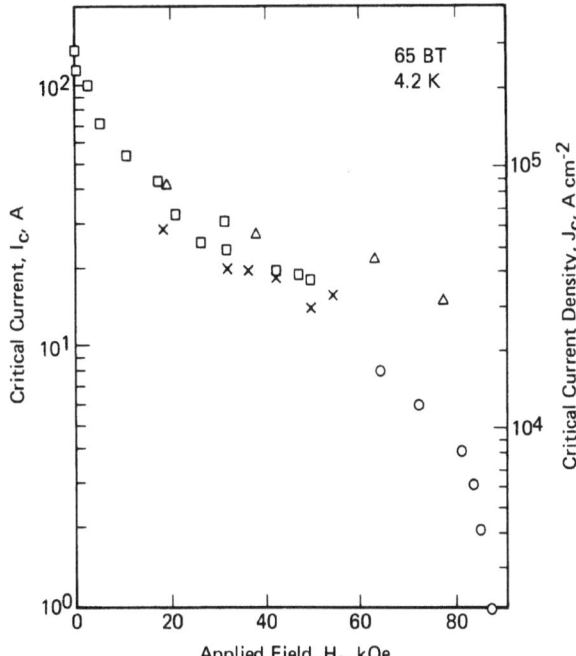

FIGURE 10-2. Critical currents (and critical current densities based on a wire diameter of 0.25 mm) of mostly not-heat-treated 65 BT wire. Depicted are the results of: (a) short-sample tests in a steady applied field, pulsed current (□); (b) short-sample tests in a pulsed applied field, steady current (○); (c) coil tests (×); (d) coil tests of 500°C-annealed wire (△) — after ALEKSEEVSKII et al [Aℓe67].

and SS 2 (Ti$_{25}$-Zr$_{25}$-Nb$_{50}$) at temperatures of 2 and 4.2 K in fields of up to about 140 kOe. ALEKSEEVSKII et al [Aℓe67] and LAZAREV et al [Laz72] then went on to intercompare the short-sample and small-coil critical current densities of these alloys. KADYKOVA [Kad73], in a brief review (not restricted to Soviet alloys) of precipitation effects and critical current density in Ti-Nb-base superconductors, has considered the properties of the three Soviet types: 35 BT, 50 BT, and 65 BT. Some typical J$_c$-data for five Soviet alloy types are given in Table 10-2. A listing of other sources of data is provided in Table 10-3.

10.6 APPLICATIONS OF SOVIET TECHNICAL ALLOYS

10.6.1 Coil Tests of T 60 and SS 2

LAZAREV et al [Laz68] have conducted coil tests, inside a larger solenoid, of the alloys T 60 and SS 2. The coils themselves, 4 mm I.D. x 30 to 40 mm long, were assembled from nine or ten 100-turn coils of Cu-plated (30 μm^t) or enamelled 0.25-mm$^\phi$ wire. In a sub-

sequent study of the two alloys, LAZAREV et al [Laz72] have tested and compared short-sample and coil properties. Lengths of 0.23-mm$^\phi$ wire, coated with Cu to thicknesses of 30-45 μm, were formed into: (a) loosely wound coils of one or two layers with a turn spacing of about 2 mm; (b) tightly wound single-layer coils; (c) tightly wound 2-layer (14 turns) to 15-layer (970 turns) coils of I.D. 0.6-1.6 cm and 0.5-2.5 cm in length.

10.6.2 Welded Joints

In the fabrication of large coils it is usually necessary to make very low resistance, or perfectly superconductive, joints. KRAINSKII and SHCHEGOLEV [Kra71] have studied the preparation and properties of welded joints in T 60, NT 50 and 65 BT monofilamentary wires. With the varnish and Cu coating removed,

TABLE 10-2 SOME CRITICAL CURRENT DENSITIES (J$_c$) OF SOVIET ALLOYS

Alloy Type	Condition	J$_c$ Presumably at 4.2 K, 10^4 A cm^{-2}	Applied Field, kOe	Reference
Ti-Nb-Type				
T 60	Def. + 1h/450°C	5	50	†
T 60	Optimized	6	50	††
Ti-Zr-Nb-Type				
SS 2	Optimized	10	50	††
35 BT	Def. + 1h/400°C	6	50	†
50 BT	Def. + 4h/400°C	2	50	†
65 BT	Def. + 4h/400°C	5	50	†

† Kadykova [Kad73]

†† Lazarev et al. [Laz72]

TABLE 10-3 CRITICAL CURRENT DENSITIES IN SOVIET TECHNICAL ALLOYS -- DATA SOURCES

Alloy Type	Condition	Procedures	Literature
SS 2	Homogenized (0-5h)/1500°C cold forged to 11x11 mm and drawn 99.96% to 0.25 mm$^\phi$. J_c measured as function of homogenization time.	Short-sample J_c in 16-kOe transverse field.	Y.N. Kunakov *et al.* (1967/70) [Kun70]
65 BT	---------------------------------------	Pulse current and steady field up to 50 kOe; steady current and pulse field up to ∿90 kOe.	N.E Alekseevskii *et al.* (1967) [Aℓe67]
65 BT	0.25 mm$^\phi$ wire, cold drawn or annealed 3h/1000°C.	Short-sample J_c(4.2K) ≳ 85 kOe.	N.E. Alekseevskii *et al.* (1968) [Aℓe68]
65 BT	Drawn from 8 mm$^\phi$ to 0.25 mm$^\phi$ (99.9%) with intermediate low-temperature anneals at 3, 1.5 and 0.8 mm$^\phi$. Aged (0.006-10h)/(100-600°C); (0-24h)/(250 and 400°C).	(a) Relative J_c *vs* aging temperature (0-600°C). (b) J_c(20kOe) *vs* aging time at 250, 400°C. (c) Influence of moderate temperature aging and enamel-baking on J_c.	N.B Gorina *et al.* (1969) [Gor69]
T 60 SS 2	---------------------------------------	Short-sample J_c(2 and 4.2K) *vs* H ≳ 145 kOe; also small-coil critical current tests.	B.G. Lazarev *et al.* (1972) [Laz72]
T 60	Heavily cold deformed to 0.25 mm$^\phi$, then heat treated 2h/350,400,500,550°C.	---------------------	B.G. Lazarev *et al.* (1974) [Laz74]
T 60	Optimally heat treated T 60 alloy with Cu coating varying in thickness from 0 to 120 μm.	Short-sample J_c(4.2K) in transverse fields of up to 60 kOe.	B.G. Lazarev and S.I. Goridov (1973) [Laz73]
T 60 35 BT 50 BT 65 BT	Deformed plus 1h/450°C. Cold worked + 1h/400°C. Cold worked + 4h/450°C. Cold worked + 4h/450°C.	---------------------	G.N. Kadykova (1973) [Kad73]

a pair of wire ends were laid across each other, spot welded, and flattened between die plates along a 5-mm length of each wire. It was shown that by suitably adjusting the operating mode of the spot welder, and the subsequent deformation pressure, current densities across the joint reasonably close to those of the un-interrupted wire were achievable. The joining of superconductors is reviewed in [Mon28.16].

10.6.3 Small Coil Properties of 65 BT

Using single-layer coils (8 mm$^\phi$ x 13 mm$^\ell$), wound from 65 BT wire subjected to varying amounts of plastic deformation and heat treatment, PREKUL and VOLKEN-STEYN [Pre67] studied the relationship between critical current density (whose 20-kOe value varied from 2.2×10^3 to 1.5×10^6 A cm^{-2}) and magnetic hysteresis. Again, using small solenoids (∿11 mm I.D. and 40-70 mm$^\ell$), ALEKSEEVSKII *et al* [Aℓe67] have compared with good results long-sample and short-sample critical current densities (cf. Fig.10-2). In addition, two of the coils in which the superconductive windings

were interleaved with a Cu winding connected to a variable resistance, provided an interesting study of the so-called "transformer method" of quench protection. ALEKSEEVSKII *et al* [Aℓe67] claimed that "data obtained from this experimental study of transformer-protection enabled the thickness of the stabilizing Cu coating to be estimated". For use in the solenoids, the 0.25-mm$^\phi$ wire was electrolytically coated with 10-20 μm of Cu. Subsequently, ALEKSEEVSKII *et al* [Aℓe68] modified a small, 35-mm bore, 55-kOe solenoid to enable the short-sample critical current densities of numerous alloy samples to be measured in fields as high as 85 kOe. Such a field was achieved within the 1-mm gap between a pair of "concentrators" -- solid cylinders of permen-dur, 27 mm in diameter and 65 mm in length, positioned along the axis of the superconducting solenoid. A similar technique was employed by BYCHKOV *et al* [Byc69] who placed the superconducting short-sample to be measured within the 1-mm gap between a pair of dyspro-sium inserts, each of diameter 9 mm, tapering to 3 mm diameter at the pole ends. The host solenoid, of bore 18 mm, length 176 mm, wound with 9981 turns of KSMI-6 cable (see introductory paragraph), was itself capable of producing a 62 kOe magnetic field.

11

TITANIUM-ZIRCONIUM-NIOBIUM TERNARY ALLOYS

The development of Ti-Zr-Nb alloy superconductors in the early 1960's was motivated by a desire to combine the advantages of Zr-Nb and Ti-Nb alloys and at the same time to avoid their disadvantages. In so doing, the following electrical properties were taken into consideration: (a) the maximum transition temperature in the Zr-Nb system of 10.9 K at about 80 at.% Nb, as compared to the 9.7 K of the Ti-Nb system, at 60-70 at.% Nb; (b) the maximum 4.2-K upper critical field in the Zr-Nb system of 95 kOe, considerably less than the 113 kOe of Ti-Nb. On the mechanical-property side, recognizing that $Zr-Nb_{50}$, for example, underwent a ductile-brittle transition at about 100°C [Doi68[b]], it was deemed advantageous to substitute some Ti for Zr in order to lower the transition point, which then dropped to room temperature with 1 at.% Ti and to about -50°C with 10 at.% Ti. As pointed out in [Mon3.24], in which the development of contemporary superconducting alloy formulations *via* a workability route is described, this search for improvements to the workability of Zr-Nb led eventually to the well-known alloy $Ti_{10}-Zr_{40}-Nb_{50}$ [Doi68[b]]. By hindsight the substitution of Zr for Ti in Ti-Nb alloys might have been expected to have led to higher paramagnetically limited upper critical fields by way of the spin-orbit-scattering mechanism referred to above (Sects.3.6.3-3.6.5, Sect.5.12.4, and Sect.7.10.2) and discussed in detail in [Mon15.00]. In fact, such an approach did *not* guide the early development of the ternary alloys.

More recently, however, with spin-orbit scattering in mind, HORIUCHI *et al* [Hor73] did substitute 6 at.% Zr for some of the Ti in $Ti-Nb_{33}$, but not stopping there went on to substitute 6 at.% Ta for some of the Nb, leading to Kobe Steel Ltd.'s commercial alloy "cryozitt" whose properties are discussed in Chapter 13.

A word on nomenclature: independently of the relative amounts of the constituent elements (either in atomic or weight percent) the alloy-formula components are herein arranged in order of: (i) column in the periodic table; (ii) atomic number. This defines the sequence Ti, Zr, Nb, which carries with it the scientifically sound implication that Ti and Zr are mutually substitutive.

The literature to be reviewed has been sorted into three major categories thereby giving rise to the three parts into which this chapter has been subdivided:

PART 1 -- international contributions to the research and development of Ti-Zr-Nb superconductors.

PART 2 -- Japanese Ti-Zr-Nb technical alloy development, an activity which is to be thought of as being juxtaposed against the Soviet work in this area, the subject of Chapter 10.

PART 3 -- a discussion of AC effects, studied during the development, again in Japan, of Ti-Zr-Nb as a conductor for AC applications.

11.1 SUPERCONDUCTIVITY AND METALLURGY IN Ti-Zr-Nb ALLOYS

In an early comparison by DOI et al [Doi66] of the critical current densities of a series of unoptimized Zr-Nb, Ti-Nb, and Ti-Zr-Nb alloys, the initial expectations for critical current density improvement were born out. In the ternary alloys, moderately high critical current densities (e.g. between 3.0×10^4 and 5.6×10^4 A cm^{-2} at 40 kOe) were maintained up to fields in the neighborhood of 80 kOe in the alloys: $Ti_{10}-Zr_{20}-Nb_{70}$, $Ti_{10}-Zr_{40}-Nb_{50}$, and $Ti_{28}-Zr_{24}-Nb_{48}$. At the same time the "best" critical current densities of $Zr-Nb_{75}$ and $Ti_{10}-Zr_{40}-Nb_{50}$ were intercompared. As indicated in Table 11-1, which lists the "optimum" critical current densities of the binary and ternary alloys, the 1965 prognosis for Ti-Zr-Nb as a replacement for Zr-Nb was very good. A justification for continuing with the development of Ti-Zr-Nb as a commercial product had thereby been established.

Detailed investigations of the ternary phase diagrams, the superconductive properties, and the relationships between superconductive and metallurgical properties, were carried out simultaneously in Japan (principally at the Central Research Laboratory of Hitachi Ltd.) and in the Soviet Union (at the Baikov Institute of Metallurgy and the University, in Moscow, and at the Physico-Technical Institute for Low Temperatures, Kharkov).

Phase relationships in the Ti-Zr-Nb system have been studied in detail by ALEKSEEVSKII et al [Ale67[a]] in the Soviet Union, and by DOI et al [Doi66[a]] in Japan. At 1050°C, and after quenching from that temperature, ternary Ti-Zr-Nb alloys are single-phase-bcc. The equilibrium phases which appear at lower temperatures are illustrated in Fig.11-1, based on the combined results of the Soviet and Japanese authors. The principal features of these diagrams are: (i) at 900°C, a two-phase-bcc region ($\beta'+\beta''$) has already begun to grow out from the Zr-Nb edge;[†] (ii) below 700°C, $\alpha+\beta$ regions grow out from the Ti and Zr corners; (iii) at 600°C, the Zr $\alpha+\beta$ corner makes connection with the growing $\beta'+\beta''$ lobe; (iv) below about 500°C, apart from a small β-zone at Nb, and a narrow α-strip along the Ti-Zr edge, most of the alloys are two-phase $\alpha+\beta$. The significance of the $\beta \rightarrow \beta'+\beta''$ immiscibility reaction, not to be confused with $\beta \rightarrow \beta'+\beta$ "phase separation" as

---

† As usual, β' is the Nb-lean phase and β'' the Nb-rich.

TABLE 11-1 COMPARISON OF VARIOUS PROPERTIES OF $Zr_{25}-Nb_{75}$ AND $Ti_{10}-Zr_{40}-Nb_{50}$ ALLOYS[†] -- Doi et al. [Doi66[c]]

Applied Field kOe	Critical Current Density[††] J_c, A cm^{-2}	
	$Zr_{25}-Nb_{75}$	$Ti_{10}-Zr_{40}-Nb_{50}$
50	7.5×10^4	$>1 \times 10^5$
60	2.9×10^4	9.2×10^4
70	almost zero	6×10^4
80	zero	3.1×10^4
90	"	6×10^3
100	"	almost zero
T_c	10.9 K	10.07 K
H_r[†††]	∿75 kOe	∿100 kOe
ρ_n 293 K	39 $\mu\Omega$ cm	53 $\mu\Omega$ cm
ρ_n 77 K	28 $\mu\Omega$ cm	42 $\mu\Omega$ cm

† Processing needed to produce results similar to these in each alloy is: 75% cw + (1h/700°C) + 93.75% cw.

†† J_c criterion: 5 μV across 3.5∿4 cm [Doi68[a], Doi68[b]]

††† H_r criterion: 100 μV across 3 cm at 5 mA [Doi68[a]]

described in Sect.9.1 and elsewhere, is considered in [Mon2.7].

As will be shown in the following review, the superconducting transition temperatures of isoelectronic sets of ternary alloys tend to be intermediate between, or a little below, those of the binary Ti-Nb and Zr-Nb end-points. For example, transition temperatures encountered along $Nb_{70}-(Ti-Zr)_{30}$ are: 10.9 K ($Nb_{70}-Zr_{30}$), followed by 10.2 K, 10.5 K, 10.0 K, 9.5 K, and 9.9 K (for $Nb_{70}-Ti_{30}$) [Doi66]; or 10.9 K ($Nb_{70}-Zr_{30}$), followed by 10.3 K, 9.5 K, 9.7 K, and 9.9 K (for $Nb_{70}-Ti_{30}$) [Sav73, p.312].

Two versions of the Ti-Zr-Nb upper critical field surfaces appear in the Soviet literature: (a) that due

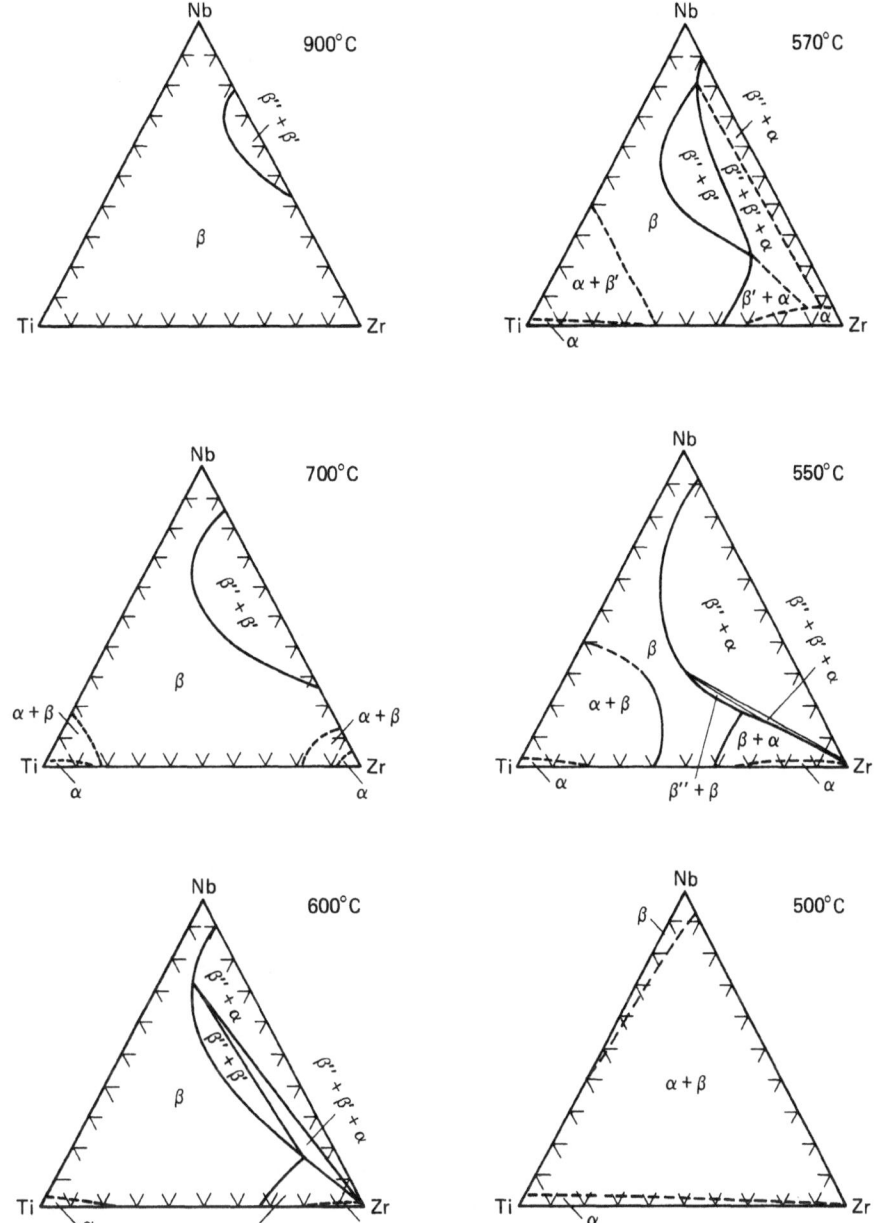

FIGURE 11-1. Ti-Zr-Nb equilibrium phase diagram (at. % linear), for temperatures between 500°C and 900°C, based on the work of DOI *et al* [Doi66[a]] (900°C, 700°C and 570°C) and ALEKSEEVSKII *et al* [Ale67[a]] (600°C, 550°C and 500°C) the latter two diagrams having been modified to take into consideration the absence of appreciable α-phase precipitation from moderate-temperature aged Ti-Nb alloys with Nb concentration greater than ~40 at. % (cf. [Nea71] [Wes80, Wes81, Wes82].

to LAZAREV *et al* [Laz68[a], Laz70], in which a low-H_{c2} trough extends from the Nb corner about half-way (the end of the data set) towards the Ti-Zr edge; *(b)* that due to ALEKSEEVSKII *et al* [Ale68[a]], in which the upper critical field profile for a line connecting Ti-Nb$_{35}$ with Zr-Nb$_{51}$ on the ternary diagram passes through a

peak of height 135.8 kOe at compositions near Ti$_{39}$-Nb$_{40}$-Zr$_{21}$. These seemingly disparate results, which also have been presented side-by-side without comment in the review-book by SAVITSKII *et al* [Sav73, p.313], will be reconciled below.

PART 1: RESEARCH AND DEVELOPMENT OF TITANIUM-ZIRCONIUM-NIOBIUM ALLOY SUPERCONDUCTORS

11.2 TRANSITION TEMPERATURES OF Ti-Zr-Nb RESEARCH ALLOYS

In a fundamental energy-gap investigation using current-tunnelling between a normal metal and superconductor, SULLIVAN and ROOS [Sul67] studied current-voltage characteristics and transition temperatures of several Ti-Zr-Nb and other alloys. An estimated gap voltage, V_g, was compared in magnitude and temperature dependence with predictions from BCS theory [Bar57]. Rather satisfactory correlations were obtained. For the three alloys investigated (against several contact metals in most cases) the values of $V_g/k_B T_c$ derived were:

$$Ti_{10}\text{-}Zr_{70}\text{-}Nb_{20}: \quad 1.02,$$

$$Ti_{10}\text{-}Zr_{50}\text{-}Nb_{40}: \quad 1.29 \pm 0.06,$$

$$Ti_{10}\text{-}Zr_{40}\text{-}Nb_{50}: \quad 1.18 \pm 0.08,$$

which may be compared with the BCS-half-gap width of 1.76 (see, for example, [Mon15.1]).

The transition temperatures of large groups of Ti-Zr-Nb alloys have been studied and tabulated by: *(a)* LAZAREV *et al* [Laz68[a]] (13 ternary alloys, in the as-cast condition and after annealing 24h/520°C and 120h/560°C, respectively); *(b)* ZWICKER *et al* [Zwi68] (5 ternary alloys, in the cast-plus-deformed, and heat-treated conditions); *(c)* SAVITSKII *et al* [Sav73, p.312] (11 ternary single-phase-bcc alloys), whose results were assembled as contours on a Ti-Zr-Nb concentration triangle. The highest binary-alloy transition temperatures were those of $Ti\text{-}Nb_{60\text{-}70}$ at 9.7 K and $Zr\text{-}Nb_{\sim 80}$ at 10.9 K. It is interesting to note that at no point on the concentration triangle for quenched alloys did the transition temperature exceed 10.9 K. In fact the contours were representative of a shallow trough extending from the vicinity of the Nb vertex towards $Nb_{30}\text{-}(Ti_{50}\text{-}Zr_{50})_{70}$. At low Nb levels the isoelectronic compositions, definable by $(Ti\text{-}Zr)\text{-}Nb_{const.}$, were approximately "isothermal" with characteristic T_c's that decreased with decreasing Nb content. The superconducting transition temperatures of some twenty-four Ti-Zr-Nb alloys have been measured by DOI *et al* [Doi66]. These results, augmented by those for five Ti-Nb alloys and five Zr-Nb alloys ob-

tained by other workers, were plotted on the composition triangle reproduced in Fig.11-2. For Nb concentrations greater than about 60 at.%, the contour arrangement differs in detail from that due to SAVITSKII *et al* especially for compositions well removed from the sides of the triangle. Such differences could be a result partly of interpretation, and partly of microstructural differences between the alloys.

Transition temperature of course responds to annealing as a consequence of the β-phase decomposition which inevitably occurs for all compositions excepting those in a restricted zone near the Nb-vertex of the composition triangle. Increases in T_c are frequently noted in response to the precipitation of Nb-enriched β-phase precipitates.

Some transition temperatures are listed in Table 11-2.

11.3 CRITICAL FIELDS OF Ti-Zr-Nb RESEARCH ALLOYS

11.3.1 The Lower Critical Field, H_{c1}

LAZAREV *et al* [Laz70[a]] have determined the lower critical fields at 4.2 K of several Ti-Zr-Nb alloys from measurements of magnetic moment (determined ballistically) *versus* applied field strength of up to 40 kOe. The alloys investigated covered a wide range of compositions (not all reported), from $Ti_{15}\text{-}Zr_{50}\text{-}Nb_{35}$ to $Ti_{48}\text{-}Zr_5\text{-}Nb_{47}$. At least one alloy, $Ti_{30}\text{-}Zr_{22}\text{-}Nb_{48}$, was measured in each of the three metallurgical conditions: *(a)* as-cast, *(b)* deformed, and *(c)* deformed-plus-annealed, in response to which H_{c1} underwent successive increases. Values of H_{c1} and H_{c1}^* (the field at maximum magnetization -- somewhat greater than H_{c1} since in irreversible material $M(H_a)$ undergoes a rounded, rather than sharp, "entry" into the mixed state, cf. Fig.7-17) were, respectively:

(a) as-cast : 600 and 1000 Oe,
(b) deformed : 850 and 2500 Oe,
(c) deformed + annealed: 1500 and 4500 Oe.

This increase in H_{c1}^* in response to thermomechanical

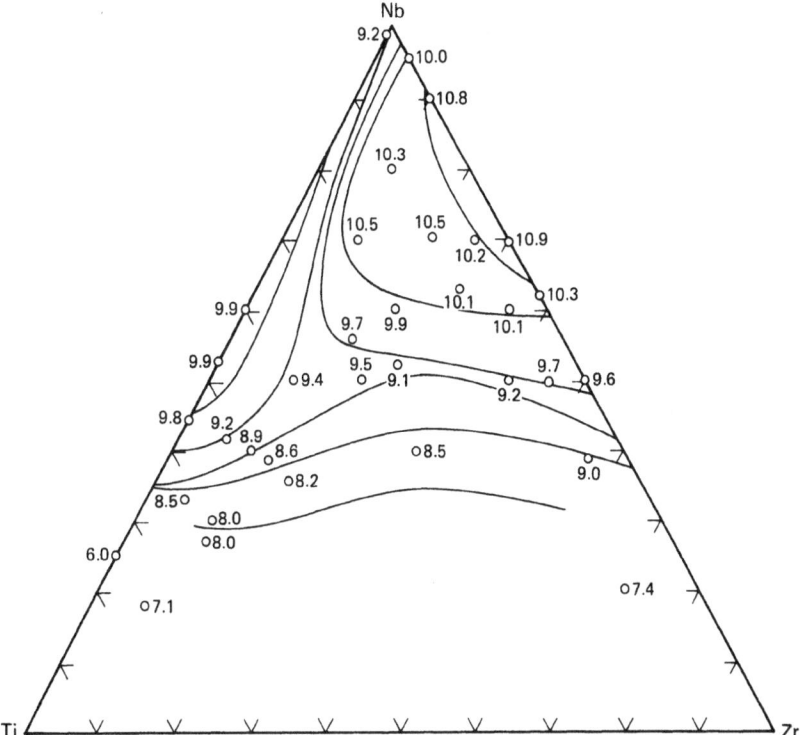

FIGURE 11-2. Ternary diagram (at. % linear) of the superconducting transition temperatures (K) of twenty-four Ti-Zr-Nb alloys as measured by DOI *et al* [Doi66] (augmented by results for several Ti-Nb and Zr-Nb alloys obtained by other workers) — after DOI *et al* [Doi66].

treatment reflected an increasing resistance of the sample to the entry of flux, in other words an increasing bulk pinning strength, consistent with which was a corresponding set of increases in J_c(4.2 K, 20 kOe), viz 10^3, 10^4, and 10^5 A cm^{-2}, respectively.

11.3.2 The Upper Critical Field, H_{c2}

In this section, as elsewhere in this chapter, attention is directed solely towards the *experimentally* determined value of the upper critical field. This being so, distinctions between H_{c2}, H_{c2}^*, H_r, H_u, useful in other contexts (cf., for example, Sect.3.6) are abandoned and the symbol H_{c2} will be frequently used to represent the experimentally determined upper critical field.

In studies of the influences of *dilute* third-element additions (mostly SM) and other metallurgical variables on the superconductive properties of Ti-Nb alloys, HELLER *et al* [Hel71[a]] and ZWICKER *et al* [Zwi70] have investigated the upper critical fields of: (a) Ti$_{65}$-Zr$_2$-Nb$_{33}$, 97.2% cold worked and annealed 10h/400°C; (b) the three alloys Ti-Zr$_4$-Nb$_{31,33,35}$

cold worked and heat treated; (c) the companion set Ti-Zr$_{2,3,4}$-Nb$_{33}$. The results of these studies suggested that it was generally disadvantageous to add a few percent of Zr to Ti-Nb$_{33}$-type alloys. Although the modulus did not change much, and the microhardness increased rather gradually at the rate of 5 kg mm^{-2} per at.% Zr, the upper critical fields were lower than those of the binary alloy.

One of the first systematic studies of upper critical field in *concentrated* Ti-Zr-Nb alloys was by RALLS and co-workers (e.g. [Ral64]) who measured critical current densities, and at the same time the resistive upper critical fields of several members of the alloy series Nb$_x$-(Ti$_{50}$-Zr$_{50}$)$_{100-x}$ which perpendicularly bisects the Nb-apex composition triangle. The alloys had been 99.92% deformed from the homogenized condition and were probably bcc, although no metallurgical studies had been undertaken. As has come to be expected for single-phase-bcc alloys of varying electron/atom ratio, H_{c2} varied smoothly with composition and passed through a maximum, $H_{c2,max.}$(4.2 K) = 90.0 kOe, at e/a = 4.4 (x = 40 in the above formula). Other investigators of the variation of H_{c2} throughout the interior of the Ti-Zr-Nb composition triangle were

TABLE 11-2 SUPERCONDUCTING TRANSITION TEMPERATURES OF SOME TITANIUM-ZIRCONIUM-NIOBIUM
ALLOYS LISTED WITHIN EACH GROUP IN ORDER OF INCREASING NIOBIUM CONTENT

Compositions, At.%			Transition Temperatures in the Conditions Listed, K			
Ti	Zr	Nb				
Zwicker et al. [Zwi68]			A	B	C	D
51	30	19	7.3	6.4	6.4	9.8
74	7	19	7.9	6.6	8.1	8.2
61	18	21	7.0	6.7		7.0
25	53	22	7.5	7.3		8.8
56	8	36	9.0	9.8		8.7
Lazarev et al. [Laz70]			E	F	G	
15	50	35	8.6	9.2	9.3	
15	44	41	8.7		9.3	
27	30	43	8.6	9.0	9.1	
48	5	47	8.7		8.7	
30	22	48	8.9	9.1	9.0	
16	32	52	9.4		9.5	
14	24	62	9.6			
15	20	65	9.8		9.7	
15	10	75	9.7			
Savitskii et al. [Sav73, p.311]			H			
--	75	25	8.3			
10	65	25	7.5			
20	55	25	7.5			
30	45	25	7.7			
40	35	25	7.7			
50	25	25	7.2			
65	10	25	7.7			
70	5	25	7.5			
74	1	25	7.4			
75	--	25	7.2			
--	50	50	10.2			
5	45	50	10.2			
10	40	50	9.8			
25	25	50	9.7			

CONDITIONS:

A - Cast + deformed (c+d)
B - c+d + 0.5h/900°C + deformed (d)
C - c+d 160h/500°C
D - c+d + 0.5h/900°C + d + 160h/500°C
E - As-cast
F - 24h/520°C
G - 120h/560°C
H - Quenched

DeSORBO et al [Des67], LAZAREV et al [Laz68[a]], and ALEKSEEVSKII et al [Aℓe68[a]]. The results of the latter two groups were summarized by SAVITSKII et al [Sav73, p.313] although no attempt at a reconciliation of the apparently divergent results was made.

In setting up a Nb-apex triangular diagram for the display of the ternary upper critical fields results, the data of RALLS for the binary systems Nb-Ti and Nb-Zr provide reliable and suitably spaced boundary values. But when *all* of the ternary data from the references cited above were transferred to a single diagram, it was found to be impossible to construct a meaningful set of contours. It was decided, therefore, to take the data one set at a time. But the three sets of contours, when independently plotted, were found to differ seriously one from the other. Strong similarities would certainly have been expected among the results of the two Soviet groups whose alloys had received rather similar heat treatments (at about 550°C in some cases). But in fact different prominent features showed up in each of the three individual diagrams. For example, LAZAREV et al claimed the existence of a valley extending from the Nb vertex towards the Ti-Zr edge, ALEKSEEVSKII et al obtained a prominent peak at position A in Fig.11-3, while the data of DeSORBO et al led to minor peaks near A and again at B. Closer examination of the three sets of data led to the following conclusions:

(i) LAZAREV: The results of LAZAREV did not include data in the vicinity of A and could therefore not be construed to include the possibility of a peak at that position.

(ii) ALEKSEEVSKII: Such *binary alloy* data as were available were some 10 and 26% too high. However, when *all* these data were reduced by 18%, no change in character of the upper critical field diagram resulted. The fact that the general features of the upper critical field diagram, and in particular the peak at A, were invariant to a sizeable systematic error in the ternary alloy H_{c2} determination, in the presence of fixed boundary values (due to RALLS), lent some credence to the reliability of the general form of the diagram.

(iii) DeSORBO: Close examination of these data revealed that the peak at B relied essentially on two data points with unusually high values. Returning to the original paper it was noted that these particular points were the only two which lay well above a general curve of H_{c2} *versus* e/a for numerous ternary and quaternary alloys. The lowering of this pair of

FIGURE 11-3. Salient features of the unmodified upper critical field results of: (a) LAZAREV et al [Laz68[a]] (a trough extending "downwards" from the Nb vertex); (b) ALEKSEEVSKII et al [Aℓe68[a]] (a peak near A); (c) DeSORBO [Des67] (peaks near A and B).

points onto the common line removed the anomalous B-peak in the triangle diagram.

After making this change in DeSORBO's data, applying a scaling correction to that of ALEKSEEVSKII, and reconnecting each of the three data sets with self-consistency in mind, three remarkably similar figures emerged. Figs.11-4, 11-5, and 11-6, possess the following features in common: *(a)* a monotonic increase in

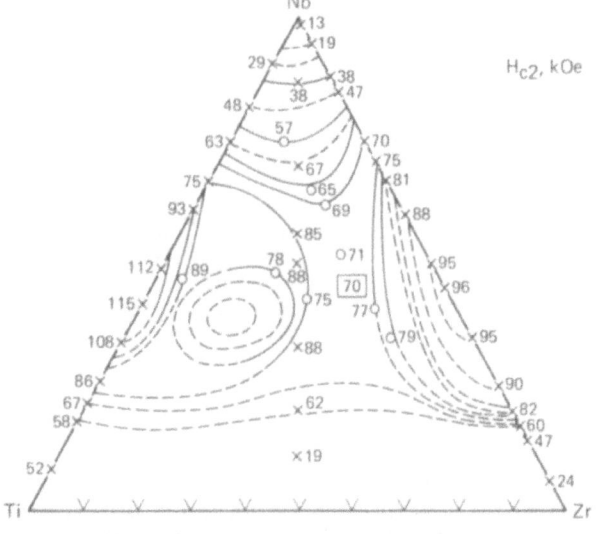

FIGURE 11-4. Ternary diagram (at. % linear) of the resistively measured upper critical fields of nine Ti-Zr-Nb alloys as determined by LAZAREV et al [Laz68[a]]. The binary boundary-value data and ternary "axial" data, all designated (x), are due to RALLS [Raℓ64].

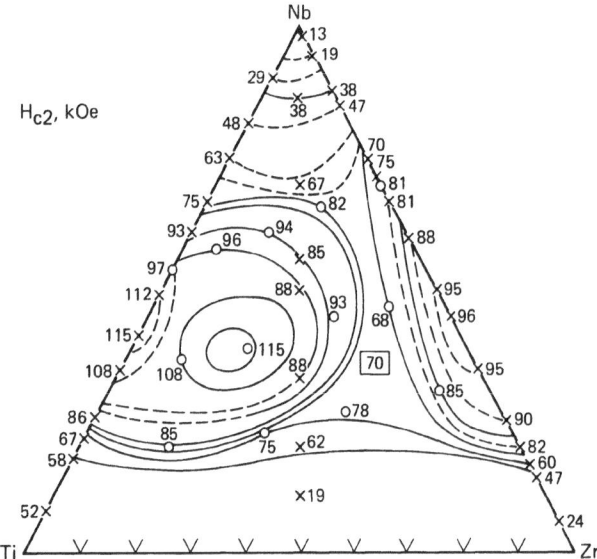

FIGURE 11-5. Ternary diagram (at. % linear) of the resistively measured upper critical fields of eleven Ti-Zr-Nb alloys as initially determined by ALEKSEEVSKII *et al* [Aℓe68ᵃ] but herein uniformly reduced by 18%. The binary boundary-value data and ternary "axial" data, all designated (x), are due to RALLS [Raℓ64].

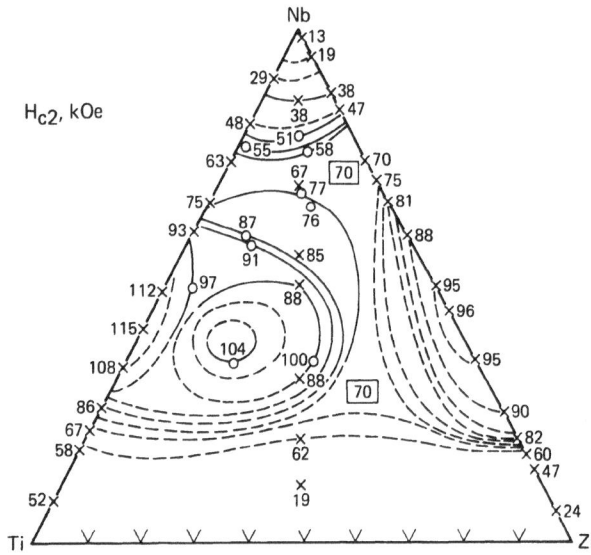

FIGURE 11-6. Ternary diagram (at. % linear) of the magnetically determined upper critical fields of ten Ti-Zr-Nb alloys as determined by DeSORBO [Des67] plotted as-measured excepting that the two values which gave rise to a local maximum in the vicinity of region B in Fig. 11-3 have, with justification, been reduced from 79.5 and 76.0 kOe to 58 and 51 kOe, respectively. The binary boundary-value data and ternary "axial" data, all designated (x), are due to RALLS [Raℓ64].

H_{c2} for isoelectronic alloys of compositions up to about Nb_{30}-$(Ti$-$Zr)_{70}$, followed by a plateau at ∿70 kOe lying between the maximum and the Nb-Zr edge, which itself rises to ∿96 kOe; *(b)* a local maximum (>100 kOe) centered about Ti_{40}-Zr_{20}-Nb_{40}. It is not certain, however, that this maximum is completely isolated; by a slight rearrangement of the contours it could be connected by a sharp ridge to a maximum on the Ti-Nb edge. The presence of such a maximum is quite reasonable, since its position can be regarded as being centered within a sequence of bcc-when-quenched alloys extending from Zr-Nb_{80} (e/a = 4.8) towards Ti-Zr_{20} (e/a = 4.0); the peak thus occurs in the vicinity of e/a = 4.4 as it does for binary Ti-Nb, binary Zr-Nb, and similar alloys.

11.4 CRITICAL CURRENT DENSITIES OF Ti-Zr-Nb RESEARCH ALLOYS

The influences of *small* additions of Zr (viz 2, 4, and 6 at.%) upon the 100-kOe J_c of cold-worked and cold-worked-plus-aged Ti-Nb_{33} have been studied by ZWICKER *et al* [Zwi70]. The best values were obtained with alloys which had been heat treated 17 or 34 h at 400°C after deformation. Under the conditions established by ZWICKER, Zr had a deleterious effect, the addition of 2 to 4 at.% of it lowering the 100-kOe J_c from a little above $1{\times}10^4$ A cm⁻² to about $5{\times}10^3$ A cm⁻². As noted above, a decrease in H_{c2} was also observed. In both cases, however, H_{c2} and J_c seem to be passing through maxima at 4 at.% Zr suggesting that improved properties lay beyond the range of the investigation. It is interesting to note in passing that one of the compositions studied by ZWICKER approximated that of the Soviet technical alloy 50 BT (cf. Table 10-1) which according to KADYKOVA and FEDOTOV [Kad72] was 42.5Ti-7.5Zr-50Nb, i.e. Ti_{59}-Zr_6-Nb_{35}. In a second series of studies, this time of *concentrated* alloys, ZWICKER *et al* [Zwi68] investigated the relative critical current densities[†] of a series of almost isoelectronic alloys (e/a ∿ 4.2) with almost constant (∿20 at.%) Nb concentration, but

---

[†] Because of the low degree of cold work the critical current densities of all alloys, including the binary control, were very low. It was hoped, presumably, that the alloys would preserve their ranking after process optimization.

with the Zr level varying from 7 to 53 at.%. The rationale for choosing the alloys $Ti_{74}-Zr_7-Nb_{19}$, $Ti_{51}-Zr_{30}-Nb_{19}$ and $Ti_{25}-Zr_{53}-Nb_{22}$ had to do with the selection of an alloy sequence characterized by e/a = 4.2 (actually 4.19, 4.19, and 4.22, respectively) in contrast to the e/a = 4.5 of the previously well-studied technical alloy, $Ti_{10}-Zr_{40}-Nb_{50}$, to be discussed below. It was concluded from the results of the J_c versus $H_a(\widetilde{<}100$ kOe) experiments performed on the above alloys, that small additions of Zr (<10 at.%) could increase the high-field critical current density beyond that of the isoelectronic Zr-free alloy $Ti-Nb_{20}$. Further substitution of Zr for Ti brought about a drop in the high-field critical current density.

In a study of the influence of alloying, normal to the isoelectronic direction in the composition triangle, RALLS [Raℓ64] investigated the critical current densities of 99.92% cold-worked wires prepared from the alloy series $Nb_x-(Ti_{50}-Zr_{50})_{100-x}$. Since companion studies were also undertaken on binary-alloy samples of Ti-Nb and Zr-Nb prepared under exactly similar conditions, it was possible for comparative evaluations to be made. The validity of deductions based on intercomparisons among unoptimized samples is of course questionable unless one can be certain that the ranking of the critical current densities will be preserved during optimization by heat treatment and additional cold work. This will tend to be the case if precipitation effects are primarily responsible for the initial ranking. Cold-worked-only alloys generally exhibit a pronounced "peak effect", a local maximum in $J_c(H_a)$ at fields just below H_{c2} -- see, for example, [Mon21.12]. The effect of process optimization is to "fill-in" the critical current density at fields lower than H_{peak} leaving $J_{c,peak}$ practically invariant, Fig.11-7. Consequently, the value of $J_{c,peak}$ can be regarded as a reasonable indicator of high-field performance and, plotted on a composition triangle, may be employed as the basis for an intercomparison of the critical current densities of Ti-Nb, Zr-Nb, and Ti-Nb-Zr. Fig.11-8(a) constructed from data listed in Table 11-3 depicts the composition dependence of $J_{c,peak}$ for the Ti-Zr-Nb ternary system. Also indicated on that figure are the locations of some U.S. technical alloys ($Ti-Nb_{30}$ and $Ti-Nb_{37}$), a Japanese alloy to be considered below (Hitachi Cable Ltd.'s $Ti_{62.5}-Zr_{2.5}-Nb_{35}$), and the Soviet alloy types 35 BT, 50 BT, and 65 BT (cf. Table 10-1).

In selecting a general-purpose technical superconducting alloy, assuming fabricability criteria have

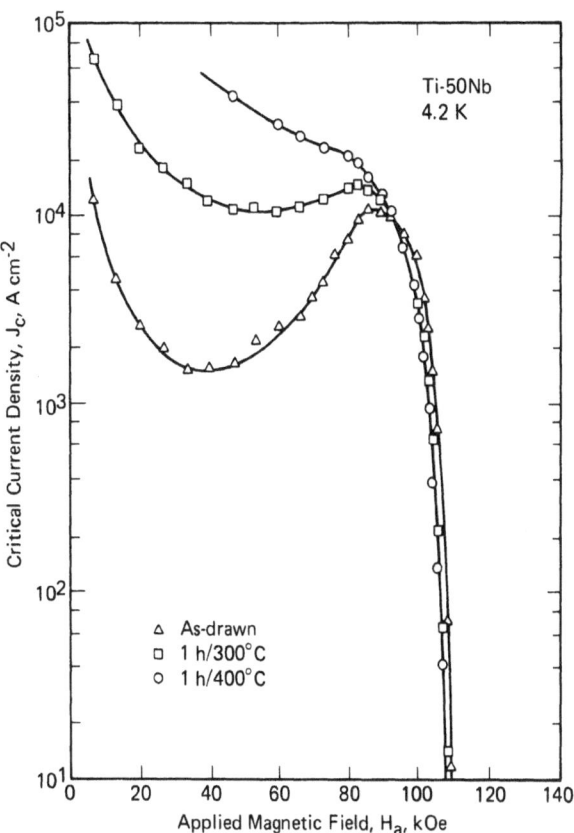

FIGURE 11-7. Typical set of J_c versus H_a characteristics (criterion: 1 μV across ~ 4 cm) for a Ti-alloy superconductor (Ti-50Nb in this case) illustrating the insensitivity of the very-high-field J_c to variation of the final processing conditions — after RAUCH [Rau68].

been met, the best alloys are those which combine large current-carrying capacities with high upper critical fields (but cf. Sect.7.32, and in particular Sect.7.32.3, for a discussion of this topic). Accordingly, a useful figure-of-merit to assist in predictions based on cold-work-only data, if only these are available, might be the product of $J_{c,peak}$ and the upper critical field. Values of these quantities based on RALL's data [Raℓ64] are presented in Table 11-3 and plotted on a composition triangle in Fig.11-8(b). As a consequence of the relatively small range of variability of the upper critical field, the effect of the H_{c2}-multiplier on the general character of the contour diagram is not strong. The diagrams show the standard U.S. alloys, the Hitachi alloy, and 65 BT to be reasonably well placed, although improvements in the latter pair of alloys may follow increases in the Nb levels. In the U.S., the Ti-Nb alloy corresponding to 35 BT, which occupied a very favorable position on the "cold-worked" diagram, was abandoned because of fabricability difficulties. The work

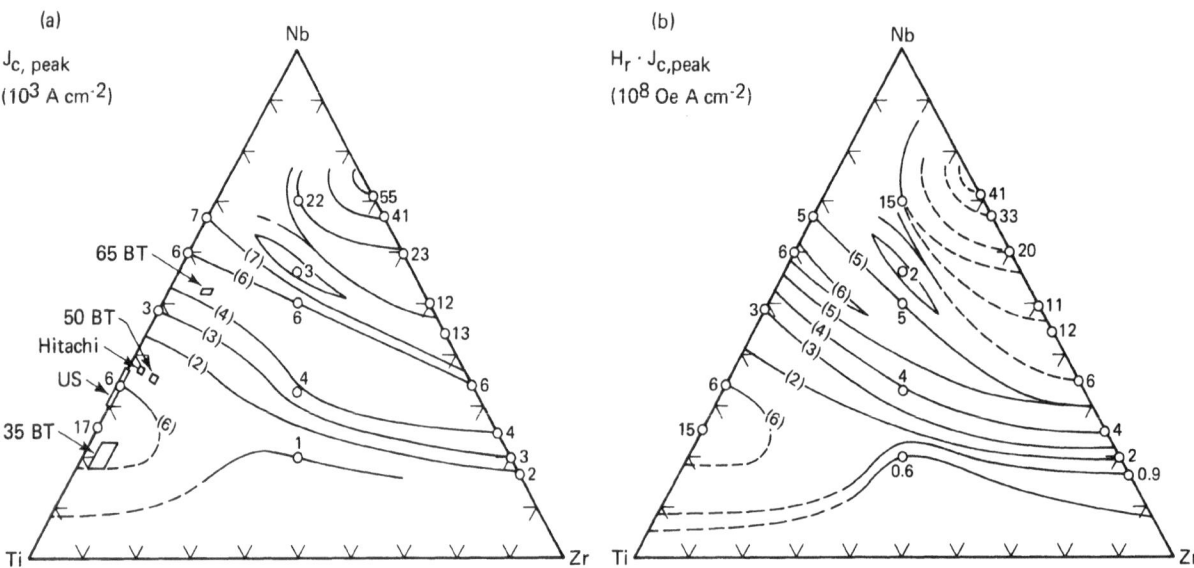

FIGURE 11-8. Ternary Ti-Zr-Nb diagrams (at. % linear) displaying: (a) the "peak" critical current densities (i.e., J_c's taken from the peak of $J_c(H_a)$ — cf. Fig. 11-7) for cold-worked (~99.9%) Ti-Zr-Nb alloys, presented in the units 10^3 A cm^{-2} — data source, RALLS [Raℓ64]; (b) the product $H_r \cdot J_{c,peak}$ in the units 10^8 Oe A cm^{-2}, where H_r is the resistive upper critical field; data from Table 11-3 (criteria: J_c, 1/4 μV cm^{-1}; H_r, 1/4 μV cm^{-1} at 1 or 10 A cm^{-2}) — data source, RALLS [Raℓ64].

of DOI et al [Doi68], who have plotted on a composition triangle the 80-kOe J_c's of fifty-seven Ti-Zr-Nb alloys (including six Ti-Nb and six Zr-Nb alloys) subjected to cold work and intermediate or final heat treatment, will be considered together with numerous other papers by the same authors in PART 2 of this chapter. The ternary critical current density plot for thermomechanically processed Ti-Zr-Nb alloys differs considerably from Fig.11-8.

RASSMANN and ILLGEN [Ras72^b] have surveyed the critical current densities and metallurgical characteristics of a series of Ti-Zr-Nb alloys whose compositions lie on a set of lines parallel to the Ti-Nb edge of the composition triangle at the levels of 5, 10, and 15 at.% Zr. With Nb contents between about 17 and 48 at.%, they embraced the compositions of 35 BT and 50 BT, and approached that of 65 BT. The alloys were tested in two metallurgical conditions: (a) cold-worked (cw) ∿99.9%, and (b) cw plus 1-h heat treated at temperatures of 200, 300, 400 and 500°C; critical current density values were quoted for an applied field of 45 kOe. In experiments with the alloy sequence:

$$Zr_{10} - \begin{cases} Ti_{72}\text{-}Nb_{18} \\ Ti_{64.5}\text{-}Nb_{25.5} \\ Ti_{57.5}\text{-}Nb_{32.5} \\ Ti_{52}\text{-}Nb_{38} \end{cases},$$

as the 1-h heat-treatment temperature was increased from 200 to 500°C in steps of 100°C, the 45-kOe J_c rose to a peak at 400°C. This is a universally encountered peak which, as pointed out in Sect.7.26.5, is a property of the J_c versus aging-temperature characteristics of both precipitation-immune and precipitation-prone previously-cold-worked Ti-Nb-base alloys. In the "as-worked" family of curves, in which the 45-kOe J_c was plotted versus Nb content, the highest critical current densities were encountered below about 20 at.% Nb, a minimum occurring near 30 at.% (cf. Fig.11-8(a)). After individual heat treatments at empirically determined optimal temperatures the picture was completely inverted, maximal 45-kOe J_c occurring in alloys whose Nb contents were in the range of about 25-30 at.%, Fig.11-9. For the alloy group in question, an absolute maximum 45-kOe J_c of 9.9×10^4 A cm^{-2} was attained by optimally heat treating $Ti_{48.5}\text{-}Zr_{15}\text{-}Nb_{36.5}$. Some other data are listed in Table 11-4.

As a result of this study it was noted that in Nb-(Ti-Zr) isoelectronic alloys with Nb content greater than about 35 at.%, the substitution for Ti of up to 15 at.% Zr resulted in improved current-carrying capacities. This is in accord with Fig.11-8(a) which shows in the region concerned that, in association with a $J_{c,peak}$ which generally increases in the Nb-direction, the iso-$J_{c,peak}$ lines "slope downwards"

TABLE 11-3 "PEAK" CRITICAL CURRENT DENSITIES ($J_{c,peak}$) AND RESISTIVE UPPER CRITICAL
FIELDS (H_r) OF COLD-WORKED ($\sim$99.9%) TITANIUM-NIOBIUM, ZIRCONIUM-NIOBIUM
AND Nb_x-$(Ti_{50}$-$Zr_{50})_{100-x}$ ALLOYS -- Ralls [Raℓ64]

Compositions, at.%[+]			$J^{++}_{c,peak}$ $A\ cm^{-2}$	H_r^{+++} kOe	$H_r \cdot J_{c,peak}$ $10^8\ Oe\ A\ cm^{-2}$
Ti	Zr	Nb			
74	--	26	1.7×10^4	86	14.6
66	--	34	5.9×10^3	108	6.4
51	--	49	2.6 "	112	2.9
40	--	60	6.0 "	92.5	5.6
33	--	67	6.8 "	75	5.1
--	83	17	1.5×10^3	60	0.9
--	80	20	2.8 "	81.5	2.3
--	75	25	4.2 "	90	3.8
--	65	35	6.3 "	95	6.0
--	55	45	1.3×10^4	95.5	12.4
--	50	50	1.2 "	95	11.4
--	40	60	2.3 "	88	20.2
--	33	67	4.1 "	81	33.2
--	29	71	5.5 "	75	41.3
40	40	20	1.0×10^3	62	0.6
33.5	33.5	33	4.4 "	88	3.9
25	25	50	5.5 "	88	4.8
22	22	56	2.7 "	85	2.3
15	15	70	2.2×10^4	67	14.7

+ To the nearest integer.

++ J_c criterion: 1 µV across 4 cm.

+++ H_r criterion: 1 µV across 4 cm at 1 or 10 A cm^{-2}.

in the Ti→Zr direction. RASSMANN and ILLGEN [Ras72[b]]
did, however, caution against an automatic substitution of Zr in this way, noting that: (a) although the
current-carrying capacity is improved, the workability
suffers: (b) the properties of a Ti-Zr-Nb alloy prepared from iodide-process Ti are no better than those
of the corresponding binary alloy prepared from commercial Kroll Ti, a starting-element with a higher
interstitial content (cf. Table 7-14, Sect.7.26.2).
In this vein RASSMANN and ILLGEN then went on to consider improvements to the current-carrying capacity of
$Ti_{60.5}$-Zr_5-$Nb_{34.5}$ in response to the addition of
3000 ppm of oxygen in the form of Nb_2O_5.

11.5 INTRODUCTION TO THE PATENT LITERATURE OF Ti-Zr-Nb ALLOY SUPERCONDUCTORS

Technically important Ti-Zr-Nb alloys arose from
the discovery that considerable improvement in the
workability of Nb-Zr followed the substitution
into it of 10 or more wt.% of Ti. It was noted for
example in a patent awarded to Imperial Metal Industries [Imp65] that, although Zr-75Nb could not be
successfully forged without pre-extrusion and was
difficult to draw, an alloy with the same Zr content
but with 15 wt.% Ti substituted for some of the Nb,

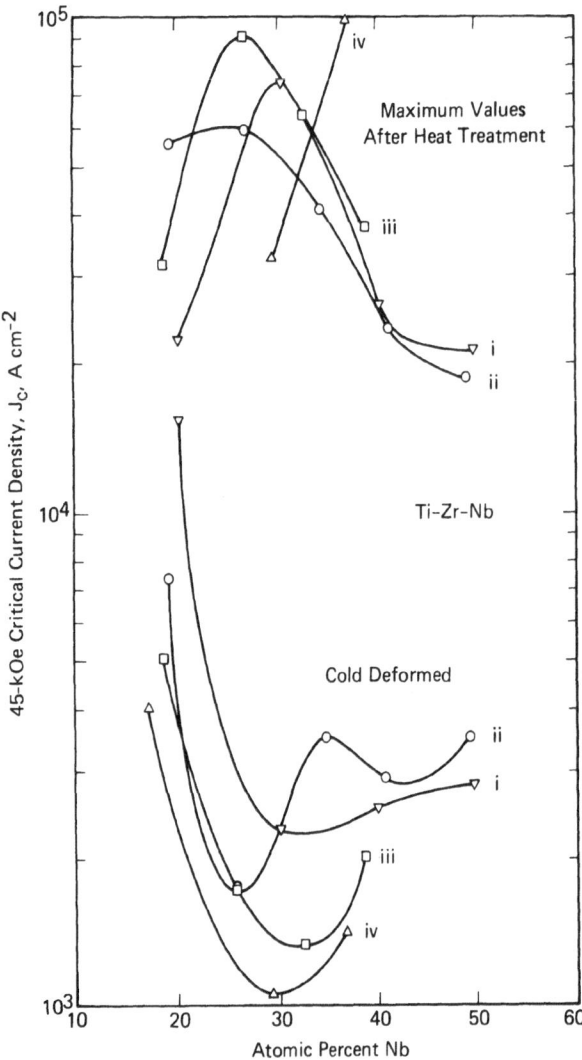

FIGURE 11-9. Critical current densities at 45 kOe of four series of Ti-Zr-Nb alloys measured in the cold-deformed ($\sim$ 99.9%) as well as heat-treated (for maximum J_c — generally 1 h/400°C) conditions plotted as a function of Nb content. The four alloy series depicted are characterized by the following fixed levels of Zr: (i) 0 at. % Zr; (ii) 5 at. % Zr; (iii) 10 at. % Zr; and (iv) 15 at. % Zr (transition criterion: 40 μV) — after RASSMANN and ILLGEN [Ras72b].

TABLE 11-4 CRITICAL CURRENT DENSITIES (J_c) OF COLD-WORKED ($\sim$99.9%) AND AGED Ti-Zr-Nb[+] AND Ti-Nb[++] ALLOYS -- Rassmann and Illgen [Ras72b]

Compositions, at.%			J_c (45 kOe)[+++]
Ti	Zr	Nb	10^4 A cm^{-2}
72	10	18	3.2
64.5	10	25.5	8.9
57.5	10	32.5	6.3
52	10	38	3.8
80	--	20	2.2
70	--	30	7.4
60	--	40	2.6
50	--	50	2.1

+   Aged 1h/400°C.

++ Aged 1h at optimal temperature (presumably $\sim$400°C).

+++ J_c criterion: 40 μV across sample.

TABLE 11-5 TERNARY ALLOY COMPOSITIONS IN WT.% [Imp65].

Ti	Zr	Nb
5-80	5-80	≥25
"	"	≥20
"	"	≥15
5-50	20-60	≥25
"	"	≥20
"	"	≥15

could be easily forged at 1000°C, and cold swaged and drawn (>95%) to a 0.25-mm$^\phi$ filament without difficulty. The lower limit for the Ti concentration was stated to be 5 wt% (where very small ingots were concerned) and 15 wt.% Ti was preferable. Awarded to Imperial Metal Industries [Imp65] were the alloys defined in Table 11-5. No superconductive data were provided.

A wide range of Ti-Zr-Nb alloys were claimed, data presented, and applications cited in U.S. Patent 3,408,604 (1963-1968) assigned to Hitachi Ltd. by DOI *et al* [Doi68b]. More-or-less the same range of alloys were also covered in French Patent 1,410,055 (1963-1965) [Hit65], both patents referring back to an application made in Japan in October, 1963. The alloys claimed in the U.S. patent can be defined by Ti$_{1-79}$-Zr$_{1-79}$-Nb$_{20-63}$, but particular attention was drawn to alloys that fell within the region on the ternary diagram defined by straight lines connecting the points: Ti_1-Zr_{36}-Nb_{63}, Ti_{10}-Zr_{27}-Nb_{63}, Ti_{47}-Zr_1-Nb_{52}, Ti_{79}-Zr_1-Nb_{20}, Ti_{68}-Zr_{12}-Nb_{20}, Ti_1-Zr_{59}-Nb_{40}, with particular attention being paid to the alloy group Ti_{1-21}-Zr_{29-53}-Nb_{37-59} within that region. Some critical temperature values were acquired as well as a large body of critical current data, which was then presented numerically on ternary diagrams for applied field values of 50, 70 and 80 kOe. The condition of the 0.25-mm$^\phi$ wire prior to critical current measurement could be simply defined as: 95% cold work plus 1h/600°C plus 93.8% cold work. Alloys which when prepared in this way exhibited 50-kOe J_c's of more than 1.0×10^5 A cm^{-2} are listed in Table 11-6. Some useful comparative I_c *versus* composition data for various slices of the composition triangle were given, some of which are reproduced in Fig.11-10. A limited amount of critical current data were also presented in the French patent [Hit65].

DOI *et al* [Doi68^b] have also introduced briefly: *(a)* alloy workability, with reference to the ductile/brittle transition commonly regarded as a property of

bcc metals and alloys; *(b)* the influences of impurity elements; *(c)* the application of insulation and coatings of metals such as Cu, Al or Ag. In addition they suggested the following uses for superconductive wire: *(i)* superconducting coils for electron microscopes and research magnets; *(ii)* saddle magnets for MHD; *(iii)* magnets for particle accelerators; *(iv)* windings for submarine propulsion motors.

Other Ti-Zr-Nb patents, less comprehensive in materials claims, were those issued to the British Central Electricity Generating Board (1964-1966)[Che65]

TABLE 11-6 COMPOSITIONS OF THE TERNARY Ti-Zr-Nb
ALLOYS STUDIED BY Doi *et al*. [Doi68^b]

Ti	Zr	Nb	(at.%)
50	15	35	
10	50	40	
15	45	40	
30	30	40	
45	15	40	
10	45	45	
15	40	45	
5	45	50	
10	40	50	
15	35	50	
30	20	50	
10	30	60	
--	30	70	
5	25	70	
10	20	70	

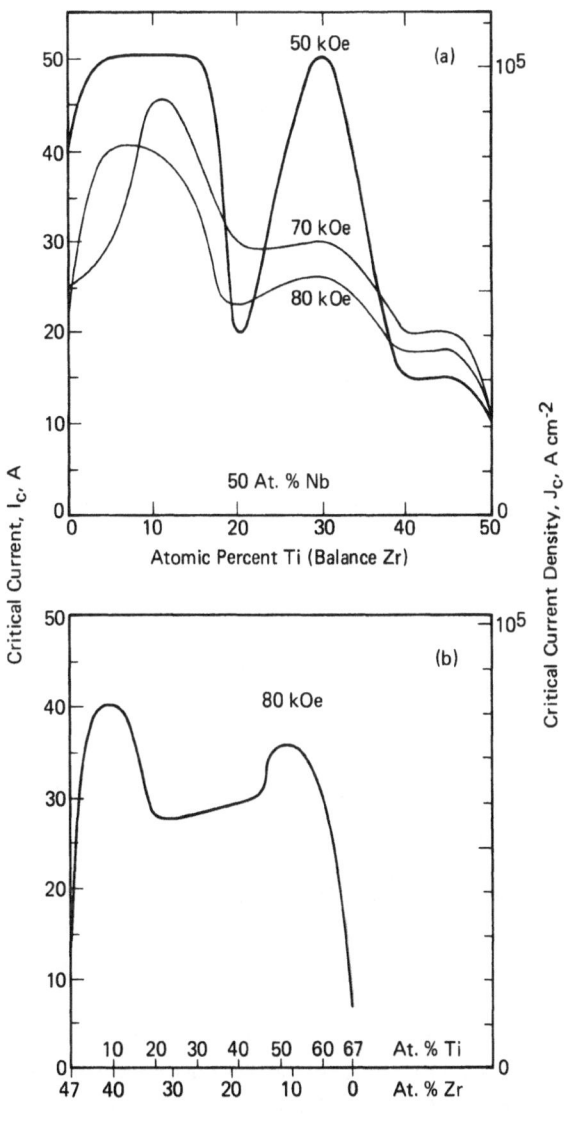

FIGURE 11-10. Critical currents and corresponding critical current densities of 0.25 mm$^\phi$ Ti-Zr-Nb wire measured at 4.2 K in the fields indicated. Depicted are data taken from sections of the ternary diagrams defined by: (a) lines connecting (50 at. % Nb, 50 at. % Zr) to (50 at. % Nb, 50 at. % Ti); (b) a line connecting (53 at. % Nb, 47 at. % Zr) to (33 at. % Nb, 67 at. % Ti) (criterion: 5 µV across 3.5 cm) — after DOI *et al* [Doi68^b].

and the Compagnie Francaise Thomson-Houston (1965-1967) [Com67[a]]. The latter also referred to the fabrication of conductor, and its cladding with metals such as Cu, Al, Ag, and stainless steel. Finally, attention is directed towards an interesting patent issued in 1973 to Hitachi Ltd. [Doi73]. In this, DOI and KUDO described equipment for the continuous coating of superconducting filaments by passing them continuously through baths of the molten elements Al, Sn, Cd, and Pb, whose melting points are commensurate with the heat treatment temperatures used in optimizing superconducting critical current density. In fact it was claimed that under such conditions, satisfactory results could be obtained even with wire that had received its normal heat treatment, and data were presented in support of that point.

PART 2: TITANIUM-ZIRCONIUM-NIOBIUM TECHNICAL ALLOY DEVELOPMENT IN JAPAN

Through the medium of some sixteen papers published between 1966 and 1970, researchers at the Central Research Laboratory of Hitachi Ltd., Tokyo, have discussed the metallurgy of Ti-Zr-Nb alloys and their thermomechanical processing to superconducting wire with technically useful current-carrying capacities. A further group of five papers (1970-1972) have dealt with the textures, heat treatment, and microstructures of rolled sheet, and critical-current anisotropies in ribbons cut from the sheet. These papers are to be discussed below. Finally, eight papers by the same group, published in 1972-77, considered several important technologically related properties of the Hitachi alloys such as: flux jumping, AC loss, and other phenomena related to alternating fields and currents. These are to be discussed in PART 3 of this chapter.

11.6 METALLURGY OF THE TECHNICAL SUPER-CONDUCTING Ti-Zr-Nb ALLOYS

The metallurgy of Ti-Zr-Nb in general having been considered in Sect.11.1, the following discussion is confined to the investigations conducted by the Hitachi laboratories. After applying optical metallography and x-ray analysis to fifty Ti-Zr-Nb alloys (including five Ti-Nb and nine Zr-Nb alloys) quenched from temperatures of 1100, 900, 800, 700, and 570°C, DOI et al [Doi66[a]] prepared a set of ternary-equilibrium phase diagrams, an adequate sampling of which contributed to Fig.11-1.

The technical superconducting Ti-Zr-Nb alloys fall into two categories: (a) those derived from the original Zr-75Nb alloy[†] in which the product of $\beta \rightarrow \beta' + \beta''$ "beta-phase immiscibility" was an important contributor to flux pinning -- these are the Zr-Nb-base or "X-type" alloys; (b) alloys based on the usual Ti-Nb$_{34}$ formulation in which the effects of thermomechanical processing, in the form of α-precipitation-decorated subbands, are responsible for flux pinning -- these Ti-Nb-base alloys have been referred to as "Z-type".

In the metallurgically based superconductive studies by the Hitachi group, emphasis was placed on precipitation as the flux-pinning agent. Whereas this may perhaps have been appropriate when dealing on a laboratory scale with the X-type alloys, it does not do justice to those based on Ti-Nb for which, according to Sects.7.30 and 9.18, precipitation is but one component of the total flux-pinning microstructure. A prerequisite for strong bulk pinning in Ti-Nb alloys is of course an optimized deformation-subband structure.

11.6.1 Precipitation from the Zr-Nb-Base (X-Type) Alloy, Ti$_{10}$-Zr$_{40}$-Nb$_{50}$

SOENO and KURODA [Soe69] have investigated the kinetics of the $\beta \rightarrow \beta' + \beta''$ immiscibility reaction in Ti$_{10}$-Zr$_{40}$-Nb$_{50}$ within the temperature range 650-800°C, presenting the results in the form of time, temperature, transformation (T-T-T) curves. The rate of α-phase precipitation in alloys previously β-decomposed

---

[†] This could also be written Zr-Nb$_{75}$, since the atomic weight of Nb is only $2\frac{1}{2}$% greater than that of Zr.

within the miscibility-gap region was also discussed with the aid of T-T-T curves. As expected, the rates of both β-phase decomposition and α-phase precipitation were increased by cold work, significant rate changes being observed as the cold-drawing deformation increased from 50 to 99%. These effects have been fully discussed with the aid of diagrams and figures [Mon2.7] and [Mon2.11].

As reported in a series of three papers, KITADA and DOI [Kit70^c, Kit70^d, Kit70^e] have conducted a detailed investigation, using optical metallography, TEM, hardness and other diagnostic measurements, of the β→β'+β" decomposition process in Ti_{10}-Zr_{40}-Nb_{50} within the temperature range 620 to about 825°C. Particular attention was given to the morphology of the lamellar-like precipitation. It was noted that the interlamellar spacing was proportional to aging temperature in the temperature range investigated and varied from about 0.5 to 2 μm [Kit70^c]. Following cold work, the β'+β" structure which resulted from the heat treatment acquired a much finer interlamellar spacing (ratio of about 10:1) [Kit70^c]; in addition, discontinuous precipitation took place, initially at dislocations. During aging at lower temperatures (400-500°C, in the α+β equilibrium-phase region) it was noted by KITADA and DOI [Kit70^e], that the first step in the process of β-decomposition was the formation of a continuous Nb-rich precipitate (i.e. β") within the grains. This was followed by α-phase precipitation (Nb-lean of course) taking place first at the grain boundaries, and then outwards from them in the form of an α+β" lamellar structure. The development of these precipitates is described quantitatively with the aid of the equilibrium-phase diagrams in [Mon2.7].

11.6.2 Precipitation from the Ti-Nb-Base (Z-Type) Alloy, Ti_{60}-Zr_5-Nb_{35}

Using transmission electron microscopy, KITADA and DOI [Kit69] have studied precipitation from the β-phase during the aging of Ti_{60}-Zr_5-Nb_{35}. A precipitate referred to as a "G.P. zone" about 50 to 200 Å in size, formed during aging at 350-500°C. Although not identified as such at the time, the size, coherency, and temperature range of its occurrence are suggestive of the β→β'+β phase-separation reaction (cf. the discussion of this in connection with Ti-TM-SM metallurgy in Sect.9.1). The precipitate so formed, a very effective flux pinner, gave way to an "intermediate phase"

(not observable at temperatures below 400°C even after prolonged aging, e.g. ∿1000 h) and α-phase precipitation at aging temperatures above 400°C.

11.7 ALLOY AND PROCESS DEVELOPMENT FOR Ti-Zr-Nb WIRE

Aided by thorough knowledge of the equilibrium and near-equilibrium microstructures acquired by means of optical metallography, replica electron microscopy, and transmission electron microscopy, researchers at the Central Research Laboratory of Hitachi Ltd. carried out a program directed towards the development of high-performance (high H_{c2} and J_c) Ti-Zr-Nb superconducting wires. Beginning with a broad survey of the critical current densities of a wide range of alloys, the program quickly became focused onto a pair of promising alloys: one of them lying near the Zr-Nb edge of the composition triangle, and the other near the Ti-Nb edge. Optimal -- or near-optimal, since the starting "billet" was of very small diameter -- thermomechanical processing sequences for the two classes of alloy were developed.

Optical metallography and replica electron microscopy were used to reveal the longitudinal drawing textures of the wires [Doi66] and to follow the coarsening of the fibrous as-drawn structure during isothermal aging [Doi67]. Optical metallography was also used to examine α-phase precipitation during the aging of a selected Z-type alloy in the α+β region [Ish68] and, together with carbon-replica electron microscopy, to study the decomposition reaction β→β'+β" during the ∿700°C aging of Ti_{10}-Zr_{40}-Nb_{50} [Doi68^a]. This work was augmented by x-ray diffraction [Doi66^b] in order to follow the resulting lattice-parameter, hence compositional, changes of the β' and β" phases. In addition, the carbon-replication technique was able to reveal the fine fibrous structure which could be developed in Ti_{10}-Zr_{40}-Nb_{50} as a result of final cold work administered after the β-decomposition intermediate anneal [Doi68^a]. With the aid of this metallurgical background, the optimization process was discussed in terms of the nature, morphology, and distribution of the flux-pinning agents: α-phase in the case of alloys represented by Ti_{60}-Zr_5-Nb_{35}, and the β'+β" lamellar structure in the case of Ti_{10}-Zr_{40}-Nb_{50}. The development of technical Ti-Zr-Nb alloys by the Hitachi group can be traced with the aid of a chronologically arranged table, Table 1-33(b), which outlines the contents of nine papers (among which considerable overlap is frequently to be found) covering the period 1966-1970.

In alloy preparation, fairly uniform starting procedures were adopted. Alloys were levitation melted and cast into ingots some 4.5 mm in diameter and 30 mm long. These were generally cold rolled in a wire mill to either 2 or 1 mm$^\phi$ and homogenized 5h/1100°C in a vacuum of 1x10^{-6} torr and water quenched. The 2- or 1-mm$^\phi$ wire rods were then subjected to various processing sequences, leading to the test product, a 0.25-mm$^\phi$ wire, upon which critical current density was measured at 4.2 K in transverse fields of up to about 90 kOe.

11.7.1 Screening Studies

In an initial series of experiments, DOI et al [Doi66] measured the critical current densities of thirty-three Ti-Zr-Nb alloys (including nine Zr-Nb and six Ti-Nb alloys) all subjected to the same thermomechanical processing sequence. This was intended as a screening study for the identification of compositions worthy of more detailed evaluation and development. Although the indiscriminate application of a common processing schedule to a wide group of alloys (presumably subdivisible into several classes) has obvious deficiencies, the one alloy which was identified by this means as having promising properties, viz Ti$_{10}$-Zr$_{40}$-Nb$_{50}$, held up to expectations during subsequent optimization studies [Doi66^b].

11.7.2 Properties of the X-Type Alloy Ti$_{10}$-Zr$_{40}$-Nb$_{50}$

In seeking an optimum thermomechanical processing schedule, DOI et al [Doi66^b] exposed Ti$_{10}$-Zr$_{40}$-Nb$_{50}$ to: (a) variation of final heat treatment temperature within 500-1100°C; (b) variation of intermediate heat treatment temperature within 570-1100°C; (c) variation of the 700°C intermediate heat treatment time; (d) variation of the degree of final cold deformation after an intermediate heat treatment of 1h/700°C and fixed overall size reduction, viz 2 mm$^\phi$ → 0.25 mm$^\phi$ (98.44%). The processing schedule leading to the best results, viz J$_c$(4.2 K) = 1x10^5 A cm^{-2} (58 kOe) and 5x10^4 A cm^{-2} (80 kOe), was:

2 mm$^\phi$ (5h/1100°C)→ 1 mm$^\phi$ (3h/700°C)→ 0.25 mm$^\phi$, (11-1)

while the second contender was:

2 mm$^\phi$ (5h/1100°C)→ 0.25 mm$^\phi$ (1h/570°C). (11-2)

11.7.3 Properties of the Z-Type Alloys Ti$_{60}$-Zr$_5$-Nb$_{35}$ and Ti$_{45}$-Zr$_{15}$-Nb$_{40}$

A limited number of Z-type alloys close to the Ti-Nb edge of the composition triangle were made by ISHIDA et al [Ish68] who intercompared the effects of various final heat treatments on the critical current densities of the Ti$_{45-70}$-Zr$_5$-Nb series of alloys. A favorable composition was Ti$_{60}$-Zr$_5$-Nb$_{35}$ which yielded J$_c$(4.2 K, 80 kOe) = 4.4x10^4 A cm^{-2} in response to:

2 mm$^\phi$ (5h/1100°C) ⟶ 0.22 mm$^\phi$ (1h/500°C).

The effect of process variation on the critical current density of an alloy containing a somewhat higher level of Zr, viz Ti$_{45}$-Zr$_{15}$-Nb$_{40}$, was also explored by DOI et al [Doi67]. Best results, viz J$_c$(4.2 K, 80 kOe) = 2.6x10^4 A cm^{-2}, were obtained with:

2 mm$^\phi$ (5h/1100°C) ⟶ 0.25 mm$^\phi$ (1h/500°C), (11-3)

while the J$_c$ versus H$_a$ behavior of the non-optimally processed alloy was used as a forum for a discussion of the peak effect, cf. [Mon21.12].

11.7.4 An Intercomparison of the Properties of X-Type Ti$_{10}$-Zr$_{40}$-Nb$_{50}$ and Z-Type Ti$_{60}$-Zr$_5$-Nb$_{35}$

In an intercomparison of the properties of selected representatives of the two types of ternary Ti-Zr-Nb alloy, DOI et al [Doi68^a] took Ti$_{60}$-Zr$_5$-Nb$_{35}$, and after applying the heat treatment defined by Eqn.(11-3), compared its properties with those of Ti$_{10}$-Zr$_{40}$-Nb$_{50}$ processed according to Eqn.(11-1). The results were as listed in Table 11-7. A set of J$_c$ versus H$_a$ curves for Ti$_{10}$-Zr$_{40}$-Nb$_{50}$ and Ti$_{60}$-Zr$_5$-Nb$_{35}$, processed according to Eqns.(11-1) and (11-3), respectively, as well as other less favorable schedules, are presented in Figs.11-11(a) and (b).

According to the table, although the X-type alloy has superior (presumably precipitation-enhanced) low-field performance, the situation is reversed at very high fields, presumably as a consequence of the domination of the high-field critical current density by the higher value of the upper critical field (cf. Sect.7.32.3). But as a result of recent experiments with binary Ti-Nb alloys, there is now no doubt that much better overall performances can be anticipated from the Z-type alloys through increasing by several

TABLE 11-7 CRITICAL TEMPERATURE (T_c) RESISTIVE CRITICAL FIELD (H_r) AND CRITICAL CURRENT DENSITY
IN TWO THERMOMECHANICALLY PROCESSED TITANIUM-ZIRCONIUM-NIOBIUM ALLOYS [Doi68[a]]

Type	Compositions, at.%			Process Schedule	T_c K	H_r(4.2K)[†] kOe	Critical Current Density[††] at 4.2 K, A cm^{-2}		
	Ti	Zr	Nb				40 kOe	80 kOe	90 kOe
X	10	40	50	Eqn (11-1)	10.3	105	3.6×10^5	5.0×10^4	1.0×10^4
Z	60	5	35	Eqn (11-3)	8.6	113	1.0×10^5	4.2×10^4	3.0×10^4

† H_r criterion: 100 µV across 3 cm at 5 mA [Doi68[a]]

†† J_c criterion: 5 µV across 3.5∼4 cm [Doi68[a], Doi68[b]]

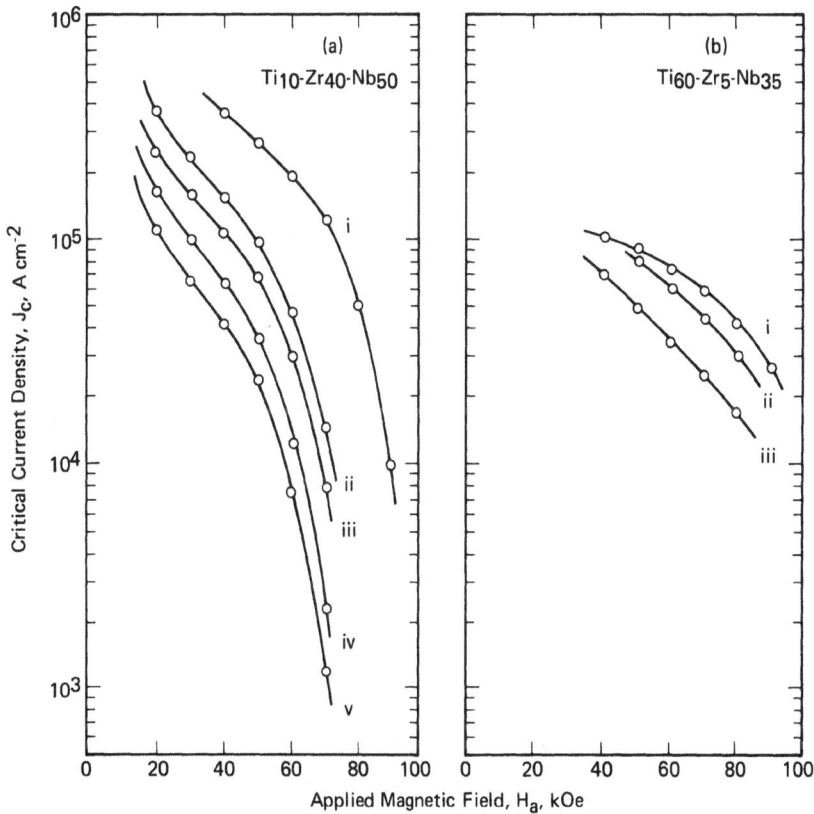

FIGURE 11-11. Intercomparison of the critical current density performances of representative X-type and Z-type Ti-Zr-Nb alloys (criterion: 5 µV across about 4 cm) — after DOI *et al* [Doi68[a]].

The alloys whose results are depicted were heat treated 5 h/1100°C (HT) at 2 mm$^\phi$ and processed according to:

(a) Ti$_{10}$-Zr$_{40}$-Nb$_{50}$

(i) 2 mm$^\phi$ (HT) ⟶1 mm$^\phi$ (3 h/700°C) ⟶ 0.25 mm$^\phi$ (as-cw)
(ii) 2 mm$^\phi$ (HT) ⟶ 0.25 mm$^\phi$ (3 h/700°C)
(iii) 2 mm$^\phi$ (HT) ⟶ 0.25 mm$^\phi$ (10 h/700°C)
(iv) 2 mm$^\phi$ (HT) ⟶ 0.25 mm$^\phi$ (50 h/700°C)
(v) 2 mm$^\phi$ (HT) ⟶ 0.25 mm$^\phi$ (100 h/700°C)

(b) Ti$_{60}$-Zr$_5$-Nb$_{35}$

(i) 2 mm$^\phi$ (HT) ⟶ 0.25 mm$^\phi$ (1 h/500°C)
(ii) 2 mm$^\phi$ (HT) ⟶ 0.25 mm$^\phi$ (6 h/500°C)
(iii) 2 mm$^\phi$ (HT) ⟶ 0.25 mm$^\phi$ (99 h/500°C)

orders of magnitude the level of cold work prior to some final-heat-treatment plus final-cold-work sequence (cf. Sect.7.28). Because of workability problems, large cold-work area reductions are not possible with the Zr-Nb-base alloys. Thus, the somewhat comparable and restrictive thermomechanical processes administered to both the Z-type and the Z-type alloys may not have rendered a fair comparison between the potentialities of these two distinctly different alloy classes.

Advantage may, however, be taken of the differing properties of these two classes of alloy in the construction of large superconducting magnets. For example, Hitachi Ltd. has supplied to the Electrotechnical Laboratory, Japan, a three-stage solenoid whose middle stage (O.D., 30 cm x I.D., 8 cm) and outer (hence low-field) stage (O.D., 55 cm x I.D., 34 cm) were wound with Cu-stabilized multifilamentary conductors prepared from the above-specified Z-type and X-type alloys, respectively.

11.8 COMPARATIVE STUDIES OF X-TYPE AND Z-TYPE ALLOY WIRES

11.8.1 The Ternary Critical-Current-Density Triangle

DOI *et al* [Doi68] completed the comprehensive alloy survey of [Doi66], as described at the beginning of Sect.11.7, by administering one of four somewhat individualized thermomechanical processing sequences to each of fifty-seven Ti-Zr-Nb alloys (including six Ti-Nb and six Zr-Nb alloys) and measuring their critical current densities. Particular attention was paid to the X-type alloy Ti_{10}-Zr_{40}-Nb_{50}, the Z-type alloy Ti_{60}-Zr_5-Nb_{35}, and the mid-range alloy Ti_{25}-Zr_{30}-Nb_{45}, but the 4.2-K, 80-kOe J_c's of all the alloys were measured and mapped on a composition triangle as in Fig.11-12. An inspection of the data led to the claim that high critical current densities were to be found on the line joining Ti-Nb_{33} to Zr-Nb_{53}. According to RASSMANN and ILLGEN [Ras72[b]], whose results were discussed in Sect.11.4, the J_c-favorable composition line was defined by:

$$[Zr] = 1.25[Nb] - 28.2 \qquad (11\text{-}4)$$

where [A] represents the concentration (in at.%) of element A in the alloy. This line, which connects Ti-Nb_{23} to Zr-Nb_{57}, is also shown in Fig.11-12.

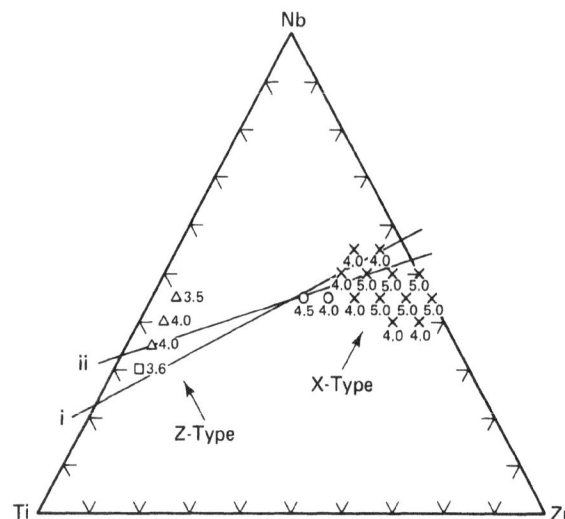

FIGURE 11-12. Ternary Ti-Zr-Nb diagram (at. % linear) representing those alloys whose 80-kOe critical current densities (indicated in the units 10^4 A cm^{-2}) are greater than 3 x 10^4 A cm^{-2} — after DOI *et al* [Doi68]. Depicted are:

X-type alloys processed according to:

2 mm$^\phi$ (5 h/1100°C) → 1 mm$^\phi$ ((3 ~ 5 h)/700°C) → 0.25 mm$^\phi$ (x)
2 mm$^\phi$ (5 h/1100°C) → 1 mm$^\phi$ ((1 ~ 10 h)/600°C) → 0.25 mm$^\phi$ (○)

Z-type alloys processed according to:

2 mm$^\phi$ (5 h/1100°C) ⟶ 0.25 mm$^\phi$ ((1 ~ 10 h)/500°C) (△)
2 mm$^\phi$ (5 h/1100°C) ⟶ 0.25 mm$^\phi$ ((1 ~ 10 h)/400°C) (□)

Also indicated are the transi of RASSMANN and ILLGEN ((i), Eqn (11-4)) and DOI *et al* ((ii), Eqn (11-5)).

With regard to this linear representation, whereas it is true that some of the important alloys cited by DOI *et al* do lie on, or close to, the line

$$[Z] = 2.35[Nb] - 77.6 \qquad (11\text{-}5)$$

implied by their statement, both lines in fact traverse a central region in the composition triangle occupied by single-phase, precipitation-free alloys, Fig.11-1, possessing relatively low critical current densities. After connecting up points of similar critical current densities and inserting a few appropriate equilibrium-phase boundaries, a meaningful congregation of high-J_c alloys into two zones is obtained, as in Fig.11-13. The occurrence of high critical current densities in Ti-Zr-Nb alloys can in this way be correlated with the existence of two composition zones: For the Zr-Nb-rich alloys aged at 700°C, the attainment of high critical current density is clearly associated with the β→β'+β" "immiscibility" reaction; for the moderate-temperature-aged Ti-Nb-rich

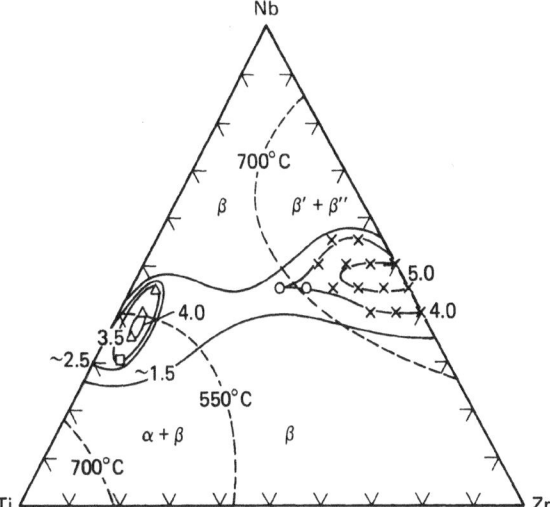

FIGURE 11-13. Ternary Ti-Zr-Nb diagram (at. % linear) displaying the critical current densities of X-type and Z-type Ti-Zr-Nb alloys (at 4.2 K, 80 kOe; units, 10^4 A cm^{-2}) — after DOI et al [Doi68] — superimposed on some relevant equilibrium transi for 700°C (after DOI et al [Doi66[a]]) and 550°C (after ALEKSEEVSKII et al [Aℓe68[a]], modified by the information that significant α-phase precipitation in Ti-Nb does not occur beyond ~ 40 at. % Nb at that temperature).

alloys, best performance is achieved with an optimal deformation-cell structure aided by the presence of α-phase precipitates when the Nb level is below about 40 at.%.

11.8.2 The Zr-Nb-Rich (X-Type Superconductor) Zone

In Fig.11-13, the right-hand 4×10^4 A cm^{-2} contour encloses a group of alloys which have received an intermediate heat treatment of (3∿5h)/700°C and is itself centered within a region of β-immiscibility. Precipitation in a typical cold-worked alloy, Ti_{10}-Zr_{40}-Nb_{50}, and its relationship to flux pinning has been discussed in, for example, [Doi68[a]]. Although the completeness of β→β'+β" decomposition increases with aging time at 700°C, and is almost complete at 10 h, the increasing grossness of the microstructure reduces its effectiveness as a flux pinner, thereby causing J_c to decrease after about 3 h of heat treatment at that temperature. In Ti_{10}-Zr_{40}-Nb_{50}, optimally processed according to Eqn.(11-1), the 3h/700°C heat treatment was responsible for the occurrence of an evenly dispersed precipitate a few thousand Å in both size and spacing [Doi68[a]]. The subsequent 93.75% final cold work contributed to the fibrous structure and, presumably, redistributed the precipitates.

11.8.3 The Ti-Nb-Rich (Z-Type Superconductor) Zone

The high-J_c alloys of the Z-type alloy group, if equilibrated below 550°C, would lie within the α+β-phase field. If quenched from the common 5h/1100°C anneal they would of course be single-phase bcc, but after deformation (98.44% in all the present examples) and 1∿10 h aging at 400 or 500°C, precipitation is expected. The alloys which yielded the highest critical current densities lay more-or-less within the purview of the "intermediate-concentration" binary Ti-Nb alloys (cf. Sect.7.26) which are characterized by an α-phase-decorated subband structure after wire drawing and short-term aging at moderate temperatures. Indeed, carbon-replica electron micrographs of cold-drawn and 1h/500°C-aged Ti_{60}-Zr_5-Nb_{35} exhibited precisely such a structure [Doi68[a]]. After 99 h of aging, the precipitates became globular particles about 1 μm in diameter and less effective as flux pinners.

11.8.4 Final Commentary

The critical current densities of alloys such as Ti_{10}-Zr_{40}-Nb_{50} and Ti_{60}-Zr_5-Nb_{35} have been studied against a background of detailed equilibrium or quasi-equilibrium microstructural information. At no time did the total cold work exceed 98.44% and when intermediate heat treatment was associated with the cold reduction, it was confined to 75% before, and 93.75% after, the heat treatment. In its predilection towards clearly recognizable precipitation, and in its lack of emphasis on the development of fine fibrous structures through heavy cold work (>99.9% is customary), particularly with regard to the Ti-Nb-base alloys, the research philosophy adopted by the Hitachi group with Ti-Zr-Nb was comparable to that used by ZWICKER and colleagues in their investigations of both binary and ternary-alloyed Ti-Nb.

Of course, in the usual research program, the smallness of the starting ingot precludes the development of the heavily-cold-worked structures which are possible in full-scale commercial processes. As a result, those alloys whose critical current densities benefit from cold work must in small-scale research and development programs always fail to attain their full current-carrying potentials. However, during the course of such investigations it has frequently been hoped, not necessarily with justification, that the critical current densities obtained in comparative studies would at least retain their ranking during scale-up.

11.9 PROPERTIES OF CONTEMPORARY COMMERCIAL Ti-Zr-Nb CONDUCTORS

Since about 1970 when it was employed in the outer section of a three-stage solenoid in order to take advantage of its superior low-field critical current density, cf. Table 11-7, production of the low-workability X-type alloy has ceased. It appears that the Z-type alloy may follow suit, since whereas in 1979, Hitachi Cable produced about two tonnes of it but none of the binary, the situation was reversed as of May 1980. During 1970 the Zr-content of the Z-type alloy was reduced to 2.5 at.%. As an alloy such as $Ti_{62.5}$-$Zr_{2.5}$-Nb_{35}, it found numerous applications in devices such as the Electrotechnical Laboratory (Japan) 1-MW MHD generator, and the Japan National Railway magnetic suspension coils.

Other variants of the basic Z-type formulation have been manufactured. Their properties have been measured by TADA, ANDO, OKA, and SHIMAMOTO [Tad80] and compared with those of Ti-Nb conductors supplied by various Japanese, British, and American manufacturers as part of a conductor evaluation program in support of the Japanese Large-Coil Program. The details of the binary-alloy tests and their results are given in

Sect.7.33. Listed in Table 11-8 are the specifications of the three Ti-Zr-Nb alloys tested by TADA *et al* [Tad80] together with those of two binary alloys, selected from Table 7-22, which have equivalent Nb contents. The critical currents and critical current densities of the alloys are presented for intercomparison in Fig.11-14, in the three parts of which the Ti-Zr-Nb alloys are compared with their binary counterparts (reckoned on a Nb-concentration basis). Figs.11-14(a) and (b) display the same crossing-over of the $J_c(H_a)$ characteristics that was previously encountered among the X-type and Z-type alloys themselves, Table 11-7, and which was attributed to the better precipitation-enhanced intermediate-field J_c of the Zr-Nb-base alloy in association with the higher H_r of the Ti-Nb-base alloy. The resistive upper critical fields of all the alloys are listed in Table 11-9. A similar pair of mechanisms is probably influencing the relative properties of Ti-Zr-Nb-(i) and -(ii) and Ti-Nb-(iv). On the other hand, Ti-Nb-(v) is superior to Ti-Zr-Nb-(iii) at all fields. The binary alloy of course has the higher H_r and consequently the better high-field performance. But with regard to intermediate-field flux pinning, the binary alloy with 38.9 at.% Nb has almost exited from the precipitation regime (cf. Fig.11-13), in which case the ternary alloy at 40 at.% Nb will certainly be outside it. Under these conditions, critical current density depends most strongly on undisclosed factors such as the extent of

TABLE 11-8 SPECIFICATIONS OF THREE TITANIUM-ZIRCONIUM-NIOBIUM CONDUCTORS AND TWO RELATED TITANIUM-NIOBIUM CONDUCTORS (from Table 7-22) TESTED BY TADA *et al.* [Tad80]

Sample Code (cf.Fig.11-14)	Compositions, at.%			Overall Diameter or Dimensions mm	Filament Diameter μm	Number of Filaments	Cu/SC Ratio	Twist Pitch mm	Manufacturer[+]
	Ti	Zr	Nb						
(i)	62.7	2.5	34.8	0.778	45.1	127			HC
(ii)	62.5	2.5	35	2.3	58.6	271	4.6	29	HC
(iii)	57.6	2.5	39.9	0.778	46.8	127			HC
(iv)	65.71	---	34.29	7.0x7.0	96	750	8.0	150	SEI
(v)	61.15	---	38.85	1.00	34.02	144	5.0	12.7	IGC

+ HC : Hitachi Cable, Ltd.
  SEI: Sumitomo Electric Industries, Ltd.
  IGC: Intermagnetics General Corporation

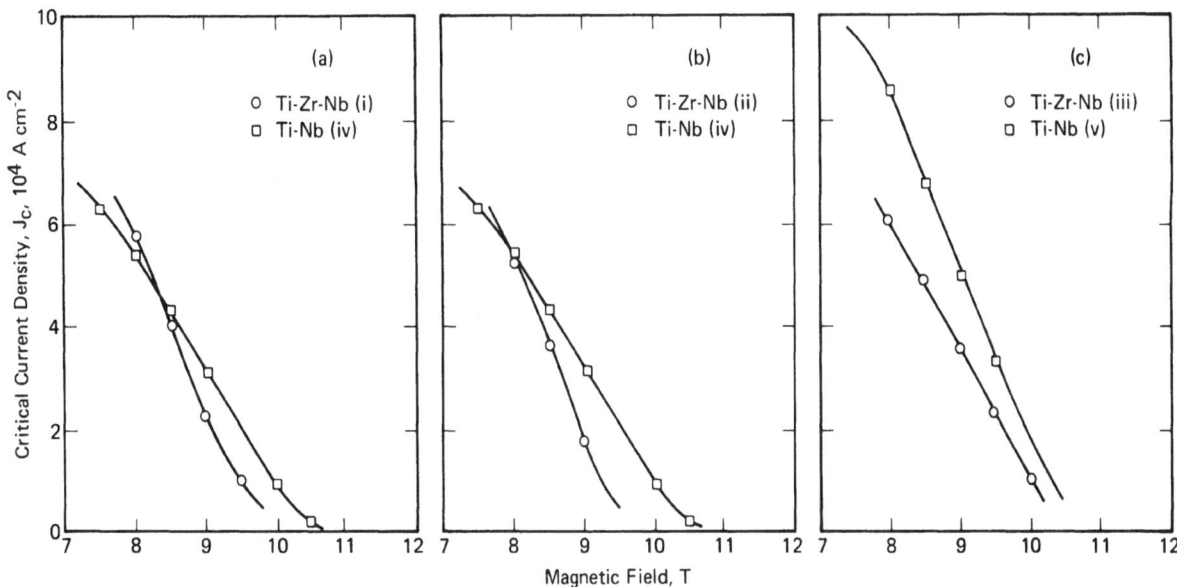

FIGURE 11-14. Intercomparison of the J_c versus H_a performances of "comparable" (Nb-concentration basis) pairs of commercial Ti-Zr-Nb and Ti-Nb multifilamentary Cu-stabilized conductors. The specifications of the conductors represented are listed in Table 11-8 (critical current criterion: $10^{-11} \Omega$ cm) — after TADA et al [Tad80].

the cold deformation and the manner in which it may have been "fine-tuned" by moderate-temperature final heat treatment and also, to some degree, on the impurity contents of the starting elements (see Table 8-12).

TABLE 11-9 UPPER CRITICAL FIELDS (H_r) OF THE ALLOYS REPRESENTED IN FIGURE 11-14 [Tad80].

Sample Code	Compositions, at.%			$H_r^\dagger$
	Ti	Zr	Nb	T
(i)	62.7	2.5	34.8	10.1
(ii)	62.5	2.5	35	10.1
(iv)	65.71		34.29	10.5
(iii)	57.6	2.5	39.9	10.5
(v)	61.15		38.85	10.7

† H_r criterion: J_c ($10^{-10}\Omega$ cm criterion) plotted linearly versus H_a was extrapolated to "zero" J_c (cf. Figs. 7-54, 11-14) -- information courtesy of E. Tada.

11.10 ALLOY AND PROCESS DEVELOPMENT FOR Ti-Zr-Nb ROLLED-RIBBON (STRIP) CONDUCTOR

11.10.1 Superconductivity in Ti-Zr-Nb Rolled Strip

KITADA, DOI and UMEZAWA, in a series of five papers published between 1970 and 1972, have discussed the magnitude and anisotropy of critical current density in heat-treated cold-rolled strips of Ti-Zr-Nb alloys. As with the wire conductor, attention was focused on the X-type and Z-type composition ranges, examples from each being: (i) the Zr-Nb-base alloy Ti_{10}-Zr_{40}-Nb_{50} and (ii) a set of Ti-Nb-base compositions Ti_{55-75}-Zr_5-Nb. The starting materials, generally in the form of solution-heat-treated strips, 1 mm thick and annealed 1h/950°C in the case of Ti_{10}-Zr_{40}-Nb_{50} and 5 mm thick, annealed 3h/1000°C, in the case of the Ti-Nb-base alloys, were cold rolled to thicknesses of 0.1 mm and 0.05 mm respectively with variable intermediate and final heat treatments, as shown in Table 1-33(c). The precipitation heat-treatment temperature for the Zr-Nb-base alloy, for reasons already given, was usually 700°C, while the Ti-Nb-base alloys were generally aged in the temperature range 350-500°C. Critical current measurements were carried out at 4.2 K on strips 2 mm wide and 15 to 20 mm long usually cut parallel to the rolling direction. Thus, if α is the angle between the current direction and

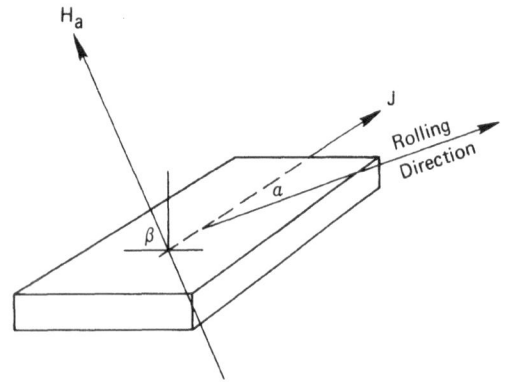

FIGURE 11-15. Diagram showing: (a) a transverse applied magnetic field, H_a, at an angle β to the rolling plane; (b) a transport current, J, at an angle α to the rolling direction. The influences of these factors on J_c have been studied by KITADA, DOI et al (see for example [Kit72]). In most studies $\alpha = 0°$, in which case the critical current densities for the $\beta = 0°$ and $\beta = 90°$ field orientations are conveniently referred to as J_c (||) and J_c ($\perp$), respectively.

the rolling direction, Fig.11-15, in most of the measurements reported, $\alpha = 0$. A second important parameter in measurements of this kind is the angle, β, between the applied magnetic field in a plane normal to the current direction, and the rolling plane. In rolled strips, J_c, which is generally quite anisotropic, is maximum when $\beta = 0$; i.e. when the applied field is parallel to the rolling plane. Using optical metallography, electron microscopy, and x-ray pole-figure determination, considerable attention was given to the metallurgical consequences of rolling both the Zr-Nb-base [Kit70[a]] and Ti-Nb-base [Kit72[a]] alloys. Critical current anisotropy was correlated with structural anisotropy as parameterized by the quotient $(a+d)/a$ where a is the thickness of "defect" layers and d that of "defect-free" layers of the laminar cold-rolled structure (clearly the rolled-strip equivalents of the cell-wall thickness and subband diameter, respectively, of wire-drawn material).

11.10.2 Properties of X-Type Ti_{10}-Zr_{40}-Nb_{50} Rolled Strip

(a) Maximum Critical Current Densities. KITADA, DOI and UMEZAWA have studied the anistropic critical current densities [Kit70] and microstructures [Kit70[a]] of cold-rolled and heat-treated Ti_{10}-Zr_{40}-Nb_{50}. The alloys were solution heat treated at either 1 mm^t or 0.2 mm^t, cold rolled to 0.1 mm^t, and either final-heat-treated or measured in the as-rolled condition. When the starting thickness was 1 mm, an intermediate

heat treatment at 700°C was administered at 0.2 mm^t. The extent of the cold work was in all cases relatively small, the steps from 1 to 0.2 mm^t and 0.2 to 0.1 mm^t representing only 80% and 50% reductions in area, respectively (assuming no change in width, which is generally true in unlubricated cold rolling). According to the results of x-ray analysis as summarized in Table 11-10, heat treatments of 1 h at 700 or 800°C decomposed the 90%-cold-rolled alloy into β', β'' and α, while an intermediate anneal of 10h/700°C after 80% cold work led to $\beta'+\beta''$, a structure which may be refined by further cold work. The highest critical current densities were achieved after the processing sequence:

$$1 \text{ mm}^t (1h/950°C) \longrightarrow 0.2 \text{ mm}^t (1\text{-}5h/700°C) \longrightarrow 0.1 \text{ mm}^t .$$

Of considerable interest is a direct comparison between the properties of more-or-less similarly processed X-type strip and wire samples. Such a comparison can best be stated in the following way, with reference to the 4.2-K, 60-kOe J_c:

$$strip: \quad \xrightarrow[80\%]{CW} (1\text{-}5h)/700°C \xrightarrow[50\%]{CW} (1.6\text{-}1.2)\times10^5 \text{ A cm}^{-2},$$

$$wire: \quad \xrightarrow[75\%]{CW} 3h/700°C \xrightarrow[93.75\%]{CW} 1.9\times10^5 \text{ A cm}^{-2}.$$

(b) Critical-Current Anisotropy. After each thermomechanical processing sequence, J_c was measured

TABLE 11-10 RESULTS OF X-RAY ANALYSIS OF THERMO-MECHANICALLY PROCESSED Ti_{10}-Zr_{40}-Nb_{50} STRIP -- Kitada et al. [Kit70]

HT-1[†]	HT-2[†]	Phases Present
None	1h/570°C	β
None	1h/700°C	β, β', β'', α
None	1h/800°C	β, β', β'', α
None	1h/1100°C	β
10h/700°C	None	β, β''

† 1 mm^t (1h/950°C) $\longrightarrow$ 0.2 mm^t (HT-1) $\longrightarrow$ 0.1 mm^t (HT-2)

under the conditions: $\alpha = 0$; $\beta = 0$ and $90°$. In the case of optimal processing it was measured under the conditions:

(i) $\alpha = 0$; $\beta = 0, 30, 45, 60,$ and $90°$,

(ii) $\beta = 0$; $\alpha = 0, 15, 30, 60,$ and $90°$.

In the latter orientation, i.e. with H_a parallel to the rolling plane, J_c was independent of the angle it made with the rolling direction. On the other hand, significant anisotropy was encountered when H_a was rotated within a plane transverse to the current (and rolling) direction. As a typical example, with $\alpha = 0$ and $H_a = 60$ kOe:

$$J_c(||)/J_c(\perp) \equiv J_c(\beta=0)/J_c(\beta=90°)$$
$$= (14 \times 10^4)/(4.3 \times 10^4) = 3.3/1 \quad .$$

Anisotropic strip conductor could conveniently be used in the center pancakes of a solenoidal magnet. But the existence of critical current anisotropy would be a distinct disadvantage, and would introduce additional complications, during the design of saddle magnets and magnets for non-uniform field configurations.

11.10.3 Properties of Z-Type $Ti_{55-75}-Zr_5-Nb$ Rolled Strip

KITADA and DOI have discussed the thermomechanically controlled critical current density [Kit70[f]], and the microstructure and critical current density anisotropy [Kit72[a]], of $Ti_{60}-Zr_5-Nb_{35}$ rolled strip and in a subsequent paper, [Kit72] have discussed the results of applying a comprehensive matrix of thermomechanical processes to a family of five alloys, $Ti_{55-75}-Zr_5-Nb$, lying close to the Ti-Nb edge of the composition triangle (and, in particular, to $Ti-Nb_{30}$). The processing sequence, uniformly applied, was:

5 mm^t (3h/1000°C) $\longrightarrow$ 0.05 mm^t (HT)

in which the heat treatment (HT) was one selected from (0.5-1000h)/(250-500°C). With regard to the laminar-structure quotient, $(a+d)/a$ referred to above, KITADA [Kit72[a]] found that for several values of the applied field, several aging temperatures, several alloy compositions, and for a wide range of aging times (0~400 h),

$J_c(||)/J_c(\perp) \propto (a+d)/a$ with a constant of proportionality close to unity, and secondly that for $Ti_{60}-Zr_5-Nb_{35}$ aged at 350°C, for example, this ratio tended towards unity at very long aging times (several hundreds of hours). As a result of a continuous refinement of the defect layers with increasing cold work, a decreased with respect to d leading to an increase in the quotient $(a+d)/a$ and with it the critical-current anisotropy. Indeed, according to KITADA [Kit72[a]], $J_c(||)/J_c(\perp)$ increased linearly according to $0.012\varepsilon + c$, where ε was the percent strain and c was some constant.

In studies of the effect of final heat treatment on the critical current density of $Ti_{60}-Zr_5-Nb_{35}$, KITADA and DOI [Kit70[f]] showed that best results could be obtained with final heat treatments of $\gtrsim$500h/300°C or $\gtrsim$100h/350°C, hence after a processing sequence such as: 5 mm^t (3h/1000°C) $\longrightarrow$0.05 mm^t (125h/350°C). Again it is interesting to compare the critical current density which results from a typical such heat treatment with that of a comparably processed cylindrical wire. The comparison can be conveniently expressed in the following way:

strip: 5 mm^t (3h/1000°C) $\xrightarrow[99\%]{CW}$ 0.05 mm^t (125h/350°C)

-- yielding 1.4×10^5 A cm^{-2}

wire: 2 mm$^\phi$ (5h/1100°C) $\xrightarrow[98.4\%]{CW}$ 0.25 mm$^\phi$ (100h/400°C)

-- yielding 9.1×10^4 A cm^{-2}

again in terms of the 4.2-K, 60-kOe J_c.

In quoting the 4.2-K, 80-kOe, J_c of $Ti_{65}-Zr_5-Nb_{30}$ it is again instructive to draw a comparison with comparably processed wire of the same composition [Ish68][Doi68]. Presented in the, by now, usual format the results of so doing are:

strip: 5 mm^t (3h/1000°C) $\xrightarrow[99\%]{CW}$ 0.05 mm^t (25h/350°C)

-- yielding 3.4×10^4 A cm^{-2}

wire: 2 mm$^\phi$ (5h/1100°C) $\xrightarrow[98.4\%]{CW}$ 0.25 mm$^\phi$ (10h/400°C)

-- yielding 3.6×10^4 A cm^{-2}.

In seeking an optimal composition for cold-rolled alloys, KITADA and Doi [Kit72] studied five alloys

TABLE 11-11   OPTIMAL (60-kOe J_C) FINAL HEAT TREATMENTS
FOR FOUR Tr-Zr-Nb ALLOY ROLLED STRIPS --
Kitada and Doi [Kit72]

| Composition, at.% | | | Final |
Ti	Zr	Nb	Heat Treatment
75	5	20	2h/400°C
70	5	25	10h/450°C
65	5	30	50h/350°C
55	5	40	≥250h/350°C

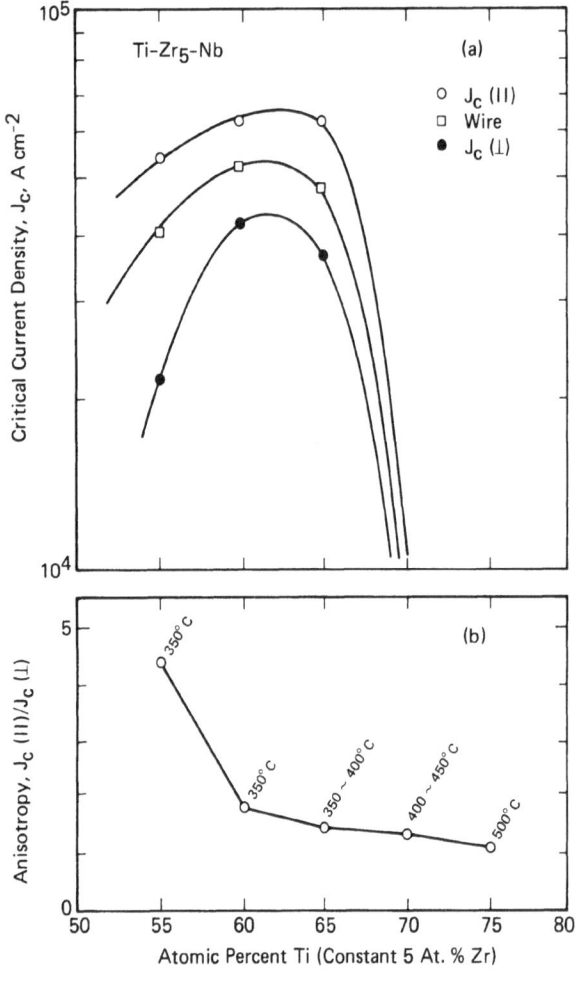

FIGURE 11-16. Maximum (i.e., "optimum") critical current densities and minimum critical-current anisotropies in thermomechanically processed Ti-Zr-Nb rolled strip. (a) 80-kOe values of J_C (∥) and J_C (⊥) as functions of Ti content in Ti-Zr$_5$-Nb strip. Wire data are inserted for comparison. (b) Minimum values of J_C (∥)/J_C (⊥) (applied field unspecified), attained in response to aging at the temperatures indicated, also as a function of Ti content in Ti-Zr$_5$-Nb strip — after KITADA and DOI [Kit72].

within the composition range Ti$_{55-75}$-Zr$_5$-Nb, subjecting each of them to a wide range of heat treatments. Optimal heat treatments for four of the alloys (as judged by the 60-kOe J_C's) are listed in Table 11-11. In order to summarize a rather large body of experimental data, KITADA and DOI [Kit72] presented graphs of $J_{C,max}$ (60 and 80 kOe) *versus* at.% Ti in the cases α = 0, β = 0 and α = 0, β = 90°, enabling the magnitudes and anisotropies of the critical current densities *(a)* to be traced as functions of composition and *(b)* to be compared with the 80-kOe J_C for optimized wire, also presented on the same graphs. The 80-kOe results are reproduced in Fig.11-16(a). They also presented graphs of the minimum anisotropy quotient for each alloy (associated with its individual aging temperature), $J_C(∥)/J_C(⊥)|_{min.}$, as a function of Ti concentration, Fig.11-16(b). From such curves it was possible to draw the following conclusions:

(a) Maximal J_C was obtained with appropriately processed alloys of the type Ti$_x$-Zr$_5$-Nb$_{95-x}$, where $60 < x < 65$.

(b) In an applied field of 80 kOe, $J_{C,wire}$ lay approximately midway between the curves for $J_C(∥)$ and $J_C(⊥)$.

(c) The anisotropy quotient itself dropped rapidly with at.% Ti, reaching values less than about 1.5 for Ti concentrations greater than about 60 at.%.

(d) The alloy with the highest 60-kOe and 80-kOe $J_C(∥)$ was Ti$_{65}$-Zr$_5$-Nb$_{30}$.

11.11 INTERCOMPARISON OF THE PROPERTIES OF Ti-Zr-Nb WIRE AND ROLLED STRIP

In considering possible technical applications of cold-rolled ribbon conductor, critical current anisotropy as well as optimal critical current density must be taken into consideration. Table 11-12 summarizes and compares the properties of rolled strip with those of drawn wire of the same composition prepared under similar conditions. In considering the X-type alloy, Ti$_{10}$-Zr$_{40}$-Nb$_{50}$, and two examples of Z-type formulations, Ti$_{60}$-Zr$_5$-Nb$_{35}$ and Ti$_{65}$-Zr$_5$-Nb$_{30}$, data have been assembled from several papers by members of the Hitachi research group.

The microstructure common to Ti$_{10}$-Zr$_{40}$-Nb$_{50}$ ribbon and wire is the cold-work-refined β'+β" decomposition product. Their critical current densities are comparable, but the anisotropy of 3:1 exhibited by the ribbon conductor probably renders it unacceptable for many applications.

TABLE 11-12 COMPARISON OF THE 60-kOe CRITICAL CURRENT DENSITIES OF TITANIUM-ZIRCONIUM-NIOBIUM STRIP AND WIRE CONDUCTOR PREPARED UNDER SIMILAR CONDITIONS

At.% Ti	Zr	Nb	Strip(s) or Wire(w)	Processing Sequence[+]	$J_c(\|\|)$ and J_c (4.2K, 60kOe) A cm^{-2}	Anisotropy (4.2K, 60kOe) $J_c(\|\|)/J_c(\perp)$	Literature
10	40	50	s	$\xrightarrow[80\%]{roll}$→(3-3/4h/700°C)$\xrightarrow[50\%]{roll}$	1.4×10^5	14/4.3=3.3	Kitada *et al.* [Kit70, Fig.6]
			w	$\xrightarrow[75\%]{draw}$→(3h/700°C)$\xrightarrow[93.75\%]{draw}$	1.9×10^5	-------	Doi *et al.* [Doi68[a], Fig.7]
60	5	35	s	$\xrightarrow[99\%]{roll}$→(125h/350°C)	1.4×10^5	14/7.9=1.8	Kitada *et al.* [Kit70[f], Fig.5]
			w	$\xrightarrow[98.44\%]{draw}$→(100h/400°C)	9.1×10^4	-------	Ishida *et al.* [Ish68, Fig.1]
65	5	30	s	$\xrightarrow[99\%]{roll}$→(25h/350°C)	1.5×10^5	15/9.3=1.6	Kitada *et al.* [Kit72, Fig.4]
			w	$\xrightarrow[98.44\%]{draw}$→(10h/400°C)	8.1×10^4	-------	Ishida *et al.* [Ish68, Fig.1]

+ Following a high-temperature anneal.

The anisotropies of the alloys with Ti contents greater than 60 at.% are less than 2:1. The $J_c(\|\|)$ of strip is considerably greater than the J_c of wire prepared under similar conditions. But since Ti-rich alloys rely on heavy cold work, in association with moderate aging, for the establishment of acceptable critical current densities, much higher values than those reported here can be expected as a result of cold work to levels of 99.9 ~ 99.99%. An extension of the joint wire-drawing/cold-rolling study into this deformation range would be particularly profitable. It is anticipated, however, that as the critical current density of the rolled strip increases with strain, so also will its anisotropy [Kit72[a]], perhaps to unacceptable levels.

PART 3: ALTERNATING-CURRENT AND -FIELD EFFECTS IN TITANIUM-ZIRCONIUM-NIOBIUM ALLOY SUPERCONDUCTORS

PART 2 of this chapter has described the DC properties of Ti-Zr-Nb technical alloys. In doing so some sixteen papers published during 1966-1970, dealing with wire development, and five papers published during 1970-72 describing ribbon conductors, were reviewed. In order to complete this survey of the commercial development of Cu/Ti-Zr-Nb composites it is necessary to consider briefly here the more recently published work dealing with their AC properties. The purpose of the research to be reviewed was the achievement of high AC critical current density and stability, minimal hysteretic AC loss in the superconductor itself, and minimal eddy-current loss in its matrix.

11.12 AC LOSS STUDIES OF TECHNICAL Ti-Zr-Nb SUPERCONDUCTORS

The goal of the investigations of AIHARA, DOI, KUDO, KURODA, and SHIIKI of Hitachi Ltd.'s Central Research Laboratory, whose results appeared as a series of eight publications during the period 1972-77 (cf. Table 1-33(d)), was the development of superconductors for AC service such as in power transmission lines, magnetically levitated trains, rotating superconducting machinery, and so on. In pursuit of these goals, the Hitachi team studied: (a) AC critical currents in zero applied field, (b) DC critical currents in AC applied fields, (c) AC current loss in zero field, and (d) magnetic hysteresis loss in the absence of transport current, as function of (i) alloy composition, (ii) wire diameter, (iii) multifilamentary twist pitch, (iv) multifilamentary composite matrix resistivity, (v) amplitude and the frequency of the applied magnetic field.

Conductor design must take into account both AC loss and stability [Eas70][Wil70, Wil77]. Thus loss measurements are *doubly* useful: not only do they yield data relating to the loss itself, but they can be used diagnostically, the "anomalous" field-amplitude dependence of the loss and other such measures of interfilamentary coupling providing, under appropriate conditions, information helpful in the choice of stability-related parameters such as filament diameter, twist pitch, and matrix resistivity. In the measurements of hysteresis loss (zero current) and AC transport-current loss (in zero field) to be considered below, He-boil-off calorimetry was used. In doing so, noninductively wound wire samples were mounted inside a nylon chamber immersed in liquid He, the gas boiled off as a result of the dissipation in the sample being conducted through a nylon pipe to a calibrated flow meter.

11.13 CRITICAL ALTERNATING CURRENT IN ZERO APPLIED MAGNETIC FIELD

The AC critical current densities and stabilities of heavily cold-worked (99.94 - 99.999% area reduction) and final-heat-treated (typically 100h/350°C) mostly Z-type bare wires and two Cu-clad (to 34 μm^t and 78 μm^t) monofilamentary composites have been measured by KUDO and SHIIKI [Kud76]. Currents of frequencies 20-500 Hz, swept at rates of 0.1-10 A sec^{-1} to more

than 700 A, were applied to 3-cm long samples; sample current and voltage signals were monitored and stored electronically. The results of such studies may be discussed in terms of three characteristic current values: (i) I_c, the DC critical current; (ii) I_{mc}, the amplitude of the critical alternating current; (iii) I_q (or "quench current"), the instantaneous value of the circuit current at which the s/n transition actually takes place. It was noted that I_{mc} could be less than I_c, and that the s/n transition occasionally took place at a current value I_q less than I_{mc}.

In well-stabilized material, all the currents are equal: $I_q = I_{mc} = I_c$. Examples of materials in which this occurred are: (a) the finest (23 μm^ϕ) bare superconducting wire and (b) the Cu-clad 78 μm^t monofilament. This relationship was also approximately true for the 34 μm^t Cu-clad monofilamentary composite wire for which $I_q \stackrel{\sim}{<} I_{mc} \stackrel{\sim}{<} I_c$. On the other hand, for heavier-gauge material, such as the 101 μm^ϕ Ti$_{60.2}$-Zr$_{2.8}$-Nb$_{37}$ bare wire, anomalous current transitions of the type represented by $I_q < I_{mc} < I_c$ occurred as a result, it was suggested, of magnetic flux jumps triggered by small temperature rises associated with the hysteretic loss.

11.14 CRITICAL DIRECT CURRENT IN A LONGITUDINAL ALTERNATING MAGNETIC FIELD

Whereas in the usual transverse-applied-field configuration the maximum superconducting transport current (critical current) is limited by the ability of flux-pinning forces to balance the Lorentz $\vec{J} \times \vec{B}$ force between the current and the magnetic induction, flux pinning plays a less important role and much larger direct currents can be supported when the applied field is directed *along* the axis of the wire. Both direct and alternating current transport in steady longitudinal applied fields, and the concepts of "force-free" and "nearly-force-free" current flow have been considered in [Mon22.00], a discussion based primarily on the work of BELANGER [Bel68], GAUTHIER [Gau76], and LeBLANC *et al* [Leb65, Leb66, Leb68].

In experiments with heavily-cold-worked Ti$_{60.2}$-Zr$_{2.8}$-Nb$_{37}$, SHIIKI and KUDO [Shi77] compared the critical direct currents in the presence of longitudinal applied DC fields (<3.5 kOe) with those in the presence of longitudinal alternating fields of amplitude, H_m (also <3.5 kOe), and frequencies of 35, 50, and 100 Hz. As the DC field was increased from zero to about 400 Oe, J_c increased from 1.25×10^6 to 2.1×10^6 A cm^{-2} and thereafter remained fairly independent of field. On the other hand, in AC fields of 35 and 100 Hz (which yielded mutually comparable results), J_c decreased monotonically with increasing field such that at 3.5 kOe it was about a factor of two lower than the corresponding DC value. It was suggested that the longitudinal current/field configuration was stable statically but not dynamically and that flux-flow dissipation was responsible for the reduced DC critical current in the AC longitudinal field.

11.15 ALTERNATING CURRENT LOSS IN ZERO APPLIED MAGNETIC FIELD

Using a He-boil-off calorimeter the AC loss associated with a sinusoidal AC transport current in zero applied magnetic field has been studied by SHIIKI et al [Shi74^c]. The Ti-Nb-base alloy $Ti_{62.5}$-$Zr_{2.5}$-Nb_{35}, in the form of a 0.25-mm$^\phi$ monofilament, both bare and after plating with Cu to thicknesses of 34, 51, 78, 126 μm , was the subject of the investigation. As pointed out in [Mon26.00] which discusses the topic in detail, AC loss is usually expressed in terms of either the power per unit volume, $\dot{Q}$, or the energy per unit volume dissipated per cycle of the current or the applied field, Q. With regard to the latter, since Q/cycle $\equiv \dot{Q}/f$, the cyclical energy loss is equivalent to the power loss per unit frequency. In what follows it is assumed that the frequencies are so low that classical skin effect need not be taken into account (cf. [Mon26.8]). The AC loss of a composite superconductor can always be expressed as the sum of the eddy-current loss of the matrix, $\dot{Q}_e$, and the hysteretic loss, $\dot{Q}_h$, of the superconductor itself. Thus, in general:

$$\dot{Q} = \dot{Q}_h + \dot{Q}_e \quad . \tag{11-6}$$

According to SHIIKI [Shi74^c], the eddy-current loss was

$$\dot{Q}_e = K \frac{t^2}{\rho} f^2 H_m^2 \qquad (W\ cm^{-3}) \tag{11-7}$$

where t and ρ were the thickness (cm) and resistivity (Ω cm), respectively, of the Cu layer and H_m was the amplitude (Oe) of the applied AC field (in this case the surface self-field $H_s = I_m/5r$) of the superconduc-

tor (of radius r cm) due to the passage of the AC transport supercurrent of amplitude I_m. K is a constant which according to the experiment had the numerical value 6.5×10^{-16}. As the Cu layer became thinner, however, departures from the quadratic field dependence became noticeable, values of the index considerably larger than 2 being encountered.

The hysteretic loss due to the self-field, H_s, of a transport current has been dealt with by LONDON [Lon63] according to whom, if the current is very much less than critical, or what is the same thing, if $H_s \ll H^*$ (the "full-penetration" field, cf. Sect.7.14.2):

$$\dot{Q}_h/f \propto H_s^3/J_c \quad . \tag{11-8}$$

With regard to the frequency dependence of the combined loss, Eqns.(11-6), (11-7), and (11-8) together show that

$$\dot{Q} = af + bf^2 \tag{11-9a}$$

or

$$\dot{Q}/f = a + bf \quad , \tag{11-9b}$$

a form which indicates that a plot of Q/cycle or $\dot{Q}/f$ *versus* f should be linear with intercept proportional to H_s^3/J_c and slope, b, related to the eddy-current loss coefficient. Such has indeed been shown to be the case, particularly for monofilamentary composites for which the question of interfilamentary coupling does not arise [Kud72][Shi74]. Secondly, with regard to the field dependence, it has already been pointed out that $\dot{Q}_e \propto H_s^\alpha$ with $\alpha \gtrsim 2$. Eqn.(11-8) has indicated that $\dot{Q}_h \propto H_s^\beta$, with $\beta = 3.0$ if the field dependence of J_c can be neglected. It is well known that J_c generally decreases with H_a (assuming an absence of peak-effect); consequently a field dependence of $\dot{Q}_h$ somewhat stronger than cubic is to be expected. As a result of the self-field studies of SHIIKI et al [Shi74^c], β was found to be equal to 3.5; values of β larger than 3.0 have also been encountered in other such experiments [Eas70].

11.16 MAGNETIC HYSTERESIS LOSS IN BARE AND STABILIZED Ti-Zr-Nb SUPERCONDUCTORS

An alternative experimental approach to the study of AC stability and AC loss is to expose a noninductively wound sample of wire to an AC magnetic field. Fig.11-17 represents a magnetic-hysteresis-aided case

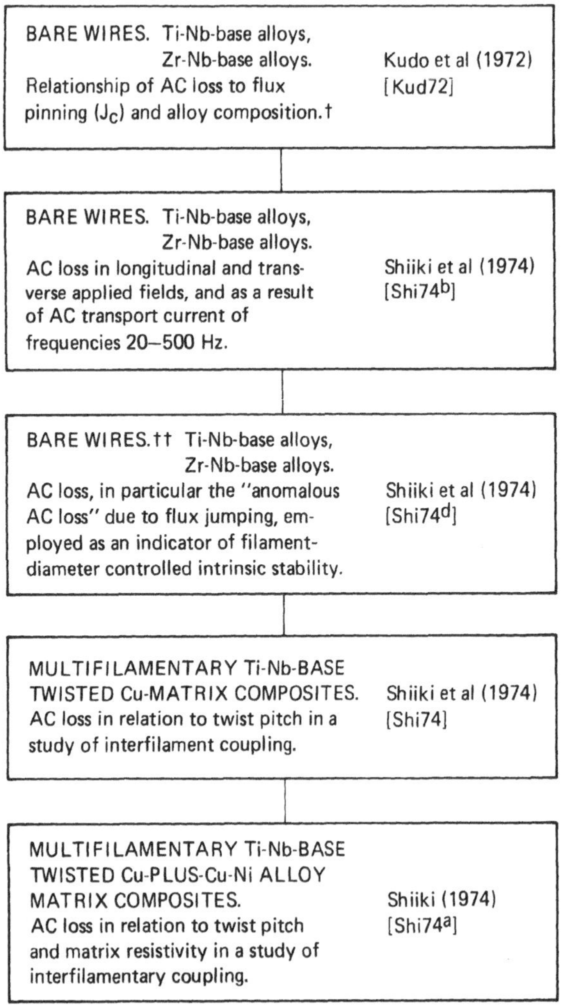

BARE WIRES. Ti-Nb-base alloys, Zr-Nb-base alloys. Relationship of AC loss to flux pinning (J_c) and alloy composition.†	Kudo et al (1972) [Kud72]
BARE WIRES. Ti-Nb-base alloys, Zr-Nb-base alloys. AC loss in longitudinal and transverse applied fields, and as a result of AC transport current of frequencies 20—500 Hz.	Shiiki et al (1974) [Shi74b]
BARE WIRES.†† Ti-Nb-base alloys, Zr-Nb-base alloys. AC loss, in particular the "anomalous AC loss" due to flux jumping, employed as an indicator of filament-diameter controlled intrinsic stability.	Shiiki et al (1974) [Shi74d]
MULTIFILAMENTARY Ti-Nb-BASE TWISTED Cu-MATRIX COMPOSITES. AC loss in relation to twist pitch in a study of interfilament coupling.	Shiiki et al (1974) [Shi74]
MULTIFILAMENTARY Ti-Nb-BASE TWISTED Cu-PLUS-Cu-Ni ALLOY MATRIX COMPOSITES. AC loss in relation to twist pitch and matrix resistivity in a study of interfilamentary coupling.	Shiiki (1974) [Shi74a]

†Also frequency dependence of AC loss in single- and multifilamentary composites.
††One Cu-clad monofilament investigated.

FIGURE 11-17. AC-loss studies in the design of twisted multifilamentary mixed-matrix Ti-Zr-Nb conductors.

study of conductor design, from bare alloy wire through through twisted multifilamentary mixed-restivity-matrix composites. Several important elements of this study are the subjects of the following discussions.

11.17 MAGNETIC HYSTERESIS LOSS IN BARE Ti-Zr-Nb ALLOY WIRE

11.17.1 Influence of Composition

Using bare heavily cold-worked and final-heat-treated wire SHIIKI and co-workers [Kud72][Shi74b, Shi74d] have studied the magnetic hysteretic loss characteristic of the superconducting alloy itself.

The composition dependences of this loss in both X-type (Zr-Nb-base) and Z-type (Ti-Nb-base) Ti-Zr-Nb alloys were measured and compared with those of a pair of representative binary alloys, $Zr-Nb_{75}$ and $Ti-Nb_{36}$. The several important results coming out of this work were: (a) the AC loss in Ti-Nb-base alloys (constant 2.5 at.% Zr) passed through a minimum as the Ti content was varied between 55 and 75 at.%; (b) AC loss in the Zr-Nb-base alloys (i) with constant 8 at.% Ti and [Zr] = 35∿45 at.% and (ii) with constant 40 at.% Zr and [Ti] = 5∿11 at.%, also passed through minima. It was therefore possible to identify a pair of minimal-AC-loss X-type and Z-type alloys viz: $Ti_8-Zr_{40}-Nb_{52}$ and $Ti_{70}-Zr_{2.5}-Nb_{27.5}$ whose AC losses (per cycle) in the frequency range considered (20-500 Hz) were about 30% lower than those of their binary counterparts. In an intercomparison of the above-mentioned binary alloys, the ternary Z-type alloy, and a slight variant of the X-type alloy, the magnitudes of the AC losses fell in the sequence:

$$Zr-Nb_{75} > Ti-Nb_{36} > Ti_{70}-Zr_{2.5}-Nb_{27.5} > Ti_6-Zr_{40}-Nb_{54}.$$

11.17.2 Influence of Wire Diameter

The applicability of early superconductors was severely limited by flux-jump instability. Since the introduction of fine multifilamentary twisted composites this has ceased to be a serious problem. Investigations of flux-jump instability in AC applied fields were undertaken by SHIIKI and KUDO [Shi74d]. Heavily cold-worked and heat-treated bare X-type and Z-type conductors were prepared in filament diameters ranging from 45 to 508 μm and exposed to AC magnetic fields of frequencies 35, 50, and 100 Hz, and surface amplitudes of up to about 3.8 kOe. At certain amplitudes, H_{fj}, depending on wire diameter, the AC loss per cycle would increase suddenly by an order of magnitude, signalling the occurrence of a flux jump. As the wire diameter decreased, H_{fj} moved to higher fields until, for example, with the alloy $Ti_{60.2}-Zr_{2.8}-Nb_{37}$ at a diameter of 45 μm the wire was intrinsically stable. It follows that the wire diameter needed for intrinsic (adiabatic) stability under *transport-current* conditions (half of the above value) would be 27 μm.

11.17.3 Hysteresis and Flux-Jump Anisotropies

In what might be regarded as an extension of several of the above studies, SHIIKI et al [Shi74b] using bare wire prepared under a wide range of cold work (99.7 - 99.998%) and final-heat-treatment conditions ($2\frac{1}{2}$h/290°C to 100h/350°C) have investigated magnetic-hysteresis and flux-jump "anisotropies". AC loss measurements were carried out: (a) with an applied AC field (35-100 Hz) either perpendicular to or parallel to the wire: (b) in zero applied field with an AC current ($\gtrsim$100 A) of frequency 20-500 Hz passing through the wire. As a result of: (i) the pronounced longitudinal versus transverse pinning center anisotropy (cf. for example Figs.7-34 and 7-35) and (ii) the tendency for the current flow to be "force-free" (cf. [Mon22.00]) when the applied field is directed along the wire, both AC loss and the flux-jump field, H_{fj}, depended on whether the field to which the current was exposed was transverse, longitudinal, or circumferential (i.e. the self-field).

11.18 AC LOSSES IN OPEN-CIRCUITED (i.e., NON-INDUCTIVELY WOUND) COMPOSITE SUPERCONDUCTORS IN TRANSVERSE MAGNETIC FIELDS

The total AC loss is the sum of contributions from the matrix (eddy-current loss) and the superconducting filaments (hysteretic loss) both of which are frequency- and field-amplitude-dependent. The situation is fully discussed in [Mon26.00], in particular [Mon26.8], and briefly summarized below.

The eddy-current losses in the "low-frequency" and "intermediate-frequency" regimes, which are separated by a characteristic frequency

$$f_c = 10^9 \rho_\perp / L_p^2 \quad , \qquad (11\text{-}10)$$

where $\rho_\perp$ is the transverse resistivity (Ω cm) of the composite and L_p is the twist pitch (cm), are given by:

$$\dot{Q}_e = \frac{H_m^2 f^2}{1 + (f/f_c)^2} \quad , \qquad (11\text{-}11)$$

where H_m is the amplitude of the applied transverse magnetic field. This very important equation shows that: (a) At low frequencies (i.e. $f < f_c$) $\dot{Q}_e \propto H_m^2 f^2$, while at intermediate frequencies ($f > f_c$) $\dot{Q}_e$ is proportional only to H_m^2. (b) Whether or not a given applied-field frequency is designated "low" or "intermediate" depends upon $\rho_\perp$ and L_p, i.e. upon the factors controlling f_c.

The hysteretic loss at low frequencies, as defined above, is the sum of the individual-filament hysteretic losses, each of which is given by:

$$\dot{Q}_h \propto H_m f \qquad . \qquad (11\text{-}12)$$

Such a linear field dependence is appropriate for hysteretic loss associated with fully-field-penetrated superconductors. As the frequency increases and interfilamentary coupling builds up, $\dot{Q}_h$ becomes larger than the sum of the individual-filament contributions. At frequencies $f > f_c$ coupling is sufficiently strong, especially towards the outside of the conductor where the relative twist pitch (pitch-to-diameter ratio) is lower, that the outer layers of the conductor tend to behave like a solid superconducting (or "current-saturated" [Car74]) cylindrical shell. From the external-field standpoint this exterior shell gives the conductor the appearance of a partially-penetrated ($H_m \ll H^*$ in the usual terminology) solid superconductor for which the proportionality

$$\dot{Q}_h \propto H_m^3 f \qquad (11\text{-}8)$$

is again appropriate. Of course the remaining uncoupled filaments in the interior of the conductor still contribute an Eqn.(11-12)-type of $H_m f$-proportional hysteretic loss.

At very high frequencies (usually in the high audio-frequency range for practical twisted composites [Mur75]) the classical skin effect takes over to confer an $f^{1/2}$ frequency dependence on both Q_e and Q_h; but if the conductor is untwisted, it has been suggested [Shi74a] that this "high-frequency" effect will tend to take place at frequencies as low as several hundred Hz.

Relying on principles embodied in Eqns.(11-8) through (11-12), SHIIKI et al [Kud72][Shi74, Shi74a] used calorimetric measurements of AC loss to investigate the influences of twist pitch and matrix resistivity on interfilamentary coupling (hence also stability) in Cu-matrix and mixed-matrix multifilamentary composite superconductors.

11.19 AC LOSS IN Ti-Zr-Nb-BASE COMPOSITE CONDUCTORS—Cu MATRIX

Using a thirty-seven-filament (23 μm$^\phi$) $Cu/(Ti_{60.0}\text{-}Zr_{2.4}\text{-}Nb_{37})$ composite conductor, SHIIKI et al [Shi74] have measured AC loss per unit length as functions of twist pitch (in particular, 2, 5, 20, 50 mm and ∞), applied-field amplitude, and applied-field frequency. Some experiments were also conducted on bare and Cu-clad-monofilamentary $Ti_{60.2}$-$Zr_{2.8}$-Nb_{37}.

In studies of the latter pair of samples, the hysteretic bare-wire loss was found to be proportional to $H_m^{3.5}f$, in agreement with arguments presented in the previous section, and the total AC loss of the mono-filament (which is of course a canonically uncoupled composite) was found to obey the relationship:

$$\dot{Q} = {}_af + {}_bf^2 \qquad (11\text{-}9a)$$

as befits a linear combination of saturated ($H_m > H^*$) hysteretic and normal eddy-current losses.

With the multifilamentary composite an opportunity existed for examining interfilamentary coupling and its response to applied-field amplitude and twist pitch.

11.19.1 Applied Field Amplitude

As indicated by Eqns.(11-8) and (11-12) the constant-frequency AC loss will undergo a pronounced change in field dependence as soon as H_m exceeds H^*, the penetration threshold. As is well known by now, $H^* = (2\pi/10)J_cd$, Eqn.(7-24a), where d is the diameter of the superconductor (a single filament in this case). It follows that H^* (hence the β of Sect.11.15) is responsive to the increase in effective filament diameter that accompanies interfilamentary coupling -- especially if this results in the establishment of the "current-saturated outer shell" [Car74] referred to above. Thus, as SHIIKI has confirmed experimentally [Shi74], a log-log plot of Q versus H_m should exhibit a d-dependent break-point as Q_h switches from H_m^3-dependence to H_m-dependence at $H_m = H^*$. Furthermore, for a given Cu-matrix conductor, this break-point should also be L_p-dependent in as much as twist controls the effective diameter of the superconducting component of the composite.

11.19.2 Twist Pitch

SHIIKI et al [Shi74] found, as a result of experiments conducted at each of the frequencies 35, 50, 100, and 250 Hz, that as a function of twist pitch the

constant-frequency AC loss passed through a maximum. This effect can be qualitatively understood with reference to Eqns.(11-10) and (11-11), which in association with Eqn.(11-6) lead to:

$$\dot{Q} = \dot{Q}_h + \frac{H_m^2 f^2}{1 + Kf^2L_p^4} \qquad (11\text{-}13)$$

where K is a constant (= $10^9 \rho_\perp)^{-2}$.

If L_p is very large, the eddy-current term vanishes and Q is simply Q_{hc}, the strongly coupled hysteretic loss of the saturated outer layer; if L_p is very small, Q_h is minimal, being that of the sum of the individual-filament losses. It is not unreasonable to expect, especially in the light of the experimental observations, that at some intermediate value of twist pitch (5∿10 mm at 50∿250 Hz for the present conductor) Q, then being the sum of the eddy-current and partially enhanced hysteretic losses, will have its maximum values.

11.20 AC LOSS IN Ti-Zr-Nb-BASE COMPOSITE CONDUCTORS—RESISTIVE (MIXED) MATRIX

In what was essentially a continuation of the studies discussed in the previous section, SHIIKI [Shi74a] has investigated the AC-loss properties of 37-filament (∿24 μm^ϕ) $Ti_{62.5}$-$Zr_{2.5}$-Nb_{37}-base composite conductors (∿0.25 mm O.D.) whose matrices, consisting of Cu and a Cu-Ni alloy in the proportions 4.7:1, 0.28:1, and 0:1, possessed measured average 12-K resistivities of about 3.85×10^{-8}, 6.88×10^{-7}, and 1.12×10^{-5} Ω cm, respectively.

Log-log plots of $\dot{Q}$ versus H_m for twist pitches of typically 2, 3, 5, 20, 50 mm and ∞, at each of the three matrix resistivities, were constructed in a search for the coupling-dependent break-points and their resonses to L_p and ρ_{matrix}.

$\dot{Q}$ was also plotted versus twist pitch (actually L_p^{-1}) as before. For the conductor with the highest-resistance matrix, the AC-loss maximum (for f∿250 Hz) occurred at a twist pitch of about 50 mm as compared to the 5∿10 mm noted above with respect to the pure-Cu-matrix conductor. If, as was suggested above, this maximum is related to the passage of $f_c = 10^9 \rho_\perp/L_p^2$ through the frequency of measurement, it should be found that, for constant $f_c = f_{meas}$:

$$\rho_\perp \propto L_p^2 \qquad (11\text{-}14)$$

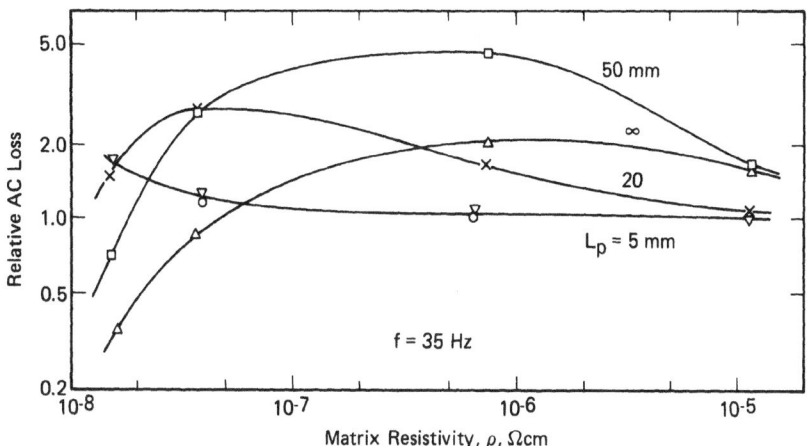

FIGURE 11-18. Total 35-Hz AC loss, normalized to that of a 2-mm-twist-pitch conductor, plotted *versus* matrix resistivities (of Cu, Cu/Cu-Ni and Cu-Ni-matrix composites) for four values of the twist pitch — after SHIIKI [Shi74[a]].

In a further examination of the interplay between $\dot{Q}$, L_p, and ρ_{matrix}, SHIIKI [Shi74[a]] plotted relative AC loss at 35 Hz *versus* matrix resistivity for conductors with twist pitches of 5, 20, 50 mm and ∞, data having been provided from measurements on the three resistive-matrix conductors just referred to, and the pure-Cu-matrix conductor of the previous section. The results, as summarized in Fig.11-18, suggest that whereas the relative AC loss for L_p = 5 mm appears to be maximizing at $\rho \cong 10^{-8}$ Ω cm, that for L_p = 50 mm maximizes at $\rho \cong 10^{-6}$ Ω cm, in good agreement with the requirements of Eqn.(11-14). The characteristic frequency, f_c, has of course separated the uncoupled-filament regime from one in which interfilamentary coupling, *via* eddy currents circulating within the matrix, is beginning to influence the AC loss.

12

TITANIUM-NIOBIUM-BASE TERNARY TRANSITION-METAL ALLOYS (EXCEPT TITANIUM-ZIRCONIUM-NIOBIUM)

This discussion of ternary Ti-Nb-base transition metal alloys commences with Ti-Hf-Nb -- Hf, being a group-IV element may, like Zr, be regarded as a substitute for Ti -- and continues with the two Ti-Nb-(group-V)TM alloys Ti-Nb-V, and Ti-Nb-Ta, in which the third solute may be regarded as substituting for Nb. In recent years Ti-Nb-Ta has been shown to be a very important system both economically and technically; accordingly, its properties have been given special emphasis in this review. The chapter concludes with

a survey of the superconducting properties of ternary alloys of Ti-Nb with the group-VI elements Cr, Mo, and W, the group-VII elements Mn and Re, and with several members of group VIII of the periodic table.

Many of the authors whose work is about to be reviewed here have expressed magnetic fields in tesla (1 T is equivalent to 10 kOe). In conformity with the original literature that convention, with a few exceptions, is adopted in this chapter.

ALLOY GROUP 1: TITANIUM-HAFNIUM-NIOBIUM AND TITANIUM-NIOBIUM-VANADIUM ALLOYS

12.1 SUPERCONDUCTIVITY IN Ti-Hf-Nb ALLOYS

12.1.1 Introduction

Hf, a group-IV element, may be regarded as substituting for Ti when added to Ti-Nb alloys. As pointed out in Chapter 11, Ti-Zr-Nb, the other alloy in this class, has in the recent past assumed considerable importance as a commercial superconducting alloy. Although presently out of favor, it has been the subject of several score of publications; by comparison, Ti-Hf-Nb has received little attention. The upper critical field of Ti_{64}-Hf_4-Nb_{32} was measured

magnetically be DeSORBO et al [Des67] as part of a comprehensive study of the systematics, in terms of an "effective electron/atom ratio", of upper critical field in Ti-Nb-base ternary and quaternary alloys. The transition temperature, critical current density, and some metallurgical quantities have been studied by BYCHKOVA et al [Byc70[a]] on homogenized and 99.9% cold-worked samples of Ti_{65}-Hf_{10}-Nb_{25} and Ti_{50}-Hf_{25}-Nb_{25}. With increasing Hf content, the transition temperature after an initial decrease tended to remain fairly constant ($T_c \stackrel{\sim}{=} 6.8$ K) within the range considered. The critical current densities were rather low, and not particularly meaningful, since no attention had been paid to either the adequacy of the cold work or to heat treatment of any kind. Both of these, and a related study,[†] were concluded during the mid 1960's. But with the recently recognized necessity of improving the properties of Ti-alloy superconductors, Ti-Hf-Nb together with several other ternary and quaternary alloys have been experiencing a revival of interest. For example, HAWKSWORTH and LARBALESTIER [Haw80, Haw81] have investigated the upper critical fields and other relevant electrical properties of three families of Hf-substituted Ti-Nb alloys. The systems referred to in [Haw80], the source of the data presented in this chapter, were:

$$Ti-Hf_{2.5}-Nb_{27.5-37.5}$$
$$Ti-Hf_5-Nb_{27-5-42.5}$$
$$\text{and } Ti-Hf_{10}-Nb_{25-35} \quad ;$$

they are of course variants of the standard Ti-50Nb ($Ti-Nb_{34}$) commercial binary alloy.

Hf is a heavier element that Ti; thus the rationale underlying this investigation, as with the comparable but more extensive study by WADA et al [Wad81] to be discussed below, was to take advantage of a supposed reduced spin-flip-scattering electron mfp in an attempt to increase the Pauli paramagnetism of the superconductive component of the mixed state, along the lines first investigated theoretically by MAKI [Mak66] and WERTHAMER et al [Wer66] and experimentally by KIM and STRNAD [Kim66] and others. The paramagnetic mixed state has been dealt with in [Mon15.00] and considered elsewhere in this book (see for example Sects.3.6.3 to 3.6.5, Sect.5.12.4, and Sects.7.10, 7.11). Although the partial relief of Pauli spin paramagnetic limitation by this mechanism is probably illusory [Orℓ79] [Bea82], the idea has at least established a starting point for empirical studies which, for reasons yet to be fully understood, have in some cases been successful in paving the way towards the development of alloys with improved upper critical fields.

+ In this regard it is interesting to note that in a patent issued in 1965 to Imperial Metal Industries of Great Britain [Imp65] it was observed that Hf, which may occasionally be associated with Zr to the extent of 2 to 3%, could be present in Ti-Zr-Nb in amounts of up to 5% without adversely affecting its superconducting properties (cf. Table 1-38).

12.1.2 Transition Temperatures of Ti-Hf-Nb Alloys

As a result of their transition-temperature measurements on the Ti-Hf-Nb alloy systems mentioned above, HAWKSWORTH and LARBALESTIER [Haw80] noted that the substitution of 2.5, 5, and 10 at.% Hf for Ti in $Ti-Nb_{35}$ decreased the transition temperature at the rate of about 0.08 K per at.% Hf. This result was at least in qualitative agreement with the earlier work of BYCHKOVA et al [Byc70[a]] referred to in the introduction.

Involving as it does, a set of Vickers hardness, critical temperature, critical field, and J_c measurements, on a series of thirty-four ternary and eight binary alloys, the recently published work of WADA et al of the Japan National Research Institute for Metals represents the most complete study yet undertaken of the superconducting properties of the Ti-Hf-Nb system [Wad81]. Critical temperatures were measured resistively on forty-two cold-rolled (80% to 0.2 mm[t]) alloys. The results, summarized in Fig.12-1, suggest the existence of a "high-T_c" zone in the vicinity of 50-60 at.% Nb and 3 at.% Hf. In general, however, a decrease in T_c accompanies the addition of Hf. For example, at a constant 40 at.% Nb the T_c's at 5 at.% Hf and 7.5 at.% Hf are, respectively, 9.0 K and 8.5 K -- a rate of decrease of 0.2 K per at.% Hf. WADA et al also studied the T_c's of the 40 at.% Nb family as a function of 24-h aging temperature [Wad81].

12.1.3 Critical Fields of Ti-Hf-Nb Alloys

As mentioned in a pair of recent publications [Haw80, Haw81], HAWKSWORTH and LARBALESTIER have performed resistive measurements at 4.2, 3, and 2 K of the upper critical fields of several sequences of

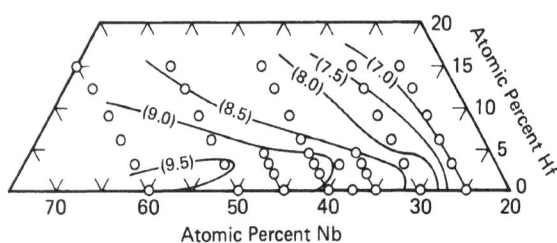

FIGURE 12-1. Superconducting transition temperatures (T_c, K) of thirty-four ternary Ti-Hf-Nb alloys and eight Ti-Nb binaries measured resistively in the cold-rolled (80%) condition — after WADA et al [Wad81].

Ti-Hf-Nb alloys[†]. The 4.2 K results of the initially reported investigation of alloys containing 2.5, 5, and 10 at.% Hf [Haw80], as extracted from the full data-set listed at the end of this chapter, are depicted in Fig.12-2(a). These alloys had been homogenized 8h/1350°C, square-rolled >75% and recrystallized 1h/875°C before being swaged, Cu-sheathed, and drawn more than 97% to wire. A subsequent report of these studies included data for alloys containing 15 and 20 at.% Hf [Haw81]. As a result of either of these reports it can be concluded that the addition of increasing amounts of Hf to Ti-Nb results in a continuous deterioration of the 4.2 K resistive critical field. On the other hand, at 2 K, some slight but practically unimportant improvement in H_r seems to result from the substitution of 2.5 and 5 at.% Hf into some Ti-Nb alloys.

WADA's study of Ti-Hf-Nb alloys [Wad81] was published simultaneously with the second paper of HAWKSWORTH and LARBALESTIER [Haw81]. In it were reported the resistive upper critical fields[††] of those cold-rolled alloys that had exhibited the highest T_c's (cf. Sect.12.1.2). The results of this work are depicted in Fig.12-2(b), according to which the maximum H_r for the complete system seems to be slightly offset from the Ti-Nb boundary, and lies within a zone de-

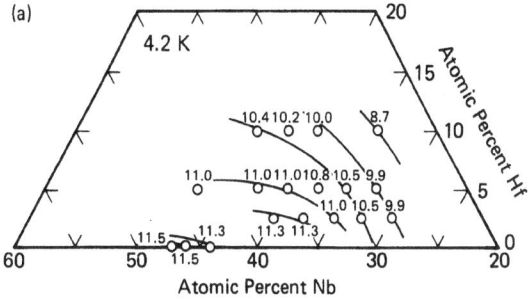

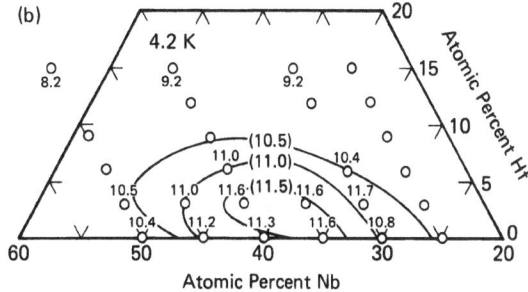

FIGURE 12-2. Resistive upper critical fields (tesla) of Ti-Hf-Nb alloys at 4.2 K. (a) Data of HAWKSWORTH and LARBALESTIER [Haw80] on > 97% cold-drawn material (criterion: first appearance of resistance ("onset") at J = 5 A cm^{-2}). (b) Data of WADA *et al* [Wad81] on 80%-cold-rolled material (criterion: mid-point of resistive transition at J = 1 A cm^{-2}).

fined by 33-42 at.% Nb, 0-3 at.% Hf. According to WADA *et al* [Wad81], the highest value of H_r(4.2 K) to be attained was 11.7 T with Ti_{57}-Hf_3-Nb_{40}. By comparison, that for the Ti-Nb_{40} control was 11.3 T, a value which agrees well with the 11.3 T (4.2 K) obtained by HAWKSWORTH and LARBALESTIER [Haw80] with Ti-Nb_{44}.

12.1.4 Critical Current Densities of Ti-Hf-Nb Alloys

WADA *et al* [Wad81] went on to measure the 4.2-K J_c's in fields of up to 8 T of those Ti-Hf-Nb alloys which had exhibited high values of H_r after being aged 24h/400°C, 24h/800°C, or after being subjected to a two-stage thermomechanical processing sequence consisting of 24h/800°C + 80% cold work + 24h/400°C. In addition, the critical current densities of several single-filament and 7-filament Cu-clad composites were measured as functions of applied field ($\gtrsim$10 T) and as functions of aging time at 400°C in fields of 3, 4.5, 6.5 and 8 T. Fig.12-3 which compares the current carrying capacities of Ti_{57}-Hf_3-Nb_{40} with that of the corresponding binary control alloy, indicates the extent to which the substitution of 3 at.% Hf for Ti in Ti-Nb_{40} improves the high-field, 4.2-K, J_c.

[†] *The H_r Criterion of HAWKSWORTH and LARBALESTIER:*

In the measurements of H_r described in [Lar80], [Haw80], and [Haw81], a circuit current of 10 mA (or 5 A cm^{-2} through the 0.5-mm$^\phi$ monofilamentary core) was employed. The transition criterion was defined as the "first appearance of resistance" [Lar80] or the "onset of normalcy -- (as distinct from) the disappearance of the superconducting state" [Haw80]. It is obviously more conservative than the "half-transition" point adopted by WADA *et al* [Wad80, Wad81]. In the authors' illustrations of this condition [Haw80][Lar81], H_r is depicted as residing close to the foot of the resistive transition to the normal state. This criterion is implied by the notation *first appearance of resistance ("onset")* or simply *"onset"*, when used in the captions and footnotes, respectively, of the relevant figures and tables. For further discussion, see [Lar81].

[††] *The H_r Criterion of WADA:* At a current density of 1 A cm^{-2}, H_r was defined as the mid-point of the transition to the normal state at 5 μV cm^{-1}.

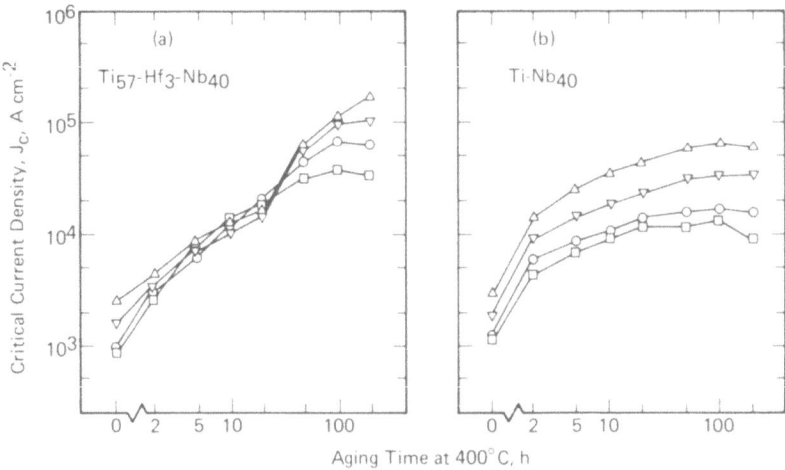

FIGURE 12-3. Critical current densities of Ti$_{57}$-Hf$_3$-Nb$_{40}$ at 4.2 K in fields of 3 T (△), 4.5 T (▽), 6.5 T (○), and 8 T (□) as functions of aging time at 400°C compared with those of the corresponding Ti-Nb (40 at. %) control alloy (criterion: 5 μV cm^{-1}) — after WADA *et al* [Wad81].

12.2 SUPERCONDUCTIVITY IN Ti-Nb-V ALLOYS

Motivated by a desire to improve on the properties of the basic Ti-Nb alloy, numerous studies have been made of the effects on T_c, H_{c2} and J_c of replacing some of the Nb with one or other of its electronic isomers, V and Ta. Progress in the development of Ti-Nb-V alloys is recorded in Table 1-35 and discussed below, where it will be shown that V substitution has a detrimental influence on both T_c and H_{c2}.

12.2.1 Transition Temperatures of Ti-Nb-V Alloys

When added to Ti-Nb alloys, V substitutes directly for the Nb to form ternary solid solutions with no unexpected anomalies in the equilibrium-phase boundaries. The results of studies of transition temperature in the Ti-Nb-V system have been reported by SAVITSKII and co-workers, first in 1966 and again in 1973 [Sav73, p.326]. In 1971 SAVITSKII and co-workers [Sav71] applied multiple-linear-regression analysis in order to express T_c as a function of ternary composition and to plot it in a three-dimensional diagram.

In studying an extensive matrix of Ti-Nb-V alloys, BELLIN *et al* [Bel69] obtained results which were in reasonably satisfactory agreement with those of the earlier workers. The influence of V on the superconducting transition temperature of Ti-Nb was expressed in the format T_c *versus* e/a for several fixed values

of the V concentrations (viz 9, 10, 20, 30 and 40 at.%). The results showed that for each fixed value of e/a, T_c decreased rapidly as V was substituted for Nb. For example, at 50 at.% Ti, the magnetically measured transition temperatures corresponding to 0, 10, 20, 30 and 40 at.% V were, respectively, 9.9, 8.8, 8.3, 7.8$_3$ and 7.8 K.

POLLOCK *et al* [Pol69] investigated the influence of neutron irradiation damage on the T_c's of a set of four Ti-Nb-V alloys. After receiving a fast *neutron* dose of 3.7x10^{19} cm^{-2}, the transition temperature was found to have dropped by about 0.3 K from an initial value of about 5.8 K. No mechanistic reason for the decrease was offered.

12.2.2 Critical Fields of Ti-Nb-V Alloys

The resistive upper critical field of an extensive matrix of some thirty-five ternary Ti-Nb-V alloys has been the subject of a phenomenological investigation by BELLIN *et al* [Bel70]. The results were summarized on a two-dimensional ternary diagram and also as families of curves representing [V] = constant, [Nb] = constant, and [Ti] = constant isoelectronic slices of the three-dimensional T_c-composition solid. Noting that the isoelectronic curves all possessed minima, BELLIN *et al* showed that the H_r maxima lay only along the binary Ti-Nb and Ti-V edges. In other words, the isoelectronic addition of V to Ti-Nb, and Nb to Ti-V, was always detrimental to H_r.

The study by RASSMANN and ILLGEN [Ras73] of the critical current densities of six Ti-Nb-V alloys is referred to below within the context of a companion investigation of Ti-Nb-Ta and other alloys.

ALLOY GROUP 2: TITANIUM-NIOBIUM-TANTALUM ALLOYS

12.3 SUPERCONDUCTIVITY IN Ti-Nb-Ta ALLOYS

The motivation for several investigations of Ti-Nb-Ta alloys -- notably the early studies by SUENAGA and RALLS [Sue69] as well as the recent work of HAWKSWORTH and LARBALESTIER [Haw80][Lar80, Lar81] -- was a desire to improve on the upper critical field of Ti-Nb. As mentioned above, the rationale was to replace some of the Nb by the heavier element Ta, in order to increase spin-orbit scattering and consequently relieve the impact of Pauli upper critical field limitation by reducing the difference between the Pauli spin susceptibilities of the superconductive and normal states. In pursuit of this goal a wide range of alloy compositions, and the effects of rather high concentrations of Ta, have been explored.

The second motivation for studying the influence of Ta on the mechanical and superconductive properties of Ti-Nb alloys was economic. It was first pointed out by CURTIS and McDONALD [Cur79] that important Nb reserves occur within the ore *pyrochlore* which in addition to other easily separable impurities contains up to 1% of Ta. If, therefore, up to about this much Ta can be tolerated in the finished Ti-Nb alloy, it becomes attractive for both economic and political reasons to utilize pyrochlore as the source of one of the starting materials. The refinement of the ores *columbite-tantalite* and *pyrochlore* is described in [Mon28.1].

The rationales outlined in the preceding two paragraphs have provided the stimuli for a series of recent studies of the effects of both "high" (i.e. up to about 10 at.%) and "low" levels, respectively, of Ta in Ti-Nb. If improved superconductive properties, unaccompanied by any serious deterioration in workability or other difficulties[+] that would render their

---

[+] According to HEMACHALEM *et al* [private communication] considerable difficulty has been experienced in obtaining *large homogeneous melts* of Ti-Nb-Ta.

use infeasible, were to follow the deliberate addition of several percent of Ta to Ti-Nb, both goals would be simultaneously satisfied. Such has more or less turned out to be the case; indeed, as will be outlined below, recent studies of Ti-Nb-Ta alloys have led to the development of an important new high-field alloy superconductor. The literature of Ti-Nb-Ta superconductivity, with particular emphasis on recent developments, is surveyed in Table 1-36.

12.4 TRANSITION TEMPERATURES OF Ti-Nb-Ta ALLOYS

In the early 1960's, during studies of Nb-Zr and the related ternary alloy Nb-Zr-Ti, REINBACH [Rei67] noted that improved properties accompanied the substitution of Ta for Zr. In a patent stemming from this work, transition temperatures as high as 13 K were claimed for Ti-45Nb-30Ta and Ti-60Nb-10Ta after cold work and heat treatment. Such surprisingly high transition temperatures for this system have never been confirmed. Major studies of Ti-Nb-Ta were conducted by BARANOV and co-workers [Bar70] (see also SAVITSKII *et al* [Sav73, p.329]) and SUENAGA and RALLS [Sue69]. Subsequently, the Soviet work was the subject of multiple regression analysis to establish a T_c *versus* composition relationship. The results of these investigations were in reasonable agreement, and showed that the superconducting transition temperatures of all isoelectronic quenched Ti-Nb-Ta alloys decreased as Ta was substituted for Nb, both in the single-phase bcc regime and in the Ti-rich corner of the ternary diagram where the alloys contained some α-phase precipitates. The composition dependence of T_c in the Ti-Nb-Ta system is displayed on a composition triangle in Fig.12-4. As an example of the influence of Ta on the transition temperature, we note that at 50 at.% Ti the magnetically determined T_c's corresponding to 0, 10, 20, 30, and 40 at.% Ta were respectively about 9.9, 9.4, 9.1, 8.9, and 8.7 K, yielding

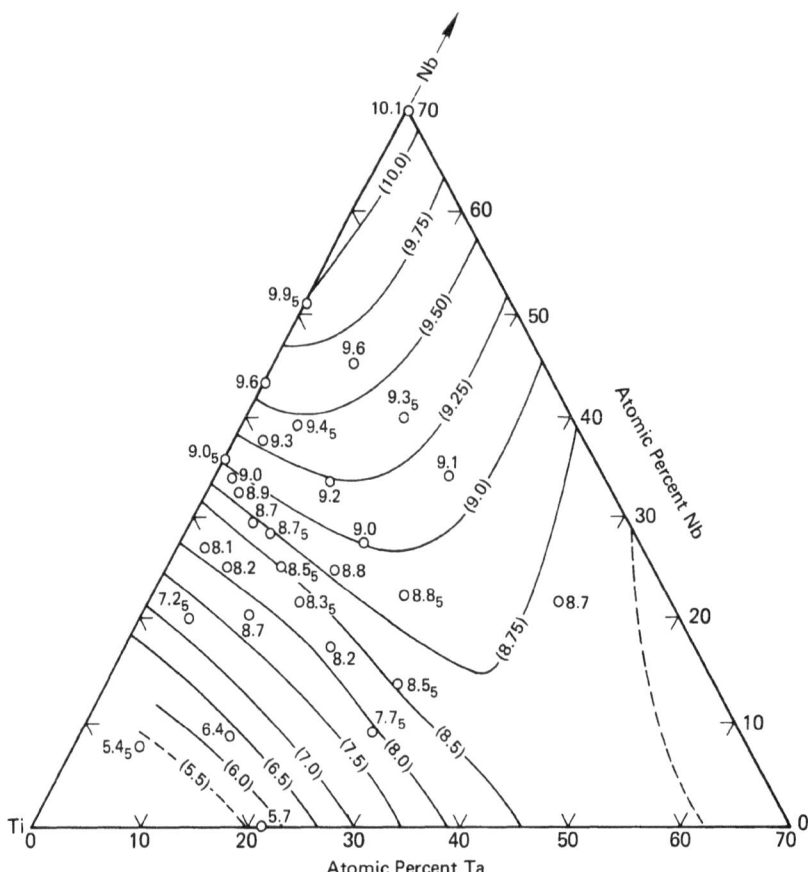

FIGURE 12-4. Superconducting transition temperature (T_c, K) of Ti-Nb-Ta alloys. The isothermals indicated are based on magnetic half-transition data — after SUENAGA and RALLS [Sue69].

a rate of decrease not quite as steep as that found in response to V additions under otherwise similar conditions. In the studies of Ti-Nb-Ta by HAWKSWORTH and LARBALESTIER [Haw80], it was shown that the substitution of 5, 10, 15, and 20 at.% Ta for Nb in Ti-Nb$_{35}$ decreased T_c at the rate of about 0.04 K per at.% Ta.

12.5 UPPER CRITICAL FIELDS OF Ti-Nb-Ta ALLOYS

Performed independently of, and slightly prior to, the Ti-Nb-V investigation reported above [Bel70] was a comparable study of Ti-Nb-Ta by SUENAGA and RALLS [Sue69], Fig.12-5, But unlike BELLIN'S inquiry, the results of which were treated phenomenologically, the work of SUENAGA and RALLS was guided by theories of the paramagnetic mixed state and, as pointed out above, the effects on the upper critical field of the competition between mixed-state and normal-state Pauli paramagnetism. Based on the concept of heavy-atom-

assisted decoupling by spin-orbit scattering of the parallel-aligned spins of the Cooper pairs, it was anticipated that with the partial reduction of Pauli limitation, H_r would increase as Ta was substituted (isoelectronically, of course) for Nb in the Ti-Nb lattice. Indeed, the addition of small amounts of Ta to Ti-Nb, at an e/a ratio corresponding to the maximum value of H_r for the binary system, increased H_r still further. Unlimited substitution is of course far from beneficial, since the effect of increased spin-orbit scattering (if such a mechanism is in fact operating within this context, cf. Sect.7.11 and [Mon15.8]) may have to compete with a decrease in the GOR'KOV-GOODMAN-MAKI-EILENBERGER dirty-limit upper critical field (see [Mon14.6]):

$$H_{c20}^* = 3.06 \times 10^4 \, \rho_n \, \gamma \, T_c \, \text{(Oe)} \quad [3\text{-}13, \, 7\text{-}10] \quad (12\text{-}1)$$

(with ρ_n in Ω cm, γ in erg cm^{-3} K^{-2}, and T_c in K) brought about by an alloying-induced decrease in T_c. Possibly as a consequence of such a competition, H_r

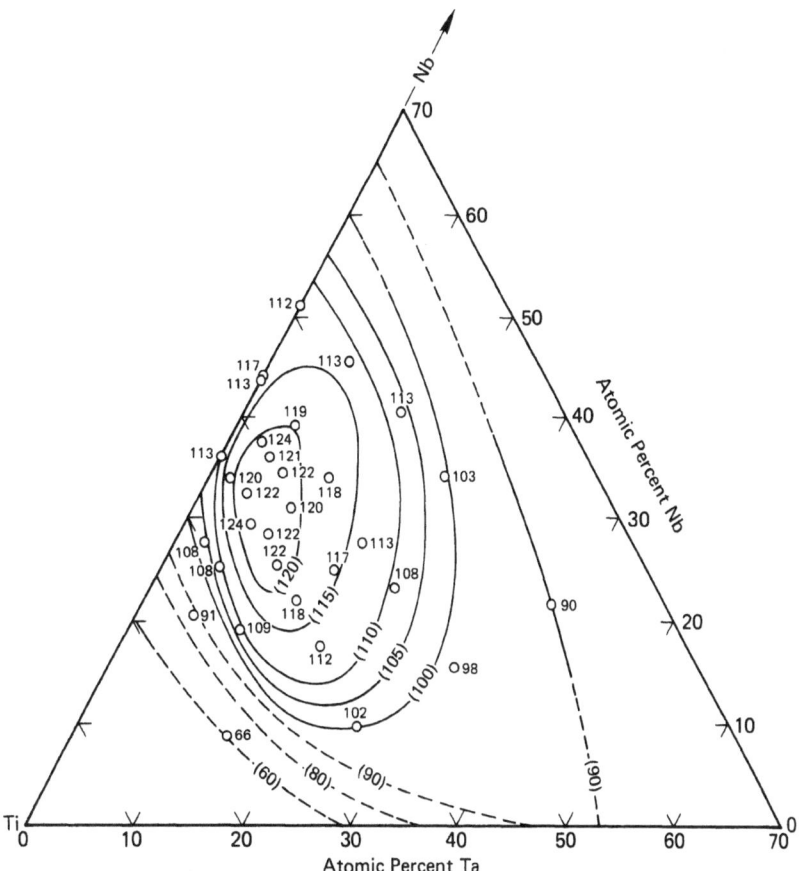

FIGURE 12-5. Upper critical fields (kOe) of Ti-Nb-Ta alloys. The data represent the resistively determined ($J \sim 30$ A cm^{-2}) end of the transition to the normal state in a pulsed magnetic field — after SUENAGA and RALLS [Sue69].

passes through a maximum. Maximum H_r for the group of some twenty-seven Ti-Nb-Ta alloys studied by SUENAGA and RALLS [Sue69] was 12.4 T at 4.2 K for a composition of about Ti_{64}-Nb_{30}-Ta_6. It is interesting to note that in an independent study, DeSORBO et al [Des67], using a magnetic measuring technique, obtained a critical field of 12.4 T in the alloy Ti_{60}-Nb_{36}-Ta_4. SUENAGA and RALLS were led to the conclusion that within a zone in the composition triangle bounded by: 2 at.% $\lesssim$ [Ta] $\lesssim$ 12 at.% and 23 at.% $\lesssim$ [Nb] $\lesssim$ 38 at.%, are to be found alloys with upper critical fields in excess of 12 T. This is some 1.0-1.5 T higher than the maximal H_{c2} exhibited by binary Ti-Nb and higher still than the maximal upper critical fields of Nb-Zr alloys.

Unfortunately, these promising conclusions were not fully substantiated by the results of the recent study of some thirty-five Ti-Nb-Ta alloys by HAWKS-WORTH and LARBALESTIER [Haw80][Lar80], see also [Lar81]. Part of the discrepancy is explicable in terms of differences in measuring techniques. In the earlier work, the normal-state end of the s/n resistive transition in a pulsed magnetic field was deliberately chosen as the criterion for assigning a value to H_r(4.2 K). The rationale underlying the selection of this H_{rn} rather than H_{rs} (the field at the start of the transition) or even the half-transition field (as favored by [Wad80, Wad81] for example) lay in the belief that, whereas H_{rs} depended to some extent on the bulk pinning strength, and consequently the metallurgical state of the sample, H_{rn} would tend to be more representative of an intrinsic alloy property. The pulsed-field experiment was performed at a current density of 30 A cm^{-2}, considerably higher than the densities favored subsequently by HAWKSWORTH et al [Haw80] (5 A cm^{-2}), HELLER et al [Hel71a] (1.3 A cm^{-2}) or WADA et al [Wad81] (1 A cm^{-2}). A comparison of the results of pulsed-field and DC-field measurements of a sample of commercial Ti-Nb wire showed that at $J = 30$ A cm^{-2}, whereas $H_{r,DC} = 11.2$ T, $H_{rs,pulse}$ and $H_{rn,pulse}$ were, respectively, about 0.25 T lower and 0.45 T higher than that value [Sue69].

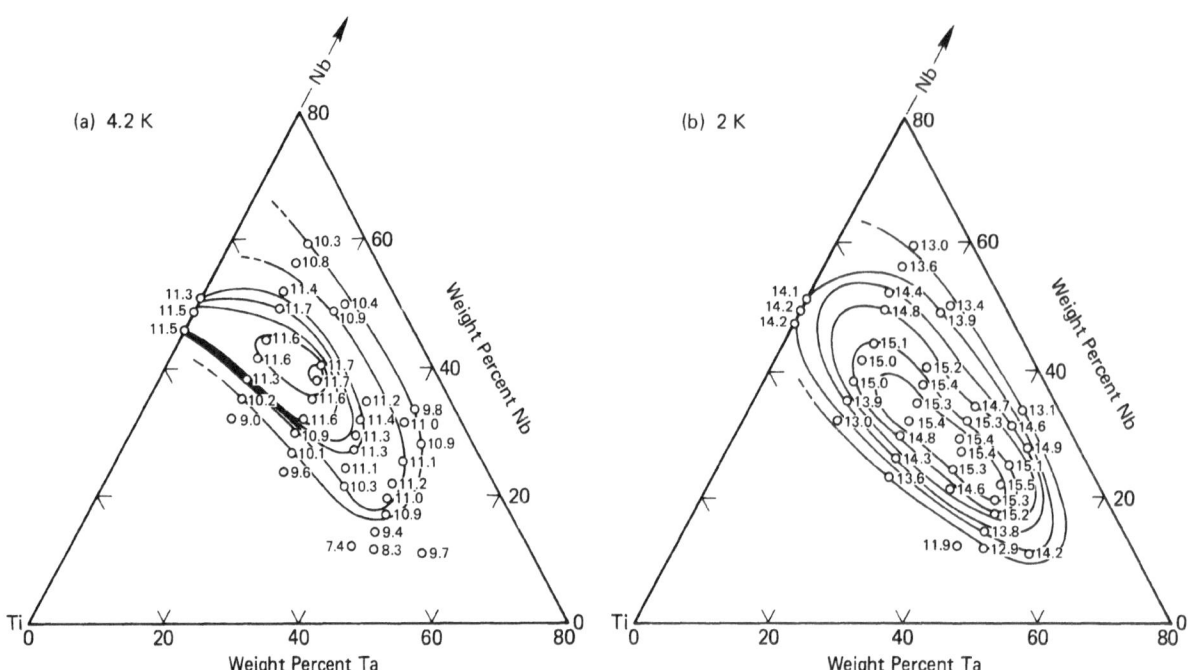

FIGURE 12-6. Resistive DC upper critical fields (tesla) of bare cold-drawn Ti-Nb-Ta alloy wires at (a) 4.2 K and (b) 2 K (criterion: resistive "onset" at J = 5 A cm^{-2}). The contours have been inserted, with considerable license, to guide the eye – data from LARBALESTIER [Lar80, Lar81].

HAWKSWORTH and LARBALESTIER whose results are summarized in Fig.12-6 from [Lar80] (cf. also Table 12-7(c) towards the end of this chapter) have, according to [Haw80], obtained maximal H_r's of 11.5-11.7 T at 4.2 K with the alloys Ti-Nb$_{30-37.5}$-Ta$_5$ and Ti-Nb$_{25-32.5}$-Ta$_{10}$ and 15.2-15.4 T at 2 K with the alloys Ti-Nb$_{25-30}$-Ta$_{10}$ and Ti-Nb$_{20-27.5}$-Ta$_{15}$. As a result of this work it was possible to conclude that, whereas Ti-Nb-Ta within the reported composition range possessed little advantage over Ti-Nb at 4.2 K ($\sim$0.1 T by direct comparison using the same DC measuring technique at 5 A cm^{-2}), at 2 K the upper critical fields of the best ternary alloys were some 1.15 T higher than those of Ti-Nb$_{36,37}$.

12.6 CRITICAL CURRENT DENSITIES OF Ti-Nb-Ta ALLOYS

Ti-Nb-Ta alloys were the subject of a German patent filed by REINBACH in 1964, assigned to Vacuumschmelze in 1967 [Rei67], and of a French patent [Mit68] and a U.S. patent (No. 3,671,226) issued to the Mitsubishi Electric Corporation in 1968 and 1972, respectively. A Netherlands patent issued to the Compagnie Francaise Thomson-Houston in 1967 [Com67[a]]

claimed processes for producing magnet wire from Ti-Nb, Ti-Zr-Nb and Ti-11Nb-25Ta, although no properties were presented. The work of REINBACH arose from an attempt to improve the workability of Ti-Zr-Nb by replacing the Zr with Ta. The claim of improved superconducting properties resulting from this substitution was not substantiated by a useful body of data. The only material measured, 99%-cold-worked Ti-60Nb-10Ta, had a 5-T J_c at 4.2 K of only 3.4x10^3 A cm^{-2} -- albeit within the range of expectation for moderately-cold-worked but un-heat-treated Ti-lean Ti-Nb and Ti-Nb-TM alloys. The Mitsubishi Electric Company has claimed a group of Ti-Nb-Ta alloys defined by Ti$_{20-80}$-Nb$_{2-80}$-Ta$_{1-80}$ and presented some comparative data (with respect to some early results for Ti-Nb$_{31}$ and Zr-Nb$_{75}$) for Ti$_{60}$-Nb$_{10,35,38}$-Ta$_{30,5,2}$. These alloys, cold reduced 99% and heat treated 1h/400°C, yielded J_c(4.2K, 4 T) = 4.3 to 8.0x10^4 A cm^{-2}.

In a study of the influence of the "β-stabilizing" elements Cu, Si, V, Ta, and Mo, on the critical current densities of Ti-Nb-base alloys, RASSMANN and ILLGEN [Ras73] have investigated six Ti-Nb-V alloys with V levels of 5, 10, 16 and 17 at.%, and four Ti-Nb-Ta alloys with Ta levels of 5 and 10 at.%, in the conditions: 99.9%-cold-deformed and deformed-plus-heat-treated (1h/500°C -- Ti-Nb-V and 1h/350°C -- Ti-Nb-Ta). Metallography was performed, and diagnosis of the phases present was aided by hardness measurements. Measured critical current densities were correlated with the microstructural conditions of the samples. In the absence of serious attempts at process optimization it is difficult to make accurate assessments of the ultimate effects of alloying additions on critical current density. If large deformation, accompanied by the usual heat treatments at temperatures near 400°C, have been applied a reasonably reliable ranking of properties can be achieved; otherwise this is not possible. An example of the results of RASSMANN and ILLGEN [Ras73] is given in Table 12-1, which shows that under the processing conditions administered, the addition of Ta (and V) was in fact *detrimental* to the low-field critical current density.

SUENAGA and RALLS [Sue69] measured the critical current density of a lightly cold-rolled ($\sim$90%) sample of $Ti_{63.8}$-$Nb_{32.5}$-$Ta_{3.7}$, aged 4h/500°C, and obtained of course very modest critical current densities such as, for example, 4.2×10^3 A cm^{-2} at 4.0 T.

In what was in effect an extension of the critical field investigations of SUENAGA and RALLS in the direction of critical current density optimization, BYCHKOV *et al* [Byc74] studied the influence of heat-treatment-induced microstructures on the critical current densities of a series of Ti-Nb-Ta alloys characterized by a fixed 40 wt.% Ti and Ta contents of 9, 10, 20, and 40 wt.%. The alloys were cold reduced 99.9% and heat treated 3 h at temperatures of 300, 350, 375, 400, 450, 500 and 600°C. The J_c-optimal heat-treatment temperature varied from 350 to 400°C depending on composition. The optimal heat treatment for Ti-40Nb-20Ta was 3h/375°C after which the 4-T J_c, for example, was 2.2×10^4 A cm^{-2} -- identical to that reported by RASSMANN and ILLGEN [Ras73] for 99.9% cold-deformed and 1h/350°C-aged Ti_{65}-Nb_{25}-Ta_{10}, but slightly lower than the values of 4.3×10^4 to 8.0×10^4 A cm^{-2} (at 4.2 K, 4 T) claimed in the Mitsubishi patent [Mit68] for Ti_{60}-$Nb_{10,35,38}$-$Ta_{30,5,2}$. TEM and x-ray diffractometry, conducted on ribbon samples, failed to reveal any traces of phase decomposition leading to the conclusion that under the thermomechanical processing conditions described, flux pinning in these alloys was governed by the dislocation-cell structure. Although, according to BYCHKOV *et al* [Byc74], Ti-60Nb treated under the same conditions as the Ti-Nb-Ta alloys yielded a *lower* critical current density, the same criticisms with regard to deviations from optimization that were levelled at the work of RASSMANN and ILLGEN must also apply to that work.

In the meantime unbeknown apparently to BYCHKOV *et al* [Byc74], although not to SUENAGA *et al* [Sue69] and RASSMANN and ILLGEN [Ras73], HASHIMOTO and coworkers [Has68] of the Mitsubishi Electric Company had published an article outlining briefly a comprehensive series of studies of the influence of solute concentration and time and temperature of heat treatment on the 4.2 K critical current density of Ti-Nb-Ta alloys. Two series of alloys were investigated (cf. Table 1-36): an "A-series" in which 2.5 to 30 at.% Ta was substituted for Nb in Ti-Nb$_{30}$ and a "B-series" in which a fixed 5 at.% Ta was present in Ti_{50-75}-Nb-Ta$_5$. After recrystallization at 900°C, the alloys were cold drawn 98.33%, 99.60% and 99.93% to a diameter of 0.25 mm. Reported with regard to the 99.93%-cold-worked alloys were the results of J_c(4.2 K) measurements as functions of: (a) composition for alloys aged

TABLE 12-1 EARLY STUDIES OF CRITICAL CURRENT DENSITY (J_c) IN TITANIUM-NIOBIUM-TANTALUM ALLOYS

Alloy	Aging After $\sim$99.9% Deformation	H_a, kOe	J_c, 10^4 A cm^{-2}
Ti_{75}-Nb_{20}-Ta_5	1h/400°C	40	12 [†]
Ti_{70}-Nb_{25}-Ta_5	1h/400°C	40	6.2[†]
Ti_{70}-Nb_{25}-Ta_5	80h/400°C	40	25 [†]
Ti_{70}-Nb_{30}	1h/500°C	50	4.2[††]
(Ti_{65}-Nb_{30}-V_5	1h/500°C	50	1.7[††])
Ti_{70}-Nb_{30}	1h/400°C	40	7.8[††]
Ti_{65}-Nb_{25}-Ta_{10}	1h/350°C	40	2.2[††]
Ti-40Nb-20Ta	3h/375°C	40	2.2[†††]

[†] Hashimoto *et al.* [Has68]

[††] Rassmann and Illgen [Ras73] (Criterion: 40 µV across sample).

[†††] Bychkov *et al.* [Byc74]

1h/400°C; (b) *temperature* up to 600°C of 1-h anneals; (c) *time* up to 40 h of 500°C anneals and up to 80 h of 400°C anneals. Finally, J_c(4.2 K) *versus* H_a($\lesssim$5.5 T) was reported for fully-annealed through to fully-cold-worked wire. The J_c(4.2 K, 4 T) of Ti_{70}-Nb_{25}-Ta_5 *versus* 1-h aging temperature maximized at the usual 400°C. In the A-series alloys, J_c(4.2 K, 4 T) peaked in the vicinity of 5-10 at.% Ta. In the B-series, J_c(4.2 K, 4 T) increased steeply and monotonically with increasing Ti content, becoming largest for Ti_{75}-Nb_{20}-Ta_5 (the end alloy of its series); at higher Ti levels than this the alloys were too brittle to be cold drawn. Some typical data are given in Table 12-1, while Fig.12-7 is a plot of J_c(4.2 K, 4 T) as function of aging time at 400 and 500°C for 99.93% cold-worked Ti_{70}-Nb_{25}-Ta_5, the composition which had been selected for detailed investigation.

The early studies of Ti-Nb-Ta seem to have been motivated primarily by a desire to improve on the superconductive properties of the basic Ti-Nb alloy. An additional incentive was soon to enter the picture. In 1970, ADAM (then at Thomson-CSF, France) and co-workers [Ada70] directed attention towards the high cost of Nb (as compared to Ti) and pointed out that the use of only partially purified Nb (in the sense that it retained significant levels of Ta) was advantageous from the standpoint, not only of superconductive properties, but also price.

In spite of these promising beginnings it was not until 1977-1978 that serious interest in the Ti-Nb-Ta system became rekindled, this time in the U.S. As indicated in the introduction to this section, the two reasons motivating the new studies were again both economic and technical: (1) It was pointed out by CURTIS and McDONALD [Cur79] to be economically advantageous to replace high-purity Nb as a starting material with one containing up to 1 wt.% Ta as "tramp" contamination; at the same time it was demonstrated that the resulting alloys had acceptable workability, slightly better tensile properties, and slightly improved J_c's in both high fields (H $\gtrsim$ 7.5 T, 4.2 K) [Cur79][†] and very high fields (H < 13.5 T, 1.9 K) [Ste78]. (2) The deliberate substitution of 10 at.% Ta for Nb in Ti-Nb_{35} resulted in an alloy with significantly improved current-carrying capacity in the very high fields-at-the-winding ($\sim$12 T) [Seg81] needed in magnets for dense-plasma containment.

CURTIS and McDONALD examined the effect of substituting 0.5, 1 and 2 wt.% Ta for Nb in a Ti-53.5Nb alloy. These three ternary alloys, as well as a reference binary alloy, were prepared and fabricated into Cu-clad monofilamentary composite wires according to the schedule summarized in Table 12-2. All of the alloys were accorded the "normal treatment" but only the 2 wt.% Ta alloy experienced the "extra cold work". The alloys themselves were intended to simulate those which would result if low levels of Ta were present in the starting Nb, and the purpose of the study was to examine the effect of such Ta substitutions on the mechanical and electrical properties. J_c(4.2 K) *versus* H (<7.5 T) measurements were conducted in three independent laboratories. Fig.12-8 and Table 12-3, which summarize the results of these tests, indicate the extents to which the critical current densities of the various alloys have been improved by the addition of Ta. The results emphasize the well-known importance of cold work and also suggest that under process optimization the ranking of the four alloys, and with it the beneficial effect of Ta additions, should be preserved. In a continuation and extension of the critical current investigation to fields of 9 and 10 T (at 4.2 and 1.9 K) and 11 through 13.5 T (at 1.9 K), STEKLY *et al* [Ste78] noted that the ternary alloys frequently continued to perform better than the Ti-Nb control.

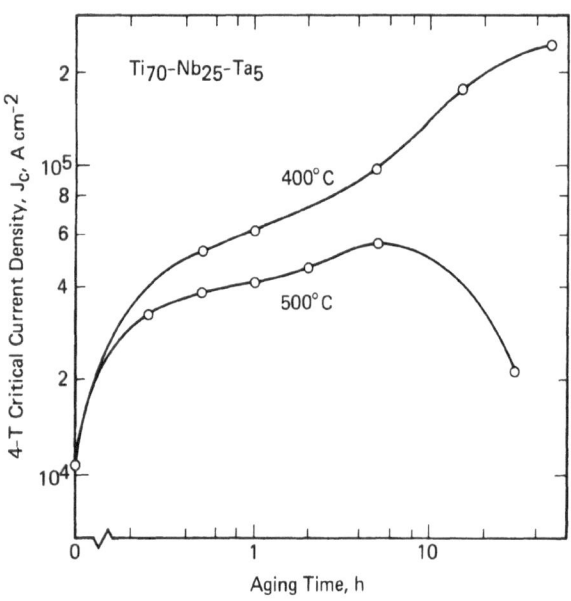

FIGURE 12-7. Critical current density at 4.2 K and 4 T of bare 99.93% cold swaged and drawn Ti_{70}-Nb_{25}-Ta_5 as a function of aging time at temperatures of 400°C and 500°C — after HASHIMOTO *et al* [Has68].

---

† The qualitative results of other such measurements are commented on in the footnote to Table 12-3.

TABLE 12-2 THERMOMECHANICAL PROCESSING OF TANTALUM-DOPED Ti-53.5Nb CONDUCTOR [Cur79]

	"Normal Treatment"		"Extra Cold Work"	
Diameter when recrystallized 2h/1600F (in.):	1.5	1.5	1.5	1.5
Diameter when vacuum annealed 1-1/2h/1475F (in.):	0.56	0.56	(not annealed)	
Cold reduced by:	99.4%	99.9%	99.9%	99.99%
to (in.):	0.045	0.018	0.045	0.018
Heat treatment:	24h/380°C	24h/380°C	24h/380°C	24h/380°C
Cold reduction by:	95%	95%	95%	95%
to (in.):	0.010	0.004	0.010	0.004
(mm):	0.25	0.10	0.25	0.10
Final heat treatment:	2h/300°C	2h/300°C	2h/300°C	2h/300°C

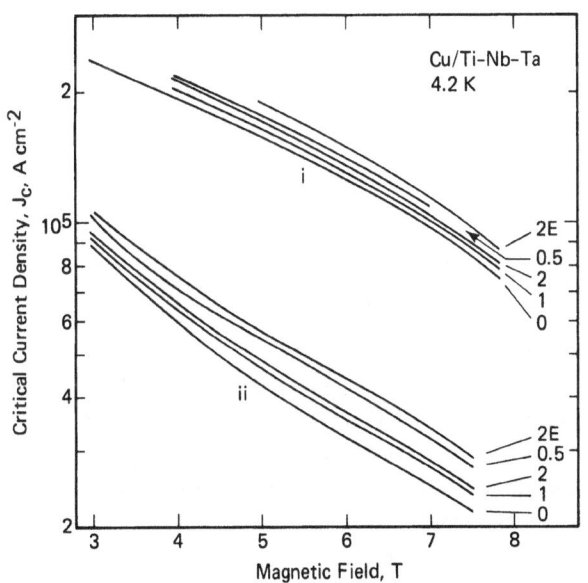

FIGURE 12-8. Critical current densities at 4.2 K (criterion: 10^{-12} Ω cm) of Ti-Nb (37 at. %) (i.e., Nb-46.5Ti) and three Ti-Nb-Ta alloys processed according to Table 12-2. The numbers refer to the nominal Ta concentrations in wt. % and E indicates the "extra cold work". Group (i): final composite wire diameter, 0.10 mm. Group (ii): final composite wire diameter, 0.25 mm — after CURTIS and McDONALD [Cur79].

TABLE 12-3 PERCENTAGE INCREASE IN CRITICAL CURRENT DENSITY AT 4.2 K AS FUNCTION OF TANTALUM CONTENT IN Ti-53.5Nb ALLOY. 0.010-in DIAMETER WIRES PROCESSED ACCORDING TO TABLE 12-2 [Cur79][†]

Applied Field T	Percent Increase in J_c			
	Nominal Added Ta Content, wt.%			
	0.5	1	2	2E[††]
3	19	2	0	18
4	16	3	2	23
5	17	1	1	25
6	17	-2	-1	22
7	27	10	15	37
7.5	25	10	13	34

† These measurements were performed on monofilamentary conductors [Cur79]. In a subsequent commercial-scale investigation, SEGAL *et al.* [Seg81[a]] concluded, with reference to tabulated data, that "Although no degradation is observed because of the tantalum additions, no improvement takes place either".

†† E indicates "extra cold work".

We have been reminded recently by LARBALESTIER [Lar80] and again by WADA et al [Wad81] that, according to KRAMER [Kra73], the high-field J_c is dominated by H_{c2}. In other words, it is unimportant how strong the local pinning forces are if the *existence* of the mixed state is threatened by an approach to the upper critical field (cf. Sect.7.32 and [Mon21.20]). Thus the search for high-H_{c2} alloys must always be the first phase of a program for the development of high-field superconductors; the second phase then consists of the optimization of $J_c(H_a)$ itself. The upper critical fields of Ti-Nb-Ta alloys have been considered above. Having been shown to possess favorable H_{c2}'s, two Ti-Nb-Ta alloys were selected for wire-fabrication and critical current density studies. As reported by SEGAL et al [Seg81], the alloys Ti-48Nb-8Ta (i.e. Ti_{62}-Nb_{35}-Ta_3) and Ti-32Nb-25Ta (i.e. Ti_{65}-Nb_{25}-Ta_{10}) were incorporated into 300-element 10-cm$^\phi$ composite billets, each with a Cu/SC-ratio of 2.3:1, and reduced to wires having diameters ranging from 0.75 to 0.25 mm. Using this material, J_c(5-13 T) was measured as a function of temperature between 2 and 4.2 K, and

J_c(1.9 K, 12 T) was measured as a function of the overall area reduction ratio which of course ranged from 1.73×10^4:1 (99.99%) to 1.6×10^5 (99.999%). It was concluded that at 2 K in a field of 12 T the critical current density of the 25 wt.% Ta (10 at.% Ta) alloy was about 10% higher than that of the 8 wt.% (3 at.%) alloy, and approximately "75% higher than that of binary Ti-Nb". It was anticipated that, as with Ti-Nb, the critical current density of the Ti-Nb-Ta alloys would be strongly dependent on cold reduction. Fig.12-9, which intercompares the effects of cold work on the critical current densities of the two classes of alloy, confirms this expectation and also demonstrates the superior performances of the Ta-bearing alloys. Of the latter, the influence of cold work on Ti-Nb-25Ta is the more pronounced; for example, with reference to the 12 T, 1.9 K data, and a reduction in wire diameter of from 0.76 to 0.25 mm [Seg81], while the critical current density of the 8 wt.% Ta (3 at.% Ta) alloy rose only 22%, that of the 25 wt.% Ta (10 at.% Ta) alloy was reported to experience an increase of 32%.

TABLE 12-4   INTERCOMPARISON OF CRITICAL CURRENT DENSITIES (J_c) OF Ti-Nb AND Ti-Nb-Ta ALLOYS

| Alloy | J_c, 10^4 A cm^{-2} | | | |
| | 1.9 K | | 4.2 K | |
	10 T	12 T	10 T	Literature
Nb-55Ti (Ti-Nb_{30})			0.505	[Ste78][Seg79]
Nb-50Ti (Ti-Nb_{34})	11.3	3.94	1.95	[Seg80]
Nb-46.5Ti (Ti-Nb_{37})	12.7	6.60	2.75 3.13	[Ste78][Seg79]
Nb-44Ti-8Ta (Ti_{62}-Nb_{35}-Ta_3)	14.5	6.76	3.27	[Seg81]
Nb-43Ti-25Ta (Ti_{65}-Nb_{25}-Ta_{10})	13.5 13.3	7.35 7.28	3.68	[Seg80] [Seg81]

J_c *Criteria:* Probably take-off current in the earlier measurements, but certainly 1 µV/cm for [Seg80, Seg81]. If so, a common criterion would accentuate the already improved relative performances of the Ta-containing alloys.

12.7 INTERCOMPARISON OF THE CRITICAL CURRENT DENSITIES OF Ti-Nb AND Ti-Nb-Ta ALLOYS

12.7.1 Comparative Data

In [Ste78] no direct intercomparisons between binary- and ternary-alloy performances, under the experimental conditions referred to above, were offered. However, with the aid of data from subsequent papers by the same research group it has been possible to synthesize some useful relationships. In Table 12-4, assembled from the work of SEGAL et al [Ste78][Seg79, Seg80, Seg81], the positions of Ti_{62}-Nb_{35}-Ta_3 and Ti_{65}-Nb_{25}-Ta_{10} vis-à-vis each other and a representative group of Ti-Nb alloys, are abundantly clear. Fig.12-9 was another such synthetic intercomparison. In comparisons of this type, the specification of measurement criterion is particularly important. Although the ternary alloys were measured at 1 µV cm^{-1} [Seg81] the earlier binary-alloy data [Ste78][Seg79] were actually take-off currents[†]. This of course does not invalidate the present comparison

---

[†] I am indebted to H.R. SEGAL [private communication, 1982] for discussing these questions.

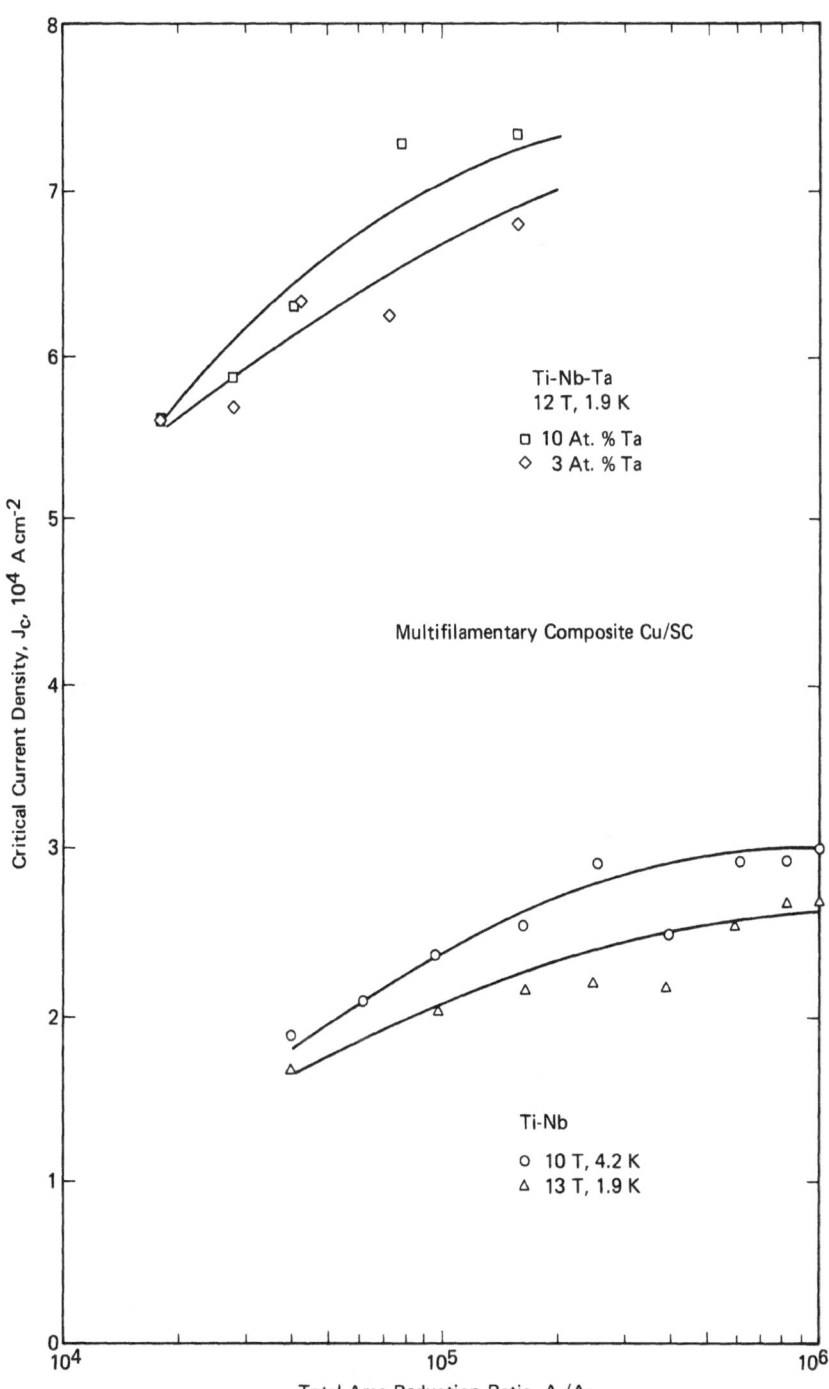

FIGURE 12-9. Intercomparison of the critical current densities of Ti-Nb (37 at. %) (i.e., Nb-46.5Ti), $Ti_{62}-Nb_{35}-Ta_3$ (i.e., Nb-44Ti-8Ta) and $Ti_{65}-Nb_{25}-Ta_{10}$ (i.e., Nb-43Ti-25Ta), at the fields and temperatures specified, *versus* the total area reduction from billet to final wire size. For the ternary alloys a 1 μV cm^{-1} sensitivity criterion was used.

Ti-Nb Conductor Details			Ti-Nb-Ta Conductor Details	
Starting billet diameter (cm)		25.4	Starting billet diameter (cm)	10
Final wire diameters (mm)		1.27-0.25	Final wire diameters (mm)	0.76-0.25
Cu/SC ratio		1.8:1	Cu/SC ratio	2.3:1
Number of filaments		2046	Number of filaments	300
Filament diameters (μm)		16-3.1		
Processing (5.1±0.1) $\cdot$ 10^3:1| |75 h/375°C| |87.1-99.5%			Processing details not specified	

— after STEKLY, SEGAL *et al* [Ste78, Seg79] (Ti-Nb) and [Seg81] (Ti-Nb-Ta).

which, if it errs at all, does so on the conservative side.

Further comparisons are made in Fig.12-10 constructed with data from [Seg81a]. In this case, although *all* curves were taken at 1 μV cm^{-1}, the reference data, for want of actual binary-alloy results, is that of Ti-53Nb-0.5Ta whose J_c in these studies was practically the same as that of its binary counterpart, Ti-53.5Nb (i.e. Ti-Nb$_{37}$), see footnote to Table 12-3.

12.7.2 Concluding Summary

As pointed out by SEGAL *et al* [Seg80], although at 5 T as well as in higher fields at 4.2 K the performance of Ti$_{65}$-Nb$_{25}$-Ta$_{10}$ is not much different from that of Ti-Nb$_{37}$ (Ti-53.5Nb), the Ta-bearing alloy exhibits a marked superiority at high fields and low temperatures. This is particularly true when the comparison is made with Ti-Nb$_{34}$ (Ti-50Nb); thus at 1.9 K in fields of 10, 12, and 13 T the critical current densities of the Ta-bearing alloy have attained values some 15, 85, and 330% higher, respectively, than those of Ti-50Nb. In fact at 1.9 K, Ti$_{65}$-Nb$_{25}$-Ta$_{10}$ is still functional in a field of 14 T in which it is able to support a transport current density of 1×10^4 A cm^{-2}.

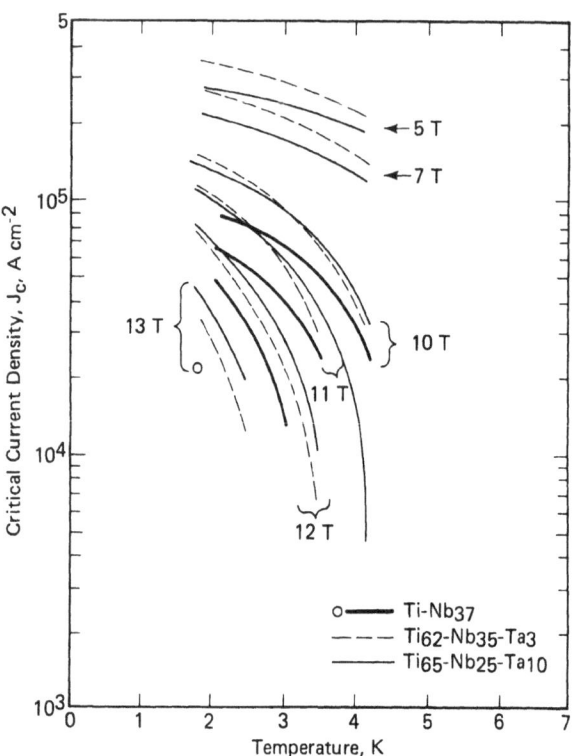

FIGURE 12-10. Intercomparison of the critical current density temperature dependences of Ti-Nb (37 at. %) (i.e., Nb-46.5Ti), Ti$_{62}$-Nb$_{35}$-Ta$_3$ (i.e., Nb-44Ti-8Ta), and Ti$_{65}$-Nb$_{25}$-Ta$_{10}$ (i.e., Nb-43Ti-25Ta). Criterion: Ti-Nb-Ta, 1 μV cm^{-1} [Seg81]. The conductors were all 0.25 mm$^\varphi$ wires conforming to the specifications listed in the previous caption — after STEKLY, SEGAL *et al* [Ste78, Seg79] (Ti-Nb) and [Seg81] (Ti-Nb-Ta).

ALLOY GROUP 3: ALLOYS OF TITANIUM-NIOBIUM WITH THE GROUPS VI THROUGH VIII TRANSITION ELEMENTS

12.8 ALLOYS OF Ti-Nb WITH THE GROUP-VI ELEMENTS Cr, Mo, AND W

The literature dealing with ternary additions of the group-VI transition elements to Ti-Nb is indexed in Table 1-37.

12.8.1 Transition Temperatures of Ti-Nb-$\begin{pmatrix} Cr \\ Mo \\ W \end{pmatrix}$ Alloys

Numerous studies have been undertaken of the superconducting transition temperatures of Ti-Nb-Cr, Ti-Nb-Mo, and Ti-Nb-W alloys, in which their responses to changes of average electron/atom ratio and heat-treatment-induced microstructures have been investi-

gated. The research was generally motivated by desires to improve upon the transition temperature and possibly some ancillary physical or mechanical properties of Ti-Nb. Another underlying purpose, in some cases, was to investigate the general applicability of average e/a as a descriptor of transition-temperature variation in multicomponent transition-metal alloy systems.

BYCHKOVA *et al* [Byc70a] (see also SAVITSKII *et al* [Sav73, p.331]) have investigated the influences of 0.4∿26 at.% Mo on the T_c of alloys with Nb concentrations within the range 25∿28 at.%; SADAGOPAN *et al* [Sad70] have presented data for seven alloys containing 5-20 at.% Mo and 20∿48 at.% Nb; COTTON *et al* [Cha73][Cot73, Cot74] have investigated a matrix of

some nineteen Ti-Nb-Mo alloys; most recently WADA *et al*
[Wad80] have conducted an extensive series of studies
on the influence of 24-h aging temperature on the
transition temperatures of several systems of Ti-Nb-Cr
and Ti-Nb-Mo alloys conforming to the general formulae:
$(Ti-Nb_{30})-Cr_{2,6,10}$, $(Ti-Nb_{50})-Cr_{2,6,10}$,
$(Ti-Nb_{75})-Cr_{2,6,10}$, $(Ti-Nb_{30})-Mo_{5,10,15}$,
$(Ti-Nb_{50})-Mo_{5,10}$, and $(Ti-Nb_{75})-Mo_{5,10}$.

Fig.12-11, which intercompares the transition
temperatures of Ti-Nb and several ternary alloy systems, shows clearly that when different alloy classes
are involved, T_c is not a universal function of e/a.
In fact, as COTTON [Cot74] has demonstrated, *families*
of curves are needed to systematize the data.
Fig.12-11, reinforced by Fig.12-12 from the work of
the above author, shows that the effect of Cr and Mo
in bcc solid solution is to severely depress the
transition temperature of Ti-Nb.

COTTON *et al* [Cot73, Cot74] have also investigated the influence on the T_c of ternary Ti-Nb-Mo alloys,
of various microstructural or precipitational effects
brought about by heat treating either the quenched or
the slow-cooled alloys in various ways. Heat treatments administered were:

(a) To Quenched Alloys:

   (i) Aging at 350°C to develop the ω+β structure.

   (ii) Annealing at 500°C to revert the ω-phase to
   bcc, thus creating a modulated β'+β structure which turned out to be stable during
   subsequent quenching.

(b) To Slow-Cooled Alloys:

   Isothermal transformation in the α+β region
   at the temperatures 750, 650 or 550°C.

The observed changes in superconducting transition
temperature were correlated with the observed or
deduced microstructural changes. For example, some
of the aged ω+β alloys exhibited double superconducting transitions, and certain of the alloys annealed
in the α+β region developed increased transition temperatures traceable to solute enrichment of the β
matrix. Effects such as these are generally observable in all Ti alloys including the binaries. In
ternary alloys a full interpretation of the results
is complicated by clustering, a phenomenon related to
interatomic affinities, and diffusional effects.

The superconducting transition temperatures of a
wide range of Ti-Nb-W alloys have been considered
phenomenologically by SAVITSKII *et al* [Sav73, p.333].
The alloy system itself is quite interesting in that

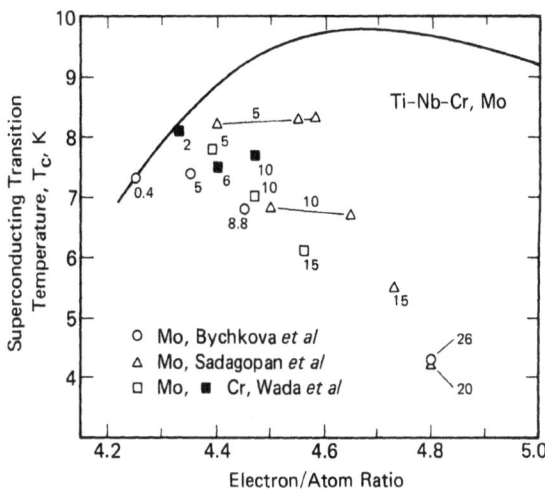

FIGURE 12-11. Superconducting transition temperatures of
Ti-Nb-Mo (and three Ti-Nb-Cr) alloys as functions of electron/atom
ratio. The numbers refer to the Cr or Mo concentrations in at. %;
the full line constructed from various sources represents the transition temperature of Ti-Nb. Ternary alloy data sources —
BYCHKOVA *et al* [Byc70[a]] [Sav73, p 333], SADOGOPAN *et al*
[Sad70], WADA *et al* [Wad80].

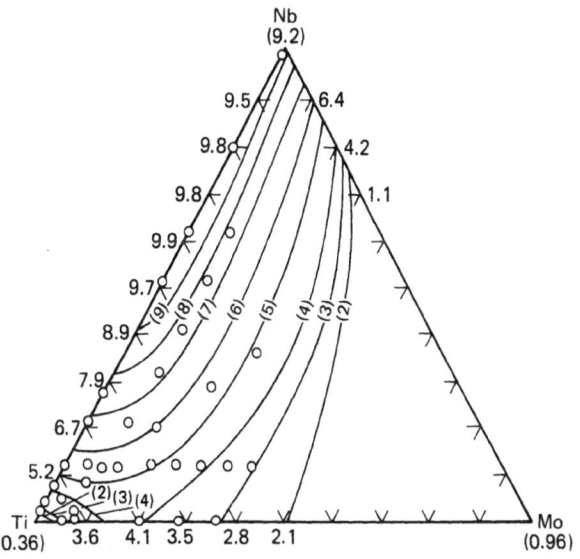

FIGURE 12-12. Ternary Ti-Nb-Mo diagram (at. % linear) indicating the superconducting transition temperatures (T_c, K) of the
quenched alloys. Temperatures corresponding to the binary alloys
(10 at. % intervals) and the ternary isothermals (integers) are indicated. The reduction of T_c with substitution of Mo into Ti-Nb is
clearly demonstrated — after COTTON *et al* [Cot74].

over certain composition and temperature ranges two
immiscible bcc solid solutions coexist. Of the alloys
considered: viz Nb-10W-(2-48)Ti, Nb-20W-(2-56)Ti and
Nb-30W-(8-30)Ti, those of compositions Nb-10W-(5-48)Ti
had the highest transition temperature, 7.5 K.

12.8.2 Critical Fields of Ti-Nb-$\binom{Cr}{Mo}_{W}$ Alloys

As part of a comprehensive study of the upper critical fields of Ti-base alloys, several dilute members of the Ti-Nb-(group-VI)TM family, viz: Ti_{65}-Nb_{33}-Cr_2, Ti_{65}-Nb_{34}-Mo_1 and Ti_{64}-Nb_{35}-W_1, have been measured by DeSORBO et al [Des67] and by HELLER et al [Hel71[a]]. The upper critical fields were found to increase in the sequence H_{c2}(-Cr) < H_{c2}(-Mo) < H_{c2}(-W) with the latter, at 11.4 T, according to HELLER et al [Hel71[a]], equalling that of binary Ti-Nb$_{33}$ itself. SADAGOPAN et al [Sad70] prepared a series of eight Ti-Nb-Mo alloys and obtained critical field data on four of them. The most extensive investigation of the influences of Cr and Mo on the H_{c2} of Ti-Nb was the recent study by WADA et al [Wad80] of the systems (Ti-Nb$_{30,50,75}$)-Cr$_{2,6,10}$ and (Ti-Nb$_{30,50,75}$)-Mo$_{5,10,15}$ referred to above. Some representative data from the above papers, plotted *versus* electron/atom ratio as in Fig.12-13, show that the presence of either Cr or Mo results in severe depressions of the upper critical field. With regard to Ti-Nb-Mo, SADAGOPAN et al

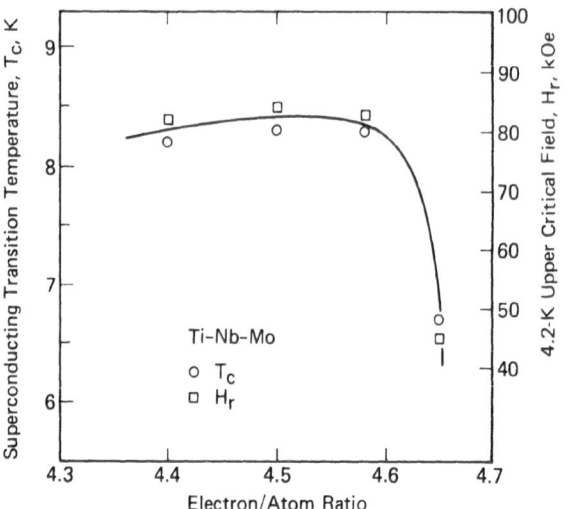

FIGURE 12-14. Variation of superconducting transition temperature (T_c) and 4.2-K resistive upper critical field (H_r) with electron/atom ratio in Ti-Nb-Mo alloys — data from SADAGOPAN et al [Sad70].

[Sad70] measured both T_c and H_r thus providing an opportunity for a comparison to be made of the ways in which each vary with electron/atom ratio. According to Fig.12-14, T_c and H_r do in fact scale, as required by the GOR'KOV-GOODMAN-MAKI-EILENBERGER relationship, Eqn.(12-1).

12.8.3 Critical Current Densities of Ti-Nb-$\binom{Cr}{Mo}$ Alloys

As part of a comprehensive study of the influence of the so-called "β-stabilizing" elements Si, Cr, Mo, V, and Ta, on the critical current density of Ti-Nb, RASSMANN and ILLGEN [Ras73] have investigated the alloys Ti-Nb$_{21-41}$-Mo$_5$ and Ti-Nb$_{10-15}$-Mo$_{10}$ in the cold-deformed (∼99.9%) and deformed-plus-aged (1h/500°C) conditions. The critical current densities of the cold-worked ternary alloys were an order of magnitude lower than those of the corresponding binary Ti-Nb alloys, and heat treatment resulted in only insignificant improvements. WADA et al [Wad80] have investigated the critical current densities of 50% cold-rolled samples of Ti$_{48}$-Nb$_{50}$-Cr$_2$ and Ti-Nb$_{30}$-Cr$_{2,6}$ as functions of 24-h aging at temperatures between 300 and 800°C, during which the usual "optimization" peak at 400°C was recorded. It was noted that the precipitation was enhanced, but on the basis of the data presented it was not possible to estimate what improvements in J_c, if any, might accompany the addition of Cr to Ti-Nb.

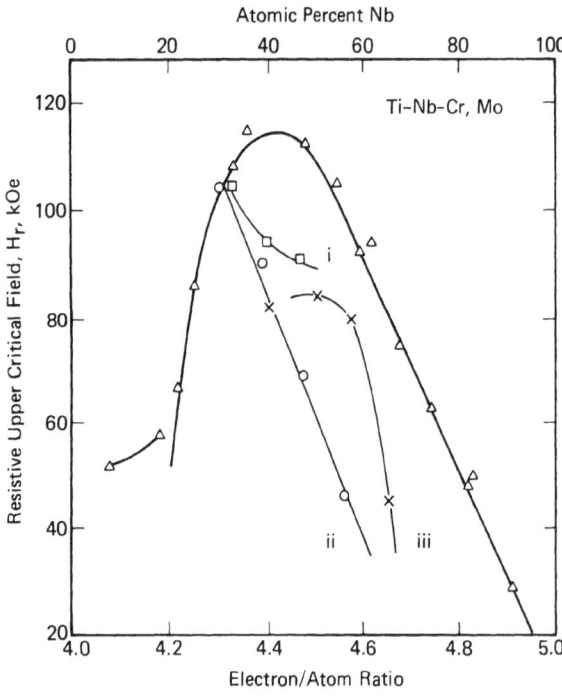

FIGURE 12-13. Influence of Cr and Mo substitutions on the upper critical fields of Ti-Nb alloys. The (△) represent Ti-Nb data from various old and new sources. Curve (i): Ti-Nb-Cr — after WADA et al [Wad80]; Curve (ii): Ti-Nb-Mo (Nb concentrations ∼ 30 at. %) — after WADA et al (○) and SADAGOPAN et al [Sad70] (x); Curve (iii): Ti-Nb-Mo (Nb concentrations 40 ∼ 48 at. %) — after SADAGOPAN et al [Sad70].

12.9 ALLOYS OF Ti-Nb WITH THE GROUP-VII ELEMENTS Mn AND Re

The influence on the upper critical field of a 2 at.% Mn addition to Ti-Nb$_{33}$ was investigated resistometrically (current density 1.3 A cm^{-2}) by HELLER et al [Hel71^a] using 97%-cold-worked and 10h/400°C-aged 1-mm$^\phi$ wire. A value of H$_r$(4.2 K) = 10.81 T was obtained, slightly lower than the 11.40 T determined for the binary Ti-Nb$_{33}$ control alloy.

The influence of Re on the superconducting transition temperature of Ti-Nb alloys has been studied by BYCHKOVA et al [Byc70^a]. The Nb concentration remained fixed at 25 at.% while the Re content took on the values 1, 5, 10 and 25 at.%. Although single phase after homogenization for 200h/1100°C followed by quenching, a hexagonal precipitate appeared during cold deformation (99.9%) to wire. A rapid decrease of T$_c$ with increasing Re concentration was noted.

12.10 ALLOYS OF Ti-Nb WITH THE GROUP-VIII ELEMENTS (Fe THROUGH Pt)

12.10.1 Transition Temperatures of Ti-Nb-(Group-VIII)TM Alloys

The influences of most of the group-VIII TM elements, viz:

Fe	Co	Ni
Ru	Rh	Pd
Os	Ir	Pt

on the transition temperature and some electrical and metallurgical properties of Ti-Nb have been investigated by ZWICKER and colleagues. In a series of early investigations, ZWICKER [Zwi65] intercompared the effects on T$_c$ of the addition of 2.5 at.% of all six of the 4d and 5d group-VIII elements. Alloys were measured in the annealed (900°C)-and-quenched condition, and the results of the additions were expressed in the format T$_c$ versus "valence-electron concentration" (meaning in this case electron/atom ratio). The electron/atom ratio is usually the same as the average group number excepting that within group VIII the numbers of electrons-per-atom associated with members of the three columns are assigned the values 8, 9 and 10, respectively. Thus, for example, the e/a of Ti$_{87.5}$-Nb$_{10}$-Ru$_{2.5}$ is 4.2, that of Ti$_{87.5}$-Nb$_{10}$-Rh$_{2.5}$ is

4.225 and so on. The results showed a severe decrease of T$_c$ with the addition of small amounts of the group-VIII elements. The substitution of Ru and Os for Nb at constant e/a = 4.20 lowered T$_c$ by about 2.5 K. Within the Pt group itself, T$_c$ increased with increasing e/a (4.2 to 2.5) but was always much lower than that of Ti-Nb$_{20}$.

An extensive investigation of the influence of Fe on the properties of several classes of Ti-Nb alloy was undertaken by NISHIMURA and ZWICKER [Nis68]. Host alloys were Ti-Nb(4.9, 10, 25, and 33.3 at.%) and the Fe additions varied from 3 to 33.3 at.%. T$_c$'s were measured on alloys in the as-quenched (from 900°C) condition and also as functions of aging time at 400°C and 500°C. Other properties investigated and measurements undertaken included electrical resistivity temperature dependence, metallography, and electron-beam microanalysis. The results for the as-quenched alloys, as summarized in Table 12-5, show that additions of Fe severely depress the transition temperature.

During aging at 500°C the following changes were noted in the transition temperatures and the microstructures of the alloys listed in Table 12-5:

ALLOY (a) - T$_c$ increased with aging time; a decomposition into α+β seemed to be taking place.

ALLOY (b) - T$_c$ increased rapidly with aging time; grain-boundary α-precipitation was noted.

ALLOY (c) - T$_c$ changed little with aging time; grain-boundary TiFe precipitation was observed.

TABLE 12-5   INFLUENCE OF Fe ADDITIONS ON THE TRANSITION TEMPERATURES (T$_c$) OF SOME Ti-Nb ALLOYS

Ti-Nb-Fe Alloy	e/a	T$_c$[+]	Ti-Nb Alloy With the Same e/a	T$_c$ of the Binary Alloy K
(a) Ti$_{86}$-Nb$_{10}$-Fe$_4$	4.26	4.2	Ti-Nb$_{26}$	6.2
(b) Ti$_{86.9}$-Nb$_{4.9}$-Fe$_{8.2}$	4.37	4.4	Ti-Nb$_{37}$	8.4
(c) Ti$_{72}$-Nb$_{25}$-Fe$_3$	4.37	7.4	Ti-Nb$_{37}$	8.4

† From the plotted data of Nishimura and Zwicker [Nis68].

12.10.2 Critical Fields of Ti-Nb-(Group-VIII)TM Alloys

The influences of small additions of Fe and Ni upon the upper critical field of Ti-Nb have been investigated by DeSORBO et al [Des67] (Fe) and HELLER et al [Hel71[a]] (Fe, Ni). DeSORBO's measurements, carried out on cold-worked wire samples, were part of a magnetic study of the manner in which H_u varied with an effective (atomic-volume corrected) electron/atom ratio. The alloy measured, Ti_{19}-Nb_{80}-Fe_1, had an H_u(4.2 K) of 4.1 T, a value which was not much lower than the 4.3 T of the "equivalent" (in conventional e/a terminology) binary alloy, Ti-Nb_{84}. HELLER et al [Hel71[a]], using a resistometric technique, investigated the upper critical fields of 97%-cold-worked plus 10h/400°C-aged wires of Ti_{65}-Nb_{33}-Fe_2 and Ti_{65}-Nb_{33}-Ni_2. Small deviations from the critical field of

Ti-Nb_{33} (H_r = 11.40 T) were noted; the effect of the Fe addition was to lower H_r to 10.54 T, and that of the Ni addition was to raise it to 11.61 T.

12.10.3 Critical Current Density of Ti-Nb-Fe

The influence of vacuum annealing for 10 h at 350, 450, 500, 600, 700 and 800°C on the J_c of 97.7%-cold-rolled Ti_{57}-Nb_{38}-Fe_5 has been investigated by PROKOSHKIN and PUZEI [Pro70]. Critical currents were measured with H_a perpendicular to the rolling plane -- the most *unfavorable* orientation from a flux-pinning standpoint (see for example, Sect.11.10) -- and the atomistics of the aging process were followed using Mössbauer spectrometry, as a result of which, evidence for Fe-rich precipitation was detected.

TABULATED DATA—SUPERCONDUCTIVITY IN TITANIUM-NIOBIUM-TRANSITION-METAL ALLOYS

TABLE 12-6   TRANSITION TEMPERATURES OF TITANIUM-NIOBIUM-TM ALLOYS

ALLOY CLASS: Ti-Nb-Group IV

(a) Transition Temperatures of Ti-Hf-Nb_{25}

Compositions, At.%			T_c
Ti	Hf	Nb	K
65	10	25	6.8
50	25	25	6.9

M.I.Bychkova et al. (1967/70)[Byc70[a]], see also E.M.Savitskii et al.[Sav73, p.325].

Condition: homogenized at 1500°C, annealed 200h/1100°C, water quenched, 99.9% cold deformed.

Magnetic measurement

(b) Transition Temperatures of Ti-Hf-Nb

Compositions, At.%			T_c
Ti	Hf	Nb	K
24	6	70	9.7, 9.8
12	18	70	9.6, 6.7
6	24	70	9.4, 6.6
30	20	50	7.5
10	40	50	7.0
56	14	30	6.8
14	56	30	5.6

E.M.Savitskii et al. [Sav76, p.174]

ALLOY CLASS: Ti-Nb-Group IV -- *continued*

(c) Transition Temperatures of Ti-Hf-Nb$_{35}$

Compositions, At.%			T_c
Ti	Hf	Nb	K
62.5	2.5	35	9.0
60	5	35	8.5
55	10	35	8.3$_5$

D.G.Hawksworth and D.C.Larbalestier [Haw80].

Condition: homogenized 8h/1350°C in gettered Ar, square rolled more than 70%, recrystallized 1h/875°C, swaged to 3 mm$^\phi$, sheathed in Cu and drawn to 1 mm$^\phi$ (core ∿0.5 mm$^\phi$).

Resistive measurement at 10 mA (J ∿ 5 A cm^{-2})

ALLOY CLASS: Ti-Nb-Group V

(d) Transition Temperatures of Ti-Nb-V

Compositions, At.%			T_c
Ti	Nb	V	K
20	50	30	8.5
30	50	20	7.8
40	40	20	7.6
40	10	50	5.3

J.T.A.Pollock *et al.* [Pol69].

Condition: Annealed 24h/1000°C, cold rolled 15:1.

Magnetic measurement

ALLOY CLASS: Ti-Nb-Group V -- *continued*

(e) Transition Temperatures of Ti-Nb$_{1-10}$-V$_{50}$ and Ti-Nb$_{25}$-V$_{5-25}$

Compositions, At.%			T_c
Ti	Nb	V	K
50	--	50	7.8
49	1	50	7.5
45	5	50	7.4
40	10	50	7.1
75	25	--	7.2
70	25	5	7.7
65	25	10	8.0
50	25	25	7.3

E.M.Savitskii *et al.* [Sav71][Sav73, p.329].

Condition: cast and cold worked 50%, homogenized 3h/1400°C at 10^{-5} to 10^{-6} Torr and quenched.

Magnetic measurement

(f) Transition Temperatures of Ti$_{70}$-Nb-Ta$_{0-30}$[†]

Compositions, At.%			T_c
Ti	Nb	Ta	K
70	30	--	8.97±0.05
70	27.5	2.5	9.86
70	25	5	8.34
70	20	10	8.45
70	15	15	7.82
70	5	25	7.72
70	--	30	7.62

Y.Hashimoto *et al.* [Has68].

Condition: Cast, remelted, recrystallized at 900°C, cold drawn 99.93% to 0.25 mm$^\phi$, aged 1h/400°C.

† For Ti-Nb-Ta results of Suenaga and Ralls [Sue69], see Fig. 12-4.

ALLOY CLASS: Ti-Nb-Group V -- *continued*

(g) Transition Temperatures of Ti-Nb-Ta

Compositions, At.%, Nominal[†]			T_c
Ti	Nb	Ta	K
75	25	--	7.2
74	25	1	7.5_5
70	25	5	7.8_5
65	25	10	8.4
50	25	25	8.6
75	24	1	7.4
75	20	5	7.1
75	15	10	6.7
75	--	25	6.0

I.I.Baranov *et al.* (1967/70)[Bar70], also
E.M.Savitskii *et al.* [Sav71][Sav73, p.330].

Condition: cast and homogenized 4h/1500°C (or as
for Ti-Nb-V, above).

Magnetic measurement

† See literature for analyzed composition.

(h) Transition Temperatures of Ti$_{65}$-Nb-Ta$_{5-20}$

Compositions, At.%			T_c
Ti	Nb	Ta	K
65	30	5	8.9
65	25	10	8.6
65	20	15	8.1_5
65	15	20	8.1

D.G.Hawksworth and D.C.Larbalestier [Haw80].

Condition: homogenized 8h/1350°C in gettered Ar,
square rolled more than 70%, recrystallized
1h/875°C, swaged to 3 mm$^\phi$, sheathed in Cu and
drawn to 1 mm$^\phi$ (core ∿0.5 mm$^\phi$).

ALLOY CLASS: Ti-Nb-Group VI

(i) Transition Temperatures of (Ti-Nb$_{30}$)-Cr$_{2,6,10}$

Compositions, At.% [†]			T_c
Ti	Nb	Cr	K
70	30	--	8.7
69	29	2	8.1
66	28	6	7.5
63	27	10	7.7

H.Wada *et al.* [Wad80].

Condition: cold rolled with intermediate heat
treatment, final annealed 5h/1000°C, cold rolled 50%.

Resistometric measurement

† Data available for the sytems (Ti-Nb$_{50}$)-Cr$_{2,6,10}$
   and (Ti-Nb$_{75}$)-Cr$_{2,6,10}$ also as function of
   24-h aging temperature to 800°C [Wad80].

(j) Transition Temperatures of Ti-Nb$_{25}$-Mo$_{1-25}$

Compositions, At.%, Nominal[†]			T_c
Ti	Nb	Mo	K
74	25	1	7.3
70	25	5	7.4
65	25	10	6.8
50	25	25	4.3

M.I.Bychkova *et al.* (1967/70)[Byc70[a]], also
E.M.Savitskii *et al.* [Sav73, p.333].

Condition: 200h/1100°C and water quenched; cold
deformed to wire 99.9%

† See literature for analyzed composition.

ALLOY CLASS: Ti-Nb-Group VI -- *continued*

(k) Transition Temperatures of Ti-Nb-Mo

Electron/Atom Ratio	At.% Nb	At.% Mo	Transition Temperature After Heat Treatment						
			A	B	C	D	E	F	G
4.12	4.09	3.42	3.81	3.42	3.62				5.28
4.20	8.01	6.00	4.78	4.17	4.88				5.55
4.20	2.02	9.03	3.84	3.10	3.83	3.60	3.40	3.54	4.17
4.22	12.85	4.46[†]	5.37	5.15	5.85	5.88			
4.22	12.86	4.49	5.46	4.29	5.18	4.87	4.45	5.18	5.95
4.30	11.85	8.85[†]	5.51	5.37	5.45				
4.40	12.00	13.95	5.42				5.29	5.26	5.15
4.40	21.96	9.03	6.49	6.40	6.34				
4.50	12.01	18.98	4.87	4.10	4.77				4.65
4.50	19.94	15.05	5.62				5.63	5.65	5.45
4.50	32.01	8.97	6.77				6.84		6.80
4.60	11.93	23.89	4.31		4.25	4.15			4.18
4.60	11.9	23.91[†]	4.22	4.35	4.61	3.94			
4.60	42.03	9.02	7.21	7.14	7.16	7.01	7.46	7.25	7.20
4.70	12.02	28.99	4.06	4.04	3.95				
4.70	28.03	20.98	4.95						
4.70	51.77	9.04	6.98	6.90	7.00		7.38	7.32	7.20
4.80	12.00	34.02	3.34	3.29	3.17				2.95
4.80	61.84	9.05	7.16	7.11	7.20				
4.90	12.00	39.03	2.75		2.82		2.87	2.71	
4.90	36.04	27.00	4.87						

W.L.Cotton *et al*. [Cot73, Cot74].

Condition:

A - Annealed 1050°C/water quench (ST+WQ)

B - Above plus 40h/350°C/water quench (WQ+ωAge)

C - Above plus 2min/550°C/water quench (ωAge+Revert)

D - Annealed 1050°C/furnace cooled (ST+FC)

E - (ST+FC) to 750°C; 40h/750°C/water quench

F - (ST+FC) to 650°C; 40h/650°C/water quench

G - (ST+FC) to 550°C; 150h/550°C/water quench

† Some oxygen contamination.

ALLOY CLASS: Ti-Nb-Group VI -- *continued*

(1) Transition Temperatures of Ti-Nb-Mo$_{5,10,15,20}$

Compositions, At.%			T_c
Ti	Nb	Mo	K
47.5	47.5	5	8.3
55.0	40.0	5	8.3
64.7	30.3	5	8.2
45.0	45.0	10	6.7
60.0	30.0	10	5.8
42.5	42.5	15	5.5
65.0	20.0	15	---
40.0	40.0	20	4.2

V.Sadagopan *et al.* [Sad70].

Condition: 24h/1300°C annealed and cold rolled to ribbon.

Magnetic measurement

(m) Transition Temperatures of (Ti-Nb$_{30}$)-Mo$_{5,10,15}$

Compositions, At.%[+]			T_c
Ti	Nb	Mo	K
70	30	--	8.7
66	29	5	7.8
63	27	10	6.9$_5$
59	26	15	6.1

H.Wada *et al.* (1980) [Wad80].

Condition: cold rolled with intermediate heat treatment, final annealed 5h/1000°C, cold rolled 50%.

Resistometric measurement

[+] Data available for the systems (Ti-Nb$_{50}$)-Mo$_{5,10}$ and (Ti-Nb$_{75}$)-Mo$_{5,10}$ (as-rolled), also as function of 24-h aging temperature to 800°C [Wad80].

ALLOY CLASS: Ti-Nb-Group VII

(n) Transition Temperatures of Ti-Nb$_{25}$-Re$_{1-25}$

Compositions, At.%			T_c
Ti	Nb	Re	K
74	25	1	7.2
70	25	5	6.8
65	25	10	5.4
50	25	25	<4.2

M.I. Bychkova *et al.* (1967/70)[Byc70[a]].

Condition: 200h/1100°C and water quenched; cold deformed to wire 99.9%.

TABLE 12-7 UPPER CRITICAL FIELDS OF TITANIUM-NIOBIUM-TM ALLOYS

ALLOY CLASS: Ti-Nb-Group IV

(a) Critical Fields of Ti-Hf-Nb[+]

Compositions, At.%			Critical Field, H_r, T		
Ti	Hf	Nb	4.2 K	3 K	2 K
60	2.5	37.5	11.2_5	13.2	14.3_5
62.5	2.5	35	11.2_5	13.1_5	14.3_5
65	2.5	32.5	11.0	12.9	14.2
67.5	2.5	30	10.5	12.6	13.9_5
70	2.5	27.5	9.8_5	12.0	13.3_5
52.5	5	42.5	10.9_5	12.8_5	14.0_5
57.5	5	37.5	11.0	13.0	14.3
60	5	35	11.0	13.1	14.5
62.5	5	32.5	10.8	12.9	14.3
65	5	30	10.5	12.7	14.2
67.5	5	27.5	9.9	12.2_5	13.6_5
55	10	35	10.4	12.7	14.1_5
57.5	10	32.5	10.2	12.6_5	14.2
60	10	30	10.0	12.5	14.2
65	10	25	8.7	11.3_5	13.2
64	4	32	11.4[++]		

+ Principally D.G.Hawksworth and D.C.Larbalestier
   [Haw80].

Condition: cold rolled >70% and recrystallized
1h/875°C, swaged to 3 mm$^\phi$, sheathed in Cu and drawn
to 1 mm$^\phi$ (core ∿0.5 mm$^\phi$).

Criterion: First appearance of resistance ("onset")
at 10 mA (i.e. at J = 5 A cm^{-2}).

++ W.DeSorbo *et al.* [Des67].

Condition: cold worked by swaging and drawing.

Magnetic measurement

ALLOY CLASS: Ti-Nb-Group V

(b) Critical Fields of Ti-Nb-V

Compositions, At.%			H_r
Ti	Nb	V	T
70	10	20	7.0
60	10	30	8.8
50	10	40	8.4
40	10	50	7.8
30	10	60	5.7
20	10	70	4.2
10	10	80	2.7
--	10	90	0.12
70	20	10	7.2
60	20	20	9.2
50	20	30	8.3
40	20	40	7.5
30	20	50	5.3
20	20	60	3.6
10	20	70	1.4
--	20	80	0.6
70	30	--	9.0
60	30	10	10.3
50	30	20	9.0
40	30	30	7.4
30	30	40	5.0
20	30	50	3.3
10	30	60	1.3
--	30	70	0
60	40	--	11.4
50	40	10	9.8
40	40	20	7.4
30	40	30	5.3
20	40	40	3.1
10	40	50	1.2
--	40	60	0

TABLE 12-7 UPPER CRITICAL FIELDS OF TITANIUM-NIOBIUM-TM ALLOYS -- *continued*

ALLOY CLASS: Ti-Nb-Group V -- *continued*

(b) Critical Fields of Ti-Nb-V -- *continued*   (c) Critical Fields of Ti-Nb-Ta[+]

Compositions, At.%			H_r
Ti	Nb	V	T
50	50	--	12.6
40	50	10	8.8
30	50	20	5.5
20	50	30	3.3
10	50	40	1.0
--	50	50	0.5
40	60	--	11.8
30	60	10	8.2
20	60	20	4.0
10	60	30	1.4
--	60	40	0.6
30	70	--	9.5
20	70	10	4.4
10	70	20	2.0
--	70	30	0.5
20	80	--	7.7
10	80	10	2.5
--	80	20	0.6
10	90	--	3.3
--	90	10	0.9

P.H.Bellin *et al.* [Beℓ70]

Condition: cast and homogenized 24h/1600°C, furnace cooled (1h), cold rolled 15:1.

Resistive measurement, 1 μV criterion.

Compositions, At.%			Critical Field, H_r, T		
Ti	Nb	Ta	4.2 K	3 K	2 K
57.5	37.5	5	11.6	13.5	14.8
62.5	32.5	5	11.6	13.7	15.1
65	30	5	11.6	13.7	15.0
67.5	27.5	5	11.2	13.4	14.9
70	25	5	10.1	12.4	13.9
57.5	32.5	10	11.7	13.7	15.1
60	30	10	11.6	13.8	15.3
62.5	27.5	10	11.5	13.7	15.2
65	25	10	11.6	13.8_5	15.4
67.5	22.5	10	10.9	13.4	14.8
70	20	10	10.1	12.5	14.3
57.5	27.5	15	11.4	13.6	15.2
60	25	15	11.3	13.7	15.3
62.5	22.5	15	11.3	13.5_5	15.4
65	20	15	11.1	13.5	15.3
67.5	17.5	15	10.2	12.8	14.6
60	36	4	12.4[++]		

[+] D.G. Hawksworth and D.C. Larbalestier [Haw80], except final entry. For *Condition* and *Criterion*, see Ti-Hf-Nb listing by these authors.

For the Ti-Nb-Ta results of Suenaga and Ralls [Sue69], see Fig. 12-5.

[++] W. DeSorbo *et al.* [Des67].

Condition: See Ti-Hf-Nb listing by these authors.

ALLOY CLASS: Ti-Nb-Group VI

(d) Critical Fields of Ti-Nb-Cr

Compositions, At.%			H_r	
Ti	Nb	Cr	T	
67	33	--	11.4_0	†
65	33	2	10.9_0	†
70	30	--	10.6	††
69	29	2	10.5	††
66	28	4	9.4	††
63	27	6	9.2	††

† W. Heller *et al.* [Heℓ71[a]].

Condition: cold reduced 97%, aged 10h/400°C.

Criterion: 1 µV across 1 cm at 10 mA (i.e., 1.3 A cm^{-2}).

†† H. Wada *et al.* [Wad80].

Condition: cold rolled with intermediate heat treatment, final annealed 5h/1000°C, cold rolled 50%.

Alloy group listed is (Ti-Nb$_{30}$)-Cr$_{2,6,10}$; data for (Ti-Nb$_{50}$)-Cr$_{2,6,10}$ and (Ti-Nb$_{75}$)-Cr$_{2,6,10}$ also are available in the original article.

Criterion: mid-point of resistive transition.

(e) Critical Fields of Ti-Nb-Mo

Compositions, At.%			$H_{r,u}$	
Ti	Nb	Mo	T	
65	34	1	11.2	†
47.5	47.5	5	8.0	††
55.0	40.0	5	8.4	††
64.7	30.3	5	8.2	††
45.0	45.0	10	4.5	††
60.0	30.0	10	----	††
42.5	42.5	15	----	††
65.0	20.0	15	----	††
40.0	40.0	20	0	††
70	30	--	10.6	†††
66	29	5	9.0	†††
63	27	10	6.9	†††
59	26	15	4.8	†††

† W. DeSorbo *et al.* [Des67].

Condition: cold worked by swaging and drawing.

Magnetic measurement

†† V. Sadagopan *et al.* [Sad70].

Condition: annealed 24h/1300°C and furnace cooled; cold rolled to ribbon.

Resistive Criterion: 1 µV across sample [Beℓ70].

††† H. Wada *et al.* [Wad80].

Condition: as for Ti-Nb-Cr above.

Alloy group listed is (Ti-Nb$_{30}$)-Mo$_{5,10,15}$; data for (Ti-Nb$_{50}$)-Mo$_{5,10}$ and (Ti-Nb$_{50}$)-Mo$_{5,10}$ also are available in the original article.

Resistive Criterion: Mid-point of the transition to the normal state.

ALLOY CLASSES: Ti-Nb-Groups V, VII, and VIII

(f) Critical Fields of Ti-Nb-TM(groups V, VII, VIII) Alloys

Group Number of Addition	Composition	$H_{r,u}$ T	Reference	
VI	$Ti_{64}-Nb_{35}-W_1$	11.4	DeSorbo	[Des67]
VII	$Ti_{65}-Nb_{33}-Mn_2$	10.8_1	Heller	[Heℓ71[a]]
VIII	$Ti_{19}-Nb_{80}-Fe_1$	4.1	DeSorbo	[Des67]
	$Ti_{65}-Nb_{33}-Fe_2$	10.5_4	Heller	[Heℓ71[a]]
	$Ti_{65}-Nb_{33}-Ni_2$	11.6_1	Heller	[Heℓ71[a]]

13

TITANIUM-NIOBIUM BASE QUATERNARY ALLOYS

Quaternary alloys with the *interstitial* elements C, N, and O have been considered in Chapter 8 and Table 1-30. Following the scheme introduced in Table 1-38, which summarizes the literature on the subject, this chapter deals with *substititional* quaternary Ti-Nb-base alloys in the three categories: *(a)* alloys of Ti-Nb with two simple-metal additions, designated Ti-Nb-SM$_1$-SM$_2$, *(b)* alloys of Ti-Nb with another transition element and a simple metal, designated Ti-Nb-TM-SM, and finally *(c)* alloys of Ti-Nb with two additional transition elements, designated Ti-Nb-TM$_1$-TM$_2$. In practice the relevant quaternary TM alloys are generally modifications of the basic Ti-Nb alloy -- which is of course a (group-IV)TM-(group-V)TM alloy -- in which substitutions have been made on one hand for Ti (by Zr and/or Hf) and on the other for Nb (by V and/or Ta). Typical alloys conforming to this specification are, for example:

$$(Ti-Zr-Hf) - Nb$$
$$\text{and } (Ti-Zr) - (Ta-Nb),$$

where the parentheses enclose elements of similar group numbers.

13.1 THE PATENT LITERATURE

An early patent assigned to Imperial Metal Industries Ltd. [Imp65] made reference to the alloys Ti-Zr-Hf-Nb and Ti-Zr-Nb-Ta not, however, from the standpoint of advocating superior superconductive properties, but rather in order to point out that contamination of the Nb by Ta, or the Zr by Hf, would not adversely affect the properties of a basic Ti-Zr-Nb alloy. A French patent assigned by CHESTER to the Central Electricity Generating Board [Che65] laid claim (in a general way, without specific supporting data) to quaternary (and ternary) alloys containing Nb and Ti or Nb and Zr.

In a series of Japanese and German patents assigned to Kobe Steel Ltd. [Hor73[b], Hor74, Hor75], an abundance of technical data supported claims that selected Ti-Zr-Nb-Ta alloys exhibited attractive superconductive properties. A preferred composition was revealed by plotting critical field *versus* Ta concentration for the alloy sequence Ti_{40-y}-Zr_{30}-Nb_{30}-Ta_y and noting that a pronounced maximum occurred for $y \sim 7$ at.% [Hor75]. In that document very modest values of upper critical field were claimed, and the reader is referred to the corresponding research article [Hor73] for more useful data. According to the latter, critical fields were determined resistometrically at a current of 5 mA -- i.e. at a current density in the 0.25-mm$^\phi$ wire of 10 A cm^{-2}. In a series of tests [Hor74] on the alloy group $[(Ti_{70}$-$Nb_{30})_{95}$-$Zr_5]_{100-y}$-Ta_y, a peak in H_r was noted for $y = 10$ at.%, corresponding to an alloy of composition (to the nearest 1 at.%) Ti_{60}-Zr_5-Nb_{25}-Ta_{10}, which is in fact fairly close to that of Kobe Steel Ltd.'s proprietary alloy "cryozitt", Ti_{61}-Zr_6-Nb_{27}-Ta_6, currently in production [Kob80, Kob80[a]].

PART 1: THE SUPERCONDUCTING TRANSITION IN TITANIUM-NIOBIUM-BASE QUATERNARY ALLOYS

13.2 INFLUENCE OF SIMPLE-METAL ADDITIONS ON THE TRANSITION TEMPERATURE

The influences of simple-metal additions on the superconducting transition temperatures of quaternary alloys based on Ti-Nb were first considered by BACH-MANN et al [Bac68] and subsequently by COTTON [Cot73]. As part of a comprehensive study of the effects of O, N and simple-metal ternary additions to Ti-Nb, BACH-MANN et al had investigated the superconducting transition temperatures of cast-plus-deformed quaternary alloys formed by adding 1 at.% of Sb (group V) plus 1 at.% of either Al, Ga or In (all of group III) to Ti-Nb$_{20}$, and of adding 2.5 at.% Sn plus 5 at.% U to Ti-Nb$_{10}$. Of interest in each case was the way in which T_c differed from that of binary Ti-Nb$_{20}$ (T_c = 6.7$_5$ K). It was noted that whereas Al, Ga and In raised T_c slightly, Sb lowered it, and that the superconducting transition temperatures of the quaternaries could be either intermediate between those of the constituent ternaries (in the case of Sb + Al) greater than either of them (Sb + In), or less than either of them (Sb + Ga).

COTTON [Cot73] has studied the effect of Al additions on the superconducting transition temperatures of Ti-Nb-Mo alloys as a function of their metallurgical conditions, viz:

(a) quenched,

(b) quenched and aged at 350°C for the purpose of developing an ω+β microstructure,

(c) quenched, aged at 350°C, and annealed 2 min/500°C in order to develop a double-bcc structure through the so-called "reversion" of ω to β-phase,

(d) isothermally transformed to α+β and furnace cooled (100-150°C per h),

(e) the above, followed by aging for 40h/750, 650 or 550°C.

It was noted: (i) that as a result of phase decomposition and solute redistribution, T_c for the Ti-Nb-base multicomponent alloys was not a universal function of electron/atom ratio; (ii) that in general the alloys were not homogeneously superconducting, and could exhibit more than one superconducting transition; (iii) that the Nb content, hence the T_c of the Ti-Nb superconducting component, could be altered by appropriate heat treatment.

13.3 TRANSITION TEMPERATURES OF QUATERNARY ALLOYS SELECTED FROM THE SCHEME

$$\underbrace{\text{GROUP IV}}\quad\underbrace{\text{GROUP V}}$$
$$\text{Ti} - (\text{Zr} - \text{Hf}) - (\text{V} - \text{Ta}) - \text{Nb}$$

Under this heading fall the studies by: (a) RAYEVSKII et al [Ray71], who has taken the basic Ti-Nb$_{30}$ alloy and substituted varying amounts of Zr and/or Hf for some of the Ti; (b) KITADA and DOI [Kit70^g], who took Ti$_{10}$-Zr$_{40}$-Nb$_{50}$ and made small substitutions of V, Ta and Mo for Nb; (c) HORIUCHI et al [Hor73], who examined a wide range of Ti-Zr-Nb alloys with 5 at.% and 10 at.% substitutions of Ta. It is instructive to represent these investigations diagrammatically as follows:

$$(\text{Ti-Zr-Nb})_X \xrightarrow{\hspace{1cm}} \text{Ta}_{5,10}, \quad \text{HORIUCHI } et\ al\ [\text{Hor73}],$$

$$(\text{Ti}_{10}\text{-Zr}_{40}\text{-Nb}_{50})-\begin{pmatrix}\text{V}\\\text{Ta}\\\text{Mo}\end{pmatrix}_X, \quad \text{KITADA and DOI } [\text{Kit70}^g],$$

$$(\text{Ti-Zr-Hf})_X \xrightarrow{\hspace{1cm}} \text{Nb}_{30}, \quad \text{RAYEVSKII } et\ al\ [\text{Ray71}].$$

Presented in this way it is clear that the first two studies involved the effects of small additions of the group-V elements (also Mo, a group-VI element) to a broadly composed Ti-Zr-Nb base, or to the frequently encountered Ti$_{10}$-Zr$_{40}$-Nb$_{50}$, and the following picture emerges:

In the ternary Ti-Zr-Nb system, according to Sect.11.2, the highest T_c's lay along the Zr-Nb and Ti-Nb edges of the Gibbs triangle, the maxima occurring near 20 at.% Zr ($T_c \sim$ 11 K) and 30-40 at.% Ti ($T_c \sim$ 9.6 K). The T_c surface itself is a trough, narrow near the Nb apex and widening out towards the Ti-Zr base.

Similar features are suggested by the (incomplete) data of HORIUCHI et al [Hor73], excepting that with the addition first of 5 at.% and then of 10 at.% Ta, the entire Ti-Zr-Nb surface (for 2h/1000°C-annealed alloys) appears to drop uniformly in two steps of up to about 0.5 K each. The superconducting transition temperatures of Ti-Zr-Nb-Ta$_{5,10}$, from the work of HORIUCHI et al [Hor73], are presented in Fig.13-1. The results of KITADA and DOI [Kit70^g] as far as they go,

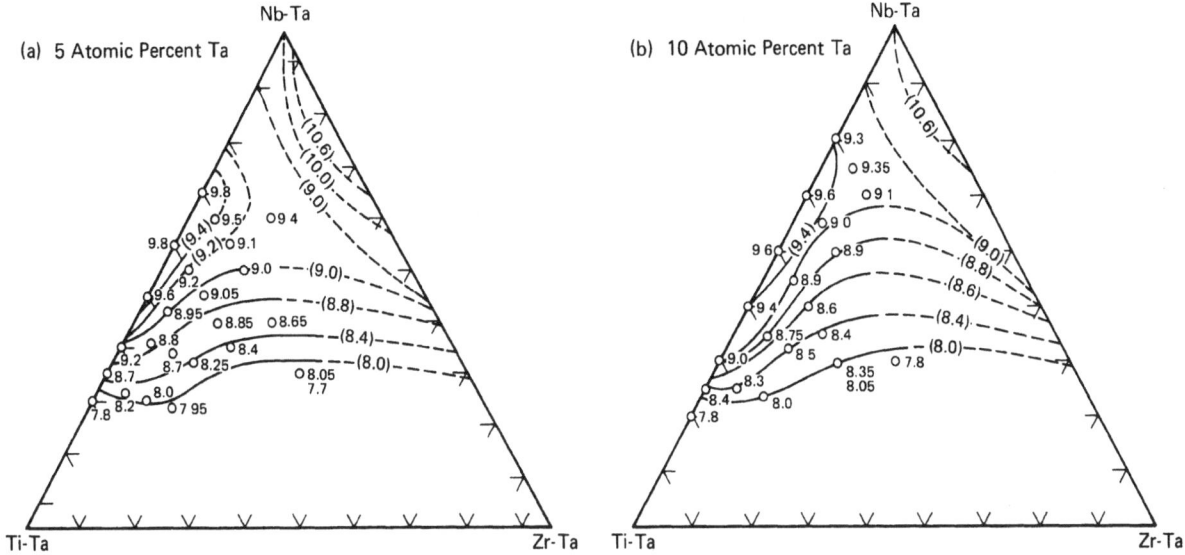

FIGURE 13-1. Pseudoternary diagrams (at. % linear) indicating the superconducting transition temperatures (T_c, K) of the quaternary alloys Ti-Zr-Nb-Ta. Two series of alloys are represented: (a) at. % Ta = const. = 5 at. %; (b) at. % Ta = const. = 10 at. %. The dashed lines which are based on binary-alloy data suggest the approximate positions of the unmeasured isothermals. The full curves (experimentally determined isothermals) are from HORIUCHI *et al* [Hor73].

are in agreement with this. As illustrated in Fig.13-2, additions of Ta, V, and Mo, decrease the T_c of the alloy-base Ti_{10}-Zr_{40}-Nb_{50} (annealed 3h/1100°C) at average rates (based on 2 at.% additions) of 0.4, 0.8_5, and 1.0 K per at.% solute, respectively.

The work of RAYEVSKII *et al* [Ray71] dealt with a (group-IV)TM-Nb_{30} type of alloy (annealed 1h/500°C),

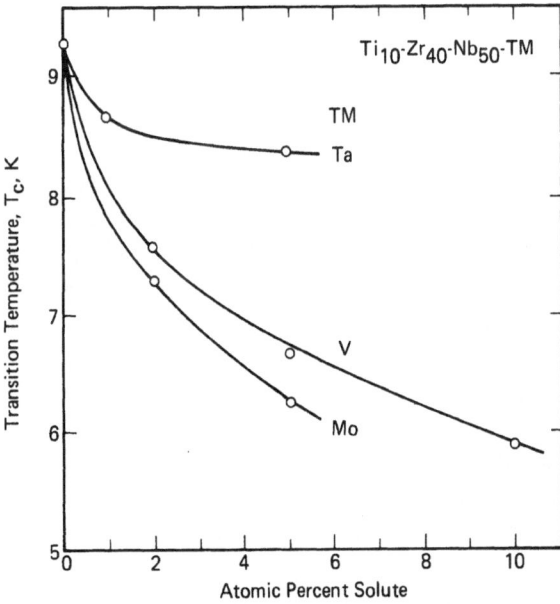

FIGURE 13-2. Influence on the superconducting transition temperature of the alloy Ti_{10}-Zr_{40}-Nb_{50} of small additions of V, Ta and Mo — after KITADA and DOI [Kit70g].

and considered the effect of changing the relative amounts of the group-IV component (i.e. the Ti, Zr, and Hf). If T_c were dominated by electron/atom ratio, no change would be expected; in practice, T_c had its largest values along the (Ti-Zr)-Nb_{30} edge (with of course $T_c \sim 7$ K and $T_c \sim 8$ K at the "Ti" and "Zr" extremities, respectively) and decreased more-or-less continuously with addition of Hf, excepting for some slight local maxima and minima. The highest quaternary-alloy T_c was 7.4 K (Ti_{14}-Zr_{42}-Hf_{14}-Nb_{30}) and the lowest was 5.8 K (Ti_{42}-Zr_{14}-Hf_{14}-Nb_{30}).

13.4 TRANSITION TEMPERATURES OF Ti-Nb-TM₁-TM₂ ALLOYS—CONCLUSION

The highest alloy transition temperatures are those of binary Zr-Nb followed by binary Ti-Nb. The intermixing of these two alloys yields the Ti-Zr-Nb system whose transition temperatures are lower than those of the binary parent alloys. Finally, the substitution of the remaining group-IV element, Hf, and the remaining group-V elements, V and Ta, and also Mo, serve to lower T_c still further. In general, only as an artifact of metallurgical heat treatment, if phase decomposition results in regions with more favorable Ti:Nb ratios, is it possible for the T_c of Ti-Nb-base alloys to be increased.

PART 2: THE CRITICAL FIELDS OF TITANIUM-NIOBIUM-BASE QUATERNARY ALLOYS

13.5 QUATERNARY-ALLOY CRITICAL FIELDS— AN OVERVIEW

The forerunners of all contemporary alloy super-conductors are of course the two (group-IV)TM-(group-V)TM alloys Zr-Nb and Ti-Nb. The combining of this pair led to the ternary Ti-Zr-Nb system dealt with in Chapter 11. Within this context it was then natural to ask if improved critical fields could be obtained by formulating multicomponent alloys through the substitution of Hf for some of the Ti or Zr, and V and/or Ta for some of the Nb. Alloys systems which have resulted from applications of this philosophy are illustrated in Table 13-1. As can be seen from the table, only a few of the possibilities have yet been explored. ·

Superconductor alloy development has as its inter-related goals the attainment of high T_c, H_{c2}, and J_c. The upper critical field may be related to the transition temperature through the usual relationship (in Oe) $H_{c20}^* = 3.06 \times 10^4 \, \rho_n \, \gamma \, T_c$ (3-13, 7-10, 12-1). The coupling with J_c is then made by recognizing that at high fields the critical current density is dominated by H_{c2}, cf. Sect.7.32. The approaches adopted during superconducting alloy development over the years, which have varied from the "systematically empirical" to the "fundamentally motivated", are exemplified by the various philosophies which have guided quaternary-alloy upper critical field studies:

(a) DeSORBO [Des67] in investigating "effective electron/atom ratio" systematics has conducted magnetic measurements of the upper critical fields of all the alloy types represented in Table 13-1, as well as those of numerous other related binary and ternary alloys whose conventional electron/atom ratios were in the range $4.2 \gtrsim e/a \gtrsim 4.8$.

(b) The starting point of the investigation conducted by RAYEVSKII *et al* [Ray68, Ray71] was the recognition that Ti-Zr-Nb itself possessed satisfactory properties coupled with the fact that, although MORTON *et al* (cf. [16] of [Ray69]) had noted that "up to 5 wt.% Hf (did) not undesirably affect the properties of (Ti-Zr-Nb)", the quaternary Ti-Zr-Hf-Nb system had not been subjected to systematic metallurgical or superconductive investigation. Accordingly, those authors set about preparing, heat treating, and examining three sets (defined by [Nb] = 30, 50, and 70 at.%) each of six quaternary alloys together with numerous related ternary and binary alloys.

(c) The work of HORIUCHI *et al* [Hor73, Hor74, Hor75] appears to have been motivated by a desire to improve on the properties of Ti-Zr-Nb, by substituting for Nb sufficient Ta to significantly increase the atomic number (Z), and consequently the spin-orbit-scattering frequency ($\propto Z^4$), in order to increase the limit placed on H_{c2} by normal-state Pauli paramagnetism. The underlying principle, proposed initially by NEURINGER and SHAPIRA [Neu66] and others, and criticized by ORLANDO *et al* [Orℓ79][Bea82], has been discussed in [Mon15.00] and outlined in Sects.3.6.3 to 3.6.5, Sect.5.12.4, and Sects.7.10, 7.11. For the same reason, apparently, HAWKSWORTH and LARBALESTIER [Haw80][Lar80] attempted to increase the upper critical field of Ti-Zr-Nb-Ta by substituting for the Zr in HORIUCHI's alloys the heavier, isoelectronic, element Hf.

The effects on H_{c2} of these various substitutions for the Ti and Nb of the basic alloy are summarized below.

TABLE 13-1   SYSTEMATICS OF QUATERNARY ALLOY CRITICAL FIELD STUDIES

Group IV			Group V			
Ti	Zr	Hf	V	Ta	Nb	References
X	X	X			X	+, ++
X	X			X	X	+, +++
X		X		X	X	+, +++*
X		X	X		X	+

+   DeSorbo *et al.* [Des67]

++   Rayevskii *et al.* [Ray69, Ray71]

+++   Horiuchi *et al.* [Hor73, Hor80]

+++*   Hawksworth *et al.* [Haw80]

13.6 THE INFLUENCE OF Hf ON THE H_{c2} OF Ti-Zr-Nb AND Ti-Nb-Ta

The results of the work of RAYEVSKII et al [Ray69, Ray71] on three "complete" sets of Ti-Zr-Hf-Nb alloys with Nb concentrations of 30, 50 and 70 at.%, can be usefully expressed in terms of the influence of Hf substitutions on the upper critical fields of Ti-Zr-Nb. In general, it can be said that the effect of appreciable amounts of Hf is deleterious. In particular for the [Nb] = 50 at.% set of alloys (annealed 2h/500°C) [Ray69] H_r, which had its maximum value of 10.0∿11.5 T in the Ti-Nb corner, was found to decrease upon the substitution of Zr for Ti (i.e. along the Ti-Nb_{50} to the Zr-Nb_{50} edge of the pseudoternary composition triangle) and to do so even more severely with the substitution of Hf (i.e. towards the Nb_{50}-Hf_{50} corner of the triangle). Similarly for the [Nb] = 30 at.% set of alloys (annealed 1h/500°C) [Ray71] it was noted that, apart from the formation of some small local troughs and ridges, substituting Hf for Ta and Zr generally lowered H_{c2}, which dropped to less than 4.5 T in the Nb_{30}-Hf_{70} corner.

The results of DeSORBO et al [Des67], interpreted from a "Hf-effect" standpoint, tended to confirm these conclusions and in particular that the substitution of 4 at.% Hf for the same amount of Zr in Ti_{53}-Zr_4-Nb_{39}-Ta_4[+] decreased H_u by about 1.1 T, and that a further increase in the Hf level to 20 at.% (plus a substitution of Zr for half of the remaining Ti, leading to the alloy Ti_{20}-Zr_{20}-Hf_{20}-Nb_{40}) reduced

[+] The new alloy was actually Ti_{50}-Hf_4-Nb_{42}-Ta_4.

TABLE 13-2 MAGNETICALLY MEASURED UPPER CRITICAL FIELDS (H_u) OF SEVERAL QUATERNARY ALLOYS -- DeSorbo et al. [Des67]

Compositions (at.%)						H_u
Group IV			Group V			
Ti	Zr	Hf	V	Nb	Ta	T
11	11	11		67		7.1
20	20	20		40		9.1
50		4		42	4	10.7
52		4	4	40		11.0
53	4			39	4	11.8

it by another 1.6 T. The magnetically measured upper critical fields of several quaternary alloys, according to DeSORBO et al [Des67], are listed in Table 13-2.

From the careful studies by HAWKSWORTH and LARBALESTIER of a large matrix of compositions [Haw80], it is possible to glean further information about the influence of Hf on the 4.2-K and 2-K upper critical fields of multicomponent Ti-Nb-base alloys. Their results on ternary alloys, as reported in Chapter 12, showed that the presence of Hf (a group-IV element) tended to decrease H_r. This tendency is repeated in the quaternary alloys wherein, as indicated in Table 13-3 and Fig.13-3, the substitution of Hf for

TABLE 13-3 RESISTIVE UPPER CRITICAL FIELDS (H_r) OF Hf-substituted Ti-Nb-Ta ALLOYS -- after Hawksworth and Larbalestier [Haw80]

Compositions, at.%				H_r[+] at temperatures specified tesla		
Ti	Hf	Nb	Ta	4.2 K	3 K	2 K
57.5	0	32.5	10	11.7	13.7	15.1
60	"	30	"	11.6	13.8	15.3
62.5	"	27.5	"	11.5	13.7	15.2
65	"	25	"	11.6	13.9	15.4
67.5	"	22.5	"	10.9	13.4	14.8
70	"	20	"	10.1	12.5	14.3
55	2.5	32.5	10	11.0	13.1	14.6
57.5	"	30	"	11.0	12.3	14.7
60	"	27.5	"	11.3	13.6	15.3
62.5	"	25	"	10.9	13.4	14.9
70	"	17.5	"	9.1	11.7	13.7
52.5	5	32.5	10	10.9	13.0	14.6
55	"	30	"	10.9	13.2	14.8
60	"	25	"	10.8	13.3	15.0
47.5	10	32.5	10	9.8	12.0	13.7
50	"	30	"	9.8	12.2	13.8
55	"	25	"	9.8	12.4	14.3
60	"	20	"	9.4	12.2	14.3
62.5	"	17.5	"	8.6	11.7	13.8

[+] Criterion: Resistive onset at 10 mA (J = 5 A cm^{-2}) -- see footnote to Sect. 12.1.3.

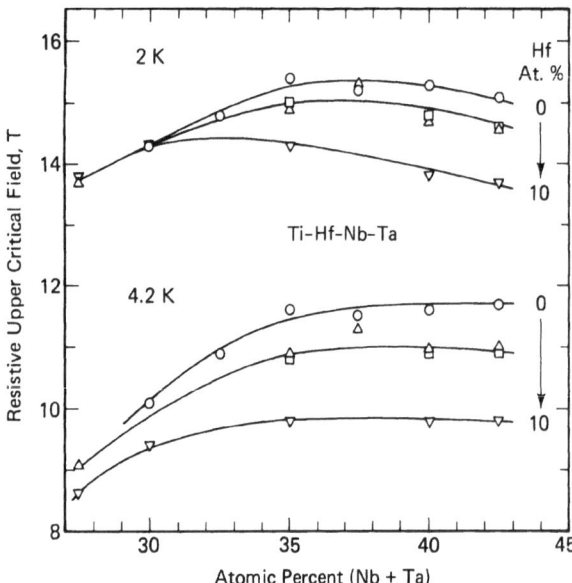

FIGURE 13-3. Resistive upper critical fields (criterion: resistive "onset" at J = 5 A cm^{-2}) of Ti-Hf-Nb-Ta (constant 10 at. % Ta) alloys plotted using the total Nb+Ta concentration as display parameter. The Hf levels depicted are 0 at. % (○), 2.5 at. % (△), 5 at. % (□) and 10 at. % (▽). H_r generally decreases with increasing Hf concentration — after HAWKSWORTH and LARBALESTIER [Haw80].

Ti at constant group-V-element (Nb + 10 at.% Ta) concentration, reduces H_r at 4.2 K and again at 2 K. Next, by comparing the result of HAWKSWORTH and LARBALESTIER [Haw80] for Ti$_{60}$-Hf$_5$-Nb$_{25}$-Ta$_{10}$ whose 4.2-K H_r was 10.8 T, with that of HORIUCHI et al [Hor73] on Ti$_{60}$-Zr$_5$-Nb$_{25}$-Ta$_{10}$ for which a 4.2-K H_r of 13.1 T was claimed, we note again that the substitution of Hf for Zr reduces the upper critical field of this rather remarkable alloy by about 2.3 T.

It may therefore be concluded that if quaternary alloys are to be considered, the inclusion of Hf is to be avoided. According to HORIUCHI et al of the Asada Fundamental Research Laboratory of Kobe Steel Ltd., the alloy Ti$_{60}$-Zr$_5$-Nb$_{25}$-Ta$_{10}$ referred to above had the highest upper critical field of any known alloy.

13.7 THE INFLUENCE OF Ta ON THE H_{c2} OF Ti-Zr-Nb

Culminating in the discovery of high-H_{c2} alloys of compositions close to that referred to above, and in particular the Kobe Steel proprietary alloy "cryo-zitt" of composition Ti$_{61}$-Zr$_6$-Nb$_{27}$-Ta$_6$, were the studies of HORIUCHI et al on what was in effect the influence of Ta-for-Nb substitution on the properties of the Ti-Zr-Nb alloys recommended at one time by

Hitachi Ltd. Thus the alloy Ti$_{61}$-Zr$_6$-Nb$_{27}$-Ta$_6$ ("cryo-zitt"), and the alloy Ti$_{60}$-Zr$_5$-Nb$_{25}$-Ta$_{10}$ referred to above, can be regarded as being derived from Hitachi Ltd.'s Ti-Nb-rich (Z-type) alloy Ti$_{60}$-Zr$_5$-Nb$_{35}$ by a replacement of some of the Nb by Ta.

Background studies consisted of measuring the properties of 5h/1000°C heat-treated 0.25-mm$^\phi$ wire samples of alloys represented roughly by Ti$_{20 \sim 65}$-Zr$_{5 \sim 35}$-Nb$_{25 \sim 60}$-Ta$_5$ and Ti$_{20 \sim 60}$-Zr$_{5 \sim 30}$-Nb$_{25 \sim 65}$-Ta$_{10}$ (i.e. containing constant 5 and 10 at.% Ta and low levels of Zr) [Hor73, Hor74], and by the following formulae for the "quaternary slices": Ti-Zr$_{32.3}$-Nb$_{32.3}$-Ta$_y$ (0 < y < 12) [Hor73], Ti-Zr$_{30}$-Nb$_{30}$-Ta$_y$ (0 < y < 15) [Hor75], and [(Ti-Nb$_{30}$)$_{95}$-Zr$_5$]$_{100-y}$-Ta$_y$ (0 < y < 25) [Hor74].

In [Hor74], data plotted on the composition triangle severely underestimated the upper critical fields of the [Ta] = 5 and 10 at.% quaternary alloys. According to [Hor73], which seems to give a more optimistic account of their potentialities, H_r(4.2 K) in the 10 at.% Ta plot of Fig.13-4(b) rises to the maximum of 13.1 T referred to above, and the [Zr] = [Nb] = 32.3 at.% (0 < [Ta] < 12 at.%) "slice" yields a maximal H_r(4.2 K) of 10.9 T at 6 at.% Ta. In [Hor75] the [Zr] = [Nb] = 30 at.% (0 < [Ta] < 14 at.%) "slice" also yields a maximal H_r(4.2 K) of 10.9 T with 7 and 8 at.% Ta, while in [Hor74] H_r(4.2 K) for the low-Zr alloy group [(Ti-Nb$_{30}$)$_{95}$-Zr$_5$]$_{100-y}$-Ta$_y$ has a maximal value of 12.85 T for y = 10 at.% (i.e. for a composition in this system of Ti$_{59.9}$-Zr$_{4.5}$-Nb$_{25.7}$-Ta$_{10}$ -- very close to that of cryozitt). The results of this early discovery are depicted in Fig.13-5. Data presented at the Eighth Symposium on the Engineering Problems of Fusion Research, November, 1979, by HORIUCHI et al [Hor80] seem to confirm the early claims.

Table 13-4, which summarizes some critical field results for the alloy systems [(Ti-Nb$_m$)$_{100-n}$-Zr$_n$]$_{100-y}$-Ta$_y$, in which m = 30, 40, 50, n = 5, 10, 15, and y = 0 to 10, shows indeed that in these systems the highest 4.2-K critical fields are to be found in [(Ti-Nb$_{30}$)$_{90,95}$-Zr$_{10,5}$]$_{100-y}$-Ta$_y$ and that, of these, the alloy with "5 at.%" Zr and 10 at.% Ta possesses the highest value, 13.1 T. The highest critical fields were generally to be found among alloys in the as-drawn 99.8% cold-reduced condition. Aging at about 380°C tended to lower H_r, and reduce its sensitivity to Ta content to such an extent that the standard deviation in H_r (average value, 11.7 T) for the twenty-four alloys measured was then only 0.25 T.

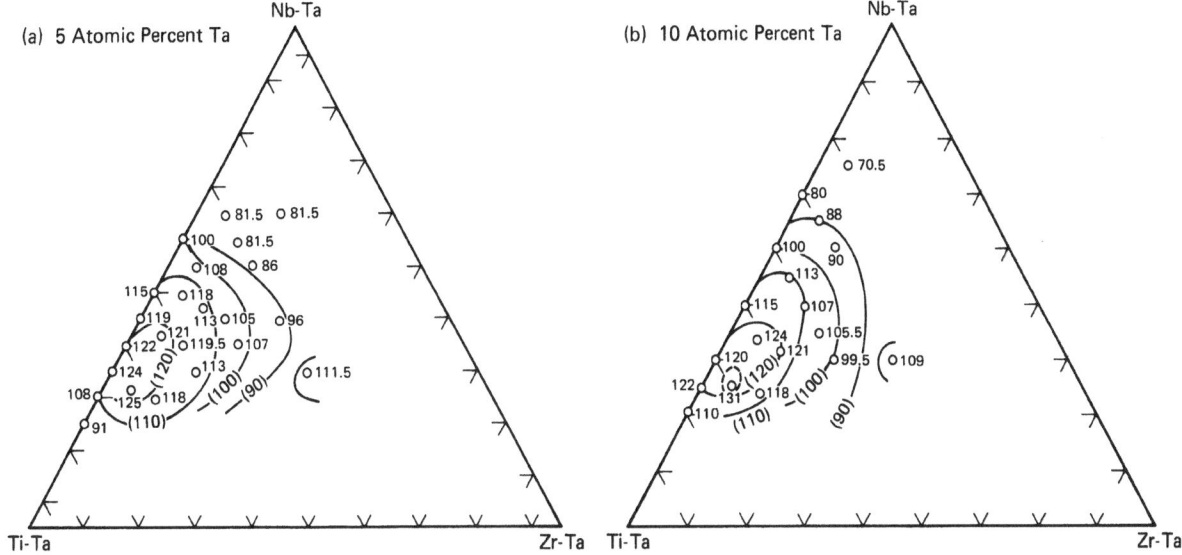

FIGURE 13-4. Pseudoternary diagrams (at. % linear) indicating the resistive upper critical fields (H_r,kOe,4.2K) of the quaternary alloys Ti-Zr-Nb-Ta. The full lines are in steps of 10 kOe. Two series of alloys are represented: (a) at. % Ta = const. = 5 at. %; (b) at. % Ta = const. = 10 at. %. Measurements were made on 0.25 mm$^\phi$ wire using a current of 5 mA (J = 10 A cm^{-2}) — after HORIUCHI et al [Hor73].

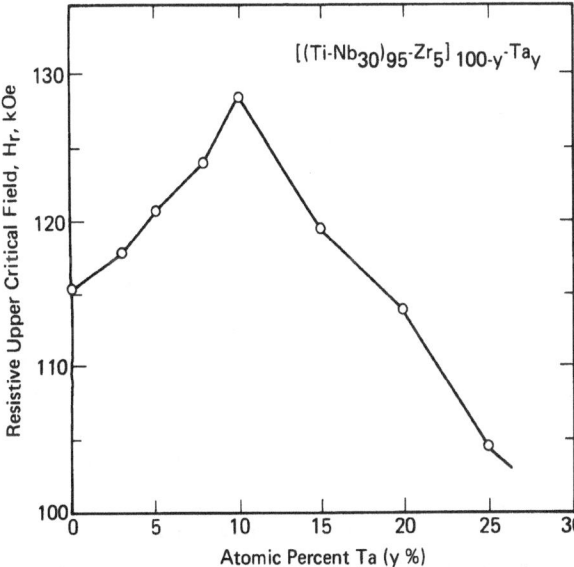

FIGURE 13-5. Resistive upper critical fields at 4.2 K (probably measured at a current density of 10 A cm^{-2} [Hor73]) for the alloy sequence [(Ti-Nb$_{30}$)$_{95}$-Zr$_5$]$_{100-y}$-Ta$_y$ plotted as function of Ta content in at. % — after HORIUCHI et al [Hor74].

TABLE 13-4   RESISTIVE UPPER CRITICAL FIELDS, H_r, OF Ta-SUBSTITUTED Ti-Zr-Nb ALLOYS[†] --- Horiuchi et al. [Hor80]

Compositions	H_r, 4.2 K[††] tesla
(Ti-Nb$_{50}$)-Zr$_5$	11.3
[(Ti-Nb$_{50}$)-Zr$_5$]-Ta$_5$	10.3
[(") "]-Ta$_7$	10.1
(Ti-Nb$_{40}$)-Zr$_5$	12.4
[(Ti-Nb$_{40}$)-Zr$_5$]-Ta$_4$	12.0
[(") "]-Ta$_6$	12.4
[(") "]-Ta$_{10}$	12.0
(Ti-Nb$_{40}$)-Zr$_{10}$	11.6
[(Ti-Nb$_{40}$)-Zr$_{10}$]-Ta$_6$	11.9
[(") "]-Ta$_{10}$	11.7
(Ti-Nb$_{40}$)-Zr$_{15}$	10.8
[(Ti-Nb$_{40}$)-Zr$_{15}$]-Ta$_5$	10.9

$(Ti-Nb_{30})-Zr_5$	11.3
$[(Ti-Nb_{30})-Zr_5]-Ta_3$	11.7
$[(\quad " \quad) \quad " \quad]-Ta_5$	12.4
$[(\quad " \quad) \quad " \quad]-Ta_7$	12.6
$[(\quad " \quad) \quad " \quad]-Ta_{10}$	13.1
$(Ti-Nb_{30})-Zr_{10}$	12.2
$[(Ti-Nb_{30})-Zr_{10}]-Ta_4$	12.2
$[(\quad " \quad) \quad " \quad]-Ta_6$	12.2
$[(\quad " \quad) \quad " \quad]-Ta_{10}$	11.9
$(Ti-Nb_{30})-Zr_{15}$	11.5
$[(Ti-Nb_{30})-Zr_{15}]-Ta_5$	11.6

† The compositions are generally of the type
$[(Ti-Nb_m)_{100-n}-Zr_n]_{100-y}-Ta_y$

†† The specimens were measured in the as-drawn 99.8% cold-reduced condition. All the aged ($\sim 380°C$) specimens tended to have the same H_r. The mean H_r for 24 aged alloys is 11.7 ± 0.25 T.

Criterion: Measurements were made on 0.25 mm$^\phi$ wire using a current of 5 mA (J = 10 A cm^{-2}) [Hor73].

PART 3: CRITICAL CURRENT DENSITY IN TITANIUM-NIOBIUM-BASE QUATERNARY ALLOYS

During the past ten years there has been a continual search for multicomponent Ti-Nb-base alloys with improved critical current densities -- especially at high magnetic fields -- as compared with those of selected Ti-Nb and Ti-Zr-Nb alloys. When considering the critical current densities of multicomponent alloys, the task of optimization is complicated by the enormous number of possible composition/heat-treatment combinations. Recently, however, attempts have been made to simplify matters by separating the compositional and heat-treatment variables. The technique used has been to seek high H_{c2} values through composition variation, under the assumption that high critical current densities can be expected to accompany subsequent process optimization of the selected compositions.

13.8 QUATERNARY ALLOY CRITICAL CURRENT DENSITIES—AN OVERVIEW

BACHMANN *et al* [Bac68] have investigated the effects of adding pairs of simple metals to a Ti-Nb$_{20}$ alloy base. The critical current densities of 1-mm$^\phi$ wire samples, cold deformed 97.2% from cast rods of the alloys Ti-Nb$_{20}$-SM$_1$-SM$_2$, where SM$_1 \equiv 1$ at.% Sb and SM$_2 \equiv 1$ at.% of either Al, Ga or In, were always greater than those of the base, suggesting that in general such additions can lead to improvements in the current-carrying capacity, but at the expense of workability.

But more important from a practical standpoint are the critical current densities of all-transition-metal Ti-Nb-base quaternaries, which fall into the following categories:

(a) Hf additions to Ti-Zr-Nb -- RAYEVSKII *et al* [Ray69, Ray71],

(b) Ta and other additions to Zr-Nb-rich alloys -- KITADA and DOI [Kit70^g],

(c) Ta additions to equiatomic Ti-Zr-Nb alloys -- HORIUCHI *et al* [Hor73^b, Hor75],

(d) Ta additions to Ti-Nb-rich Ti-Zr-Nb alloys -- HORIUCHI *et al* [Hor74].

13.9 THE INFLUENCE OF Hf ON THE CRITICAL CURRENT DENSITY OF Ti-Zr-Nb

RAYEVSKII *et al* [Ray69, Ray71] have considered the effect on three series of Ti-Zr-Nb alloys, conveniently represented by the scheme Ti-Zr-Nb$_{30}$, 50, and 70, of replacing some of the Ti or the Zr by Hf. With so many variables the data may be arranged for presentation in numerous ways (e.g. with [Nb] = constant, and J_c plotted *versus* H for various Ti levels, etc.) and can give rise to a complicated discussion of the influences of alloying on critical current density and its field dependence. The critical current data can be presented most clearly on sets of compositional triangles -- one set for each Nb level (viz 30, 50 and 70 at.%) -- each set representing several values of the magnetic field (e.g. 4, 5, 6, and 7 T). Based on such presentations, the influence of Hf content on the critical current density of cold-worked and final-heat-treated Ti-Zr-Hf-Nb alloys may be summarized as in Table 13-5 and Fig.13-6.

Since the low upper critical fields of Hf-Nb and nearby alloys lead to the occurrence of large normal regions in the composition triangle at fields above 6 T, the quaternary Ti-Zr-Hf-Nb alloys are useful only at low to intermediate fields ($\lesssim$5 T). None of the quaternary alloys has a higher critical current density than binary Zr-Nb itself. The advantage of Ti-Nb or Ti-Zr-Nb over binary Zr-Nb is its higher upper critical field. The decrease of critical field with

TABLE 13-5 CRITICAL CURRENT DENSITY IN RESPONSE TO VARIATION OF HAFNIUM CONTENT IN TITANIUM-ZIRCONIUM-HAFNIUM-NIOBIUM ALLOYS -- AFTER RAYEVSKII *et al.* [Ray69, Ray71]

Nb Concentration and Final Heat Treatment	40 kOe	50 kOe	60 kOe	70 kOe
30 at.% (1h/500°C)	A small pocket in the Hf-Nb corner is nonsuperconducting.	The Hf-Nb corner is also normal.		A large normal triangle with apex at the Hf-Nb corner covers one-third of the diagram.
	J_c is maximal in the Zr-Nb corner and minimal in the Ti-Nb corner. A ridge of diminishing height connects the Zr-Nb corner to the opposite edge at about Ti$_{20}$-Hf$_{50}$-Nb$_{30}$.	J_c is maximal in the Zr-Nb corner and lies on a weak ridge projecting towards the opposite edge at about Ti$_{42}$-Hf$_{28}$-Nb$_{30}$.		J_c is maximal in the Zr-Nb corner and along most of the Ti-Zr-Nb edge.

The existence of the ridge demonstrates that it is beneficial to substitute Hf for Ti at constant Zr (and of course, Nb), Fig. 13-6(a).

	40 kOe	60 kOe	70 kOe	80 kOe
50 at.% (2h/500°C)	J_c is very small in the Hf-Nb corner.	A small pocket in the Hf-Nb corner is normal.	A normal zone extends from the Hf-Nb corner.	A large normal triangle with apex at the Hf-Nb corner covers half of the diagram.
	J_c is maximal in the Zr-Nb corner and, at corresponding Zr levels, is larger along the Ti-Zr-Nb edge.			

The tilt acquired by the iso-J_c lines increases with increasing H$_a$ until at 80 kOe they are almost parallel to the Ti-Zr-Nb edge. Starting with Zr$_{50}$-Nb$_{50}$, replacing Zr by Ti and Ti by Hf, lowers J_c, Fig. 13-6(b).

	40 kOe	50 kOe	60 kOe	70 kOe
70 at.% (3h/550°C)	J_c is maximal in the Zr-Nb corner and equally small in the Ti-Nb and Hf-Nb corners.		A small pocket in the Hf-Nb corner is normal.	A large normal triangle with apex at the Hf-Nb corner covers two-thirds of the diagram.
	Apart from minor undulations, J_c decreases fairly uniformly from the Zr-Nb corner to the Ti-Hf-Nb edge.		At corresponding Zr levels J_c is larger along the Ti-Zr-Nb edge.	
			The tilt acquired by the iso-J_c lines increases with increasing H$_a$ until at 70 kOe they tend to be parallel to the Ti-Zr-Nb edge, Fig. 13-6(b).	

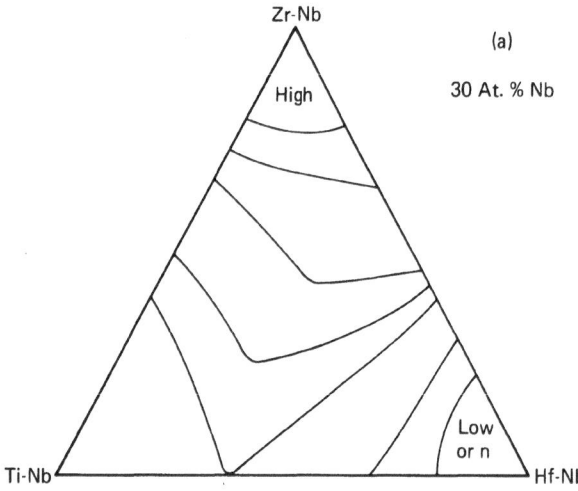

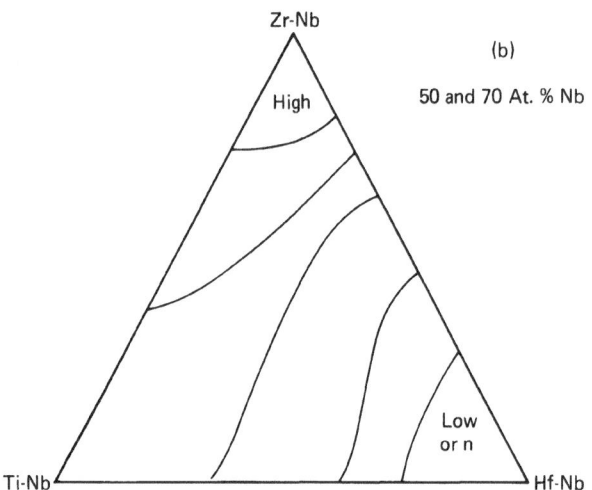

FIGURE 13-6. General features of the J_c-composition relationships in Ti-Zr-Hf-Nb alloys. The relative magnitudes (or existences) of critical current density are indicated qualitatively by the notations: 'hi', 'lo', and n (i.e., nonsuperconducting) — after RAYEVSKII *et al* [Ray69, Ray71], see also Table 13-5.

Hf addition is accompanied by the additional penalty of reduced intermediate-field critical current density.

13.10 THE INFLUENCE OF Ta ON THE CRITICAL CURRENT DENSITY OF Ti-Zr-Nb

13.10.1 Ta and Other Additions to Zr-Nb-Rich Ti-Zr-Nb

KITADA and DOI [Kit70[g]] have applied process optimization in the form of intermediate heat treatments at 600 and 700°C (the optimal temperature range for this class of alloy -- cf. Sect.11.1) and final cold work to Ti_{10}-Zr_{40}-Nb_{50} alloyed with a few at.% of either V, Mo or Ta (whose solubilities in that base

in the relevant temperature range are, respectively, about 5, 5 and 2 at.%). No advantages accrued, the highest critical current density, that of $Ti_{9.5}$-Zr_{38}-$Nb_{47.5}$-Ta_5, being about equal to that of the base alloy.

13.10.2 Ta Additions to Equiatomic Ti-Zr-Nb

The alloy Ti_{33}-Zr_{30}-Nb_{30}-Ta_7, introduced in 1973 by HORIUCHI *et al* [Hor73[a], Hor73[b]], see also [Hor75], can be regarded as being derived from those considered above by an equalization of the Ti and Zr levels at fixed group-V element content. Process optimization consisted of varying the first heat-treatment time and temperature, and the second heat-treatment (550°C) time prior to a final 90% cold reduction. No direct comparisons of performance with other technical Ti-Nb or Ti-Zr-Nb alloys were made, but detailed information on the effects of first heat-treatment conditions and second heat-treatment time on J_c(6 T, 4.2 K) was presented. A maximal critical current density of about 2×10^5 A cm^{-2} (6 T, 4.2 K) was obtained in response to:

$$(2h/850°C) \xrightarrow[87\%]{CW} (3h/650°C) \xrightarrow[90\%]{CW} (5h/550°C) \xrightarrow[90\%]{CW} \square$$

13.10.3 Ta Additions to Ti-Nb-Rich Ti-Zr-Nb

HORIUCHI *et al* were the first to recognize the advantages which derive from the substitution of 5 at.% Ta and 10 at.% Ta, respectively, for Nb in Ti-Nb-rich Ti-Zr-Nb alloys such as, for example, Ti_{55}-Zr_{15}-Nb_{30}. In their development program, the superconductor group at the Asada Research Laboratory of Kobe Steel Ltd. sought an alloy with the best combination of superconductive and metallurgical properties (including workability and tensile strength of the final product). The result was Ti_{61}-Zr_6-Nb_{27}-Ta_6, which under the name "cryozitt" appeared on the market in 1978. The first conductors to become available commercially possessed up to 361 filaments each of diameter about 50 μm. Conductor outside dimensions were up to 2.0 mm$^\phi$ or 1.3x2.6 mm^2, with Cu/SC ratios of either 2.5:1 or 4.0:1 [Kob78]. By 1980, the product line had been extended to include a 3721-filament conductor of outside dimensions 1.5x3.0 mm^2 with a 5-T current-carrying capacity of 1800 A [Kob80].

As a measure of the workability of the quaternary alloys, it was claimed that drawing to an area reduction of 99.9999% was possible without any intermediate annealing [Hor81]. Thus filament diameters as small as 2-5 μm were supposed to be achievable enabling conductors of up to 200,000 filaments to be easily fabricated. In general, unless a special application[†] demanded a very small filament diameter, such large area reductions by cold work were found to be neither necessary nor desirable for this particular alloy. Initially the multifilamentary wire was prepared only by drawing; in a typical production process a 7.4-mm$^\phi$ alloy rod would be jacketed in a Cu tube of O.D. 9.5 to 12.1 mm, depending on the Cu/SC ratio required, and cold drawn using in succession a 100-ton draw bench and a 45-ton draw bench. The single-core product would then be rebundled, again inserted into a Cu tube, and the drawing process repeated. More recently, multifilamentary rods have been produced by one or two stages of conventional extrusion followed by cold drawing. In the first stage of this process single-core billets are reduced from 230 to 76 ~ 50mm$^\phi$ using a 3000-ton extrusion press. The product is then transferred to the draw benches prior to rebundling, jacketing and extrusion as before. The reduction ratios achievable by the conventional extrusion just referred to are obviously within the range 9:1 to 21:1. In the subsequent rod and wire drawing operations, after proper selection of die angle, lubricant, and drawing speed, it was found possible to achieve area reductions of 25 to 15% in each pass while at the same time maintaining uniform deformation and avoiding centerbursting.

It is planned to take advantage of the relatively low cold-work requirements of the quaternary alloy (see below) coupled with the large area-reduction-ratio capabilities of hydrostatic extrusion to eliminate a large number of drawing steps and by so doing to lower production costs [Hor80^a].

13.11 PROPERTIES OF THE TECHNICAL QUATERNARY ALLOY Ti$_{61}$-Zr$_6$-Nb$_{27}$-Ta$_6$

This discussion of quaternary-TM alloys concludes with a review of a series of detailed studies, recently completed, of stress effects, process optimization, and flux pinning in the important technical superconductor Ti$_{61}$-Zr$_6$-Nb$_{27}$-Ta$_6$, whose preparation and proper-

---

† To reduce hysteretic loss in AC conductors, small
  filament diameters are required [Wil70][Mon27.10].

---

ties were introduced in the previous section. The geometries and specifications of four representative quaternary-alloy-base multifilamentary conductors are presented in Fig.13-7 and Table 13-6, respectively.

13.11.1 Stress Effects

As indicated above, superconductive materials development must have as its goal the optimization of both electrical *and* mechanical properties. With regard to the latter, HORIUCHI *et al* [Kob80] have measured 4.2-K tensile strengths, as functions of Cu/SC ratio, of conductors of various external diameters. In the results of these tests strict adherence to the rule of mixtures was noticeable. Critical current degradation in response to static stress (with repeated loading) [Kob78, Kob80, Kob80^a][Hor80] and dynamic stress [Kob78, Kob80, Kob80^a] were also investigated. Stress effects with particular reference to Cu/Ti-Nb composites have been discussed at length in [Mon29.00]. As for Cu-clad Ti$_{61}$-Zr$_6$-Nb$_{27}$-Ta$_6$ composites, HORIUCHI *et al* noted that under static stresses of 30%, 50% and 70% of $\sigma_B{}^†$, the 4.2-K J$_c$ degradations experienced by a 61-filament, 2:1 Cu/SC-ratio, conductor were 0, 2 and 5%, respectively, and furthermore that at a stress level of 64% of $\sigma_{B,4.2K}$, no increase in degradation of the critical current (i.e. no deterioration of current-carrying capacity) was detected during repeated applications (of up to at least 100 times) of the static load. In dynamic tensile tests of a 61-filament 0.9:1 Cu/SC-ratio conductor, at various rates of from 0.1 to 2.0 cm min^{-1}, the 5-T J$_c$ never dropped below about 0.98 of its unstressed value provided that the stress level was kept below 70% of $\sigma_{B,4.2K}$.

13.11.2 Optimization Studies

During the development of the quaternary Ti-Zr-Nb-Ta superconductor, numerous studies were undertaken of the effects on J$_c$ of: *(a)* various levels of total cold reduction; *(b)* aging time at a given temperature prior to final cold work; *(c)* various levels of final cold work after cold reduction and aging [Kob80^a][Hor80^a].

---

† σ_B (4.2 K) for the 61-filament, Cu/SC = 2:1,
  conductor is 10.8$_5$x10^8 N m^{-2}.

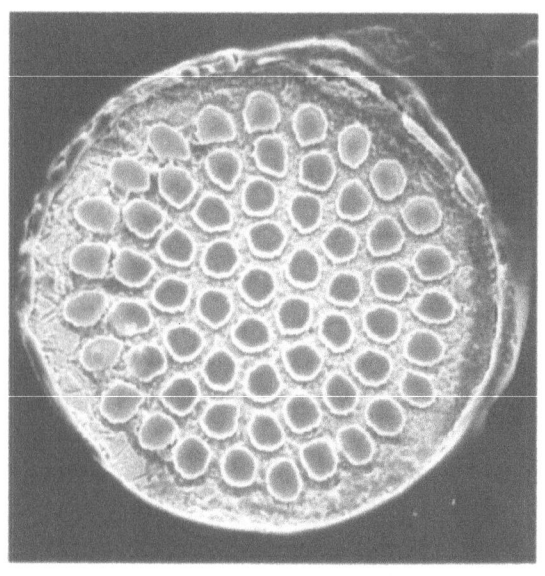
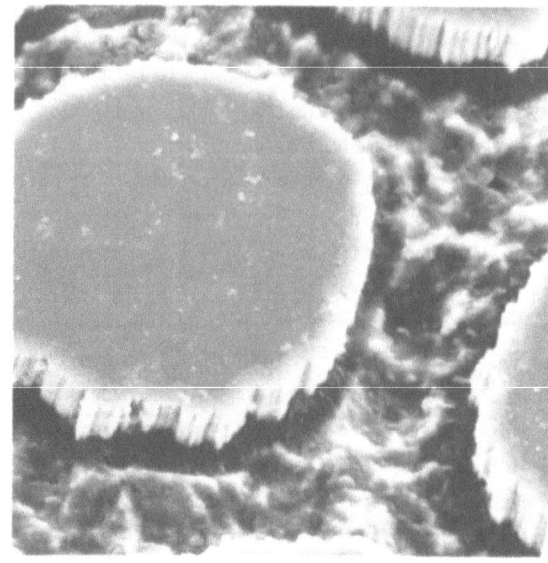

(a)

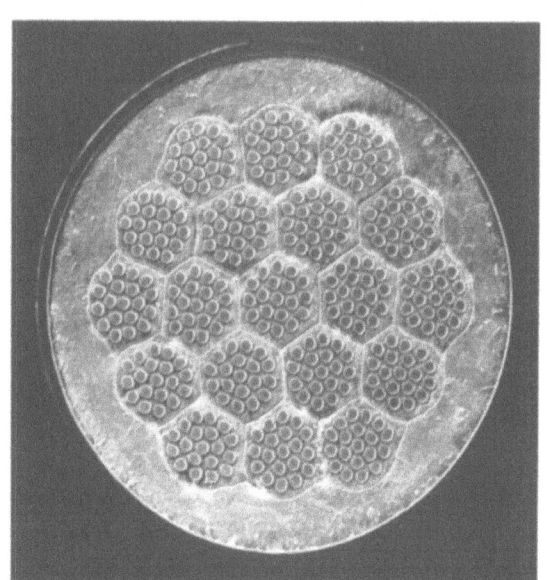
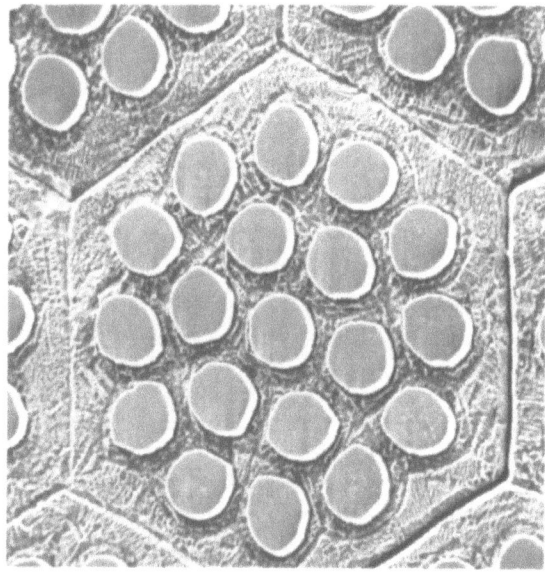

(b)

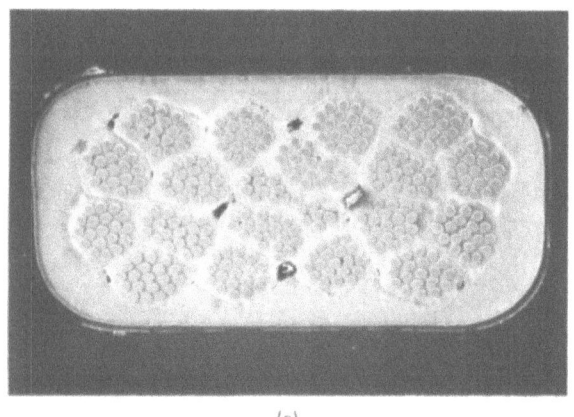

(c)

FIGURE 13-7. Structures of the quaternary-alloy-base composites listed in Table 13-6: (a) a 61-filament composite [Kob78]; (b) a 361-filament composite intended for a ship-propulsion magnet; (c) another 361-filament composite; (d) a 3721-filament composite, intended for an energy-storage magnet, before and after cold drawing to a rectangular cross-section. Conductor samples and one diagram from group-(d) supplied by T. Horiuchi, Kobe Steel, Ltd.; samples mounted by R. D. Smith, polished and etched (40% HNO_3) by K. L. Hammond, and SEM microphotographed by C. R. Barnes and A. Skidmore, all of Battelle.

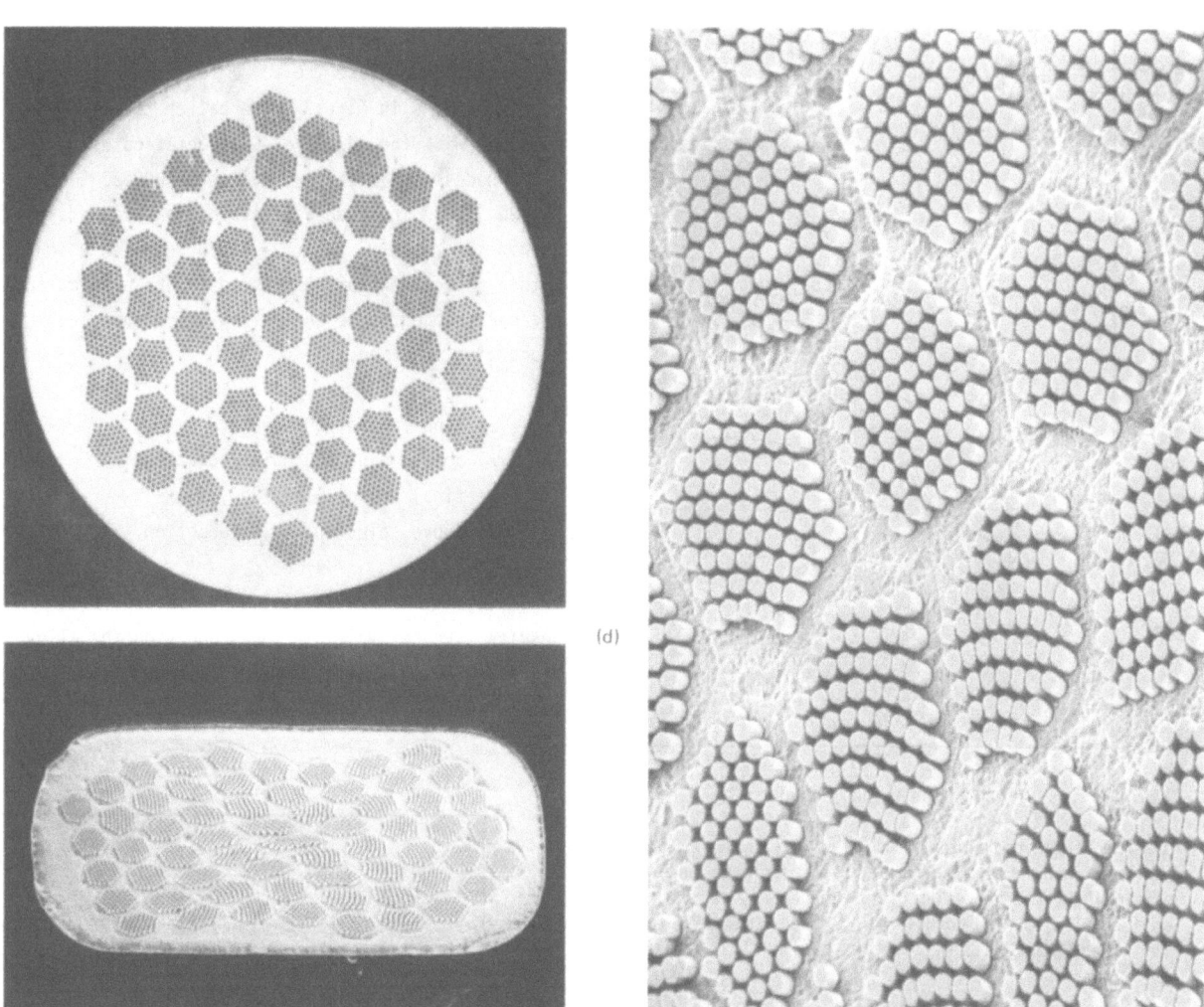

(d)

TABLE 13-6 SPECIFICATIONS[†] OF THE QUATERNARY ALLOY SUPERCONDUCTORS OF FIGURE 13-7

Fig. Code	Conductor O.D. mm	No. of Fils.	Fil. Diam. μm	Cu/SC Ratio	Twist Pitch mm	Critical Current Density 10^5 A cm^{-2}				5-T Critical Current[††] A
						3T	4T	5T	6T	
(a)	0.37	61	27	2.0	10	3.3	2.9	2.4	1.8	86
(b)	1.8	19x19	43	3.8	35	2.9	2.4	1.8	1.3	930
(c)	1.6x3.1[†††]	19x19	60	3.2	45	2.5	2.0	1.6	1.2	1600
(d)	1.5x3.0[†††*]	61x61	18	3.2	35	3.2	2.7	2.1	1.4	2100

† Most of the data supplied for this table by T. Horiuchi [Hor81].

†† I_c is defined as the current associated with a sample resistivity of $1x10^{-11}$ Ω cm.

††† Includes Formvar coating.

†††* Also depicted in Fig.13-7(d), upper left, is a partially processed conductor; diam., 30.5 mm.

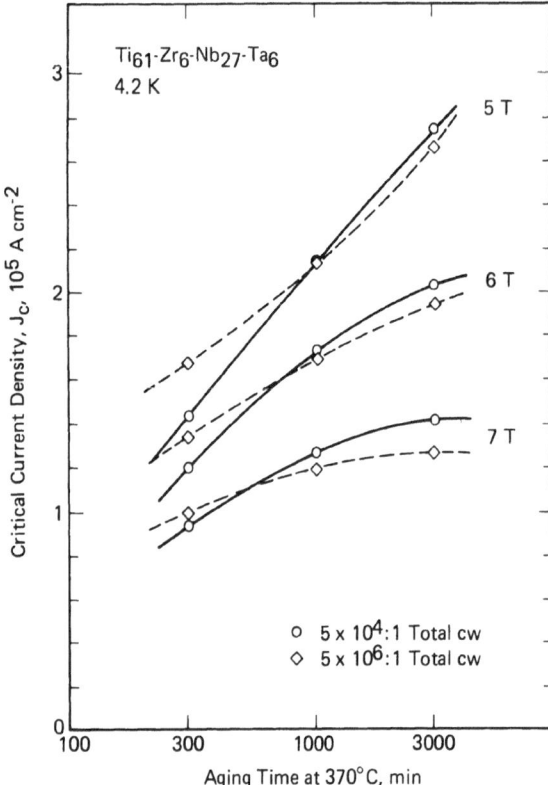

FIGURE 13-8. Critical current densities of $Ti_{61}\text{-}Zr_6\text{-}Nb_{27}\text{-}Ta_6$ at 4.2 K as functions of aging time at 370°C for three values of magnetic field, and two values of total cold-work area reduction (criterion: 1×10^{-11} Ω cm [Hor81]) – after HORIUCHI *et al* [Hor80] and [Kob80^a].

Fig.13-8 depicts the critical current densities of alloys which have been subjected to two levels of total cold work (99.998% and 99.99998%, respectively) as functions of aging time at 370°C [Kob80^a]. The results indicate that what is judged to be an optimal aging time depends on the field to which the conductor is exposed during the critical current measurement. Higher fields call for shorter aging times; thus in process scheduling, the choice of aging time must be made after the conductor's design-field has been selected. The figure shows that for *these* alloys: *(a)* the lower level of cold work yields the higher critical current densities in strong fields, *(b)* J_c optimization with respect to fields such as 7 T necessitates 370°C heat-treatment times of at least 50 h.

Fig.13-9 emphasizes the importance of post-heat-treatment cold reduction on the critical current density. For high-field optimization of this alloy the figure indicates that at least 47% final cold work is needed [Kob80^a]. Figure 5 of [Hor80] suggested

that the 7-T optimal final cold reduction was about 60%, but in that case the *total* cold work was not specified. As in Fig.13-8 the better results were obtained with the lower level of total cold work.

Process optimization studies by HORIUCHI *et al* [Kob80^a] also included a series of measurements of $J_c(4.2$ K) *versus* H_a (up to 7 T) for samples which, within the context of 99.998% total area reduction, had been subjected to the following set of final processing treatments:

(a) final aging, 50h/340°C,
(b) above, plus 40% cold reduction,
(c) final aging, 50h/370°C,
(d) above, plus 40% cold reduction.

Of these, treatment-*(d)* yielded by far the best results. For example, in that condition the alloy possessed 4.2-K J_c's of 2.7×10^5, 2.0×10^5, and 1.4×10^5 A cm^{-2} in fields of 5, 6 and 7 T, respectively.

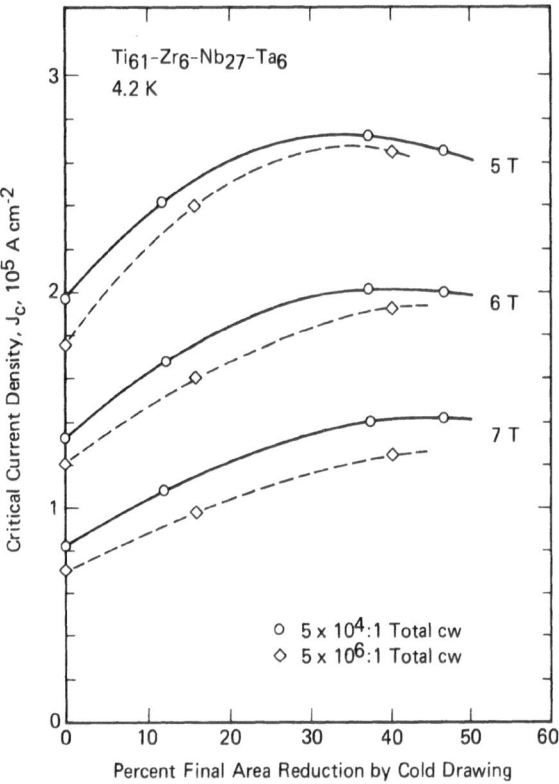

FIGURE 13-9. Critical current densities of $Ti_{61}\text{-}Zr_6\text{-}Nb_{27}\text{-}Ta_6$ at 4.2 K as functions of final area reduction by cold drawing following an aging heat treatment of 50 h/370°C, for three values of the applied magnetic field and two values of total cold-work area reduction (criterion: 1×10^{-11} Ω cm [Hor81]) – after HORIUCHI *et al* [Hor80] and [Kob80^a].

Of course in any new alloy development program it is necessary to compare performance data with those of pre-existing materials. In evaluating the total cold-work requirements of the quaternary alloy, $Ti-Nb_{37}$ was taken as the reference. Fig.13-10 intercompares the 4.2-K, 7-T critical current densities of $Ti_{61}-Zr_6-Nb_{27}-Ta_6$ and $Ti-Nb_{37}$ as functions of total cold work [Kob80[a]]. It turns out that in the quaternary alloy, J_c develops its maximum value at lower levels of cold work and remains fairly constant thereafter, while in $Ti-Nb_{37}$, J_c is slow to approach its maximum value which it attains only after relatively heavy cold work. HORIUCHI et al [Kob80[a]] have intercompared the thermomechanical processing requirements of $Ti-Nb_{37}$ and $Ti_{61}-Zr_6-Nb_{27}-Ta_6$ as they apply to the attainment of a 4.2-K J_c of 1.2×10^7 A cm^{-2} at 7 T. Their conclusions, as outlined in Table 13-7, can be summarized simply by pointing out that the quaternary alloy requires much less cold work and a much shorter heat-treatment time than does the binary reference.

Comparisons between the performances of $Ti_{61}-Zr_6-Nb_{27}-Ta_6$ and several commercial binary $Ti-Nb_{31-37}$ alloys have been made by HAWKSWORTH and LARBALESTIER [Haw80[a]]. Their results for 4.2 K and $\sim$2 K are depicted in Fig.13-11, from which it can be concluded that although the 4.2-K J_c of the quaternary alloy is inferior to that of $Ti-Nb_{34}$ throughout the entire applied-field range, and to that of $Ti-Nb_{37}$ above about 7 T, at 2 K and in fields of less than about 8 T, $Ti_{61}-Zr_6-Nb_{27}-Ta_6$ is superior to both $Ti-Nb_{34}$ (i.e. Nb-50.5Ti) and $Ti-Nb_{37}$ (i.e. Nb-46.5Ti).

TABLE 13-7   THERMOMECHANICAL PROCESSING FOR OBTAINING 4.2-K, 7-T, J_c's OF ABOUT 1.2×10^5 A cm^{-2} [Kob80[a]]

Alloy	Cold Area Reduction	Aging Time at 370°C
$Ti-Nb_{37}$	99.9997% (164 mm $\rightarrow$ 0.3 mm$^\phi$)	50 $\sim$ 150 h
$Ti_{61}-Zr_6-Nb_{27}-Ta_6$	99.98% (21 $\rightarrow$ 0.3 mm$^\phi$)	15 $\sim$ 50 h

13.11.3 Flux Pinning and the Scaling Laws

(a) Flux Pinning. As considered in detail elsewhere (cf. Sect.7.30 and [Mon21.13] et seq), in the presence of a transverse magnetic field, transport current relies for its support on the pinning of that field by a suitable arrangement of lattice defects and precipitate particles -- so-called subbands and precipitates.

In comparative studies of the binary and quaternary alloys considered above, HORIUCHI et al [Hor80] (see also [Osa80]) used TEM and low-angle x-ray diffraction to monitor the influence of aging for various times at 380°C on the size and distribution of the precipitate particles. Data from the diffraction experiment, in the form of Guinier radii (a measure of precipitate size) and relative interparticle distances, were plotted versus aging times of up to 565 h for alloy wires of several external diameters; i.e. wires which had been subjected to various amounts of previous cold work. The area reductions administered (as percentages and as ratios) and the corresponding filament diameters for the two classes of alloy studied were:

$Ti_{61}-Zr_6-Nb_{47}-Ta_6$: 99.994% ($1.67 \times 10^4$:1)   40 μm$^\phi$

$Ti-Nb_{36}$: 99.96 % (2.50×10^3:1) 110 μm$^\phi$
99.986% (7.14×10^3:1)   60 μm$^\phi$
99.994% (1.67×10^4:1)   40 μm$^\phi$

In these experiments, the precipitate particle sizes

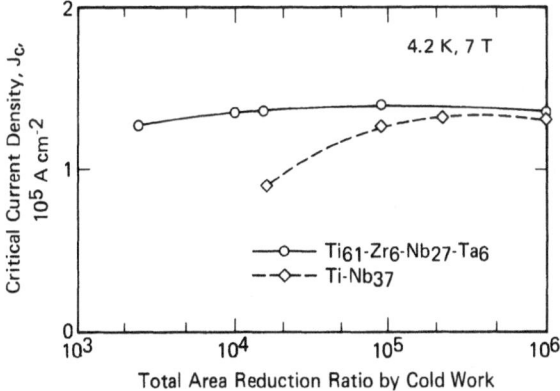

FIGURE 13-10. Critical current densities at 4.2 K and 7 T as functions of total area reduction by cold work; a comparison of the properties of $Ti_{61}-Zr_6-Nb_{27}-Ta_6$ and Ti-Nb (37 at. %) (criterion: 1×10^{-11} Ω cm [Hor81]) — after [Kob80[a]].

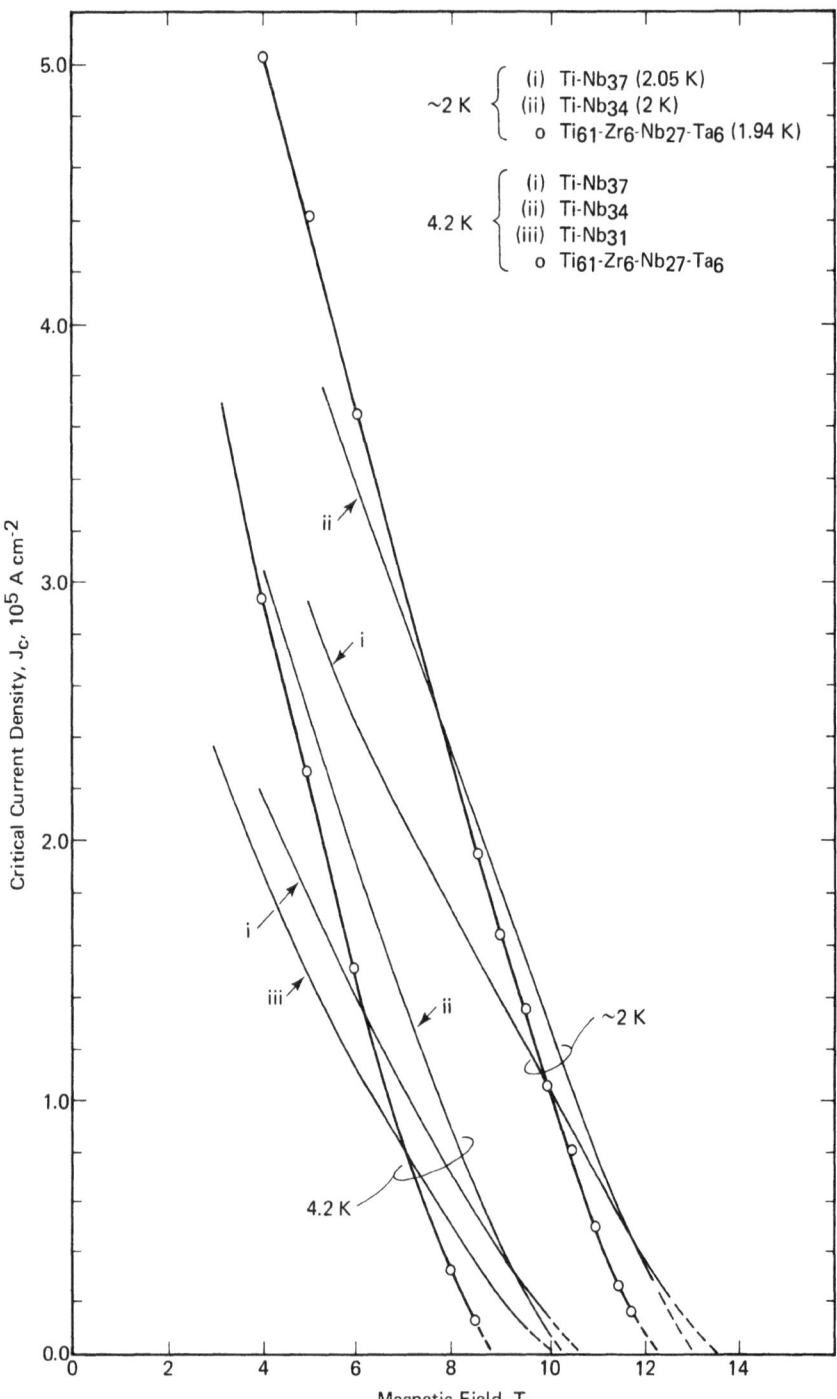

FIGURE 13-11. Intercomparison of the critical current densities of Ti_{61}-Zr_6-Nb_{27}-Ta_6 and Ti-Nb composite conductors at 4.2 K and specified temperatures near 2 K (criterion: 10^{-12} Ω cm) — after HAWKSWORTH and LARBALESTIER [Haw80a].

increased most rapidly to saturation in the most heavily cold-worked wires, in accordance with expectations (cf. [Mon2.11]). The saturation diameters of the particles, identified as α-phase by TEM, were slightly larger in the quaternary alloy (75 Å) than in the binary (65 Å). Nevertheless, in both cases the precipitates must be regarded as "very fine" in that their diameters were of the order of a coherence length (cf. [Mon11.1]). As compared with Ti-Nb_{36} (40 μm^ϕ and 110 μm^ϕ), the quaternary alloy possessed the smallest relative interparticle spacing; its excellent critical current performance was attributed to the high density of very fine α-phase precipitate particles.

(b) The "Scaling-Law" Representation of $J_c(T)$ versus H_a Data.

In studies of the theory of flux pinning it has been noted that the bulk pinning force $F_p = J_c H_a$ can be expressed in the form

$$F_p \propto H_{c2}^m(t)\, f[h(t)] \quad , \qquad (13\text{-}1)$$

where $t \equiv T/T_c$, $h(t) \equiv H_a/H_{c2}(t)$, and m is a whole- or half-integer. A typical example would be

$$F_p \propto H_{c2}^{3/2}(t)\, h^{1/2}(1-h) \quad . \qquad (13\text{-}2)$$

By evaluating $H_{c2}^m(t)$ at any particular value of h (say h_1, but usually h_{max}) it can be eliminated from the proportionality, leading in general to

$$F_p/F_{p,1} = f(h)/f(h_1) \quad , \qquad (13\text{-}3)$$

where the RHS represents a unique curve for a given specimen. The existence of such a "scaling relationship" as it is called, cf. [Mon21.10], relies on:
(a) the fact that the only polynomial factor in the pinning force function is $(H_{c2}-H_a)$ (or perhaps some power of it) enabling H_{c2} to be brought to the outside; *(b)* the axiom that any function f(x) plotted *versus* x is independent of whatever x itself is a function of.

Whether or not the form of the pinning function, $F_p(h,t)$, has fundamental significance as an indicator of the nature of the dominant flux-pinning mechanism operative within a given alloy sample, the existence of the scaling relationship that has been shown to derive from it has been claimed to be useful from a design-engineering standpoint [Lar80]. An excellent example of the operation of the scaling principle is provided by Fig.13-12, from the work of HAWKSWORTH and LARBALESTIER [Haw80[a]]. A design engineer needs to know the J_c of a conductor as a function of H_a and T. The scaling principle, when it holds, eliminates the need for complete sets of two-dimensional $J_c(H_a)$ curves or a three-dimensional diagram. It enables

the J_c-H_a-T data to be embodied within a universal curve such as Fig.13-12 if augmented by information on the temperature dependence of $F_{p,max}$ (obtained from J_c *versus* H_a data near $H_r/2$) and the temperature dependence of H_r [Lar80].

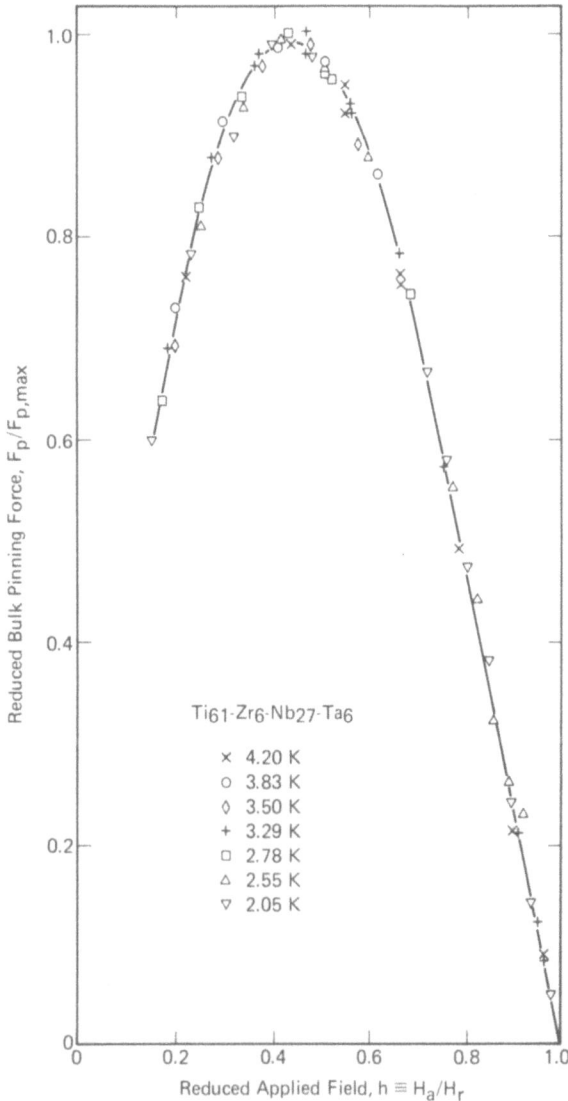

FIGURE 3-12. A plot of normalized pinning force $F_p/F_{p,max}$ (where $F_p \equiv J_cH_a$) *versus* reduced applied field, $h \equiv H_a/H_r$ in this case, for the alloy Ti_{61}-Zr_6-Nb_{27}-Ta_6 at the seven temperatures listed. An excellent example of pinning-force "scaling" provided by HAWKSWORTH and LARBALESTIER [Haw80[a]].

14

AMORPHOUS TITANIUM ALLOY SUPERCONDUCTORS

Although studies of amorphous superconducting alloys have over the years been conducted in the spirit of pure scientific inquiry, the goal of some of the more recent investigations has been the economical production of continuous lengths of superconducting ribbon possessing ductility and strength superior to those of existing technical superconductors and at least comparable superconducting properties. It should be stated at the outset that, as yet, these goals have not been achieved and the subject is under active investigation. In the first systematic study of superconductivity in amorphous metals, that conducted in the early 1950's by BUCKEL and HILSCH Table 14-1, in which the simple metals Al, Zn, In, Sn, Hg, Tl, Pb, and the alloys Sn-Bi(0-100%) were vacuum deposited onto cryocooled substrates, it was noted that all amorphous films except Pb, for which the ratio $T_{c,am}/T_{c,cryst}$ equalled 1.00 and Hg for which it equalled 0.94, possessed T_c's higher than those of the bulk. The superconducting properties of amorphous 4d and 5d TM elements and alloys remained unknown until 1973 when the work of COLLVER and HAMMOND [Coℓ73[b]] revealed that their T_c's could be either greater than or less than those of the corresponding crystalline TM elements or binary alloys depending on their "positions" (expressed in terms of the usual electron/atom ratio) along the 4d and 5d rows, respectively, of the periodic table. On the basis of this work, although the prognosis for useful superconducting amorphous alloys with electron/atom

ratios less than about 5 (that of Nb) was not good, the future of alloys based on Mo appeared to be quite promising. The initial studies of amorphous superconductivity were generally carried out on cryodeposited films which usually crystallized upon warming to room temperature. Stabilization by some chemical means is necessary if the amorphous structure is to exist at room temperature and above. This is the basis for the metallic-glass group of amorphous alloys which are generally prepared by rapid liquid quenching onto a room-temperature substrate. The first results from a metallic-glass superconductor prepared in this way were reported in 1975 by JOHNSON *et al*, Table 14-1.

The chapter commences with a summary in tabular form of the recent literature of amorphous and glassy metal superconductivity, after which some representative T_c *versus* e/a curves for crystalline and amorphous transition metal alloys are described and briefly discussed. Ti alloys had not been studied by COLLVER and HAMMOND [Coℓ73[b]] and in fact seem to have remained neglected until quite recently when their properties in metallic-glass-ribbon form were investigated by MASUMOTO and colleagues at Tôhoku University. The recent results of this group are the basis for the present review. Accordingly the chapter continues with a description of the influence of metallurgical factors such as cold work, aging, and crystallization on T_c, and concludes with a discussion of the magnetic-field dependence of J_c and the influence of heat-treatment-induced precipitation on it.

TABLE 14-1 INTRODUCTION TO SUPERCONDUCTIVITY IN AMORPHOUS METALS AND ALLOYS

Element or Alloy	Solidification Process	Superconductive Property Measured	Literature
Al, Zn, In, Sn, Hg, Tl, Pb	Vapor deposited onto 4K substrate	T_c	W.Buckel and R.Hilsch Z.Phys. <u>138</u> 109 (1954)
Sn-Bi	Vapor deposited onto 4K substrate	T_c	W.Buckel and R.Hilsch Z.Phys. <u>146</u> 27 (1956)
4d and 5d TM_1-TM_2 alloys	Vapor deposited onto 4.2K substrate	T_c	M.M.Collver and R.H.Hammond Phys.Rev.Lett. <u>30</u> 92 (1973)
La-Au	Melt-quenched splat	T_c, $H_{c_2}(T)$	W.L.Johnson et al. Phys.Rev. B, <u>11</u> 150 (1975)
Au-La Nb-Ni Nb-Rh Pd-Zr	Melt-quenched splat	T_c, $H_{c_2}(T)$	W.L.Johnson and S.J.Poon J.Appl.Phys. <u>46</u> 1787 (1975)
Zr-Rh	Melt-quenched splat	T_c	K.Togano and K.Tachikawa J.Appl.Phys. <u>46</u> 3609 (1975)
Zr-Pd	Melt-quenched ribbon	T_c	J.E.Graebner et al. Phys.Rev.Lett. <u>39</u> 1480 (1977)
Be-Zr	Melt-quenched ribbon	T_c, $H_{c_2}(T)$	R.Hasegawa and L.E.Tanner Phys.Rev B, <u>16</u> 3925 (1977)
La-Ga	Melt-quenched splat	T_c, $H_{c_2}(T)$	W.H.Shull et al. Phys.Rev.B, <u>18</u> 3263 (1978)
Zr-Rh La-Au	Melt-quenched splat	T_c, $H_{c_2}(T)$	W.L.Johnson et al. Phys.Rev.B, <u>17</u> 2884 (1978)
$(Mo-Ru)_{80}$-P_{20} $(Mo-Re)_{80}$-P_{10}-B_{10} Mo_{80}-P_{10}-B_{10}	Melt-quenched splat	T_c, $H_{c_2}(T)$	W.L.Johnson et al. Phys.Rev.B, <u>18</u> 206 (1978)
Zr-Rh $(Mo-Ru)_{80}$-P_{20} Mo_{80}-P_{10}-B_{10}	Melt-quenched splat	T_c, $H_{c_1}(T)$, $H_{c_2}(T)$	E.R.Domb and W.L.Johnson J. Low Temp.Phys. <u>33</u> 29 (1978)
Mo-Si-B W-Si-B	Melt-quenched ribbon	T_c	A.Inoue et al. Scripta Met. <u>14</u> 235 (1980)
Nb-Si Nb_{80-x}-X_x-Si_{20} Nb_{80}-Si_{20-x}-Y_x with X = Zr,V,Ta,Mo,W Y = C,B,Ge	Melt-quenched ribbon	T_c, $H_{c_2}(T)$, J_c	T.Masumoto et al. Trans. JIM <u>21</u> 115 (1980)

ALLOY GROUP 1: AMORPHOUS AND GLASSY METALS

14.1 STABILITY AND PROPERTIES OF AMORPHOUS ALLOYS

Whether intrinsically stable, or stabilized in the presence of interstitial oxygen from residual air in the vacuum enclosure (cf. [Col73[b]] and [Joh79]), amorphous elemental metallic films prepared by vacuum deposition are usually only cryostable, and generally crystallize upon warming up to room temperature. On the other hand metallic glasses, although they may exhibit aging effects over large periods of time (many months) at room temperature, are usually amorphous-stable to several hundreds of degrees celsius. Metallic glasses, unlike amorphous elements, share with oxide glasses a glass transition followed (with increasing temperature) by crystallization, both of which are detectable calorimetrically through the technique of differential thermal analysis (DTA).

In selecting candidate compositions for possible metallic-glass stabilization it is usual to take advantage of one or other of the following three empirical principles:

(i) The presence of a deep eutectic in the phase diagram, indicative of a large negative heat of formation of the liquid alloy, is considered to be an indicator of compositional conditions favorable to metallic glass existence.

(ii) In addition to the foregoing, and not necessarily exclusive of it, it is possible to group elements into classes in such a way that metallic glass stability is achievable in binary systems composed of elements selected one from each of almost any pair of such classes. Fig.14-1 serves to illustrate this statement, and shows that metallic-glass candidate alloys fall into the following categories (see caption, Fig.14-1):

(a)	RE/ETM - Nbl.M.	e.g. La-Au
(b)	ETM - LTM	e.g. Zr-Rh,Pd
(c)	LTM - Mtld.	e.g. Ru-P
(d)	Nbl.M. - Mtld.	e.g. Au-Si
(e)	RE - LTM	e.g. Sm-Co

(iii) Finally, and again in overlap with the above, "metalloids" such as P and Si seem eminently capable of filling the holes in the Bernal random-close-packed structure and thereby stabilizing the amorphous phase of the host. The commonly encountered "80-20" metal-

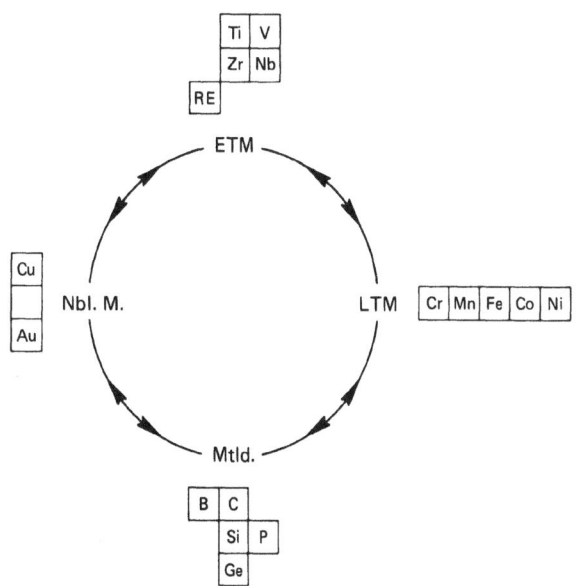

FIGURE 14-1. Cyclic diagram for the selection of metallic-glass binary alloy constituents. RE = rare earth metal, ETM = early transition metal, LTM = late transition metal, Mtld = metalloid, Nbl.M. = noble metal.

metalloid combinations (such as Fe_{80}-$(P-C)_{20}$ and so on) are a result of this design philosophy.

In turning to metallic glasses as candidates for technical superconducting alloys, it has been hoped that advantage could be taken of their following general properties: *(a)* the readiness with which metallic glasses can be "spun" into continuous ribbons directly from the melt; *(b)* the high tensile strengths of the resulting ribbons, which are usually greater than those of the parent crystalline alloys, coupled with their excellent bend ductilities (e.g. they can be permanently folded through 180°); *(c)* the excellent short-range chemical homogeneities of the as-quenched ribbons which, even if they are not ultimately required in glassy form are amenable to fine-grain crystallization and the support of an even distribution of fine precipitate particles.

14.2 AMORPHOUS AND GLASSY ALLOY SUPERCONDUCTORS

Table 14-1 provides an introduction to the recent literature of superconductivity in amorphous and glassy alloys and indicates that the amorphous Ti-Nb-Si alloys whose superconducting properties are the basis of this chapter, far from being isolated curiosities, form an integral part of a continuous series of investigations into the properties and technical potentialities of superconducting metallic-glass ribbon. By so doing, Table 14-1 serves as the staging point for the discussions to follow. This is not the place to review the general subject of superconductivity in amorphous simple-metal and transition-metal alloys, which in any case has already been accomplished by BERGMANN [Ber76] (simple metals) and JOHNSON [Joh79] (transition metals). In dealing with the systematics of glassy-TM-alloy superconductivity, reference is always made to the, by now, classical experiments of COLLVER and HAMMOND [Col73[b]] who studied the T_c's of vapor-cryodeposited films of 4d- and 5d-binary-TM alloys and came to the remarkable conclusion that T_c *versus* electron/atom ratio no longer exhibited the double-peaked curve characteristic of crystalline binary-TM alloys but instead increased and decreased again, with increasing e/a, in a rather structureless way with maxima situated at e/a(4d) = 6.4 and e/a(5d) = 6.8. The 4d- results are illustrated somewhat schematically in Fig.14-2. The purpose of drawing attention to them is because these curves establish a kind of standard, or benchmark, against which the transition temperatures of other amorphous and subsequently recrystallized alloys may be compared. Reference to the curves of Fig.14-2 would not be complete without at least a brief comment on the origins of the two distinctly different T_c *versus* e/a characteristics, full treatment of which would of course be a discussion of the microscopic mechanisms of superconductivity along the lines of the recent reformulation by ALLEN and DYNES [All75] of the McMILLAN theory [Mcm68] as described, for example, in [Mon12.2].

14.3 TRANSITION TEMPERATURES OF AMORPHOUS SUPERCONDUCTORS

According to the strong-coupling theory of McMILLAN [Mcm68], which is perfectly adequate for the present purpose:

$$T_c = \frac{<\omega^2>^{1/2}}{1.20} \exp\left[\frac{-1.04(1+\lambda)}{\lambda - \mu^*(1+0.62\lambda)}\right] \quad [2\text{-}1] \ (14\text{-}1)$$

where $\mu^* = 0.1$ is a Coulomb pseudopotential for

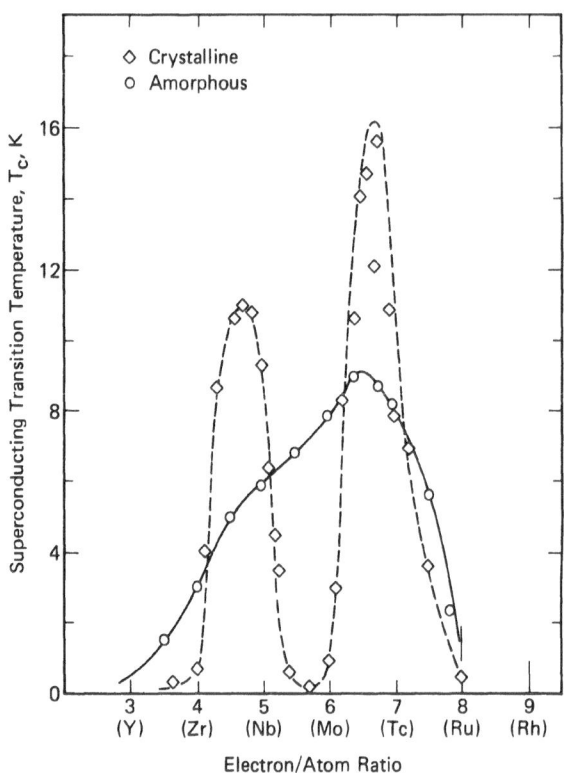

FIGURE 14-2. Superconducting transition temperature (T_c,K) plotted as function of average group number or electron/atom ratio for 4d-transition metals and their alloys in the crystalline form ($\diamond$) (cf. Fig. 2-2) and as amorphous vapor-cryodeposited films ($\circ$) (COLLVER and HAMMOND [Col73[b]]) — after JOHNSON [Joh79].

electron-electron interaction, $<\omega^2>^{1/2}$ is the root-mean square phonon "frequency" (in the form of a temperature), and $\lambda(\stackrel{\sim}{<}1.5)$ is the dimensionless electron-phonon coupling constant which has been shown to be rigorously represented by

$$\lambda = \frac{n(E_F)<I^2>}{M<\omega^2>} \quad , \quad (14\text{-}2)$$

where $n(E_F)$ is the Fermi density-of-states, $<I^2>$ is the average over the Fermi surface of the electron-ion (phonon) matrix elements, and M is the atomic mass. For a full discussion of superconducting transition temperature involving the use of these and the following equations [Mon12.00] is recommended.

Initially, McMILLAN recognized that for a series of bcc transition metals $n(E_F)<I^2>$ was practically constant, and suggested that the variation of λ (hence T_c) along a series of transition metals and their binary alloys was dominated by the variation of lattice stiffness (as expressed *via* the $M<\omega^2>$ factor).

Subsequently, HOPFIELD [Hop71] recast Eqn.(14-2) in the form:

$$\lambda = \frac{\eta}{M<\omega^2>} \qquad (14\text{-}3)$$

where η was defined as a parameter descriptive of *individual atomic* rather than band properties. Comparable approaches were also adopted by BARAŠIC' *et al* [Bar70[a]] and BENNEMANN and GARLAND [Ben72]. The fact that η was supposed to vary smoothly along an alloy series as did the T_c *versus* e/a data of COLLVER and HAMMOND as depicted in Fig.14-2 led the authors to suggest that they had detected evidence in support of the existence and variation of some "atomic-like" local parameter. However, during the early 1970's when studies such as COLLVER-HAMMOND were conducted, the sharing of the responsibility for T_c variation between $n(E_F)$ and $M<\omega^2>$ was still in an unsatisfactory state; as a matter of fact at the very beginning, McMILLAN, while a proponent of the $M<\omega^2>$ approach, had already admitted that elastic softness was a consequence of a high density-of-states. It remained for VARMA and DYNES [Var76] to come to grips with the causality of the relationship between $n(E_F)$ and $M<\omega^2>$. They pointed out that for numerous A15 compounds, $\lambda/n(E_F)$ was constant within close limits in spite of strong compensatory individual variations in both λ and $n(E_F)$. A similar result for groups of transition-metal alloys had already been demonstrated by HULM and BLAUGHER [Huℓ72]. It followed from Eqn.(14-2) then that $<I^2>/M<\omega^2>$ was constant. Combining this with the initial observation of the approximate constancy of $n(E_F)<I^2>$ led immediately to

$$n(E_F) \stackrel{\sim}{\propto} (M<\omega^2>)^{-1} \qquad (14\text{-}4)$$

in accordance with: *(a)* the initial comment by McMILLAN; *(b)* the frequently observed inverse scaling between γ (the electronic specific heat coefficient)

and θ_D, both quantities obtainable from low-temperature-calorimetric data.

All three quantities on the right-hand side of Eqn.(14-2) are thus interrelated, and it becomes a matter of judgment aided by external evidence to decide which, if any, is to be the independent variable. VARMA and DYNES [Var76] decided that $n(E_F)$, as an electronic property of the equilibrium zero-K lattice, was the independent variable while the other quantities, being associated with perturbations of that lattice, were derivatives of the electronic structure.

Returning to the experimental curves of Fig.14-2, there is ample evidence in the form of curves of γ *versus* electron/atom ratio [Hei66] to support the conclusion that $T_{c,cryst}$ follows the variation of $n(E_F)_{cryst}$. By the same token we anticipate that $T_{c,am}$ is a copy of $n(E_F)_{am}$. It is not difficult to justify the existence of a featureless "single-band" $n(E_F)$ function for amorphous metals and alloys. An analogy could be drawn with liquid Ge which is metallic, and in which structural disorder has "washed out" the gap between the erstwhile valence and conduction bands. In a recent argument, bolstered by relevant experimental evidence, JOHNSON [Joh79] has pointed out that d-bands in amorphous transition metals should be comparable in width to those of the corresponding crystalline metals but devoid of structure on account of the lack of local structural order, consequently (the d-band contribution to) $n(E_F)$ should exhibit a single broad maximum.[+] It is believed that the amorphous curve" in Fig.14-2 is an example of this effect.

[+] It would be useful to have some other experimental density-of-states data, such as electronic specific heat and magnetic susceptibility results, to support this assertion (P. B. ALLEN [private communication, 1982]).

ALLOY GROUP 2: GLASSY TITANIUM ALLOYS

14.4 PHASE STABILITY AND MECHANICAL PROPERTIES OF GLASSY Ti-Nb-Si ALLOYS

Using a combination of the "eutectic trough" and "metal-metalloid" glassy alloy design philosophies referred to in Sect.14.1, MASUMOTO and his group at the Research Institute for Iron, Steel and Other Metals (Tôhoku University) selected Si as the metallic-glass stabilizer for Ti-Nb, and explored the glass stabilities of ternary alloys lying in the eutectic trough centered on a line connecting the Ti-Si$_{14}$ eutectic (temperature, 1313°C) to the Nb-Si$_{18}$ eutectic (temperature, 1943°C) [Ino80]. The composition ranges explored, whether or not they were glass-stable and, if so, their crystallization temperatures, are indicated in Fig.14-3. According to these studies, an amorphous phase without any traces of crystallinity (according to TEM) is stable in Ti-Nb-Si within compositional limits defined by: 0-43 at.% Nb, 13-21 at.% Si. Ternary Ti-Nb-Si compositions favored for study by the Tôhoku group were: Ti$_{85-x}$-Nb$_x$-Si$_{15}$ (x = 15, 25, 30, 35, 40) and Ti$_{56}$-Nb$_{30}$-Si$_{14}$ [Ino80, Ino80b][Mas80a]. Also prepared, for a brief investigation of their transition temperatures, were the related (isoelectronic) alloys Ti$_{85}$-V$_x$-Si$_{15}$ (x = 5, 10, 20, 30) [Mas80a]. Using as basis a ternary system with constant 30 at.% Nb, quaternary amorphous alloys with B, C, Ge and Mo were also prepared. The particular compositions selected for investigation were [Ino80a] [Mas80a]:

Ti$_{57}$-Nb$_{30}$-Si$_{10}$-B$_3$

(Ti-Nb)$_{85}$-Si$_{12}$-B$_3$

Ti$_{55}$-Nb$_{30}$-Si$_{15-x}$-M$_x$ (M = B, C, and Ge)

Ti$_{55-x}$-Nb$_{30}$-Si$_{15}$-M$_x$ (M = Mo, Ru, Rh, Pd, and Ir).

Finally with the B and Si levels fixed, the Ti:Nb ratio was varied in a study of the properties of (Ti-Nb)$_{85}$-Si$_{12}$-B$_3$ [Ino80a].

The mechanical properties of the glassy alloys achieved their anticipated superiority over those of the crystalline binary bases. Thus the Vickers hardness, H$_v$, of Ti$_{55}$-Nb$_{30}$-Si$_{15}$ was 580 kg mm^{-2} (57x10^8 N m^{-2}) while its ultimate tensile strength was 20.1x10^8 N m^{-2} [Ino80], values which may be com-

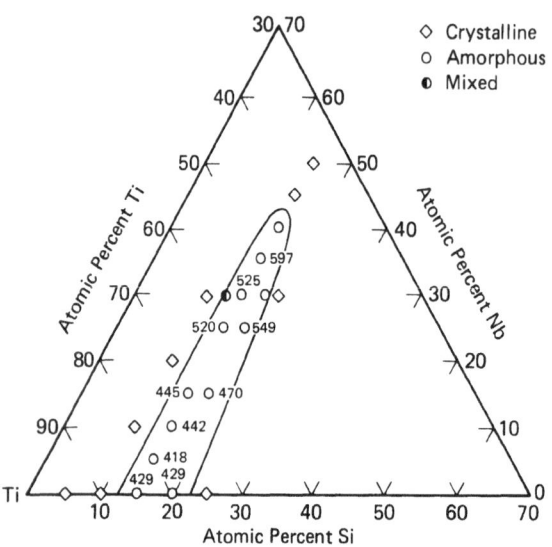

FIGURE 14-3. Composition range favorable to the formation of the amorphous phase in the Ti-Nb-Si system. The numerals indicate crystallization temperatures in °C — after INOUE et al [Ino80b].

pared with H$_v$ = 293 kg mm^{-2} (28.7x10^8 N m^{-2}) and a yield strength of 10.3x10^8 N m^{-2} for thermomechanically processed Ti-Nb$_{30}$ rod [Rea78] (cf. [Mon3.14]). The hardness-to-strength ratio of the metallic glass was 2.83, that of the crystalline alloy was 2.56, both indicative of the "non-plastic/elastic", high yield-strength/modulus, condition of Marsh-type hardness, cf. [Mon3.6]. The above hardness and strength levels were of course maintained in the B-containing quaternary alloys [Ino80a]. At the completion of a tensile test, fracture occurred at the usual angle of 45-55° to the tensile axis and the fracture surface carried (a) regions of smooth shear-slip and (b) the usual vein-like evidence for plastic-instability. As has come to be expected for melt-quenched ribbons, the Ti-Nb-Si alloys exhibited good 180°-bend ductility.

14.5 AGING AND CRYSTALLIZATION OF GLASSY Ti-Nb-Si ALLOYS

During room-temperature or moderate-temperature aging of metallic-glass alloys, various manifestations of "stress relaxation" or changes of the atomic pair correlation function are not uncommon. In monitoring the superconducting transition temperature of Ti-Nb-Si alloys (in particular Ti_{55}-Nb_{30}-Si_{15}) during 1-h aging at temperatures of up to 500°C [Ino80] continuing to 900°C [Mas80[a]], evidence was obtained for two different types of microstructural change: *(a)* an aging effect near 400°C, the immediate precursor to crystallization which approached completion more and more closely as the 1-h aging temperature approached 525°C, the crystallization temperature (cf. Fig.14-3); *(b)* crystallization itself.

14.5.1 Aging

A metallurgical explanation for the superconductive anomaly occurring near 400°C was sought with the aid of 200 kV TEM [Ino80]. It was reported that, although the structure remained amorphous during 1-h aging at that temperature, changes in the bright-field images from SAD patterns were observed, indicative of changes taking place in the "microscopic state of the amorphous phase".

14.5.2 Crystallization

During the crystallization of Ti-V-Si [Mas80[a]] and Ti-Nb-Si [Ino80][Mas80[a]] an interesting series of events took place, either as the sample's temperature was steadily increased with time as in DTA, or as it was isothermally aged at some elevated temperature. The types of metastable crystalline phases which first appeared can best be appreciated in the results of crystallization experiments on the low-Nb (e.g. 5 and 15 at.%) members of the Ti_{85-x}-Nb_x-Si_{15} series of glasses [Ino80[b]]: *(a)* During slow steady heating at say ∼0.1°C per sec, the first crystalline precipitate to form (>418°C, x = 5; >445°C, x = 15) is a supersaturated bcc solid solution; at a slightly higher temperature (>480°C, x = 5; >490°C, x = 15) the bct intermetallic compound Nb_3Si separates out. It can be seen that the precipitation of the bcc solid solution is deferred to higher temperatures as the Nb concentration increases. *(b)* During the isothermal aging of Ti_{70}-Nb_{15}-Si_{15} at 490°C, the first phase to crystallize out is the bcc solid solution, which after 10 min is a fine globular or elliptical precipitate of average diameter 150 Å. Subsequently in the remaining amorphous phase a transformation to bct Nb_3Si in the form of a heavily defected cylindrical precipitate of

average size 400 μm (four orders of magnitude larger than the previous precipitate) takes place. After 1 h at a higher aging temperature, say 560°C, the supersaturated bcc solid solution gives way to α-phase, still in association with bct Nb_3Si. As the Nb concentration increases, the stability of the supersaturated β-Ti-Nb also increases; thus in Ti_{50}-Nb_{35}-Si_{15} it is still present after 1-h aging at 650°C, 700°C and 750°C. On the other hand, the bct Nb_3Si which forms under these conditions is very much finer and appears as particles of diameter 500, 900 and 2000 Å, respectively, imbedded in the β-matrix [Ino80[b]]. The paper indicated that the precipitates were evenly dispersed rather than clustered; the spacing of the 9000 Å precipitates was about equal to their diameter.

In these detailed studies, as reported by INOUE *et al* [Ino80[b]], no A15 Nb_3Si was detected. In the Ti-V-Si series of alloys, according to MASUMOTO and INOUE [Mas80[a]], the intermetallic compound which precipitated out was hexagonal-Ti_5Si_3. The description of the aged product of Ti-Nb-Si as β-Ti-Nb plus Nb_3Si is of course an oversimplification, since not enough Nb is present in the entire alloy to accommodate all the Si present in the form of the stoichiometric compound. There seems moreover, in the light of the results for Ti-V-Si, no reason to exclude the possibility of Ti_5Si_3 precipitation during the aging of amorphous Ti-Nb-Si.

14.6 TRANSITION TEMPERATURES OF AMORPHOUS Ti ALLOYS

14.6.1 Composition Dependence in the Ternary Alloys

In the alloy system Ti_{85-x}-Nb_x-Si_{15} the transition temperature *increased* at the average rate of 0.33 K per 10 at.% Nb as x was raised from 15 to 40. In the system Ti_{70-x}-Nb_{30}-Si_x an increase in the Si level from x = 14 to 18 *reduced* T_c at the average rate of 0.75 K per 10 at.% Si [Ino80].

In Fig.14-4 the transition temperatures of amorphous Ti-Nb_x-Si_{15} (x = 15, 30, 40) plotted *versus* the conventional electron/atom ratios of the transition-metal components (equal, respectively, in this case to 4.15, 4.30 and 4.40) [Ino80[a]][Mas80[a]] display remarkably satisfactory agreement with the associated COLLVER-HAMMOND curve, whose validity with respect to metallic-glass alloys is further confirmed by the results for a series of Nb-base alloys from the work

of MASUMOTO *et al* [Mas80] which are also included in
the figure. The slight departures from COLLVER-
HAMMOND that do occur (positive for the Ti-base alloys
and negative for the Nb-base alloys) were discussed by
INOUE *et al* [Ino80]. Further agreement with earlier
results, this time for *crystalline* transition-metal
alloys, is also exhibited in Fig.14-4 [Ino80, Ino80[b]];
after crystallization, the three alloys referred to
yielded transition temperatures which fell nicely on
the double-peaked T_c *versus* e/a curve previously
established for 4d-TM binary alloys.

14.6.2 Composition Dependence in the Quaternary Alloys

According to INOUE *et al* [Ino80[a]] to whom the
data depicted in Fig.14-5 are due: *(a)* substitutions
of C or Ge for Si in Ti_{55}-Nb_{30}-Si_{15} reduce T_c at the
average rate (calculated at the lowest T_c's achieved)
of about 2.5 K per 10 at.%; *(b)* substitutions of Mo
for Ti in the same basic ternary reduce T_c at the rate
of 3.0 K per 10 at.% Mo (both this and the previous

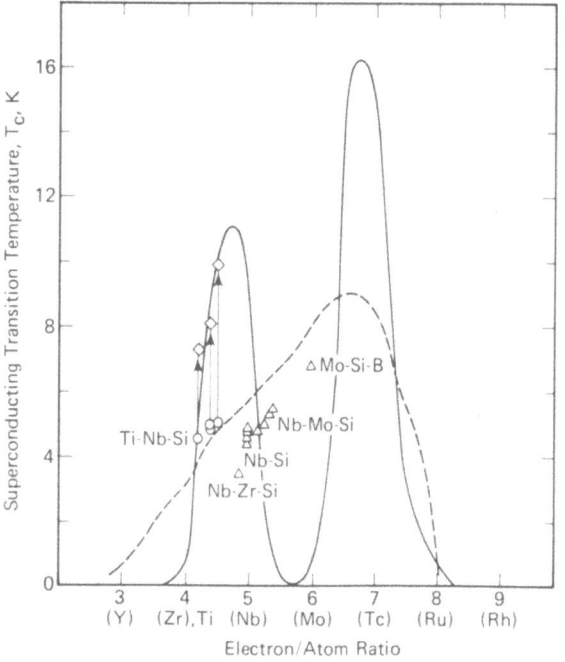

FIGURE 14-4. Superconducting transition temperatures (T_c) of
amorphous Ti-Nb-Si alloys (○) (and other amorphous alloys
(△) — from [Mas80]) juxtaposed against the COLLVER-
HAMMOND [Col73[b]] curve (---) for vapor-cryoquenched amorphous 4d-TM alloy films. Upon crystallization, the T_c's of the
Ti-Nb-Si alloys (◇) rise to meet the double-peaked curve (—)
for crystalline 4d-TM alloys (cf. in both cases, Fig. 14-2) — after
INOUE *et al* [Ino80].

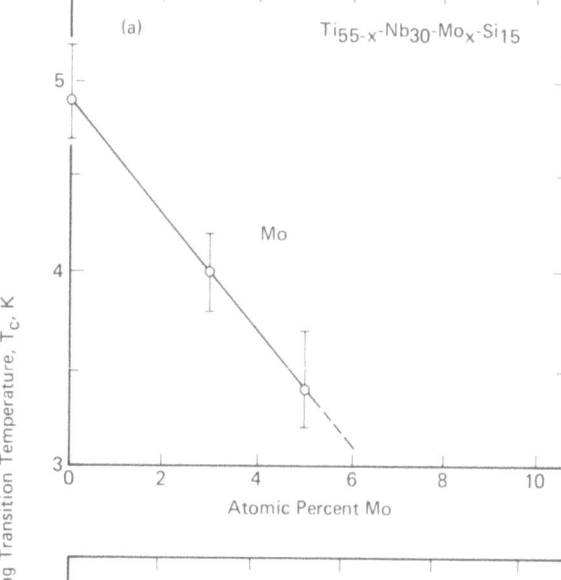

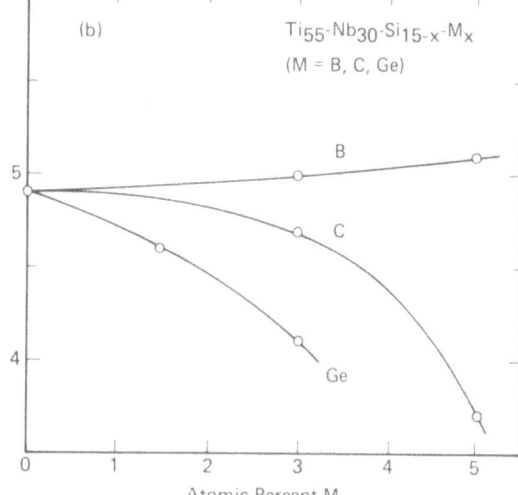

FIGURE 14-5. Influence of fourth-element substitutions on
the superconducting transition temperature of amorphous
Ti_{55}-Nb_{30}-Si_{15}. Depicted are the effects of: (a) substituting Mo
for Ti; (b) substituting B, C or Ge for Si — after INOUE *et al*
[Ino80[a]].

effects are an order of magnitude stronger than those
encountered when the Si or Nb contents of the ternary
alloy are varied); *(c)* substitutions of Ru, Rh, Pd or
Ir for Ti also bring about reductions in the T_c, but
at unspecified rates; *(d)* substitutions of B for Si in
Ti_{55}-Nb_{30}-Si_{15} *increase* T_c slightly, the rate being
about 0.38 K per 10 at.% B.

Finally, taking Ti_{55}-Nb_{30}-Si_{12}-B_3 (whose T_c,
according to the above, was higher than that of the
B-free ternary) as prototype, a series of alloys with
fixed Si and B levels but variable Ti:Nb ratio was
prepared. The transition temperatures across the

series Ti_{85}-Nb_x-Si_{12}-B_3 ($x = 20$, 30, 40, 50 and 60)
were, respectively, 4.5, 5.0, 5.4, 4.0 and <3.5 K.
Similar studies were carried out with the
$(Ti-Nb)_{85}$-Si_{10}-B_5 system whose maximum T_c was 5.5 K,
again at a composition corresponding to 40 at.% Nb.

14.6.3 Influence of Cold Deformation

Cold rolling to about 15% reduction in thickness
had little effect on the transition temperatures of
Ti_{70}-Nb_{15}-Si_{15} and Ti_{55}-Nb_{30}-Si_{15}, but by 40% reduction in thickness the transition temperatures, initial-
ly 4.6 K and 4.9 K, respectively, had dropped below
4.3 K, Fig.14-6. However, as also indicated in that
figure, the T_c could be partially restored to its
starting value by aging for 1h/300°C [Ino80].

14.6.4 Influence of One-Hour Heat Treatment

The influence on the transition temperature of
Ti_{55}-Nb_{30}-Si_{15} of 1-h aging at temperatures up to
900°C has been reported in papers by MASUMOTO *et al*
(up to 500°C [Ino80][Mas80[a]], continuing to 900°C
[Ino80[b]][Mas80[a]]). The latter authors have also considered the annealing of Ti_{85-x}-V_x-Si_{15} ($x = 5$, 10,
20, 30) and Ti_{85-x}-Nb_x-Si_{15} ($x = 10$, 15, 40) within
the temperature range 400∿900°C.

The results are easily explicable, with particu-
lar reference to the behavior of Ti_{55}-Nb_{30}-Si_{15} as
depicted in Fig.14-7, in terms of the aging and precipitational effects outlined above. The figure shows
that during 1-h aging, T_c remains practically constant
up to about 227°C, then begins to decrease at about
277°C as the glassy phase undergoes some microscopic
change. At about 400°C, T_c passes through a minimum
as precipitation of the β-phase commences, an effect
which is accompanied by a reduction in the ductility.
Upon further increase of the temperature to 800°C, T_c
increases rapidly reflecting the completion of the
precipitation of the metastable β-Ti-Nb phase with its
intrinsically high T_c. Particularly noticeable in all
the Ti-V-Si alloys and in the Ti_{85-x}-Nb_x-Si_{15}
($x = 10$, 15) alloys is the appearance of a *maximum* in
T_c induced by the β→α decomposition of the solid-
solution phase which begins to take place at sufficiently high temperatures. The α-Ti-Nb phase, of
course, has a much lower transition temperature than
the corresponding β-phase (see, for example, Fig.7-3).

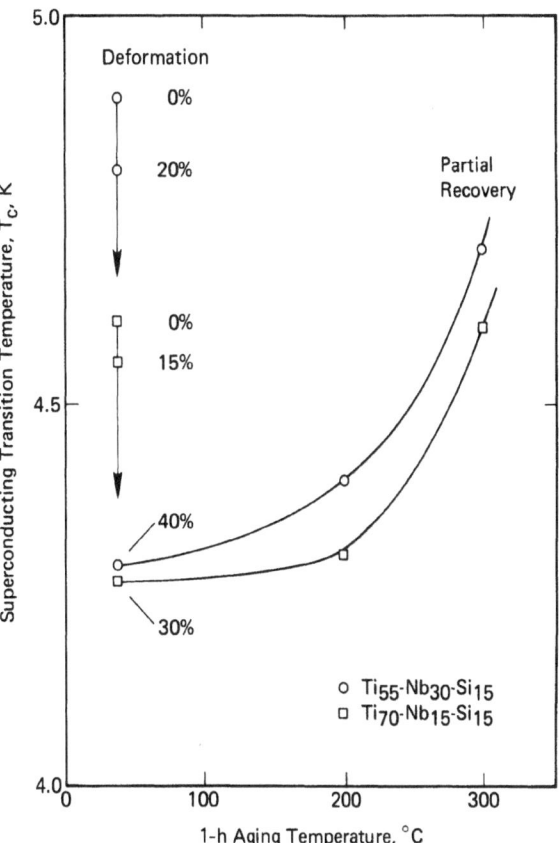

FIGURE 14-6. Influence of cold rolling of up to 30∿40% followed by 1-h aging at temperatures of up to 300°C on the superconducting transition temperatures of amorphous Ti_{70}-Nb_{15}-
Si_{15} and Ti_{55}-Nb_{30}-Si_{15} ribbons — after INOUE *et al* [Ino80].

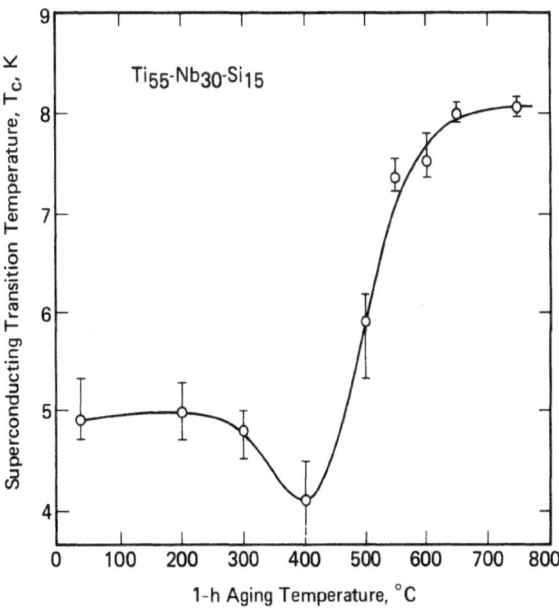

FIGURE 14-7. Influence of 1-h aging and annealing at temperatures of up to 750°C on the superconducting transition temperature of Ti_{55}-Nb_{30}-Si_{15} (note: brittleness is setting in at 400°C;
crystallization at 525°C) — after INOUE *et al* [Ino80], also
MASUMOTO and INOUE [Mas80[a]].

The maximum transition temperatures attained in this way in the $Ti_{85-x}-Nb_x-Si_{15}$ group of alloys were: (a) 4.8 K, x = 10; (b) 7.3 K, x = 15; (c) 8.1 K, x = 30; (d) 9.9 K, x = 40 [Ino80[b]]. An interesting result to come out of this study, but one that is to be expected considering the opportunities that exist for shifts in the composition of any of the component phases away from the average composition, was that the maximum T_c's obtained were significantly different from those of binary Ti-Nb alloys with the corresponding nominal Nb contents, viz: (a) 5 K, $Ti-Nb_{10}$; (b) 6 K, $Ti-Nb_{15}$; (c) 7.3 K, $Ti-Nb_{30}$; (d) 8.8 K, $Ti-Nb_{40}$ (cf. Fig.7-3).

14.7 CRITICAL FIELDS OF AMORPHOUS Ti ALLOYS

The critical fields, determined resistometrically at 4.2 K, and other relevant properties of some Ti-Nb-Si alloys, are listed in Table 14-2. It is not possible to speculate on the reason for the discrepancies in critical field between the ternary and quaternary alloys. On the theoretical side it is well known that upper critical field is proportional to the product $\rho_n \gamma T_c$, Eqn.(7-10), where ρ_n is the residual resistivity and γ is the electronic specific heat co-

efficient -- which itself is roughly proportional to the Fermi density-of-states, and hence to T_c, cf. [Mon8.00][Mon10.00]. It follows, therefore, that the measured resistive upper critical field, H_r, should be approximately proportional to $\rho_n T_c^2$. For the amorphous alloys listed in Table 14-2, this product is about 4000 $\mu\Omega$ cm K^2. For crystalline $Ti-Nb_{39}$ wire, ρ_n = 63 $\mu\Omega$ cm [Ber63[a]] and for a similar alloy, $T_c \sim 8.6$ K; the product $\rho_n T_c^2$ is, therefore, 4700 $\mu\Omega$ cm K^2, not much different from the amorphous value. This finding, as far as it goes, suggests that the upper critical fields of the amorphous Ti-Nb-base alloys are not to be regarded as anomalous, in spite of the fact that they appear to be low, and should be amenable to calculation using standard theory (cf. [Dom78] for a further discussion of this subject). That the H_r's listed are generally low (e.g. 62 kOe in $Ti_{55}-Nb_{30}-Si_{12}-B_3$) must therefore be attributable to low values of γ (yet to be measured in these systems) with which the relatively low T_c's (as compared to those of the crystallized alloys, Fig.14-4) must be associated.

With regard to the temperature dependence of H_r, it was shown in Sect.7.32.3 that:

$$-dH_r/dT = 4.50 \times 10^4 \, \rho_n \gamma \qquad (7-36)$$

TABLE 14-2   ROOM-TEMPERATURE ELECTRICAL RESISTIVITY (ρ), SUPERCONDUCTING TRANSITION TEMPERATURE (T_c), LOWER AND UPPER CRITICAL FIELDS (H_{c1} and H_{c2}) AND TEMPERATURE DEPENDENCE OF $H_{c2}(dH_{c2}/dT)$ FOR TERNARY [Ino80] AND QUATERNARY [Ino80[a]] AMORPHOUS TITANIUM-NIOBIUM-SILICON-BASE ALLOYS LISTED IN ORDER OF INCREASING NIOBIUM CONTENT

Composition, At.%				ρ_n $\mu\Omega$ cm	T_c K	H_{c1} kOe	H_r[+] kOe	dH_{c2}/dT[++] kOe K^{-1}
Ti	Nb	Si	B					
70	15	15		170	4.6			
60	25	15		180	4.8			
55	30	15		170	4.9	5.9	34	75
55	30	12	3		5.0	14	62	
56	30	14		170	5.0			
50	35	15		160	5.0			
45	40	15		150	5.1	6.4	39	60
45	40	12	3		5.4	11	77	

† Resistive onset.

†† Estimated values.

from which in the case of Ti-Nb$_{34}$ it followed that:

$$-dH_r/dT = 36.5 \text{ kOe K}^{-1} \quad , \qquad (7\text{-}37a)$$

while the experimental value was shown to be 26.4 kOe K^{-1} [Haw80]. Compared with either 36.5 or 26.4 kOe K^{-1}, the values of $\sim$70 kOe K^{-1} listed in Table 14-2 seems to be anomalously high. That $\rho_{am}/\rho_{cryst} \sim 2$ would be sufficient to explain the discrepancy had it not already been implied in the preceding paragraph that $\gamma_{am}/\gamma_{cryst} \sim 0.5$. Further discussion of the upper critical field temperature dependences of these alloys is to be found in the original article [Ino80] and elsewhere [Mas80, Mas80^a], where it is shown that the dH_r/dT's in alloys such as (Nb-Mo)$_{80}$-Si$_{20}$, Nb$_{80}$-(Si-M)$_{20}$ (M = C and Ge) and W$_{70}$-Si$_{20}$-B$_{10}$ are typically 27 kOe K^{-1}. For a general discussion of dH_{c2}/dT in amorphous alloys, [Dom78] is recommended.

14.8 CRITICAL CURRENT DENSITIES OF AMORPHOUS Ti ALLOYS

During 1980, MASUMOTO *et al* reported on the critical current densities of several Ti-Nb-Si-base metallic glass ribbons both in the as-quenched form and after heat treatment. The measurements were carried out at 4.2 K with the applied field, H_a, perpendicular to the current direction; no mention was made of the relative field direction in the transverse plane, or of the possibility of $J_c(H_a)$ anisotropy in that plane as is generally found to be present in rolled ribbon (cf., for example, Sect.11.10). A representative set of data are presented in Table 14-3.

14.8.1 As-Quenched Metallic-Glass Ribbon

As pointed out by INOUE *et al* [Ino80, Ino80^a], the critical current densities of the as-quenched amorphous alloys are small. Even in zero applied field, that of Ti$_{45}$-Nb$_{40}$-Si$_{12}$-B$_3$ is only $4.5 \sim 4.8 \times 10^2$ A cm^{-2} [Ino80^a], and with increasing field J_c decreases rapidly, in this case to half its initial value in a field of 40 kOe. The low critical current densities are attributable to the extreme homogeneity of the material which offers little opportunity for flux pinning excepting at the edges of the ribbon. The as-quenched alloys, therefore, have no technical value as current-carrying superconductors, but may have device applications in situations where relatively large upper critical fields and low flux-flow viscosities are desirable attributes.

The critical current density responds readily to heat treatment as the products of glass decomposition

TABLE 14-3 CRITICAL CURRENT DENSITIES (J_c) AS FUNCTION OF APPLIED TRANSVERSE MAGNETIC FIELD (H_a) IN AS-QUENCHED AND HEAT-TREATED TITANIUM-NIOBIUM-BASE METALLIC-GLASS RIBBONS

Composition, At.%				Condition	H_a kOe	J_c [+] A cm^{-2}	References
Ti	Nb	Si	B				
55	30	15		As-quenched	17	1.5×10^2	[Ino80]
55	30	12	3	As-quenched	40	2.2×10^2	[Ino80^a]
55	30	4	11	Amorphous + β	56	1.0×10^3	[Mas80^a]
57	30	10	3	Amorphous + β	34	1.0×10^3	"
70	15	15		Aged 1h/600°C	70	5.0×10^3	[Ino80^b]
55	30	15		Aged 1.5h/650°C	70	1.0×10^4	"
50	35	15		Aged 1h/700°C	70	4.7×10^4	"

[+] J_c criterion: 1 μV across 2.5 cm [Mas80][Ino80, Ino80^b] or 5 cm [Ino80^a].

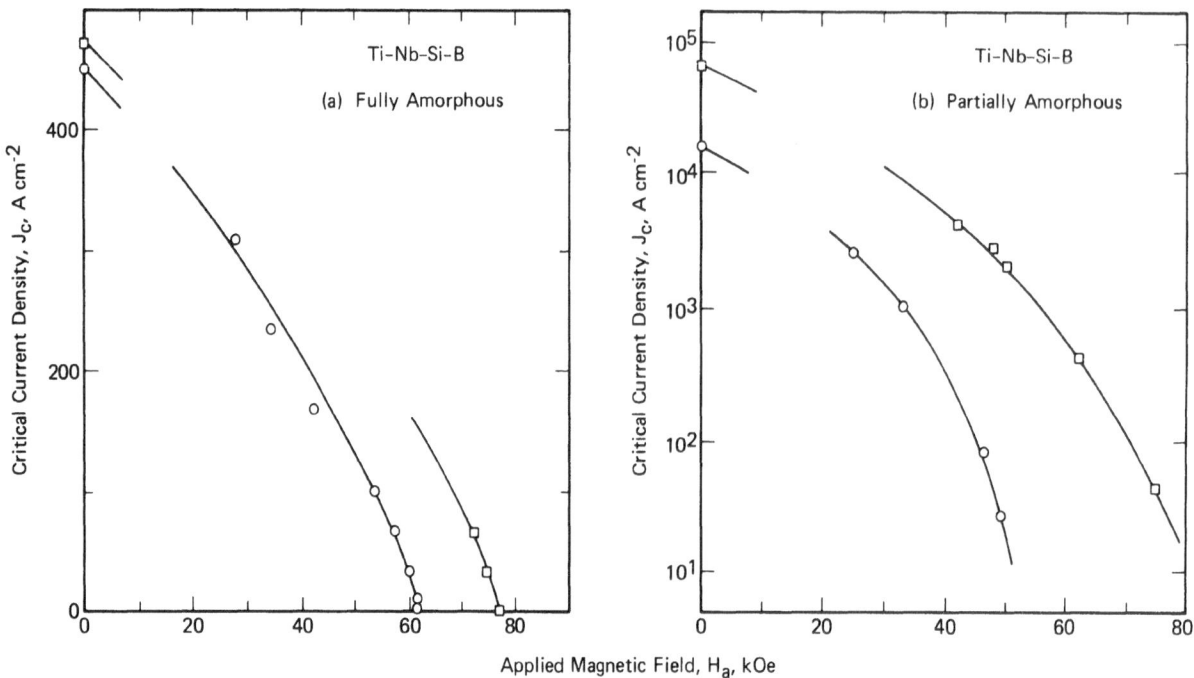

FIGURE 14-8. Critical current densities at 4.2 K *versus* applied magnetic field for: (a) as-quenched amorphous ribbons of Ti_{55}-Nb_{30}-Si_{12}-B_3 (○) and Ti_{45}-Nb_{40}-Si_{12}-B_3 (□) — after INOUE *et al* [Ino80a]; (b) partially crystallized (i.e., β-Ti-Nb precipitates in an amorphous matrix) Ti_{57}-Nb_{30}-Si_{10}-B_3 (○) and Ti_{55}-Nb_{30}-Si_4-B_{11} (□) — after MASUMOTO and INOUE [Mas80a].

provide flux-pinning centers. Fig.14-8 compares $J_c(H_a)$ for a typical as-quenched ribbon with that for one which has been subjected to moderate heat treatment such that the β-Ti-Nb phase has precipitated out within the amorphous matrix -- the first stages of recrystallization, cf. Sect.14.5.2. In this condition, referred to as "duplex structured" [Ino80a], the ribbon exhibits the improved current carrying capacity without loss of ductility. If higher critical current densities are to be obtained (at the expense of ductility) full crystallization is necessary.

14.8.2 Alloy Ribbons Crystallized from the Amorphous Phase

According to Table 14-3, the highest critical current densities in the glassy alloys are attainable only after complete crystallization to the β-Ti-Nb phase condition. The most complete recent study of the effects of crystallization on the superconducting properties of the present alloys is that of INOUE *et al* [Ino80b], who pointed out that even within the context of crystallization, the optimal critical current densities were achieved only under closely controlled heat treatment conditions. Thus with Ti_{50}-Nb_{35}-Si_{15}, for example, the sample is "underaged" at 1h/650°C and "overaged" at 1h/750°C. The best overall $J_c(H_a)$ performance was obtained after a heat treatment of 1h/700°C which, as pointed out in Sect.14.5.2, was responsible for producing an even dispersion (particle separation $\lesssim$1000 Å) of 900 Å Nb_3Si precipitates in a β-Ti-Nb matrix.

REFERENCES

A

[Abr57]   ABRIKOSOV, A.A., "On the Magnetic Properties of Superconductors of the Second Group", Sov. Phys. JETP, 5 1174-82 (1957), [Transl. of Zh. Eksp. Teor. Fiz., 32 1442-52 (1957)].

[Ada70]   ADAM, E., DOSDAT, J.P. and BLANC, J.M., "Composite Filamentary Superconductors" (Proc. of the 3rd Int. Cryogenic Engineering Conf., Berlin, Germany (May 1970)), Iliffe Science and Technology Publications Ltd., Guilford, Surrey, UK (1970) pp. 333-8.

[Aga74]   AGARWAL, K.L., "Low Temperature Calorimetry of Transition Metal Alloys", University of New Orleans, LA, Ph.D. Thesis (1974).

[Aka75]   AKACHI, T., KIM, D.F. and KIM, Y.B., "Flux-Flow Resistance Minima in Type-II Superconductors", Low Temperature Physics - LT14 (Proc. 14th Int. Conf., Otaniemi, Finland (Aug. 1975)) Vol.2, ed. by M. Krusius and M. Vuorio, American Elsevier Publishing Co., NY (1975) pp. 477-80.

[Aℓb76]   ALBERT, H. and PFEIFFER, I., "Temperaturabhängigkeit der Festigkeitseigenschaften des Hochfeldsupraleiters NbTi50", Z. Metallkde., 67 356-60 (1976).

[Aℓe67]   ALEKSEEVSKII, N.E., DUBROVIN, A.V., MIKHAILOV, N.N., SOKOLOV, V.I. and FEDOTOV, L.N., "Principal Properties of Superconducting Wire Made from an Alloy of the 65 BT Type in Samples and Solenoids", Sov. Phys. Dokl., 11 993-6 (1967), [Transl. of Dok. Akad. Nauk SSSR, 171 566-9 (1966)].

[Aℓe67ᵃ]   ALEKSEEVSKII, N.E., IVANOV, O.S., RAEVSKII, I.I. and STEPANOV, N.V., "Constitution Diagram of the System Niobium-Titanium-Zirconium and Superconducting Properties of the Alloys", Phys. Met. Metallogr. (USSR), 23 No.1, 28-35 (1967), [Transl. of Fiz. Met. Metalloved., 23 28-36 (1967)].

[Aℓe68]   ALEKSEEVSKII, N.E., HLASNIK, I. and DUBROVIN, A.V., "Some Features of the Transition due to Destruction of Superconductivity by a Current in Superconducting Alloys", Sov. Phys. JETP, 27 47-50 (1968), [Transl. of Zh. Eksp. Teor. Fiz., 54 84-90 (1968)].

[Aℓe68ᵃ] ALEKSEEVSKII, N.E., IVANOV, O.S., RAEVSKII, I.I. and STEPANOV, N.V., "Critical Magnetic Fields of Niobium-Titanium-Zirconium Alloys", Sov. Phys. Dokl., 12 898-900 (1968), [Transl. of Dok. Akad. Nauk SSSR, 176 305-7 (1967)].

[Aℓℓ75] ALLEN, P.B. and DYNES, R.C., "Transition Temperature of Strong-Coupled Superconductors Reanalyzed", Phys. Rev. B, 12 905-22 (1975).

[And62] ANDERSON, P.W., "Theory of Flux Creep in Hard Superconductors", Phys. Rev. Lett., 9 309-11 (1962).

[Arn74] ARNDT, R. and EBELING, R., "Einfluss von Gefügeparametern auf die Stromfähigkeit von Niob-Titan-Supraleitern", Z. Metallkde., 65 364-73 (1974).

[Asa65] ASAYAMA, K. and MASUDA, Y., "Nuclear Spin-Lattice Relaxation in Type II Superconductors", J. Phys. Soc. Jpn., 20 1290-1 (1965).

[Ass68] ASSOCIATED ELECTRICAL INDUSTRIES LTD., "Procédé pour Améliorer les Propriétés Supra-Conductrices d'un Alliage de Niobium et de Titane", French Patent 1,512,971 (Feb. 1968).

[Ass68ᵃ] ASSOCIATED ELECTRICAL INDUSTRIES LTD., "Alliages de Titane Superconducteurs", French Patent 1,517,216 (Mar. 1968).

[Ato66] ATOMICS INTERNATIONAL, "Investigation of Supercurrent Instabilities in Type II Superconductors", Atomics International, Canoga Park, CA, Quart. Progr. Report No.1, NASA-CR-76060, AI-66-87, N66-29908 (May 1966).

[Ato67] ATOMICS INTERNATIONAL, "Investigation of Supercurrent Instabilities in Type II Superconductors", Atomics International, Canoga Park, CA, Final Report, NASA-CR-86651, AI-67-83, N67-31510 (May 1967).

[Aut60] AUTLER, S.H., "Possible Applications of Superconducting Solenoids to Plasma Containment and Energy Storage", Bull. Am. Phys. Soc., 5 367 (1960).

B

[Bac68] BACHMANN, D., HEDRICH, D., KALSCH, E., RÖSCHEL, E., RAUB, E. and ZWICKER, U., "Über den Einfluss von Zusätzen auf Verarbeitbarkeit, Gefüge, Elektrische und Supraleitende Eigenschaften der Legierungen TiNb33 and TiNb40", Z. Metallkde., 59 426-32 (1968).

[Bag59] BAGARIATSKII, Yu.A., NOSOVA, G.I. and TAGUNOVA, T. V., "Factors in the Formation of Metastable Phases in Titanium-Base Alloys", Sov. Phys. Dokl., 3 1014-8 (1959), [Transl. of Dok. Akad. Nauk SSSR, 122 593-6 (1958)].

[Bak69] BAKER, C. and SUTTON, J., "Correlation of Superconducting and Metallurgical Properties of a Ti-20 at.% Nb Alloy", Phil. Mag., 19 1223-55 (1969).

[Bak70] BAKER, C., "Effect of Heat Treatment and Nitrogen Addition on the Critical Current Density of a Worked Niobium 44 wt.% Titanium Superconducting Alloy", J. Mater. Sci., 5 40-52 (1970).

[Bak71]   BAKER, C., "The Shape-Memory Effect in a Titanium-35wt.-%Niobium Alloy", Met. Sci. J., $\underline{5}$ 92-100 (1971).

[Baℓ72]   BALCERZAK, A.T. and SASS, S.L., "The Formation of the ω Phase in Ti-Nb Alloys", Met. Trans., $\underline{3}$ 1601-5 (1972).

[Bar50]   BARDEEN, J., "Zero-Point Vibrations and Superconductivity", Phys. Rev., $\underline{79}$ 167-8 (1950).

[Bar57]   BARDEEN,J., COOPER, L.N. and SCHRIEFFER, J.R., "Theory of Superconductivity", Phys. Rev., $\underline{108}$ 1175-1204 (1957).

[Bar61]   BARDEEN, J. and SCHRIEFFER, J.R., "Recent Developments in Superconductivity", in Progress in Low Temperature Physics, ed. by C.J. Gorter, North-Holland Publishing Co. (1961) pp. 277-8.

[Bar65]   BARDEEN, J. and STEPHEN, M.J., "Theory of the Motion of Vortices in Superconductors", Phys. Rev., $\underline{140A}$ 1197-207 (1965).

[Bar66]   BARNES, L.J. and HAKE, R.R., "Specific Heat and Magnetization of a Pauli-Paramagnetic Superconductor", Ann. Acad. Sci. Fenn., $\underline{A\text{-}VI}$ 78-84 (1966).

[Bar67]   BARNES, L.J. and HAKE, R.R., "Calorimetric Evidence for Pauli-Paramagnetic Superconductivity", Phys. Rev., $\underline{153}$ 435-7 (1967).

[Bar67[a]] BARANOV, I.A., VASIL'EV, N.G., KARASIK, V.R., and SHMULEVICH, R.S., "Influence of Additions of Carbon, Oxygen, Nitrogen and Hydrogen on the Critical Temperatures of Nb-Ti Alloys", Low Temperature Physics and Chemistry (Proc. 10th Int. Conf., Moscow, USSR (Aug.-Sept. 1966)), Vol.IIB Superconductivity, ed. by N.V. Zavaritskii et al, Viniti, Moscow, USSR (1967) pp. 97-100.

[Bar70]   BARANOV, I.A., SHMULEVICH, R.S., SYTNIKOV, V.A., KARASIK, V.R. and VASIL'EV, N.G., "Effect of Microinhomogeneities on the Transformation Temperature of Superconducting Alloys", in Physics and Metallurgy of Superconductors, ed. by E.M. Savitskii and V.V. Baron, Consultants Bureau, NY (Plenum Publishing Corp.) (1970) pp. 94-7; (Proc. of the Second and Third Conf. on Metallurgy, Physical Chemistry and Metal Physics of Superconductors, Moscow, USSR (May 1965 and May 1966); first published in Metalloved., Fiz.-Khim., Metallofiz. Sverkhprovodnikov, Izdatelstvo Nauka, Moscow, USSR (1967) pp. 82-6.

[Bar70[a]] BARAŠIĆ, S., LABBE, J. and FRIEDEL, J., "Tight Binding and Transition-Metal Superconductivity", Phys. Rev. Lett., $\underline{25}$ 919-22 (1970).

[Bat64]   BATT, R.H., "Apparatus for Heat Capacity Measurements in the Range 0.24 to 20°K; Applications to Some Transition-Metal Superconductors and Ferromagnetic-Superconductors", University of California, CA, Ph.D. Thesis (1964).

[Bea62]   BEAN, C.P., "Magnetization of Hard Superconductors", Phys. Rev. Lett., $\underline{8}$ 250-3 (1962).

[Bea64]   BEAN, C.P., "Magnetization of High-Field Superconductors", Rev. Mod. Phys., $\underline{36}$ 31-9 (1964).

[Bea69]   BEASLEY, M.R., LABUSCH, R. and WEBB, W.W., "Flux Creep in Type-II Superconductors", Phys. Rev. $\underline{181}$ 682-700 (1969).

[Bea82] BEASLEY, M.R., "New Perspectives on the Physics of High-Field Superconductors", Advances in Cryogenic Engineering (Materials), 28 345-60 (1982).

[Beℓ68] BELANGER, B.C., "The Behavior of Electric Currents and Magnetic Fields in Nonideal Type II Superconductors: Longitudinal Case", University of So. California, CA, Ph. D. Thesis (1968).

[Beℓ69] BELLIN, P.H., SADAGOPAN, V. and GATOS, H.C., "Ternary Superconducting Alloys of the Titanium-Niobium-Vanadium System: Transition Temperature Variation", J. Appl. Phys., 40 3982-4 (1969).

[Beℓ70] BELLIN, P.H., GATOS, H.C. and SADAGOPAN, V., "Critical Field in the Ti-Nb-V System", J. Appl. Phys., 41 2057-9 (1970).

[Ben72] BENNEMANN, K.H. and GARLAND, J.W., "Theory for Superconductivity in d-Band Metals", in Superconductivity in d- and f-Band Metals (Proc. AIP Conf., Rochester, NY (Oct. 1971)), ed. by D.H. Douglass, AIP Conf. Proc. No.4, AIP, NY (1972).

[Ber62] BERLINCOURT, T.G. and HAKE, R.R., "Upper Critical Fields of Transition Metal Alloy Superconductors", Phys. Rev. Lett., 9 293-5 (1962).

[Ber63] BERLINCOURT, T.G., "Pulsed Magnetic Field Studies of Superconducting Transition Metal Alloys at High and Low Current Densities", Low Temperature Physics - LT8 (Proc. 8th Int. Conf., London, UK (Sept. 1962)), ed. by R.O. Davies, Butterworths (1963) pp. 338-41.

[Ber63[a]] BERLINCOURT, T.G. and HAKE, R.R., "Superconductivity at High Magnetic Fields", Phys. Rev., 131 140-57 (1963).

[Ber64] BERKL, E.W. and WEYL, R., "Erwärmungseffekte bei Impulsmessungen an harten Supraleitern", Z. Angew. Phys., 16 415-9 (1964).

[Ber76] BERGMANN, G., "Amorphous Metals and their Superconductivity", Phys. Reports, 27C 160-85 (1976).

[Bes79] BEST, K.J., GENEVEY, D., HILLMANN, H., KREMPASKY, L., POLAK, M. and TURCK, B., "Anisotropy of the Critical Current in Solid Solution Superconductor NbTi", IEEE Trans. Magn., MAG-15 395-7 (1979).

[Bes79[a]] BEST, K.J., GENEVEY, D., HILLMANN, H., KREMPASKY, L., POLAK, M. and TURCK, B., "Anisotropy of Optimized and Not Optimized Technical NbTi Superconductors", IEEE Trans. Magn., MAG-15 765-7 (1979).

[Bet66] BETTERTON, J.O., Jr., "Electronic Properties of Metals and Alloys", Oak Ridge National Lab., TN, Metals and Ceramics Div., Annual Progr. Report ORNL-3970, N67-25347 (1966) pp. 25-32.

[Bet68] BETTERTON, J.O., Jr. and SCARBROUGH, J.O., "Low-Temperature Specific Heats of Zr-Ti, Zr-Hf and Zr-Sc Alloys", Phys. Rev., 168 715-25 (1968).

[Bid70] BIDAULT, M. and DOSDAT, J., "Problèmes Liés à la Production des Alliages Supraconducteurs à Base de NbTi", Rev. Phys. Appl., 5 505-11 (1970).

[Bℓa61] BLAUGHER, R.D., CHANDRASEKHAR, B.S. and HULM, J.K., "Superconducting Titanium-Molybdenum Alloys", Phys. Chem. Solids, 21 252-5 (1961).

[Bℓa63] BLAUGHER, R.D. and JOINER, W.C.H., "Superconductivity in the Ti-Ta System", Bull. Am. Phys. Soc., 8 192 (1963).

[Bℓa65]   BLAUGHER, R.D., "The Temperature Dependence of the Lower Critical Field for Some Non-Ideal Type II Superconductors", Phys. Lett., 14 181-2 (1965).

[Bℓa68]   BLACKBURN, M.J. and WILLIAMS, J.C., "Phase Transformations in Ti-Mo and Ti-V Alloys", Trans. Met. Soc. AIME, 242 2461-9 (1968).

[Boy78]   BOYD, J.D., Private Communication.

[Bra65]   BRANDT, N.B. and GINZBURG, N.I., "Effect of High Pressure on the Superconducting Properties of Metals", Sov. Phys. Usp., 8 202-23 (1965), [Transl. of Usp. Fiz. Nauk, 85 485-521 (1965)].

[Bra66]   BRANDT, N.B. and GINZBURG, N.I., "Investigation of the Effects of Hydrostatic Pressure and Plastic Deformation on the Superconducting Properties of Titanium", Sov. Phys. JETP, 22 1167-71 (1966), [Transl. of Zh. Eksp. Teor. Fiz., 49 1706-14 (1965)].

[Bra67]   BRAMMER, W.G., Jr. and RHODES, C.G., "Determination of Omega Phase Morphology in Ti-35%Nb by Transmission Electron Microscopy", Phil. Mag., 16 477-86 (1967).

[Bra72]   BRAND, R.A., "Paramagnetic Critical State: Pinning in a Series of Type II Superconducting Alloys", Cornell University, Ithaca, NY, Ph.D. Thesis (1972).

[Bra75]   BRAND, R.A., "Pinning in the Paramagnetic Limit", Low Temperature Physics - LT14 (Proc. 14th Int. Conf., Otaniemi, Finland (Aug. 1975)), ed. by M. Krusius and M. Vuorio, American Elsevier Publishing Co., NY (1975) pp. 485-8.

[Bre73]   BRECHNA, H., Superconducting Magnet Systems, Springer-Verlag, NY, etc. (1973).

[Bro64]   BROWN, A.R.G., CLARK, D., EASTABROOK, J. and JEPSON, K.S., "The Titanium-Niobium System", Nature, 201 914-5 (1964).

[Buc59]   BUCHER, E., BUSCH, G. and MÜLLER, J., "Supraleitung in Legierungen des Molybdäns mit Titan und Vanadium", Helv. Phys. Acta., 32 318-20 (1959).

[Buc61]   BUCHER, E. and MÜLLER, J., "Supraleitung in hexagonalen Ti-V- and Ti-Nb-Legierungen", Helv. Phys. Acta., 34 410-13 (1961).

[Buc62]   BUCKEL, W., DUMMER, G. and GEY, W., "Supraleitung und Gitterstruktur von Titan-Rhodium-Legierungen", Z. Angew. Phys., 14 703-6 (1962).

[Buc63]   BUCHER, E., HEINIGER, F. and MÜLLER, J., "Spezifische Wärme und Supraleitung von isoelektronischen Ubergangsmetallen", Helv. Phys. Acta., 36 806 (1963).

[Buc64]   BUCHER, E., HEINIGER, F., MUHEIM, J. and MÜLLER, J., "Superconductivity and Electronic Properties of Transition Metal Alloys", Rev. Mod. Phys., 36 146-9 (1964).

[Buc65]   BUCHER, E., HEINIGER, F. and MÜLLER, J., "Anomalies in the Superconducting Transition of HCP Titanium Alloys", Low Temperature Physics - LT9 (Proc. 9th Int. Conf., Columbus, OH (Sept. 1964)), ed. by J.G. Daunt et al, Plenum Press, NY (1965) pp. 482-6.

[Buy70]   BUYNOV, N.N., VOZILKIN, V.A. and RAKIN, V.G., "Structural Study of the Superconductive Alloy 65 BT", Phys. Met. Metallogr. (USSR), 29 No.5,115-9 (1970), [Transl. of Fiz. Met. Metalloved., 29 1005-9 (1970)].

[Byc69]   BYCHKOV, Yu.F., VERESHCHAGIN, V.G., KARASIK, V.R. and KURGANOV, G.B., "Critical Currents in a Superconducting Alloy with a Rigid Vortex Lattice", Sov. Phys. JETP, 29 276-81 (1969), [Transl. of Zh. Eksp. Teor. Fiz., 56 505-15 (1969)].

[Byc70]   BYCHKOVA, M.I., BARON, V.V. and SAVITSKII, E.M., "Effect of Heat Treatment on the Superconducting Properties of Nb-Ti Alloys", in Physics and Metallurgy of Superconductors, ed. by E.M. Savitskii and V.V. Baron, Consultants Bureau, NY (Plenum Publishing Corp.) (1970) pp. 53-60; (Proc. of the Second and Third Conf. on Metallurgy, Physical Chemistry and Metal Physics of Superconductors, Moscow, USSR (May 1965 and May 1966)); first published in Metallovd., Fiz.-Kim., Metallofiz. Sverkhprovodnikov, Izdatelstvo Nauka, Moscow, USSR (1967) pp. 64-8.

[Byc70[a]]   BYCHKOVA, M.I., BARON, V.V. and SAVITSKII, E.M., "Three-Component Alloys Based on the Nb-Ti System", in Physics and Metallurgy of Superconductors, ed. by E.M. Savitskii and V.V. Baron, Consultants Bureau, NY (Plenum Publishing Corp.) (1970) pp. 90-3; (Proc. of the Second and Third Conf. on Metallurgy, Physical Chemistry and Metal Physics of Superconductors, Moscow, USSR (May 1965 and May 1966)); first published in Metalloved., Fiz.-Khim., Metallofiz. Sverkhprovodnikov, Izdatelstvo Nauka, Moscow, USSR (1967) pp. 79-82.

[Byc74]   BYCHKOV, Yu.F., VOZILKIN, V.A. and UZLOV, V.Yu., "Influence of Structure on the Critical Current Density of Titanium-Niobium-Tantalum Alloys", Phys. Met. Metallogr. (USSR), 38 No.2,61-8 (1974), [Transl. of Fiz. Met. Metalloved., 38 295-302 (1974)].

C

[Cam68]   CAMPBELL, A.M., EVETTS, J.E. and DEW-HUGHES, D., "Pinning of Flux Vortices in Type II Superconductors", Phil. Mag., 18 313-43 (1968).

[Cam72]   CAMPBELL, A.M. and EVETTS, J.E., Critical Currents in Superconductors, Taylor and Francis Ltd., London, UK; Barnes and Noble Books, NY (1972).

[Cap63]   CAPE, J.A., "Superconductivity and Localized Magnetic States in Ti-Mn Alloys", Phys. Rev., 132 1486-92 (1963).

[Cap66]   CAPE, J.A., "Pauli-Spin and Spin-Orbit Effects in the Magnetization of a Superconducting Ti-Mo Alloy", Phys. Rev., 148 257-63 (1966).

[Car74]   CARR, W.J., Jr., "A.C. Loss in a Twisted Filamentary Superconducting Wire. I", J. Appl. Phys., 45 929-34 (1974).

[Cha62]   CHANDRASEKHAR, B.S., "A Note on the Maximum Critical Field of High-Field Superconductors", Appl. Phys. Lett., 1 7-8 (1962).

[Cha63]   CHANDRASEKHAR, B.S., HULM, J.K. and JONES, C.K., "Temperature Dependence of the Upper Critical Field in Some Niobium Solid Solution Alloys", Phys. Lett., 5 18-20 (1963).

[Cha70] CHARLESWORTH, J.P. and MADSEN, P.E., "Effect of Heat Treatment on the Superconducting Critical Current of Cold-Worked Titanium-45 at.% Niobium: Part I", Atomic Energy Research Establishment, Harwell, UK, Report No. AERE-R-6534 (Oct. 1970).

[Cha73] CHANDRASEKARAN, V., COTTON, W., NARASIMHAN, S., TAGGART, R. and POLONIS, D.H., "Study of Phase Transformation and Superconductivity", University of Washington, Seattle, WA, Annual Progr. Report No. RLO-2225-T-13-19 (Oct. 1973).

[Cha74] CHANDRASEKARAN, V., TAGGART, R. and POLONIS, D.H., "The Influence of Constitution and Microstructure on the Temperature Coefficient of Resistivity in Ti-Base Alloys", J. Mater. Sci., 9 961-8 (1974).

[Che60] CHENG, C.H., WEI, C.T. and BECK, P.A., "Low Temperature Specific Heat of Body-Centered Cubic Alloys of 3d Transition Elements", Phys. Rev., 120 426-36 (1960).

[Che62] CHENG, C.H., GUPTA, K.P., van REUTH, E.C. and BECK, P.A., "Low-Temperature Specific Heat of Body-Centered Cubic Ti-V Alloys", Phys. Rev., 126 2030-3 (1962).

[Che65] CHESTER, P.F., "Perfectionnements aux Électro-aimants et aux Matériaux pour leur Enroulement", French Patent 1,452,977 (Oct. 1965).

[Chi70] CHIKABA, J., "Effects of Thermal Insulation on Flux Jumps in Nb-50%Ti Rods", Cryogenics, 10 306-13 (1970).

[Cℓe68] CLEM, J.R., "Local Temperature-Gradient Contribution to Flux-Flow Viscosity in Superconductors", Phys. Rev. Lett., 20 735-8 (1968).

[Cℓo62] CLOGSTON, A.M., "Upper Limit for the Critical Field in Hard Superconductors", Phys. Rev. Lett., 9 266-7 (1962).

[Cof67] COFFEY, H.T., "Distribution of Magnetic Fields and Currents in Type II Superconductors", Cryogenics, 7 73-7 (1967).

[Cof68] COFFEY, H.T., "Modified London Model for Type-II Superconductors", Phys. Rev., 166 447-56 (1968).

[Coℓ66] COLLING, D.A., RALLS, K.M. and WULFF, J., "High-Field Superconductivity of Tantalum-Titanium Alloys", Trans. TMS-AIME, 236 1218-23 (1966).

[Coℓ66^a] COLLING, D.A., RALLS, K.M. and WULFF, J., "Superconducting Transition Temperature of Solid-Solution Ta-Ti Alloys", J. Appl. Phys., 37 4750-2 (1966).

[Coℓ69] COLLINGS, E.W. AND HO, J.C., "Enhancement of Superconducting Transition Temperatures in Martensitic Ti-Mo Alloys", Phys. Lett., 29A 306-7 (1969).

[Coℓ70] COLLINGS, E.W. AND HO, J.C., "Influence of Microstructure on the Superconductivity of a Dilute Ti-Mo Alloy", Phys. Rev. B, 1 4289-94 (1970).

[Coℓ71] COLLINGS, E.W. and BOYD, J.D. and HO, J.C., "Enhancement of the Superconducting Transition Temperature in Deformation-Induced Structures", Low Temperature Physics - LT12 (Proc. 12th Int. Conf., Kyoto, Japan (Sept. 1970)), ed. by E. Kanda, Academic Press of Japan (1971) p. 316.

[Co*ℓ*71ᵃ] COLLINGS, E.W. and HO, J.C., "Empirical Determination of the Electronic Specific Heat Enhancement Factor in Transition Metal Alloy Superconductors", Phys. Status Solidi (b), 43 K123-7 (1971).

[Co*ℓ*72] COLLINGS, E.W., HO, J.C. and JAFFEE, R.I., "Superconducting Transition Temperature, Lattice Instability and Electron-to-Atom Ratio of Transition-Metal Binary Solid Solutions", Phys. Rev. B, 5 4435-49 (1972).

[Co*ℓ*72ᵃ] COLLINGS, E.W., "An Experimental Study of Density of States Related Properties in Ti-Mo Alloys", Proc. Michigan State University Summer School on Alloys (Aug.-Sept. 1972), ed. by W.M. Hartman, P.A. Schroeder and C.L. Foiles (1972) pp. 236-69.

[Co*ℓ*73] COLLINGS, E.W. and HO, J.C., "Density of States of Transition Metal Binary Alloys in the Electron-to-Atom Ratio Range 4.0 to 6.0", in Proc. 3rd Materials Research Symposium, "Electronic Density of States", ed. by L.H. Bennett, National Bureau of Standards (US), Spec. Publication 323 (1973) pp. 587-96.

[Co*ℓ*73ᵃ] COLLINGS, E.W., HO, J.C. and JAFFEE, R.I., "Physics of Titanium Alloys-II: Fermi Density-of-States Properties and Phase Stability of Ti-Al and Ti-Mo", in Titanium Science and Technology (Proc. 2nd Int. Conf. on Titanium (May 1972)) Vol.2, ed. by R.I. Jaffee and H.M. Burte, Plenum Press, NY (1973) pp. 831-42.

[Co*ℓ*73ᵇ] COLLVER, M.M. and HAMMOND, R.H., "Superconductivity in "Amorphous" Transition-Metal Alloy Films", Phys. Rev. Lett., 30 92-5 (1973).

[Co*ℓ*74] COLLINGS, E.W., "Anomalous Electrical Resistivity, BCC Phase Stability and Superconductivity in Titanium-Vanadium Alloys", Phys. Rev. B, 9 3989-99 (1974).

[Co*ℓ*75] COLLINGS, E.W. and GEGEL, H.L., "Physical Principles of Solid Solution Strengthening in Alloys", in Physics of Solid Solution Strengthening, ed. by E.W. Collings and H.L. Gegel, Plenum Press (1975) pp. 147-82.

[Co*ℓ*75ᵃ] COLLINGS, E.W. and HO, J.C., "Superconducting Transition Temperature in Martensitic Titanium-Base Transition Metal Binary Alloys", J. Less-Common Metals, 41 157-64 (1975).

[Co*ℓ*75ᵇ] COLLINGS, E.W., HO, J.C. and UPTON, P.E., "Low-Temperature Calorimetric Studies of Superconductivity and Microstructure in Titanium-Vanadium Alloys", J. Less-Common Metals, 42 285-301 (1975).

[Co*ℓ*76] COLLINGS, E.W. and HO, J.C., "Solute-Induced Lattice Stability as it Relates to Superconductivity in Titanium-Molybdenum Alloys", Solid State Comm., 18 1493-5 (1976).

[Co*ℓ*76ᵃ] COLLINGS, E.W. and SMITH, R.D., "The Magnetic Susceptibility of Titanium-Niobium Alloys", J. Less-Common Metals, 48 187-98 (1976).

[Co*ℓ*77] COLLING, D.A., DeWINTER, T.A., McDONALD, W.K. and TURNER, W.C., "Superconducting Performance of Production NbTi Alloys", IEEE Trans. Magn., MAG-13 848-51 (1971).

[Coℓ78]   COLLINGS, E.W., "Anomalous Electrical Resistivity and Magnetic Susceptibility Temperature Dependences in Ti-V Alloys Exhibiting Reversible Soft-Phonon-Induced Structural Inhomogeneities", In Electrical Transport and Optical Properties of Inhomogeneous Media, ed. by J.C. Garland and D.B. Tanner, AIP Conf. Proc. No. 40, AIP, NY (1978) pp. 410-5.

[Coℓ78^a]   COLLINGS, E.W. and WHITE, J.J., "Deformation- and Solute-Induced Microstructural Effects in the Superconductivity of Transition-Metal Alloys", in Transition Metals 1977, ed. by M.J.G. Lee, J.M. Perz and E. Fawcett, The Institute of Physics, Conf. Ser. No. 39 (1978) pp. 645-9.

[Coℓ80]   COLLINGS, E.W., "The Metal Physics of Titanium Alloys", in Titanium '80 Science and Technology (Proc. 4th Int. Conf. on Titanium, Kyoto, Japan (May 1980)), ed. by H. Kimura and O. Izumi, published by The Metallurgical Society AIME, PA (1980) pp. 77-132.

[Coℓ83]   COLLINGS, E.W., Previously unpublished results.

[Com67]   COMEY, K.R., Jr., "The Superconductivity of a Titanium-Niobium Alloy", Massachusetts Institute of Technology, Cambridge, MA, M.S. Thesis (1967).

[Com67^a]   COMPAGNIE FRANCAISE THOMSON-HOUSTON, "Superconductive Wire", Netherlands Patent 6,614,471 (Apr. 1967).

[Cot73]   COTTON, W.L., "Phase Transformation and Superconductivity in the Titanium-Base Ternary Alloy Systems Ti-Nb-Mo and Ti-Nb-Al", University of Washington, WA, M.S. Thesis (TN 7 TH 23104) unpublished.

[Cot74]   COTTON, W.L., TAGGART, R. and POLONIS, D.H., "The Superconducting Transition Behavior of Ternary Alloys of Titanium", Scr. Met., 8 329-35 (1974).

[Cou67]   COURTNEY, T.H. and WULFF, J., "Quarternary Solid Solution Superconductors", Phys. Lett., 25A 477-9 (1967).

[Cou69]   COURTNEY, T.H. and WULFF, J., "Omega Phase Formation in Superconducting Ti Alloys", Mater. Sci. Eng., 4 93-7 (1969).

[Cur79]   CURTIS, C.W. and McDONALD, W.K., "Suitability of NbTi Containing up to 2 wt.% Ta for Use in Fabrication of Superconductors", IEEE Trans. Magn., MAG-15 768-71 (1979); see also Final Report to ERDA on Contr. No. RL-E-77-0189.

[Cur81]   CURTIS, C.W., Private Communication.

D

[Dan69]   DANNER, S. and DUMMER, G., "Kalorimetrische Bestimmung der Übergangstemperatur zur Supraleitung von Verdunnten Titan-Rhodium-Legierungen", Z. Phys., 222 243-52 (1969).

[Dau49]   DAUNT, J.G. and HEER, C.V., "Some Properties of Superconductors Below 1 Degree K. I. Titanium", Phys. Rev., 76 715-7 (1949).

[Deg74] DEGTYAREVA, V.F., KARIMOV, Yu.S. and RABINKIN, A.G., "Superconductivity and Magnetic Susceptibility of the α and ω Modifications of Titanium and Zirconium", Sov. Phys. Solid State, 15 2293-4 (1974), [Transl. of Fiz. Tverd. Tela (Leningrad), 15 3436-8 (1973)].

[Deh31] de HAAS, W.J. and van ALPHEN, P.M., "Resistance of Graphite, Thorium, Titanium and Titanium-Zirconium Below 20.4 K", K. Akad. Amsterdam, Proc. 34 70-4 (1931), Comm. No. 212e from the Phys. Lab. Leiden.

[Des63] DeSORBO, W., "Size Factor and Superconducting Properties of Some Transition Metal Solutions", Phys. Rev., 130 2177-87 (1963).

[Des65] DeSORBO, W., "Solute Size and Valence Effect in Some Superconducting Alloys of Transition Elements", Phys. Rev., 140 A914-9 (1965).

[Des67] DeSORBO, W., LAWRENCE, P.E. and HEALY, W.A., "Probable Upper Limit of H_{c2} for Some Higher-Order Ductile Transition-Metal Superconducting Alloys", J. Appl. Phys., 38 903-4 (1967).

[Die62] DIETRICH, I., WEYL, R. and ZWICKER, U., "Untersuchungen an Drähten aus supraleitenden Legierungen der Systeme Niob-Titan und Niob-Zirkonium", Z. Metallkde., 53 721-8 (1962).

[Die64] DIETRICH, I. and WEYL, R., "Dependence of the Critical Current and the Magnetization on the Rise-Time of the Magnetic Fields in Superconducting Niobium-Titanium", Phys. Lett., 13 274-5 (1964).

[Doi66] DOI, T., MITANI, M. and UMEZAWA, T., "Superconducting Properties of Nb-Zr-Ti Ternary Alloys (Studies of Hard Superconductor, I)", Nippon Kinzoku Gakkaishi, 30 133-8 (1966).

[Doi66[a]] DOI, T., ISHIDA, H. and UMEZAWA, T., "Study of Nb-Zr-Ti Phase Diagram (Studies of Hard Superconductor, II)", Nippon Kinzoku Gakkaishi, 30 139-45 (1966).

[Doi66[b]] DOI, T., ISHIDA, H. and UMEZAWA, T., "Effect of Cold Working and Heat Treatment on Field Critical Current Characteristics of Nb-25at.%Zr and Nb-40at.%Zr-10at.%Ti Alloys (1) (Studies of Hard Superconductor, III)", Nippon Kinzoku Gakkaishi, 30 213-9 (1966).

[Doi66[c]] DOI, T., ISHIDA, H. and UMEZAWA, T., "Effect of Cold Working and Heat Treatment on the Critical Current Characteristics of Nb-25at.%Zr and Nb-40at.%Zr-10at.%Ti Alloys (2) (Studies of Hard Superconductor, IV)", Nippon Kinzoku Gakkaishi, 30 220-5 (1966).

[Doi67] DOI, T., ISHIDA, F. and KAWABE, U., "Peak Effect in Superconducting Nb-15at.%Zr-45at.%Ti Alloys", J. Appl. Phys., 38 3811-2 (1967).

[Doi68] DOI, T., ISHIDA, F., KAWABE, U. and KITADA, M., "Effect of Precipitation on the Critical Current Density of Nb-40at.%Zr-10at.%Ti Alloy in a Magnetic Field", Nippon Kinzoku Gakkaishi, 32 886-92 (1968).

[Doi68[a]] DOI, T., ISHIDA, F., KAWABE, U. and KITADA, M., "Critical Current of Superconducting Nb(Cb)-Zr-Ti Alloys in High Magnetic Field", Trans. TMS-AIME, 242 1793-800 (1968).

[Doi68[b]] DOI, T., MITANI, M., KOBAYASHI, M. and OHARA, H., "Superconducting Alloys and Apparatus for Generating Superconducting Magnetic Field", U.S. Patent 3,408,604 (Oct. 1968).

[Doi73] DOI, T. and KUDO, M., "Method of Producing Superconducting Strips", U.S. Patent 3,710,844 (Jan. 1973).

[Dom78] DOMB, E.R. and JOHNSON, W.L., "The Upper and Lower Critical Fields of High-κ Amorphous Superconductors", J. Low Temp. Phys., $\underline{33}$ 29-40 (1978).

[Dum68] DUMMER, G. and OFTEDAL, E., "Supraleitung und Spezifische Wärme von Titan-Rhodium-Legierungen", Z. Phys., $\underline{208}$ 238-48 (1968).

[Duw53] DUWEZ, P., "The Martensite Transformation in Titanium Binary Alloys", Trans. Am. Soc. Met., $\underline{45}$ 934-40 (1953).

E

[Eas70] EASTHAM, A.R. and RHODES, R.G., "Alternating Transport Current Losses in Superconducting Wires", Cryogenic Engineering (Proc. 3rd International Conf., Berlin, Germany (May 1970)), Iliffe Science and Technology Publications Ltd., Guilford, Surrey, UK (1970) pp. 167-70.

[Eas71] EASTON, D.S., SCARBROUGH, J.O. and BETTERTON, J.O., Jr., "Effect of Heat Treatment and Impurities on the Current Density of Transition Metal Superconductors", Oak Ridge National Lab., TN, Report No. ORNL-4703, N72-11656 (June 1971).

[Edg64] EDGECUMBE, J., ROSNER, L.G. and ANDERSON, D.E., "Preparation and Properties of Thin-Film Hard Superconductors", J. Appl. Phys., $\underline{35}$ 2198-202 (1964).

[Efi65] EFIMOV, Yu.V., BARON, V.V. and SAVITSKII, E.M., "Superconducting Properties of Vanadium-Titanium", in Metal Science and Metal Physics of Superconductors, Joint Publications Research Service, Washington, DC, Transl. No. JPRS-36575, TT 66-33006 (1966) pp. 48-55; (Proc. of the First Conf. on Metal Science and Metal Physics of Superconductors (May 1964)); first published in Metallovedeniye i Metallofizika Sverkhprovodnikov, Trudy I Soveschchaniya po Metallovedeniyu i Metallofizike Sverkhprovodnikov, 25-27 Maya (1964), Izdatelvsto Nauka, 1965.

[Efi70] EFIMOV, Yu.V., BARON, V.V. and SAVITSKII, E.M., "Effect of Heat Treatment on the Superconducting Properties of V-Ti Alloys", in Physics and Metallurgy of Superconductors, ed. by E.M. Savitskii and V.V. Baron, Consultants Bureau, NY (Plenum Publishing Corp.) (1970) pp. 69-73; (Proc. of the Second and Third Conf. on Metallurgy, Physical Chemistry and Metal Physics of Superconductors, Moscow, USSR (May 1965 and May 1966)); first published in Metalloved., Fiz.-Khim., Metallofiz. Sverkhprovodnikov, Izdatelstvo Nauka, Moscow, USSR (1967) pp. 64-8.

[Efi70[a]] EFIMOV, Yu.V., BARON, V.V. and SAVITSKII, E.M., "Superconducting and Mechanical Properties of Three-Component Alloys Based on the V-Ti System", in Physics and Metallurgy of Superconductors, ed. by E.M. Savitskii and V.V. Baron, Consultants Bureau, NY (Plenum Publishing Corp.) (1970) pp. 98-101; (Proc. of the Second and Third Conf. on Metallurgy, Physical Chemistry and Metal Physics of Superconductors, Moscow, USSR (May 1965 and May 1966)); first published in Metalloved., Fiz.-Khim., Metallofiz. Sverkhprovodnikov, Izdatelstvo Nauka, Moscow, USSR (1967) pp. 86-9.

[Eil67] EILENBERGER, G., "Determination of $\kappa_1(T)$ and $\kappa_2(T)$ for Type-II Superconductors with Arbitrary Impurity Concentration", Phys. Rev., $\underline{153}$ 584-98 (1967).

[Eis54] EISENSTEIN, J., "Superconducting Elements", Rev. Mod. Phys., 26 277-91 (1954).

[Eℓb63] EL BINDARI, A. and LITVAK, M.M., "The Upper Critical Field of Nb-Zr and Nb-Ti Alloys", Avco-Everett Res. Lab., Everett, MA, Report No. BSD-TDR-63-62, AMP-101, N63-14201 (Jan. 1963).

[Eℓb63[a]] EL BINDARI, A. and LITVAK, M.M., "Critical Current and Field in Non-Ideal Superconductors', Avco-Everett Res. Lab., Everett, MA, Report No. BSD-TDR-63-148, Res. Report 162, N63-22202 (July 1963).

[Eℓb64] EL BINDARI, A. and LITVAK, M.M., "Critical Current and Field in Non-Ideal Superconductors", Rev. Mod. Phys., 36 98-102 (1964).

F

[Faℓ63] FALGE, R.L., Jr., "Superconductivity of Titanium", Phys. Rev. Lett., 11 248-50 (1963).

[Fie67] FIETZ, W.A., "Reversible and Irreversible Magnetic Properties of Some Type-II Alloy Superconductors", Cornell University, Ithaca, NY, Ph.D. Thesis (1967).

[Fie67[a]] FIETZ, W.A. and WEBB, W.W., "Magnetic Properties of Some Type-II Alloy Superconductors Near the Upper Critical Field", Phys. Rev., 161 423-33 (1967).

[Fie69] FIETZ, W.A. and WEBB, W.W., "Hysteresis in Superconducting Alloys -- Temperature and Field Dependence of Dislocation Pinning in Niobium Alloys", Phys. Rev., 178 657-67 (1969); Erratum Phys. Rev., 185 862 (1969).

[Fon81] FONER, S., and SCHWARTZ, B.B., (eds) Superconductor Materials Science -- Metallurgy, Fabrication, and Applications, NATO Advanced Study Institute Series, Series B: Physics Volume 68, Plenum Press (1981).

[Fre69] FREYHARDT, H., "Pinning of Fluxoids by Dislocations in Niobium Single Crystals, II: Effect of Temperature and Magnetic Field", Z. Metallkde., 60 409-12 (1969).

[Fri64] FRIEDEL, J., "Dislocations", Pergamon Press, Oxford (also Addison Wesley, MA) (1964); transl. of "Les Dislocations" published in 1956.

[Fro50] FRÖLICH, H., "Theory of the Superconducting State, I: The Ground State at the Absolute Zero of Temperature", Phys. Rev., 79 845-56 (1950).

G

[Gan66] GANGULY, B.N., UPADHYAYA, U.N. and SINHA, K.P., "Indirect Interaction Involving Impurity States in Superconductors", Phys. Rev., 146 317-21 (1966).

[Gan66[a]] GANDOLFO, D.A. and HARPER, C.M., "Study of Properties of High-Field Superconductors at Elevated Temperatures", Radio Corp. of America, Camden, NJ, Final Report No. NASA-CR-80154, N67-12913 (Oct. 1966).

[Gan67] GANDOLFO, D.A., "Motion of Quantized Flux Lines in Hard Superconductors in the Presence of Time-Varying Magnetic Fields", Temple University, Philadelphia, PA, Ph.D. (1967).

[Gan69] GANDOLFO, D.A., DUBECK, L. and ROTHWARF, F., "Steady-State Flux Jumping in Superconducting Niobium Titanium Tubes in Superimposed A.C. and D.C. Magnetic Fields", J. Appl. Phys., 40 2066-70 (1969).

[Gar72] GARLAND, J.W and BENNEMANN, K.H., "Theory for the Pressure Dependence of T_c for Narrow-Band Superconductors", in Superconductivity in d- and f-Band Metals (Proc. AIP Conf., Rochester, NY (Oct. 1971)), ed. by D.H. Douglass, AIP Conf. Proc. No.4, AIP, NY (1972) pp. 255-92.

[Gau67] GAUSTER, W.F., EFFERSON, K.R., HENDRICKS, J.B., KROEGER, D.M. and LUBELL, M.S., "Magnetic Flux Flow and Superconductor Stabilization", Oak Ridge National Lab., TN, Report No. ORNL-TM-2075 (Dec. 1967).

[Gau70] GAUSTER, W.F., IHARA, S. and LUBELL, M.S., "Magnetic Flux Flow and Superconductor Stabilization", Oak Ridge National Lab., TN, Report No. ORNL-TM-2980 (May 1970).

[Gau76] GAUTHIER, R., "A.C. Losses in Type-II Superconductors and Related Phenomena", University of Ottawa, Canada, Ph.D. Thesis (1976).

[Gin56] GINZBURG, V.L., "Some Remarks Concerning the Macroscopic Theory of Superconductivity" Sov. Phys. JETP, 3 621-3 (1956), [Transl. of Zh. Eksp. Teor. Fiz., 30 593-5 (1956)].

[Goo62] GOODMAN, B.B., "The Magnetic Behavior of Superconductors of Negative Surface Energy", IBM J. Res. Dev., 6 63-7 (1962).

[Gop66] GOPAL, E.S.R., Specific Heats at Low Temperatures, Plenum Press, NY (1966).

[Gor60] GOR'KOV, L.P., "The Critical Supercooling Field in Superconductivity Theory", Sov. Phys. JETP, 10 593-9 (1960), [Transl. of Zh. Eksp. Teor. Fiz., 37 833-43 (1959)].

[Gor69] GORINA, N.B., GRUZNOV, Yu.A., IL'ICHEV, A.I., LOSINA, T.N., MATORIN, V.I., MIKHAILOV, N.N., PROKOSHIN, A.F., SOKOLOV, V.I. and FEDOTOV, L.N., "Physico-Technological Research on Superconducting Alloy 65 BT", in Precision Alloys, ed. by N.N. Potapov, Joint Publications Research Service, Washington, DC (1969) pp. 18-30, [Transl. of Pretsizionnyye Splavy, Moscow, USSR (Aug. 1968)].

[Gri68] GRIGORYEV, A.M., LAYNER, D.I., PAKHOMOV, V.Ya., RAYEVSKAYA, M.V. and SOKOLOVSKAYA, E.M., "Critical Currents of Some Superconductor Nb-Zr-Ti Alloys", Russian Metallurgy (Metall) No. 3 149-51 (1968), [Transl. of Izv. Akad. Nauk. SSSR (Metal) No. 3 215-8 (1968)].

H

[Haa75] HAASEN, P. and FREYHARDT, H.C. (eds.), Proceedings of the International Discussion Meeting on Flux Pinning in Superconductors, Akademie der Wissenschaften in Göttingen, Goltze-Druck, Göttingen, 1975.

[Hac67] HACKETT, W.H., Jr., MAXWELL, E. and KIM, Y.B., "Microwave Flux-Flow Dissipation in Paramagnetically Limited Ti-V Alloys", Phys. Lett., 24A 663-4 (1967); [see also: Massachusetts Institute of Technology, Cambridge, MA, National Magnet Lab., Report No. AFOSR-67-1741 (May 19670].

[Hak61]   HAKE, R.R., "Specific Heats of Some Cubic Superconducting Titanium-Molybdenum Alloys Between 1.1 and 4.3 Degrees K", Phys. Rev., 123 1986-94 (1961).

[Hak61[a]]   HAKE, R.R., LESLIE, D.H. and BERLINCOURT, T.G., "Electrical Resistivity, Hall Effect and Superconductivity of Some BCC Titanium-Molybdenum Alloys", J. Phys. Chem. Solids, 20 177-81 (1961).

[Hak62]   HAKE, R.R., BERLINCOURT, T.G. and LESLIE, D.H., "High-Field Superconductivity in Some BCC Ti-Mo and Nb-Zr Alloys", IBM J. Res. Dev., 6 119-21 (1962).

[Hak62[a]]   HAKE, R.R., BERLINCOURT, T.G. and LESLIE, D.H., "High-Field Superconducting Characteristics of Some Ductile Transition Metal Alloys", in Superconductors, ed. by M. Tanenbaum and W.V. Wright, Interscience Publishers, NY (1962) pp. 53-9.

[Hak62[b]]   HAKE, R.R., LESLIE, D.H. and BERLINCOURT, T.G., "Low-Temperature Resistivity Minima and Negative Magnetoresistivities in Some Dilute Superconducting Ti Alloys", Phys. Rev., 127, 170-9 (1962).

[Hak63]   HAKE, R.R. and LESLIE, D.H., "High-Field Superconducting Properties of Ti-Mo Alloys", J. Appl. Phys., 34 270-6 (1963).

[Hak63[a]]   HAKE, R.R., LESLIE, D.H. and RHODES, C.G., "Giant Anisotropy in the High-Field Critical Currents of Cold-Rolled Transition Metal Alloy Superconductors", Low Temperature Physics - LT8 (Proc. 8th Int. Conf., London, UK (1962)), ed. by R.O. Davies, Butterworths (1963) pp. 342-4.

[Hak64]   HAKE, R.R. and CAPE, J.A., "Calorimetric Investigation of Localized Magnetic Moments and Superconductivity in Some Alloys of Titanium with Manganese and Cobalt", Phys. Rev., 135 A1151-60 (1964).

[Hak65]   HAKE, R.R., "Mixed-State Paramagnetism in High-Field Type-II Superconductors", Phys. Rev. Lett., 15 865-8 (1965).

[Hak67]   HAKE, R.R., "Magnetization and Resistive Behavior of Pauli-Paramagnetic Superconductors", Low Temperature Physics and Chemistry (Proc. 10th Int. Conf., Moscow, USSR (Aug.-Sept. 1966)), Vol.IIA, Superconductivity, ed. by N.V. Zavaritskii et al, Viniti, Moscow, USSR (1967) pp. 480-9.

[Hak67[a]]   HAKE, R.R., "Upper-Critical-Field Limits for Bulk Type-II Superconductors", Appl. Phys. Lett., 10 189-92 (1967); Erratum Appl. Phys. Lett., 15 107 (1969).

[Hak67[b]]   HAKE, R.R., "Paramagnetic Superconductivity in Extreme Type-II Superconductors", Phys. Rev., 158 356-76 (1967).

[Hak68]   HAKE, R.R., "Mixed-State Hall Effect in an Extreme Type-II Superconductor", Phys. Rev., 168 442-4 (1968).

[Hak69]   HAKE, R.R., "Thermodynamics of Type-I and Type-II Superconductors", J. Appl. Phys., 40 5148-69 (1969).

[Hak69[a]]   HAKE, R.R., "Evidence for Fluctuation Superconductivity in Bulk-Type-II Superconductors", Phys. Rev. Lett., 23 1105-8 (1969).

[Hak70]   HAKE, R.R., "Apparent High-Magnetic-Field Fluctuation Superconductivity in the 2-3T_c Range", Phys. Lett., 32A 143-4 (1970).

[Hak71]   HAKE, R.R., "High-Field Fluctuation Superconductivity in a Bulk Extreme-Type-II Superconductor",
Physica (Utrecht), 55 311-6 (1971).

[Hak75]   HAKE, R.R., MONTGOMERY, A.G. and LUE, J.W., "Low-Temperature Anomalies in the Normal-State
Resistivity of Some Disordered, Superconducting, Transition-Metal Alloys", Low Temperature Physics -
LT14 (Proc. 14th Int. Conf., Otaniemi, Finland (Aug. 1975) Vol.3, ed. by M. Krusius and M. Vuorio,
American Elsevier Publishing Co., NY (1975) pp. 122-5.

[Ham72]   HAMPSHIRE, R.G. and TAYLOR, M.T., "Critical Supercurrents and the Pinning of Vortices in Commercial
Nb-60 at.%Ti", J. Phys. F: Metal Phys., 2 89-106 (1972).

[Han51]   HANSEN, M., KAMEN, E.L., KESSLER, H.D. and McPHERSON, D.J., "Systems Titanium-Molybdenum and
Titanium Columbium", Trans. TMS-AIME, 191 881-8 (1951).

[Han66]   HANCOX, R., "The Paramagnetic Limit in Type-II Superconductors", Culham Lab., UK Atomic Energy
Authority, Abingdon, UK, Report No. CLM-R 60 (1966).

[Has68]   HASHIMOTO, Y., TANAKA, M., HIRATA, I., ISHIHARA, K. and KOMATA, T., "Superconducting Properties of
Titanium-Niobium-Tantalum Ternary Alloys", Cryogenic Engineering (Proc. 1st Int. Conf., Tokyo and
Kyoto, Japan (Apr. 1967)), Heywood Temple Industrial Publications Ltd., London, UK (1968) pp. 150-1.

[Has73]   HASSING, R.F., HAKE, R.R. and BARNES, L.J., "Magnetic-Field-Induced One-Dimensional Behavior in the
Specific-Heat Transition in Dirty Bulk Superconductors", Phys. Rev. Lett., 30 6-9 (1973).

[Hat68]   HATT, B.A. and RIVLIN, V.G., "Phase Transformation in Superconducting Ti-Nb Alloys", J. Phys. D:
Applied Phys., 1 1145-9 (1968).

[Haw80]   HAWKSWORTH, D.G. and LARBALESTIER, D.C., "Enhanced Values of B_{c2} in Nb-Ti Ternary and Quaternary
Alloys", Advances in Cryogenic Engineering (Materials), 26 479-86 (1980).

[Haw80[a]] HAWKSWORTH, D.G. and LARBALESTIER, D.C., "The High Field J_c and Scaling Behavior in Nb-Ti and Alloyed
Nb-Ti Superconductors", in Proc. of the 8th Symposium on Engineering Problems of Fusion Research,
Pt.I, San Francisco, CA, 13-16 Nov. 1979 (IEEE 1980) pp. 249-54.

[Haw81]   HAWKSWORTH, D.G. and LARBALESTIER, D.C., "Further Investigations of the Upper Critical Field and the
High Field Critical Current Density in Nb-Ti and its Alloys", IEEE Trans. Magn., MAG-17 49-52 (1981).

[Haw81[a]] HAWKSWORTH, D.G. and LARBALESTIER, D.C., "The Upper Critical Field and Critical Current Density of
Niobium-Titanium and Niobium-Titanium Based Alloys", Cryogenic Engineering/Int. Cryogenic Materials
Conf., San Diego, CA, Aug. 10-14 (1981) unpublished.

[Hec65]   HECHT, R. and HARPER, C.M., "Study of Properties of High-Field Superconductors at Elevated
Temperatures", Radio Corp. of America, Camden, NJ, Defense Electronic Products, Report No. NASA-CR-
67157, N65-34464 (July 1965).

[Hed64]   HEDGCOCK, F.T. and MUTO, Y., "Low Temperature Magnetoresistance in Magnesium and Aluminum Containing
Small Concentrations of Manganese and Iron", Phys. Rev., 134A, 1593-9 (1964).

[Hei64]   HEINIGER, F. and MÜLLER, J., "Bulk Superconductivity in Dilute Hexagonal Titanium Alloys", Phys. Rev., 134 1407-9 (1964).

[Hei66]   HEINIGER, F., BUCHER, E. and MÜLLER, J., "Low Temperature Specific Heat of Transition Metals and Alloys", Phys. Kondens. Materie, 5 243-84 (1966).

[Heℓ64]   HELFAND, E. and WERTHAMER, N.R., "Temperature and Purity Dependence of the Superconducting Critical Field, H_{c2}", Phys. Rev. Lett., 13 686-90 (1964).

[Heℓ66]   HELFAND, E. and WERTHAMER, N.R., "Temperature and Purity Dependence of the Superconducting Critical Field, H_{c2}, II", Phys. Rev., 147 288-94 (1966).

[Heℓ71]   HELLER, W., "Über den Einfluss von Germaniumzusätzen auf supraleitende Eigenschaften und Umwandlungsvorgänge in technischen Titan-Niob-Legierungen", Universität Erlangen-Nürnberg, West Germany, D.Eng. Dissertation (1971).

[Heℓ71[a]]   HELLER, W., LÖHBERG, R. and BABISKIN, J., "Einfluss von Zusätzen dritter Elemente auf das kritische Magnetfeld von Titan-Niob-Legierungen", J. Less-Common Metals, 24 265-76 (1971).

[Heℓ72]   HELLER, W., LÖHBERG, R. and ZWICKER, U., "Einfluss der Verankerungskräfte für Flussfäden auf die kritische Stromdichte in Magnetfeldern bis 10 Tesla bei supraleitenden Titan-Niob-Legierungen", Z. Metallkde., 63 735-9 (1972).

[Hic68]   HICKMAN, B.S., "Precipitation of the Omega Phase in Titanium-Vanadium Alloys", J. Inst. Met., 69 330-7 (1968).

[Hic69]   HICKMAN, B.S., "The Formation of Omega Phase in Titanium and Zirconium Alloys: A Review", J. Mater. Sci., 4 554-63 (1969).

[Hic69[a]]   HICKMAN, B.S., "Omega Phase Precipitation in Alloys of Titanium with Transition Metals", Trans. TMS-AIME, 245 1329-35 (1969).

[Hiℓ69]   HILLMANN, H., "Technologie der harten Supraleiter", Z. Metallkde., 60 157-65 (1969).

[Hiℓ72]   HILLMANN, H. and HAUCK, D., "Relationship Between Defect Structure, Flux-Line Lattice and Metallurgical Treatment in Highfield Superconductor NbTi", Applied Superconductivity Conf., Annapolis, MD (May 1972), IEEE Pub. No. 72CHO682-5-TABSC (1972) pp. 429-33.

[Hiℓ73]   HILLMANN, H. "Entwicklung harter Supraleiter, vorzugsweise am Beispiel Nb-Ti, Teil I", Metall (Berlin) 27 797-808 (1973); [see also: "Teil II" pp. 977-982].

[Hiℓ73[a]]   HILLMANN, H., "Werkstoffe für supraleitende Wechselfeldmagnete mit Ummagnetisierungszeiten der Grossenordnung Sekunde", Vacuumschmelze GmbH, Hanau, Forschungsbericht T 73-03 (April 1973).

[Hiℓ74]   HILLMANN, H., "Ausscheidungen und Flussverankerung bei Hochfeldsupraleitern aus Niobtitan", Siemens Forsch.u. Entwickl. Ber., 3 197-204 (1974).

[Hiℓ75]   HILLMANN, H., "Über die Entwicklung von Hochfeldsupraleitern", Z. Metallkde., 66 69-73 (1975).

[Hiℓ79]   HILLMANN, H., BEST, K.J., HOEFLICH, H., RUDOLPH, J., PFEIFFER, I. and WEBER, H., "Hochfeldsupraleiter aus NbTi mit Stromfahigkeiten über 3000 A fur die Anwendung in Magnetsystem mit Wechselfeldkomponenten", Vacuumschemelze GmbH, Hanau, Forschungsbericht NT 1022 (Apr. 1979).

[Hit65]   HITACHI LTD., "Superconductor Materials", French Patent 1,410,055 (Sept. 1965).

[Ho69]   HO, J.C. and COLLINGS, E.W., "The Influence of ω-Phase on the Superconductivity of Ti-Mo Alloys", Phys. Lett., 29A 206-7 (1969).

[Ho70]   HO,J.C., BOYD, J.D. and COLLINGS, E.W., "Bulk Superconductivity in a Non-Homogeneous Ti-Mo(10 at.%) Alloy", Low Temperature Physics - LT12 (Proc. 12th Int. Conf., Kyoto, Japan (Sept. 1970)), ed. by E. Kanda, Academic Press of Japan (1971) p. 366.

[Ho71]   HO, J.C. and COLLINGS, E.W., "Enhancement of the Superconducting Transition Temperatures of Ti-Mo (5,7 at.%) Alloys by Mechanical Deformation", J. Appl. Phys., 42 5144-50 (1971).

[Ho72]   HO, J.C. and COLLINGS, E.W., "Anomalous Electrical Resistivity in Titanium-Molybdenum Alloys", Phys. Rev. B, 6 3727-38 (1972).

[Ho73]   HO, J.C. and COLLINGS, E.W., "Physics of Titanium Alloys-I: Alloying and Microstructural Effects in the Superconductivity of Ti-Mo", in Titanium Science and Technology (Proc. 2nd Int. Conf. on Titanium (May 1972)) Vol.2, ed. by R.I. Jaffee and H.M. Burte, Plenum Press, NY (1973), pp. 815-30.

[Ho73[a]]   HO, J.C. and COLLINGS, E.W., "Calorimetric Studies of Superconductive Proximity Effects in a Two-Phase Ti-Fe(7.5 at.%) Alloy", Low Temperature Physics - LT13 (Proc. 13th Int. Conf., Boulder, CO (Aug. 1972)) Vol.3, ed. by K.D. Timmerhaus, W.J. O'Sullivan and E.F. Hammel, Plenum Press (1973), pp. 403-7.

[Hop71]   HOPFIELD, J.J., "On the Systematics of High T_c in Transition Metal Materials", Physica (Utrecht), 55 41-9 (1971).

[Hor73]   HORIUCHI, T., MONJU, Y. and NAGAI, N., "Superconducting Transition Temperatures and Resistive Critical Fields of Superconducting Ti-Nb-Zr-Ta Alloys", Nippon Kinzoku Gakkaishi, 37 882-7 (1973).

[Hor73[a]]   HORIUCHI, T., MONJU, Y., TATARA, I. and NAGAI, N., "Phase Transformations of Ti-30Nb-30Zr-7Ta Superconducting Alloy", Nippon Kinzoku Gakkaishi, 37 1057-64 (1973).

[Hor73[b]]   HORIUCHI, T., NOSHIJU, Y., TATA, R. et al, "Precipitation Heat-Treatment of a Superconducting Ti-Nb-Zr-Ta Alloy", Japanese Patent No. 73-85,408 (Nov. 1973).

[Hor74]   HORIUCHI, T., MONJU, Y. and NAGAI, N., "Supraleitende Legierung", German Patent No. 2,350,199 (July 1974).

[Hor75]   HORIUCHI, T., MONJU, Y., TATARA, I. and NAGAI, N., "Verfahren zur Behandlung einer supraleitenden Legierung", German Patent No. 2,347,400 (Apr. 1975).

[Hor80]   HORIUCHI, T., MATSUMOTO, K. and MONJU, Y., "Effect of Ta and Zr Additions to Ti-Nb Alloys on Superconducting Properties", in Proceedings of the 8th Symposium on Engineering Problems of Fusion Research, Pt.I, San Francisco, CA, 13-16 Nov. 1979 (IEEE 1980) pp. 274-7.

[Hor80[a]] HORIUCHI, T., FUKUTSUKA, T., MATSUMOTO, K. and MONJU, Y., "Consideration on the Production System of High Field Superconductors", in Workshop on High Field Superconducting Materials for Fusion -- US-Japan Cooperation Program, Tokyo (1980).

[Hor81] HORIUCHI, T., Private Communication.

[Hul61] HULM, J.K. and BLAUGHER, R.D., "Superconducting Solid Solution Alloys of the Transition Elements", Phys. Rev., 123 1569-80 (1961).

[Hul72] HULM, J.K. and BLAUGHER, R.D., "Transition-Metal Superconductors -- Experimental Survey", in Superconductivity in d- and f-Band Metals (Proc. AIP Conference, Rochester, NY (Oct. 1971)), ed. by D.H. Douglass, AIP Conf. Proc. No.4, AIP, NY (1972) pp. 1-16.

[Hwa79] HWANG, K.F. and LARBALESTIER, D.C., "Generalized Critical Current Density of Commercial Nb46.5, Nb50 and Nb53w/o Ti Multifilamentary Superconductors", IEEE Trans. Magn., MAG-15, 400-3 (1979).

I

[Ike77] IKEBE, M., NAKAGAWA, S., HIRAGA, K. and MUTO, Y., "Anomalous Phonon Thermal Resistivity in Superconducting $Ti_{55}Nb_{45}$ Alloys", Solid State Commun., 23 189-92 (1977).

[Img61] IMGRAM , A.G., WILLIAMS, D.N., WOOD, R.A., OGDEN, H.R. and JAFFEE, R.I., "Metallurgical and Mechanical Characteristics of High-Purity Titanium-Base Alloys", Battelle Memorial Institute, Report 59-595-Part II (Mar. 1961).

[Imp65] IMPERIAL METAL INDUSTRIES (KYNOCH) LTD., "Superconducting Alloys", Belgium Patent No. 659,033 (July 1965).

[Ino80] INOUE, A., KIMURA, H.M., MASUMOTO, T., SURYANARAYANA, C. and HOSHI, A., "Superconductivity of Ductile Ti-Nb-Si Amorphous Alloys", J. Appl. Phys., 51 5475-82 (1980).

[Ino80[a]] INOUE, A., MASUMOTO, T., SURYANARAYANA, C. and HOSHI, A., "Superconductivity of Ductile Titanium-Niobium-Based Amorphous Alloys", 4th Int. Conf. on Liquid and Amorphous Metals Proceedings; J. Phys. Colloq. (France), 41 C8/758-61 (1980).

[Ino80[b]] INOUE, A., SURYANARAYANA, C., MASUMOTO, T. and HOSHI, A., "Superconductivity of Ti-Nb-Si Alloys Crystallized from the Amorphous State", Sci. Report Res. Inst. Tohoku University, 28 182-94 (1980).

[Ish68] ISHIDA, F., DOI, T. and UMEZAWA, T., "The Effect of Precipitation on the Critical Current Density of Nb-5at.%Zr-60at.%Ti Alloy in Magnetic Field", Nippon Kinzoku Gakkaishi, 32 893-7 (1968).

[Ish70] ISHIDA, F., DOI, T., AIYAMA, Y. and TOMIYAMA, S., "Superconductivity of Nb-Ti-Si Alloy", Cryogenic Eng. (Tokyo), 5 33-9 (1970).

[Iwa69] IWASA, Y., "Magnetization of Single-Core, Multi-Strand and Twisted Multi-Strand Superconducting Composite Wires", Appl. Phys. Lett., 14 200-1 (1969).

J

[Jen65]   JENSEN, M.A., MATTHIAS, B.T. and ANDRES, K., "Electron Density and Electronic Properties in Noble-Metal Transition Elements", Science, $\underline{150}$ 1448-50 (1965).

[Jep70]   JEPSON, K.S., BROWN, A.R.G. and GRAY, J.A., "The Effect of Cooling Rate on the Beta Transformation in Titanium-Niobium and Titanium-Aluminium Alloys", in The Science, Technology and Application of Titanium (Proc. 1st Int. Conf. on Titanium (May 1968)), ed. by R.I. Jaffee and N.E. Promisel, Pergamon Press (1970) pp. 677-90.

[Joh79]   JOHNSON, W.L., "Superconductivity and Electronic Properties of Amorphous Transition Metal Alloys", J. Appl. Phys., $\underline{50}$ 1557-63 (1979).

[Jon64]   JONES, C.K., HULM, J.K. and CHANDRASEKHAR, B.S., "Upper Critical Field of Solid Solution Alloys of the Transition Elements", Rev. Mod. Phys., $\underline{36}$ 74-6 (1964).

[Jun76]   JUNOD, A., FLUKIGER, R. and MÜLLER, J., "Superconductivity and Specific Heat in Titanium Base A15 Alloys", J. Phys. Chem. Solids, $\underline{37}$ 27-31 (1976).

K

[Kad72]   KADYKOVA, G.N. and FEDOTOV, L.N., "Structure of Superconductive Alloys 35 BT, 50 BT and 65 BT", Phys. Met. Metallogr. (USSR), $\underline{33}$ No.4, 32-8 (1972), [Transl. of Fiz. Met. Metalloved., $\underline{33}$ 708-14 (1972)].

[Kad73]   KADYKOVA, G.N., "Properties and Structure of Superconducting Ti-Nb Alloys", Met. Sci. Heat Treat. (USSR), $\underline{15}$ Nos. 1-2, 118-22 (1973), [Transl. of Metalloved. Term. Obr. Metallov., No.2, 28-32 (1973)].

[Kad74]   KADYKOVA, G.N., KIRSHENINA, I.I. and FEDOTOV, L.N., "Influence of Iron and Zirconium on the Structure and Resistivity of the Alloy 35 BT", Phys. Met. Metallogr. (USSR), $\underline{37}$ No.3, 94-9 (1974), [Transl. of Fiz. Met. Metalloved., $\underline{37}$ 554-9 (1974)].

[Ker73]   KERKER, G. and BENNEMANN, K.H., "Theory for Superconductivity in Amorphous Transition Metals", Z. Phys., $\underline{264}$ 15-20 (1973).

[Kim62]   KIM, Y.B., HEMPSTEAD, C.F. and STRNAD, A.R., "Critical Persistent Currents in Hard Superconductors", Phys. Rev. Lett., $\underline{9}$ 306-9 (1962).

[Kim63]   KIM, Y.B., HEMPSTEAD, C.F. and STRNAD, A.R., "Flux Creep in Hard Superconductors", Phys. Rev., $\underline{131}$ 2486-95 (1963).

[Kim64]   KIM, Y.B., HEMPSTEAD, C.F. and STRNAD, A.R., "Resistive States of Hard Superconductors", Rev. Mod. Phys., $\underline{36}$ 43-5 (1964).

[Kim65]   KIM, Y.B., HEMPSTEAD, C.F. and STRNAD, A.R., "Flux-Flow Resistance in Type-II Superconductors", Phys. Rev., $\underline{139}$ A1163-73 (1965).

[Kim66] KIM, Y.B. and STRNAD, A.R., "Temperature Dependence of Upper Critical Fields of Type II Superconductors", Quantum Fluids (Proc. of the Sussex University Symposium (Aug. 1965)), ed. by D.F. Brewer, North Holland Publishing Co., Amsterdam and John Wiley and Sons, Inc., NY (1966) pp. 68-73.

[Kim67] KIM, Y.B., "Transport Phenomena in Type-II Superconductors", Low Temperature Physics and Chemistry (Proc. 10th Int. Conf., Moscow, USSR (Aug.-Sept. 1966)), Vol.IIA, Superconductivity, ed. by N.V. Zavaritskii *et al*, Viniti, Moscow, USSR (1967) pp. 43-58.

[Kim69] KIM, Y.B. and STEPHEN, M.J., "Flux Flow and Irreversible Effects", in Superconductivity, Vol.2, ed. by R.D. Parks, Marcel Dekker Inc., NY (1969) pp. 1107-62.

[Kim75] KIM, K.S. and KIM, Y.B., "Flux-Flow Resistance of Paramagnetically-Limited Type-II Superconductors", Low Temperature Physics - LT14 (Proc. 14th Int. Conf., Otaniemi, Finland (Aug. 1975)) Vol.2, ed. by M. Krusius and M. Vuorio, American Elsevier Publishing Co., NY (1975) pp. 473-6.

[Kit69] KITADA, M. and DOI, T., "Precipitation Phenomena of Nb-60Ti-5Zr Superconducting Alloy", Nippon Kinzoku Gakkaishi, 33 1115-21 (1969).

[Kit70] KITADA, M., DOI, T. and UMEZAWA, T., "Applied Magnetic Field-Critical Current Density Characteristics for Superconducting Nb-40Zr-10Ti Alloy Sheets", Nippon Kinzoku Gakkaishi, 34 1-4 (1970).

[Kit70a] KITADA, M. and DOI, T., "Relation Between the Anisotropy of Applied Magnetic Field-Critical Current Density Characteristics and Microstructures for Superconducting Nb-40Zr-10Ti Alloy Sheets", Nippon Kinzoku Gakkaishi, 34 5-10 (1970).

[Kit70b] KITADA, M. and DOI, T., "Effect of Oxygen on Superconducting Characteristics of Nb-40Zr-10Ti Alloy", Nippon Kinzoku Gakkaishi, 34 11-16 (1970).

[Kit70c] KITADA, M. and DOI, T., "Discontinuous Precipitation of Solution Treated Nb-40Zr-10Ti Superconducting Alloy", Nippon Kinzoku Gakkaishi, 34 361-5 (1970).

[Kit70d] KITADA, M. and DOI, T., "Discontinuous Precipitation of Cold Worked Nb-40Zr-10Ti Superconducting Alloy", Nippon Kinzoku Gakkaishi, 34 365-9 (1970).

[Kit70e] KITADA, M. and DOI, T., "Precipitation and Superconducting Properties of Nb-40Zr-10Ti Alloy", Nippon Kinzoku Gakkaishi, 34 369-74 (1970).

[Kit70f] KITADA, M. and DOI, T., "Applied Magnetic Field *versus* Critical Current Density Characteristics of Superconducting Nb-60Ti-5Zr Alloy Sheets", Nippon Kinzoku Gakkaishi, 34 1075-81 (1970).

[Kit70g] KITADA, M. and DOI, T., "Effects of V, Mo or Ta Additions on Superconducting Properties of Nb-40Zr-10Ti Alloy", Nippon Kinzoku Gakkaishi, 34 1082-6 (1970).

[Kit72] KITADA, M. and DOI, T., "Applied Magnetic Field *versus* Critical Current Density Characteristics of Superconducting Nb-Ti-Zr and Nb-Ti-Ge Sheets", Nippon Kinzoku Gakkaishi, 36 891-6 (1972).

[Kit72[a]] KITADA, M., "Relation Between the Applied Magnetic Field *vs* Critical Current Density Characteristics and the Microstructures for Superconducting Nb-Ti-Zr Sheets", Nippon Kinzoku Gakkaishi, 36 1064-9 (1972).

[Kit73] KITADA, M., "Superconductivity of Nb-Ti-Ge Alloys", Nippon Kinzoku Gakkaishi, 37 104-9 (1973).

[Kne62] KNEIP, G.D., Jr., BETTERTON, J.O., Jr., EASTON, D.S. and SCARBROUGH, J.O., "Increased Critical Currents in Nb-Zr Superconductors from Precipitation-Induced Defects", in High Magnetic Fields (Proc. Int. Conf., Cambridge, MA (1961)), MIT Press & John Wiley and Sons Inc. NY (1962) pp. 603-8.

[Kob78] KOBE STEEL LTD., "High Critical Current Density High Tensile Strength Superconductor", Technical Data, Kobe Steel Ltd., Asada Research Lab. (Oct. 1978).

[Kob80] KOBE STEEL LTD., "High Critical Current Density High Tensile Strength Superconductor", Technical Data, Kobe Steel Ltd., Asada Research Lab. (May 1980).

[Kob80[a]] KOBE STEEL LTD., "Critical Current Characteristics of Ti-Nb-Zr-Ta Superconductors", Technical Data, Kobe Steel Ltd., Asada Research Lab. (May 1980).

[Koc76] KOCH, C.C. and SCARBROUGH, J.O., "The Influence of Yttrium on the Superconducting Properties of Nb-Ti Alloys", Metals and Ceramics Div., Oak Ridge National Lab., Annual Progress Report for period ending June 30, 1976, p. 48.

[Koc76[a]] KOCH, C.C., "Superconductivity in the Technetium-Titanium Alloy System", J. Less-Common Met., 44 177-82 (1976).

[Koc77] KOCH, C.C. and EASTON, D.S., "A Review of Mechanical Behavior and Stress Effects in Hard Superconductors", Cryogenics, 17 391-413 (1977).

[Kou70] KOUL, M.K. and BREEDIS, J.F., "Phase Transformation in Beta Isomorphous Titanium Alloys", Acta Metall., 18 579-88 (1970).

[Kra67] KRAMER, D. and RHODES, C.G., "Omega Phase Precipitation and Superconducting Critical Transport Currents In Titanium-22 at. pct. Niobium (Columbium)", Trans. TMS-AIME, 239 1612-5 (1967).

[Kra71] KRAINSKII, I.S. and SHCHEGOLEV, I.F., "Properties of Welded Joints of Wires Made of the Alloys Nb+Zr+Ti and Nb+Ti", Instrum. Exp. Tech. (USSR), 14 1551-5 (1971), [Transl. of Prib. Tekh. Eksp., No.5 242-4 (1971)].

[Kra73] KRAMER, E.J., "Scaling Laws for Flux Pinning in Hard Superconductors", J. Appl. Phys., 44 1360-70 (1973).

[Kro66] KROEGER, D.M., "Flux Motion in High-Field Type-II Superconducting Cylinders", Vanderbilt University, Nashville, TN, Ph.D. Thesis (1966).

[Kud72] KUDO, M., AIHARA, K., KURODA, K. and DOI, T., "A.C. Loss of Superconducting Nb-Ti-Zr Ternary Alloys", Cryogenic Engineering (Proc. 4th Int. Conf., Eindhoven, Netherlands (May 1972)), IPC Science and Technology Press Ltd., Guilford, UK (1972) pp. 143-5.

[Kud76]   KUDO, M. and SHIIKI, K., "Anomalous Critical Alternating Currents of Superconducting Wires", Jpn. J. Appl. Phys., $\underline{15}$ 2445-8 (1976).

[Kun61]   KUNZLER, J.E., BUEHLER, E., HSU, F.S.L., MATTHIAS, B.T. and WAHL, C., "Production of Magnetic Fields Exceeding 15 Kilogauss by a Superconducting Solenoid", J. Appl. Phys., $\underline{32}$ 325-6 (1961).

[Kun70]   KUNAKOV, Ya.N., KACHUR, Ye.V., and PAKHOMOV, V.Ya., "Effect of Various Factors on the Superconducting Properties of Nb-Ti Alloys", in Physics and Metallurgy of Superconductors, ed. by E.M. Savitskii and V.V. Baron, Consultants Bureau, NY (Plenum Publishing Corp.) (1970) pp. 64-8; (Proc. of the Second and Third Conf. on Metallurgy, Physical Chemistry and Metal Physics of Superconductors, Moscow, USSR (May 1965 and May 1966)); first published in Metalloved., Fiz.-Khim., Metallofiz. Sverkhprovodnikov, Izdatelstvo Nauka, Moscow, USSR (1967) pp. 59-63.

[Kwa66]   KWASNITZA, K. and RUPP, G., "Measurement of Critical Field H_{c3} and Critical Surface Current in Superconducting V-Ti Alloys Up to 30 kOe", Phys. Lett., $\underline{23}$ 40-2 (1966).

L

[Lab67]   LABUSCH, R., "Elastische Konstanten des Flussfadengitters in Supraleitern zweiter Art", Phys. Status Solidi, $\underline{19}$ 715-9 (1967).

[Lab69]   LABUSCH, R., "Elastic Constants of the Fluxoid Lattice Near the Upper Critical Field", Phys. Status Solidi, $\underline{32}$ 439-41 (1969).

[Lab69[a]]   LABUSCH, R., "Calculations of the Critical Field Gradient in Type-II Superconductors", Crystal Lattice Defects, $\underline{1}$ 1-16 (1969).

[Lar80]   LARBALESTIER, D.C., "Nb-Ti Alloy Superconductors -- Present Status and Potential for Improvement", Advances in Cryogenic Engineering (Materials), $\underline{26}$ 10-36 (1980).

[Lar81]   LARBALESTIER, D.C., "Niobium-Titanium Superconducting Alloys", in Superconductor Materials Science, Fabrication and Applications" (Proc. NATO Advanced Study Institute on the Science and Technology of Superconducting Materials, Sintra, Portugal (1980)), ed. by S. Foner and B.B. Schwartz, Plenum Press (NATO Study Institute Ser.B: Physics, Vol.68) NY (1981) pp. 133-99.

[Laz64]   LAZAREV, B.G., KHORENKO, V.K., KORNIENKO, L.A., KRIVKO, A.I., MATSAKOVA, A.A. and OVCHARENKO, O.N., "Layered and Filamentlike Structure of Superconducting Nb-Zr and Nb-Ti Alloys", Sov. Phys. JETP, $\underline{18}$ 1417-9 (1964), [Transl. of Zh. Eksp. Teor. Fiz., $\underline{45}$ 2068-9 (1963)].

[Laz68]   LAZAREV, B.G., LAZAREVA, L.S. and GORIDOV, S.I., "Behavior of the Critical Current in Wires of Superconducting Niobium Alloys as a Function of Magnetic Field and Temperature", Sov. Phys. Dokl., $\underline{12}$ 1150-2 (1968), [Transl. of Dok. Akad. Nauk SSSR, $\underline{177}$ 1310-2 (1967)].

[Laz68[a]]   LAZAREV, B.G., OVCHARENKO, O.N., MATSAKOVA, A.A. and VOLOTSKAYA, V.G., "Superconducting Properties of Niobium-Base Alloys", Sov. Phys. JETP, $\underline{27}$ 549-52 (1968), [Transl. of Zh. Eksp. Teor. Fiz., $\underline{54}$ 1031-6 (1968)].

[Laz70]   LAZAREV, B.G., OVCHARENKO, O.N., MATSAKOVA, A.A. and VOLOTSKAYA, V.G., "Superconducting Properties of Niobium-Base Alloys", in Physics and Metallurgy of Superconductors, ed. by E.M. Savitskii and V.V. Baron, Consultants Bureau, NY (Plenum Publishing Corp.) (1970) pp. 86-9; (Proc. of the Second and Third Conf. on Metallurgy, Physical Chemistry and Metal Physics of Superconductors, Moscow, USSR (May 1965 and May 1966)); first published in Metalloved., Fiz.-Khim., Metallofiz. Sverkhprovodnikov, Izdatelstvo Nauka, Moscow, USSR (1967) pp. 76-8.

[Laz70^a]   LAZAREV, B.G., OVCHARENKO, O.N. and MATSAKOVA, A.A., "Critical Magnetic Fields of High-Field Superconductors", in Physics and Metallurgy of Superconductors, ed. by E.M. Savitskii and V.V. Baron, Consultants Bureau, NY (Plenum Publishing Corp.) (1970) pp. 114-6; Proc. of the Second and Third Conf. on Metallurgy, Physical Chemistry and Metal Physics of Superconductors, Moscow, USSR (May 1965 and May 1966).

[Laz72]   LAZAREV, B.G., LAZAREVA, L.S. and GORIDOV, S.I., "Critical Currents in Superconductive Wires of Deformed Alloys Based on Niobium", Sov. Phys. Dokl., 17 265-7 (1972), [Transl. of Dok. Akad. Nauk SSSR, 203 329-31 (1972)].

[Laz73]   LAZAREV, B.G. and GORIDOV, S.I., "Critical Current Value for a Superconductor Niobium-Alloy Wire as a Function of its Copper Coating Thickness", Sov. Phys. Dokl., 17 902-3 (1973), [Transl. of Dok. Akad. Nauk SSSR, 206 85-6 (1972)].

[Laz74]   LAZAREV, B.G., KOGAN, V.S., MARTYNOV, I.S. and SERYUGIN, A.L., "Electron Microscopic Structural Analysis of a Wire of the Superconductive Alloy 60 T", Phys. Met. Metallogr. (USSR), 35 No.1, 215-8 (1974), [Transl. of Fiz. Met. Metalloved., 35 221-4 (1973)].

[Leb65]   LeBLANC, M.A.R., BELANGER, B.C. and FIELDING, R.M., "Paramagnetic Helical Current Flow in Type-II Superconductors", Phys. Rev. Lett., 14 704-7 (1965).

[Leb66]   LeBLANC, M.A.R., "Pattern of Current Flow in Nonideal Type-II Superconductors in Longitudinal Magnetic Fields, I", Phys. Rev., 143 220-3 (1966).

[Leb68]   LeBLANC, M.A.R. and CHANG, C.T.M., "Dependence of the Pinning Force on Magnetic Fields in Type II Superconductors", Solid State Commun., 6 679-83 (1968).

[Lep72]   LEPPER, R., WOLFF, E.G. and MILLS, G.J., "A.C. Permeability Studies of Ternary Alloys at Cryogenic Temperatures", Applied Superconductivity Conf., Annapolis, MD (May 1972), IEEE Pub. No. 72CH0682-5-TABSC (1972) pp. 461-7.

[Loh70]   LÖHBERG, R., HELLER, W. and ZWICKER, U., "Investigations of Superconducting Ti-Nb Alloys with Additions of Third Elements", Cryogenic Engineering (Proc. of 3rd Int. Conf., Berlin, Germany (May 1970)), Iliffe Science and Technology Publications Ltd., Guilford, Surrey, UK (1970) pp. 325-32.

[Loh71]   LÖHBERG, R., "Uber den Einfluss von Kupferzusätzen auf die supraleitenden Eigenschaften und Phasenumwandlungen von technischen Titan-Niob-Legierungen", Universität Erlangen-Nürnberg, West Germany, D.Eng. Dissertation (1971).

[Loh73]   LÖHBERG, R., HELLER, W. and ZWICKER, U., "Influence of Pinning Forces of Titanium-Niobium Base Alloys on the Critical Current up to 10 Tesla", in Titanium Science and Technology (Proc. 2nd Int. Conf. on Titanium (May 1972)) Vol.2, ed. by R.I. Jaffee and H.M. Burte, Plenum Press, NY (1973) pp. 859-70.

[Lon63] LONDON, H., "Alternating Current Losses in Superconductors of the Second Kind", Phys. Lett., 6 162-5 (1963).

[Lub66] LUBELL, M.S. and WIPF, S.L., "Magnetic Diffusivity in a Type II Superconductor", J. Appl. Phys., 37 1012-4 (1966).

[Lub71] LUBELL, M.S. and KERNOHAN, R.H., "Comparison of Calculated and Measured Lower Critical Field for Some Nb-Ti Alloys", J. Phys. Chem. Solids, 32, 1531-9 (1971); first published as Oak Ridge National Lab., TN, Report No. ORNL-TM-3206 and NASA George C. Marshall Space Flight Center, Report No. NASA-CR-117493, N71-21595 (Oct. 1970).

[Lue75] LUE, J.W., MONTGOMERY, A.G. and HAKE, R.R., "Fluctuation Superconductivity at High Magnetic Fields", Phys. Rev. B, 11 3393-6 (1975).

[Luh69] LUHMAN, T.S., TAGGART, R. and POLONIS, D.H., "Correlation of Superconducting Properties with the Beta to Omega Phase Transformation in Ti-Cr Alloys", Scr. Met., 3 777-82 (1969).

[Luh70] LUHMAN, T.S., "Superconductivity and Constitution of Titanium Base Transition Metal Alloys", University of Washington, Seattle, WA, Ph.D. Thesis (1970).

[Luh70a] LUHMAN, T.S., TAGGART, R. and POLONIS, D.H., "The Effects of Step Quenching and Aging on the Superconducting Transition in Beta Stabilized Ti-Cr Alloys", Scr. Met., 4 611-5 (1970).

[Luh71] LUHMAN, T.S., TAGGART, R. and POLONIS, D.H., "The Effect of Omega Phase Reversion on the Superconducting Transition in Titanium-Base Alloys", Scr. Met., 5 81-6 (1971).

[Luh72] LUHMAN, T.S., TAGGART, R. and POLONIS, D.H., "Magnetic Hysteresis Studies of Superconducting Beta Stabilized Titanium Alloys", Scr. Met., 6 1055-60 (1972).

[Lus75] LUSTFELD, H., "Electron-Phonon Matrix Elements and Coupling Constant λ in Transition-Metal Alloys", Solid State Commun., 17 437-9 (1975).

M

[Mae67] MAEDA, S., DOI, T., ISHIDA, F. and KAWABE, U., "Superconducting Properties of Nb-Zr-Ti Ternary Alloys", Low Temperature Physics and Chemistry (Proc. 10th Int. Conf., Moscow, USSR (Aug.-Sept. 1966)), Vol.IIB, Superconductivity, ed. by N.V. Zavaritskii *et al*, Viniti, Moscow, USSR (1967) pp. 63-9.

[Mak64] MAKI, K., "The Magnetic Properties of Superconducting Alloys, I", Physics, 1 21-30 (1964).

[Mak64a] MAKI, K., "The Magnetic Properties of Superconducting Alloys, II", Physics, 1 127-43 (1964).

[Mak64b] MAKI, K. and TSUNETO, T., "Pauli Paramagnetism and Superconducting State", Prog. Theor. Phys., 31 945-56 (1964).

[Mak65] MAKI, K. and TSUZUKI, T., "Magnetic Properties of Intrinsic London Superconductors", Phys. Rev., 139 A868-77 (1965).

[Mak66]   MAKI, K., "Effect of Pauli Paramagnetism on Magnetic Properties of High-Field Superconductors", Phys. Rev., 148 362-9 (1966).

[Mak68]   MAKI, K., "Resistive States in High-Field Type-II Superconductors", Phys. Rev., 169 381-7 (1968).

[Mas80]   MASUMOTO, T., INOUE, A., SAKAI, S., KIMURA, H. and HOSHI, A., "Superconductivity of Ductile Nb-Based Amorphous Alloys", Trans. JIM, 21 115-122 (1980).

[Mas80^a]   MASUMOTO, T. and INOUE, A., "Amorphous Superconducting Alloys Produced by Liquid Quenching", in Workshop on High Field Superconducting Materials for Fusion -- US-Japan Cooperation Program, Tokyo (1980).

[Mat57]   MATTHIAS, B.T., "Superconductivity in the Periodic System", in Progress in Low Temperature Physics, Vol.II, ed. by C.J. Gorter, North-Holland Publishing Co., Amsterdam, Holland (1957) pp. 139-50.

[Mat58]   MATTHIAS, B.T., SUHL, H. and CORENZWIT, E., "Spin Exchange in Superconductors", Phys. Rev. Lett., 1 92-5 (1958).

[Mat59]   MATTHIAS, B.T., COMPTON, V.B., SUHL, H. and CORENZWIT, E., "Ferromagnetic Solutes in Superconductors", Phys. Rev., 115 1597-8 (1959).

[Mat62]   MATTHIAS, B.T., "Superconductivity and Ferromagnetism", IBM J. Res. Dev., 6 250-5 (1962).

[Mat63]   MATTHIAS, B.T., "Superconductivity and Its Relation to Transition Elements", Low Temperature Physics - LT8 (Proc. 8th Int. Conf., London, UK (Sept. 1962)), ed. by R.O. Davies, Butterworths (1963) pp. 135-8.

[Mat63^a]   MATTHIAS, B.T., GEBALLE, T.H. and COMPTON, V.B., "Superconductivity", Rev. Mod. Phys., 35 1-22, 414 (1963).

[Mat65]   MATTHIAS, B.T., "Superconducting Device Consisting of a Niobium-Titanium Composition", U.S. Patent No. 3,167,692 (Jan. 1965).

[Mat70]   MATTHIAS, B.T., "Superconductivity and the Periodic System", Amer. Scientist, 58 80-3 (1970).

[Mci67]   McINTURFF, A.D., CHASE, G.G., WHETSTONE, C.N., BOOM, R.W., BRECHNA, H. and HALDEMANN, W., "Size Effect and Critical Transport Current in Titanium (22 at.% Niobium)", J. Appl. Phys., 38 524-6 (1967).

[Mci73]   McINTURFF, A.D. and CHASE, G.G., "Effect of Metallurgical History on Critical Current Density in NbTi Alloys", J. Appl. Phys., 44 2378-84 (1973).

[Mcm68]   McMILLAN, W.L., "Transition Temperature of Strong-Coupled Superconductors", Phys. Rev., 167, 331-44 (1968).

[Mea65]   MEADEN, G.T., Electrical Resistance of Metals, Plenum Press, NY (1965).

[Mei30]   MEISSNER, W., "Messungen mit Hilfe von flüssigem Helium, VI: Die Übergangskurve zur Supraleitfähigkeit für Titan", Z. Phys., 60 181-3 (1930).

[Mei32] MEISSNER, W., FRANZ, H. and WESTERHOFF, H., "Messungen mit Hilfe von flüssigem Helium, XV: Ba, In, Tl, C and Ti", Ann. Phys. (Leipzig), 13 555-63 (1932).

[Men35] MENDELSSOHN, K., "A Discussion on Superconducting and other Low Temperature Phenomena", Proc. Roy. Soc. London, Ser. A:, A152 34-41 (1935).

[Men64] MENDELSSOHN, K., "On Different Types of Superconductivity", Rev. Mod. Phys., 36 50-1 (1964).

[Men71] MENDIRATTA, M.G., LUTJERING, G., and WEISSMAN, S., "Strength Increase in Ti 35Wt Pct Nb Through Step-Aging", Met. Trans., 2 2599-2605 (1971).

[Mit68] MITSUBISHI ELECTRIC CORPORATION, "Alliages Superconducteurs", French Patent 1,512,769 (Feb. 1968).

[Moℓ65] MOLCHANOVA, E.K., Phase Diagrams of Titanium Alloys, ed. by S.G. Glazunov, Israel Program for Scientific Translations, Jerusaleum (1965), [Transl. of Atlass Diagramm Sostoyaniya Titanovykh Splavov, Izdatel'stvo "Machinostroenie", Moskva (1964)].

[Mon00.00] MONOGRAPH entitled "Applied Superconductivity, Metallurgy and Physics of Titanium Alloys", by E.W. COLLINGS, Plenum Press (1983) International Cryogenic Monograph Series. Citations such as [Mon25.18] refer to Chap.25, Sect.25.18; those in the format [Mon25.00] indicate the entire chapter.

[Moo68] MOON, J.R., "A Possible Source of Instability in Niobium Alloy Superconductors", Phil. Mag., 18 229-36 (1968).

[Mor59] MOREL, P., "Calcul Semi-Empirique de la Temperature de Transition des Superconducteurs et de l'Effet de la Pression sur la Transition", J. Phys. Chem. Solids, 10 277-85 (1959).

[Mor63] MORIN, F.J. and MAITA, J.P., "Specific Heats of Transition Metal Superconductors", Phys. Rev., 129 1115-20 (1963).

[Mor77] MORTON, N., JAMES, B.W., WOSTENHOLM, G.H., TAYLOR, I.A. and TAYLOR, G.M., "The Lattice Thermal Conductivity of Superconducting Vanadium-Titanium Alloys", Cryogenics, 17 447-50 (1977).

[Mur75] MURPHY, J.H., WALKER, M.S. and CARR, W.J., Jr., "Alternating Field Losses in a Rectangular Multifilamentary NbTi Superconductor", IEEE Trans. Magn., MAG-11 313-6 (1975).

N

[Nar66] NARLIKAR, A.V. and DEW-HUGHES, D., "Superconductivity in Deformed Niobium Alloys", J. Mater. Sci., 1 317-35 (1966).

[Nar71] NARAYANAN, G.H., LUHMAN, T.S., ARCHBOLD, T.F., TAGGART, R. and POLONIS, D.H., "A Phase Separation Reaction in a Binary Titanium-Chromium Alloy", Metallography, 4 343-58 (1971).

[Nea71] NEAL, D.F., BARBER, A.C., WOOLCOCK, A. and GIDLEY, J.A.F., "Structure and Superconducting Properties of Nb44% Ti Wire", Acta Metall., 19 143-9 (1971).

[Nea80] NEAL, D.F., Private Communication.

[Net60]   NETZEL, R.G., "Superconducting Transition Temperatures of Titanium and Titanium Isotopes", University of Wisconsin, Madison, WI, Ph.D. Thesis (1960).

[Neu66]   NEURINGER, L.J. and SHAPIRA, Y., "Effect of Spin-Orbit Scattering on the Upper Critical Field of High-Field Superconductors", Phys. Rev. Lett., $\underline{17}$ 81-4 (1966).

[Nis68]   NISHIMURA, T. and ZWICKER, U., "Supraleitung und Phasenumwandlung bei Titanlegierungen, III: Zusätze von Eisen und Niob", Z. Metallkde., $\underline{59}$ 69-73 (1968).

O

[Oℓi74]   OLIVEI, A., "Composite Filamentary Superconductors Made from the Melt", Cryogenic Engineering [Proc. 5th Int. Conf., Kyoto, Japan (May 1974)], ed. by K. Mendelssohn, IPC Science and Technology Press, Sussex, UK (1974) pp. 494-9.

[Orℓ79]   ORLANDO, T.P., McNIFF, E.J., Jr., FONER, S. and BEASLEY, M.R., "Critical Fields, Pauli Paramagnetic Limiting and Material Parameters of Nb_3Sn and V_3Si", Phys- Rev. B, $\underline{9}$ 4545-61 (1979).

[Osa80]   OSAMURA, K., MATSUBARA, E., MIYATANI, T., MURAKAMI, Y., HORIUCHI, T., and MONJU, Y., "Effect of Cold Working on Precipitation Behaviour in Superconducting Ti-Nb Alloys", Phil. Mag. A., $\underline{42}$ 575-89 (1980).

P

[Pak66]   PAKHOMOV, V.Ya., KUNAKOV, Ya.N., KACHUR, Ye.V. and LAYNER, D.I., "Effect of Microheterogeneity on the Critical Currents of Superconductive Alloys', Phys. Met. Metallogr. (USSR), $\underline{22}$ No.4, 181-2 (1966), [Transl. of Fiz. Met. Metalloved., $\underline{22}$ 640 (1966)].

[Par69]   PARKS, R.D., Superconductivity, Marcel Dekker Inc., NY (1969).

[Pfe68]   PFEIFFER, I. and HILLMANN, H., "Der Einfluss der Struktur auf die Supraleitungseigenschaften von NbTi 50 and NbTi 65", Acta Metall., $\underline{16}$ 1429-39 (1968).

[Pin58]   PINES, D., "Superconductivity in the Periodic System", Phys. Rev., $\underline{109}$ 280-7 (1958).

[Poℓ69]   POLLOCK, J.T.A., SADAGOPAN, V. and GATOS, H.C., "Effect of Fast Neutron Irradiation of the Superconducting Properties of Some Nb(Cb)-Ti-V Alloys", Trans. TMS-AIME, $\underline{245}$ 2350-1 (1969).

[Poℓ69[a]]   POLONIS, D.H., "A Study of Phase Transformations and Superconductivity", University of Washington, Seattle, WA, Progr. Report No. 10 and Report No. RLO-1375-18 (Oct. 1969).

[Poℓ70]   POLONIS, D.H., "A Study of Phase Transformations and Superconductivity", University of Washington, Seattle, WA, Annual Progr. Report No. RLO-2225-T-13-6 (Oct. 1970).

[Poℓ71]   POLONIS, D.H., "Study of Phase Transformations and Superconductivity", University of Washington, Seattle, WA, Annual Progr. Report No. RLO-2225-T-13-9, N72-24786 (Oct. 1971).

[Pre67]   PREKUL, A.F. and VOLKENSHTEIN, N.V., "Some Peculiarities of the Magnetization Curves of Heterogeneous Superconductors of the II Kind", Phys. Met. Metallogr. (USSR), $\underline{23}$ No.4, 177-8 (1967), [Transl. of Fiz. Met. Metalloved., $\underline{23}$ 741-3 (1967)].

[Pre67ᵃ]   PREKUL, A.F., "Magnetic Properties of Rigid Superconductors', Phys. Met. Metallogr. (USSR), 24 No. 2, 59-65 (1967), [Transl. of Fiz. Met. Metalloved., 24 260-7 (1967)].

[Pre73]   PREKUL, A.F., RASSOKHIN, V.A. and VOLKENSHTEIN, N.V., "Experimental Confirmation of the Influence of Localized Spin Fluctuations (LSF) on Superconductivity", JETP Lett., 17 252-3 (1973), [Transl. of Pis'ma Zh. Eksp. Teor. Fiz., 17 354-6 (1973)].

[Pre74]   PREKUL, A.F., RASSOKHIN, V.A. and VOLKENSHTEIN, N.V., "Effect of Spin Fluctuations on the Superconducting and Normal Properties of Alloys of Ti Containing V, Nb or Ta", Sov. Phys. JETP, 40 1134-6 (1974), [Transl. of Zh. Eksp. Teor. Fiz., 67 2286-92 (1974)].

[Pre76]   PREKUL, A.F., SHCHERBAKOV, A.S. and VOLKENSHTEIN, N.V., "Resistivity and Anomalous Superconducting Transition in $Ti_{1-x}Fe_x$ Alloys ($0 < x \leq 0.2$)", Sov. J. Low Temp. Phys., 2 684-6 (1976), [Transl. of Fiz. Nizk Temp., 2 1399-404 (1976)].

[Pro70]   PROKOSHKIN, A.F. and PUZEI, I.M., "Connection Between the Superconducting Properties of the Ti-Nb-Fe Alloy and the Parameters of the NGR Spectra", JETP Lett., 11 336-9 (1970), [Transl. of Pis'ma Zh. Eksp. Teor. Fiz., 11 493-7 (1970)].

R

[Rad66]   RADEBAUGH, R. and KEESOM, P.H., "Low-Temperature Thermodynamic Properties of Vanadium, I: Superconducting and Normal States", Phys. Rev., 149 209-12 (1966).

[Raℓ64]   RALLS, K.M., "Effect of Metallurgical Structure on the Superconductivity of Nb-Zr and Nb-Ti Alloys", Massachusetts Institute of Technology, Cambridge, MA, Ph.D. Thesis (1964).

[Raℓ66]   RALLS, K.M., "Internal Surface Superconductivity: The Relationship Between Resistive Critical Field and Bulk Upper Critical Field", Phys. Lett., 23 29-31 (1966).

[Ras72]   RASSMANN, G. and ILLGEN, L., "Zum Zusammenhang zwischen Gefüge und kritischer Stromdichte bei supraleitenden binären Titan-Niob-Legierungen", Neue Hütte, 17 321-8 (1972).

[Ras72ᵃ]   RASSMANN, G. and ILLGEN, L., "Zum Zusammenhang zwischen Gefüge und kritischer Stromdichte bei supraleitenden Titan-Niob Legierungen mit Zusätzen von α-Stabilisatoren", Neue Hütte, 17 547-53 (1972).

[Ras72ᵇ]   RASSMANN, G. and ILLGEN, L., "Zum Zusammenhang zwischen Gefüge und kritischer Stromdichte bei supraleitenden Titan-Niob-Legierungen mit Zusätzen von neutralen Elementen", Neue Hütte, 17 718-23 (1972).

[Ras73]   RASSMANN, G. and ILLGEN, L., "Zum Zusammenhang zwischen Gefüge und kritischer Stromdichte bei supraleitenden Titan-Niob-Legierungen mit Zusätzen von β-Stabilisatoren", Neue Hütte, 18 33-40 (1973).

[Rau63]   RAUB, Ch.J. and ANDERSEN, C.A., "Über die Supraleitfähigkeit von Ti- und Zr-Rh-Legierungen", Z. Phys., 175 105-14 (1963).

[Rau64]   RAUB, Ch., "Supraleitfähigkeit der Edelmetalle und ihrer Legierungen", Z. Metallkde., 55 195-9 (1964).

[Rau64ᵃ]   RAUB, Ch.J. and ZWICKER, U., "Supraleitung und Phasenumwandlung bei Titanlegierungen, I: Zusätze von Vanadium + Aluminum, Niob und Molybdän", Z. Metallkde., 55 711-5 (1964).

[Rau64b] RAUB, Ch.J., "Über die Supraleitfähigkeit ternärer β-Titanlegierungen", Z. Phys., 178 216-20 (1964).

[Rau64c] RAUB, Ch.J. and HULL, G.W., Jr., "Superconductivity of Solid Solutions of Ti and Zr with Co, Rh and Ir", Phys. Rev., 133 A932-4 (1964).

[Rau65] RAUB, Ch.J. and ZWICKER, U., "Superconductivity of α-Titanium Solid Solutions with Vanadium, Niobium and Tantalum", Phys. Rev., 137 A142-3 (1965).

[Rau66] RAUB, Ch.J., RÖSCHEL, E. and ZWICKER, U., "Supraleitung und Phasenumwandlung bei Titanlegierungen, II: Zusätze von Chrom", Z. Metallkde., 57 288-95 (1966).

[Rau67] RAUB, E., RAUB, Ch.J., RÖSCHEL, E., COMPTON, V.B., GEBALLE, T.H. and MATTHIAS, B.T., "The α-Ti-Fe Solid Solution and Its Superconducting Properties", J. Less-Common Metals, 12 36-40 (1967).

[Rau68] RAUCH, G.C., "Aging Phenomena and Superconductivity in Niobium-Titanium Alloys", Massachusetts Institute of Technology, Cambridge, MA, Ph.D. Thesis (1968).

[Rau68a] RAUCH, G.C., COURTNEY, T.H. and WULFF, J., "Aging in Nb(Cb)-Ti-O Superconductors, with Appendix", Trans. TMS-AIME, 242 2263-70 (1968).

[Ray69] RAYEVSKII, I.I., STEPANOV, N.V., SKRYABINA, M.A., DUBROVIN, A.V., ALEKSEEVSKII, N.Ye. and IVANOV, O.S., "Phase Structure and Superconductive Properties of Alloys on the System Niobium-Titanium-Zirconium-Hafnium", Phys. Met. Metallogr. (USSR), 27 No.2,42-55 (1969), [Transl. of Fiz. Met. Metalloved., 27 235-49 (1969).

[Ray71] RAYEVSKII, I.I., STEPANOV, N.V., DUBROVIN, A.V., IVANOV, O.S. and ALEKSEEVSKII, N.Ye., "Superconductive Properties of Niobium-Titanium-Zirconium-Hafnium Alloys", Phys. Met. Metallogr. (USSR), 31 No.1, 71-6 (1971), [Transl. of Fiz. Met. Metalloved., 31 72-8 (1971)].

[Rea78] READ, D.T., "Metallurgical Effects in Niobium-Titanium Alloys", Cryogenics, 18 579-84 (1978).

[Rei67] REINBACH, R., "Verwendung einer Niob-Tantal-Titan-Legierung zur Herstellung von Supraleitern", German Patent 1,237,786 (Mar. 1967).

[Reu66] REUTER, F.W., RALLS, K.M. and WULFF, J., "Microstructure and Superconductivity of a 44.7 At. Pct. Niobium (Columbium)-54.3 At. Pct. Titanium Alloy Containing Oxygen", Trans. TMS-AIME, 236 1143-51 (1966).

[Rey66] REYNOLDS, W.T., "Superconductive Alloys", U.S. Patent No. 3,268,373 (Aug. 1966).

[Ric65] RICKAYZEN, G., Theory of Superconductivity, John Wiley and Sons, NY (1965).

[Ric69] RICKETTS, R.L., "Correlation of Structure and Superconducting Properties in Niobium-Titanium Quaternary Alloys", Massachusetts Institute of Technology, Cambridge, MA, Ph.D. Thesis (1969).

[Ric70] RICKETTS, R.L., COURTNEY, T.H., SHEPARD, L.A. and WULFF, J., "Correlation of Structure and Superconducting Properties in Nb(Cb)-Ti Quaternary Alloys", Met. Trans., 1 1537-44 (1970).

[Rod70] RODRIGUEZ-GONZALEZ, F.A., "Application of Neutron Activation Analysis to the Study of Interstitial Solid Solution of Oxygen in Niobium-Titanium Alloys", Univ. of Texas, Austin, TX, Ph.D. Thesis (1970).

[Rud64] RUDMAN, P.S., "An X-Ray Diffuse-Scattering Study on the Nb-Ti B.C.C. Solution", Acta. Metall., 12 1381-8 (1964).

S

[Sad70] SADAGOPAN, V., GATOS, H.C. and OLSON, G., "Superconducting Properties of bcc Alloys in the Ti-Nb-Mo System", J. Appl. Phys., 41 1874-5 (1970).

[Sai63] SAINT-JAMES, D. and de GENNES, P.G., "Onset of Superconductivity in Decreasing Fields", Phys. Lett., 7 306-8 (1963).

[Saℓ64] SALTER, L.C., Jr., "Investigation of Current Degradation Phenomenon in Superconducting Solenoids", Atomics International, Canoga Park, CA, First Quart. Prog. Report, Contr. No. NAS 8-5356 to NASA (Oct. 1964).

[Saℓ65] SALTER, L.C., Jr., "Investigation of Current Degradation Phenomenon in Superconducting Solenoids", Atomics International, Canoga Park, CA, Second Quart. Prog. Report, Contr. No. NAS 8-5356 to NASA, N65-35450 (Jan. 1965).

[Saℓ65[a]] SALTER, L.C., Jr., Investigation of Current Degradation Phenomenon in Superconducting Solenoids", Atomics International, Canoga Park, CA, Third Quart. Prog. Report, Contr. No. NAS 8-5356 to NASA, N66-27537 (1965).

[Saℓ66] SALTER, L.C., Jr., "Investigation of Current Degradation Phenomenon in Superconducting Solenoids", Atomics International, Canoga Park, CA, Summary Report, Contr. No. NAS 8-5356 to NASA, N66-23800 (Jan. 1966).

[Sas72] SASS, S.L., "The Structure and Decomposition of Zr and Ti B.C.C. Solid Solutions", J. Less-Common Metals, 28 157-73 (1972).

[Sav68] SAVITSKII, E.M., BARON, V.V. and EFIMOV, Yu.V., "Superconducting Alloy", Soviet Patent 223,357 (Aug. 1968).

[Sav71] SAVITSKII, E.M., BARON, V.V., EFIMOV, Yu.V., BYCHKOVA, M.I. and KOZLOVA, N.D., "Use of Experimental Planning to Study Superconductor Systems", Dokl. Phys. Chem., 196 176-9 (1971), [Transl. of Dok. Akad. Nauk SSSR, 196 No.5,1145-8 (1971)].

[Sav71[a]] SAVITSKII, E.M., EFIMOV, Yu.V. and KOZLOVA, N.D., "Structure and Properties of V-Ta-Ti Alloys", Russian Metallurgy (Metally), No.4 149-53 (1971), [Abridged Transl. of Izv. Akad. Nauk SSSR (Metal.), No.4 214-9 (1971)].

[Sav73] SAVITSKII, E.M., BARON, V.V., EFIMOV, Yu.V., BYCHKOVA, M.I. and MYZENKOVA, L.F., Superconducting Materials, Plenum Press, NY (1973).

[Sav76] SAVITSKII, E.M., EFIMOV, Yu.V., KOZLOVA, N.D., MIKHAILOV, B.P., MYZENKOVA, L.F. and DORON'KIN, E.D., Superconducting Materials, "Metallurgy" Publishing House, Moscow, USSR (1976).

[Sch59]   SCHAWLOW, A.L. and DEVLIN, G.E., "Effect of the Energy Gap on the Penetration Depth of Superconductors", Phys. Rev., 113 120-6 (1959).

[Sch73]   SCHMIDT, P.H., "Superconductivity of Transition Metal Thin Films Deposited by Noble Gas Ion Beam Sputtering", J. Vac. Sci. Technol., 10 611-5 (1973).

[Sch81]   SCHOPOHL, N. and SCHARNBERG, K., "Upper Critical Fields in the Presence of Electron-Spin and Spin-Orbit Effects", Physics 107B 293-4 (1981).

[Seg79]   SEGAL, H.R., HEMACHALAM, K., de WINTER, T.A., and STEKLY, Z.J.J., "Development of NbTi Conductor for High Field Applications", IEEE Trans. Magn., MAG-15 807-9 (1979).

[Seg80]   SEGAL, H.R., HRYCAJ, T.M., STEKLY, Z.J.J., de WINTER, T.A. and HEMACHALAM, K., "NbTi-Based Conductors for Use in 12 Tesla Toroidal Field Coils", in Proc. of the 8th Symposium on Engineering Problems of Fusion Research, Pt. I, San Francisco, CA, 13-16 Nov. 1979 (IEEE 1980) pp. 255-9.

[Seg81]   SEGAL, H.R., de WINTER, T.A., STEKLY, Z.J.J. and HEMACHALAM, K., "The Use of NbTiTa as a High Field Superconducting Alloy", IEEE Trans. Magn., MAG-17 53-6 (1981).

[Seg81^a]   SEGAL, H.R., STEKLY, Z.J.J., and de WINTER, T.A., "Current Densities of Commercial NbTi-Based Alloys for High Field Applications", [Proc. of the 7th Int. Conf. on Magnet Technology, Karlsruhe, W. Germany (1981)], IEEE Trans. Magn. MAG-17 1645-8 (1981).

[Sek63]   SEKULA, S.T., BOOM, R.W. and BERGERON, C.J., "Longitudinal Critical Currents in Cold-Drawn Superconducting Alloys", Appl. Phys. Lett., 2 102-4 (1963).

[Sek67]   SEKULA, S.T., KERNOHAN, R.H. and LOVE, G.R., "Superconducting Properties of Technetium", Phys. Rev., 155, 364-9 (1967).

[Sha62]   SHAPOVAL, E.A., "The Upper Critical Field of Superconducting Alloys", Sov. Phys. JETP, 14 628-32 (1962), [Transl. of Zh. Eksp. Teor. Fiz., 41 877-85 (1961)].

[Sha65]   SHAPIRA, Y. and NEURINGER, L.J., "Upper Critical Fields of Nb-Ti Alloys -- Evidence for the Influence of Pauli Paramagnetism", Phys. Rev., 140 A1638-44 (1965).

[Shc73]   SHCHETKIN, I.S. and KHARCHENKO, T.N., "Superconductivity and Electron Structure of a Solid Solution of Titanium in Niobium", Sov. Phys. JETP, 37 491-3 (1973), [Transl. of Zh. Eksp. Teor. Fiz., 64 964-9 (1973)].

[Shi65]   SHIBUYA, Y. and AOMINE, T., "Experimental Upper Critical Field for Ti-V Alloys", J. Phys. Soc. Jpn., 20 2302 (1965).

[Shi67]   SHIBUYA, Y. and AOMINE, T., "Upper Critical Fields and Magnetization Curves for Ti-V Alloys", Low Temperature Physics and Chemistry (Proc. 10th Int. Conf., Moscow, USSR (Aug.-Sept. 1966)), Vol.IIA, Superconductivity, ed. by N.V. Zavaritskii *et al*, Viniti, Moscow, USSR (1967) pp. 490-5.

[Shi74]   SHIIKI, K., AIHARA, K. and KUDO, M., "A.C. Loss and Twisting Effect in Superconducting Composite Conductor", Jpn. J. Appl. Phys., 13 345-50 (1974).

[Shi74a]   SHIIKI, K., "Effect of Matrix Resistivity on Twisting in Superconducting Composite Conductors", Jpn. J. Apply. Phys., 13 1875-80 (1974).

[Shi74b]   SHIIKI, K. and AIHARA, K., "Anisotropic Behaviours in Non-ideal Type-2 Superconducting Alloy Wires", Jpn. J. Appl. Phys., 13 1881-7 (1974).

[Shi74c]   SHIIKI, K., AIHARA, K., KUDO, M. and IRIE, F., "A.C. Losses in Single-Core Superconducting Composite Conductors", Cryogenics, 14 343-5 (1974).

[Shi74d]   SHIIKI, K. and KUDO, M., "Anomalous Hysteresis Loss in Superconducting Wire Due to Flux Jump", J. Appl. Phys., 45 4071-5 (1974).

[Shi77]   SHIIKI, K. and KUDO, M., "Longitudinal Critical Currents in Type II Superconducting Wires", Jpn. J. Appl. Phys., 16 1241-4 (1977).

[Sho40]   SHOENBERG, D., "Properties of Some Superconductors", Proc. Camb. Phil. Soc., 36 84-93 (1940).

[Sik82]   SIKKA, S.K., VOHRA, Y.K., and CHIDAMBARAM, R., "Omega Phase in Materials", Progr. Mater. Sci., 27 245-310 (1982).

[Sin68]   SINHA, A.K., "Low Temperature Specific Heat of B.C.C. Titanium Molybdenum Alloys", J. Phys. Chem. Solids, 29 749-54 (1968).

[Smi52]   SMITH, T.S. and DAUNT, J.G., "Some Properties of Superconductors Below 1 Degree K, III: Zr, Hf, Cd and Ti", Phys. Rev., 88 1172-6 (1952).

[Smi53]   SMITH, T.S., GAGER, W.B. and DAUNT, J.G., "Some Properties of Superconductors Below 1 Degree K", Phys. Rev., 89 654 (1953).

[Soe69]   SOENO, K. and KURODA, T., "Kinetics of Beta-Phase Decomposition and the Precipitation of Alpha-Zirconium in Nb-40at.%Zr-10at.%Ti Superconducting Alloy", Nippon Kinzoku Gakkaishi, 33 791-5 (1969).

[Spi71]   SPITZER, H.J., "Preparation and Superconducting Properties of Thin Films of Some Binary Transition Metal Alloy Systems", J. Vac. Sci. Technol., 8 184-7 (1971).

[Spi74]   SPITZER, H.J., "Superconducting Thin Films in the Niobium-Vanadium and Titanium-Tantalum Systems", Army Mobility Equipment Research and Development Center, Fort Belvoir, VA, Tech. Report No. USAMERDC-2088, DDC AD776 949 (Feb. 1974).

[Spi74a]   SPITZER, H.J., "Type II Film Superconductors for Power Applications", Army Mobility Equipment Research and Development Center, Fort Belvoir, VA, Report No. USAMERDC-2101, DDC AD785 534 (May 1974).

[Spi74b]   SPITZER, H.J., "Superconducting Properties of Some Vanadium-Rich Titanium-Vanadium Alloy Thin Films", Low Temperature Physics - LT13 (Proc. 13th Int. Conf., Boulder, CO (Aug. 1972)), ed. by K.D. Timmerhaus, W.J. O'Sullivan and E.F. Hammel, Plenum Press, NY (1974) pp. 485-9.

[Ste53]   STEELE, M.C. and HEIN, R.A., "Superconductivity of Titanium", Phys. Rev., 92 243-7 (1953).

[Ste78]   STEKLY, Z.J.J., SEGAL, H.R., HEMACHALAM, K., de WINTER, T.A. and COLLING, D.A., "Development of NbTi Conductors for 10T-14T Operation", Magnetic Corp. of America, Final Report, Dept. of Energy Contr. No. EG-77-C-02-4140 (Sept. 1978).

[Str64]   STRONGIN, M., MAXWELL, E., and REED, T.B., "A.C. Susceptibility Measurements on Transition Metal
            Superconductors Containing Rare Earth and Ferromagnetic Metal Solutes", Rev. Mod. Phys., 36 164-8
            (1964).

[Str68]   STRONGIN, M., KAMMERER, O.F., CROW, J.E., PARKS, R.D., DOUGLASS, D.H., Jr. and JENSEN, M.A., "Enhanced
            Superconductivity in Layered Metallic Films", Phys. Rev. Lett., 21 1320-3 (1968).

[Sue68]   SUENAGA, M., OLSON, J. and RALLS, K., Lawrence Radiation Lab., University of California, Berkeley, CA,
            Report No. UCRL-18043 (Mar. 1968) pp. 141-4.

[Sue69]   SUENAGA, M. and RALLS, K.M., "Some Superconducting Properties of Ti-Nb-Ta Ternary Alloys", J. Appl.
            Phys., 40 4457-63 (1969).

[Suk68]   SUKHAREVSKII, B.Ya. and ALAPHINA, A.V., "Some Features of the Temperature Dependence of the Specific
            Heat of a Niobium-Titanium Alloy at the Transition to the Superconducting State", Sov. Phys. JETP,
            27 897-9 (1968), [Transl. of Zh. Eksp. Teor. Fiz., 54 1675-9 (1968)].

[Suk71]   SUKHAREVSKII, B.Ya., SHCHETKIN, I.S. and FAL'KO, I.I., "Investigation of the Superconducting State of
            Solid Solutions of the Niobium-Titanium System", Sov. Phys. JETP, 33 152-55 (1971), [Transl. of Zh.
            Eksp. Teor. Fiz., 60 277-82 (1971)].

[Suℓ67]   SULLIVAN, D.B. and ROOS, C.E., "Energy-Gap Measurements in Type-II Superconductors by a New Method",
            Phys. Rev. Lett., 18 212-5 (1967).

[Sut66]   SUTTON, J. and BAKER, C., "Effect of Precipitates on the Critical Current of a Ti-32 Wt.%Nb
            Superconductor", Phys. Lett., 21 601-2 (1966).

[Swa68]   SWARTZ, P.S. and BEAN, C.P., "A Model for Magnetic Instabilities in Hard Superconductors: The
            Adiabatic Critical State", J. Appl. Phys., 39 4991-8 (1968).

[Swi59]   SWIHART, J.C., "Isotope Effect in the Bardeen-Cooper-Schrieffer and Bogoliubov Theories of
            Superconductivity", Phys. Rev., 116 45-52 (1959).

[Syc69]   SYCHEV, V.V., ZENKEVICH, V.B., ANDRIANOV, V.V. and BAEV, V.P., "Influence of an Alternating Magnetic
            Field on Current Flowing Through a Superconductor of the Second Kind", JETP Lett., 9 31-3 (1969),
            [Transl. of Pis'ma Zh. Eksp. Teor. Fiz., 9 53-7 (1969)].

T

[Tad80]   TADA, E., ANDO, T., OKA, K. and SHIMAMOTO, S., "Superconducting Characteristics of Nb-Ti Alloy in
            High Magnetic Field", Tokai Research Establishment, JEARI, Report No. JAERI-M 8785 (Feb. 1980).

U

[Upt72]   UPTON, P.E., "Low Temperature Specific Heats of Titanium-Vanadium Alloys", Wichita State University,
            KS, M.S. Thesis (1972) unpublished.

V

[Val65] VALLIER, J.C., "Le Retablissement de l'Etat Normal dans divers Alliages Supraconducteurs soumis a des Champs Magnetiques Variables", Phys. Lett., 19 83-4 (1965).

[Var76] VARMA, C.M. and DYNES, R.C., "Empirical Relations in Transition Metal Superconductivity", in Superconductivity in d- and f-Band Metals (Proc. AIP Conf., Rochester, NY (Oct. 1971)), ed. by D.H. Douglass, Plenum Press, NY (1976) pp. 507-33.

[Ver76] VERKIN, B.I., PARKHOMENKO, T.A., PUSTOVALOV, V.V. and STARTSEV, V.I., "Low-Temperature Plasticity of Superconducting Materials of Niobium-Titanium Alloys", Sov. Phys. Dokl., 21 276-7 (1976), [Transl. of Dok. Akad. Nauk SSSR, 228 586-9 (1976)].

[Ver80] VERKIN, B.I., Private Communication.

[Vet65] VETRANO, J.B. and BOOM, R.W., "High Critical Current Superconducting Titanium-Niobium Alloy", J. Appl. Phys., 36 1179-80 (1965).

[Vet68] VETRANO, J.B., GUTHRIE, G.L., KISSINGER, H.E., BRIMHALL, J.L. and MASTEL, B., "Superconductivity Critical Current Densities in Ti-V Alloys", J. Appl. Phys., 39 2524-8 (1968).

[Voz68] VOZILKIN, V.A., PREKUL, A.F., RAKIN, V.G., VOLKENSHTEIN, N.V. and BUYNOV, N.N., "Study of the Dependence of the Superconductive Properties on the Structure of Titanium with 47 wt.% Niobium", Phys. Met. Metallogr. (USSR), 26 No.4,77-84 (1968), [Transl. of Fiz. Met. Metalloved., 26 655-63 (1968)].

W

[Wad80] WADA, H., TACHIKAWA, K. and ROSE, R.M., "A Metallurgical Study on Superconducting Ti-Nb Binary and Ternary Alloys", Titanium '80 Science and Technology (Proc. of the 4th Int. Conf. on Ti, Kyoto, Japan (May 1980)), ed. by H. Kimura and O. Izumi, pub. by The Metallurgical Society, AIME, PA (1980) pp. 745-53.

[Wad81] WADA, H., TACHIKAWA, K. and KATO, T., "Superconducting Properties of Ti-Nb-Hf Alloys", IEEE Trans. Magn., MAG-17 61-4 (1981).

[Wat74] WATANABE, M., "Superconductive Materials", Reito, 49 245-9 (1974).

[Web48] WEBBER, R.T. and REYNOLDS, J.M., "Incipient Superconductivity in Titanium", Phys. Rev., 73 640 (1948).

[Wei68] WEIJSENFELD, C.H., "A Core Model Explaining Hall Effect and Resistivity in the Mixed State of Type II Superconductors", Phys. Lett., 28A 362-3 (1968).

[Wer66] WERTHAMER, N.R., HELFAND, E. and HOHENBERG, P.C., "Temperature and Purity Dependence of the Superconducting Critical Field, H_{c2}, III: Electron Spin and Spin-Orbit Effects", Phys. Rev., 147 295-302 (1966).

[Wes66] WESTINGHOUSE ELECTRIC CORPORATION, "Superconductive Alloys", British Patent 1,019,888 (Feb. 1966).

[Wes67] WESTINGHOUSE ELECTRIC CORPORATION, "Superconductive Alloys", British Patent 1,089,786 (Nov. 1967).

[Wes80] WEST, A.W. and LARBALESTIER, D.C., "Transmission Electron Microscopy of Commercial Filamentary Nb-Ti Superconducting Composites", Advances in Cryogenic Engineering (Materials), 26 471-8 (1980).

[Wes81] WEST, A.W. and LARBALESTIER, D.C., "Microstructure Superconducting Property Relationships in a Fermilab Nb-46.5w/oTi Filamentary Superconducting Composite", IEEE Trans. Magn., MAG-17 65-68 (1981).

[Wes82] WEST, A.W. and LARBALESTIER, D.C., "Alpha-Titanium Precipitation in Niobium-Titanium", Advances in Cryogenic Engineering (Materials), 337-44 (1982).

[Wey65] WEYL, R. and DIETRICH, I., "Degradation of Type II Superconductors", Cryogenics, 5 9-11 (1965).

[Whi74] WHITE, J.J., "Estimation of Confidence Intervals for the Nonlinear Parameters of a Least-Squares Fit. Application to Cholesteryl Nonanoate Data", Appl. Phys., 5 57-61 (1974).

[Whi76] WHITE, J.J. and COLLINGS, E.W., "Analysis of Calorimetrically Observed Superconducting Transition-Temperature Enhancement in Ti-Mo(5 at.%)-Based Alloys", in Magnetism and Magnetic Materials - 1976, ed. by J.J. Becker and G.H. Lander, AIP Conf. Proc. No. 34, AIP, NY (1976) pp. 75-7.

[Whi78] WHITE, J.J. and COLLINGS, E.W., "Characterization of Aged ω-Phase Precipitation in Ti-V (19 at.%) from Analysis of Rounded Superconducting Transition Calorimetry Experiments", in Electrical Transport and Optical Properties of Inhomogeneous Media (Proc. AIP Conf., Columbus, OH (1977)), ed. by J.C. Garland and D.B. Tanner, AIP Conf. Proc. No. 40, AIP, NY (1978) p. 408.

[Wil69] WILLIAMS, J.C. and BLACKBURN, M.J., "The Influence of Misfit on the Morphology and Stability of the Omega Phase in Titanium-Transition Metal Alloys", Trans. TMS-AIME, 245 2352-5 (1969).

[Wil70] WILSON, M.N., WALTERS, C.R., LEWIN, J.D., SMITH, P.F. and SPURWAY, A.H., "Experimental and Theoretical Studies of Filamentary Superconducting Composites", J. Phys. D: Appl. Phys., 3 1517-83 (1970).

[Wil71] WILLIAMS, J.C., HICKMAN, B.S. and LESLIE, D.H., "The Effect of Ternary Additions on the Decomposition of Metastable ß-Phase Titanium Alloys", Met. Trans., 2 477-84 (1971).

[Wil73] WILLIAMS, J.C., "Critical Review: Kinetics and Phase Transformations", in Titanium Science and Technology (Proc. 2nd Int. Conf. on Titanium (May 1972)) Vol.3, ed. by R.I. Jaffee and H.M. Burte, Plenum Press, NY (1973) pp. 1433-94.

[Wil75] WILLBRAND, J. and SCHLUMP, W., "Einfluss von Ausscheidungsdichte und Teilchengrösse auf die Stromtragfähigkeit von NbTi-Supraleitern", Z. Metallkde., 66 714-9 (1975).

[Wil75a] WILLBRAND, J., ARNDT, R., EBELING, R. and MOHS, R., "Optimierung supraleitender NbTi Legierungen", Fried. Krupp GmbH, Krupp Forschungsinstitut, Essen, Forschungsbericht T 75-35 (Nov. 1975).

[Wil77] WILSON, M.N., "Stabilization of Superconductors for Use in Magnets", IEEE Trans. Magn., MAG-13, 440-6 (1977).

[Wiℓ78] WILLIAMS, J.C., "Precipitation in Titanium-Base Alloys", in Precipitation Processes in Solids, ed. by K.C. Russell and H.I. Aaronson, pub. by The Metallurgical Society AIME, PA (1978) pp. 191-221.

[Wiℓ80] WILLBRAND, J., Private Communication.

[Wip65] WIPF, S.L. and LUBELL, M.S., "Flux Jumping in Nb-25%Zr Under Nearly Adiabatic Conditions", Phys. Lett., 16 103-5 (1965).

[Wip67] WIPF, S.L., "Magnetic Instabilities in Type-II Superconductors", Phys. Rev., 161 404-16 (1967).

[Wit72] WITCOMB, M.J. and NARLIKAR, A.V., "Magnetic Irreversibility and Microstructure of a Nb-40at.%Ti Alloy", Phys. Stat. Sol. (a), 11 311-8 (1972).

[Wit73] WITCOMB, M.J. and DEW-HUGHES, D., "Superconductivity of Heat-Treated Nb-65at.%Ti Alloy", J. Mater. Sci., 8 1383-400 (1973).

[Woℓ63] WOLGAST, R.C., HERNANDEZ, H.P., ARON, P.R., HITCHCOCK, H.C. and SOLOMON, K.A., "Superconducting Critical Currents in Wire Samples and Some Experimental Coils", Advances in Cryogenic Engineering, 8 601-7 (1963).

[Woℓ73] WOLFF, E.G., LEPPER, R. and MILLS, G.J., "Relationships Between Microstructure, Superconductivity and Mechanical Properties of Ti-6Al-4V" in Titanium Science and Technology (Proc. 2nd Int. Conf. on Titanium (May 1972)) Vol.2, ed. by R.I. Jaffee and H.M. Burte, Plenum Press, NY (1973) pp. 843-58.

Y

[You64] YOUNG, F.J. and SCHENK, H.L., "Critical Alternating Currents in Superconductors", J. Appl. Phys., 35 980-1 (1964).

Z

[Zwi63] ZWICKER, U., "Supraleitung von Titan und seinen Legierungen", Z. Metallkde., 54 477-83 (1963).

[Zwi63[a]] ZWICKER, U., "Procédé de Fabrication de Fils et Rubans Supraconducteurs", Belgian Patent 633,765 (Nov. 1963).

[Zwi64] ZWICKER, U., "Metallkundliche und technologische Probleme bei harten Supraleitern auf Niob- und Titanbasis", Metall., 18 941-8 (1964).

[Zwi65] ZWICKER, U., "Einflusse von Platinmetallen auf technologische Eigenschaften und Leitfahigkeit von Titan-Niob-Legierungen", Z. Metallkde., 56 222-8 (1965).

[Zwi68] ZWICKER, U., MEIER, T. and RÖSCHEL, E., "Verarbeitbarkeit, Phasengleichgewichte, Elektrische und Supraleitfähigkeit im System Niob-Titan-Zirkonium", J. Less-Common Metals, 14 253-68 (1968).

[Zwi70]   ZWICKER, U., LÖHBERG, R. and HELLER, W., "Metallkundliche Probleme und Supraleitung bei Legierungen auf Basis Titan-Niob, die als Werkstoffe für die Herstellung von supraleitenden Magneten dienen können", Z. Metallkde., 61 836-47 (1970).

[Zwi74]   ZWICKER, U., Titan und Titanlegierungen, Springer-Verlag, Berlin, Heidelberg and NY (1974).

AUTHOR INDEX

Although primarily an AUTHOR index, this is also a NAME index in that it does list some names which, in the text, are unaccompanied by references to the literature. An example of this is "LANDAU", a name which is implied each time the anagram "GLAG" (for GINZBURG, LANDAU, ABRIKOSOV, and GOR'KOV) is cited. Names such as EILENBERGER may be associated with specific literature citations (e.g. on p.226) and also in a more general way with un-referenced expressions or procedures (e.g. on p.384). All contributors to the literature under review in this book are listed. The format used,

JONES, A.B., (Smith), 123, (456), [789], ... ,

conveys the following information: *(i)* The first name is, of course, that of the author or co-author being listed. *(ii)* Co-author names in n-author (n>2) papers seldom appear explicitly in the text, generally being subsumed under one or more of the *et al's* which will occur on the pages listed. To assist in locating the contributions in such cases, the names of frequently associated, or representative, first authors may also be given -- e.g. (Smith) in the above example. *(iii)* The bare numbers in the sequence following the name(s) correspond, as usual, to pages containing references within the *TEXT*. *(iv)* The round brackets direct the reader to *FIGURE CAPTIONS*, the square brackets to *DATA TABLES*. For any page, round- and square-bracketed references pre-empt unbracketed ones but not each other.

A

ABRIKOSOV, A.A., 131, 136, 139, 141, 145, 173, 188, 191, 200, 221, 222, 225-8, 230, 231, 284, 291.

ADAM, E., 77, 388.

AGARWAL, K.L., 11, 22, 24, 27, 29, 30, 32, 153, 154, [155], 156, [157], [158], 159, 162, [164], 165, 166, [167], 168, 171, 217.

AIHARA, K., (Kudo), 98, 99, 100, 373, 374, 375, 376.

AIYAMA, Y., (Ishida), 81, 322.

AKACHI, T., 18, 144.

ALAPHINA, A.V., 54, 55, (132), 218.

ALBERT, H., 287.

ALEKSEEVSKII, N.E., 86, 90, 114, 115, 342, 344, (345), 346, 348, (349), (353), (354), (365), 406-9, 412, 413, (414).

ALLEN, P.B., 426, 427.

ANDERSEN, C.A., 6, 46, 123, (124), 128, 201, 202, [203], 205.

ANDERSON, D.E., 19, 62, 68, [146], 283.

ANDERSON, P.W., 145, 234.

ANDO, T., (Tada), 72, 95, 292, (295), (296), (297), (298), 366, (367), [367].

ANDRES, K., (Jensen), 58, 131, 169, 170.

ANDRIANOV, V.V., (Sychev), 87, 342.

AOMINE, T., 14, 17, 139.

110, 112, 116, 209, 210, [212], 255-7, 259, 263,
[264], 266, [267], 269-71, 273, [275], 277, 348,
350, [352], 379, 380, 382, 383, 392, (393), 395,
[396], [397], [398], [400].

C

CAMPBELL, A.M., 274, 291.

CAPE, J.A., 23, 25, 28, 29, 31, 40, 51, 153, 156,
[157], (158), [158], [159], 161-3, 166, [167],
168, 170, [171], 172, 186, (187), 189, (190),
(191), 208.

CARR, W.J., Jr., 375.

CASIMIR, H.B.G., 162.

CHANDRASEKHAR, B.S., 37, 40, 56, 136, (138), 140, 147,
173, 182, [194], 195, 221, 224, 225.

CHANDRASEKARAN, V., 23, 111, 134, 200, 392.

CHANG, C.T.M., (LeBlanc), 372.

CHARLESWORTH, J.P., 66, 257, 259, 263, 273, 284.

CHASE, G.G., 64, 67, 219, (220), 256, 257, 259, 262,
[264], 265, 266, [267], 269-71, 274, [275], [278].

CHENG, C.H., 8, 14, 132, 133, 136, 137, 139, [148],
149.

CHESTER, P.F., 101, 119, 243, 359, 405.

CHIDAMBARAM, R., (Sikka), 247.

CHIKABA, J., 238.

CLARK, D., (Brown), 215.

CLEM, J.R., 144.

CLOGSTON, A.M., 136, 137, (138), 139-41, 144, 147, 173,
174, 188, 221, 226, 228.

COFFEY, H.T., (238), 274.

COLLING, D.A., 33-5, 52, 69, 70, 72, 107, 172-4, 176,
(177), [178], [179], [180], 211, 255-9, 278,
(285), [289], 290, (291), 388, [390], (391), 392.

COLLINGS, E.W., 6, 11, 12, 21, 25, 37-9, 42, 49, 128,
129, 131, (132), (134), 135, 136, 146, 147, (148),
[148], 149, (158), 161, 162, [164], [165], 167,
182, (183), 184, (185), (186), (193), [194], 195,
200, 201, 207, (208), [211], 212, 217, 218, 247,
[316].

COLLVER, M.M., 121, 423-5, (426), 427, 429, 430.

COMEY, K.R., Jr., 75, 306, 307, 311.

COMPTON, V.B., (Matthias), 22, 24, 26, 27, 29, 30,
44-6, 132, 133, 153, [155], 156, [159], (160),
161, 163, [164], [165], 166, [167], 168, 198,
[199], 201, [203], 205, 217.

COOPER, L.N., (Bardeen), 126, 129, 132, 133, 182, 218,
222.

CORENZWIT, E., (Matthias), 8, 22, 24, 26, 29, 30, 44-6,
132, 133, 153, [155], 156, [159], (160), 161,
[164], [165], 166, [167], 168, 198, [199], 201,
[203], 205, 217.

COTTON, W.L., 80, 111, 114, 320, [321], 327, 392,
(393), [399], 406.

COURTNEY, T., 65, 75, 76, 79, 84, 245, 256, 258, 306,
(307), (308), [309], 311, 313, [314].

CROW, J.E., (Strongin), 183.

CURTIS, C.W., 69, 107, 257, 259, 383, 388, (389),
[389].

D

DANNER, S., 48, 201, [203], 205.

DAUNT, J.G., 2, 3, 5, 31, 123, 125, [130].

de GENNES, P.G., 142.

DEGTYAREVA, V.F., 5, 7, 128, [130].

de HAAS, W.J., 2, [130].

DeSORBO, W., 9, 31, 34, 51, 53, 58, 90, 103, 106, 110,
112, 114, 132, [148], 149, 169, 174, [179], 209,
(216), 217, 223, (353), (354), 379, 385, 394, 396,
[401], [402], [403], [404], 408, [409].

DEVLIN, G.E., 124.

DEW-HUGHES, D., 63, 76, 244, 274, 306, 309, 310.

de WINTER, T.A., (Stekly), 69, 70, 72, 107, 108, 239,
240, 255-9, 278, (285), [289], 290, (291), 388,
[390], (391), (392).

DIETRICH, I., 56, 61, 62, 231, 232, 243, 256-9, 277.

DOI, T., 78, 81, 84, 89, 93-8, 101, 102, 115, 248,
[306], 312, 322, 333, 334, (336), 343, 344, 347,
[348], (349), 350, (351), 356, 358, (359), 360-2,
(363), [363], (364), (365), (368), 369, (370),
[371], 373-5, 406, (407), 412, 414.

DOMB, E.R., 174, 424, 432, 433.

DORON'KIN, E.D., (Savitskii), 82, 103, [396].

DOSDAT, J.P., (Adam), (Bidault), 77, 306, (312), [313],
388.

DOUGLASS, D.H., Jr., (Strongin), 183.

DUBEK, L. (Gandolfo), 232.

DUBROVIN, A.V., (Rayevskii), 86, 114, 115, 342, 344,
(345), 346, 406-9, 412, 413, (414).

DUMMER, G., 6, 46-8, 123, (124), 128, 201, 202, [203],
205.

DUWEZ, P., (215).

DYNES, R.C., 426, 427.

E

EASTABROOK, J., (Brown), (215).

EASTHAM, A.R., 372.

EASTON, D.S., 76, (215), 342.

EBELING, R., 68, 82, 84, 250, (251), [252], 257, 259, 279, [280], (281), 285, 287, 325, 329, (330), [330], 331, 333, 335, (337), (338), (339).

EDGECUMBE, J., 19, 62, 68, [146], 283.

EFFERSON, K.R., (Gauster), (236).

EFIMOV, Yu.V., 10, 19, 20, 49, 50, 82, 89, 92, 103, 105, 106, 110, 112, 116, 132, 135, 145, 149, [151], 207, 209-11, [212], 348-50, [352], 382, 383, 392, (393), [396], [397], [398].

EILENBERGER, G., 226, 384, 394.

EISENSTEIN, J., 3, 123.

EL BINDARI, A., 13, 56, 136, 139, [150], 151, 188, 224-6.

EVETTS, J.E., (Campbell), 274, 291.

F

FAL'KO, I.I., (Sukharevskii), 54, 218.

FALGE, R.L., Jr., 4, 24, 26, [124], [130], 156, [157], 159, 165.

FEDOTOV, L.N., (Alekseevskii), 86-8, 342-4, (345), 346, 354.

FERRELL, R.A., 200.

FIELDING, R.M., (LeBlanc), 372.

FIETZ, W.A., 58, 220, 226, 274.

FLUKIGER, R., 47, 48, 203, [204], 205.

FONER, S., (Orlando), 141, 175, 182, 222, 228, 230, 248, 380, 408.

FRANZ, H., (Meissner), 2, [130].

FREYHARDT, H.C., 274, 291.

FRIEDEL, J., 284, 427.

FRÖLICH, H., 129.

FUKUTSUKA, T., (Horiuchi), 118, 415.

G

GAGER, W.B., (Smith), 3, 5, 31, 123, 125, [130].

GANDOLFO, D.A., 57, 231, 232, (233), 234, 238.

GANGULY, B.N., 27, 160.

GARLAND, J.W., 126, [127], 128, 427.

GATOS, H.C., 10, 54, 59, 105, 111, 132, [148], 149, 217, 382, 384, 392, (393), (394), [397], [400],

[402], [403].

GAUSTER, W.F., 235, (236).

GAUTHIER, R., 372.

GEBALLE, T.H., 27, 133, 161, 163, [164], 165, 199, 219.

GEGEL, H.L., 49, 207, [211], 212, 217.

GENEVEY, D., (Best), 70, 252, 288.

GEY, W., (Buckel), 6, 46, 123, (124), 128, 201.

GIDLEY, J.A.F., (Neal), 66, 215, 239, 246, 248, 249, (254), 257, 259, 269, 271, [272], 273, 274, [275], 283, (284), 285, (286), 338, (349).

GINZBURG, N.I., 5, 126, [130].

GINZBURG, V.L., 131, 136, 138, 139, 141, 173, 188, 189, 191, 200, 220, 221, 225-8, 230, 231.

GLADSTONE, G., (127).

GOODMAN, B.B., 13, 137, 145, 173, 221, 232, 384, 394.

GOPAL, E.S.R., 157.

GOR'KOV, L.P., 131, 136-9, 141, 173, 174, 187, 188, 191, 200, 221, 225-8, 230, 231, 384, 394.

GORIDOV, S.I., (Lazarev), 67, 86, 88, 257, 342, 344, [345], 346.

GORINA, N.B., 87, 342, 344, 346.

GORTER, C.J., 162.

GRAEBNER, J.E., 424.

GRAY, J.A., (Jepson), 247.

GRIGORYEV, A.M., 91.

GRUZNOV, Yu.A., (Gorina), 87, 342, 344, 346.

GUPTA, K.P., (Cheng), 8, 14, 132, 133, 136, 137, 139, [148], 149.

GUTHRIE, G.L., (Vetrano), 20, 145, [146], [151], 259, [275].

H

HAASEN, P., 291.

HACKETT, W.H., Jr., 17, 18, 144.

HAKE, R.R., 13, 15-7, 19, 20, 23, 25, 28-31, 34, 35, 37, 40-3, 45, 50, 51, 53, 56, 59, 61, 136, 137, (139), 140, 141, (142), 144, 145, [146], [150], 151, 153, (154), [155], 156, [157], [158], [159], 160, 163, [164], 165, 166, [167], 168, 170, [171], 172, 173, (174), 175, [179], [180], 182, 186, (187), 188, (189), (190), (191), (192), (193), [194], [195], [196], 200, 202, 208, 210, 217, 219, 220-3, 225, 226, (227), 230, 232, (239), 242, 255, 258-60, 344, 432.

HALDEMANN, W., (McInturff), 64.

HAMMOND, R.H., 121, 423-5, (426), 427, 429, 430.

HAMPSHIRE, R.G., 66, 257, 259, 263, 274.

I

SUBJECT INDEX

This is both a general and a topical index. Although some specially important references to SUPERCONDUCTING TRANSITION TEMPERATURE, LOWER, THERMODYNAMIC, and UPPER CRITICAL FIELDS, CRITICAL CURRENT DENSITY, and FLUX PINNING are called out in the general listing, all text-discussions of these topics are fully referenced in sub-indexes entitled (in alphabetical order): "**CRITICAL CURRENT DENSITY**, Factors which Influence it in Titanium Alloys", "**SUPERCONDUCTING TRANSITION TEMPERATURE**, Discussion of its Response to Alloying", "**UPPER CRITICAL FIELD**, Discussion of its Responses to Alloying and Temperature", and "**UPPER CRITICAL FIELD**, General Discussion". For the reader wishing to retrieve *DATA* from the book, three sub-indexes of **Tabulated and Plotted Data** are also provided under the headings: "**CRITICAL CURRENT DENSITY**, Tabulated and Plotted Data", "**SUPERCONDUCTING TRANSITION TEMPERATURE**, Tabulated and Plotted Data", and "**UPPER CRITICAL FIELD**, Tabulated and Plotted Data".

It should also be borne in mind that the TABLE OF CONTENTS provides another means whereby information in this book, sorted according to alloy composition, can be readily retrieved.

-- Ti-Zr-Nb: 343, 344, 360, 361.

-- Ti-Nb-Mo-Al: 406.

AGING OF METALLIC GLASS ALLOYS,

-- in general, at room temperature: 425.

-- of Ti-Nb-Si, 1-h at various temperatures: 429, 431, 432.

-- "stress relaxation" at the atomic level: 428.

AGING, Optimal Temperature in Relation to,

-- α-phase precipitation: 259, 266, 307, 308.

-- enhancement or "refinement" of the subband (deformation cell) structure: 259, 283, 308.

-- H_r maximization: 224, 332.

-- J_c maximization: 269, 270, 277, 283, 290, 307, 311, 312, 322, 329, 344, 356, 388, 394.

-- T_c maximization: 326.

-- ω-phase and α-phase precipitate formation and subband refinement, under common aging conditions: 259, 274, 283.

-- ω-precipitation: 259, 266, 283, 307, 308, 313.

ALLOY CATEGORIZATION, see CATEGORIZATION OF Ti-Nb.

ALLOY DESIGN PHILOSOPHY, Metallic Glasses: 425.

ALLOY DESIGN PHILOSOPHY, Superconductors,

-- for high H_{c2}, see HEAVY ELEMENT SUBSTITUTION.

-- for high H_{c2}, the "TWELVE-TESLA" PROGRAM: 239.

-- for high J_c: 291, 292, 412.

ALLOY DEVELOPMENT, Multicomponent, philosophies of for high critical parameters: 408.

AMORPHOUS ALLOYS: 175, 423-34.

AMORPHOUS Ti: 127, 128.

ANISOTROPIES IN (COLD) ROLLED RIBBON (STRIP), see COLD-ROLLED RIBBON.

ANOMALOUS COMPOSITION DEPENDENCES OF T_c,

-- in Ti-Mn, Ti-Fe, and Ti-Co: 133,

-- see also MAGNETIC INTERACTION MECHANISM FOR SUPERCONDUCTIVITY.

ANOMALOUS (INCOMPLETE) CALORIMETRIC SUPERCONDUCTING TRANSITION,

-- in Ti-Co: 166.

-- in Ti-Cr: 153, 154.

-- in Ti-Fe: 160-2, 164.

ANOMALOUS RESISTIVITY, see RESISTIVITY, ANOMALOUS and RESISTIVITY MINIMUM.

α-PHASE PRECIPITATE PARTICLE DIAMETERS, Data: 252, 261, 266, 273, 280, 313, 338, 339, 365, 420.

α-PHASE PRECIPITATION, Matrix Enrichment Associated with, see SOLUTE ENRICHMENT THROUGH PHASE DECOMPOSITION.

α-PHASE PROPERTIES AND PRECIPITATION,

-- at β'/β interfaces: 318.

-- at carbide needles: 300.

-- at grain boundaries: 361, 395.

-- at subband boundaries: 266, 334.

-- catalyzed by Cu addition to Ti-Nb: 328.

-- catalyzed by oxygen additions to Ti-Nb: 308, 311.

-- catalyzed by Si addition to Ti-Nb: 323.

-- catalyzed by simple metal additions in general to Ti-Nb: 318.

-- catalyzed by Y additions to Ti-Nb: 323.

-- conversion from ω-phase, non-occurrence of: 318.

-- deformability during wire drawing: 287.

-- from the $\beta'+\beta$-phase regime: 318.

-- from the $\omega+\beta$-phase regime: 318.

-- globular, after long-time aging: 365.

-- influence on solute composition of the β-matrix, see SOLUTE ENRICHMENT/PARTITIONING.

-- in relation to subband structure: 251, 253, 266, 280, 282, 284-8, 304, 338.

-- in Soviet alloys: 343, 344.

-- in Ti-Nb, compositional range of occurrence: 248, 249.

-- in Ti-Nb, during aging after prior deformation: 266.

-- in Ti-Nb in general: 246, 248-53, 266, 273.

-- in Ti-Nb-Mo: 393.

-- in Ti-Nb-O: 308.

-- in Ti-Nb-Si metallic glass ribbon: 431.

-- in Ti-TM alloys in general: 317, 318.

-- in Ti-Zr-Nb-Ta: 419-21.

-- modified by Cu and Ge additions to Ti-Nb: 339.

-- morphology of in Ti-Nb-Cu and Ti-Nb-Ge: 338.

-- nucleation and growth mechanism: 261, 266.

-- number density of: 240, 252, 266, 286, 287.

-- optimal aging temperature for, see AGING.

-- redistribution under final cold deformation: 253, 278, 287, 288.

-- refinement by Si additions to Ti-Nb: 323.

-- relative flux-pinning effectiveness *versus* ω-phase: 145, 308, 312.

-- sluggishness in high-concentration Ti-Nb alloys: 282.

-- STEM/EDAX investigations of: 249, 287.

-- stimulated by cold work: 248, 262, 278, 361, 420.

-- within β' grains: 318.

α-PHASE STABILIZATION,

-- by dissolved oxygen: 173, 308.

-- by simple metals: 183, 318, 319.

APPLICATIONS OF ALLOY SUPERCONDUCTORS: x, 239, 359, 364, 366, 372.

APPLICATIONS OF ALLOY SUPERCONDUCTORS, RIBBON: 369.

ATOMIC NUMBER, Z, Influence on Spin-Orbit Scattering: 141, 230, 408, see also HEAVY ELEMENT SUBSTITUTION.

B

BARDEEN-STEPHEN MODEL: 143, 190, 235,
    see also FLUX FLOW,
    see also HALL EFFECT IN THE MIXED STATE.
BCS,
-- deviation function, Δ: 174, 218.
-- energy gap: 133, 218, 350.
-- interaction parameter, V: 129, 132, 182.
-- theory of superconductivity: 129, 133.
BEAN MODEL, Critical State: 232, 233.
BEAN MODEL, Flux Pinning, see FILAMENTARY MESH (SPONGE
    OR MENDELSSOHN-BEAN) MODEL.
BERNAL RANDOM-CLOSE-PACKED STRUCTURE: 425.
BILLET PROCESSING: 289, 415.
β'-PHASE AND β''-PHASE, Definitions: 348.
β-PHASE IMMISCIBILITY,
-- in Nb-Zr and Nb-Zr-rich alloys: 248, 313.
-- in Ti-Zr-Nb: 313, 348, 349, 360, 361, 364, 365, 368,
    370.
-- in Ti-Zr-Nb, kinetics of: 360, 361.
-- morphology (lamellar) of the decomposition product:
    361.
-- reaction enhanced by the presence of oxygen: 313.
β-PHASE REVERSION,
-- in Ti-Cr: 155.
-- in Ti-V: 135, 144.
-- in Ti-TM alloys in general: 318.
-- in Ti-Nb-Mo: 393.
-- in Ti-Nb-Mo-Al: 406.
-- related to β-phase separation: 318.
β-PHASE SEPARATION,
-- catalyst for α-phase precipitation: 318.
-- compared to G.P. zone formation: 343.
-- contrasted with β-phase immiscibility: 348.
-- contrasted with ω-phase precipitation: 318.
-- in Ti-Cr: 155.
-- in Ti-Nb: 248, 262, 318.
-- in Ti-TM alloys in general: 317, 318.
-- in Ti-V: 135.
-- in Ti-Zr-Nb: 344, 361
-- related to β-phase reversion: 318.
β-PHASE STABILIZATION BY TRANSITION METALS: 318, 328,
    386, 394.
β-Ti, Lattice Parameter of: 202.

C

CABLING OF HIGH-CURRENT CONDUCTOR: 290.
CALORIMETRIC MEASUREMENT, of
-- Ti-Cr: 153.
-- Ti-Fe: 132, 161-3.
-- Ti-Hf: 171.
-- Ti-Ir: 203.
-- Ti-Mn: 156, 157.
-- Ti-Mo: 132, 182-6.
-- Ti-Nb: 132, 217, 218.
-- Ti-Pt: 203.
-- Ti-Rh: 201, 202.
-- Ti-V: 132, 134, 135.
-- Ti-Zr: 170.
-- Ti-Mo-Al: 207, 208.
-- Ti-V-Cr: 210.
-- unalloyed Ti: 125, 132.
CALORIMETRIC MEASUREMENT OF THE MIXED STATE, in Ti-Mo:
    186, 189.
CALORIMETRIC MEASUREMENT OF THE SUPERCONDUCTING
    TRANSITION,
-- advantages, as a measuring technique for T_c: 124,
    166.
-- as a step towards the determination of H_c: 136, 137.
-- curve-fitting procedures applied to the results of:
    134, 184.
-- relative height of the "jump" at T_c: 134, 162, 164,
    166, 167, 183-6.
-- see also ANOMALOUS (INCOMPLETE) CALORIMETRIC SUPER-
    CONDUCTING TRANSITION.
-- see also ROUNDED CALORIMETRIC TRANSITION.
CALORIMETRIC SUPERCONDUCTING TRANSITION, ANOMALOUS,
    see ANOMALOUS (INCOMPLETE) CALORIMETRIC SUPERCON-
    DUCTING TRANSITION.
CALORIMETRIC TRANSITION ROUNDING, see ROUNDED CALORI-
    METRIC TRANSITION.
CATEGORIZATION OF Ti-Nb ALLOYS, Three Important Concen-
    tration Ranges: 262, 274, 302, 307.
CLADDING MATERIAL: 360,
    see also COPPER.
CLUSTERING EFFECTS, see SOLUTE CLUSTERING, and
    MAGNETIC CLUSTERING, respectively.
COATING, LIQUID-METAL: 360.
COHERENCE LENGTH,
-- in relation to electronic mean free path (dirtiness):
    221.
-- in relation to precipitate spacing in Ti-Nb and
    Ti-Zr-Nb-Ta: 225, 420.

CRITICAL CURRENT DENSITY, Factors which Influence it in Titanium Alloys, see also Table of Contents,

-- additions of Ag to Ti-Nb: 324.

      Al to Ti-Nb: 322.

      boron to Ti-Nb: 299.

      carbon to Ti-Nb: 300, 301.

      Cr to Ti-Nb: 394.

      Cu to Ti-Nb: 327-31.

      Cu and Ge to Ti-Nb, a relative assessment: 337-9.

      Fe to Ti-Nb: 396.

      Ge to Ti-Nb: 333-7.

      Hf to Ti-Nb: 381, 382.

      Hf to Ti-Zr-Nb: 413.

      Mo to Ti-Nb: 394.

      nitrogen to Ti-Nb: 301-5.

      oxygen to Ti-Nb: 305-13.

      oxygen to Ti-Nb, influence on α-phase and ω-phase precipitation: 307.

      oxygen to Ti-Zr-Nb: 357.

      Si to Ti-Nb: 322, 323.

      simple metal to Ti-Nb: 317-39.

      Sn to Ti-Nb: 324.

      Ta to Ti-Nb: 386-92.

      Ta to Ti-Zr-Nb: 414-21.

      Y to Ti-Nb: 323, 324.

      Zr to Ti-Nb: 341-5, 354-71.

-- aged microstructure (following quenching or cooling) in Ti-Nb: 260, 261.

-- aged microstructure (following cold deformation) in Ti-Nb: 262-76.

-- aging temperature, high-concentration D//A Ti-Nb: 271.

-- aging temperature, intermediate-concentration D//A Ti-Nb: 269.

-- aging temperature, low-concentration D//A Ti-Nb: 265.

-- aging time, high-concentration D//A Ti-Nb: 271, 272.

-- aging time, intermediate-concentration D//A Ti-Nb: 270.

-- aging time, low-concentration D//A Ti-Nb: 265.

-- aging time/temperature optimization: 145, 262, 265, 269, 270, 274, 311, 336, 362, 363, 418.

-- anisotropy of cold-rolled ribbon conductors: 151, 175, 192, 260, 367-70.

-- α-phase precipitates, see also α-PHASE PROPERTIES AND PRECIPITATION.

-- α-phase precipitates, influence in ribbon conductors: 335, 336.

-- α-phase precipitates in relation to subbands: 280, 282, 284-8, 308, 338.

-- α-phase precipitates in Ti-V: 145.

-- α-phase precipitates in Ti-Nb and Ti-Nb-base alloys: 261, 266, 267, 270, 278, 280, 281, 285, 286, 307, 308, 311, 313, 323, 328, 334, 335, 339, 343, 361, 365, 368, 419, 420.

-- α-phase *vis-à-vis* ω-phase, relative flux-pinning efficacies: 145, 308, 312.

-- β-phase immiscibility in Ti-Zr-Nb: 313, 360, 361, 364, 365, 368, 370.

-- β-phase separation in Ti-Nb and Ti-Nb-base alloys: 262, 312, 344, 361.

-- carbide precipitates: 300.

-- cold deformation, aging time, aging temperature, relative importance of, in D//A Ti-Nb: 265.

-- cold deformation of high-concentration D//A Ti-Nb: 271.

-- cold deformation of intermediate-concentration D//A Ti-Nb: 269.

-- cold deformation of low-concentration D//A Ti-Nb: 264, 265.

-- cold deformation of Ti-Nb, effectiveness of total *vis-à-vis* final: 289, 290, 311.

-- cold deformation of Ti-Nb, optimal final: 287-90.

-- cold deformation of Ti-Nb, *vis-à-vis* precipitation: 273, 274, 284-7.

-- cold deformation of Ti-Nb and Ti-Nb-base alloys, total area reduction: 289, 290, 389-91, 415, 418, 419.

-- cold deformation of Ti-Ta: 175.

-- cold rolling induced critical current anisotropy: 151, 175, 192, 260, 367-70.

-- cold rolling of Ti-Mo: 191, 192.

-- cold rolling of Ti-Nb: 260, 262.

-- cold rolling of Ti-Ta: 175.

-- cold rolling of Ti-V: 145.

-- composition, alloy: 276, 285, 292, 358, 364, 365.

-- coring of starting ingot or billet: 245, 268, 341-3.

-- dislocations (also dislocation networks); 254, 309, 328,
    see also subband structure.

-- filamentary microstructures: 145, 191, 232, 260, 344.

-- heat treatment of Ti-Ta: 176.

-- heat treatment, multiple intermediate, of Ti-Nb: 280, 281, 287.

-- heat treatment, intermediate, of Ti-Nb, single *vis-à-vis* multiple: 290.

-- impurities present in the starting Ti: 268, 269, 310, 367.

-- intermetallic compound precipitates: 323, 328, 334, 339, 434,
  see also INTERMETALLIC (INCLUDING "INTERSTITIAL") COMPOUND PRECIPITATION.
-- interstitial element effects on Ti-Nb: 176, 211, 267, 268, 299-314.
-- interstitial-element effects vis-à-vis associated host-alloy microstructural effects: 299, 313.
-- interstitials, dissolved: 304, 313.
-- interstitials, precipitated: 300, 304.
-- Kroll-process Ti as compared to iodide-process Ti: 268, 269, 300, 304, 310, 311, 357.
-- martensitic platelets in Ti-Nb: 260, 261.
-- microstructural anisotropy due to cold rolling: 191, 335, 336.
-- microstructure, amorphous: 433.
-- microstructure, cold deformed/aged/final deformed (D//A//D): 276-8, 362, 364, 368, 371, 389, 418.
-- microstructure, cold deformed plus aged (D//A): 262-76, 355, 358, 362-4, 369, 371, 381, 414.
-- microstructure, cold rolled: 151, 175, 192, 260.
-- microstructure, multiple intermediate heat treated/final deformed [(A-D-A)//D]: 279-82, 330, 336, 337.
-- microstructure, quenched plus cold deformed: 260.
-- microstructure, recrystallized: 260.
-- microstructure, recrystallized from amorphous state: 434.
-- microstructure, recrystallized plus aged (R//A): 260, 261.
-- microstructure, recrystallized plus cold deformed: 260.
-- optimal aging temperature: 266, 270, 277, 290, 307, 308, 322, 324, 334, 388, 394.
-- optimal aging time/temperature: 145, 262, 265, 269, 270, 274, 311, 336, 362, 363, 418.
-- optimal final cold deformation: 282, 287-90, 418, 419.
-- optimal sequences of deformation and aging (D//A): 277, 278, 369, 371, 414, 415, 418, 419.
-- optimal subband structure: 251, 254, 280, 286, 338.
-- orientation of applied magnetic field with respect to current direction: 146, 192.
-- percolation of current along deformation-induced filamentary pathways: 145, 191, 232, 260, 344.
-- phase separation products in Ti-Nb and Ti-Nb-base alloys: 262, 312, 344, 361.
-- precipitate number-density: 240, 252, 286.
-- precipitate structures in Ti-Ta: 176.
-- precipitate structures in Ti-V: 145.

-- precipitates, interstitial elements: 300, 304.
-- precipitates, carbide: 300.
-- precipitates, grain boundary or cell-wall: 266, 282, 334.
-- precipitates, intermetallic compound: 323, 328, 334, 434,
  see also INTERMETALLIC (INCLUDING "INTERSTITIAL") COMPOUND PRECIPITATION.
-- precipitates, morphology of: 338, 365.
-- precipitates, nitride: 304.
-- precipitates, oxide: 309, 311, 313.
-- precipitates, see α-phase precipitates,
  see also α-PHASE PROPERTIES AND PRECIPITATION.
-- precipitates, see ω-phase precipitates,
  see also ω-PHASE PRECIPITATION AND PROPERTIES.
-- precipitation in β-immiscible systems: 313, 360, 361, 364, 365, 368, 370.
-- precipitation of β-phase in an amorphous matrix: 434.
-- precipitation of β' in phase-separation systems: 262, 312, 344, 361.
-- precipitation vis-à-vis cold deformation in Ti-Nb: 273, 274.
-- rare earth additions to Ti-Nb: 313, 314.
-- recrystallized microstructure in Ti-Nb: 260.
-- ribbon conductors, precipitation in: 335, 336.
-- rolled ribbon structure in Ti-Nb: 260, 262.
  Ti-Ta: 175.
  Ti-V: 145.
  Ti-Nb-Ge: 334-6.
  Ti-Zr-Nb: 367-71.
-- segregation, see solute partitioning and/or solute segregation,
  see also SOLUTE SEGREGATION.
-- shearing of rolled strips: 175, 192.
-- simple-metal effects vis-à-vis associated host-alloy microstructural effects: 317, 323.
-- solute concentration in D//A Ti-Nb: 262-76.
-- solute concentration in Ti-Nb alloys: 276, 285.
-- solute concentration in Ti-Zr-Nb alloys: 358, 364, 365.
-- solute partitioning between precipitate and matrix: 261, 328, 331.
-- solute segregation effects: 270, 273, 341-3.
-- sputter deposition of Ti-Nb: 282, 283.
-- sputter deposition of Ti-Ta: 176, 177.
-- sputter deposition of Ti-V: 146.
-- subbands in relation to α-phase precipitates: 251, 253, 266, 280, 282, 284-8, 304, 338.

-- subband structure: 250-4, 269, 273, 283, 284, 287, 304, 310, 334, 337-9.

-- subband structure, absence of in amorphous alloys: 433.

-- subband structure, thermal recovery or disintegration of: 278, 308, 338.

-- temperature: 291, 392, 420.

-- upper critical field (at high fields): 192, 231, 239, 242, 276, 291, 292, 355, 362, 366, 367, 390, 408.

-- vacancies: 328.

-- ω-phase precipitates,
    see also ω-PHASE PRECIPITATION AND PROPERTIES.

-- ω-phase precipitates in Ti-Nb and Ti-Nb-base alloys: 261, 266, 267, 270, 307, 308, 311, 313, 328, 334, 343.

-- ω-phase precipitates in Ti-V: 145.

-- ω-phase vis-à-vis α-phase, relative flux-pinning efficacies: 145, 308, 312.

CRITICAL CURRENT DENSITY, LONGITUDINAL APPLIED FIELD, see FORCE-FREE AND NEARLY-FORCE-FREE CURRENT FLOW.

CRITICAL CURRENT DENSITY, Magnetic Measurement of: 232, 233.

CRITICAL CURRENT DENSITY MEASUREMENT CRITERIA,

-- two examples: 390.

-- see captions/footnotes to J_c-relevant figures/tables.

CRITICAL CURRENT DENSITY MEASUREMENT, Pulse Method: 244.

CRITICAL CURRENT DENSITY, Tabulated and Plotted Data, see also Table of Contents,

BINARY ALLOYS

-- Soviet technical alloys, tabulated data: 345.

-- Ti-Mo rolled strip, $J_c(||)$ and $J_c(\perp)$ plotted *versus* applied field: 192.

-- Ti-Mo rolled strip, $J_c(||)$ and $J_c(\perp)$, tabulated data: 197.

-- Ti-Nb, α-phase precipitation, influence on $J_c(5T)$, plotted and tabulated data: 286, 287.

-- Ti-Nb commercial composites, plotted *versus* applied field: 295-8.

-- Ti-Nb composite conductors, $J_c(2K$ and $4K)$ plotted *versus* applied field: 291.

-- Ti-Nb composite conductors, response to total and final cold deformation, plotted and tabulated data: 288, 289.

-- Ti-Nb, intercomparison between binary alloy and Ti-Nb-Ta, plotted *versus* total cold work area reduction, also *versus* temperature: 391, 392.

-- Ti-Nb, intercomparison between binary alloy and Ti-Zr-Nb-Ta, plotted *versus* applied field: 420.

-- Ti-Nb, intercomparison between $J_c(H_a)$'s of sixteen commercial composite conductors: 295-8.

-- Ti-Nb, intercomparison of the influences of Kroll-process and iodide-process Ti, plotted *versus* applied field: 269.

-- Ti-Nb, intercomparison with Ti-Nb-Cu and Ti-Nb-Ge, plotted *versus* applied field: 338.

-- Ti-Nb, intercomparison with Ti-Zr-Nb-Ta, plotted *versus* applied field for two temperatures: 420.

-- Ti-Nb, $J_c(4.5T)$ plotted *versus* 1-h aging temperature: 270, 277.

-- Ti-Nb, $J_c(5T)$ plotted *versus* 1-h aging temperature: 307.

-- Ti-Nb, $J_c(8T)$ plotted *versus* 24-h aging temperature: 322.

-- Ti-Nb, $J_c(4.5T)$ plotted *versus* 1-h aging temperature, various levels of cold work and intermediate heat treatments: 270, 277.

-- Ti-Nb, $J_c(3-8T)$ plotted *versus* aging time at 400°C: 382.

-- Ti-Nb, $J_c(5T)$ plotted *versus* reciprocal subband diameter: 284, 287.

-- Ti-Nb, $J_c(5-13T)$ plotted *versus* temperature: 392.

-- Ti-Nb, $J_c(7T)$ plotted *versus* total cold work area reduction: 419.

-- Ti-Nb, $J_c(10$ and $13T)$ plotted *versus* total cold work area reduction ratio: 391.

-- Ti-Nb, $J_c(3$ and $5T)$ tabulated and plotted *versus* at.% Nb: 267, 275.

-- Ti-Nb, $J_c(5T)$ tabulated as function of 1-h aging temperature: 272.

-- Ti-Nb, $J_c(5T)$ tabulated as function of cold deformation: 272.

-- Ti-Nb, $J_c(3T)$ tabulated as function of final processing conditions: 278.

-- Ti-Nb, $J_c(5T)$ tabulated as function of intermediate heat treatment: 280.

-- Ti-Nb, "peak effect" demonstration: 355.

-- Ti-Nb, plotted *versus* applied field, as function of final processing conditions, in demonstration of the "peak effect": 355.

-- Ti-Nb, plotted *versus* applied field, cold-deformed wire: 265.

-- Ti-Nb, plotted *versus* applied field for numerous intermediate heat-treatment conditions: 281.

-- Ti-Nb, tabulated as function of applied field, response of composite conductors to process optimization: 282.

-- Ti-Nb, tabulated data, intermediate-concentration research alloys, various heat treatments, influence of aging: 267, 268.

-- Ti-Nb, tabulated data, low-concentration research alloys: 264.

-- Ti-Nb, Soviet technical alloy T 60, tabulated data: 345.

-- Ti-Nb, subband diameter, influence on J_c(5T), plotted and tabulated data: 284, 286, 287.

-- Ti-Ta cold-drawn wire, tabulated data: 181.

-- Ti-Ta, plotted *versus* applied field: 177.

-- Ti-Ta, plotted *versus* applied field, intercomparison between processed wire and sputtered film: 177.

-- Ti-Ta processed wire, tabulated data: 180.

-- Ti-Ta rolled strip, tabulated data: 180.

-- Ti-Ta sputtered films, plotted *versus* applied field: 177.

-- Ti-Ta sputtered films, tabulated data: 181.

-- Ti-V processed wire, tabulated data: 151.

-- Ti-V rolled strip, tabulated data, "longitudinal" and "transverse" J_c, numerous compositions and several applied fields: 146, 151.

-- Ti-V sputtered films, tabulated data, "longitudinal" and "transverse" J_c, numerous compositions and several applied fields: 146.

TERNARY ALLOYS

-- amorphous Ti-Nb-Si, tabulated data: 433.

-- Ti-Hf-Nb, J_c(3-8T) plotted *versus* aging time at 400°C: 382.

-- Ti-Nb-Al, intercomparison between Al-alloyed Ti-Nb and a binary control, plotted data: 322.

-- Ti-Nb-Al, J_c(8T) plotted *versus* 24-h aging temperature: 322.

-- Ti-Nb-Cu, intercomparison between Cu-alloyed Ti-Nb and a binary control, tabulated data: 329.

-- Ti-Nb-Cu, intercomparison with Ti-Nb and Ti-Nb-Ge, plotted *versus* applied field: 338.

-- Ti-Nb-Cu, plotted *versus* applied field for numerous intermediate heat-treatment conditions, also for two fields: 330.

-- Ti-Nb-Cu, tabulated data, influence of aging: 329.

-- Ti-Nb-Ge, intercomparison between wire and rolled strip, plotted *versus* applied field: 336.

-- Ti-Nb-Ge, intercomparison with Ti-Nb and Ti-Nb-Cu, plotted *versus* applied field: 338.

-- Ti-Nb-Ge, plotted *versus* applied field, intercomparison between wire and rolled strip: 336.

-- Ti-Nb-Ge, plotted *versus* applied field, response of composite conductor to intermediate heat treatment: 337.

-- Ti-Nb-Ge, plotted *versus* applied field, response to final heat treatment: 336.

-- Ti-Nb-N, tabulated data, intercomparison of J_c(5T) between nitrided and un-nitrided alloys: 303.

-- Ti-Nb-O, intercomparison between oxidized and un-oxidized alloys, plotted and tabulated data: 307, 312-4.

-- Ti-Nb-O, J_c(5T) plotted *versus* 1-h aging temperature: 307.

-- Ti-Nb-O, plotted and tabulated data, influence of aging: 307, 309, 310.

-- Ti-Nb-Ta, intercomparison between ternary alloys and binary Ti-Nb, tabulated and plotted data: 390-2.

-- Ti-Nb-Ta, J_c(4T) plotted *versus* aging time at 400 and 500°C: 388.

-- Ti-Nb-Ta, J_c(5-13T) plotted *versus* temperature: 392.

-- Ti-Nb-Ta, J_c(12T) plotted *versus* total area reduction ratio: 391.

-- Ti-Nb-Ta, plotted *versus* applied field, also tabulated data, composite conductors under two processing conditions: 389.

-- Ti-Nb-Ta, tabulated data: 390.

-- Ti-Nb-Ta, tabulated data, early results: 387.

-- Ti-Zr-Nb commercial superconducting composites, plotted *versus* applied field: 367.

-- Ti-Zr-Nb, intercomparison between wire and rolled strip, anisotropy of ribbon, tabulated data: 371.

-- Ti-Zr-Nb, intercomparison between wire and rolled strip, plotted *versus* applied field: 370.

-- Ti-Zr-Nb, intercomparison with Ti-Nb within the context of commercial composite conductors, plotted *versus* applied field: 367.

-- Ti-Zr-Nb, J_c at "peak" plotted on composition triangle, also tabulated data: 356, 357.

-- Ti-Zr-Nb, J_c(8T) plotted on composition triangle: 364, 365.

-- Ti-Zr-Nb, J_c(4.5T) plotted *versus* at.% Nb, influence of 1h/400°C aging: 358.

-- Ti-Zr-Nb, J_c(4.5T) plotted *versus* at.% Nb, intercomparison of cold-worked (cw) and cw+aged alloys: 358.

-- Ti-Zr-Nb, J_c(5, 7, and 8T) plotted *versus* composition, patented alloys: 359.

-- Ti-Zr-Nb, J_c(4.5T) tabulated as function of Nb concentration, comparison with Ti-Nb: 358.

-- Ti-Zr-Nb, J_c(4, 8, and 9T) tabulated for representative X-type and Z-type alloys: 363.

-- Ti-Zr-Nb, plotted *versus* applied field, comparison between wire and ribbon: 370.

DENSITY OF STATES, see ELECTRONIC DENSITY OF STATES.

DEVIATION FUNCTION, BCS: 174, 218.

DIAGNOSTIC VALUE OF SUPERCONDUCTIVITY MEASUREMENT: 135, 154, 155, 208, 255.

DIFFUSION,
-- solute-solvent in relation to ω-precipitation: 245.
-- transition metals in β-Ti: 167.

DIFFUSION OF MAGNETIC FLUX: 234.

DIMENSIONALITY, Superconductive Effects Related to: 145, 186.

DISLOCATION(S),
-- cell structures in Ti-Nb: 232, 244, 245, 254.
-- flux pinning by, early studies of: 309.
-- influence on H_{c2}: 223.
-- influence on normal-state resistivity: 223.
-- loops: 344.
-- migration to cell boundaries during aging: 274, 283.
-- networks of, influence on J_c: 328.
-- tangles, Ti enrichment at: 273.

DISLOCATION CELLS, see SUBBAND STRUCTURE.

DISLOCATION LOOPS, see COHERENCY STRAIN.

DISORDER,
-- as a factor in the T_c of unalloyed Ti films: 129.
-- influence on the T_c-composition (e/a) dependence of TM_1-TM_2 alloys: 427.
-- influence on the T_c of deformed Ti-Mo: 183.
-- influence on the T_c of unalloyed Ti: 127, 128.

DISTRIBUTION OF TRANSITION TEMPERATURES: 162, 217, see also ROUNDED TRANSITION and DOUBLE TRANSITION.

DOUBLE TRANSITION, in Phase Decomposed Ti-Nb-Mo and Ti-Nb-Mo-Al: 393, 406.

DUCTILE-BRITTLE TRANSITION: 359.

E

EDDY-CURRENT LOSS, AC Current,
-- frequency dependence of: 373.
-- influence of Cu-layer thickness: 373.
-- self-field amplitude dependence of: 373.

EDDY-CURRENT LOSS, AC Applied Field,
-- critical frequency, f_c, for: 375.
-- field-amplitude dependence of: 375, 376.
-- frequency dependence of: 375, 376.
-- "low-" and "high-frequency" regimes of: 375.
-- matrix resistivity, influence of: 375-7.
-- twist pitch, relationship to f_c and eddy-current loss: 375-7.

ELASTIC-VORTEX-LATTICE THEORY: 142, 239, 274.

ELECTRICAL RESISTIVITY, see RESISTIVITY.

ELECTRON/ATOM RATIO (e/a),
-- as a display parameter for H_{c2}, H_r, or H_u: 223, 239, 394.
-- as a display parameter for T_c: 127, 132, 169, 199, 217, 382, 392-5, 406, 426.
-- effective value of DeSORBO *et al*: 132, 169, 174, 209, 217, 223, 379, 396.
-- T_c dependence on, crystalline TM_1-TM_2 alloys: 127.
-- T_c dependence on, crystalline *vis-à-vis* amorphous cryodeposited TM_1-TM_2 alloys: 426.

ELECTRON DENSITY, as a Display Parameter for T_c: 131, 170, 211.

ELECTRON-ELECTRON INTERACTION, Coulomb Pseudopotential, μ^*: 128, 426.

ELECTRONIC DENSITY OF STATES AT THE FERMI LEVEL, $n(E_F)$,
-- electron-phonon renormalization of: 141, 170.
-- proportionality to the electronic specific heat coefficient, γ: 126, 135, 170, 182, 222, 427.
-- relationship to electron-phonon coupling constant (McMILLAN theory): 128, 427.
-- relationship to T_c (BCS theory, etc.): 129, 132, 135, 170, 173, 182, 427.

ELECTRONIC MEAN FREE PATH,
-- as a "dirtiness" parameter: 221.
-- in relation to atomic spacing: 230.
-- in relation to coherence length: 221.
-- in relation to spin-orbit mean free path: 140.

ELECTRONIC SPECIFIC HEAT, At or Below T_c,
-- GORTER-CASIMIR temperature dependence: 162.
-- "jump" at T_c: 162, 164, 166, 167, 183-6.

ELECTRONIC SPECIFIC HEAT COEFFICIENT,
-- relationship to $n(E_F)$: 126, 135, 170, 173, 182, 222, 427.
-- relationship to T_c: 127, 135, 170, 173, 182, 183, 217, 222, 432, 433.

ELECTRON-PHONON INTERACTION,
-- coupling constant, λ: 128, 182, 426, 427.
-- coupling constant, the η-parameter: 427.
-- density of states renormalization due to: 141, 170, 175, 182, 230.
-- pairing potential, V: 129, 132, 182.
-- see also INTERACTIONS RESPONSIBLE FOR SUPERCONDUCTIVITY.

ELECTRON SCATTERING,
-- influence on nonparamagnetic upper critical field temperature dependence: 226.
-- isotropic and anisotropic (p-wave): 226.
-- mean free path, ℓ: 140, 221.
-- normal relaxation time, τ: 141.

ENERGY GAP, BCS: 133, 218, 350.

ENRICHMENT EFFECTS THROUGH PHASE DECOMPOSITION, see
    SOLUTE ENRICHMENT.
ENRICHMENT, SOLUTE, see SOLUTE ENRICHMENT.
EUTECTIC COMPOSITION, Significance in Metallic-Glass
    Stability: 425, 428.
EXCESS LOW TEMPERATURE SPECIFIC HEAT, As a Result of:
    (i) nuclear hyperfine interaction; *(ii)* superpara-
    magnetic clustering; *(iii)* localized spin ordering:
    157.

F

FERMI DENSITY OF STATES, see ELECTRONIC DENSITY OF
    STATES AT THE FERMI LEVEL, $n(E_F)$.
FERMI SURFACE,
-- area of: 137, 221.
-- Γ-point on: 171.
"FERROMAGNETIC SOLUTES" AND SUPERCONDUCTIVITY: 160,
    166, 199,
    see also MAGNETIC INTERACTION MECHANISM FOR SUPER-
    CONDUCTIVITY.
FIELD CONCENTRATORS, Permendur and Dysprosium: 346.
FIELD-INDUCED TRANSITION BROADENING: 186.
FILAMENTARY INCLUSIONS OR PATHWAYS,
-- as a supercurrent "pinning" mechanism: 145, 191,
    232, 260, 344.
-- in cold-rolled Ti-Nb: 260.
-- in dilute β-Ti-Rh: 123, 124, 128.
-- in general: 166, 201.
-- in Ti-Co: 166.
-- in Ti-Mn: 158.
-- in Ti-V: 133, 134.
FILAMENTARY MESH (SPONGE OR MENDELSSOHN-BEAN) MODEL:
    145, 232, 260, 344.
FILM BOILING: 236.
FILMS, see THIN FILMS.
FINAL COLD DEFORMATION, see COLD DEFORMATION, FINAL.
FLATTENED WIRE,
-- in final cold deformation optimization studies: 288.
-- microstructure of: 253.
FLUCTUATIONS, Related to ω-Phase: 147, 163, 200, 218,
    247.
FLUCTUATIONS, SPIN: 154, 157, 164, 173, 193, 200, 218.
FLUCTUATION SUPERCONDUCTIVITY: 145, 163, 192, 193, 200,
    210.
FLUX BUNDLES,
-- creep of: 231.
-- reference to pinning of: 145.

FLUX CREEP,
-- as a diffusion phenomenon: 234.
-- general discussion: 231-4.
-- in relation to resistive H_{c2} measurement criteria:
    242.
-- in relation to J_c measurement criteria: 234.
-- resistivity associated with: 234.
FLUX FLOW,
-- BARDEEN-STEPHEN model of: 143, 235.
-- channelling effects associated with: 190.
-- general discussion: 143, 190, 231, 232, 234-6.
-- resistivity associated with, see FLUX-FLOW
    RESISTIVITY.
-- velocity of: 189, 234, 235.
-- viscosity associated with: 234, 235.
-- see also HALL EFFECT IN THE MIXED STATE.
FLUX-FLOW RESISTIVITY,
-- DC measurement of: 144, 235.
-- KIM *et al*/BARDEEN-STEPHEN (vortex core) mechanism:
    143, 235.
-- microwave surface impedance measurement of: 144.
-- minimum in: 144.
-- relationship to Hall effect: 189, 190.
-- relationship to H_{c2}^{*}: 143, 144, 235.
FLUX JUMPING: 231, 232, 236-8.
FLUX JUMPING, as Function of,
-- magnitude of applied field: 238.
-- rate of change of applied field: 237, 238.
-- sample diameter: 237, 238.
FLUX-JUMP INSTABILITY, Factors which Influence it: 237,
    238.
FLUX LATTICE *versus* CRYSTAL LATTICE, Mechanical
    Analogy: 231, 236, 237.
FLUX LATTICE SOFTENING NEAR H_{c2}: 192, 231, 239, 291,
    390,
    see also UPPER CRITICAL FIELD, -- control of J_c.
FLUXOID,
-- driving force: 234.
-- elementary pinning force: 234.
-- motion: 190, 233-8.
FLUXOID CORES,
-- dimensions (coherence length): 225.
-- role in flux-flow resistivity: 143, 235.
-- role in mixed-state Hall effect: 190.
FLUX PINNING,
-- ABRIKOSOV-ANDERSON model: 145.
-- bulk pinning force density *versus* Nb concentration:
    285.
-- bulk pinning force density *versus* subband diameter:
    284.

I

IMMISCIBILITY, SOLID STATE (As Distinct from "β-Phase Immiscibility"),
-- W in Ti-Nb: 393.
-- Y in Ti-Nb: 323.
IMPURITIES,
-- in thin films produced by ion-beam bombardment: 129.
-- influence on J_c of Ti-Nb: 245.
-- influence on precipitation kinetics: 173, 245, 268.
-- influence on T_c in general: 123, 161, 170.
-- interstitial, in Ti-Nb-Cu: 329.
-- interstitial, in Ti-Nb-Ge: 334, 335.
-- oxygen, in commercial Ar and He: 309.
-- properties of Ta as a tramp impurity in Nb, history and current situation: 388, 389, 405.
-- their contribution to magnetic irreversibility: 125, 137.
-- transition metal, influence on the T_c of unalloyed Ti: 125.
IMPURITIES, INTERSTITIAL, see INTERSTITIAL ELEMENTS.
INDIRECT EXCHANGE (s-d) INTERACTION: 133.
INGOT HOMOGENEITY, Influence on J_c of Resulting Wire: 245, 268, 334, 341-3.
IN-SITU PROCESSING: 325.
INSTABILITY, Flux Jump, in relation to,
-- applied field sweep rate: 237.
-- sample diameter: 237, 238.
INTERACTION, Coulomb Pseudopotential, μ^*, for Electron-Electron: 128, 426.
INTERACTIONS, Indirect Exchange or s-d: 133.
INTERACTIONS RESPONSIBLE FOR SUPERCONDUCTIVITY,
-- electron-phonon: 129, 133, 160, 182.
-- magnetic (MATTHIAS): 133, 134, 142, 153, 160-2, 166, 199, 201.
-- other (GANGULY): 160.
-- see also THEORIES OF SUPERCONDUCTIVITY.
INTERCOMPARISONS OF TECHNICAL SUPERCONDUCTING ALLOYS, see CRITICAL CURRENT DENSITY, Technical Superconducting Alloys, Intercomparisons.
INTERFILAMENTARY COUPLING, see EDDY-CURRENT LOSS.
INTERMEDIATE HEAT TREATMENT, see HEAT TREATMENT, INTERMEDIATE.
"INTERMEDIATE PHASE" PRECIPITATE: 343, 361.
INTERMETALLIC (INCLUDING "INTERSTITIAL") COMPOUND PRECIPITATION,
-- carbides: 300.
-- detected in Ti-Nb-Fe using Mössbauer spectrometry: 396.
-- Nb_3Si, during metallic glass aging: 429, 434.

-- nitrides: 304.
-- non-occurrence during moderate-temperature aging of Ti-Nb-Cu and Ti-Nb-Ge: 339.
-- oxides, including TiO_x, M_2O_3: 309-11, 313.
-- silicides: 323, 434.
-- $TiCr_2$: 154.
-- Ti_2Cu: 318, 328.
-- TiFe: 395.
-- Ti_5Ge_3: 318, 332-4.
-- Ti_2Ni: 168.
-- Ti_5Si_3 during metallic glass aging: 429.
INTERSTITIAL ELEMENTS, Effects of B, C, N, and O in General, see Chapter 8.
INTERSTITIAL ELEMENTS (IMPURITIES),
-- as stabilizers of amorphous metal films: 425.
-- cell size refinement due to: 273.
-- in commercial and "pure" Nb: 300, 316.
-- in commercial and "pure" Ti: 300, 316.
-- influence on J_c: 176, 211.
-- see also Chapter 8.
-- see also IODIDE Ti.
-- see also KROLL-PROCESS Ti.
IODIDE Ti,
-- comparison with Kroll-process Ti as a starting material for conductor processing: 173, 268, 269, 300, 304, 310, 311, 357.
-- impurities in: 268, 316.
-- *versus* Kroll process, influence on T_c: 173.
IRON, As an Impurity in Unalloyed Ti: 124.
ISOTOPE EFFECT: 129.

J

J_c, see CRITICAL CURRENT DENSITY.
JOINTS: 345.

K

KIM *et al*/BARDEEN-STEPHEN MODEL, see BARDEEN-STEPHEN MODEL, see also FLUX FLOW, see also HALL EFFECT IN THE MIXED STATE.
KONDO EFFECT: 163.
KROLL-PROCESS Ti,
-- as a starting material for conductor processing: 300, 304.
-- comparison with iodide Ti as a starting material for conductor processing: 173, 268, 269, 300, 304, 310, 311, 357.

-- impurities in: 176, 245, 268, 310, 311, 316.

-- imprites in, influence on J_c: 268, 269, 300, 310, 311.

-- *versus* iodide process, influence on T_c: 173.

L

LAMINAR MODEL (GOODMAN): 232, 260.

LATTICE DISPLACEMENT WAVES: 318,
    see also SOFT PHONONS.

LATTICE SOFTNESS,

-- T_c enhancement due to: 183.

-- see also SOFT PHONONS.

LATTICE STABILITY, Relationship to T_c: 217.

LATTICE STIFFNESS, Parameterized by $M\langle\omega^2\rangle$: 426.

LOCALIZED MAGNETIC MOMENT, see MAGNETIC MOMENT,
    Localized.

LOCALIZED SPIN FLUCTUATIONS, see SPIN FLUCTUATIONS.

LONGITUDINAL AC AND DC APPLIED FIELDS, see LONGITUDINAL
    APPLIED FIELD.

LONGITUDINAL AND TRANSVERSE COMPOSITIONAL MODULATION,
    Resulting from Solute Segregation: 245, 249, 250.

LONGITUDINAL APPLIED FIELD,

-- critical current in, see FORCE-FREE AND NEARLY
    FORCE-FREE CURRENT FLOW.

-- in AC loss studies: 374, 375.

-- in stability studies: 244.

LOWER CRITICAL FIELD, General Discussion,

-- determination from flux-flow resistivity data: 230.

-- experimental relationship to H_{c2}: 230.

-- indirect measurement of: 231.

-- MAKI temperature dependence of: 221.

-- Ti-Nb alloys: 230, 231.

-- Ti-V alloys: 136, 137.

-- Ti-Zr-Nb research alloys: 350, 351.

-- see also MIXED STATE, -- "threshold field", H_{c1}^*.

LOWER CRITICAL FIELD, Tabulated Data,

-- amorphous Ti-Nb-Si and Ti-Nb-Si-B: 432.

-- Ti-Mo, (16 at.%), tabulated as function of temperature, 1.25-4.11 K: 195.

-- Ti-Nb, tabulated for three compositions: 231.

-- Ti-V, tabulated for three compositions: 137, 150.

-- Ti-Zr-Nb: 350.

LOW TEMPERATURE SPECIFIC HEAT, see CALORIMETRIC
    MEASUREMENT, see also SPECIFIC HEAT "JUMP" AT T_c.

M

MAGNETIC CLUSTERING (Superparamagnetism): 157-209.

MAGNETIC FIELD "CONCENTRATORS": 346.

MAGNETIC HYSTERESIS LOSS, AC Applied Field,

-- field-amplitude dependence of: 375, 376.

-- frequency dependence of: 375, 376.

-- in relation to sample (filament) diameter: 374, 415,
    see also STABILITY OR INSTABILITY, Flux Jump.

-- influence of individual-filament diameter and twist
    pitch: 376.

-- Ti-Zr-Nb alloys, influence of composition: 374.

MAGNETIC HYSTERESIS LOSS, AC Current,

-- frequency dependence of: 373.

-- self-field-amplitude dependence of: 373.

MAGNETIC HYSTERESIS, Related to Flux Pinning and
    Critical Current Density: 232, 244.

MAGNETIC INTERACTION MECHANISM FOR SUPERCONDUCTIVITY:
    133, 134, 142, 153, 160-2, 166, 199, 201.

MAGNETIC IRREVERSIBILITY (Hysteresis): 137, 144, 187,
    210, 232, 238, 244.

MAGNETIC MEASUREMENT OF T_c,

-- flux exclusion by more than ten percent of the
    superconducting second phase: 166.

-- flux trapping, effect of: 124.

-- in association with adiabatic-demagnetization
    refrigeration, possible problems: 124, 125.

-- internal shielding by higher T_c inclusions: 124,
    158, 161.

-- perturbation of by second-phase material: 124, 158,
    161.

-- SCHAWLOW-DEVLIN resonance-frequency-shift technique,
    use of: 124.

-- Ti-Cr: 154.

-- Ti-Fe: 162.

-- Ti-Fe (dilute): 156.

-- Ti-Mn: 158.

-- Ti-Mn (dilute): 156.

-- Ti-Mo: 181, 183.

-- Ti-Nb: 216, 217.

-- Ti-Tc: 197.

-- Ti-V: 131, 135.

-- Ti-Al-V: 208.

-- Ti-Rh-TM: 210.

-- unalloyed Ti: 123, 124.

MAGNETIC MOMENT, CURIE-WEISS, see PARAMAGNETISM.

MAGNETIC MOMENT, Fluctuating: 157.

MAGNETIC MOMENT, Localized,

-- as suppressor of superconductivity: 156, 160, 170.

-- fluctuating, see SPIN FLUCTUATIONS.

-- Hall effect in: 189, 190.
-- paramagnetism in: 187, 210, 221, 222, 226, 384.
-- "threshold field", H_{c1}^{*}: 137, 350, 351.

N

NEGATIVE MAGNETORESISTANCE: 163, 164, 193, 200, 210.
NEUTRON-ACTIVATION ANALYSIS, For Oxygen Analysis: 305.
NEUTRON-IRRADIATION DAMAGE, Influence on the T_c of Ti-Nb-V: 382.
NICKEL, As a Noble-Like Solute in Ti: 168.
NIOBIUM ENRICHMENT, see SOLUTE ENRICHMENT.
NONPARAMAGNETIC EILENBERGER THEORY: 226.
NONPARAMAGNETIC HW THEORY: 226.
NONPARAMAGNETIC MAKI THEORY: 188, 223, 226.
NORMAL-STATE RESISTIVITY, ρ_n, see RESISTIVITY, NORMAL STATE.
NUCLEAR HYPERFINE INTERACTION, Possible Source of Low Temperature Calorimetric Anomaly: 157.
NUCLEAR SPIN-LATTICE RELAXATION MEASUREMENT: 133.
NUCLEATE BOILING: 236.

O

OMEGA PHASE, see ω-PHASE (under Z).
OPTIMIZATION STUDIES,
-- as in the elimination of the "peak effect": 355.
-- cold deformation (subbands) *vis-à-vis* precipitation: 284-7.
-- in relation to intended operating field strength: 282, 418.
-- precipitate-redistribution philosophy: 279.
-- precipitate/subband morphology philosophy: 279.
-- Ti-Nb: 276-82, 283-92.
-- Ti-Nb-Cu: 329-31.
-- Ti-Nb-Ge: 334-7.
-- Ti-Nb-Ta: 390-2.
-- Ti-Zr-Nb: 367-71.
-- Ti-Zr-Nb, The Optimal Composition Lines of DOI *et al* and RASSMANN and ILLGEN: 364.
-- Ti-Zr-Nb-Ta: 414-21.
ORDER (THERMODYNAMIC), of the Transition at H_{c2}.
-- general discussion: 141, 144-8, 186, 187.
-- influence on electron-phonon-enhanced CLOGSTON field: 230.
ORDERING OF LOCALIZED SPINS: 157.
ORES, Nb-bearing: 383.

OVERALL COLD DEFORMATION, *versus* Final Cold Deformation in Ti-Nb: 288-90.
OXYGEN CONTAMINATION,
-- in Ti-Ta: 173.
-- in Ti-Nb-base alloys in general: Chapter 8.
-- in Ti-Nb-Mo: 399.

P

PAIRING POTENTIAL, BCS Electron-Phonon, V: 129, 132, 182.
PARACONDUCTIVITY: 200, 207.
PARAMAGNETIC GINZBURG-LANDAU PARAMETER: 230.
PARAMAGNETIC LIMITATION,
-- conditions for absence of: 140.
-- general discussion: 136, 138-42, 147, 173-5, 188, 189, 221, 226-30, 292, 380, 383, 408.
-- "weak" and "strong": 228.
PARAMAGNETIC LIMITATION MIXED PARAMETER, β: 189, 228.
PARAMAGNETIC LIMITATION PARAMETER, α,
-- as calculated from superconductive-state and normal-state measurables, respectively: 140.
-- general discussion: 140-2, 189, 227-30.
PARAMAGNETISM, MIXED STATE,
-- example of: 187.
-- possible influence on H_{c1}: 231.
PARAMAGNETISM, NORMAL STATE,
-- Curie-Weiss: 153, 157, 160, 163, 164, 166, 200, 209.
-- Pauli: 136, 138, 141, 221, 228, 380, 383, 384.
PARTITIONING, SOLUTE, see SOLUTE SEGREGATION and/or SOLUTE ENRICHMENT.
PATENT LITERATURE,
-- Ti-Nb: 242, 243.
-- Ti-Ta: 36.
-- Ti-Nb-Intl: 74, 77, 211, 300, 301, 303, 304.
-- Ti-Nb-Ta: 109, 386, 387.
-- Ti-Zr-Nb: 101, 102, 357-60.
-- Ti-Nb-Hf-Intl: 78, 211.
-- Ti-Nb-Ta-Intl: 78, 211.
-- Ti-Zr-Hf-Nb: 119, 380, 405.
-- Ti-Zr-Nb-Ta: 119, 120, 405.
PEAK EFFECT,
-- in cold-rolled Nb-Sc: 175.
-- in cold-rolled Ti-Mo: 192.
-- in Ti-Nb: 260, 284, 355.
-- in Ti-Nb-Ge: 336.
-- in Ti-Nb-Intl: 303.
-- in Ti-Zr-Nb: 362.

QUENCH-RATE, Influence on Quenched Microstructure: 169, 170, 246, 347.

QUENCHED MICROSTRUCTURES,

-- influence of quench rate and prior annealing temperature on: 247.

-- Ti-Mo: 184, 185.

-- Ti-Nb: 246, 247.

-- Ti-V: 134.

-- Ti-Nb-Si: 428.

-- Ti-Nb-SM in general: 317.

R

RECOVERY CURRENT: 236.

REFRIGERATION, Adiabatic Demagnetization: 124, 125.

RELAXATION TIME,

-- spin-orbit: 141, 189, 228.

-- transport scattering: 141, 228.

RESIDUAL RESISTIVITY, see RESISTIVITY, NORMAL-STATE, ρ_n.

RESISTIVE DETERMINATION of Minority Component T_c: 201.

RESISTIVE UPPER CRITICAL FIELD MEASUREMENT CRITERIA: 224, 242.

RESISTIVITY, ANOMALOUS: 135, 147, 164, 193, 210, 218.

RESISTIVITY, Composite Conductor Matrix: 376, 377, see also EDDY-CURRENT LOSS.

RESISTIVITY, FLUX CREEP: 234.

RESISTIVITY, FLUX FLOW,

-- in determination of H_{c2}^*: 143, 144, 235.

-- KIM et al/BARDEEN-STEPHEN model: 143, 235.

-- measurement of: 144, 235.

-- temperature dependence of: 144.

RESISTIVITY MINIMUM: 153, 154, 157, 163, 164, 166.

RESISTIVITY, NORMAL-STATE, ρ_n,

-- calculation of α, a paramagnetic limitation parameter: 140, 228.

-- in Ti-Fe: 163.

-- in Ti-Mn: 157.

-- in Ti-Mo: 192, 193.

-- in Ti-Ta: 173.

-- in Ti-V: 137, 146, 147.

-- in Ti-Nb-Si,B metallic glass alloys: 432.

-- influence of ω-phase on: 147.

-- MATTHIESSON's rule, departure from: 147.

-- relation to flux-flow resistivity: 143.

REVERSION REACTION, see β-PHASE REVERSION.

RIBBON, see COLD-ROLLED RIBBON.

RIBBON, see MELT-SPUN RIBBON.

ROLL-FLATTENED WIRE, see FLATTENED WIRE.

ROUNDED CALORIMETRIC TRANSITION,

-- analysis of in Ti-Mo: 183, 184.

-- curve-fitting procedures applied to: 134, 184.

-- in general: 134, 162, 166, 217, 218.

-- see also DOUBLE TRANSITION.

-- see also PREMATURE TRANSITION.

RULE OF MIXTURES, for Tensile Strength: 415.

S

SCALING LAW FOR FLUX PINNING, Definition and Discussion: 421.

SCATTERING RELAXATION TIME, τ: 141.

SEGREGATION, SOLUTE, see SOLUTE SEGREGATION.

SKIN EFFECT IN AC LOSS: 375.

SMALL COIL, see COIL.

SOFT PHONONS: 134, 147, 164, 183, 193, 200, 210.

SOFTENING OF THE FLUX LATTICE NEAR H_{c2},

-- implication for the design of high-field conductors: 231, 239, 291, 390.

-- related to the peak effect: 192.

-- see also UPPER CRITICAL FIELD -- control of J_c.

SOLUTE CLUSTERING,

-- in Ti-Nb: 248, 312, 318.

-- in Ti-TM alloys in general: 318.

-- in Ti-V: 135, 144.

-- in Ti-Nb-Mo: 393.

-- related to β-phase separation: 318.

-- related to isothermal-ω precipitation: 318.

SOLUTE DIFFUSION in Ti: 167, 245.

SOLUTE ENRICHMENT THROUGH PHASE DECOMPOSITION,

-- influence on H_{c2}: 224, 239, 240, 319, 327, 333.

-- influence on J_c: 261, 328, 331.

-- influence on T_c: 219, 224, 320, 326, 332, 350, 393, 406, 407.

SOLUTE PARTITIONING, see SOLUTE SEGREGATION and/or SOLUTE ENRICHMENT.

SOLUTE REDISTRIBUTION, see SOLUTE ENRICHMENT.

SOLUTE SEGREGATION, see also SOLUTE ENRICHMENT,

-- due to phase decomposition in Ti-Nb-Mo and Ti-Nb-Mo-Al: 393, 406.

-- due to residual coring or clustering in Nb-rich Ti-Nb alloys: 312.

-- influence on α-phase precipitation: 250.

-- influence on flux-jump instability: 245.

-- influence on ω-phase precipitation in alloys of high average solute concentrations: 270.

-- ingot inhomogeneity, influence on the J_c of Ti-Nb-base alloys: 245, 268, 334, 341-3.

-- unalloyed Ti, influence of impurities, tabulated
data: 124, 157.
-- unalloyed Ti, pressure dependence, tabulated data:
127.
-- unalloyed Ti, tabulated data, various samples: 130.
-- unalloyed Ti, volume dependence, tabulated data:
127.

TERNARY ALLOYS

-- amorphous Ti-Nb-Si, plotted *versus* e/a ratio: 426.
-- amorphous Ti-Nb-Si, plotted *versus* 1-h aging
temperature: 431.
-- amorphous Ti-Nb-Si, response to deformation: 431.
-- amorphous Ti-Nb-Si, tabulated for various Nb
concentrations: 432.
-- intercomparison between amorphous Ti-Nb-Si and
other amorphous superconductors: 430.
-- intercomparison between crystalline and amorphous
Ti-Nb-Si: 430.
-- intercomparison between T_c and H_r plotted *versus*
e/a ratio for Ti-Nb-Mo: 394.
-- Ti-Hf-Nb, plotted on composition triangle: 380.
-- Ti-Hf-Nb, tabulated for numerous Hf and Nb concentrations: 396, 397.
-- Ti-Mo-Al, plotted *versus* at.% Al: 208.
-- Ti-Mo-Al, tabulated for various Mo and Al concen-
trations: 211.
-- Ti-Nb-Ag, tabulated data: 319.
-- Ti-Nb-Al, tabulated for numerous Nb and Al concen-
trations: 319, 321.
-- Ti-Nb-Au, tabulated data: 325.
-- Ti-Nb-C, tabulated as function of carbon concen-
tration: 315.
-- Ti-Nb-Cr, plotted *versus* e/a ratio: 393.
-- Ti-Nb-Cr, tabulated for three Cr concentrations:
398.
-- Ti-Nb-Cu, tabulated data: 319.
-- Ti-Nb-Fe, tabulated for three alloys: 395.
-- Ti-Nb-Ga, tabulated data: 319.
-- Ti-Nb-Ge, tabulated data: 319.
-- Ti-Nb-In, tabulated data: 319.
-- Ti-Nb-Mo, plotted on composition triangle: 393.
-- Ti-Nb-Mo, plotted *versus* e/a ratio: 393, 394.
-- Ti-Nb-Mo, tabulated for numerous Nb and Mo concen-
trations: 398-400.
-- Ti-Nb-N, tabulated as function of nitrogen concen-
tration: 315.
-- Ti-Nb-O, rate of change with oxygen content: 305.
-- Ti-Nb-O, tabulated as function of oxygen concen-
tration: 315.
-- Ti-Nb-Pb, tabulated data: 319, 325.

-- Ti-Nb-Re, tabulated for four Re concentrations: 400.
-- Ti-Nb-Sb, tabulated for three Sb concentrations:
319, 325.
-- Ti-Nb-Sn, tabulated data: 319.
-- Ti-Nb-Ta, plotted on composition triangle: 384.
-- Ti-Nb-Ta, tabulated for numerous Nb and Ta concen-
trations: 397, 398.
-- Ti-Nb-U, tabulated data: 319, 325.
-- Ti-Nb-V, tabulated for numerous Nb and V concen-
trations: 397.
-- Ti-Rh-TM (all TM's), plotted *versus* e/a ratio: 210.
-- Ti-V-Mo, tabulated for three Mo concentrations: 212.
-- Ti-V-Ta, tabulated for four Ta concentrations: 212.
-- Ti-V-Zr, tabulated for three Zr concentrations: 212.
-- Ti-Zr-Nb, plotted on composition triangle: 351.
-- Ti-Zr-Nb, tabulated for a pair of representative
X- and Z-type alloys: 363.
-- Ti-Zr-Nb, tabulated for numerous Zr and Nb concen-
trations: 352.

QUATERNARY ALLOYS

-- amorphous Ti-Nb-Si-B, plotted *versus* at.% B: 430.
-- amorphous Ti-Nb-Si-B, tabulated for two compositions: 432.
-- amorphous Ti-Nb-Si-C, plotted *versus* at.% C: 430.
-- amorphous Ti-Nb-Si-Ge, plotted *versus* at.% Ge: 430.
-- amorphous Ti-Nb-Si-Mo, plotted *versus* at.% Mo: 430.
-- Ti-Nb-Sb-Al, tabulated data: 325.
-- Ti-Nb-Sb-Ga, tabulated data: 325.
-- Ti-Nb-Sb-In, tabulated data: 325.
-- Ti-Nb-Sn-U, tabulated data: 325.
-- Ti-Zr-Nb-Mo, plotted *versus* at.% Mo: 407.
-- Ti-Zr-Nb-O, tabulated for three oxygen concentra-
tions: 306.
-- Ti-Zr-Nb-Ta(5 at.%), plotted on pseudoternary
composition triangle: 407.
-- Ti-Zr-Nb-Ta(10 at.%), plotted on pseudoternary
composition triangle: 407.
-- Ti-Zr-Nb-Ta, plotted *versus* at.% Ta: 407.
-- Ti-Zr-Nb-V, plotted *versus* at.% V: 407.
SUPERCONDUCTIVE FLUCTUATIONS (Fluctuation Superconduc-
tivity),
-- general discussion: 145, 163, 192, 193, 200, 210.
-- magnetic field quenching of: 193, 200.
SUPERCONDUCTIVITY, Mechanisms for, see INTERACTIONS
RESPONSIBLE FOR SUPERCONDUCTIVITY, see also
THEORIES OF SUPERCONDUCTIVITY.
SUPERPARAMAGNETISM: 157, 209.
SURFACE CRITICAL FIELD,
-- discussion of: 136, 142, 143, 193.
-- measurement of in Ti-V: 142.

-- relationship to H_{c2}: 142, 143.

-- Ti-V, tabulated for four compositions: 143.

SURFACE SUPERCONDUCTIVITY, see SURFACE CRITICAL FIELD.

SYNTHETIC (BEAN) SUPERCONDUCTOR: 344.

T

TAKE-OFF (QUENCH) CURRENT: 236, 264, 372, 390.

TANTALUM AS AN IMPURITY IN Nb, History and Current
   Situation: 388, 389, 405.

T_c, see SUPERCONDUCTING TRANSITION TEMPERATURE.

TENSILE PROPERTIES,

-- influence of Al on Ti-Nb: 320.

-- influence of Ta on Ti-Nb: 388.

-- of metallic glass relative to parent crystal: 425.

-- of Ti-Zr-Nb-Ta: 415.

TEXTURES, see DEFORMATION TEXTURES.

THEORIES OF SUPERCONDUCTIVITY,

-- BCS electron-phonon: 129, 133, 160.

-- search for alternatives to BCS, stimulated by
      departures from the isotope-effect law: 129.

-- spin pairing as opposed to momentum pairing: 142.

-- two-band model: 218.

-- two-fluid model: 162.

-- see also INTERACTIONS RESPONSIBLE FOR SUPER-
      CONDUCTIVITY.

THERMAL CONDUCTIVITY (RESISTIVITY), Anomalous: 147,
   219.

THERMAL EXPANSION OF Ti: 202.

THERMODYNAMIC CRITICAL FIELD,

-- BCS expression for: 136, 222.

-- relationship to H_{c1} and H_{c2}: 138.

-- temperature dependence of: 136, 225, 241.

-- Ti-V, calculated from calorimetric results: 137.

-- unalloyed Ti, temperature dependence of: 125, 130.

THERMODYNAMIC ORDER, see ORDER (THERMODYNAMIC).

THIN FILMS, see also SPUTTERED FILMS,

-- crystal structures assumed by Ti, Zr, Mo, and W
      films: 129.

-- enhanced T_c in: 129.

-- general comment: 172.

-- of amorphous simple metals, cryodeposited: 423.

-- of amorphous TM_1-TM_2 alloys, cryodeposited: 423,
      426, 430.

-- of unalloyed Ti produced by ion-beam bombardment:
      129.

THIRD CRITICAL FIELD, see SURFACE CRITICAL FIELD.

THRESHOLD FIELD, MIXED STATE, H_{c1}^*: 137, 350, 351.

Ti-Zr, AS A "HIGH-T_c" SUBSTITUTE FOR Ti, Application
   to Certain Solute-Effect Studies: 170, 208, 209.

TRANSITION, GLASS: 425.

TRANSITION ROUNDING, see ROUNDED TRANSITION,
   see also DOUBLE TRANSITION.

TRANSITION TEMPERATURE, SUPERCONDUCTING, see
   SUPERCONDUCTING TRANSITION TEMPERATURE.

TRANSMISSION ELECTRON MICROSCOPY, Results and/or
   Discussion with Regard to,

-- Ti-Mo: 182.

-- Ti-Nb: 249-54, 272, 273, 278, 287.

-- Ti-Nb-Cu: 338, 339.

-- Ti-Nb-Ge: 338, 339.

-- Ti-Nb-O: 311, 313.

-- Ti-Zr-Nb: 361, 368.

-- Ti-Zr-Nb-Ta: 419.

TRANSVERSE VOLTAGE, Hall-Effect Related: 190.

T-T-T CURVES, Discussion: 360, 361.

TUBE-MAGNETIZATION EXPERIMENTS: 233, 234.

"TWELVE-TESLA" PROGRAM: 239.

TWIST PITCH,

-- in relation to AC loss: 375-7.

-- in relation to stability: 238.

-- see also EDDY-CURRENT LOSS.

U

UP-QUENCHING,

-- definition of: 135.

-- effect of on Ti-Cr: 154, 155.

-- effect of on Ti-V: 135.

UPPER CRITICAL FIELD, Discussion of its Responses to
   Alloying and Temperature, see also Table of Contents,
   BINARY ALLOYS

-- alloying in general: 139, 222.

-- Ti-Mo, general discussion: 186.

-- Ti-Nb, composition dependence: 239, 242.

-- Ti-Nb, technical composites, temperature
      dependence: 241.

-- Ti-Nb, temperature dependence: 240, 241.

-- Ti-Ta, composition dependence, spin-orbit
      effects: 173-5.

-- Ti-Ta, Pauli paramagnetism and spin-orbit
      scattering: 174, 175.

-- Ti-V, Pauli paramagnetism and spin-orbit
      scattering: 140, 141.

-- Ti-V, temperature- and composition-dependence:
      137-9.

UPPER CRITICAL FIELD RESISTIVE MEASUREMENT CRITERIA,

-- influence on apparent H_{c2}, discussion: 385.

-- specification of, two examples: 381.

-- variously mentioned: 150, 224, 241, 242, 381, 385.

UPPER CRITICAL FIELD, Specially Important Properties,
and Relationships,

-- competition between the influences of ρ_n and γ or
γT_c: 224, 384.

-- control of J_c: 192, 231, 239, 242, 276, 291, 292,
355, 362, 366, 390, 408.

-- experimental relationship to H_{c1}: 230.

-- flux-flow resistivity measurement of: 143, 144, 235.

-- highest for any known alloy: 410.

-- inverse correlation between H_r and ρ_n in Ti-Nb and
Ti-Nb-Intl alloys, plotted *versus* 1-h aging
temperature: 224.

-- relationship to ρ_n, γ, and T_c: 137, 217, 223, 224,
384, 394, 408, 432.

UPPER CRITICAL FIELD, Tabulated and Plotted Data,
see also Table of Contents,

BINARY ALLOYS

-- Ti-Mo, determinable from plotted magnetization
results: 187.

-- Ti-Mo, tabulated as function of composition,
6.25-50.0 at.% Mo: 195.

-- Ti-Mo(16 at.%), tabulated as function of temperature, 1.18-4.25 K: 196.

-- Ti-Nb, plotted intercomparison between H_r, H_{c2}^*,
and H_p *versus* temperature: 229.

-- Ti-Nb, plotted intercomparison between $H_r(1.2K)$,
H_{c20}^{*n}, H_{c20}^{*s}, H_{c20}^{WHH}, and H_{p0} *versus* at.% Nb: 229.

-- Ti-Nb, plotted *versus* at.% Nb, H_r at 1.2 K and
4.2 K: 239.

-- Ti-Nb, plotted *versus* at.% Nb, $H_r(4.2K)$ at three
measuring current densities: 242.

-- Ti-40Nb, plotted *versus* 1-h aging temperature: 224.

-- Ti-Nb, tabulated intercomparison between H_{c20}^{*n},
H_{c20}^{*s}, and H_{c20}^{WHH} for four compositions: 229.

-- Ti-Nb, tabulated intercomparison between H_{p0}, H_{c20}^{*n},
and H_{c20}^{*s} for four compositions: 227.

-- Ti-Nb technical composite conductors, plotted
versus temperature and (temperature)^2: 241.

-- Ti-Nb technical composite conductors, plotted
versus wt.% Nb, H_r at 4.2 K: 240.

-- Ti-Ta, plotted *versus* at.% Ta: 174.

-- Ti-Ta, tabulated as function of composition,
10.0-97.7 at.% Ta: 179.

-- Ti-V, plotted *versus* at.% V: 139.

-- Ti-V, tabulated as function of composition,
15-97 at.% V: 150.

-- Ti-V, tabulated for four compositions: 143.

TERNARY ALLOYS

-- amorphous Ti-Nb-Si, tabulated for several compositions: 432.

-- Ti-Hf-Nb, plotted on composition triangle, H_r at
4.2 K: 381.

-- Ti-Hf-Nb, tabulated for numerous Hf and Nb concentrations and three temperatures: 401.

-- Ti-Nb-Ag, tabulated data: 319, 333.

-- Ti-Nb-Al, tabulated data: 319.

-- Ti-Nb-Cr, plotted *versus* e/a ratio: 394.

-- Ti-Nb-Cr, tabulated for several compositions: 403.

-- Ti-Nb-Cu, tabulated data: 319.

-- Ti-Nb-Fe, tabulated for two Fe concentrations: 404.

-- Ti-Nb-Ga, tabulated data: 319.

-- Ti-Nb-Ge, plotted *versus* aging time at 400°C and
500°C, H_r at 4.2 K: 333.

-- Ti-Nb-Ge, plotted *versus* 16-h aging temperature,
H_r at 4.2 K: 332.

-- Ti-Nb-Ge, tabulated data: 319, 333.

-- Ti-Nb-In, tabulated data: 319, 333.

-- Ti-Nb-Mn, tabulated data: 404.

-- Ti-Nb-Mo, plotted *versus* e/a ratio: 394.

-- Ti-Nb-Mo, tabulated for several compositions: 403.

-- Ti-Nb-Ni, tabulated data: 404.

-- Ti-Nb-Sn, tabulated data: 319, 333.

-- Ti-Nb-Ta, plotted on composition triangle, H_r at
2 and 4.2 K: 385, 386.

-- Ti-Nb-Ta, tabulated for numerous Nb and Ta concentrations and three temperatures: 402.

-- Ti-Nb-V, tabulated for numerous Nb and V concentrations: 401, 402.

-- Ti-Nb-W, tabulated data: 404.

-- Ti-Zr-Nb, plotted on composition triangle: 353, 354.

-- Ti-Zr-Nb, tabulated for a pair of representative
X- and Z-type alloys: 363.

-- Ti-Zr-Nb, tabulated for numerous Zr and Nb concentrations: 357.

-- Ti-Zr-Nb, tabulated for three technical composite
conductors: 367.

QUATERNARY ALLOYS

-- amorphous Ti-Nb-Si-B, tabulated for several compositions: 432.

-- Ti-Hf-Nb-Ta, plotted *versus* at.% (Nb+Ta): 410.

-- Ti-Hf-Nb-Ta, tabulated for numerous compositions
and three temperatures: 409.

-- Ti-Hf-V-Nb, tabulated data: 409.

-- Ti-Zr-Hf-Nb, tabulated data: 409.

-- Ti-Zr-Nb-Ta(5 at.%), plotted on pseudoternary
composition triangle: 411.

-- in Ti-V, T_c: 134, 135, 144, 147.

-- in Ti-Nb-Cu: 328.

-- in Ti-Nb-Mo: 393.

-- in Ti-Nb-O alloys: 307, 308.

-- in Ti-Nb-Mo-Al: 406.

-- influence on thermal conductivity: 147, 219.

-- isothermal in Ti-TM alloys in general: 317, 318.

-- kinetics of formation in Ti-Nb: 245, 248.

-- magnetic character of, in Ti-Mn: 158.

-- normal-state resistivity: 147.

-- nucleation and growth: 245.

-- optimal aging temperature for: see AGING.

-- precipitation in concentrated alloys after extremely
      long-time aging: 248, 270.

-- relative pinning effectiveness *versus* α-phase: 145,
      308, 312.

-- solute diffusion associated with: 245.

-- unalloyed Ti, T_c: 128.

-- under hydrostatic pressure: 128.

ω-PHASE REVERSION, see β-PHASE REVERSION.